Lecture Notes in Computer Science   7666

Commenced Publication in 1973
Founding and Former Series Editors:
Gerhard Goos, Juris Hartmanis, and Jan van Leeuwen

Tingwen Huang Zhigang Zeng
Chuandong Li Chi Sing Leung (Eds.)

Neural
Information Processing

19th International Conference, ICONIP 2012
Doha, Qatar, November 12-15, 2012
Proceedings, Part IV

Springer

Volume Editors

Tingwen Huang
Texas A&M University at Qatar, Education City
P.O. Box 23874, Doha, Qatar
E-mail: tingwen.huang@qatar.tamu.edu

Zhigang Zeng
Huazhong University of Science and Technology
Department of Control Science and Engineering
1037 Luoyu Road, Wuhan, Hubei 430074, China
E-mail: zgzeng@gmail.com

Chuandong Li
Chongqing University, College of Computer Science
174 Shazhengjie Street, Chongqing 400044, China
E-mail: licd@cqu.edu.cn

Chi Sing Leung
City University of Hong Kong, Department of Electronic Engineering
83 Tat Chee Avenue, Kowloon, Hong Kong, China
E-mail: eeleungc@cityu.edu.hk

ISSN 0302-9743   e-ISSN 1611-3349
ISBN 978-3-642-34477-0   e-ISBN 978-3-642-34478-7
DOI 10.1007/978-3-642-34478-7
Springer Heidelberg Dordrecht London New York

Library of Congress Control Number: 2012949896

CR Subject Classification (1998): F.1, I.2, I.4-5, H.3-4, G.3, J.3, C.1.3, C.3

LNCS Sublibrary: SL 1 – Theoretical Computer Science and General Issues

Typesetting: Camera-ready by author, data conversion by Scientific Publishing Services, Chennai, India

Printed on acid-free paper

Springer is part of Springer Science+Business Media (www.springer.com)

Preface

This volume is part of the five-volume proceedings of the 19th International Conference on Neural Information Processing (ICONIP 2012), which was held in Doha, Qatar, during November 12–15, 2012. ICONIP is the annual conference of the Asia Pacific Neural Network Assembly (APNNA). This series of conferences has been held annually since 1994 and has become one of the premier international conferences in the areas of neural networks.

Over the past few decades, the neural information processing community has witnessed tremendous efforts and developments from all aspects of neural information processing research. These include theoretical foundations, architectures and network organizations, modeling and simulation, empirical study, as well as a wide range of applications across different domains. Recent developments in science and technology, including neuroscience, computer science, cognitive science, nano-technologies, and engineering design, among others, have provided significant new understandings and technological solutions to move neural information processing research toward the development of complex, large-scale, and networked brain-like intelligent systems. This long-term goal can only be achieved with continuous efforts from the community to seriously investigate different issues of the neural information processing and related fields. To this end, ICONIP 2012 provided a powerful platform for the community to share their latest research results, to discuss critical future research directions, to stimulate innovative research ideas, as well as to facilitate multidisciplinary collaborations worldwide.

ICONIP 2012 received tremendous submissions authored by scholars coming from 60 countries and regions across six continents. Based on a rigorous peer-review process, where each submission was evaluated by at least two reviewers, about 400 high-quality papers were selected for publication in the prestigious series of *Lecture Notes in Computer Science*. These papers cover all major topics of theoretical research, empirical study, and applications of neural information processing research. In addition to the contributed papers, the ICONIP 2012 technical program included 14 keynote and plenary speeches by Majid Ahmadi (University of Windsor, Canada), Shun-ichi Amari (RIKEN Brain Science Institute, Japan), Guanrong Chen (City University of Hong Kong, Hong Kong), Leon Chua (University of California at Berkeley, USA), Robert Desimone (Massachusetts Institute of Technology, USA), Stephen Grossberg (Boston University, USA), Michael I. Jordan (University of California at Berkeley, USA), Nikola Kasabov (Auckland University of Technology, New Zealand), Juergen Kurths (University of Potsdam, Germany), Erkki Oja (Aalto University, Finland), Marios M. Polycarpou (University of Cyprus, Cyprus), Leszek Rutkowski (Technical University of Czestochowa, Poland), Ron Sun (Rensselaer Polytechnic Institute, USA), and Jun Wang (Chinese University of Hong Kong, Hong Kong). The

ICONIP technical program included two panels. One was on "Challenges and Promises in Computational Intelligence" with panelists: Shun-ichi Amari, Leon Chua, Robert Desimone, Stephen Grossberg and Michael I. Jordan; the other one was on "How to Write Better Technical Papers for International Journals in Computational Intelligence" with panelists: Derong Liu (University of Illinois of Chicago, USA), Michel Verleysen (Université catholique de Louvain, Belgium), Deliang Wang (Ohio State University, USA), and Xin Yao (University of Birmingham, UK). The ICONIP 2012 technical program was enriched by 16 special sessions and "The 5th International Workshop on Data Mining and Cybersecurity." We highly appreciate all the organizers of special sessions and workshop for their tremendous efforts and strong support.

Our conference would not have been successful without the generous patronage of our sponsors. We are most grateful to our platinum sponsor: *United Development Company PSC (UDC)*; gold sponsors: *Qatar Petrochemical Company, ExxonMobil* and *Qatar Petroleum*; organizers/sponsors: *Texas A&M University at Qatar* and *Asia Pacific Neural Network Assembly*. We would also like to express our sincere thanks to the IEEE Computational Intelligence Society, International Neural Network Society, European Neural Network Society, and Japanese Neural Network Society for technical sponsorship.

We would also like to sincerely thank Honorary Conference Chair Mark Weichold, Honorary Chair of the Advisory Committee Shun-ichi Amari, the members of the Advisory Committee, the APNNA Governing Board and past presidents for their guidance, the Organizing Chairs Rudolph Lorentz and Khalid Qaraqe, the members of the Organizing Committee, Special Sessions Chairs, Publication Committee and Publicity Chairs, for all their great efforts and time in organizing such an event. We would also like to take this opportunity to express our deepest gratitude to the members of the Program Committee and all reviewers for their professional review of the papers. Their expertise guaranteed the high quality of the technical program of the ICONIP 2012!

We would like to express our special thanks to Web manager Wenwen Shen for her tremendous efforts in maintaining the conference website, the publication team including Gang Bào, Huanqiong Chen, Ling Chen, Dai Yu, Xing He, Junjian Huang, Chaobei Li, Cheng Lian, Jiangtao Qi, Wenwen Shen, Shiping Wen, Ailong Wu, Jian Xiao, Wei Yao, and Wei Zhang for spending much time to check the accepted papers, and the logistics team including Hala El-Dakak, Rob Hinton, Geeta Megchiani, Carol Nader, and Susan Rozario for their strong support in many aspects of the local logistics.

Furthermore, we would also like to thank Springer for publishing the proceedings in the prestigious series of *Lecture Notes in Computer Science*. We would, moreover, like to express our heartfelt appreciation to the keynote, plenary, panel, and invited speakers for their vision and discussions on the latest research developments in the field as well as critical future research directions, opportunities, and challenges. Finally, we would like to thank all the speakers, authors, and participants for their great contribution and support that made ICONIP 2012 a huge success.

November 2012

Tingwen Huang
Zhigang Zeng
Chuandong Li
Chi Sing Leung

Organization

Honorary Conference Chair

Mark Weichold   Texas A&M University at Qatar, Qatar

General Chair

Tingwen Huang   Texas A&M University at Qatar, Qatar

Program Chairs

Andrew Leung   City University of Hong Kong, Hong Kong
Chuandong Li   Chongqing University, China
Zhigang Zeng   Huazhong University of Science and Technology,
                                 China

Advisory Committee

Honorary Chair

Shun-ichi Amari   RIKEN Brain Science Institute, Japan

Members

Majid Ahmadi   University of Windsor, Canada
Sabri Arik   Istanbul University, Turkey
Salim Bouzerdoum   University of Wollongong, Australia
Jinde Cao   Southeast University, China
Jonathan H. Chan   King Mongkut's University of Technology, Thailand
Guanrong Chen   City University of Hong Kong, Hong Kong
Tianping Chen   Fudan University, China
Kenji Doya   Okinawa Institute of Science and Technology, Japan
Wlodzislaw Duch   Nicolaus Copernicus University, Poland
Ford Lumban Gaol   Bina Nusantara University, Indonesia
Tom Gedeon   Australian National University, Australia
Stephen Grossberg   Boston University, USA
Haibo He   University of Rhode Island, USA
Akira Hirose   University of Tokyo, Japan
Nikola Kasabov   Auckland University of Technology, New Zealand

Irwin King	The Chinese University of Hong Kong, Hong Kong
James Kwow	Hong Kong University of Science and Technology, Hong Kong
Soo-Young Lee	Advanced Institute of Science and Technology, Korea
Xiaofeng Liao	Chongqing University, China
Chee Peng Lim	Universiti Sains Malaysia, Malaysia
Derong Liu	University of Illinois at Chicago, USA
Bao-Liang Lu	Shanghai Jiao Tong University, China
John MacIntyre	University of Sunderland, UK
Erkki Oja	Helsinki University of Technology, Finland
Nikhil R. Pal	Indian Statistical Institute, India
Marios M. Polycarpou	University of Cyprus, Cyprus
Leszek Rutkowski	Czestochowa University of Technology, Poland
Noboru Ohnishi	Nagoya University, Japan
Ron Sun	Rensselaer Polytechnic Institute, USA
Ko Sakai	University of Tsukuba, Japan
Shiro Usui	RIKEN, Japan
Xin Yao	University of Birmingham, UK
DeLiang Wang	Ohio State University, USA
Jun Wang	Chinese University of Hong Kong, Hong Kong
Li-Po Wang	Nanyang Technological University, Singapore
Rubin Wang	East China University of Science and Technology, China
Zidong Wang	Brunel University, UK
Huaguang Zhang	Northeastern University, China

Organizing Committee

Chairs

Rudolph Lorentz	Texas A&M University at Qatar, Qatar
Khalid Qaraqe	Texas A&M University at Qatar, Qatar

Members

Hassan Bazzi	Texas A&M University at Qatar, Qatar
Hala El-Dakak	Texas A&M University at Qatar, Qatar
Mohamed Elgindi	Texas A&M University at Qatar, Qatar
Jihad Mohamad Jaam	Qatar University, Qatar
Samia Jones	Texas A&M University at Qatar, Qatar
Uvais Ahmed Qidwai	Qatar University, Qatar
Paul Schumacher	Texas A&M University at Qatar, Qatar

Special Sessions Chairs

Zijian Diao	Ohio University, USA
Hassab Elgawi Osman	The University of Tokyo, Japan
Paul Pang	Unitec Institute of Technology, New Zealand

Publicity Chairs

Mehdi Roopaei   Shiraz University, Iran
Enchin Serpedin   Texas A&M University,USA
Maolin Tang   Queensland University of Technology, Australia

Program Committee Members

Sabri Arik   Chi Sing Leung
Emili Balaguer Ballester   Tieshan Li
Gang Bao   Bin Li
Matthew Casey   Yangmin Li
Li Chai   Bo Li
Jonathan Chan   Ruihai Li
Mou Chen   Hai Li
Yangquan Chen   Xiaodi Li
Mingcong Deng   Lizhi Liao
Ji-Xiang Du   Chee-Peng Lim
El-Sayed El-Alfy   Ju Liu
Osman Elgawi   Honghai Liu
Peter Erdi   Jing Liu
Wai-Keung Fung   C.K. Loo
Yang Gao   Luis Martínez López
Erol Gelenbe   Wenlian Lu
Nistor Grozavu   Yanhong Luo
Ping Guo   Jinwen Ma
Fei Han   Mufti Mahmud
Hanlin He   Jacek Mańdziuk
Shan He   Muhammad Naufal Bin Mansor
Bin He   Yan Meng
Jinglu Hu   Xiaobing Nie
He Huang   Sid-Ali Ouadfeul
Kaizhu Hunag   Seiichi Ozawa
Jihad Mohamad Jaam   Shaoning Paul Pang
Minghui Jiang   Anhhuy Phan
Hu Junhao   Uvais Qidwai
John Keane   Ruiyang Qiu
Sungshin Kim   Hendrik Richter
Irwin King   Mehdi Roopaei
Sid Kulkarni   Thomas A. Runkler
H.K. Kwan   Miguel Angel Fernández Sanjuán
James Kwok   Ruhul Sarker
Wk Lai   Naoyuki Sato
James Lam   Qiankun Song
Soo-Young Lee   Jochen Steil

Publications Committee Members

Platinum Sponsor

Gold Sponsors

Table of Contents – Part IV

Session 4: Applications

Neural Network Learning
for Blind Source Separation with Application
in Dam Safety Monitoring

Theodor Dan Popescu

National Institute for Research and Development in Informatics,
8-10 Averescu Avenue, 011455 Bucharest, Romania
pope@ici.ro
http://www.ici.ro/ici/homepage/thpopescu.html

Abstract. Usually, dam monitoring systems are based on both boundary conditions (temperature, rainfall, water level, etc.) and structural responses. Statistical analysis tools are widely used to determine eventual unwanted behaviors. The main drawback of this approach is that the structural response quantities are related to the external loads using analytical functions, whose parameters do not have physical meaning. In this paper a new approach to solve this problem, based on a neural network learning rule for Blind Source Separation (BSS), to find out the contributions of the dam external loads is presented and applied in a case study for a concrete dam.

Keywords: Blind Source Separation, Neural network learning, Dam safety monitoring, Case study.

1 Introduction

Usually, dam monitoring systems are based on both boundary conditions (temperature, rainfall, water level, etc.) and structural responses (i.e. displacements, rotations, pore pressures, etc.). Statistical analysis tools are widely used to compare the current response of the dam with a whole set of recorded data, in order to determine an eventual unwanted behavior. The main drawback of this approach is that the structural response quantities are related to the external loads using analytical functions, whose parameters do not have physical meaning. Another option consists in using the structural identification technique, based on finite element models of the structure, that can be adopted to obtain an estimate of true physical parameters.

De Sortis and Paoliani, [1], discuss and compare two different procedures: a statistical approach and a structural identification technique. In [2], Leger and Leclerc present frequency domain solution procedures to develop the HTT (hydrostatic, temperature, time) statistical model to interpret concrete dams-recorded pendulum displacements. The diagnostic analysis of concrete dams based on seasonal hydrostatic loading is examined in [3] by Ardito, Maier and

T. Huang et al. (Eds.): ICONIP 2012, Part IV, LNCS 7666, pp. 1–8, 2012.

Massalongo. An integrated system, decisional support based on multisensorial information fusion provided by supervisor sensors from dams and hydropower plans, related to meteorological and geophysical factors, is presented by Calarasu and colleagues in [4].

The approach presented in this paper is a new one and makes use of Blind Source Separation (BSS) , [5], to find out the contributions due to the external loads: air temperature and hydrostatic pressure, to structure deformation and to identify irreversible component in structural response, based on a neural network learning rule. The key objective of BSS is to retrieve the source signals without resorting to any a priori information about the source signals and the transmission channel.

2   Dam Displacements Monitoring

Usually, the analysis of the dams behavior deals with few and reliable measures, such as air mean daily temperature (T), reservoir water level (Q), and the displacements of the dam, (D), with respect to the foundation of each buttress. The last choice was made in order to take into account just the contribution of the structure deformation to the crest displacement.

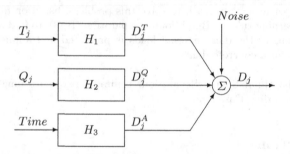

Fig. 1. Arch dam physical model

The experience in the field (see [1], [6], etc) gives the values of the displacement, D_j, where j is the time step, measured by pendulum instruments, as the sum of three terms: the first is due to air temperature change, D_j^T, and the second is related to the hydrostatic pressure, D_j^Q; the third term takes into account unexpected behavior of the dam, in the following called the irreversible component, or the trend line, D_j^A (see Fig. 1). The expressions for each terms are given in [1].

$$D_j = D_j^T + D_j^Q + D_j^A \tag{1}$$

The irreversible or trend line component, D_j^A, corresponds to the evolution in time of the dam behavior. It can be amortized (strengthened) or amplified (deteriorated). The reversible component, D_j^Q, corresponds to the hydrostatic pressure effect of the reservoir water level, while the reversible component, D_j^T, depends on the distribution of temperatures and precipitations.

The objective of the dams monitoring, by BSS, is to separate the external loads of the dam: air temperature and hydrostatic pressure, and the time effect on the dam, or the components D_j^T, Dj^Q, D_j^A, mentioned above, without a priori knowledge of the generator phenomena or of the propagation environment, and by using only the raw displacement measures of the dam.

3 Blind Source Separation

3.1 Problem Formulation

BSS deals with the problem of recovering multiple independent sources from their mixtures. BSS is closely related to the Independent Component Analysis (ICA), [5]. ICA is one method, perhaps the most widely used, for performing BSS. BSS has become a mature field of research with many technological applications.

The simple model for BSS assumes the existence of n independent signals $s_1(t), \ldots, s_n(t)$ and the observation of as many mixtures $x_1(t), \ldots, x_n(t)$, these mixtures being linear and instantaneous, i.e.

$$x_i(t) = \sum_{j=1}^{n} a_{ij} s_j(t) \tag{2}$$

for each $i = 1, \ldots, n$. This is compactly represented by the mixing equation

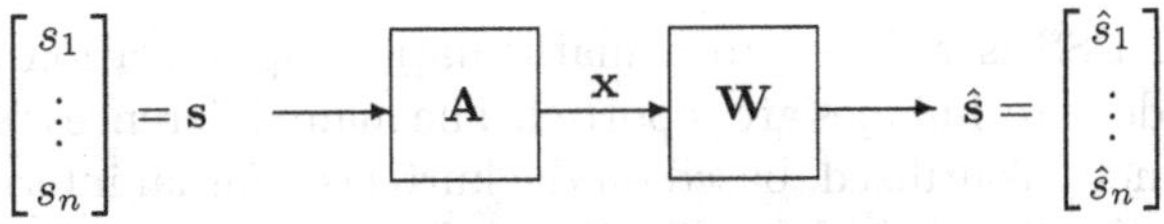

$$\begin{bmatrix} s_1 \\ \vdots \\ s_n \end{bmatrix} = \mathbf{s} \longrightarrow \boxed{\mathbf{A}} \xrightarrow{\mathbf{x}} \boxed{\mathbf{W}} \longrightarrow \hat{\mathbf{s}} = \begin{bmatrix} \hat{s}_1 \\ \vdots \\ \hat{s}_n \end{bmatrix}$$

Fig. 2. Mixing and separating. Unobserved variables: **s**; observations: **x**; estimated source components: $\hat{\mathbf{s}}$

$$\mathbf{x}(t) = \mathbf{A}\mathbf{s}(t) \tag{3}$$

where $\mathbf{s}(t) = [s_1(t), \ldots, s_n(t)]^T$ is an $n \times 1$ column vector collecting the source signals, while vector $\mathbf{x}(t)$ collects the n observed signals and the square $n \times n$ "mixing matrix" $\mathbf{A}$ contains the mixture coefficients (see Fig. 2).

In the case of convoluted mixtures the model has the following form:

$$x_i(t) = \sum_{j=1}^{n} \sum_{\tau=0}^{P} a_{ij\tau} s_j(t - \tau) \tag{4}$$

for each $i = 1, \ldots, n$, or compactly

$$\mathbf{x}(t) = \sum_{\tau=0}^{P} \mathbf{A}(\tau)\mathbf{s}(t - \tau) \tag{5}$$

In many applications, it would be more realistic to assume that there is some noise in the measurements, which would mean adding a noise term in the model:

$$\mathbf{y}(t) = \mathbf{A}\mathbf{s}(t) \tag{6}$$
$$\mathbf{x}(t) = \mathbf{y}(t) + \mathbf{n}(t)$$

BSS consists in recovering the source vector $\mathbf{s}(t)$ using only the observed data $\mathbf{x}(t)$, the assumption of independence between the entries of the input vector $\mathbf{s}(t)$ and possible some a priori information about the probability distribution of the inputs. It can be formulated as the computation of an $n \times n$ "separating matrix" $\mathbf{W}$ whose output $\hat{\mathbf{s}}(t)$ is an estimate of the vector $\mathbf{s}(t)$ of the source signals, and has the form:

$$\hat{\mathbf{s}}(t) = \mathbf{W}\mathbf{x}(t) \tag{7}$$

in the case of an instantaneous mixture and

$$\hat{\mathbf{s}}(t) = \sum_{\tau=0}^{Q} \mathbf{W}(\tau)\mathbf{x}(t - \tau) \tag{8}$$

in the case of an convolved mixture.

In our analysis it will be used an instantaneous mixture model of the sources, which is specific for a dam structure.

3.2 Algorithms

The problem of BSS is reduced to a mathematical optimization problem, for which a multitude of techniques are reported. The main differences rest on the varieties of cost functions utilized, based on the kurtosis, mutual information, cross power-spectra, negentropy and log-likelihood. In many cases these approaches are the result of different formalisms, and can be shown to be mathematically equivalent, [5].

When the signals are temporal coherent, it is possible to solve BSS problem using only the second-order statistics. If the signals are temporal white or have identical normalized spectral densities, without any information on a priori source distributions, the solution will need higher-order statistics. If the source signal distributions are known, the problem could be solved by maximum likelihood method. The following algorithms could be used in the case of instantaneous mixture, for vibration signal analysis: SOBI (Second Order Blind Identification), which uses second-order statistics, [7], JADE (Joint Approximate Diagonalization of Eigen-matrices), using 4th order cumulants, [8], and FastICA (Fixed-Point Algorithm), [9]. In the case of convolutive mixture the algorithm proposed by Parra and Spence, [10], could be used.

4   Fixed-Point Algorithm

In this section we give the conceptual description of a well-known fast algorithm for ICA used for blind source separation and feature extraction, [9], to be applied in dam safety monitoring. A neural network learning rule is transformed in a fixed-point iteration, providing an algorithm that is very simple, does not depend on any user-defined parameters, and is fast to converge to the most accurate solution allowed by the data.

The idea of the algorithm is to use a very simple, yet highly efficient, *fixed-point iteration scheme* for finding the local extrema of the kurtosis of a linear combination of the observed variables. It is well-known, [11] that finding the local extrema of kurtosis is equivalent to estimating the non-Gaussian independent components. Most suggested solutions to the ICA problem use the forth-order cumulant or *kurtosis* of the signals.

Assume that we have collected a sample of the sphered (or prewhited) random vector $\mathbf{x}_w(t)$, which in the case of blind source separation is a collection of linear mixtures of independent source signals. The derivation of the *fixed-point algorithm for ICA* is given in [9]:

Step 1. Take a random initial vector $\mathbf{w}(0)$ of norm 1. Let $k = 1$.
Step 2. Let $\mathbf{w}(k) = E\{\mathbf{x}_w(t)(\mathbf{w}(k-1)^T\mathbf{x}_w(t))^3\} - 3\mathbf{w}(k-1)$. The expectation can be estimated using a large sample of $\mathbf{x}_w(t)$ vectors.
Step 3. Divide $\mathbf{w}(k)$ by its norm.
Step 4. If $|\mathbf{w}(k)^T\mathbf{w}(t-1)|$ is not close enough to 1, let $k = k+1$ şi and go back to step 2. Otherwise, output the vector $\mathbf{w}(k)$.

The final vector $\mathbf{w}(k)$ given by the algorithm equals one of the columns of the (orthogonal) mixing matrix $\mathbf{B}$. In the case of blind source separation, this means that $\mathbf{w}(k)$ separates one of the non-Gaussian source signals in the sense that $\mathbf{w}(k)^T\mathbf{x}_w(t), t = 1, 2, \ldots$ equals one of the source signals.

A remarkable property of the algorithm is that a very small number of iterations seems to be enough to obtain the maximal accuracy allowed by the sample data, due to the cubic convergence of the algorithm.

To estimate n independent components, we run the algorithm n times. To ensure that we estimate each time a different independent component, we only need to add a simple *orthogonalizing projection* inside the loop. Recall that the columns of the mixing matrix $\mathbf{B}$ are orthogonal because of sphering. Thus we can estimate the independent components one by one by projecting the current solution $\mathbf{w}(k)$ on the space orthogonal to the columns of the mixing matrix $\mathbf{B}$ previous found. Define the matrix $\bar{\mathbf{B}}$ as a matrix whose columns are the previously found columns of $\mathbf{B}$. The add the projection operation in the beginning of **Step 3**:

$$\textbf{Step 3.} \quad \mathbf{w}(k) = \mathbf{w}(k) - \bar{\mathbf{B}}\bar{\mathbf{B}}^T\mathbf{w}(k)$$

and divide $\mathbf{w}(k)$ by its norm.

Also the initial random vector should be projected this way before starting the iterations. To prevent estimation errors in $\bar{\mathbf{B}}$ from deteriorating the estimate $\mathbf{w}(k)$, this projection step can be omitted after the first few iterations: once the solution $\mathbf{w}(k)$ has entered the basin of attraction of one of the fixed points, it will stay there and converge to that fixed point.

The algorithm can be used in a *semi-adaptive manner*, [9], to avoid the storage of large amounts of data. This can be accomplished by computing the expectation $E\{\mathbf{x}_w(t)(\mathbf{w}(k-1)^T\mathbf{x}_w(t))^3\}$ by an on-line algorithm for N consecutive sample points, keeping $\mathbf{w}(k-1)$ fixed, and updating the vector $\mathbf{w}(k)$ after the average over all N sample points has been computed.

5   Case Study - External Loads Separation

The objective of this case study was to separate the external loads of the dam: air temperature and hydrostatic pressure, and the time effect on the dam, or the components D_j^T, Dj^Q, D_j^A, without a priori knowledge of the generator phenomena or of the propagation environment, and using only the values of the displacement, D_j, measured by pendulum on x and y axes. The application was dedicated to Vidraru dam, Romania, for a period of 1200 days, [12]. The data are measured daily, at 5 different measurement points, for x and y directions. The evolution of the dam displacements is given in Fig. 3 and Fig. 4.

For these displacements, the ICA algorithm has been used to determine the 3 independent sources, which can be assimilated with the seasonal component (temperature), *Source*1, hydrostatic pressure component (reservoir water level), *Source*2, and irreversible component, *Source*3.

The original sources, the real values of the temperature and the reservoir water level, and the corresponding estimated sources are given in Fig. 5. It can be noted that there are strong similarities between the estimated sources (seasonal and hydrostatic components) and original sources (temperature and reservoir water level).

The blind source separation provided also the instantaneous mixing model of the sources, so it will be possible to see how the sources are reflected in the

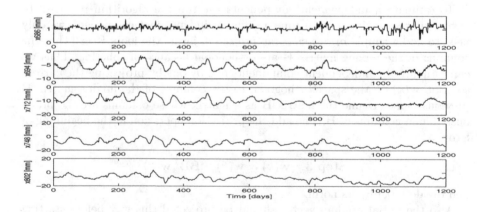

Fig. 3. Displacements for x axis at different measurement points

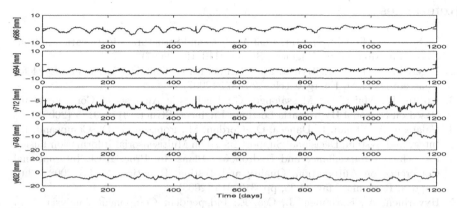

Fig. 4. Displacements for y axis at different measurement points

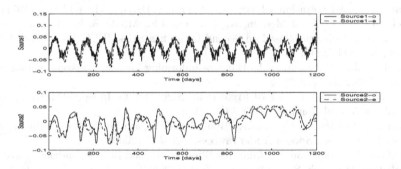

Fig. 5. Original (Source1-o, Source2-o) and estimated (Source1-e, Source2-e) sources

components D_j^T, Dj^Q, D_j^A, and finally in D_j. So, the individual contributions of the sources in the displacement measurements can be estimated. This means recovering the responses of the system to every single source, simultaneously, as if all other sources were switched off. It can be shown that this approach provides the source contributions with the exact scaling factors, thus solving the difficulty related in Subsection 3.2. This approach is refers as Blind Component Separation (BCS) in order to stress the difference with BSS.

6   Conclusions

The paper presented a new approach in dam safety monitoring, based on a neural network learning rule for blind source separation. The conceptual description of the algorithm has been given and a case study having as object the separation of the external loads and the time effect on the dam, has been presented, without a priori knowledge of the generator phenomena or of the propagation environment, and using only the displacements of the dam.

References

1. De Sortis, A., Poliani, P.: Statistical Analysis and Structural Identification in Concrete Dam Monitoring. Engineer. Struct., 110–120 (2007)
2. Leger, P., Leclerc, M.: Hydrostatic, Temperature, Time-Displacement Model for Concrete Dams. J. Engrg. Mech., 267–277 (2007)
3. Ardito, R., Maier, G., Massaiongo, G.: Diagnostic Analysis of Concrete Dams Based on Seasonal Hydrostatic Loading. Engineer. Struct., 3176–3185 (2008)
4. Calarasu, A., Stoian, I., Dancea, O., Gordan, M., Popescu, T. D., Campeanu, R.: Integrated System, Decisional Support Based on Multisensorial Information Fusion for Behavior Surveillance and Hydropower Plans. In: Proc. IEEE International Conference on Automation, Quality and Testing, Robotics, AQTR 2008, Cluj-Napoca, Romania, May 22-25, pp. 312–315 (2008)
5. Hyvärinen, A., Karhunen, J., Oja, E.: Independent Component Analysis. John Wiley & Sons (2001)
6. Mazenot, P.: Methode Generale d'Interpretation des Mesures de Surveillance des Barrages en Exploitation a Electricite de France. Divis. Techn.e Gener. (1971)
7. Belouchrani, A., Abed Meraim, K., Cardoso, J.F., Moulines, E.: A Blind Source Separation Technique Based on Second Order Statistics. IEEE Trans. Signal Process., 434–444 (1997)
8. Cardoso, J.F., Souloumiac, A.: Blind Beamforming for Non Gaussian Signals. IEE Proc. F, 362–370 (1993)
9. Hyvärinen, A., Oja, E.: A Fast Fixed-Point Algorithm for Independent Component Analysis. Neural Comput., 1483–1492 (1997)
10. Parra, L., Spence, C.: Convolutive Blind Source Separation of Non-Stationary Signals. IEEE Trans. Speech Audio Process., 320–327 (2000)
11. Delfosse, N., Loubaton, P.: Adaptive Blind Separation of Independent Sources: A Deflation Approach. Signal Process., 59–83 (1995)
12. Popescu, T.D.: A New Approach for Dam Monitoring and Surveillance Using Blind Source Separation. Int. J. Innov. Comput. Inform. Control, 3811–3824 (2011)

Improved BTC Algorithm for Gray Scale Images Using K-Means Quad Clustering

Jayamol Mathews[1], Madhu S. Nair[1,*], and Liza Jo[2]

[1] Department of Computer Science, University of Kerala, Kariavattom
Thiruvananthapuram – 695581, Kerala, India
[2] Philips Electronics India Ltd, Bangalore, India
{jayamolm,liza.jose}@gmail.com, madhu_s_nair2001@yahoo.com

Abstract. With images replacing textual and audio in most technologies, the volume of image data used in everyday life is very large. It is thus important to make the image file sizes smaller, both for storage and file transfer. Block Truncation Coding (BTC) is a lossy moment preserving quantization method for compressing digital gray level images. Even though this method retains the visual quality of the reconstructed image it shows some artifacts like staircase effect, etc. near the edges. A set of advanced BTC variants reported in literature were analyzed and it was found that though the compression efficiency is increased, the quality of the image has to be improved. An Improved Block Truncation Coding using k-means Quad Clustering (IBTC-KQ) is proposed in this paper to overcome the above mentioned drawbacks. A new approach of BTC to preserve the first order moments of homogeneous pixels in a block is presented. Each block of the input image is segmented into quad-clusters using k-means clustering algorithm so that homogeneous pixels are grouped into the same cluster. The block is then encoded by means of the pixel values in each cluster. Experimental analysis shows an improvement in the visual quality of the reconstructed image with high Peak Signal-to-Noise Ratio (PSNR) values compared to the conventional BTC and other modified BTC methods.

Keywords: Image compression, Block Truncation Coding, Image clustering, k-means clustering.

1 Introduction

With the continuing growth of modern communication technology demand for image transmission and storage is increasing rapidly. To improve the efficiency of transmission and storage of images, image compression is needed to save space and time. A good image compression algorithm should achieve reduction in the number of bits to represent the image, while preserving its quality [1-3].

BTC is a simple and fast lossy compression technique which involves less computational complexity. The basic idea of BTC [4] is to perform moment preserving quantization for blocks of pixels. The input image is divided into non-

---

*Corresponding author.

T. Huang et al. (Eds.): ICONIP 2012, Part IV, LNCS 7666, pp. 9–17, 2012.
© Springer-Verlag Berlin Heidelberg 2012

overlapping blocks of pixels of sizes 4x4, 8x8 and so on. Each block is coded individually into bit planes consisting of 0's and 1's. Each block is coded by the values of the block mean, standard deviation and the bit plane. Since BTC produces a set of bitmap, mean and standard deviation to represent a block, it gives a compression ratio of 4 and hence the bit rate is 2 bits per pixel for a 4x4 block. This method provides good compression without much degradation on the reconstructed image. But it shows some artifacts like staircase effect or raggedness near the edges. Due to its low complexity and easy implementation, BTC has gained wide interest in its further development and application for image compression. To improve the quality of the reconstructed image and for better compression efficiency several modifications of BTC have been done during the last many years.

Absolute Moment Block Truncation Coding (AMBTC) [5] preserves the higher mean and lower mean of each of the blocks and use this quantity to quantize output. AMBTC provides better image quality than image compression using BTC. Moreover, the AMBTC is quite faster compared to BTC. Cheng and Tsai [6] propose an algorithm for image compression based on the application of the moment preserving edge detection. The algorithm is computationally faster as it offers simple analytical formulae to compute the parameters of the edge feature in an image block. Reconstructed images are of good quality in accordance with human perception. Desai et al. [7] propose an edge and mean-based compression algorithm that produces good quality images at very low bit rates. The algorithm represents the image in terms of its binary edge map, mean information, and the intensity information on both sides of the edges. Amarunnishad et al. [8] propose an improved BTC image compression using a Fuzzy Complement Edge Operator (YIFCEO) [9] while shows an improvement of visual quality of reconstructed images compared to the conventional BTC. This method is based on replacement of bit blocks obtained using conventional BTC with the fuzzy logical bit block (LBB) such that the sample mean and standard deviation in each image block are preserved. Enhanced Block Truncation Coding (EBTC) [10] is an improved compression algorithm for gray scale image to reduce the correlation and spatial redundancy between pixels of an image. This method also maintains the compression ratio and quality of an image. Futuristic Algorithm for Gray Scale Image based on Enhanced Block Truncation Coding (FEBTC) [11] has greater PSNR value than AMBTC and EBTC without degradation of bit rate. Hence the improvements on BTC continue to reduce the low bit rate and computational complexity by keeping the image quality to acceptable limits.

An improved Block Truncation Coding Algorithm using k-means Quad Clustering (IBTC-KQ) is proposed in this paper. In this method, instead of the bi-clustering technique used in the conventional BTC, quad clusters are formed for each block using k-means clustering algorithm [12] so that similar pixel values come under the same cluster. And by getting the means of the pixel values of each cluster, the reconstructed image is generated. Therefore, 2 bits are required to represent a pixel in each cluster. Even though the bit rate is high compared to conventional BTC, it increases the visual quality with high PSNR values. It also reduces the ruggedness and other artifacts near the edges. By increasing the block size, compression ratio can be increased and hence bit rate can be reduced. Still PSNR value remains high when compared with conventional BTC and other modified BTC's.

The organization of this paper is as follows. The proposed algorithm is explained in section 2, detailed analysis of the algorithm in section 3, performance measures in section 4, experimental analysis in section 5 and finally conclusion in section 6.

2 Proposed Method (IBTC-KQ)

The proposed method IBTC-KQ segments the blocks of pixels into quad-clusters and by preserving the first order moment of each cluster, the reconstructed image is generated. The algorithm for this method is as follows.

Step 1: Input a gray scale image of size M×N pixels and the block size k by which the image is to be divided into non-overlapped blocks.

Step 2: Divide the image into blocks, each of size k×k, value of k can be 4, 8, 16, and so on. Each block, W is represented as,

$$W = \begin{bmatrix} w_1 & w_2 \cdots & w_k \\ \vdots & \ddots & \vdots \\ & \cdots & w_{k^2} \end{bmatrix}$$

Step 3: Segment the block into 4 clusters (c_0, c_1, c_2, c_3) using k-means algorithm so that similar pixels are grouped into the same cluster.

$$c_{i \ (i=0 \ to \ 3)} = \{w_j \mid w_j \in W \ and \ w_j \ closure \ to \ centroid_i\}$$

Step 4: Compute the mean (μ_0, μ_1, μ_2, μ_3) of the pixel values corresponding to each cluster using equation (1).

$$\mu_{j \ (j = 0 \ to \ 3)} = \frac{1}{m}\sum_{i=1}^{m} x_i \tag{1}$$

where x_i is the intensity value of the pixel in each cluster and m is the total number of pixels in each cluster.

Step 5: Based on these 4 clusters the bit map B is generated.

$$B = \begin{bmatrix} b_1 & b_2 \cdots & b_k \\ \vdots & \ddots & \vdots \\ & \cdots & b_{k^2} \end{bmatrix} \quad where \ b_j = \begin{cases} 00, if \ w_j \in c_0 \\ 01, if \ w_j \in c_1 \\ 10, if \ w_j \in c_2 \\ 11, if \ w_j \in c_3 \end{cases}$$

Therefore, 2 bits per pixel are required to represent each b_j.

Step 6: Repeat the steps 3 to 5 for each block. The resultant bit map represents the encoded image.

For the reconstruction of the image, the bit map and the four means (μ_0, μ_1, μ_2, μ_3) are transmitted to the decoder. The decoding procedure is as follows:

Step 1: Divide the bit map into k×k blocks.

$$B = \begin{bmatrix} b_1 & b_2 \cdots & b_k \\ \vdots & \ddots & \vdots \\ & \cdots & b_{k^2} \end{bmatrix}$$

Step 2: Decode bitmap block with the four means (μ_0, μ_1, μ_2, μ_3) in such a way that the elements assigned 00 are replaced with μ_0, elements assigned 01 are replaced with μ_1, elements assigned 10 are replaced with μ_2 and elements assigned 11 are replaced with μ_3. Then the decoded image block Z can be represented as,

$$Z = \begin{bmatrix} z_1 & z_2 & \cdots & z_k \\ \vdots & & \ddots & \vdots \\ & & \cdots & z_{k^2} \end{bmatrix} \qquad \text{where} \quad z_i = \begin{cases} \mu_0, b_j = 00 \\ \mu_1, b_j = 01 \\ \mu_2, b_j = 10 \\ \mu_3, b_j = 11 \end{cases}$$

Step 3: Repeat step 2 for each block and the resultant matrix represents the reconstructed image.

3   Detailed Analysis

In this section, the conventional BTC algorithm and the proposed algorithm IBTC-KQ are analyzed in detail by taking a 4×4 block of a test image 'cameraman'. The detailed analysis of BTC and IBTC-KQ algorithms are illustrated in *Table 1*.

Input block $\qquad\qquad W = \begin{bmatrix} 156 & 159 & 158 & 155 \\ 160 & 154 & 157 & 158 \\ 156 & 159 & 158 & 155 \\ 160 & 154 & 157 & 158 \end{bmatrix}$

Table 1. Detailed analysis of BTC and IBTC-KQ algorithms

BTC Algorithm [13]	IBTC-KQ Algorithm
• Calculate the mean μ and standard deviation σ of pixel values of W. $\quad \mu = 157.1250, \sigma = 0.9860$ • The compressed bit map is obtained by $$B = \begin{cases} 1, w_i \geq \mu \\ 0, w_i < \mu \end{cases}$$ $$\therefore B = \begin{bmatrix} 0 & 1 & 1 & 0 \\ 1 & 0 & 0 & 1 \\ 0 & 1 & 1 & 0 \\ 1 & 0 & 0 & 1 \end{bmatrix}$$ • The bit map B, μ and σ are transmitted to the decoder. • Calculate H and L. $\quad \therefore H = 158.1250, L = 156.1250$ • The reconstructed image Z can be obtained by replacing the element 1 in B with H and element 0 with L. $$\therefore Z = \begin{bmatrix} 156 & 158 & 158 & 156 \\ 158 & 156 & 156 & 158 \\ 156 & 158 & 158 & 156 \\ 158 & 156 & 156 & 158 \end{bmatrix}$$ • Mean Square Error (MSE) = 1.375	• Segment the block W into 4 clusters using k-means clustering algorithm. $\quad c_0 = \{156 \ \ 156 \ \ 157 \ \ 157\}$ $\quad c_1 = \{160 \ \ 160\}$ $\quad c_2 = \{154 \ \ 154 \ \ 155 \ \ 155\}$ $\quad c_3 = \{159 \ \ 159 \ \ 158 \ \ 158 \ \ 158 \ \ 158\}$ • The compressed bit map obtained is, $$B = \begin{bmatrix} 00 & 11 & 11 & 10 \\ 01 & 10 & 00 & 11 \\ 00 & 11 & 11 & 10 \\ 01 & 10 & 00 & 11 \end{bmatrix}$$ • Calculate the mean of the pixel values in each cluster. $\quad \mu_0 = 157 \quad \mu_1 = 160 \quad \mu_2 = 155 \quad \mu_3 = 158$ • The bit map B and $\mu_0, \mu_1, \mu_2, \mu_3$ are transmitted to the decoder. • The reconstructed image Z can be obtained by replacing the element 00 in B with μ_0, element 01 with μ_1, element 10 with μ_2, and element 11 with μ_3. $$\therefore Z = \begin{bmatrix} 157 & 158 & 158 & 155 \\ 160 & 155 & 157 & 158 \\ 157 & 158 & 158 & 155 \\ 160 & 155 & 157 & 158 \end{bmatrix}$$ • Mean Square Error (MSE) = 0.375

4   Performance Measures

Once an image compression system has been designed and implemented, it is important to be able to evaluate its performance based on some image quality measures. This evaluation should be done in a way to be able to compare results against other image compression techniques. In this work we focus on measures such as Peak Signal to Noise Ratio (PSNR), Compression ratio (CR) and Structural Similarity Index (SSIM) [14, 15].

Peak Signal to Noise Ratio (PSNR) - The PSNR is most commonly used as a measure of quality of reconstruction of lossy compression. PSNR is a qualitative measure based on the mean-square-error (MSE) of the reconstructed image. MSE gives the difference between the original image and the reconstructed image and is calculated using equation (2).

$$MSE = \frac{1}{MN} \sum_{i=1}^{M} \sum_{j=1}^{N} [y(i,j) - x(i,j)]^2 \qquad (2)$$

The PSNR is the quality of the reconstructed image and is the inverse of MSE. If the reconstructed image is close to the original image, then MSE is small and PSNR takes a large value. PSNR is dimensionless and is expressed in decibel. PSNR can be calculated using equation (3).

$$PSNR = 10 \log \left[\frac{L^2}{MSE} \right] \qquad (3)$$

where L is the dynamic range of the pixel values (255 for 8-bit grayscale images).

Compression Ratio (CR) - The performance of image compression schemes can be specified in terms of compression efficiency. Compression efficiency is measured by the compression ratio or by the bit rate. Compression ratio is the ratio of the size of original image to the size of the compressed image and the bit rate is the number of bits per pixel required by the compressed image. Compression ratio (CR) can be calculated using equation (4).

$$CR = \frac{size\ of\ the\ original\ image}{size\ of\ the\ compressed\ image} \qquad (4)$$

Structural Similarity Index (SSIM) - The structural similarity (SSIM) index [16] is a method for measuring the similarity between two images. SSIM can be defined as a function of three components luminance, contrast and structure and each component can be calculated separately using equations 5, 6 & 7 respectively.

Luminance change, $l(x,y) = \frac{2\mu_x\mu_y + c_1}{\mu_x^2 + \mu_y^2 + c_1}$ $\qquad (5)$

Contrast change, $c(x,y) = \frac{2\sigma_x\sigma_y + c_2}{\sigma_x^2 + \sigma_y^2 + c_2}$ $\qquad (6)$

Structural change, $s(x,y) = \frac{\sigma_{xy} + c_3}{\sigma_x\sigma_y + c_3}$ $\qquad (7)$

Then SSIM can be calculated using equation 8.

$$SSIM(x,y) = l(x,y) \cdot c(x,y) \cdot s(x,y) \qquad (8)$$

where, x represents the original image, y represents the reconstructed image and

$\mu_x = average\ of\ x,\ \ \mu_y = average\ of\ y$

$\sigma_x = variance\ of\ x,\ \ \sigma_y = variance\ of\ y,\ \ \sigma_{xy} = covariance\ of\ x\ and\ y.$

$c_1\ and\ c_2$ are two variables to stabilize the division with weak denominator

$c_1 = (k_1L)^2,\ c_2 = (k_2L)^2, c_3 = \frac{c_2}{2},\ k_1 = 0.001,\ k_2 = 0.002\ by\ default$

$L\ is\ the\ dynamic\ range\ of\ the\ pixel\ values\ (2^{\#bits\ per\ pixel} - 1)$

The resultant SSIM index is a decimal value between -1 and 1 and in the case of two identical sets of data value of SSIM becomes 1.

5   Experimental Analysis

Performance of the proposed method IBTC-KQ has been evaluated for a set of test images of different sizes, viz., 'cameraman', 'barb', 'baboon', 'goldhill', 'lena', and 'sar'. IBTC-KQ is compared with conventional BTC and AMBTC. Table 2 shows the comparative performance results of BTC, AMBTC and IBTC-KQ. The performance is measured based on three parameters PSNR, SSIM and CR.

Table 2. Comparative performance results of BTC, AMBTC and IBTC-KQ algorithms

Image	Method	Block size – 4			Block size – 8		
		PSNR	SSIM	CR	PSNR	SSIM	CR
cameraman (256×256)	BTC	24.2787	0.8336	4	22.0288	0.7351	6.4
	AMBTC	29.1585	0.9427	4	26.6107	0.9006	6.4
	IBTC-KQ	36.7714	0.9890	2	33.6339	0.9754	3.2
barb (512×512)	BTC	24.8451	0.7600	4	22.8007	0.6393	6.4
	AMBTC	29.3866	0.9257	4	27.2603	0.8665	6.4
	IBTC-KQ	36.3729	0.9847	2	33.5212	0.9632	3.2
baboon (512×512)	BTC	22.9667	0.6909	4	21.2916	0.5399	6.4
	AMBTC	26.9828	0.8869	4	25.1846	0.8276	6.4
	IBTC-KQ	33.8605	0.9777	2	31.2925	0.9550	3.2
goldhill (512×512)	BTC	28.5057	0.7903	4	25.6795	0.6614	6.4
	AMBTC	32.8605	0.9203	4	29.9261	0.8526	6.4
	IBTC-KQ	39.9867	0.9840	2	36.1776	0.9599	3.2
lena (512×512)	BTC	29.0602	0.8566	4	25.7395	0.7457	6.4
	AMBTC	33.2365	0.9413	4	29.9340	0.8818	6.4
	IBTC-KQ	40.3478	0.9874	2	36.4511	0.9664	3.2
sar (1472×1472)	BTC	20.4338	0.6717	4	18.6995	0.5305	6.4
	AMBTC	23.2078	0.8563	4	21.5774	0.7990	6.4
	IBTC-KQ	30.3053	0.9732	2	27.7087	0.9483	3.2

From *Table 2*, it is seen that performance of the proposed method IBTC-KQ is better than BTC and AMBTC algorithms on the basis of the three performance measures. For all the test images with 4×4 and 8×8 blocks, PSNR and SSIM values are high in comparison with BTC and AMBTC. It shows an enhancement in the visual quality of the reconstructed image. Compression ratio calculated by IBTC-KQ is half that of the BTC and AMBTC, but even then the visual quality of the image is maintained. To compromise with the compression ratio, 8×8 block can be considered. By increasing the block size, compression ratio can be increased and even then the PSNR value is high compared to that of the other BTC methods with 4×4 block. This is shown by the shaded region in *Table 2*.

In BTC and AMBTC, bi-clustering is done and in the bit plane block, the changes in the bit values indicate edge positions and these are determined by the block threshold. Because of the in-built fuzziness present in the image, these edge positions may or may not be accurate. This is reflected in the visual quality of the image. In IBTC-KQ, since quantization is done based on quad clustering and it preserve the first order moment of each cluster, the mean square error between the original image and the decoded image is very low when compared with the conventional BTC, which is illustrated in *Table 1*. Hence the visual quality of the reconstructed image even at the edges is considerably enhanced by IBTC-KQ.

Test images are shown in *Figure 1(a) - 4(a)* and the reconstructed images using 4×4 block by the BTC, AMBTC and IBTC-KQ algorithms are shown in *Figure 1(b) – 4(b), 1(c) –4(c)* and *1(d) – 4(d)* respectively.

    (a)   (b)   (c)   (d)

Fig. 1. (a) Original Image'cameraman'; Reconstructed images (b) BTC,(c)AMBTC,(d)IBTC-KQ

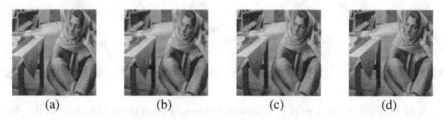

    (a)   (b)   (c)   (d)

Fig. 2. (a) Original Image 'barb'; Reconstructed images (b) BTC,(c) AMBTC, (d) IBTC-KQ

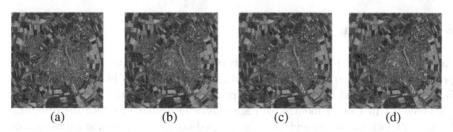

    (a)   (b)   (c)   (d)

Fig. 3. (a) Original Image 'sar'; Reconstructed images (b) BTC, (c) AMBTC, (d) IBTC-KQ

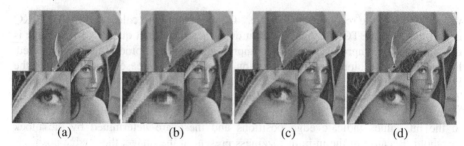

(a)   (b)   (c)   (d)

Fig. 4. (a)Original Image 'lena512'; Reconstructed images (b)BTC, (c) AMBTC, (d) IBTC-KQ

The above four figures show the improvement in the visual quality of the images from *(b)* through *(d)*. In *Figure 4,* a small portion of the image is marked and its zoomed effect is separately shown in the image itself. Here we can see the staircase effect and raggedness near the edges in 4*(b)*, i.e., using BTC and in 4*(c)* using AMBTC, though the raggedness is reduced still there is some distortions near the edges. But in *Figure 4(d)*, our method corresponds to perceptually high-quality reconstructed image with minimum distortions near the edges.

Figure 5(d) shows the performance of IBTC-KQ using 8×8 block. The visual quality of the reconstructed image is still increased when compared with 4×4 block BTC and AMBTC algorithms. From *Table 2*, it is also seen that the compression ratio is increased from 2 to 3.2 for the 8×8 block. Since this method involves less number of simple computations, the computational complexity is also very less when compared with BTC method.

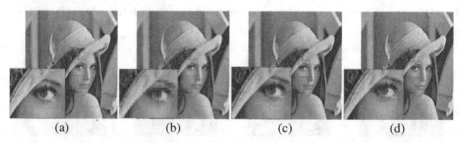

(a)   (b)   (c)   (d)

Fig. 5. (a) Original Image 'lena512'; Reconstructed images using (b) 4×4 block BTC, (c) 4×4 block AMBTC, (d) 8×8 block IBTC-KQ

6   Conclusion

An improved BTC algorithm for enhancing the conventional BTC algorithm is proposed. This method uses k-means quad clustering instead of bi-clustering in BTC. Because of this quad clustering similar pixels come under the same cluster and it preserves the first order moment of each cluster in decoding. Hence the difference between the original image and the reconstructed image is very less and therefore the MSE also is less. A set of images with different textures and edges has been tested and found that the reconstructed images show a better visual quality than BTC and other modified BTC's. The test results also show the performance of the proposed method based on the parameters PSNR, SSIM and CR. The results show that the PSNR and SSIM values are very high when compared to BTC, even when the block size is increased. It gives a better enhancement in the visual quality of the reconstructed images even at the edges. Also the computational complexity of this method is very less when compared with BTC making it suitable for real time transmission. Future work is focused on the modification of IBTC-KQ for the compression of color images.

References

1. Gonzalez, R.C., Woods, R.E.: Digital Image Processing, 3rd edn. Prentice Hall (2008)
2. Khalid, S.: Introduction to Data Compression, 3rd edn. (2005)
3. Baxes, G.A.: Digital Image Processing – Principles and Applications, pp. 179–179. John Wiley & Sons (1994)
4. Delp, E.J., Mitchell, O.R.: Image Compression Using Block Truncation Coding. IEEE Trans. Commun. 27(9), 1335–1342 (1979)
5. Lema, M.D., Mitchell, O.R.: Absolute Moment Block Truncation Coding and Its Application to Color Image. IEEE Trans. Commun. 32, 1148–1157 (1984)
6. Cheng, S.C., Tsai, W.H.: Image Compression by Moment-Preserving Edge Detection. Pattern Recogn. 27, 1439–1449 (1994)
7. Desai, U.Y., Mizuki, M.M., Masaki, I., Horn, B.K.P.: Edge and Mean Based Compression. MIT Artif. Intell. Lab. AI Memo 1584 (1996)
8. Amarunnishad, T.M., Govindan, V.K., Abraham, T.M.: Improving BTC Image Compression Using a Fuzzy Complement Edge Operator. Signal Process. Lett. 88, 2989–2997 (2008)
9. Amarunnishad, T.M., Govindan, V.K., Abraham, T.M.: A Fuzzy Complement Edge Operator. In: IEEE Proceedings of the Fourteenth International Conference on Advanced Computing and Communications, Mangalore, Karnataka, India (2006)
10. Kumar, A., Singh, P.: Enhanced Block Truncation Coding for Gray Scale Image. Int. J. Comput. Techn. Appl. 2(3), 525–530 (2011)
11. Kumar, A., Singh, P.: Futuristic Algorithm for Gray Scale Image based on Enhanced Block Truncation Coding. Int. J. Comput. Inform. Syst. 2, 53–60 (2011)
12. Kanungo, T., Mount, D.M., Netanyahu, N., Piatko, C., Silverman, R., Wu, A.Y.: An efficient k-means clustering algorithm: Analysis and implementation. In: Proceeding IEEE Conference of Computer Vision and Pattern Recognition, pp. 881–892 (2002)
13. Doaa, M., Fatma, A.: Image Compression Using Block Truncation Coding. Cyber J.: Multidiscipl. J. Sci. Techn. J. Sel. Areas Telecom. (2011)
14. Eskicioglu, A.M., Fisher, P.S.: Image Quality Measures and Their Performance. IEEE Trans. Commun. 34, 2959–2965 (1995)
15. Yamsang, N., Udomhunsakul, S.: Image Quality Scale (IQS) for Compressed Images Quality Measurement. In: Proceedings of the International Multiconference of Engineers and Computer Scientists, vol. 1, pp. 789–794 (2009)
16. Wang, Z., Bovik, A.C., Sheikh, H.R., Simoncelli, E.P.: Image Quality Assessment: from Error Measurement to Structural Similarity. IEEE Trans. Image Process. 13 (2004)

Optimization of a Neural Network for Computer Vision Based Fall Detection with Fixed-Point Arithmetic

Christoph Sulzbachner[1,*], Martin Humenberger[1],
Ágoston Srp[2], and Ferenc Vajda[2]

[1] Austrian Institute of Technology
Donau-City-Strasse 1, 1220 Vienna, Austria
[2] Budapest University of Technology and Economics
1117 Budapest, Magyar tudósok krt. 2., Hungary
{christoph.sulzbachner,martin.humenberger}@ait.ac.at,
{srp.agoston,vajda}@iit.bme.hu
http://www.ait.ac.at,https://www.iit.bme.hu

Abstract. This paper presents an optimized implementation of a neural network for fall detection using a Silicon Retina stereo vision sensor. A Silicon Retina sensor is a bio-inspired optical sensor with special characteristics as it does not capture images, but only detects variations of intensity in a scene. The data processing unit consists of an event-based stereo matcher processed on a field programmable gate array (FPGA), and a neural network that is processed on a digital signal processor (DSP). The initial network used double-precision floating point arithmetic; the optimized version uses fixed-point arithmetic as it should be processed on a low performance embedded system. We focus on the performance optimization techniques for the DSP that have a major impact on the run-time performance of the neural network. In summary, we achieved a speedup of 48 for multiplication, 39.5 for additions, and 194 for the transfer functions and, thus, realized an embedded real-time fall detection system.

Keywords: Neural network, Fall detection, Fixed-point arithmetic, Embedded systems.

1 Introduction

According to the regional population projection statistics of the European Commission [1], the median age of the population in 2030 is projected between 34.2 and 57.0 years, while in 2008, the range was between 32.9 and 47.8 years. The population aged over 65 years is expected to increase in a range from 10.4% to 37.3%. The major cause of injuries of people aged 60 and above are falls [2]. To detect fall situations, robust detection systems are required to ensure safety of

---

* Corresponding author.

T. Huang et al. (Eds.): ICONIP 2012, Part IV, LNCS 7666, pp. 18–26, 2012.
© Springer-Verlag Berlin Heidelberg 2012

older people. This paper presents an approach for a fall detection system for an embedded systems using stereo vision and a neural network.

The paper is organized as follows: Section 2 gives a brief overview about existing work on the Silicon Retina technology, and state-of-the-art fall detection techniques. Section 3 gives details about our approach using a neural network for fall detection. Section 4 gives an overview of the embedded platform and shows the performance optimization techniques that had a major impact on the run-time performance. Section 5 shows results of the optimization, and section 6 concludes the paper and gives an outlook of further research.

2 Related Work

2.1 Silicon Retina Technology

The Silicon Retina technology is a bio-inspired vision sensor that uses an address-event-representation (AER) concept for data representation. AER was proposed in 1991 by Sivilotti [3] as a method for exchanging neural information within biological systems. Later, AER has been modified for exchanging asynchronous data streams. Each time a variation of intensity is recognized by the sensor, an event E is emitted. An event E is a tuple consisting of the coordinates x and y representing the position where the variation occurred, a timestamp t indicating a precise time information, and the polarity s that indicates the direction of the change.

The technology goes back to Fukushima et al. [4] in 1970, who first implemented a model of a retina. Later, Mead and Mahowald [5] in 1988 developed a first silicon based retina. The work of Lichtsteiner et al. [6,7] shows recent developments in bio-inspired sensor systems that established the basis for this type of sensor. Recent developments by Posch et al. [8] have higher optical and time resolution.

Derived from the AER concept, timed 3D events (T3DE) additionally include depth information (z coordinate). The stereo matching algorithm is an area-based normalized sum of absolute differences (NSAD) approach, that processes aggregated grayscale images and delivers T3DE as output. To be independent of the mounting position of the system, the results are transformed to world coordinates. Details can be found in our previous work [9].

2.2 Fall Detection

There are various methods and approaches for fall detection. Most commercially available systems are based on special equipment such as wearable components or components integrated in clothes. Here, we will give a brief overview about recently introduced methods in research. Fu et al. [10] uses a horizontal asynchronous temporal contrast sensor and uses normalized Y velocity to determine falls. Wu et al. [11] analyzed the velocity characteristics during falls and activities of daily living (ADL) and discovered differences during falls. This approach can be used with wearable devices such as accelerometers. Anderson [12] uses the bounding box of a body to distinguish falls from ADL. He claims that all falls ending in a specified position will be detected. Li et al. [13] proposed to use context information and body posture as a means of identifying falls. Both Nyan et al. [14] and Juang et al. [15] tried similar approaches, where Nyan uses the tilt angle of the body, while Juang used the silhouette to identify features.

3   Neural Network for Fall Detection

Due to several reasons, such as privacy on the one hand and the high temporal resolution on the other hand, a stereo setup of two Silicon Retina sensors is well suited for passive, computer vision based fall detection. Our approach analysis the observed area to detect falls in real-time by extracting specific features, all calculated out of the 3D point cloud generated by the stereo vision system.

An analysis of hundreds of recorded fall scenarios showed that the following features are most promising: position, velocity and acceleration of both the center of gravity (cog) and the highest point (hp), a 3D bounding cylinder (bc), its height to diameter ratio, and a measure for the motion in the scene within a specific time frame (event-rate). The velocity is determined by tracking the cog, and the acceleration is the change of velocity with time.

The raw features are not suitable for being directly processed with the neural network. Thus, pre-processing to a suitable format is required, where the raw input from each sampling is put into a tapped delay line and the whole content of the tapped delay line is given to the network as input. For successful fall detection, the temporal sensitivity is needed which necessitates to use dynamic networks. The most promising solution is a focused time delay neural network (FTDNN), a type of feed forward neural networks.

Experiments show that the key to detect falls is to analyze a time interval of about 2s with a temporal resolution of 10ms, resulting in 200 sample vectors. A sample vector contains all features with a total size of 33 elements per feature, see table 1. Thus, the resulting original input vector size would be $33 \times 200 = 6600$, which cannot be handled on a low performance embedded system.

Thus, we use two approaches for reducing the input vector: Firstly, instead of providing the whole original input vector to the network, we sample the vector by averaging the history. To ensure fast detection of falls, the 10 most recent samples are fully retained. The next three group of 10 samples, four groups of 20 samples, and two groups of 40 samples are averaged from elapsed events. Thus, we reduced the input vector size to 19. Secondly, we reduced the size of a feature vector from 33 to 14 elements by eliminating the x, y coordinates and the timestamp t (table 1) because the z coordinate has the most influence during a fall and the time information is systematically included in the 2s interval. Hence, the total size of the input vector is reduced to $14 \times 19 = 266$.

Table 1. Original and optimized feature vector: x, y, z are world coordinates, t is the timestamp, m is magnitude, d is diameter of cylinder, h is height of cylinder

feature	original elements	size	opt. elements	opt. size
position cog	x, y, z, t	4	z	1
velocity cog	x, y, z, t	4	z, m	2
acceleration cog	x, y, z, t	4	z, m	2
position hp	x, y, z, t	4	z	1
velocity hp	x, y, z, t	4	z, m	2
acceleration hp	x, y, z, t	4	z, m	2
BC	x, y, z, t, d, z_{low}, z_{high}	7	h, d	2
BC ratio	BC(h)/BC(d)	1	BC(h)/BC(d)	1
event-rate		1		1
total		33		14

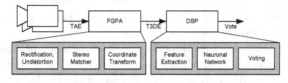

Fig. 1. System overview

4 Performance Optimization

The processing platform is based on both a field programmable gate array (FPGA) and a digital signal processor (DSP). Figure 1 shows a block diagram of the embedded system. Both sensors continually capture the area to observe, and the FPGA acquires and pre-processes the captured data (T3E) by a stereo matcher [9]. The stereo algorithm is an area-based normalized sum of absolute differences (NSAD) approach processed on greyscale pseudo-frames which are generated by aggregating the event information from the sensors. The overall FPGA utilization requires 32k equivalent logic cells and 2MBit Block RAM for a typical system configuration. The resulting depth information (T3DE) is sent to the DSP in chunks and stored in the internal memory for feature computation. Due to the size of the network, several components need to be reloaded to internal memory using DMA transfers. The used Blackfin DSP from Analog Devices [16] is a 16-/32-bit mixed embedded processor with an internal memory of 132kB running at $600MHz$. It is restricted to two 16-bit MAC or four 8-bit ALU plus two load/store plus two pointer updates per cycle. Thus, floating point or 32-bit integer arithmetic does not allow exploitation of the architecture.

4.1 Floating- and Fixed-Point Arithmetic

To represent fractional numbers in computer systems, two common approaches are floating-point and fixed-point representation. The IEEE Standard for

Floating-Point Arithmetic (IEEE 754) last revised in 2008 [17] specifies interchange and arithmetic formats, and methods for binary and decimal floating-point arithmetic in computer programming environments. The standard separates four formats, basic and extended, each consisting of single and double precision. The floating data consists of three fields: a sign field s, an exponent field E with a minimum and maximum exponent E_{min} and E_{max} encoding the exponent offset, and a fraction field b encoding the significant. A floating-point number is represented as

$$x_{float} = (-1)^s 2^E (b_0.b_1 b_2 ... b_{p-1}).$$

Depending on the format, the standard specified different parameters such as 8-bit exponent and 23-bit fraction for basic single precision, and 11-bit exponent and 52-bit fraction for basic double precision. The extended formats and special values are not further covered in this document.

Computing float point operations such as additions or multiplications are performance critical to computer systems without a floating-point instruction set. Fixed-point architectures such as fixed-point digital signal processors emulate floating-point operations in software [18].

In contrast, in fixed-point arithmetic the same number of digits is used to represent every value and the binary point is fixed on a specific position depending on the actual fixed-point format. Figure 2 shows the representation of fixed-point numbers, where $[b_{m+n}..b_n]$ represents the whole part and $b_{n-1}...b_0$ represents the fractional part.

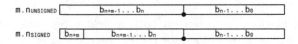

Fig. 2. Fixed-point layout

Unsigned fixed-point rationales have a size of $N_{unsigned} = m + n$ and are represented by

$$x_{unsigned,m.n} = \frac{1}{2^n} \sum_{i=0}^{N_{unsigned}-1} 2^i b_i$$

with a dynamic range of $[0..2^m - 2^{-n}]$ and a resolution of 2^{-n}. Signed two's complements fixed-point rationales have a size of $N_{signed} = m + n + 1$ and are represented by

$$x_{signed,m.n} = \frac{1}{2^n} [-2^{N_{signed}-1} b_{N_{signed}-1} + \sum_{i=0}^{N_{signed}-2} 2^i b_i]$$

with a dynamic range of $[-2^m..2^m - 2^{-n}]$ and a resolution of 2^{-n}. Arithmetic operations use the fixed-point instruction set of a computer system. Different operations such as multiplications and division change the format of the result.

4.2   Optimization of the Neuron Model

An artificial neuron is a mathematical function representing the behavior of a biological neuron. It is defined as

$$y_k = \varphi\left(\sum_{j=0}^{m} w_{kj}x_j + bias\right)$$

where $m + 1$ are the number of inputs, x_j are the input signals, and w_{kj} are the weights. To specify the property of the artificial neuron, different types of transfer functions φ exist. The introduced neural network uses the logsig, tansig, and purelin functions.

The logsig function is defined as $logsig(n) = 1/(1 + \exp(-n))$ and the tansig function is defined as $2/(1 + \exp(-2n)) - 1$. The purelin function is a linear transfer function that needs no special optimization.

Optimizing $\exp(n)$ with Series Expansion. Series expansion for $\exp(n)$ is very time consuming when a high accuracy is required. Series expansion requires the computation of the factorial function which can be computed previously and the power function which needs to be computed at each function call. Higher numbers of expansion also require larger data type sizes, which are cost intensive to compute. The optimization arithmetic mean difference $\overline{y}$ is defined as

$$\overline{y} = \frac{1}{len} \sum_{x=min}^{max} \left| \exp(x) - \sum_{n=0}^{m} \frac{x^n}{n!} \right|$$

where m represents the number of series, min and max define the x range, and len is the number of samples in the range. For $x < min$ and $x > max$, the result is saturated by 0 and 1. Evaluations shows that the network is very sensitive to the accuracy of the logsig function as series expansions with $m > 15$ were required to achieve $\overline{y} < 10^{-3}$, resulting in a vast of multiplications and additions. Thus, this optimization approach is not applicable.

Optimizing with Fixed-Point Lookup Tables and Linear Interpolation. Another approach is based on lookup tables in signed fixed-point 3.12 format storing the values of the transfer function, then to use linear interpolation for computing the result for $x_{min} \leq x \leq x_{max}$, and finally saturating the residual values. Figure 3 shows the absolute error of the transfer functions optimized with 3.12 signed fixed-point and linear interpolation.

Table 2 shows the runtime performance of the single-precision and double-precision floating-point, and fixed-point version of the transfer functions.

4.3   Optimization of the Matrix Multiplications and Additions

Computing the neural network requires number crunching of 64168 multiplications that cannot be reduced. The optimized implementation of the neural network uses fixed-point representation. Table 3 shows the runtime of arithmetic operations for the functional behavior implementations and the optimized version.

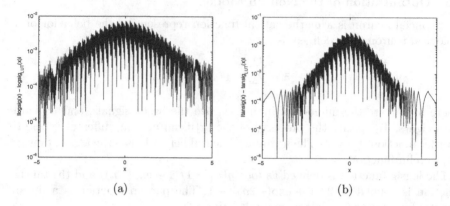

(a)   (b)

Fig. 3. Absolute mean error of optimized version of transfer functions logsig (a) and tansig (b) compared with double precision floating-point for $x_{min} = -5$ and $x_{max} = 5$

Table 2. Runtime performance of transfer functions for double-precision, single-precision floating point and fixed-point arithmetic

function	range	t_{double} [cycles]	t_{single} [cycles]	t_{opt} [cycles]
logsig(x)	$x < x_{min} \lor x > x_{max}$	8050	3165	16
logsig(x)	$x_{min} \leq x \leq x_{max}$	8230	3282	43
tansig(x)	$x < x_{min} \lor x > x_{max}$	3271	3308	16
tansig(x)	$x_{min} \leq x \leq x_{max}$	8469	3425	43

5   Results

Processing the complete neural network including the feature computation, feature assembler, input sampler, and classifier requires $1552.84\mu s$ for only internal memory usage. Figure 4 shows a comparison of the original double-precision floating-point and the optimized fixed-point arithmetic version of the fall probability (output of the neural network). Based on both probabilities, fall and non

Table 3. Runtime performance of multiplication operation for double-precision, single-precision floating point and fixed-point arithmetic

function	t_{double} [cycles]	t_{single} [cycles]	t_{opt} [cycles]
$a * b$	192	95	4
a/b	1512	250	53
$a + b$	158	120	4
$a - b$	199	148	4

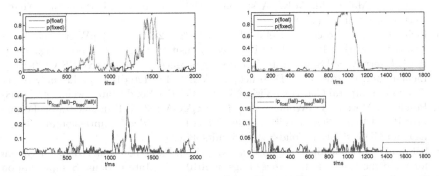

(a) person trying to climb over bed mesh and falling down   (b) person trying to stand up from chair and falling down

Fig. 4. Comparison of double-precision floating-point to fixed-point version of neural network for different scenarios (a) and (b)

fall, a voting strategy decides whether a fall occurred or not. For a detailed evaluation of the voting strategies see our previous work [19].

6   Conclusion and Future Work

In our work, we presented optimization approaches for fitting a neural network to a low performance embedded system by simplifying the feature vector, sampling the input vector, optimizing the whole computation in fixed-point arithmetic and optimizing the transfer functions. With this approach, multiplications and addition have been boosted at a factor of 48 and 39.5. The transfer functions were boosted by an average factor of 194.

Next, we will train the fixed-point neural network again with the modified configuration and try to add additional functionality such as gesture detection.

Acknowledgment. This work is supported by the AAL JP project Grant CARE "aal-2008-1-078". The authors would like to thank all CARE participants working on the success of the project.

References

1. Giannakouris, K.: Regional population projections EUROPOP 2008: Most EU regions face older population profile in 2030. European Union (2010)
2. Bauer, R., Steiner, M.: Injuries in the European Union - Statistics Summary 2005-2007. European Union (2009)
3. Sivilotti, M.: Wiring consideration in analog vlsi systems with application to field programmable networks. Phd-thesis, California Institute of Technology (1991)
4. Fukushima, K., Yamaguchi, Y., Yasuda, M., Nagata, S.: An Electronic Model of the Retina. Proceedings of the IEEE 58 (1970)

5. Mead, C., Mahowald, M.K.: A silicon model of early visual processing. Neural Networks Journal 1 (1988)
6. Lichtsteiner, P., Posch, C., Delbruck, T.: A 128×128 120dB 30mW Asynchronous Vision Sensor that Responds to Relative Intensity Change. In: Proceedings of the IEEE International Solid-State Circuits Conference, SanFrancisco, USA (2006)
7. Lichtsteiner, P., Posch, C., Delbruck, T.: A 128×128 120 dB 15 μs Latency Asynchronous Temporal Constrast Vision Sensor. IEEE J. Solid-State Circuits 43 (2008)
8. Posch, C., Matolin, D., Wohlgenannt, R.: A QVGA 143 dB Dynamic Range Frame-Free PWM Image Sensor With Lossless Pixel-Level Video Compression and Time-Domain CDS. IEEE J. Solid-State Circuits 46 (2011)
9. Schraml, S., Belbachir, A., Milosevic, N., Schön, P.: Dynamic stereo vision system for real-time tracking. In: Proceedings of 2010 IEEE International Symposium on Circuits and Systems, ISCAS (2010)
10. Fu, Z., Delbruck, T., Lichtsteiner, P., Culurciello, E.: An Address-Event Fall Detector for Assisted Living Applications. IEEE Trans. Biomed. Circuits Syst. 2 (2008)
11. Wu, G.: Distinguishing fall activities from normal activities by velocity characteristics. J. Biomech. 11 (2000)
12. Anderson, D., Keller, J.M., Skubic, M., Chen, X., He, Z.: Recognizing Falls from Silhouettes. In: 28th Annual International Conference of the IEEE Engineering in Medicine and Biology Society (2006)
13. Li, Q., Zhou, G., Stankovic, J.: Accurate, Fast Fall Detection Using Posture and Context Information. In: Proceedings of the 6th ACM Conference on Embedded Network Sensor Systems (2008)
14. Nyan, M.N., Tay, F.E.H., Mah, M.Z.E.: Application of motion analysis system in pre-impact fall detection. J. Biomech. 10 (2008)
15. Juang, C.F., Chang, C.M.: Human Body Posture Classification by a Neural Fuzzy Network and Home Care System Application. IEEE Trans. Syst., Man, Cybern. A, Syst., Humans 6 (2007)
16. Analog Devices: ADSP-BF537 Blackfin Processor Hardware Reference. Analog Devices (2005)
17. IEEE Std 754-2008: IEEE Standard for Floating-Point Arithmetic. IEEE Standard (2008)
18. Analog Devices: Fast Floating-Point Arithmetic Emulation on Blackfin Processors. Engineer-to-Engineer Note EE-185 (2007)
19. Humenberger, M., Schraml, S., Sulzbachner, C., Srp, A., Vajda, F.: Embedded Fall Detection with a neuronal Network and Bio-Inspired Stereo Vision. In: 2012 IEEE International Conference on Computer Vision and Pattern Recognition Workshops, CVPR Workshops (2012)

Customer Relationship Management
Using Partial Focus Feature Reduction

Yan Tu and Zijiang Yang[*]

School of Information Technology, York University, Toronto, Canada
zyang@yorku.ca

Abstract. Effective data mining solutions have for long been anticipated in Customer Relationship Management (CRM) to accurately predict customer behavior, but in a lot of research works we have observed sub-optimal CRM classification models due to inferior data quality inherent to CRM data set. This paper is proposed to present our new classification framework, termed Partial Focus Feature Reduction, poised to resolve CRM data set with Reduced Dimensionality using a collection of efficient data preprocessing techniques characterizing a specially tailored modality grouping method to significantly improve feature relevancy as well as reducing the cardinality of the features to reduce computational cost. The resulting model yields very good performance result on a large complicated real-world CRM data set that is much better than ones from complex models developed by renowned data mining practitioners despite all data anomalies.

Keywords: Customer relationship Management, Feature reduction, Classification, Data mining.

1   Introduction

Customer Relationship Management (CRM) is "the strategic use of information, processes, technology, and people to manage the customer's relationship with your company (Marketing, Sales, Services, and Support) across the whole customer life cycle" [1]. It is widely recognized nowadays that CRM represents one vital business function that generates long term profit by developing harmonious relationship with customers. The technological advancement has enabled new approaches – notably data mining – to be applied for finding the best CRM strategies. Ngai, Xiu and Chau [2] argue analytical CRM as a sub category of CRM, where data mining can play an essential role in analyzing customer data. Their paper also gives a manifest of data mining researches from year of 2000 to 2006. It is clear that the potential of data mining techniques on all aspects of CRM is being extensively studied. However, there are still problems jeopardizing the success of data mining applications for CRM. Some problems are more specific to CRM domain. For example, Privacy-Preserving Data Mining (PPDM) is the discipline commonly associated with CRM data mining projects data that offer data transformation, through which information that can

---

[*] Corresponding author.

T. Huang et al. (Eds.): ICONIP 2012, Part IV, LNCS 7666, pp. 27–35, 2012.

potentially identify individuals is altered for privacy protection. Various techniques are proposed to address the data privacy issue, from naïve approaches like data scrambling to advanced algorithmic approach such as data reconstruction [3] and feature set partitioning [4]. Other problems, on the other hand, stand for the natural challenges of data mining. Usually, the classification tasks involve imbalanced classification of identifying scarce prospective customers from the large population; the gathering of customer information for analysis is non-standardized process and the quality of the data collected cannot be guaranteed. A good CRM classification model should be capable of addressing an amalgamation of these issues while still having a reasonable degree of generality across the entire population.

However, there are factors imposed by industry nature that constitute major challenges for building high performance CRM classification models in the real-world application. Data quality is a salient issue for CRM classification practitioners in that various types of data anomaly largely complicate the data preparation and classification processes. Data imbalance is the most detrimental and yet pervasive data anomaly from which classification methods suffer most as the interesting customers for most CRM function are extremely rare in the entire population. A related concept reflecting this reality is the classical Purchase Funnel Model developed by St. Elmo Lewis in the late 1800's [5], where only small percentage of targeted customers pass through each phase of the Attention-Interest-Desire-Action (AIDA) model. The same phenomenon is observed in many concepts closely related to CRM such as the conversion rate internet marketing. Another complication arising from the cascading types of data anomaly is that no generally accepted data mining classification procedure can be established since it is hard to find one methodology that addresses all common data mining problems possessed by CRM data. With the ever increasing importance of CRM in every industry domain, CRM classification practitioners demand a standardized framework with streamlined data mining processes capable of delivering satisfactory result for general CRM data with varying quality and attribute; its workflow needs to be scalable so that data sets of different scale can be processed in the same fashion.

This paper proposes an enhanced customer relationship management classification framework with partial focus feature reduction to significantly improve feature relevancy as well as reduce the cardinality of the features to reduce computational cost. The resulting model yields very good performance result on a large complicated real-world CRM data set. The rest of the paper is organized as follows. Section 2 gives a brief literature review. Section 3 provides the methodology utilized in this paper. Section 4 discusses the sample, presents and discusses the result. Finally, the conclusions are presented.

2   Literature Review

The data set used in this research work is the KDD Cup 2009 challenge, targeting a CRM marketing problem. An official disclosure of detail of the KDD Cup 2009 contest can be found in [6]. The objective is to build classifiers to predict three target indicators: the propensity of customer defection (churn), buy new products or services (appetency) and buy additional services (up-selling). There are two versions of the data set corresponding to two different tasks – the original set and a scaled down scrambled set – available for modeling. Both versions have training and testing set containing 50,000 instances each. The small challenge is chosen for this research due to hardware limitation.

The IBM Research Lab is the definite winner of the contest by achieving the best overall score in both tasks. The solution from IBM Research Lab is detailed in [7]. This research is based on the FAST TRACK task which is the challenge for the result of original data set submitted within five days of the training label's release. A wide variety of data mining classifiers are selected to generate base classifiers, including Decision Trees, Logistic Regression, SVM, Naïve Bayes, K-nearest Neighbour and others. A pool of 500-1000 individual classifier model is constructed for each target indicator, and the ensemble is built by a greedy forward stepwise search which starts with the base individual classifier with the best performance and adds the base classifier that yields the most performance increase at each step. The performance is judged by the target metric AUC%, and is validated via a hill-climbing set not used to train the base classifiers. The data preprocessing is done via a quite standard approach; features with missing numeric values are imputed using the mean value, and missing nominal values are considered as a separate value. For the sake of classifiers incapable of nominal features, additional features are constructed to represent the nominal attributes. Data cleansing is also applied to normalize feature values by its range and to remove redundant features that contain constant values or are scrambled duplicates of other features. Moreover, an attempt is made to construct even more features based on observations of discrepancies of feature correlation between different measurements, which yields positive impact on both the performance of individual classifiers and the ensemble. For the SLOW TRACK, another attempt to induce additional features and additional cross-validation folds further boosts the performance by a significant amount.

ID Analytics – the Second Prize Winner of the FAST TRACK – presented their solution in [8]. Similarly, the solution of the classification task takes advantage of an ensemble of bagged boosting trees to achieve optimal performance. Compared to the work of IBM Research Lab, ID Analytics' solution dedicates more effort to the preprocessing rather than a meticulous ensemble selection. A histogram analysis has been applied into the data set to first establish the equality of distribution of the training and testing samples. The result also reveals the skewed distribution and hinted artificial encoding of some of the numeric features. Moreover, the binned values of the features show many constant-valued features with only one entry, which are then removed from the problem feature space. Discretization is performed on 10 most correlated numeric features with the target indicators. Grouping of scarcely populated value ranges of both discretized numeric features and nominal features ensure that the effect of outliers is smoothed out. Eventually, a wrapper feature selection is applied on the remaining features to filter out the most informative features based on Information Gain via the TreeNet classifier. It is worth noting that a down-sampling on the population of negative class has been applied to address the extreme data imbalance.

The SLOW TRACK task involves the results of the original data set past the FAST TRACK's deadline and those of the small data set. Substantially fewer contestants have devoted to the research of the small data set. Some teams such as ID Analytics did attempt the small data set in addition to the large one, and discovered strong evidence of data scrambling by the features of the large data set during analysis. Most of the results for small data set are generally worse than for large data set.

Daria Sorokina from the School of Computer Science at Carnegie Mellon University has a fully developed research on the KDD Cup 2009's small data set [9]. She has applied a classification method called the Additive Grove which is proposed by herself in her work. The Additive Groves is an ensemble of bagged additive models of regression tree [10], and in this case has been proven the best model on the Appetency target indicator of the small data set. In addition, a bagged decision tree ensemble is used to perform feature selection in the preprocessing. Unfortunately, Additive Groves' excellent performance does not carry over to the other two target indicators. The classifier performs much worse on the Churn target indicator and has scored an overall AUC% of 0.80171, a much lower result than those from the winning teams using the large data set discussed earlier. The reason for the inferior performance, according to the author herself, is partially due to insufficient preprocessing causing the trees to overfit on nominal features with too many values. In contrast, a group of data mining course students from RWTH Aachen University of Germany has gone extra length on preprocessing [11]. In particular, they have divided selected features into categories based on Missing Value Ratio – a measure of degree of how much the values of a certain feature is missing – and applied different imputation techniques to fill in missing values. Additional features based on correlation with the target indicators are also generated during the process. The combined feature set is then filtered by Information Gain Ratio feature selection. The team's best classification model is a Logistic Model Tree with AUC Split Criterion, which yields an overall 0.8081 AUC%. This number is better than what Sorokina has achieved, yet similarly low compared to the best results of the overall task.

3   Methodology

In the proposed classification framework, a data mining workflow has been developed to exploit the reduced dimensionality. Firstly, a new supervised binning method is introduced, called the Modality Grouping with Partial Focus. This special binning method targets the nominal features in the data set other than the numeric ones, and will be used to bin every value of a target nominal feature that has instances belonging to the minority class, and merges other values into a surrogate value. For a nominal feature A_i with k value categories, a value category j either retains its value or is converted into the surrogate value category if the value it represents is absent in the reduced dimension:

$$A'_{ij} = \begin{cases} A_{ij}, & A_{ij} \in A_i^+ \\ A_{is}, & A_{ij} \notin A_i^+ \text{ or } |A_{ij}| = 0 \end{cases} \qquad 1 < i < n,\ 1 < j < k$$

Here A'_{ij} is the equivalent nominal feature in the new data set after the modality grouping, and A is the surrogate value category for feature A_i. In this way, the modality grouping can efficiently smooth out most noise values and outliers relative to the target minority class while reducing the computation complexity of data processing. The same process is applied to the numeric features of the data set after feature discretization. To further ensure that the most informative features are acquired, all the resulting features are binary-encoded into a feature pool and are then put against an Information Gain ranking [12] to retain the collection of the most relevant features according to a cut off threshold. At the end of the workflow, the classification framework can employ one from the set of classification algorithms highly compatible with the preprocessed data. We coin the methodology Partial Focus Feature Reduction.

Moreover, missing values in the data set can be effectively addressed in this classification framework. Rather than using imputation methods, the missing values for both nominal and discretized numeric features are labeled to another surrogate value. The conversion not only considers additional information from the missing values, but eliminates possible bias caused by value imputation. All these data cleansing/transformation measures for dealing with data anomalies in Partial Focus Feature Reduction will be proven later to work very well with the classification data set used in this research.

The methodology implementation is primarily based on the popular WEKA data mining package [13], aided with custom developed WEKA modules and external Java applications. The WEKA Data Mining Package is open-source software developed by the researchers of University of Waikato. WEKA offers a wide collection of data mining algorithms for classification and clustering as well as a variety of data mining utilities for data preprocessing and feature selection, etc. The source code for all WEKA components is available, and most of the time the reference to the scholarly papers on which the specific data mining technique bases on is included for most of its components. The development is all Java based, and coding is done within the Eclipse IDE.

4   Results and Discussion

The classification performance for this data set is measured by the Area under ROC Curve (AUC%). The small version of the KDD Cup 2009 challenge is used in this paper, which embraces the most typical problems to contemporary CRM classification projects including, but not limited to, the following:

- Large number (50,000) of data set instances
- Large number of missing values (about 60%)
- Large number of features (230 for the chosen SMALL data set)
- Nominal features with high cardinality (large amount of feature values)
- Imbalanced class (less than 10% of the data set instances belong to positive class)
- Very noisy data

More importantly is that this data set exhibits apparent reduced dimensionality for the minority class instances. Compared to the full feature space, the cardinality for most sub feature spaces for minority class instances is only 20% the magnitude compared to the overall. Nonetheless, this trait is not regarded valuable by other researchers and has not been treated in all published papers involving this data set.

After the initial data scan, the features that are either empty or too sparse are removed. Over 77% of the numeric features are sparse features concentrated in the top 20% range and 20% in the bottom 20% range, with only the rest 3% in the middle. For nominal features, 30% of features are in the upper 60% range with 15% in the top 20% range. Based on the observation, the removal threshold is set at 80%, that is, any features having over 80% missing rate will be removed.

The numeric features are discretized to allow for further preprocessing. The WEKA filter (supervised) Discretize is applied to perform the task, which uses Fayyad and Irani's Minimum Description Length (MDL) based Decision Tree method to partition the continuous range [14].

A data filter will extract a collection of all instances of the minority class and create a new data set definition based on it. A data transformer will then convert the original data set, preserving all the values present in the new definition, and merge all the redundant ones into a surrogate value "OTHER". The numeric features are discretized in the previous step to enable processing in this modality grouping task.

As the last preprocessing task, all the original features are encoded into binary features which will later be ranked based on entropy score. After the feature ranking process, the final version of preprocessed data sets with the most relevant features for each target features are obtained.

In this research, a Jackknife-based (also called "leave X% out" approach) renders a 50/50 split on the original training set – referred to as the base set from this point forward – as the new training and testing set to train and evaluate our classifiers. The AUC result of the selected algorithms' 10-fold cross validation is shown in Figure 1 in which the results are clustered by scores of each task and the overall average.

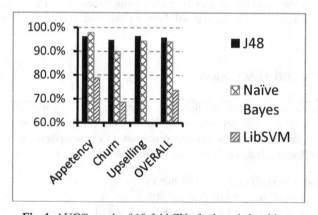

Fig. 1. AUC% result of 10-fold CV of selected algorithms

Judged by AUC and also other metrics, J48 has become the best performing single classifier which will be used for further classification scenarios. Also exciting is that the preliminary result of more than 95% average AUC, which is already significant, compared to those from other researchers. Since there is no access to the testing data at this moment, we resort to the official ranking of the contest for inference of the generated model's competitiveness. On the leaderboard for the KDD Cup 2009 SLOW TRACK [15], it is able to identify several winners of the tasks by their scores and team names. When examining the comparative performance of results from the participants on both training and testing data set, it is clear that most of the participants' models already have lower performance on the training data than obtained by the J48 in this research, except ones from the grand winner of IBM Research Lab which clearly overfits the training data by having all 100% AUC across all three tasks. Assuming the models perform averagely well on the testing data, it will then become the best score for the SLOW TRACK task!

Provided the classification results from above, J48 will be used to augment the bagging ensemble classifier to produce an ensemble classification result. The rationale for choosing bagging algorithm for ensemble classification in this research is that it provides the most cost-efficient yet robust ensemble solution for this particular data set and hardware environment. WEKA has implemented a variety of other ensemble classifiers by default, such as the AdaBoostM1 and Stacking for the equally famous boosting and stacking algorithms. However, they cost way more processing time than bagging to build models on data set of the same size, and occasionally crash the WEKA's memory heap with over-stretching parameter settings. Moreover, the performance of these two ensemble classifiers has shown notable fluctuation as the classifier parameters change during classification. The stacking classifier in particular, suffers varying classification predicaments like data imbalance bias while having different configurations for the base/meta classifiers. It is believed that the performance of these ensemble classifiers cannot match that of bagging without significant effort in parameter tuning. On the other hand, the bagging algorithm can be fairly easily without tampering too much the classifier parameters while still capable of producing models comparable or even superior to other ensemble classifiers.

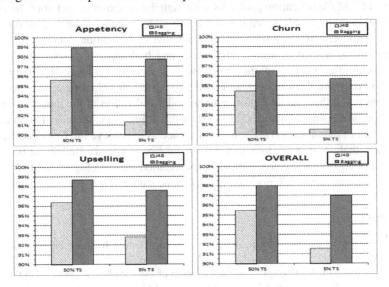

Fig. 2. Comparison of AUC% of J48 and Bagging Ensemble

In this research, the bagging implementation of WEKA is used for ensemble classification, with the bootstrap sample size set to 100% of original data set and number of bootstrap iteration set to 10. A similar approach as in the last section is adopted which repeats the process with different training set size, again generated with three random seeds of 1, 10 and 100. Just curious of the best/worst case of degree of improvement from ensemble classification, only the sizing options of 5% and 50% will be incorporated, whose results are compared side by side with that of the original J48 classifier. The mean AUC% measures of the three versions for the three target features and the overall average are shown in Figure 2. Once more an improvement over the previous result is observed. The performance of bagging ensemble exceeds that of the J48 classifiers by far, and the difference is especially obvious with the 5% training set option. Overall, the AUC% of bagging ensemble with the 5% training set option is 0.970, which is 0.55 higher than J48's 0.915; this is actually expected behavior since the bootstrap process can largely compensate the inadequate size of the small training set. Another reason for the performance improvement is that the J48 decision tree is an unstable classifier that can generate sufficiently different decision boundaries upon even small changes in training parameters, a characteristic desired for base models of ensemble classifications [16]. This is equivalent to saying, however, that performance improvement observed with this ensemble J48 may not carry over to ensembles with other base classifiers which do not possess such characteristics. Nevertheless, it is certain that ensemble classification is a viable way to maximize the benefit of this classification framework.

5   Conclusion

Partial Focus Feature Reduction, a classification framework that effectively resolves real-world CRM classification problems with high data imbalance and poor data quality is developed. In the classification phase, the effectiveness of the proposed methodology is validated by learning a real-world CRM classification data set perceiving the array of common data mining challenges and building both a J48 and bagging ensemble classifier that beat all the competitors' models by a considerable margin.

There is yet more work to be done in order to perfect this classification framework. It has been noted that one limitation may be under-utilization of the numeric features with the preprocessing techniques rendered by this framework; there is also no definitive standard of classification algorithms of choice as to achieve the best classification result. Nonetheless, it is obvious that the Partial Focus Feature Reduction possesses the specialty absent in any conventional classification methodology to address imbalanced classification and inferior data quality in the studied data set.

Reference

1. Kincaid, J.W.: Customer Relationship Management: Getting It Right. Prentice Hall PTR (2003)
2. Ngai, E.W.T., Li, X., Chau, D.C.K.: Application of Data Mining Techniques in Customer Relationship Management: A Literature Review and Classification. Expert Syst. Appl. 36, 2592–2602 (2009)

3. Zhu, D., Li, X., Wu, S.: Identity Disclosure Protection: A Data Reconstruction Approach for Privacy-preserving Data Mining. Dec. Supp. Syst. 48, 133–140 (2009)
4. Matatov, N., Rokach, L., Maimon, O.: Privacy-Preserving Data Mining: A Feature Set Partitioning Approach. Inform. Sci. 180, 2696–2720 (2010)
5. Barry, T.: The Development of the Hierarchy of Effects: An Historical Perspective. Current Issues Research Advert., 251–295 (1987)
6. Guyon, I., Lemaire, V., Boullé, M., Dror, G., Vogel, D.: Analysis of the KDD Cup 2009: Fast Scoring on a Large Orange Customer Database. In: JMLR W&CP, vol. 7, pp. 1–22 (2009)
7. IBM Research: Winning the KDD Cup Orange Challenge with Ensemble Selection. In: W&CP, vol. 7 (2009)
8. Xie, J., Rojkova, V., Pal, S., Coggeshall, S.: A Combination of Boosting and Bagging for KDD Cup 2009 - Fast Scoring on a Large Database. In: JMLR W&CP, vol. 7 (2009)
9. Sorokina, D.: Application of Additive Groves Ensemble with Multiple Counts Feature Evaluation to KDD Cup 2009 Small Data Set. In: JMLR W&CP, vol. 7, pp. 101–109 (2009)
10. Sorokina, D., Caruana, R., Riedewald, M.: Additive Groves of Regression Trees. In: Proceedings of the 18th European Conference on Machine Learning (2007)
11. Doetsch, P., et al.: Logistic Model Trees with AUC Split Criterion for the KDD Cup 2009 Small Challenge. In: JMLR W&CP, vol. 7, pp. 77–88 (2009)
12. Quinlan, J.R.: Induction of Decision Trees. Mach. Learn., 81–106 (1986)
13. Hall, M., et al.: The WEKA Data Mining Software: An Update. SIGKDD Exploration 11 (2009)
14. Fayyad, U.M., Irani, K.B.: Multi-interval Discretization of Continuousvalued Attributes for Classification Learning. In: Proceedings of the International Joint Conference on Uncertainty in AI, pp. 1022–1027 (1993)
15. The KDD Cup 2009 Results (Online) (2009),
    http://www.kddcuporange.com/results.php?ds=small
16. Polikar, R.: Ensemble Based Systems in Decision Making. IEEE Circuits Syst. Magazine 6, 21–45 (2006)

An Automated System for the Grading of Diabetic Maculopathy in Fundus Images

Muhammad Usman Akram*, Mahmood Akhtar, and M. Younus Javed

Department of Computer Engineering
College of Electrical & Mechanical Engineering
National University of Sciences & Technology, Pakistan
usmakram@gmail.com, mahmood@unswalumni.com,
myjaved@yahoo.com

Abstract. Computer aided diagnosis systems are very popular now days as they assist doctors in early detection of the disease. Diabetic maculopathy is one such disease which affects the retina of the diabetic patients. It affects the central vision of the person and causes blindness in severe cases. In this paper, an automated system for the grading of diabetic maculopathy has been developed, that will assist the ophthalmologists in early detection of the disease. Here, we propose a novel computerized method for the grading of diabetic maculopathy in fundus images. Our proposed system comprises of preprocessing of retinal image followed by macula and exudate regions detection. This is followed by feature extractor module for the formulation of feature set. SVM classifier is then used to grade the diabetic maculopathy. The publicly available fundus image database MESSIDOR has been used for the validation of our algorithm. The results of our proposed system have been compared with other methods in the literature in terms of sensitivity and specificity. Our system gives higher values of sensitivity and specificity as compared to others on the same database.

Keywords: Computer aided diagnosis systems, Diabetic maculopathy, Fundus images, SVM classifier.

1 Introduction

Over the years, medical imaging has become a significant part in early detection of various diseases. It is the fastest growing area within medicine and research at present and plays a central role in developing cost effective health care systems. Diabetes is one of the chronic disease all over the world. Diabetic retinopathy (DR) is a condition where diabetes starts effecting the human retina. There are several stages of diabetic retinopathy namely non proliferative DR, proliferative DR and diabetic maculopathy (commonly known as macular edema) [1].

Macula is the central portion of the retina which is usually the darkest portion and is rich in cones [1]. Macula is accountable for the clear, sharp and detailed

---

* Corresponding author.

T. Huang et al. (Eds.): ICONIP 2012, Part IV, LNCS 7666, pp. 36–43, 2012.

vision. When the damaged blood vessels in the retina leaks out and the fluid gets deposited near macula, then it leads to distorted central vision. Exudates are the yellow color deposits of protein present in the retina, and maculopathy occurs when exudates affect the central vision. The ophthalmologists grade maculopathy into two stages i.e. Non Clinically Significant Macular Edema (Non-CSME) and Clinically Significant Macular Edema (CSME) [2]. Non-CSME is a mild form of maculopathy where there are no symptoms of the disease. Because in Non-CSME, the location of exudates are at a distance from fovea, so the central vision is not affected. CSME is the severe form of maculopathy, in which the exudates leak out and get deposited very close to or on fovea, affecting central vision of the eye [2]. Figure 1 shows the retinal images having different types of macular edema.

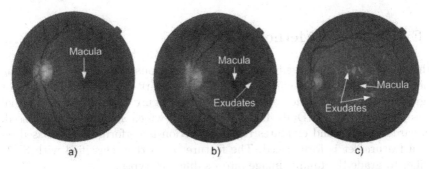

Fig. 1. Stages of diabetic maculopathy: a) Healthy retinal image, b) Non-CSME retinal image, c) CSME retinal image

Irrespective of diabetic retinopathy, long term diabetic patients have chances of developing diabetic maculopathy. Automated detection of diabetic maculopathy is vital for the early cure of the disease. There are various computerized methods in the literature which are useful for the detection of diabetic maculopathy. In [3], diabetic maculopathy is graded by location of exudates in marked region of macula in fundus image. Exudates are detected using clustering and mathematical morphological techniques. The method is tested on local dataset and the sensitivity and specificity are found to be 95.6% and 96.15% respectively. [4] proposed an intelligent diagnostic system for diabetic maculopathy using fundus images. The feed forward artificial neural network is used for classification. They have stated a sensitivity value of 95% and specificity value of 100%. In [5], marker controlled watershed transformation is used for exudates feature extraction and diabetic macular edema classification. The exudates from the fundus image are extracted, and their location along with marked macular regions is utilized for the classification of macular edema into different stages. The method is tested on MESSIDOR database and the sensitivity is found to be 80.9% and specificity is 90.2%. Deepak et. al [6] proposed a method for automatic assessment of macular edema using supervised learning approach to capture the global characteristics in fundus images. Disease severity is assessed using a rotational asymmetry metric (motion pattern,) by examining the symmetry of macular region. The method is tested on publicly available databases like diaretdb0, diaretdb1, MESSIDOR and DMED. The accuracy for the maculopathy detection is found to be 81%. [7] presented a method for classification of exudative maculopathy. This technique uses FCM clustering and artificial neural networks. The authors have reported sensitivity of 92% and specificity of 82% on some local dataset.

This paper is organized in four sections. Section 2 consists of systematic overview of our proposed methodology for the grading of diabetic maculopathy. This section also explains the detailed proposed system and its various modules. Experimental results and analysis are given in section 3, followed by conclusion in section 4.

2   Experimental Methodology

In this section, our proposed method and its various stages for the grading of maculopathy are explained in detail. The flowchart of proposed method for the grading of diabetic maculopathy is shown in figure 2. In our work, retinal images present in MESSIDOR database are used, which are then preprocessed. Afterwards, macula and exudates region detection is performed, and based on them a feature set is formulated. The feature set is then classified with SVM classifier to grade the fundus image into its different types.

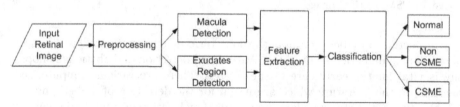

Fig. 2. Flow diagram of our proposed method

2.1   MESSIDOR Database

The retinal images present in MESSIDOR database has been used in our study. This publicly available database has been established to facilitate computer aided DR lesions detection. The database is collected using TopCon TRC NW6 Non-Mydriatic fundus camera with 45° FOV and resolutions of 1440×960, 2240×1488 or 2304×1536 with 8 bits per color plane. It contains total 1200 images which are divided into three sets of 400 images and each set is further divided into 4 parts to facilitate thorough testing. Each set contains an Excel file with medical findings which can be use for testing purposes.

2.2   Preprocessing

The acquired retinal image contains extra background pixels which are not required for further processing and add more time in overall processing. The purpose of preprocessing is to differentiate between background and foreground and eliminate background and noisy pixels. A mean and variance based method for background estimation, and ratio of Hue and Intensity channel for noise detection are used in preprocessing. The details of these methods are given in [8].

2.3   Exudates Detection

Exudates are the bright lesions which appear on the surface of retina if the leaking blood contains fats and proteins along with water. Their occurrence is a main threat to vision especially when they occur near or on macula. The presence of Optic Disc (OD) makes it difficult for automated system to detect exudates with high accuracy. The proposed system detects and removes OD pixels for accurate detection of exudates. Followings steps are used for exudate detection [10]:

- Take preprocessed image as an input and apply morphological closing to remove the effect of blood vessels and dark lesions
- Apply adaptive contrast enhancement technique to improve the contrast of exudates on retinal surface
- Create filter bank given in equation 1 based on Gabor kernel and convolve it with contrast enhanced image to further enhance the bright lesions [15]

$$G_{FB} = \frac{1}{\sqrt{\pi r \sigma}} e^{-\frac{1}{2}[(\frac{d_1}{\sigma})^2 + (\frac{d_2}{\sigma})^2]}(d_1(cos\Omega + \iota sin\Omega)) \tag{1}$$

where σ, Ω and r are the standard deviations of Gaussian, spatial frequency and aspect ratio respectively θ is the orientation of filter and $d_1 = xcos\theta + ysin\theta$ and $d_2 = -xsin\theta + ycos\theta$ [15].

- Create binary map containing candidate exudate regions by applying adaptive threshold value T which is calculated using OTSU algorithm [9]
- Detect OD using averaging and Hough transform given in [11] and remove all OD pixels from binary map.

2.4   Macula Detection

Macula detection is an important module for developing the computerized system for the grading of diabetic maculopathy. It is the macular area of the eye that is affected in diabetic maculopathy upsetting the central vision of the eye and in severe cases leading to blindness. The technique which we have used for macula detection is described in [13]. In this technique macula is first localized with the help of localized OD and enhanced blood vessels [12]. Finally macula is detected by taking the distance from the center of optic disk along with enhanced blood vessels image to locate the darkest pixel in this region, and making clusters of these pixels. The largest cluster formed is macula [13]. Figure 3 shows the outputs of different modules, i.e. preprocessing, exudate and macula detection.

Fig. 3. a) Original retinal image; b) Preprocessing mask; c) OD detection; d) Filter bank response; e)Binary map for exudates; f)Macula detection

2.5   Feature Extraction and Classification

The binary map generated in exudate detection phase may contain spurious and non exudate regions. A feature set for each object in binary map consisting of area, mean intensity value, energy and mean value of filter bank response is created to find true exudates. Exudates are used to grade the risk of macular edema. Table 1 shows the three diabetic maculopathy grading conditions which have been used while designing MESSIDOR database. We have used Support Vector Machine (SVM) classifier to grade the input image. The complete feature vector containing features for exudates and location of macula is passed to SVM, where it grades the test image into three categories as defined in table-1.

Table 1. Conditions for grading of diabetic maculopathy [16]

Grade	Condition	Class
0	No exudate present	Normal
1	A few exudates present and distance between macula and exudates > one papilla diameter	Non CSME
2	Exudates present and distance between macula and exudates ≤ one papilla diameter	CSME

Support Vector Machine (SVM) separates the exudates and non exudates regions from each other with maximum margin by using a separating hyperplane. Let the separating hyperplane be defined by $x \cdot w + b = 0$, where w is its normal. For linearly separable data labeled x_i, y_i, $x_i \in R^{N_d}$, $y_i = \{-1, 1\}$, $i = 1, ..., N$, the optimum boundary chosen with maximum margin criterion is found by minimizing the objective function using equation 2.

$$E = \| w \|^2, \tag{2}$$

subject to $(x_i \cdot w + b)y_i \geq 1$, for all i. We apply linear SVM for classification of exudates, hence the inequality in equation 2 doesn't hold in that case. The new objective function is defined in equation 3.

$$E = \frac{1}{2} \| w \|^2 + C \sum_i L(\xi_i), \tag{3}$$

subject to $(x_i \cdot w + b)y_i \geq 1 - \xi_i$, for all i.

$C \sum_i L(\xi_i)$ is the empirical risk associated with misclassified cases where L is a cost function and C is the parameter that minimizes the risk against maximizing the SVM margin.

The linear cost function is robust to outliers hence equation 1 is generalized by taking $L(\xi_i) = \xi_i$ in equation 4.

$$\alpha^* = \max_\alpha(\sum_i \alpha_i + \sum_{i,j} \alpha_i \alpha_j y_i y_j x_i . x_j), \tag{4}$$

subject to $0 \leq \alpha_i \leq C$ and $\sum_i \alpha_i y_i = 0$ in which $\alpha = \{\alpha_1, ..., \alpha_i\}$ is the set of Lagrange multipliers. The optimum decision boundary ω_0 which is a linear combination of all vectors is given in equation 5.

$$\omega_0 = \Sigma_i(\alpha_i y_i x_i) \tag{5}$$

This decision boundary is then used to classify candidate object into exudate and non exudate region. The final output of SVM depends on macular coordinates and their distance from exudates if present.

3   Experimental Results

The proposed system is tested and evaluated properly to check the validity of proposed method. The SVM grades the images into different categories depending on the number and position of lesions. Figure 4 shows different images classified as normal, Non-CSME and CSME by the classifier.

Fig. 4. Maculopathy detection. 1st row: Normal images; 2nd row: Non-CSME or grade 1 images; 3rd row: CSME or grade 2 images.

We have used sensitivity, specificity and accuracy as the figures of merit for performance evaluation. Sensitivity and specificity are the true positive and true negative rates respectively and accuracy is the ratio of truly classified images to the total number of images. Table-2 shows the comparison of our proposed system with existing methods in the literature in terms of sensitivity, specificity and accuracy. The results show that our proposed method achieved high values of sensitivity, specificity and accuracy as compared to other methods using MESSIDOR database. Furthermore, the results from other methods are comparable with our method as we are using a large dataset of images than [3], [4] and [7].

Table 2. Comparison of our proposed method with existing techniques

Author	Technique	Database	Sensitivity	Specificity	Accuracy%
Siddalingaswamy et. al. [3]	Clustering and morphology	Local dataset	95.6%	96.15%	-
Nayak et. al. [4]	Feed forward ANN	Local dataset	95%	100%	-
Lim et. al. [5]	Marker controlled watershed transform	MESSIDOR	80.9%	90.2%	-
Deepak et. al. [6]	Rotational Asymmetric Motion Pattern	MESSIDOR	95%	90%	-
		DMED	100%	74%	-
Osareh et. al [7]	FCM Clustering and ANN	Local dataset	92%	82%	-
Aquino et. al. [14]	Image Processing	MESSIDOR	-	-	96.51%
Proposed Method	Filter bank and SVM	MESSIDOR	92.6%	97.8%	97.3%

4   Conclusion

In this paper, we have proposed a method for developing computerized system for the grading of diabetic maculopathy. Our proposed system consists of preprocessing, exudates region detection followed by macula detection. The exudate detection stage creates a binary map of candidate regions. The SVM based classifier first detected true exudate regions based on feature set then using coordinates of macula and the distance of exudates from macula, the classifier grades the input image into three categories. The success of computerized diagnostic system mainly depends upon three factors such as sensitivity, specificity and accuracy of the system. The results showed that our system has higher values of sensitivity and specificity on MESSIDOR database as compared to the values in literature, hence making our system significant for the screening purposes.

References

1. Causes and Risk Factors. Diabetic Retinopathy. United States National Library of Medicine (2009)
2. Iwasaki, M., Inomara, H.: Relation Between Superficial Capillaries and Fovea Structures in the Human Retina. J. Invest. Ophthalm. & Visual 27, 1698–1705 (1986)

3. Siddalingaswamy, P.C., Prabhu, K.G.: Automatic Grading of Diabetic Maculopathy Severity Levels. In: Proceedings of 2010 International Conference on Systems in Medicine and Biology, pp. 331–334 (2010)
4. Nayak, J., Bhat, P.S., Acharya, U.R.: Automatic Identification of Diabetic Maculopathy Stages Using Fundus Images. J. Med. Engin. & Techn. 33(12), 119–129 (2009)
5. Lim, S.T., Zaki, W.M.D.W., Hussain, A., Lim, S.L., Kusalavan, S.: Automatic Classification of Diabetic Macular Edema in Digital Fundus Images. In: IEEE Colloquium on Humanities, Science and Engineering (CHUSER), pp. 265–269 (2011)
6. Deepak, K.S., Sivaswamy, J.: Automatic Assessment of Macular Edema From Color Retinal Images. IEEE Trans. Med. Imag. 31(3), 766–776 (2012)
7. Osareh, A., Mirmehdi, M., Thomas, B., Markham, R.: Automatic Recognition of Exudative Maculopathy using Fuzzy C-means Clustering and Neural Networks. In: Proceedings of Medical Image Understanding and Analysis Conference, pp. 49–52 (2001)
8. Tariq, A., Akram, M.U.: An Automated System for Colored Retinal Image Background and Noise Segmentation. In: IEEE Symposium on Industrial Electronics and Applications, pp. 405–409 (2010)
9. Gonzalez, R.C., Woods, R.E.: Digital Image Processing, 2nd edn. Prentice Hall (2002)
10. Akram, M.U., Tariq, A., Anjum, M.A., Javed, M.Y.: Automated Detection of Exudates in Colored Retinal Images for Diagnosis of Diabetic Retinopathy. OSA J. Appl. Opt. 51(20), 4858–4866 (2012)
11. Usman Akram, M., Khan, A., Iqbal, K., Butt, W.H.: Retinal Images: Optic Disk Localization and Detection. In: Campilho, A., Kamel, M. (eds.) ICIAR 2010, Part II. LNCS, vol. 6112, pp. 40–49. Springer, Heidelberg (2010)
12. Akram, M.U., Khan, S.A.: Multilayered Thresholding-based Blood Vessel Segmentation for Screening of Diabetic Retinopathy. Engin. Comput. (2012), doi:10.1007/s00366-011-0253-7
13. Mubbashar, M., Usman, A., Akram, M.U.: Automated System for Macula Detection in Digital Retinal Images. In: International Conference on Information and Communication Technologies, pp. 1–5 (2011)
14. Aquino, A., Gegundez, M.E., Marin, D.: Automated Optic Disc Detection in Retinal Images of Patients with Diabetic Retinopathy and Risk of Macular Edema. Int. J. Biolog. Life Sci. 8(2), 87–92 (2012)
15. Sung, J., Bang, S.Y., Choi, S.: A Bayesian Network Classifier and Hierarchical Gabor Features for Handwritten Numeral Recognition. Pattern Recogn. Lett. (2005)
16. MESSIDOR Database: http://messidor.crihan.fr/index-en.php

An Iterative Method for a Class
of Generalized Global Dynamical System
Involving Fuzzy Mappings in Hilbert Spaces

Yun-zhi Zou[1], Xin-kun Wu[2], Wen-bin Zhang[3,*], and Chang-yin Sun[1]

[1] School of Automation, Southeast University, Nanjing, 210096, P.R. China
[2] College of Mathematics, Sichuan University, Chengdu, 610064, P.R. China
[3] College of Statistics and Mathematics,
Yunnan University of Finance and Economics, Kunming, 650221, P.R. China
wbzhangyn@126.com

Abstract. This paper presents a class of generalized global dynamical system involving (H, η) set-valued monotone mappings and a set-valued function induced by a closed fuzzy mapping in Hilbert spaces. By using the resolvent operator technique and Nadler fixed-point theorem, we prove the equilibrium point set is not empty and closed. Furthermore, we develop a new iterative scheme which generates a Cauchy sequence strongly converging to an equilibrium point.

Keywords: Generalized dynamical system, Variational inequality, equilibrium, Fuzzy mapping, Iterative method, Resolvent operator.

1 Introduction

Projective dynamical systems have been studied extensively due to their enormous applications arised in the fields of economics, optimization, control theory, mechanics, physical equilibrium analysis, linear and nonlinear programming and so on. For details, we refer to [1–4] and the references therein.

In 1993, Dupuis and Nagurney [1] introduced and studied the following class of dynamics given by solutions to a differential equation with a discontinuous right-hand side, namely local projected dynamical systems as follows

$$\frac{dx}{dt} = \lim_{\rho \to 0} \frac{P_K(x - \rho N(x)) - x}{\rho}. \tag{1}$$

They also proved that the critical points of the equation are the same as the solutions to a variational inequality problem: to find an $x \in R^n$, such that

$$(N(x), y - x) \geq 0 \text{ for all } y \in R^n.$$

Recently Friesz, Xia and Vincent [3] analyzed the global asymptotic stability behavior of a class of global dynamical system

$$\frac{dx}{dt} = P_K(x - \rho N(x)) - x. \tag{2}$$

---

* Corresponding author.

T. Huang et al. (Eds.): ICONIP 2012, Part IV, LNCS 7666, pp. 44–51, 2012.
© Springer-Verlag Berlin Heidelberg 2012

In 2007, by using the resolvent operator technique, Zou, Huang and Lee [5] considered a class of generalized global dynamical system in Hilbert Space $\mathcal{H}$: to find absolutely continuous functions $x\,(\cdot)$ from $[0, J] \to \mathcal{H}$ such that

$$\begin{cases} \frac{dx}{dt} \in R_{M,\rho}^{H,\eta}\,(g\,(x) - \rho N\,(x)) - g\,(x) & \text{for } a.a.t \in [0, J] \\ x\,(0) = b, \end{cases} \tag{3}$$

where M, and N are set-valued mapping $\mathcal{H} \to 2^{\mathcal{H}}$, and $\eta : \mathcal{H} \to \mathcal{H}$, is a singled valued mapping. They proved that the set of the equilibrium points of [5] is nonempty and closed. However, they did not give a way to find any equilibrium points and this is exactly one of the two major motivations of this paper.

On the other hand, in 1965, Zadeh [6] introduced the concept of fuzzy sets, which became a cornerstone of modern fuzzy mathematics. In particular, fuzzy mappings have been of many authors' interest. To explore properties of the generalized globle dynamic system involving fuzzy mapping is another motivation of this paper.

In this paper, we consider a class of generalized projective dynamical system in Hilbert spaces involving (H, η)-mappings and a set-valued mapping induced by a fuzzy mapping. Also, we use similar techniques as in [7] to obtain similar results as in [5]. Moreover a new iterative method for finding an equilibrium point is proposed and the strong convergence of the iterative sequence is proved.

2   Preliminaries

Let $\mathcal{H}$ be a real Hilbert space endowed with a norm $||\cdot||$ and inner product $(\cdot, \cdot)$. Let $F\,(\mathcal{H})$ be a collection of all fuzzy sets over H. A mapping $F : \mathcal{H} \to \mathcal{F}\,(\mathcal{H})$ is said to be a fuzzy mapping, if for each $x \in H$, $F\,(x)$ is (denote it by F_x in the sequel) is a fuzzy set on H and $F_x\,(y)$ is the membership function of y in F_x.

A fuzzy mapping $F : \mathcal{H} \to F\,(x)$ is said to be closed if for each $x \in \mathcal{H}$, the function $y \to F_x\,(y)$ is upper semicontinuous, i.e., for any given net $\{y_\alpha\} \subset \mathcal{H}$ satisfying $y_\alpha \to y_0 \in \mathcal{H}$, $\limsup\limits_{\alpha} F_x\,(y_\alpha) \leq F_x\,(y_0)$. For $B \in \mathcal{F}\,(\mathcal{H})$ and $\lambda \in [0, 1]$, the set $(B)_\lambda = \{x \in B | B(x) \geq \lambda\}$ is called a λ-cut set of B. Suppose that $\alpha : \mathcal{H} \to [0, 1]$ is a real valued function. It is known that $(F_x)_{\alpha(x)}$ is a closed subset of $\mathcal{H}$ if F is a closed fuzzy mapping over $\mathcal{H}$. Let $C\,(\mathcal{H})$ denote all the closed subsets of $\mathcal{H}$. Let $E : \mathcal{H} \to F\,(\mathcal{H})$ be a closed fuzzy mappings and $\alpha : \mathcal{H} \to [0, 1]$ be a real valued function, then for each $x \in H$, we have $(E_x)_{\alpha(x)}$. Therefore we can define a set-valued mappings, $\tilde{E} : \mathcal{H} \to C(\mathcal{H})$ by $\tilde{E}\,(x) = (E_x)_{\alpha(x)}$. In this paper, we say that the multi-valued mapping $\tilde{E}$ is induced by the fuzzy mapping E.

Let $\eta : \mathcal{H} \times \mathcal{H} \to \mathcal{H}$ and $g, H : \mathcal{H} \to \mathcal{H}$ be single-valued mappings and let $E : \mathcal{H} \to F\,(\mathcal{H})$ be fuzzy mappings, Let $\alpha : \mathcal{H} \to [0, 1]$ be a given function. We will consider the following generalized dynamical system, find absolutely continuous functions $x\,(\cdot) : [0, J] \to R \cup \{+\infty\}$ such that

$$\frac{dx}{dt} \in R_{A,\rho}^{H,\eta}(g\,(x) - \rho \tilde{E}\,(x)) - g\,(x) \quad \text{for } a.a.t \in [0, J], \tag{*}$$

where $\tilde{E}$ is the set valued mapping $\mathcal{H} \to C(\mathcal{H})$ induced by the fuzzy mapping E.

We start with some basic definitions.

Definition 1. A point x^* is said to be an equilibrium point of global dynamical system $(*)$, if x^* satisfies the following inclusion

$$0 \in R_{A,\rho}^{H,\eta}(g(x^*) - \rho\tilde{E}(x^*)) - g(x^*).$$

Definition 2. A mapping $H : \mathcal{H} \to \mathcal{H}$ is said to be

(i) α-strongly monotone with respect to first argument if there exists some $\alpha > 0$ such that

$$(H(x) - H(y), x - y) \geq \alpha\|x - y\|^2, \quad \forall x, y \in K;$$

(ii) ξ-Lipschitz continuous if there exists a constant $\xi \geq 0$ such that

$$\|H(x) - H(y)\| \leq \xi\|x - y\|, \quad \forall x, y \in K.$$

Definition 3. A set-valued mapping $T : \mathcal{H} \to 2^{\mathcal{H}}$ is said to be ξ-Lipschitz continuous if there exists a constant $\xi > 0$ such that

$$M(T(x), T(y)) \leq \xi\|x - y\|, \quad \forall x, y \in K,$$

where $M(\cdot, \cdot)$ is the Hausdorff metric on $C(\mathcal{H})$ defined by $M : 2^H \times 2^H \to R \cup \{+\infty\}$

$$M(\Gamma, \Lambda) := \max \left\{ \sup_{u \in \Gamma} dist(u|\Lambda), \sup_{v \in \Lambda} dist(v|\Gamma) \right\}.$$

Definition 4. Let $\eta : \mathcal{H} \times \mathcal{H} \to \mathcal{H}$ and $H : \mathcal{H} \to \mathcal{H}$ be two single valued mappings and $A : \mathcal{H} \to 2^{\mathcal{H}}$ be a set-valued mapping. A is said to be

(i) monotone if $(x - y, u - v) \geq 0$ for all $u, v \in \mathcal{H}$, $x \in Au$, and $y \in Av$;
(ii) η-monotone if $(x - y, \eta(u, v)) \geq 0$ for all $u, v \in \mathcal{H}$, $x \in Au$, and $y \in Av$;
(iii) (H, η)-monotone if A is η-monotone and $(H + \lambda A)(\mathcal{H}) = \mathcal{H}$ for all $\lambda > 0$.

Example 1. If f is a proper convex function from $\mathcal{R}^2$ to $R \cup \{+\infty\}$ defined by

$$f : \mathbf{x} = \begin{pmatrix} x \\ y \end{pmatrix} \to \sqrt{x^2 + y^2}$$

and then the subdifferential of such a function f at any point $\mathbf{x}$ in $\mathcal{R}^2$

$$\partial f(\mathbf{x}) = \{\mathbf{v} \in \mathcal{R}^2 | f(\mathbf{y}) \geq f(\mathbf{x}) + \mathbf{v} \cdot (\mathbf{y} - \mathbf{x}), \forall \mathbf{y} \in \mathcal{R}^2\}$$

is a maximal monotone set-valued mapping. Moreover, if H is the identity mapping and $\eta(x, y) = x - y$, then ∂f is also an (H, η)-monotone set-valued mapping.

Lemma 1. ([8]) Let $\eta : \mathcal{H} \times \mathcal{H} \to \mathcal{H}$ be a single-valued operator, $H : \mathcal{H} \to \mathcal{H}$ be a strictly η-monotone operator and $A : \mathcal{H} \to 2^{\mathcal{H}}$ be an (H, η)-monotone operator. Then, the operator $(H + \lambda A)^{-1}$ is single-valued.

By this lemma, we can define the resolvent operator $R_{A,\lambda}^{H,\eta}$ as follows.

Definition 5. ([8]) Let $\eta : \mathcal{H} \times \mathcal{H} \to \mathcal{H}$ be a single-valued operator, $H : \mathcal{H} \to \mathcal{H}$ be strictly η-monotone operator and $A : \mathcal{H} \to 2^{\mathcal{H}}$ be an (H, η)-monotone operator. The resolvent operator $R_{A,\lambda}^{H,\eta}$ is defined by

$$R_{A,\lambda}^{H,\eta}(u) = (H + \lambda A)^{-1}(u), \quad \forall u \in \mathcal{H}.$$

Lemma 2. ([8]) Let $\eta : \mathcal{H} \times \mathcal{H} \to \mathcal{H}$ be a single-valued Lipschitz continuous operator with constant τ. Let $H : \mathcal{H} \to \mathcal{H}$ be stongly η-monotone operator with constant r and $A : \mathcal{H} \to 2^{\mathcal{H}}$ be an (H, η)-monotone operator. The resolvent operator $R_{A,\lambda}^{H,\eta}$ is Lipschitz continuous with constant $\frac{\tau}{r}$.

3 Existence of the Equilibrium Points

In this section, we prove that the equilibrium points set of the generalized global set-valued dynamical system $(*)$ is also nonempty and closed.

Theorem 1. Let $E : \mathcal{H} \to F(\mathcal{H})$ be a closed fuzzy mapping. Let $\tilde{E} : \mathcal{H} \to C(\mathcal{H})$ be μ-Lipschitz continuous, $g : \mathcal{H} \to \mathcal{H}$ be ϵ-Lipschitz continuous and β-strongly monotone, $\eta : \mathcal{H} \times \mathcal{H} \to \mathcal{H}$ be a single-valued τ-Lipschitz continuous operator. Let $a : \mathcal{H} \to [0, 1]$ be a single-valued function and $H : \mathcal{H} \to \mathcal{H}$ be strictly η-monotone operator with constant r and $A : \mathcal{H} \to 2^{\mathcal{H}}$ be an (H, η)-monotone operator. If

$$\sqrt{1 + \epsilon^2 - 2\beta} + \tau r^{-1}(\epsilon + \rho\mu) < 1,$$

then the equilibrium points set of the generalized global dynamical system $(*)$ is nonempty and closed.

Proof. Let

$$T(x) = x - g(x) + R_{A,\rho}^{H,\eta}(g(x) - \rho\tilde{E}(x)), \quad \forall x \in K.$$

Then $T : \mathcal{H} \to C(\mathcal{H})$. From Definition 2.1, it is easy to know that x^* is an equilibrium point of global dynamical system $(*)$ if and only if x^* is a fixed point of T in $\mathcal{H}$, i.e.,

$$x^* \in T(x^*) = x^* - g(x^*) + R_{A,\rho}^{H,\eta}(g(x^*) - \rho\tilde{E}(x^*)).$$

Therefore, the equilibrium points set of (2.1) is the same as the fixed pints set of T.

We first prove that $F(T)$ is nonempty. In fact, for any $x, y \in \mathcal{H}$ and $a_1 \in T(x)$, there exists $u \in \tilde{E}(x)$ such that

$$a_1 = x - g(x) + R_{A,\rho}^{H,\eta}(g(x) - \rho u). \tag{1}$$

Since $u \in \tilde{E}(x)$, and $\tilde{E} : \mathcal{H} \to C(\mathcal{H})$, it follows from Nadler [9] that there exists $v \in \tilde{E}(y)$ such that

$$\|u - v\| \leq M(\tilde{E}(x), \tilde{E}(y)). \tag{2}$$

Let

$$a_2 = y - g(y) + R_{A,\rho}^{H,\eta}(g(y) - \rho v). \tag{3}$$

Then $a_2 \in T(y)$. From (1) to (3), we have

$$\|a_1 - a_2\| = \|x - y - (g(x) - g(y)) + R_{A,\rho}^{H,\eta}(g(x) - \rho u) - R_{A,\rho}^{H,\eta}(g(y) - \rho v))\|$$
$$\leq \|x - y - (g(x) - g(y))\| + \|R_{A,\rho}^{H,\eta}(g(x) - \rho u) - R_{A,\rho}^{H,\eta}(g(y) - \rho v))\|. \tag{4}$$

Since g is ϵ-Lipschitz continuous and β-strongly monotone,

$$\|x - y - (g(x) - g(y))\|^2 \leq (1 + \epsilon^2 - 2\beta)\|x - y\|^2. \tag{5}$$

From Lemma 2, $R_{A,\rho}^{H,\eta}$ is Lipchitz continuous, we have

$$\|R_{A,\rho}^{H,\eta}(g(x) - \rho u) - R_{A,\rho}^{H,\eta}(g(y) - \rho v))\| \leq \tau r^{-1}(\|g(x) - g(y)\| + \rho\|u - v\|)$$
$$\leq \tau r^{-1}(\epsilon\|x - y\| + \rho\|u - v\|). \tag{6}$$

From the selection of v and the Lipschitz continuity of $\tilde{E}$,

$$\|u - v\| \leq M(\tilde{E}(x), \tilde{E}(y)) \leq \mu\|x - y\|. \tag{7}$$

In light of (4)-(7), we have

$$\|a_1 - a_2\| \leq (\sqrt{1 + \epsilon^2 - 2\beta} + \tau r^{-1}(\epsilon + \rho\mu))\|x - y\| = L\|x - y\|, \tag{8}$$

where $L = \sqrt{1 + \epsilon^2 - 2\beta} + \tau r^{-1}(\epsilon + \rho\mu)$. Now (3.8) implies that

$$d(a_1, T(y)) = \inf_{a_2 \in T(y)} \|a_1 - a_2\| \leq L\|x - y\|. \tag{9}$$

Since $a_1 \in T(x)$ is arbitrary, we have

$$\sup_{a_1 \in T(x)} d(a_1, T(y)) \leq L\|x - y\|. \tag{10}$$

Similarly, we can prove

$$\sup_{a_2 \in T(y)} d(T(x), a_2) \leq L\|x - y\|. \tag{11}$$

From (10), (11), and the definition of the Hausdorff metric $M\left(\cdot,\cdot\right)$ on $C(\mathcal{H})$, we have

$$M(T(x),T(y)) \leq L\|x-y\|, \forall x,y \in K. \tag{12}$$

Now the assumption of the theorem implies that $L < 1$ and so $T(x)$ is a set-valued contractive mapping. By the fixed-point theorem of Nadler[9], there is x^* such that $x^* \in T\left(x^*\right)$, and thus x^* is an equilibrium point of $(*)$. This means that $F(T)$ is nonempty.

Now we prove that $F(T)$ is closed. Let $\{x_n\} \subset F(T)$, and $x_n \to x_0 (n \to \infty)$. Then $x_n \in T(x_n)$ and (12) implies that

$$M(T(x_n),T(x_0)) \leq L\|x_n - x_0\|.$$

Thus,

$$d(x_0,T(x_0)) \leq \|x_0 - x_n\| + d(x_n,T(x_n)) + M(T(x_n),T(x_0))$$
$$\leq (1+L)\|x_n - x_0\| \to 0, \quad \text{as} \quad n \to \infty.$$

It follows that $x_0 \in F(T)$ and so $F(T)$ is closed. This completes the proof.

4   A New Iterative Method for Finding an Equilibrium Point

Assume all the notations are as stated in the previous sections. Under the same assumptions as in Theorem 3.1, this section provides a new iterative method for finding an equilibrium point for (2.1).

Algorithm 1. Step 0. Let $\rho > 0$ be a constant. Choose $x_0 \in int(dom(\tilde{E}))$ and choose $u_0 \in \tilde{E}(x_0)$. Set $n = 0$.

Step 1 Let

$$x_{n+1} = x_n - g\left(x_n\right) + R_{A,\rho}^{H,\eta}\left(g\left(x_n\right) - \rho u_n\right). \tag{13}$$

Step 2 Choose an $u_{n+1} \in \tilde{E}\left(x_{n+1}\right)$ satisfying

$$\|u_{n+1} - u_n\| \leq M(\tilde{E}\left(x_{n+1}\right),\tilde{E}\left(x_n\right)),$$

Step 3 If x_{n+1} satisfies a required accuracy, then stop, otherwise, $n := n+1$, go to step 1.

Theorem 2. If all the assumptions in Theorem 3.1 hold, then the iterative sequence $\{x_n\}$ generated by Algorithm 1 strongly converges to an equilibrium point for $(*)$.

Proof. By 13, we have

$$\|x_{n+1} - x_n\| = \| -g\left(x_n\right) + R_{A,\rho}^{H,\eta}\left(g\left(x_n\right) - \rho u_n\right) \|$$

Let $T_n = -g\left(x_n\right) + R_{A,\rho}^{H,\eta}\left(g\left(x_n\right) - \rho u_n\right) = x_{n+1} - x_n.$

Then,

$$\|T_n\| = \|T_{n-1} + T_n - T_{n-1}\|$$
$$\leq \|(x_n - x_{n-1}) - (g(x_n) - g(x_{n-1}))\|$$
$$+ \|R_{A,\rho}^{H,\eta}(g(x_n) - \rho u_n) - R_{A,\rho}^{H,\eta}(g(x_{n-1}) - \rho u_{n-1}). \tag{14}$$

For the first term in (14)

$$\|(x_n - x_{n-1} - (g(x_n) - g(x_{n-1}))\|^2 \leq (1 - 2\beta + \epsilon^2)\|x_n - x_{n-1}\|^2. \tag{15}$$

For the second term in (14)

$$\|R_{A,\rho}^{H,\eta}(g(x_n) - \rho u_n) - R_{A,\rho}^{H,\eta}(g(x_{n-1}) - \rho u_{n-1})\|$$
$$\leq \tau r^{-1}(\|\rho u_n - \rho u_{n-1}\| + \epsilon\|x_n - x_{n-1}\|)$$
$$\leq \tau r^{-1}(\rho M(\tilde{E}(x_n), \tilde{E}(x_{n-1})t) + \epsilon\|x_n - x_{n-1}\|)$$
$$\leq \tau r^{-1}(\rho\mu + \epsilon)\|x_n - x_{n-1}\|. \tag{16}$$

Therefore, in light of (14) to (16), we have

$$\|x_{n+1} - x_n\| \leq (\sqrt{(1 - 2\beta + \epsilon^2)} + \tau r^{-1}(\rho\mu + \epsilon))\|x_n - x_{n-1}\|.$$

Let $k := \sqrt{(1 - 2\beta + \epsilon^2)} + \tau r^{-1}(\rho\mu + \epsilon)$ and $0 < k < 1$ by the assumptions of Theorem 1, then we conclude that $\{x_n\}$ is a Cauchy sequence and therefore $\{x_n\}$ must strongly converge to a point x^*. From the construction of Algorithm 1, it is obviously that $x^* \in F(T)$, and therefore is an equilibrium point for (*). This completes the proof.

A demonstration example is given below.

Example 2. Now let $A = \partial\phi$, where $\phi : \mathcal{H} \to \mathcal{R} \cup \{+\infty\}$ is a lower semicontinuous and $\eta = x - y$. Furthermore, let H be the identity mapping and ϕ is the indicator function of K, then $R_{A,\lambda}^{H,\eta} = P_K$, the projection of X onto K (see [8]). Moreover, if $F_x(y)$ is the characteristic function, and if $\tilde{E} = N$ is a single-valued mapping, then (*) could further reduce to the global projective dynamical system

$$\begin{cases} \frac{dx(t)}{dt} = P_K(g(x) - \rho N(x(t))) - g(x) \\ x(0) = x_0 \end{cases}$$

which was studied by Friesz et al. [2], Xia and Vincent [3]. Now, Let $g(x) = x$ and K is the unit ball in R^4, and

$$N = \begin{pmatrix} 15 & 3.4 & 5.7 & -1 \\ 9.2 & 14 & 3 & -0.5 \\ -8 & 2.1 & 12 & 1 \\ 2 & -3 & 7 & 15 \end{pmatrix}$$

Obviously, the equilibrium point is $(0, 0, 0, 0)$. Now choose the initial point $x_0 = (-1.5, 1, -1, 1.7)^T$ and $\rho = 1/189$, the sequence generated by the algorithm 1 is

$x(t) = (x1, x2, x3, x4)$. The following figure is the simulation of the sequences, and it does converge to the equilibrium point.

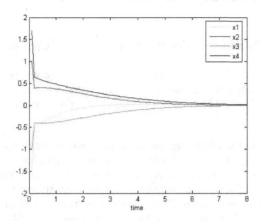

Remark. Compared with [7, 5], a more generalized monotone operator is considered in this paper and an iterative algorithm is constructed to find an equilibrium point. Also the strong convergence of the generated iterative sequence is proved.

References

1. Dupuis, P., Nagurney, A.: Dynamical Systems and Variational Inequalities. Ann. Oper. Res. 44, 19–42 (1993)
2. Friesz, T.L., Bernstein, D.H., Mehta, N.J., Tobin, R.L., Ganjlizadeh, S.: Day-to-Day Dynamic Network Disequilibria and Idealized Traveler Information Systems. Oper. Res. 42, 1120–1136 (1994)
3. Xia, Y.S., Vincent, T.L.: On the Stability of Global Projected Dynamical Systems. J. Optim. Theory Appl. 106, 129–150 (2000)
4. Zhang, D., Nagurney, A.: On the Stability of Projected Dynamical Systems. J. Optim. Theory Appl. 85, 97–124 (1995)
5. Zou, Y.Z., Huang, N.J., Lee, B.S.: A New Class of Generalized Global Set-Valued Dynamical Systems Involving (H, η) -monotone operators in Hilbert Spaces. Nonlinear Analysis Forum 12(2), 191–193 (2007)
6. Zadeh, L.A.: Fuzzy Sets. Inform. Control 8, 338–353 (1965)
7. Zou, Y., Huang, N.: An Iterative Method for Quasi-Variational-Like Inclusions with Fuzzy Mappings. In: Wang, G.-Y., Peters, J.F., Skowron, A., Yao, Y. (eds.) RSKT 2006. LNCS (LNAI), vol. 4062, pp. 349–356. Springer, Heidelberg (2006)
8. Fang, Y.P., Huang, N.J., Tompson, H.B.: A New System of Variational Inclusions with (H, η)-Monotone Operators in Hilbert Spaces. Comput. Math. Appl. 49(2-3), 365–374 (2005)
9. Nadler, S.B.: Multivalued Contraction Mappings. Pacific J. Math. 30, 475–485 (1969)

Estimation of Missing Precipitation Records Using Modular Artificial Neural Networks

Jesada Kajornrit, Kok Wai Wong, and Chun Che Fung

School of Information Technology, Murdoch University
South Street, Murdoch, Western Australia, 6150
j_kajornrit@hotmail.com, {k.wong,l.fung}@murdoch.edu.au

Abstract. Estimation of missing precipitation records is one of the important tasks in hydrological study. The completeness of precipitation data leads to more accurate results from the hydrological models. This study proposes the use of modular artificial neural networks to estimate missing monthly rainfall data in the northeast region of Thailand. The simultaneous rainfall data from neighboring control stations are used to estimate missing rainfall data at the target station. The proposed method uses two artificial neural networks to learn the generalized relationship of rainfall recorded in dry and wet periods. Inverse distance weighting method and optimized weight of subspace reconstruction method are used to aggregate the final estimation value from both networks. The experimental results showed that modular artificial neural networks provided a higher accuracy than single artificial neural network and other conventional methods in terms of mean absolute error.

Keywords: Missing precipitation records, Modular artificial neural networks, Northeast region of Thailand, Inverse distance weighting method, Optimized weight of subspace reconstruction method.

1   Introduction

Precipitation data are one of the most important variables used in hydrological modeling in the assessment of streamflow and rainfall-runoff. These models fundamentally require the complete and reliable rainfall data records [1]. Normally, ground-based observations are the primary sources of rainfall data. A large number of rain gauge stations are installed throughout the study area to record the rainfall. However, in practice, rainfall records often contain missing data values due to malfunctioning of the equipment and/or other conditions. Such imperfect rainfall record could affect the performance of the hydrological models. Therefore, estimating missing rainfall data is an important task in hydrological modeling [2]. This study proposes the use of Modular Artificial Neural Networks (MANN) to estimate missing monthly rainfall data. This paper is organized as follows: Section 2 describes some of the related works. Section 3 illustrates four case studies and the dataset being used. Section 4 describes the details of MANN used in this study. Section 5 shows the experimental results and an analysis of the outcomes. Finally, a conclusion is presented in Section 6.

T. Huang et al. (Eds.): ICONIP 2012, Part IV, LNCS 7666, pp. 52–59, 2012.

2   Related Works

In the last decade, many studies have been dedicated to address the missing rainfall data problem. Teegavarapu et al. [2] examined Inverse Distance Weighting Method (IDWM) and its variants to estimate the missing precipitation data. They suggested several ways to improve IDWM by defining some parameters and surrogate measures for distance used in IDWM. They concluded that using correlation coefficient as weight for revised IDWM and Artificial Neural Network (ANN) yielded better accuracy. Later, Teegavarapu et al. [3] improved Ordinary Kriging (OK) by using ANN to create semivariogram instead of using a prior definition of a mathematical function. This revised technique was used to estimate the missing precipitation data. The results showed that the use of ANN with OK had more advantages than the original OK. Nevertheless, Teegavarapu et al. [4] purposed a fixed functional set genetic algorithm method to derive the optimal functional forms for estimating the missing precipitation data. The method used genetic algorithm and non-linear optimization formulation to obtain functional form and its coefficients. The proposed method was compared with IDWM and Correlation Coefficient Weighting Method (CCWM). Their method showed improvement to IDWM and CCWM in term of root mean square error. Kim et al. [1] applied Regression Tree (RT) and ANN to construct missing precipitation data. Regression tree was used to create the list of influenced stations. These stations were then used to estimate the missing precipitation data by ANN models. The result showed that the use of RT + ANN provided better estimation than the use of RT or ANN alone. Piazza et al. [5] compared various spatial interpolation methods to create a serially complete monthly precipitation time series. Their study suggested that the best estimation result could be derived from the use of a combination method called residual kriging in which the residual from linear regression are interpolated by ordinary kriging method. Another comparison work is Kajornrit et al. [6]. They compared several spatial interpolation methods to estimate missing rainfall data in the northeast region of Thailand. They suggested the use of statistics of dataset as a guideline to select the appropriate estimation techniques. All works mentioned above used the single model to estimate missing rainfall data. Since the nature of rainfall data could be grouped into dry and wet period, the use of modular models may improve the estimation accuracy. Therefore, this study proposes the use of modular artificial neural networks to perform this task.

3   Four Case Studies and Dataset

The case study area selected sites in the northeast region of Thailand as illustrated in Figure 1. In this study, four rainfall stations are assumed to have missing rainfall data records (*target station*). The simultaneous rainfall data from neighboring stations (*control station*) are used to estimate the missing data at target station. Many researchers have recommended the use of three or four closest stations for application of IDWM [2]. This suggestion related to the work of Eischeid [7], which showed that inclusion of more than four stations does not significantly improve the interpolation and may in fact degrade the estimate.

This study selected three closest control stations to estimate the missing data at the target station. An additional reason to select only three control stations is due to the availability of data. Since the dataset contains a few real missing data, the data records that have missing data must be removed. The number of available data records decreases when the number of control stations increases. Thus, the use of three control stations is deemed to be an appropriate selection for this study. However, it does not necessarily mean it is the best.

The rainfall data range from 1981 to 2001. The data from 1981 to 1998 are used to calibrate the models, and data from 1999 to 2001 are used to validate the developed models. Since there are a few real missing data records in control stations in the earlier period, such records have been removed. After removing missing records from calibration data, the proportion between validation and calibration data falls between 18 to 20 percents approximately. To validate the models, Mean Absolute Error (MAE) is adopted as given in equation (1).

$$MAE = \sum_{i=1}^{m} |Oi - Pi|/m. \tag{1}$$

where O_i and P_i is the observed the estimated value respectively, m is the number of missing data.

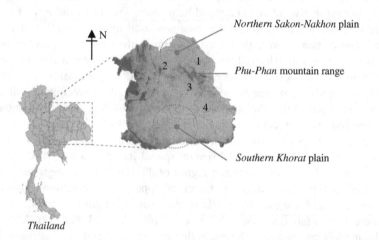

Fig. 1. Four selected case study sites in the northeast region of Thailand, case 1: ST356010, case 2: ST381010, case 3: ST388002, case 4: ST407005. Case 1 and Case 3 sites are located over and under the *Phu-Phan* mountains range. Case 2 sites are located in the *Northern Sakon-Nakhon* plain and Case 4 sites are located in the *Southern Khorat* plain.

4   The Modular Artificial Neural Networks

Figure 2 shows an overview of the proposed model. The proposed methodology could be divided into two steps. The first step is to partition the data and create the estimation modules. In this step, the training data are clustered into different groups and then the data in each group are used to train an ANN. The second step is to create an aggregation module. The function of this module is to finalize the decision value from those networks. In this study two aggregation methods are introduced. Both methods are based on the concept of Tobler's first law, *"Everything is related to everything else, but near things are more related than distant things"* [8].

In the first step, since the nature of rainfall data could be divided into dry and wet period, the proposed method partitions the data into two clusters according to the seasons. All the input-output pairs are then clustered by using Fuzzy C-Mean (FCM) clustering technique. Once the two training data are prepared, supervised neural network are used to capture the relationship between these input-output pairs. Among several types of supervised neural network, Back-Propagation Neural Network (BPNN) has been widely used in hydrological study. In this study one hidden layer BPNN is used to learn from the training data. The numbers of input node, hidden node and output node are three, four and one respectively. The transfer functional used in the model is sigmoid function.

The second step is to create an aggregation module. This study proposed two aggregation methods, *Inverse Distance Weighting Method* (MANN-IDWM) and *Optimized Weight of Subspace Reconstruction Method* (MANN-OWSR). In the first method, the final decision output should be closer to the decision output from closer ANN than farther ANN. The distance between the data point and the center of clusters are used to weight the final decision value from both ANNs inversely. The mathematic formula of MANN-IDWM is

$$z_0 = [z_1 \frac{1}{d_1^k} + z_2 \frac{1}{d_2^k}] / [\frac{1}{d_1^k} + \frac{1}{d_2^k}] . \tag{2}$$

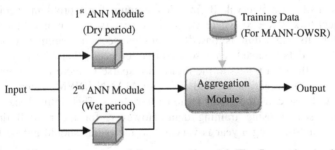

Fig. 2. The architectural overview of the proposed model. The first and second ANN captures the relationship of rainfall in the dry and wet period respectively. In the aggregation module, the training data are used only for the MANN-OWSR model.

where Z_1 is predicted value from ANN$_1$ and Z_2 is the predicted value from ANN$_2$, d_1 and d_2 are the distance between the data point to the centroid of cluster 1 and cluster 2 respectively. k is the power parameter. The optimized k parameter can be found from training data.

In the second method, the optimized weight of subspace reconstruction method, based on the idea that if the weight assigned to a data point in order to weight final decision value from both ANNs is optimal, this weight value should also be optimal for the nearest data points in the same manner. Assume δ to be a small region around an input vector Z_s and a set of data points $\{Z_1, Z_2,...,Z_k\}$ to be the data points around the region δ in which

$$\|Z_i - Z_s\| \ll \delta . \tag{3}$$

If the weight applied to Z_s is the optimal weight. That weight should be the optimal value for all the points in the region. So, if the weight applies to all the points in that region is optimal, the error of equation shown below should be minimal.

$$\varepsilon = \frac{1}{k}\sum_{i=1}^{k}(z_i' - z_i)^2 . \tag{4}$$

where ε is mean square error, k = number of data point in the region δ, z' is predicted value from MANN and z is the observed value. Considering this case study, the final decision value comes from two ANNs. The final estimated value is $z' = \alpha z_d' + \beta z_w'$ and $\alpha + \beta = 1$ Then

$$z' = \alpha z_d' + (1 - \alpha)z_w' . \tag{5}$$

where z' is final predicted results and α is weight applied. Replace equation (5) into equation (4). Then

$$\varepsilon = \frac{1}{k}\sum_{i=1}^{k}((\alpha z_{di}' + (1 - \alpha)z_{wi}') - z_i)^2 . \tag{6}$$

The equation (6) is the cost function that we have to minimize in order to find the optimal value of α. Then the problem is to optimize one variable equation. In order to optimize the cost function, this study uses a MATLAB function call "*fminbnd*" to minimize MSE. The function "*fminbnd*" is used to find the minimum of the single variable function of a fix interval. It finds a minimum for a problem specified by $\min_x f(x)$, subject to $x_1 < x < x_2$ where x_1, x, x_2 are scalars and $f(x)$ is a function that returns a scalar. Its algorithm is based on golden section search and parabolic interpolation. More details have been described in references [9] and [10].

In the case that the data points in the region are sparse or there is no point in the defined region, the aggregation method will use MANN-IDWM instead. Another consideration is the size of the small region (or radius). This optimal size can be found by direct search using training data. However, for the rainfall data, the distribution of rainfall among a year is varying, so the radius should not be fixed in the input space.

Taking the distribution of data in Figure 3 into account, one can see that the distribution of data is concentrated near origin and spread out in all dimension. Thus, the proposed method partition input space into three regions. Region 1 begins from the origin to center of the first cluster. Region 2 is between center of first and second cluster. Region 3 is the area outside center of cluster 2. In each region, the training data have been used to investigate the appropriate radius by direct searching.

5   Experimental Results

To evaluate the accuracy of the developed models, the rainfall data from 1999 to 2001 are assumed to be missing data records and they needed to be estimated. The proposed models have been compared with the Inverse Distance Weighting Method (IDWM), the Correlation Coefficient Weighting Method (CCWM) and Artificial Neural Network (ANN). Table 1 shows the results of evaluation.

In IDWM, the optimized power parameter k could be defined by considering MAE of data in the calibration period when increasing power parameter. It was found that the optimized power parameters are 0.8, 4.5, 2.8 and 0 for case 1 to case 4 respectively. In CCWM, the correlation coefficient of rainfall data between each control stations and target station in calibration period are used in this method. The network architecture of the ANN is the same as the architecture used in the MANN method.

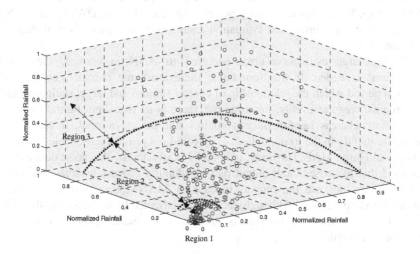

Fig. 3. An example of the distribution of rainfall data in the input space (TS356010)

Table 1. Mean Absolute Error (MAE) of validation data

Models	ST356010	ST381010	ST388002	ST407005
IDWM	245.02	267.32	500.46	399.10
CCWM	261.05	285.38	484.92	399.02
ANN	244.59	258.13	487.95	481.11
MANN - IDWM	232.04	230.83	456.25	387.52
MANN - OWSR	212.91	228.40	448.36	389.87

In case 1 (ST356010), CCWM gave the highest estimation error. IDWM and ANN showed no different in the accuracy. MANN-IDWM provided an improvement from ANN and other conventional method up to 5 percents. In turn, MANN-OWSR provided significantly improvement from MANN-IDWM to almost 8 percents. This case study pointed out that the proposed methods, especially MANN-OWSR can improve the performance of ANN and other conventional models.

In case 2 (ST381010), CCWM provided the lowest accuracy. IDWM provided better estimation than CCWM, and ANN provided better estimation than IDWM.

MANN-OWSR showed a slight improvement over MANN-IDWM. However, both models showed better result than CCWM, IDWM and ANN of up to 13 percents approximately.

In case 3 (ST388002), IDWM provided the lowest accuracy whereas CCWM and ANN provided almost similar performance. MANN-IDWM improved the estimation of the ANN by 6.50 percents and MANN-OWSR improved the estimation of ANN by 8 percents. In this case study, MANN-OWSR again showed better estimation results than MANN-IDWM.

In case 4 (ST407005), IDWM and CCWM provided almost similar estimation results whereas ANN showed very high estimation error in this case study. However, both MANN-IDWM and MANN-OWSR still provided lower estimation error than IDWM and CCWM. This case study pointed out that MANN-IDWM and MANN-OWSR could still perform good estimation results even though ANN provided high estimation error.

Since the large estimation error occurred to ANN in case 4, then, more investigation is needed. It was found that there are some rainfall records in the calibration period which the relationship of control stations and target station could be considered as irregular events; For example, there is an overshoot rainfall record at target station whereas rainfall data at control stations are normal. If such record occurred frequently in the training data, ANN could not provide reasonable estimation and thus yield large MAE. However, only ANN is affected by this noise data because ANN used this record as input-output pair in adapting process whereas the IDWM and CCWM do not use the output. In case of MANN, these irregular data are separated into two datasets. Although one ANN is affected by this data, another ANN is not. Therefore, when the final decision value is evaluated from both ANN, the irregular effect is reduced.

6   Conclusion

This study proposed the use of modular artificial neural networks to estimate the missing monthly precipitation records. The proposed models use fuzzy c-mean clustering technique to partition the data into dry and wet period according to the nature of the data. Back-propagation neural networks have been used to capture the relationship of rainfall in each period. In the aggregation module, this study used an inverse distance weighting method and an optimized weight of subspace reconstruction method to form the final decision value. Four case studies in the northeast region of Thailand have been used to test the proposed models. The simultaneous rainfall records from three nearest control stations were used to estimate the missing rainfall record at the target station. The experimental results reported so far have showed that the use of modular artificial neural network can improve the performance of single artificial neural network and other conventional method to estimate the missing rainfall data. Furthermore, modular artificial neural networks are more tolerant to irregular data than single artificial neural networks.

References

1. Kim, J., Pachepsky, Y.A.: Reconstructing Missing Daily Precipitation Data using Regression Trees and Artificial Neural Networks for SWAT Streamflow Simulation. J. Hydrol. 394, 305–314 (2010)
2. Teegavarapu, R.S.V., Chandramouli, V.: Improved Weighting Methods, Deterministic and Stochastic Data-driven Models for Estimation of Missing Precipitation Records. J. Hydrol. 312, 191–206 (2005)
3. Teegavarapu, R.S.V.: Use of Universal Function Approximation in Variance-dependent Surface Interpolation Method: An Application in Hydrology. J. Hydrol. 332, 16–29 (2007)
4. Teegavarapu, R.S.V., Tufail, M., Ormsbee, L.: Optimal Functional Forms for Estimation of Missing Precipitation Data. J. Hydrol. 374, 106–115 (2009)
5. Piazza, A.D., Conti, F.L., Noto, L.V., Viola, F., Loggia, G.L.: Comparative Analysis of Different Techniques for Spatial Interpolation of Rainfall Data to Create a Serially Complete Monthly Time Series of Precipitation for Sicily, Italy. Int. J. Appl. Earth Obs. Geoinf. 13, 396–408 (2011)
6. Kajornrit, J., Wong, K.W., Fung, C.C.: Estimation of Missing Rainfall Data in Northeast Region of Thailand using Spatial Interpolation Methods. AJIIPS 13(1), 21–30 (2011)
7. Eischeid, J.K., Pasteris, P.A., Diaz, H.F., Plantico, M.S., Lott, N.J.: Creating a Serially Complete, National Daily Time Series of Temperature and Precipitation for the Western United States. J. Appl. Meteorol. 39, 1580–1591 (1999)
8. Miller, H.J.: Tobler's First Law and Spatial Analysis. A. Assoc. Am. Geog. 94(2), 284–289 (2004)
9. Forsythe, G.E., Malcolm, M.A., Moler, C.B.: Computer Methods for Mathematical Computations. Prentice-Hall (1976)
10. Brent, R.P.: Algorithms for Minimization without Derivatives. Prentice-Hall, New Jersey (1973)

Office Employees Authentication
Based on E-exam Techniques

Ameer H. Morad

Al-Khwarizmi Engineering College, University of Baghdad-Iraq
ameer_morad@yahoo.com

Abstract. The paper presents a method for protection office door to open only for authentic employees. Firstly, the employee enter his ID into remote computer through keypad fixed out door, then the computer automatically reply by asking the employee a few questions appear on the keypad screen. To open the door, the employee must answer all of questions within certain and short period of the time. The questions are been selected randomly using e-exam technique from a bank of questions which have been designed based on invariant private information related with employee. The paper is aimed to use a simple and accurate method for employee authentication comparing with traditional office security techniques.

Keywords: Person authentication, E-exam, Random generation, Door locking system.

1 Introduction

Historically the access control system to determine who is allowed to enter through the door was partially accomplished through keys and locks. When a door is locked only someone with a key can enter through the door, the keys can be easily copied or transferred to an unauthorized person.

Electronic person authentication technology [1] is used to replace mechanical key. Traditionally there are different types of person authentication, such a, password, ID, card key, smart card, and Biometric [2].

Biometric is the most secure and convenient authentication tool. The biometric door lock uses measure individual unique or behavioral characteristic [3-8] such as fingerprint, signature, voice, hand geometry, eye, and facial verification. Even biometrics can't be borrowed, stolen, or forgotten but authentication systems based on biometrics are safer from the following:

- Ease of use: some biometric devices are difficult to handle unless there is proper training.
- Error incidence: Time and environmental conditions may affect the accuracy of biometric data. For instance, biometrics may change as an individual becomes old. Environmental conditions may either alter the biometric directly (if a finger is cut and scarred) or interfere with the data collection (background noise when using a voice biometric).

T. Huang et al. (Eds.): ICONIP 2012, Part IV, LNCS 7666, pp. 60–65, 2012.

- Complexity and cost: to extract biometric information needs preprocessing techniques. Overhead processing.

Our project aim to design protection system for office based on E-exam techniques uses ID code with biometric. Firstly, the employee enter his special (ID) into remote computer through keypad fixed out door, then the computer automatically reply by generating randomly a few questions appear on the keypad screen. To open the door, the employee must answer all of questions within certain and short period of the time. When the door is open, the computer records time of the entering. At time the employee needs to leave the office, he must only enter his (ID) number into the computer through keypad fixed indoor, the computer records the time of leaving. The questions are selected randomly from a bank of question are designed based on invariant private information related with employee himself. The objectives of our project are use simple and accurate idea, ID code with biometric.

2   Proposed Structure and Design of the System

In this study, we proposed a security system contains door locking system using ID code with E-exam technique as illustrate in figure (1).

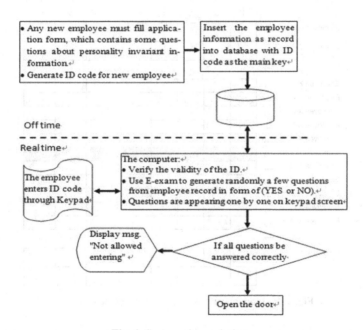

Fig. 1. System phases design

The system is implemented in two phases, in the first phase, we design application form contains 50 questions.The questions are invariant bio information related with office employees such as family name, name of higher school he/she graduated from, name and date of his/her birthday, name of his/her older brother, and so on. All employees must fill the application form with the clear answer known to him. These information are used to design a database table, each record in the database represent the answers of questions for one employee with main key represent by ID code that the system generate it randomly.

The database is linked to e-examining software, to generate ten questions in form YES/NO randomly from the employee record which is linked to the main key (ID code) for the employee. This code is magnetized on the employee card.

The questions generation method from the database is illustrate in fig. 2, suppose that the issue employee is define by employee X. Selection of one of fifty questions (Q_S) is based on random uniform distribution method according to equation (1), and for creation of questions (Q_C), usessame method with window of ±5 to specified employee row according to equation (2).

$$Q_S = INT\ (1+50*R) \tag{1}$$

$$Q_C = INT\ ((ID-5)+10*R) \tag{2}$$

where R is common random number, $R \in [0, 1]$, and INT ( ) is integer value.

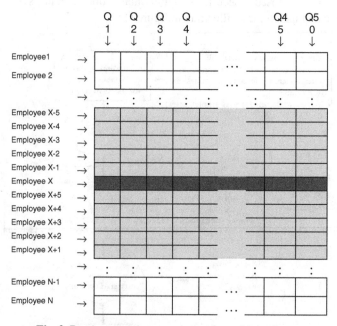

Fig. 2. Random questions generation from the database

In the second phase, represent by real time work, to open the door for any person who wants to enter the office, firstly he/she must feeds the employee card to card reader, then the system extracts the ID code from the card, as illustrate in figure (3), and verify it is valid or not, if it is valid the system use it to specify the employee record in the database. The system is automatically replying by generate randomly ten different questions from employee record according to equations (1) and (2). The questions are displaying one by one on the keypad screen, and the person who stands out of door must answer all of them. The door is be open if and only if the employee answer all ten questions correctly within short period of the time, otherwise the door continue closed and the system inform the person that he/she not allowed to enter by sending message appear on the keypad screen.

The answers of question are test for correctness by compare one by one with data-base record fields which is specified by entered ID code.

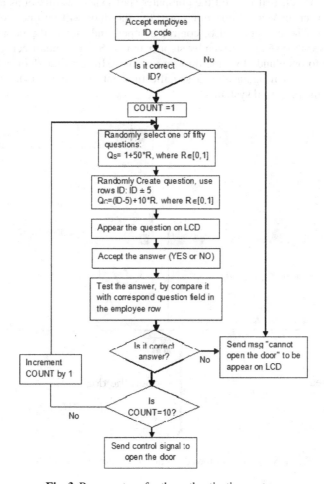

Fig. 3. Process steps for the authentication system

The system used hardware as well as software. The hardware components are interface card reader and keypad, USB connections and connecting cables etc. In addition we have used actuator (stepper motor for this purpose) to open the lock controlled by output signal from computer after employee authentication is verified.

3   Implementation and Results

The system has being implemented using MATLAB program, and use Microsoft office excess to construct the database. The database contains employees bio information uses unique ID code to indicate employee record. Also, the database contains the record of the check-in and check-out of the user. User must have ID code, if the user does not have any previous record registered to the database, the door will not be open thus unauthorized entries will be avoided. So firstly, the computer tests the user ID, it is valid or not. If it is valid the computer start generates questions and test answers. If the user answer all question correctly, the door will be open, otherwise it remain closed. The computer also, control of open and closed the door. The door along with locking system is driven by stepper motor. Stepper motor acts as actuator, which is able to open and close the door in real-time. In real time door is open automatically and closes it again after a specific time interval. Figure(4), illustrate the simulation of the proposed system.

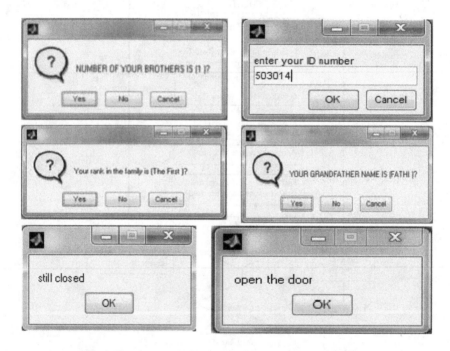

Fig. 4. Simulation of the authentication system use MATLAB

4   Conclusion

In this work we have successfully implemented security system which can be apply to record attendance in office with high reliability and more security than traditional and present techniques. Once the user information (ID code) matched with any ID code stored in central database system, then the system ask the user 10 randomly question.

The user must answer all questions correctly to allow him to enter the office. The system is also able to maintain the record of a user such as how many time and what time user check-in and time of the entering and date.

There many benefits gains form this authentication method comparing with traditional techniques (fingerprint, eye print, ..) such as:

- Most of office organization, already have its employee's database, and only it needs to update by add some bio information to face the requirement of this work.
- Not need to use extra software or equipment, only card reader with keypad.
- The system is more reliable because it based on static bio information, rather than images or sound signals which are used in other techniques.
- Easy to check the absent of any employees, also the cashier and the accounting department will use this database to determine the amount of employee work's hour to determine exactly employee's salary.
- Even the processing time is not large, the time of answering the questions may be use as important factor to verify that the person stand on the door is right person or not.

References

1. http://www.peterindia.net/BiometricsView.html
2. Andrijchuk, V.A., Kuritnyk, I.P., Kasyanchuk, M.M., Karpinski, M.P.: Modern Algorithms and Methods of the Person Biometric Identification. IEEE Intelligent Data Acquisition and Advanced Computing Systems, Technology and Applications, 403–406 (2005) ISBN: 0780394453
3. Zhang, J.H., Liu, X.J., Chen, B.: The Design and Implementation of ID Authentication System Based On Fingerprint Identification. In: Fourth International Conference on Intelligent Computation Technology and Automation, vol. 2, pp. 1217–1220 (2011)
4. Wang, F.L., Zhang, Y.Y.: Study and Design oh Intelligent Authentication System Based on Fingerprint Identification. In: Second International Symposium on Knowledge Acquisition and Modeling, vol. 3, pp. 170–173 (2009)
5. Miroslav, B., Petra, K., Tomislav, F.: Basic On-line Handwriting Signature features for Person Biometric authentication. In: Proceedings of the 34th International Convention MIPRO, pp. 1458–1463 (2011)
6. Eshwarappa, M.N., Latte, M.V.: Multimode Biometric Person Authentication using Speech, Signature and Handwriting Features. International Journal of Advanced Computer Sciences and Applications Special +Issue, Artificial +Intelligent, 77–86 (2011)
7. Chibelushi, C.C., Mason, J.S.D., Deravi, N.: Integrated Person Identification Using Voice and Facial Features. IEE Colloquium on Image Processing for Security Applications (Digest No.: 1997/074), 4/1–4/5 (1997)
8. Qu, Z.Y., Wang, T., Zheng, R.D., Liu, Y.M.: An Identity Authentication System Based on Face Biometrics and Eye Location. In: International Conference on Measuring Technology and Mechatronics Automation, vol. 1, pp. 479–483 (2010)

Obtaining Single Document Summaries
Using Latent Dirichlet Allocation

Karthik Nagesh* and M. Narasimha Murty

Dept of Computer Science and Automation, Indian Institute of Science,
Bangalore-560012, India
{karthik.n,mnm}@csa.iisc.ernet.in

Abstract. In this paper, we present a novel approach that makes use of
topic models based on Latent Dirichlet allocation(LDA) for generating
single document summaries. Our approach is distinguished from other
LDA based approaches in that we identify the summary topics which
best describe a given document and only extract sentences from those
paragraphs within the document which are highly correlated given the
summary topics. This ensures that our summaries always highlight the
crux of the document without paying any attention to the grammar
and the structure of the documents. Finally, we evaluate our summaries
on the DUC 2002 Single document summarization data corpus using
ROUGE measures. Our summaries had higher ROUGE values and better
semantic similarity with the documents than the DUC summaries.

Keywords: Single Document Summaries, Latent Dirichlet Allocation,
SVM, Naïve Bayes Classifier.

1   Introduction

Generating summaries for large document corpora is an arduous task as it requires specialized human effort. The preciseness of a summary is limited to the
understanding of a document or the document domain by the human summarizer. This has led to a large amount of research in the field of automatic text
summarization to generate precise and concise summaries.

Text summarization can be classified into abstractive summarization and extractive summarization. Abstractive summarization involves understanding and
identifying the main concepts of the document and generating the summary by
using well formed sentences with this knowledge. Construction of such a system relies on the machine's ability to "understand" the natural language of the
document with all its ambiguities.

An extractive summarization method consists of identifying important sentences or paragraphs in a document and concatenating them to form a meaningful summary. Extractive summarization techniques exploit statistical, linguistic
and positional properties of sentences or paragraphs for ranking the same. In

---

* Corresponding author.

T. Huang et al. (Eds.): ICONIP 2012, Part IV, LNCS 7666, pp. 66–74, 2012.
© Springer-Verlag Berlin Heidelberg 2012

this paper we present a new extractive summarization technique on LDA(Latent Dirichlet allocation) to identify the summary sentences for generating single document summaries.

2 Related Work

The earliest known works on extractive summarization constructed summaries by identifying important sentences in the original document. Properties such as presence of high frequency words [4], location of sentences [3], etc were used to identify these important sentences. However such methods did not capture the semantics of the document. Thus, a different approach [5] using HMM's which considered the local dependencies of the extracted sentences tried to address this problem.

Our approach makes use of Latent Dirichlet Allocation(LDA) for extracting sentences which reflect the core "theme" of the document. According to Blei [1], the LDA topics are synonymous to the themes of the document. Most of the works on LDA based summarization are with regard to the multi-document summarization problem. To the best of our knowledge, only a few significant methods are available which deal with using LDA for single document summarization. Chang et al. [2] have proposed a new model for extractive summarization called SLDA (Sentence-based LDA) designed for query based summarization. This model assumes that each sentence in a document is influenced by a single theme or topic and ranking of sentences is according to their query likelihood calculated by the SLDA parameters.

Our approach exploits the class specific properties of the document namely the class specific topics. This is because a given document can be summarized in different ways based on the document class. We first do pre-processing on the data to select appropriate features for every class in the corpus, such that the vocabulary words of each class have least entropy in the class. Then we generate a document word matrix separately for every class and provide it as input to LDA for generating class-wise topics. These class-wise topics are post processed as explained in later sections and the summary topics for each document are obtained. Using these summary topics, the paragraphs are clustered and sentences extracted from such clusters are ranked based upon the importance of the paragraph of their origination, the importance of that cluster within the summary topic and the summary topic to which the cluster belongs to. Thus our summaries capture the central theme of the document.

3 Processing Data and LDA Topics for Summarization

The summary topics of a document D are the ones that highlight the main content of the document. For our summaries to effectively capture the semantics of the document, non-summary topics should have negligible probability within the documents. We choose appropriate asymmetric Dirichlet priors [7] for LDA such that only a few topics have high probabilities in the document-topic distribution.

3.1   Pre-processing Data

Feature Selection Using Naïve Bayes and SVM. Firstly, the stop words are removed by using the Zipf's law. Let this set of words be called V_{wsw}. Next vocabulary words with high Mutual information[8] "I(U;C)" scores for a given class C are selected(where U depicts the presence or absence of the word in C). This ensures that our vocabulary words have least entropy for a given class. Next we employ Naïve Bayes and SVM methods for feature selection. The features generated by this step generate two different summaries by our algorithm. The Naïve Bayes method involves training a classifier on two classes at a time using features selected from previous steps and then, only selecting the high probability features within each class. In the SVM method, we learn the "w" vector for two classes at a time and then segregate features into positive class and negative class features based upon their sign in the "w" vector. We then rank the features based on their absolute weights and select the high ranking features for every class.

3.2   Post-processing LDA Topics

Within a given topic t_k, the words are ranked on the basis of decreasing order of their probabilities in t_k and the top 70% of the words are retained. Let this set of words in each topic t_k be called HPt_k. Each word in HPt_k is considered as a vector of its topic distributions. The words within each HPt_k are clustered using this vector representation to generate semantically similar clusters. The set HPt_k is replaced by the cluster representatives instead of the words where the probability of each cluster is the sum of the probabilities of its components. This proves to be useful in reducing the dimensionality of the topics by consolidating the probabilities of semantically similar words into clusters. Now, semantically similar words can be assigned the same probability(their cluster probability), for a given topic.

4   Summarization Algorithm

"A paragraph is a distinct section of a piece of writing, usually dealing with a single theme or topic."[10] Our approach is based on this fundamental definition of a paragraph. The semantics of a sentence is dependent on its local environment or nearby sentences, its meaning is ambiguous without context. Thus our approach differs from [2] by considering the paragraph as an entire document which defines a topic, whereas SLDA [2] considers the sentence as a single document defining a topic.

The steps of our algorithm are explained (in the same order) in detail in the following subsections.

4.1   Building Paragraph Similarity Matrix for Every Topic

The words in the set V_{wsw} are arranged in the decreasing order of their frequencies in the corpus and partitioned into blocks of size L. Thus let vector

---

Algorithm 1. Get Similarity matrix

Require: Paras $\in D$, t_k, W
Ensure: Similarity matrix $\Rightarrow SMt_k$
 1: **for** each $para_i \in$ Paras **do**
 2:     Initialize P_i vector of length(V_{t_k}) to 0
 3:     **for** x = 1 to length(V_{t_k}) **do**
 4:         $g' \leftarrow V_{t_k}[x]$
 5:         **for** each $w \in$ words in $para_i$ **do**
 6:             **if** $w \in g'$ **then**
 7:                 $P_i[x]=W[x]$
 8:                 break {Even if one word of block g' is present in P_i, then that corresponding x entry is made non zero.}
 9:             **end if**
10:         **end for**
11:     **end for**
12: **end for**{The vector P_i is a modified representation of a paragraph such that it contains the W(g) of the corresponding g if even one of the words of g' are in P_i }
13: **for** $i=1$ to length(P) **do**
14:     **for** $j=1$ to $j \leq i$ **do**
15:         $SMt_k[i][j]$=cosine-similarity($P[i]$,$P[j]$) {where, $P[i]$ and $P[j]$ are vectors}
16:     **end for**
17: **end for**{Two paragraphs are considered exactly same if they have atleast one word in g', $\forall g' \in V_{t_k}$}

$G=\{$ V_{wsw} partitions of size L$\}$. We now define a vector W(G) which contains weights $\forall g \in G$ such that:

$$W(g) = \frac{1}{\sum_{words \in g} term frequency} \qquad (1)$$

Now given a topic t_k, we construct a vector V_{t_k} of same dimensions as G such that $V_{t_k}=\{g'|g'=g\bigcap HPt_k, \forall g \in G\}$. The weight vector for V_{t_k} is same as W(g). We make use of V_{t_k} in Algorithm 1 which gives the paragraph similarity matrix SMt_k for topic t_k. Paras is the set of all paragraphs in the document D, t_k is the topic under consideration and W(G) is the weight vector.

4.2   Finding Summary Topics

For each topic t_k in set of all topics, we sum the entries in the Similarity matrix SMt_k to get the values $TW[k]$ of a vector TW. We rank the topics based on the decreasing order of their $TW[k]$ values and the top 70% of the topics are chosen for each document. These constitute the summary topics for document D.

4.3   Paragraph Clustering

Let the set of summary topics of document D be $SummaryTopics_D$. For every topic $t_i \in SummaryTopics_D$ we find the paragraph clusters by applying the Single-Link clustering algorithm by considering the entries in the corresponding

SMt_i as the distances between paragraphs P_j and P_k. The clusters of topic t_i are stored in the array $Clusterst_i$. The corresponding weights of the clusters in $Clusterst_i$ are stored in the array CWt_i. The cluster weights are got by adding the corresponding $SMt_i[j][k]$ value into a cluster when a paragraph P_j and P_k are added into the same cluster. Each entry in the array CWt_i is a measure of the closeness of the paragraphs within that cluster entry.

4.4 Sentence Ranking

For every $t_i \in SummaryTopics_D$ of document D, we can find the weight of the sentence S_i by using (2)

$$P(S_i|t_j) =$$
$$\sum_{Cl \in Clusterst_j} \sum_{\{Pa \in Cl\}} \prod_{\{w_{i_p} \in Pa \& w_{i_p} \in HPt_j\}} P(w_{i_p}|Pa)P(Pa|Cl)P(Cl|t_j)P(t_j),$$
(2)

where w_{i_p} is the pth word of the sentence S_i. Equation (2) is a direct result of the application of the Bayes theorem. The value of the entities $P(w_i| Pa)$, $P(Pa | Cl)$ and $P(Cl |t_j)$ are direct. The value of $P(t_j)$ is got by normalizing the array TW calculated in Sect.4.2. The entry $TW[j]$ is $P(t_j)$. All the candidate sentences are arranged in the decreasing order of their weights and the top sentences are extracted until we encounter our word limit. However one disadvantage of (2) is that long sentences are penalized due to the multiplication of the probabilities. So on the same lines as Arora et al [6], we also replace the multiplication of $P(w_i| Para)$ with summation. Now in order to ensure that length of the sentence does not play a significant role in determining the weight of the sentence, (2) is modified as (3).

$$P(S_i|t_j) =$$
$$\frac{\sum_{Cl \in Clusterst_j} \sum_{\{Pa \in Cl\}} \sum_{\{w_{i_p} \in Pa \& w_{i_p} \in HPt_j\}} P(w_{i_p}|Pa)P(Pa|Cl)P(Cl|t_j)P(t_j)}{length(S_i)}$$
(3)

5 Evaluation and Results

We evaluated our algorithm on the DUC 2002 single document summarization corpus. There were a total of 60 sets of documents with 10 documents in each set. The corpus comprises of four classes distributed across the 60 sets.The Single Document Summarization task requires the generating of a summary not exceeding 100 words for every given document. To evaluate the automatic summaries, two human generated summaries are provided in the DUC corpus for every single document. We adopted the ROUGE-N [9] measure for evaluation. Specifically our summaries were evaluated using the ROUGE-1,2,3 and 4 measures. Given a document D, we shall refer to the summaries generated for D

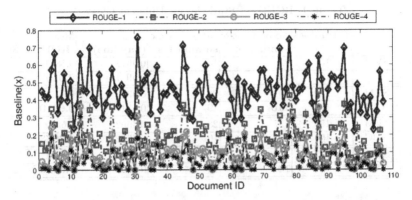

Fig. 1. Comparison of Baseline(D) values for ROUGE-1,2,3,4 values

using our algorithm as **Auto_Summ(D)** and the DUC provided summaries as **DUC_Summ$_x$(D)**, where x $\in \{1, 2\}$ refers to either of the human summaries provided.

For every document D, we fix the **baseline(D)** as maximum ROUGE value got by evaluating $DUC\_Summ_1(D)$ and $DUC\_Summ_2(D)$ against each other. This follows from the assumption that the baseline is the maximum ROGUE value that the human generated summaries can attain over one another. Figure 1 shows the comparison between the baseline ROUGE-1,2,3,4 values for documents of a single class. It was observed that the ROUGE-1 values are the highest and outscore the ROUGE-2 by a factor of 2. The corresponding graph for all four classes is more or less the same. This clearly shows that the DUC summaries are more word oriented than phrase oriented. Thus for our evaluation we only consider the ROUGE-1 values.

Our algorithm was run on documents pertaining to a single class at a time. The Auto_Summ(D) of every document is analysed using the ROGUE-1 measure against the two DUC_Summ(D) that are provided. The maximum ROUGE-1 value that is found for Auto_Summ(D) of the document D against the two DUC_Summ(D) is taken as the **ROUGE_score(Auto_Summ(D))** as in [9].

We define △ROUGE_Gain(D)=ROUGE_score(Auto_Summ(D)) - baseline(D). If △ROUGE_Gain(D) $\geq$ 0, then we classify the corresponding Auto_Summ(D) as **Better summary** than the two DUC_Summ(D). Else if △ROUGE_Gain(D) < 0, we classify such summaries as **Worse summaries**. Table 1 shows the statistics with respect to the evaluation of our Auto_Summ(D) for all classes of documents in the DUC data corpus. The column "Method" refers to the method used in feature selection. The results of the corpus clearly show that Auto_Summ(D) have consistently higher ROUGE values than the DUC_Summ(D) for both the SVM as well as the NBC feature selection methods. Another interesting observation is that for every class, NBC summaries have higher number of Better summaries and Mean(ROUGE_Score) than the SVM method. This could be because NBC focusses on features related to document clusters within the class. However SVM focusses on features of the support vectors at the boundaries.

Table 1. ROUGE-1 Statistics for Different Classes

Class	Method	Total docs	No of **Better** summaries	No of **Worse** summaries	Mean (baseline)	Mean (**ROUGE_Score**)
1	SVM	107	92	15	0.4911	0.5831
1	NBC	107	**96**	11	0.4911	**0.6107**
2	SVM	129	101	28	0.4316	0.5722
2	NBC	129	**110**	19	0.4316	**0.6010**
3	SVM	155	127	28	0.4662	0.5822
3	NBC	155	**145**	10	0.4662	**0.6073**
4	SVM	147	115	32	0.4504	0.5831
4	NBC	147	**129**	18	0.4504	**0.6016**

Thus NBC summaries tend to be more descriptive with higher ROUGE values than the discriminative summaries of SVM method.

A major outcome of summarization based on labelled corpus is that, the summaries could be used to train a classifier instead of the documents. We motivate this interesting direction by looking at whether the Auto_Summ(D) are semantically more similar to D than the DUC_Summ(D). For the Auto_Summ(D) generated using SVM based feature selection, we train a SVM classifier on the set of all documents pertaining to two classes at a time. For every document D of the two classes, using the "w" and "b" value learnt by the SVM model, we calculate the value $Weight(\mathbf{X}) = w^T X + b$, where X can be Auto_Summ, DUC_Summ and D. Let the corresponding weights be Weight(D) for D, Weight(Auto_Summ(D)) for Auto_Summ(D) and Weight(DUC_Summ(D)) for DUC_Summ(D). To measure the semantic difference between the summaries and the document, we define Error(X) = abs(Weight(X) - Weight(D)), where X can be Auto_Summ(D) or DUC_Summ(D). In the case of DUC summaries, Weight(DUC_Summ(D)) is chosen as that summary from the two whose weight minimizes Error(DUC_Summ(D)).

To be able to substitute the summaries instead of D, it is required that Error(Summary) should be as small as possible. This would imply that the summary will behave in the same way as D. We try to verify that our Auto_Summ have lesser error compared to DUC_Summ. This would mean that Auto_Summ is semantically more similar to D than DUC_Summ. For evaluating the semantic similarities we define a quantity ΔSVM as

$$\Delta SVM(D) = Error(DUC\_Summ(D)) - Error(Auto\_Summ(D))$$

Similarly, for the Naïve Bayes approach we train a NBC model using the original documents. Now for every document, the posterior probability, given its parent class is found. Similary, for both Auto_Summ and DUC_Summ also the posterior probabilities are found. We define Error(X)=abs(P(Class|X) - P(Class| D)), where X is Auto_Summ(D) or DUC_Summ(D). As in the case above, the summary with the minimum Error is deemed more semantic. We define the value Δ NBC(D) for document (D) as

$$\Delta NBC(D) = Error(DUC\_Summ(D)) - Error(Auto\_Summ(D))$$

Table 2. Semantic similarity comparisons between DUC_Summ and Auto_Summ

Class	Method	Total docs	No of Semantically better summaries	No of Semantically worse summaries	Mean (Error (DUC_Summ))	Mean(Error (Auto_Summ))
1	SVM	107	73	34	0.4312	**0.214**
1	NBC	107	90	17	0.056	**0.0139**
2	SVM	129	93	36	0.4111	**0.3648**
2	NBC	129	92	37	0.0913	**0.0237**
3	SVM	155	103	52	0.5394	**0.4041**
3	NBC	155	132	23	0.0357	**0.0154**
4	SVM	147	105	42	0.5277	**0.4033**
4	NBC	147	115	32	0.0061	**0.0030**

Now for $\Delta NBC(D) \geq 0$ ($\Delta SVM(D) \geq 0$) we define the Auto_Summ(D) as **Semantically better**, else Auto_Summ(D) is defined as **Semantically worse** for the NBC based and SVM based methods respectively. The semantic similarity of the Auto_Summ(D) for all D in the DUC corpus is found out class wise and the statistics are presented in Table 2.

It is observed that the number of Semantically better summaries is higher in all classes of the DUC corpus. The Mean(Error(Auto_Summ)) are also consistently lesser than the corresponding Mean(Error(DUC_Summ)) for all the classes.

6   Conclusion and Future Work

The results of Table 1 prove that our summaries are better than the DUC summaries. Majority of the documents in every class have higher ROUGE_Scores than the baseline. The last column of the table gives the Mean(ROUGE_Score) which is consistently higher than the Mean(Baseline). This means our summaries in general have a higher rouge score than their individual baselines. The NBC method consistently outperforms the SVM method. This clearly explains our claim that NBC summaries are more descriptive that SVM summaries.

The results of Table 2 prove that our Auto_Summ are semantically more similar to the documents compared to the human generated summaries of the DUC corpus. This means that our algorithm is able to effectively capture the semantics of the document for a given class better than a human summarizer. Measures like ΔSVM and ΔNBC were chosen to compare the difference in the Error between DUC_Summ and Auto_Summ. This was done to discover whether our summaries could be substituted for the documents for learning the classifier. The Mean(Error(Auto_Summ)) was lesser than the Mean(Error(DUC_Summ)) for all classes. Thus, according to the results in Table 2 we can conclude that instead of the DUC_Summ our Auto_Summ could be used for training as they appear to be more semantically closer to the documents.

From this work, we get a few interesting directions in which this work could be extended. Firstly, LDA can be used for abstractive summarization or for phrase extraction from the documents, which is another task with the DUC 2002 data corpus. The usage of summaries in place of documents for classifier learning can be looked at in more detail with the help of topic models.

References

1. Blei, D.: Probabilistic Topic Models. Commun. ACM 55(4), 77–84 (2012)
2. Changm, Y., Chien, J.: Latent Dirichlet Learning for Document Summarization. In: ICASSP, pp. 1689–1692 (2009)
3. Edmundson, H.P.: New Methods in Automatic Extracting. J. ACM 16(2), 264–285 (1969)
4. Luhn, H.P.: The Automatic Creation of Literature Abstracts. Presented at IRE National Convention (1958)
5. Conroy, J.M., O'leary, D.P.: Text Summarization via Hidden Markov Models. In: SIGIR, pp. 406–407 (2001)
6. Arora, R., Ravindran, B.: Latent Dirichlet Allocation Based Multi-Document Summarization. In: AND, pp. 91–97 (2008)
7. Hanna, W., David, M., Andrew, M.: Rethinking LDA: Why Priors Matter. In: NIPS, pp. 1973–1981 (2009)
8. Manning, C.D., Raghavan, P., Schütze, H.: Introduction to Information Retrieval, 1st edn. Cambridge University Press (2008)
9. Lin, C.Y.: Rouge: A package for Automatic Evaluation of Summaries. In: Proceedings of the Workshop on Text Summarization Branches Out, WAS 2004 (2004)
10. Simpson, J., Weiner, E.: Oxford English Dictionary, 2nd edn. Oxford University Press (1989)

Object Recognition Using Sparse Representation of Overcomplete Dictionary

Chu-Kiong Loo and Ali Memariani

Faculty of Computer Science and Information Technology
University of Malaya, Kuala Lumpur, Malaysia
ckloo.um@um.edu.my,
ali_memariani@siswa.um.edu.my

Abstract. Research in computational neuroscience via Functional magnetic resonance imaging (fMRI) argued that recognition of objects in mammalian brain follows a sparse representation of responses to bar-like structures. We considered different scales and orientations of Gabor wavelets to form a dic-tionary. While previous works in the literature used greedy pursuit based meth-ods for sparse coding, this work takes advantage of a locally competitive algo-rithm (LCA) which calculates more regular sparse coefficients by combining the interactions of artificial neurons. Moreover proposed learning algorithm can be implemented in parallel processing which makes it efficient for real-time applications. A synergetic neural network is used to form a prototype template, representing general characteristic of a class. A classification experiment is performed based on multi-template matching.

Keywords: Object recognition, Sparse coding, Dictionary learning.

1   Introduction

Categorization models in object recognition suffer from poor accuracy, rooted in wide range of required flexibilities. Models are not only supposed to handle the variation of one object (e.g. posture, occlusion, illumination, etc) but also generalize between different samples of a class. Since human brain can effectively categorize ob-jects, recent categorization methods are inspired by biological findings of computa-tional neuroscience [1, 2].

Effect of lateral geniculate nucleus (LGN) in human brain can be formulated as a high-pass filtering [3]. Process of images in receptive fields (V1) is more sensitive on

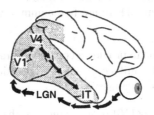

Fig. 1. Recognition process in brain [4]

T. Huang et al. (Eds.): ICONIP 2012, Part IV, LNCS 7666, pp. 75–82, 2012.
© Springer-Verlag Berlin Heidelberg 2012

bar-like structures. Responses of V1 are combined together by extrastriate visual areas and passed to inferotemporal cortex (IT) for recognition tasks [4].

1.1 Dictionary Learning

Representing an image with a few elementary functions is widely used in image processing and computer vision. Determining image component is useful to remove the noise. Also decomposition is used for compression by simplifying image representation. In computer vision decomposition is a tool for feature extraction. An elementary function is called basis and set of bases functions is a dictionary. In early models choice of dictionary elements was subject to orthogonality condition. A complete representation of image is a linear combination of bases in the dictionary, derived by projection of image into bases. However, poor quality of representation in complete solutions resulted in relaxation of orthogonality condition and applying overcomplete dictionaries. Due to useful mathematical characteristics obtained by orthogonality (e.g. computing decomposition coefficients with projection), overcomplete dictionaries are still meant to be partially orthogonal. A common approach is to use an orthogonal subset of a large dictionary containing all possible elements.

Dictionary learning is applied to provide sparsity which was inspired by response of mammalian brain's simple sell to stimuli in receptive fields of visual cortex. [2] Early works applied gradient descent to train the dictionary. Bayesian approaches have been applied for MAP estimation of the dictionary. [5]

Textons are developed as a mathematical representation of basic image objects. [6] First images are coded by a dictionary of Gabor and Laplacian of Gaussian elements; Responses to the dictionary elements is Combined by transformed component analysis. Furthermore, sparse approximation helps to find a more general object models in terms of scale and posture. [7]

Active basis model [8] provides a biological deformable template using Gabor wavelets as dictionary elements. They also proposed a shared sketch algorithm (SSA) inspired by AdaBoost. SSA is based on matching pursuit which greedily selects a wavelet element in each iteration. SSA is tested on number of classes of objects in different location, scale and orientation.

1.2 Sparse Coding

Assuming an image (Io) its sparse approximation I is, $I = \sum_{m=1}^{M} a_m \phi_m$. Optimal sparse coding tries to minimize the number of nonzero coefficients a_m, which is an NP-hard optimization problem. There are tree common approaches to obtain a near optimal sparse solution:

1. Changing objective function to provide a convex programming definition
2. Greedily selecting elements until a convergence condition is satisfied
3. Sparse coding using neural systems

Basis pursuit (BP) algorithm by [10] is an example of the first approach that replaces $\|a\|_0$ with $\|a\|_1$. Donoho and Elad (2003) derived that BP finds the optimal solution, if $\|a\|_0 < min_{m \neq n} \frac{1}{2}[1 + \frac{1}{\langle \phi_m, \phi_n \rangle}]$, where $\langle \phi_m, \phi_n \rangle$ denotes the response of element m to element n in the dictionary. In presence of noise, [10] developed basis pursuit denoising (BPDN) which uses Lagrange method: $Min_a (\|I - \sum_m a_m \phi_m\|^2 + \lambda \|a\|_1)$, where λ is the Lagrange coefficient.

Matching pursuit (MP) algorithm by [11] follows the second approach. An approximation is initialized with original image $r_0 = I_0$, MP iteratively finds a subset of dictionary elements $\theta_i i=1,\ldots,$ k as follows:

$$\theta_i = argmax_m |< r_{i-1}, \phi_m >| \tag{1}$$

$$a_i = < r_{i-1}, \phi_{\theta_i} > \tag{2}$$

$$r_i = r_{i-1} - a_m \phi_{\theta_i} \tag{3}$$

1.3   Sparse Coding Using Neural Systems

Since computational neurophysiologic works argued that response of visual cortex (V1) can be described as a sparse representation of a dictionary [2], some sparse coding algorithm is developed based on the interactions of artificial neurons. Most of them use an iterative approach with a thresholding policy to obtain converged sparse coefficients.

Locally competitive algorithm [12] corresponds each element of the dictionary with a neuron. Internal states of neurons are initialized with zero. Firstly a few neurons that have the highest response with I_0 become active. During the iterations internal state of neurons is charged with respect to their response to I_0. Active neurons suppress the internal state of other nodes by sending a signal proportional to the response of the receiving neurons with corresponding active neuron. This leads to a have a semi orthogonal subset of dictionary elements. A threshold function derives sparse coefficients. Objective function of LCA is the mean square error combined by a cost function to enforce sparsity constraint.

$$Min_a \left(\left\| I_0 - \sum_m a_m \phi_m \right\|^2 + \lambda \sum_m C_m (a_m) \right) \tag{4}$$

$$C_m (\alpha, \lambda) = \frac{(1-\alpha)^2 \lambda}{2} + \alpha \|a_m\| \tag{5}$$

where α and λ determine the threshold policy (e.g. ideal, sigmoidal, etc).

2   Sparse Dictionary Learning with LCA

2D Gabor function centered at (x_0, y_0) is:

$$G(x, y) = \frac{1}{2\pi \sigma_x \sigma_y} e^{-\left[\frac{(x-x_0)^2}{\sigma_x^2} + \frac{(y-y_0)^2}{\sigma_y^2} \right]} e^{i[\xi_0 x + v_0 y]} \tag{6}$$

where (ξ_0, υ_0) is optimal spatial frequency. Using wavelet transform, a Gabor function can be rotated, dilated or translated. A general form of Gabor wavelet function [13] is as the following:

$$GW(x, y, \omega, \theta) = \frac{\omega}{\sqrt{2\pi}\kappa} e^{-\frac{\omega^2}{8\kappa^2}(4(x\cos\theta + y\sin\theta)^2 + (-x\sin\theta + y\cos\theta)^2)} \cdot \left[e^{i\omega(x\cos\theta + y\sin\theta)} - e^{-\frac{\kappa^2}{2}} \right]$$

(7)

where ω is the radial frequency and θ is the wavelet orientation. κ is a constant representing bandwidth frequency. Approximation of $\kappa \approx \pi$ and $\kappa \approx 2.5$ are common for 1 and 1.5 octave bandwidth (ϕ) respectively. Generally κ is:

$$\kappa = \sqrt{2\ln 2}\left(\frac{2^\phi + 1}{2^\phi - 1}\right)$$

(8)

A family of Gabor wavelets can be formed with different scale and orientation by setting (ω, θ) values. This set of Gabor wavelets is called a dictionary. Image information can be encoded by taking the response of each element in the dictionary.

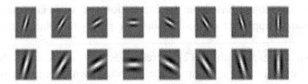

Fig. 2. A family of Gabor Wavelets

2.1   Feature Extraction

A dictionary of Gabor wavelets, including n orientations and m scales is in the form of, $GW_j (\theta, \omega)$, $j = 1, \ldots, m \times n$. where, $\theta \in \left\{\frac{k\pi}{n}, k = 0, \ldots, n - 1\right\}$ and $\omega = \left\{\frac{\sqrt{2}}{i}, i = 1, \ldots, m\right\}$.

Gabor wavelet features are able to represent the shape of an object considering small variance in posture and size and location. However general structure of the shape is considered to be maintained during the recognition process. Response (convolution) to each element provides shape information along θ and ω.

$$B = \langle GW, I \rangle = \sum\sum GW(x_0 - x, y_0 - y : \omega_0, \theta_0) I(x, y)$$

(9)

2.2   Learning Method

Suppose a dictionary of Gabor wavelets is constructed as $\{GW_j, j = 1, \ldots, n\}$ and all dictionary elements are in the same bounding box. In order to learn the shape of an object, locally competitive algorithm (LCA) is used to select a subset of responses which their combination fits in with learning images $\{I_i, i = 1, 2, \ldots, m\}$ with a minimum error. Selection is done for all pixel values among elements of dictionary. For each training image I_i, learning method has the following steps:

(1) Compute response of Ii with all the elements in the dictionary.

$b_j = \langle GW_j, I_i \rangle$,(Set t = 0 and $u_j(0) = 0$, for j = 1, ..., n.)

(2) Determine active nodes by activity thresholding.
(3) For each pixel calculate internal state of element j, $u_j(t)$.

$$\dot{u}_j(t) = \frac{1}{\tau}\left[b_j(t) - u_j(t) - \sum_{j \neq k}\Phi_{j,k}.a_k(t)\right]$$

(10)

$$\Phi_{j,k} = <GW_j.GW_k>$$

(11)

(4) Compute sparse coefficients $a_j(t)$ for $u_j(t)$.
(5) If $a_j(t-1) - a_j(t) > \delta$, t ← t+1 and go to step 2, otherwise finish.

Limit of T as $\lambda \to \infty$ is called ideal thresholding function. $T_{(0,\infty,\lambda)}(.)$ is hard tresholding function and $T_{(1,\infty,\lambda)}(.)$ is soft tresholding function.

Result of learning method is in complex format. Magnitude of a learned sample is similar to an edge detection output. Choice of real or complex values depends on type of classifier. Complex values show better result on projection based classifiers. Sparse approximation of the image is obtained by reconstruction: $\hat{I} = \sum_{j=1}^{M} a_j GW_j$

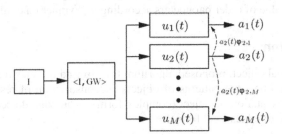

Fig. 3. LCA structure[12]

2.3   Synergetic Neural Network

Synergetic neural network developed by Haken describes pattern recognition process in the human brain. A common approach to combine learned samples is averaging feature values. However averaging is not flexible to change of orientations. Thus such templates have unclear boundaries. One way to deal with inflexibility is to use learning object in the same view which will restrict the classification task. A melting algorithm is proposed by [14] to combine objects in deferent pose. Suppose a learned sample object I_i' consists of n pixel values. I_i' is reshape to a column vector vi and normalized so that: $\sum_{j=1}^{n} v_{ij} = 0$, $\sum_{j=1}^{n} v_{ij}^2 = 1$.

Adjoint prototype matrix V+ is: $V^+ = (V^T V)^{-1}$, Where V is the set of all learned samples vi= 1, ..., m. and each column satisfies the orthonormality condition: $v_k^+ v_j = \delta_{ij}$, for all j and k. Where δ_{ij} is the Kronecker delta. For a test sample q, order parameters represent the match between test sample and each class. Order parameter for class k is derived as, $\varepsilon_k = v_k^+.q$, $k = 1, ..., m$.

However, melting fails to generalize the learning patterns and suffers from pseudo inverse overfitting. [15] used a most probable optimum Design (MPOD) penalty function to enhance the generalization and classify face object in terms of their pose. This modification of synergetic melt combines a number of similar object patterns into a template which could be used for classification. Accordingly synergic template is:

$$V_p^+ = E(V^TV + p_1O + p_2I)^{-1}V^T \tag{12}$$

I is an identity matrix, O is a unitary matrix. p_1 and p_2 are penalty coefficients. E is an enhanced identity matrix; each element of E is a row vector of size j as the following:

$$E = \begin{bmatrix} e_1^{n(1)} & e_0^{n(2)} & \cdots & e_0^{n(M)} \\ e_0^{n(1)} & e_1^{n(2)} & \cdots & e_0^{n(M)} \\ \vdots & \vdots & \ddots & e_0^{n(M)} \\ e_0^{n(1)} & e_0^{n(2)} & \cdots & e_1^{n(M)} \end{bmatrix}, e_0^j = (0,...,0), e_1^j = (1,...,1), \tag{13}$$

Obtaining a template for each class of objects, order parameter can represent the match of a test image with corresponding template. Image is classified as the class with maximum value of order parameters according to "winner takes all" [16].

3   Discussion

For each individual object, proposed algorithm tries to find the most relevant features that can represent the edge patterns of objects. Reconstruction of result would be a compressed representation of object (complex form) or an edge detected image with different intensity of edges (real form).

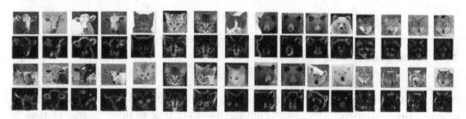

Fig. 4. Learned samples of four classes (real form)

3.1   Classification

Proposed algorithm is used to classify four classes of animals[1]. Each image used in classification contains an animal object. Image objects are obtained from different

---

[1] Dataset can be accessed at
http://www.stat.ucla.edu/~ywu/AB/templates.html

animals in variety of postures, and shapes. Result of classification is a tag for each image representing the corresponding class. For each class 10 images are selected randomly to form the test set. Remaining images are used for training. Using cross validation classification is performed over all the images. This result is compared to Active basis model (ABM).

Table 1. Classification result

	Bear	Cat	Cow	Wolf
Proposed Algorithm	86%	91%	77%	93%
ABM	87%	100%	76%	60%
Total number of images	60	70	60	50

Learning method can effectively detect the edge patterns and represent the main components of objects. This leads to have more distinguishable object definitions (specifically between classes of Bear and Wolf) rather than shared sketching used in ABM. However, we consider simple projection for template matching that could be improved by EM structures.

4   Conclusion

We used LCA to enforce sparsity on a dictionary of Gabor wavelets. According to the literature LCA has not been used on image data. We also consider multi-scale Gabor wavelet elements. Sparse coefficients are localized so that curve-like patterns could also be detected. Average time of learning for each image is less than three seconds on a laptop with 2.1 GHz CPU and 3GB RAM; Regarding the parallel structure of the learning method, implementation could be optimized via parallel processing which is essential for real-time applications.

References

1. Daugman, J.G.: Two-dimensional spectral analysis of cortical receptive field profiles. Vision Research 20(10), 847–856 (1980)
2. Olshausen, B.A., Field, D.J.: Emergence of simple-cell receptive field properties by learning a sparse code for natural images. Nature 381(6583), 607–609 (1996)
3. Braccini, C., et al.: A model of the early stages of the human visual system: Functional and topological transformations performed in the peripheral visual field. Biological Cybernetics 44(1), 47–58 (1982)
4. Riesenhuber, M., Poggio, T.: Neural mechanisms of object recognition. Current Opinion in Neurobiology 12(2), 162–168 (2002)
5. Kreutz-Delgado, K., et al.: Dictionary Learning Algorithms for Sparse Representation. Neural Computation 15(2), 349–396 (2003)
6. Zhu, S.C., et al.: What are textons? International Journal of Computer Vision 62(1-2), 121–143 (2005)

7. Figueiredo, M.A.T.: Adaptive sparseness for supervised learning. IEEE Transactions on Pattern Analysis and Machine Intelligence 25(9), 1150–1159 (2003)

8. Wu, Y.N., et al.: Learning Active Basis Model for Object Detection and Recognition. International Journal of Computer Vision 90(2), 198–235 (2010)

9. Wu, T., Zhu, S.C.: A Numerical Study of the Bottom-Up and Top-Down Inference Processes in And-Or Graphs. International Journal of Computer Vision 93(2), 226–252 (2011)

10. Chen, S.S.B., Donoho, D.L., Saunders, M.A.: Atomic decomposition by basis pursuit. SIAM Review 43(1), 129–159 (2001)

11. Mallat, S.G., Zhang, Z.F.: Matching pursuits with time-frequency dictionaries. IEEE Transactions on Signal Processing 41(12), 3397–3415 (1993)

12. Rozell, C.J., et al.: Sparse Coding via Thresholding and Local Competition in Neural Circuits. Neural Computation 20(10), 2526–2563 (2008)

13. Tai Sing, L.: Image representation using 2D Gabor wavelets. IEEE Transactions on Pattern Analysis and Machine Intelligence 18(10), 959–971 (1996)

14. Hogg, T., Rees, D., Talhami, H.: Three-dimensional pose from two-dimensional images: a novel approach using synergetic networks. In: IEEE International Conference on Neural Network (1995)

15. Lee, G.C., Loo, C.K.: Facial pose estimation using modified synergetic computer. In: Second World Congress on Nature and Biologically Inspired Computing, NaBIC (2010)

16. Wagner, T., et al.: Using a synergetic computer in an industrial classification problem. In: Proceedings of the International Conference on Artificial Neural Nets and Genetic Algorithms, pp. 206–212 (1993)

PEAQ Compatible Audio Quality Estimation Using Computational Auditory Model

Jia Zheng, Mengyao Zhu, Junwei He, and Xiaoqing Yu

School of Communication & Information Engineering,
Shanghai University, Shanghai, China
zhumengyao@shu.edu.cn

Abstract. This paper proposed an improved objective audio quality estimation system compatible with PEAQ (Perceptual Evaluation of Audio Quality). Based on the computational auditory model, we used a novel psychoacoustic model to assess the quality of highly impaired audio. We also applied the robust linear MOA (Least-squares Weight Vector algorithm) and MinmaxMOA (Minmax-Optimized MOV Selection algorithm) to cognitive model of the estimation system. Compared to the PEAQ advanced version, the proposed estimation system has a considerable improvement in performance both in terms of the correlation and MSE (Mean Square Error). By combining the computational auditory model and PEAQ, our estimation system can be applied to the quality assessment of highly impaired audio.

Keywords: Audio quality, Computational, Auditory, Estimation.

1 Introduction

Objective audio quality assessment is an interdisciplinary research area about psychoacoustic and cognitive science. Over the years, the research has been focused on the cognitive aspect. However, the psychoacoustic model bridges the gap between physical representations of sound and subjective hearing sensations. The investigation on it also deserves attention.

In 1996, Torsten Dau put forward a basic computational auditory model of the "effective" signal processing[1], which has been proved to be closely to human ear acoustic transformation. The computational auditory model was later applied in [2]for objective speech quality measurement. As PEAQ[3] has been the only available standardized method for the purpose of audio quality assessment[4], it has been studied by many researchers. The psychoacoustic models of PEAQ are respectively a lower complexity "Basic Version" with FFT based ear model and a more accurate "Advanced Version" with FFT based ear model and Filter bank based ear model[4].However, the PEAQ has been proved to be unable to measure the quality of highly impaired audio[5].To overcome the limitation of PEAQ, a new method called PEMO_Q[6] was proposed by Rainer Huber. It is important to note that PEMO_Q adopted a computational auditory model[1, 7]. Furthermore, ZIYUAN GUO also used

T. Huang et al. (Eds.): ICONIP 2012, Part IV, LNCS 7666, pp. 83–90, 2012.

the computational auditory model in the non-intrusive objective audio quality assessment system for utilizing spectro-temporal modulation[8]. Recently, the computational auditory models have been updated several times and the latest one is the CASP (Computational Auditory Signal-processing and Perception) [9] model. Based on the CASP models, we proposed a novel psychoacoustic model applied to the audio quality assessment system. The cognitive algorithm used in PEAQ is ANN (Artificial Neural Network). The weighting value of ANN is fixed and trained, which can not be used universal when a little change in the metric. For this problem, MOA (Least-squares Weight Vector algorithm) and MinmaxMOA (Minmax-Optimized MOV Selection algorithm)[10] are put forward successively. Since they are linear mapping algorithms and show good robustness for perceptual quality assessment, we adopted them in proposed PEAQ compatible audio quality estimation system.

The proposed audio quality assessment system is completely compatible with the traditional PEAQ advanced version. But the correlation of the SDG (Subjective Difference Grade) and ODG (Objective Difference Grade) of proposed system increased sharply. At the same time, it can assess the quality of highly impaired audio. In this regard, it is an available choice that PEAQ adopt the improvement approach for evaluating audio quality more efficiently.

2   Audio Quality Estimation Compatible with PEAQ

Similar to our previous work[11], we keep the main frame of PEAQ advanced version and just make improvement on the psychoacoustic model and adopt the linear cognitive algorithms. The improved computational auditory model is applied to the proposed estimation system instead of the filterbank based ear model of PEAQ. FFT based ear model and the calculation method of MOVs (Model Output Variables)[3] are as the same with the PEAQ advanced version. Figure 1 shows the comparison of proposed assessment system with PEAQ advanced version. We can see that the proposed assessment system is compatible with the PEAQ advanced version.

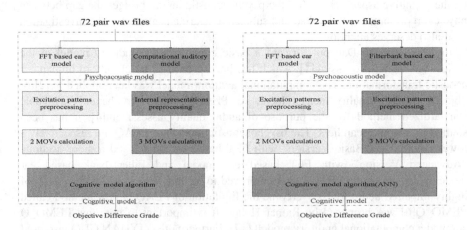

Fig. 1. Comparison of proposed assessment system (left) with PEAQ advanced version (right)

3   Proposed Psychoacoustic Model with Computational Auditory Model

The whole psychoacoustic model is shown in the Figure 2. Based on the psychoacoustic model of PEAQ advanced version, the reference and test signals are processed simultaneously by FFT based ear model and modified computational auditory model.

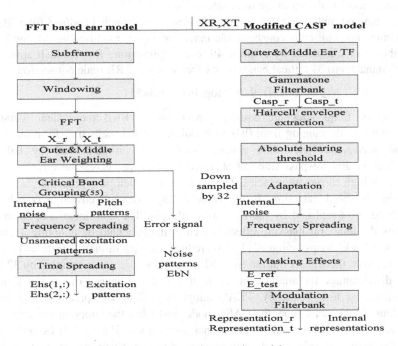

Fig. 2. Proposed psychoacoustic model

3.1   FFT Based Ear Model

The inputs of the FFT based ear model are aligned reference and test signals with sampling rate of 48KHz, they are cut into frames of 2048 samples with an overlap of 1024 samples. Through Hann window and short term FFT, the frequency domain audio signals are grouped into 55 critical bands by a frequency to Bark scale conversion,

$$bark=B(f)=7asinh(f/650) \tag{1}$$

To compensate the influence of the noise internally generated by the human ear, the internal noise is added to get the "pitch patterns" before the frequency spreading. The outputs of the frequency spreading are the so-called "unsmeared excitation patterns"[3]. Finally, we can get the "excitation patterns" which can be used to calculate the MOVs after time spreading.

3.2   Improved Computational Auditory Model

The improved computational auditory model is based on the CASP model by adding frequency spreading and masking effect. The CASP model is the latest computational auditory model, representing major changes at the peripheral and more central stages of processing in the basic computational auditory model by Dau. We plan to apply the CASP model in proposed psychoacoustic model and begin with the basic Dau96 model by adding frequency spreading and masking effect. The block diagram of the auditory model is shown in the right side of Figure 2.

The out and middle ear processing is modeled by a simple bandpass filter. Basilar-membrane and hair cells constitute the inner ear. In our modified model, the basilar-membrane is stimulated by a linear 4th order gammatone filterbank. It split up the audio signals into 34 critical bands by the frequency to ERB scale conversion,

$$erb=E(f)=9.265*\log(1+f/228.8455) \tag{2}$$

The center frequencies are equally spaced on an ERB (equivalent rectangular bandwidth) scale ranging from 0Hz to 8000Hz. The hair cells transform mechanical sound waves into electronic neural signals. It is implemented by half-wave rectification and lowpass filtering at 1 KHz, which preserves the envelope of the signal for high carrier frequencies. The value of the envelope is processed by absolute hearing threshold detection. If the value of the extracted envelope is below the threshold, it is replaced by the threshold value[6]. The non-linear adaptation stage models the temporal masking effect, which is implemented by adopting a chain of five feedback loops circuits[1]. To reduce computational effort and storage consumption, the signals out put from adaptation part are down sampled by 32 times. It is disadvantage for analysis of the signals that the energy of out-put signals is centralized at low frequency. PEAQ adopt the frequency spreading to solve the problem. At the same time, the CASP model just takes the temporal masking effects into account but ignores the frequency masking effects. Hence, it is essential to add the frequency spreading and frequency masking effects processing stage in the modified CASP model.

Other quality assessment auditory models besides PEAQ do not pay much attention on the analytical investigation of cortex, just considering the working mechanism of peripheral ear, middle ear and inner ear. In the computational auditory model, we take the 8Hz lowpass modulation filtering into account corresponding to the cortex of the human auditory. The outputs of the modulation filtering stage are "internal representations" expanding the processing on modulation frequency domain besides the temporal and frequency domain.

4   Cognitive Model

After the frame by frame processing of FFT based ear model, the "excitation patterns" related to audio quality are calculated. Through excitation patterns preprocessing and

MOVs calculation, FFT based ear model generates 2 MOVs, Segmental NMRB (Noise-to-Mask Ratio) and EHSB (Harmonic Structure of the Error). For computational auditory model, we take the "internal representations" as the "excitation patterns" and use the similar MOVs calculation methods of PEAQ to get another 3 MOVs: RmsModDiffA, RmsNoiseLoudAsymA and AvgLinDistA[3]. According to the linear MOA and MinmaxMOA algorithms, the 5 MOVs are mapped to ODG. In MOA linear algorithm, the weighs are training according to the equation:

$$Aw=p \qquad (3)$$

The p is the average subjective score for inputs 72 pair audio sequences. We take 2/3 part of the audio to training, the p is a 48*1 vector, representing the 48 subjective scores of the training audio. A is a 48*5 matrix. The weight vector w is calculated by the equation (4).

$$w=(A^T A)^{-1} A^T p \qquad (4)$$

Finally, the objective quality score is:

$$q=w^T m \qquad (5)$$

The MinmaxMOA algorithm is based on MOA algorithm by only implementing Minimax-Optimal MOVs selection[10]. The MinmaxMOA optimization algorithm is implemented as follows:

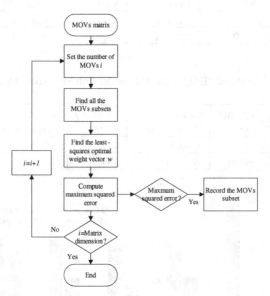

Fig. 3. Flow chart of MinmaxMOA optimization algorithm

5   Experiment Results and Discussion

In our experiment, we used eight kinds of mono audio with subjective scores[12] obtained by MUSHRA as audio database. 2/3 part of the audio files (the corresponding frame loss rate is 10% and 30%) is used to train and the remaining 1/3 part to test (the corresponding frame loss rate is 20%). For comparison, the correlation and MSE between SDG and ODG of PEAQ were obtained as benchmark. Instead of ANN, we improved PEAQ by using the linear MOA and MinmaxMOA algorithm. On that basis, we used the modified CASP model as the computational auditory model of the psychoacoustical stage as well as respectively adopted the linear MOA and MinmaxMOA algorithm as the cognitive model algorithms, obtaining a novel audio quality assessment system described in section 2. The comparison results are shown in Table1 and Table 2. In the tables, the prefixes of each method present the different psychoacoustical processing stages. The Ba_PEAQ and Ad_PEAQ means the experiment adopt the psychoacoustic model of PEAQ basic and advanced version while the CASP means the computational auditory model. At the same time, the suffixes of each method present the different cognitive stages. The ANN means the original artificial neural network algorithm used in PEAQ. The MOA and MinmaxMOA respectively means the linear prediction MOA algorithm and Minmax-optimal MOV selection algorithm.

Table 1. Results of PEAQ and CASP model with MOA algorithm

Methods	Ba_PEAQ_ ANN	Ba_PEAQ_ MOA	Ad_PEAQ_ ANN	Ad_PEAQ_ MOA	**CASP_ MOA**
Correlation	0.124747	0.326631	0.244129	0.316974	**0.576802**
MSE	7.550451	0.621434	0.596015	0.712967	**0.671301**

Table 2. Results of PEAQ and CASP model with MinmaxMOA algorithm

Methods	Ba_PEAQ _ANN	Ba_PEAQ_ MinmaxMOA	Ad_PEAQ _ ANN	Ad_PEAQ_ MinmaxMOA	**CASP_ MinmaxMOA**
Correlation	0.124747	0.379999	0.244129	0.101446	**0.431274**
MSE	7.550451	0.666900	0.596015	0.707361	**0.677081**

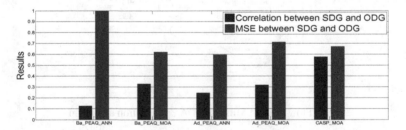

Fig. 4. Histogram of PEAQ and CASP model with MOA algorithm

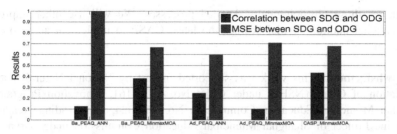

Fig. 5. Histogram of PEAQ and CASP model with MinmaxMOA algorithm

From Table 1, comparing ANN and MOA algorithms used in PEAQ basic and advanced version, we can see that the correlation between SDG and ODG using MOA linear cognitive algorithm is improved, but the MSE is raised a little at the same time. It means the linear MOA algorithm can increase the closeness of estimation at a cost of slightly increase the error of evaluation. When compare the Ad_PEAQ_MOA and CASP_MOA, it is worth to notice that the correlation is increased sharply by 81.97%. Furthermore, the MSE is reduced by 5.84%. In this regard, the computational auditory model has been well proved to be accuracy and effective in objective audio quality assessment.

From Table 2, comparing Ad_PEAQ_ANN and Ad_PEAQ_MinmaxMOA, the correlation is decreased by 58.45% while the MSE is raised by 18.68%. The bad performance is caused by selecting very few MOVs when calculate the final quality score, so the information related to audio quality reduced and led to the inaccuracy of assessment. However, it does not affect the accuracy of the proposed objective audio quality assessment system. When compare the Ad_PEAQ_MinmaxMOA and CASP_MinmaxMOA, the correlation is also increased sharply by 325.13% with MSE reduced by 4.28%. It is proved again that the audio quality assessment using computational auditory model is effective and reasonable.

Combined the results in Table 1 and Table 2, it shows that the proposed method gives better results over existing quality measures and correlate well with subjective quality measures. In particular, the audio used in our experiment is highly impaired audio, the correlation and MSE of PEAQ performs badly while the results of improved system using modified CASP model is better. It indicates that the improved audio quality assessment system can overcome the weakness of PEAQ for measuring the quality of highly impaired audio as well.

6   Conclusion and Future Work

We developed a PEAQ compatible audio quality evaluation system. The performance is highly improved by using the effective computational auditory model. Furthermore, our proposed objective evaluation system is compatible with PEAQ advanced version and can predict the quality of highly impaired audio, the PEAQ can be further refined by using the computational auditory model. In this paper, the basic computational auditory model has been proved to be closely to human ear acoustic transformation, but ignores those physiologically detailed processes. We will use other improved computational auditory model besides the latest CASP to implement audio quality evaluation.

Acknowledgement. This work was supported in part by National Natural Science Foundation of China (NSFC) under Grant No. 61001161, Innovation Program of Shanghai Municipal Education Commission No.12YZ024, and Leading Academic Discipline Project of Shanghai Municipal Education Commission under Grant No. J50104.

References

1. Dau, T., Kollmeier, B., Kohlrausch, A.: A quantitative model of the effective signal processing in the auditory system. I. Model structure. The Journal of the Acoustical Society of America 99, 3615–3622 (1996)
2. Hansen, M., Kollmeier, B.: Using a quantitative psychoacoustical signal representation for objective speech quality measurement. In: IEEE International Conference on Acoustics, Speech, and Signal Processing, ICASSP 1997, vol. 2, pp. 1387–1390 (1997)
3. ITU-R recommendation BS.1387-1. Methods for Objective Measurements of Perceived Audio Quality. ITU-R Recommendation BS.1387-1 (2001)
4. Campbell, D., Jones, E., Glavin, M.: Audio quality assessment techniques-A review, and recent developments. Signal Process 89, 1489–1500 (2009)
5. Creusere, C.D.: Understanding perceptual distortion in MPEG scalable audio coding. IEEE Transactions on Speech and Audio Processing 13, 422–431 (2005)
6. Huber, R., Kollmeier, B.: PEMO-Q:A New Method for Objective Audio Quality Assessment Using a Model of Auditory Perception. IEEE Transactions on Audio, Speech, and Language Processing 14, 1902–1911 (2006)
7. Dau, T., Kollmeier, B., Kohlrausch, A.: Modeling auditory processing of amplitude modulation. I. Detection and masking with narrow-band carriers. The Journal of the Acoustical Society of America 102, 2892–2905 (1997)
8. Guo, Z.: Objective Audio Quality Assessment Based on Spectro-Temporal Modulation Analysis. Master's Degree Project of Stockholm, Sweden (2011)
9. Morten, L.J., Stephan, D.E., Torsten, D.: A computational model of human auditory signal processing and perception. The Journal of the Acoustical Society of America 124, 422–438 (2008)
10. Creusere, C.D., Kallakuri, K.D., Vanam, R.: An Objective Metric of Human Subjective Audio Quality Optimized for a Wide Range of Audio Fidelities. IEEE Transactions on Audio, Speech, and Language Processing 16, 129–136 (2008)
11. Zhu, M.Y.: Zheng. J., Yu, X.Q., Wan, W.G.: Audio Quality Assessment Improvement via Circular and Flexible Overlap. In: 2011 IEEE International Symposium on Multimedia (ISM), pp. 47–52 (2011)
12. Zhu, M.Y., Zheng, J., Yu, X.Q., Wan, W.G.: Streaming audio packet loss concealment based on sinusoidal frequency estimation in MDCT domain. IEEE Transactions on Consumer Electronics 56, 811–819 (2010)

A System for Offline Character Recognition Using Auto-encoder Networks

Sagar Dewan[1] and Srinivasa Chakravarthy[2]

[1] Department of Electrical Engineering
Indian Institute of Technology, Madras, Chennai 600036, India
[2] Department of Biotechnology
Indian Institute of Technology, Madras, Chennai 600036, India
sagar.dewan@outlook.com, srinivasa.chakravarthy@gmail.com

Abstract. We present a technique of using Deep Neural Networks (DNNs) for offline character recognition of Telugu characters. We construct DNNs by stacking Auto-encoders that are trained in a greedy layer-wise fashion in an unsupervised manner. We then perform supervised fine-tuning to train the entire network. We provide results on Consonant and Vowel Modifier Datasets using two and three hidden layer DNNs. We also construct an ensemble classifier to increase the classification performance further. We observe 94.25% accuracy for the two hidden layer network on Consonant data and 94.1% on Vowel Modifier Dataset which increases to 95.4% for Consonant and 94.8% for Vowel Modifier Dataset after combining classifiers to form an ensemble classifier of 4 different two hidden layer networks.

Keywords: Autoencoder Neural Networks, ANN, Deep Networks, Offline Telugu Character Recognition.

1 Introduction

Handwriting recognition is a complex problem since every human has a different style of writing and bears a different character model in mind. We aim at building an offline character recognition system for Telugu, a language spoken in the southern part of India. There is very limited literature available in the area of offline character recognition of Indian scripts. (Pal & Chaudhuri, 2004) provides a detailed review of offline recognition methodologies in OCR development as well as work done on Indian scripts (Pal et al., 2012; Rajashekararadhya & Ranjan, 2008; Banashree et al.,2007).

Telugu is one of the prominent languages in India having more than 62 million speakers. There are 16 vowels, 36 Consonants in this language. A single character or Akshara in Telugu consists of none, one, or multiple consonants (C) and a vowel (V), expressed as C^*V. In the present study, we consider only CV type Telugu characters. Character recognition in Telugu is more complicated than in isolated English characters for the following reasons:

T. Huang et al. (Eds.): ICONIP 2012, Part IV, LNCS 7666, pp. 91–99, 2012.

- Compared to English, most Indian languages have more basic characters and also large number composite characters. Particularly Telugu script has over 5,000 composite characters.
- Though most Indian scripts (Urdu is an exception) are written in left-to-right direction, information is also distributed vertically in multiple tiers. In case of Telugu, information is typically organized in three tiers and in rare situations even in four tiers.

1.1   Literature Survey

(Pal et al., 2012) report results for offline handwritten character recognition of Telugu, but consider only the pure vowels (V) and consonants (C), without considering CV combinations. Furthermore, they consider very small data sets (about 20-30 samples on the whole). Pal et al (2007) describe offline recognition of Telugu numerals but not of regular characters. Moreover, there is very little work on offline recognition of Telugu script.

Although artificial neural networks (ANNs) showed great initial potential as adaptive, universal learning systems, practical difficulties arise in training networks with many (>3) layers, - the so-called deep neural networks (DNNs) – seemed to seriously hamper the full realization of that potential.   However, recent breakthroughs in development of rapid training algorithms for DNN opened up immense opportunities in machine learning applications (Hinton et al, 2006).   In this paper we use newly proposed methodology of training DNNs with greedy layer-wise training (Bengio et al., 2007). The approach involves stacking auto-encoders layer by layer as compared to the conventional approach of initializing the weight values randomly and training the entire network in parallel. DNNs have recently shown to perform better than many state of the art techniques (Erhan et al., 2010). In this paper, we present a multi-classifier DNN-based system for offline HCR of Telugu script. The outline of the paper is as follows: Section 2 describes acquisition of Telugu character data used for training and its preprocessing. Section 3 presents the architecture of the model used and the relevant equations. Performance results are described in Section 4. A discussion of the entire study is presented in the final section.

2   Data Acquisition and Pre-processing

2.1   Data

In the present preliminary study, we collect online data and then convert it to offline data, since online data offers many conveniences in data collection and preprocessing, compared to directly handling offline data. We used the tablet G-Note 7000 to collect handwriting samples. During data collection, each writer was asked to write a list of words in different fields of an electronic form, such that individual words are written in separate rectangles, enabling easy segmentation of words. The words are further segmented into characters in a semi-automated fashion.

Fig. 1. Sample Character Image   **Fig. 2.** Sample Image after Pre-processing

2.2   Pre-processing of Images

Isolated characters obtained from data collection are converted into 28X28 images. The character image is first registered/centered appropriately. To this end we make use of available online information related to the character. The first stroke that is written in a Telugu character is almost always the "main" stroke around which other strokes in the character are organized. We register the character image such that the mid-point of the bottom span of the main stroke is at the point Vo = (18, 14) in image coordinates (Fig. 1). Vo serves as some sort of an origin for the character. Once the origin is fixed, the two dimensions of the character are scaled (preserving the aspect ratio) so that the character fits the image frame (of size 28X28) tightly at least in one dimension (Fig. 2).

3   Class Imbalance Problem

Since the data was collected from several sources, we found a great disparity in the number of samples present in various classes. In such a situation, it is well-known that machine learning algorithms tend to over-train on the majority class (Guo, et al 2008). We expanded the number of samples in under-represented classes by shifting the images corresponding to such classes by 2 to 4 pixels in top/down/left/right directions.

The recognition problem addressed in this study considers CV type characters, in which both consonant (24 classes) and vowel modifier (15 classes) information is present in the same image. It is challenging to separate consonant and vowel modifier image from the same character. We train separate networks for classifying consonants and vowel modifiers. Accordingly, the training data for Consonants includes 21,600 training images (=900 X 24) and 7200 test images (= 300 X 24). Similarly, the Vowel Modifier Dataset consists of the training set of 13500 images (= 900 X 15) and test dataset size of 4500 (=300 X 15).

4   Classification Using Auto-encoder Neural Networks

4.1   Sparse Auto-encoders

Auto-encoders are 3-layered neural networks that model the function y=x. The idea is to extract reduced dimensionality features from the data points by restricting the feature space to the number of hidden nodes (Ng, 2011).

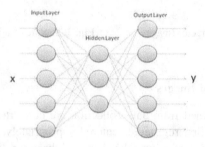

Fig. 3. Sample Auto-encoder Neural Network

To extract interesting features from the Auto-encoder, we impose a sparsity constraint on the hidden layer, which makes the output vector of the hidden layer sparse, i.e. making sure that most of the nodes in the hidden layer are inactive most of the time. We define a sparsity parameter ρ (ρ~0), and restrict the average activation of every node in the hidden layer (averaged over all training examples) to be close to ρ. We do so by calculating ρ(i) corresponding to the activation of ith node in the hidden layer such that:

$$\frac{1}{n}\Sigma_{j \in 1:n}\, ac(i;j) \;=\; \rho(i) \tag{1}$$

where ac(i;j) corresponds to the activation of hidden node 'i' to training example 'j', n corresponds to the number of training examples and 'm' corresponds to the number of hidden layer nodes. We force $\rho(i)$ to approach ρ by calculating the KL Divergence and minimizing it.

$$KL = \Sigma_{j \in 1:m}\, \rho * log\left(\frac{\rho}{\rho(j)}\right) + (1 - \rho) * log\left(\frac{1-\rho}{1-\rho(j)}\right) \tag{2}$$

To minimize the KL divergence term, we add it to the net cost function (Ng, 2011)

$$Cost \;=\; \left\| d - y(x) \right\|^2 + \beta * KL + \lambda * \Sigma \left\| w \right\|^2 \tag{3}$$

Where y(x) is the predicted output by the network, d is the desired output, β is the weight given to the sparsity term and λ is the regularization (weight decay) parameter.

4.2   Optimization

We use limited memory Broyden–Fletcher–Goldfarb–Shanno (L-BFGS) optimization algorithm to implement back-propagation in training autoencoders, using Mark

Schmidt's minfunc package (Schmidt, 2012). minFunc uses a quasi-Newton strategy, where limited-memory BFGS updates with Shanno-Phua scaling are used in computing the step direction. In the line search, cubic interpolation is used to generate trial values, and the method switches to an Armijo back-tracking line search on iterations where the objective function enters a region where the parameters do not produce a real valued output.

4.3   Greedy Layer-Wise Stacking Auto-encoders

Greedy layer-wise pre-training overcomes the challenges of deep learning by introducing a prior in the weights to the supervised fine-tuning training procedure. The parameters are restricted to a relatively small volume of parameter space within a local basin of attraction generated by supervised fine-tuning cost function (Erhan et al., 2010). Auto-encoders can be stacked to form a deep network by feeding the latent representation of layer below as input to the current layer. The unsupervised pre-training is done one layer at a time. Each layer is trained by minimizing the reconstruction of its input. Once the first k layers are trained, we can train the $(k+1)^{th}$ layer because we can now compute the output of the k'th layer and thus build an independent autoencoder that reconstructs that output. Once all layers are pre-trained, the network goes through a second stage of training called fine-tuning. Here we consider supervised fine-tuning where we minimize prediction error on a supervised task. To this end, we first add a logistic regression layer that acts as a classification layer for the network. We then train the entire network as a simple feed-forward neural network using standard back-propagation algorithm.

4.4   Ensemble Classifier

We create an ensemble classifier by combining various classifiers with least correlation in the performance on the validation set. We trained 10 different Deep Networks of different sizes on the training set. We ranked the networks on the basis of their performance on the validation set. We selected 4 networks among these 10 with maximum performance and least correlation amongst them. This was carried out by using a greedy approach wherein the performance of all the 10 networks on the validation set was calculated and we took all possible combinations of 4 networks and carried out majority voting to find the performance of the ensemble classifier over the validation set. The ensemble that gave the best performance over the validation set was choosen.

5   Results

We performed classification on Telugu dataset of Consonants and Vowel Modifiers, consisting of 2 classes and 15 classes respectively. We hereby present the results obtained after fine tuning the parameters using five–fold cross validation technique.

Parameters Used:

Weight decay parameter (λ) = 0.001, Weight of the sparsity penalty term (β) = 4, Sparsity parameter (ρ) =0.1

First two Hidden Layers use sigmoid activation functions

The Output Layer uses soft-max function to perform classification

The training was performed in batch mode and fine tuning back propagation with maximum number of iterations=2000.

Table 1. Parameters used for Consonant and Vowel Modifier dataset

Type	Train Size	Testing Size	Validation Size	Classes
C	21600	6000	1200	24
VM	13500	3750	750	15

5.1   Network Performance with Consonant Dataset

Performance Using Two and Three Hidden Layer Networks

The trained autoencoder can be visualized by plotting the extracted feature vectors as images. Fig.4 depicts the feature weights of the network with hidden layer size (24X24, 20X20). It represents the weights from the input layer to the first hidden layer, with each patch having a similar size as the input image and having total number of images same as the total number of hidden nodes, where each image corresponds to a node in the first hidden layer.

Fig. 4. Feature weights of the two Layer Network(24X24, 20X20)

Table 2. Performance using 2 Hidden Layers

Hidden Layer1	Hidden Layer2	Test Accuracy
24X24	20X20	94.25
20X20	14X14	91.5
14X14	7X7	89.5
30X30	28X28	94.25
20X20	16X16	92.3
14X14	10X10	90.8
18X18	10X10	91.25

Table 3. Performance using three Hidden Layers

Hidden Layer 1	Hidden Layer 2	Hidden Layer 3	Test Accuracy
24X24	**20X20**	**16X16**	**94.8**
24X24	14X14	10X10	91.75
20X20	16X16	10X10	92.25
20X20	10X10	7X7	90.5

5.2   Network Performance with Vowel Modifier Dataset

Performance Using Two and Three Hidden Layer Networks
The trained auto-encoder can be visualized in a similar fashion as we did for the consonant dataset.

Fig. 5. Feature weights of the two Hidden Layer Network (24X24,20X20)

Table 4. Performance using 2 Hidden Layers

Hidden Layer1	Hidden Layer2	Test Accuracy
24X24	20X20	94.1
20X20	14X14	90.8
14X14	7X7	88.7
30X30	28X28	94.2
20X20	16X16	91.2
14X14	10X10	89.6
18X18	10X10	90.2

Table 5. Performance using 3 Hidden Layers

Hidden Layer 1	Hidden Layer 2	Hidden Layer 3	Test Accuracy
24X24	**20X20**	**16X16**	**94.3**
24X24	14X14	10X10	91.3
20X20	16X16	10X10	92.0
20X20	10X10	7X7	89.8

5.3   Performance with Ensemble Classifier

We get a performance of 94.8% using the ensemble classifier that combines 4 networks (each two hidden layer) selected among 10 different networks. The selected 4 networks are with sizes {24X24, 20X20}, {20X20, 14X14}, {20X20, 16X16}, {14X14, 10X10}.

6   Discussion

We present a technique using Deep Networks for offline Telugu character recognition. We experiment with various parameters and various layer sizes for the network. After analyzing the results we can make certain observations such as:

- An increase in the test performance as we increase the number of nodes in the hidden layers
- A slight increase in the test performance on using 3 hidden layers opposed to two, i.e. from 94.25% to 94.8% for the Consonant dataset and from 94.1% to 94.3% in Vowel Modifier Dataset

One possible explanation for three hidden layer network to give better performance than a two hidden layer network is the increased non-linearity introduced by the extra layer. Similarly, increasing the number of nodes in each layer plays a significant role in determining the network's performance. We also observe that after a certain layer size, the performance becomes stagnant and does not increase any further. We find the hidden layers sizes of 24*24 and 20*20 to be optimal.

We can thereby conclude that DNNs essentially can be trained to give higher performance provided they are:

- Not trained using traditional back-propagation approach
- Using large datasets for training to prevent overtraining a highly non-linear network
- Use the dataset balanced over all labels, to prevent overtraining the network on one particular label

Lastly, we propose an approach to combine various classifiers to form an ensemble classifier based on correlation in the error patterns on the validation set. This approach seems to increase the classifier performance from 94.25% to 95.4% for consonant dataset and 94.1% to 94.8% for Vowel Modifier dataset. Our future efforts will be aimed at experimenting with Stacking De-noising auto-encoders (Vincent, 2010), which have recently shown to extract robust features from the dataset and give better performance.

Acknowledgements. The authors acknowledge the support of the Department of Information Technology, India.

References

1. Banashree, N., Andhre, D., Vasanta, R.: OCR for script identification of Hindi (Devnagari) numerals using error diffusion Halftoning. In: Proceedings of World Algorithm with Neural Classifier, pp. 46–50 (2007)
2. Bengio, Y., Lamblin, P., Popovici, D.: Greedy layer-wise training of deep networks. In: Advances in Neural Information Processing Systems (2007)
3. Erhan, D., Bengio, Y., Courville, A.: Why does unsupervised pre-training help deep learning? Journal of Machine Learning 11, 625–660 (2010)

4. Guo, X., Yin, Y., Dong, C., Yang, G.: On the class imbalance problem. In: Fourth International Conference on Natural Computation, pp. 192–201 (2008)
5. Hinton, G.E., Osindero, S., Teh, Y.-W.: A fast learning algorithm for deep belief nets. Neural Computation 18(7), 1527–1554 (2008)
6. Ng, A.: Ufldl Tutorial on Neural Networks. Ufldl Tutorial on Neural Networks (2011) retrieved from,
   `http://ufldl.stanford.edu/wiki/index.php/UFLDL_Tutorial`
7. Pal, U., Chaudhuri, B.B.: Indian script character recognition: a survey. Pattern Recognition 37(9), 1887–1899 (2004)
8. Pal, U., Jayadevan, R., Sharma, N.: Handwriting Recognition in Indian Regional Scripts. ACM Transactions on Asian Language Information Processing 11(1), 1–35 (2012)
9. Rajashekararadhya, S.V., Vanaja Ranjan, P.: Neural network based handwritten numeral recognition of Kannada and Telugu scripts. In: IEEE Region 10 Conference, pp. 1–5 (2008)
10. Schmidt, M.: minFunc.,
   `http://www.di.ens.fr/~mschmidt/Software/minFunc.html`
11. Vincent, P.: Stacked Denoising Autoencoders: Learning Useful Representations in a Deep Network with a Local Denoising Criterion. Journal of Machine Learning Research 11(5), 3371–3408 (2010), doi:10.1111/1467-8535.00290

TrafficS: A Behavior-Based Network Traffic Classification Benchmark System with Traffic Sampling Functionality

Xiaoyan Yan[1], Bo Liang[1], Tao Ban[2], Shanqing Guo[1], and Liming Wang[3]

[1] School of Computer Science and Technology, Shandong University, China
[2] National Institute of Information and Communications Technology, Japan
[3] DNSLAB, China Internet Network Information Center, Beijing, China
bantao@nict.go.jp, guoshanqing@sdu.edu.cn, wlm@cnnic.cn

Abstract. In recent years, there have been many methods proposed to perform network traffic classification based on application protocols. Still, there is a pressing need for a practical tool to benchmark the performance of these approaches in real-world high-performance network environments. In this paper, based on rigorous requirements analysis on real-world environments, we present a real-time traffic classification benchmark system, termed TrafficS, which aims at easy performance-evaluation between different intelligent methods. TrafficS is not only extensible to incorporate multiple traffic classification engines but supports different packet/stream sampling techniques as well. Furthermore, it could provide users a comprehensive means to perceive the difference between inspected methods in various aspects.

Keywords: Network traffic classification, high-performance network.

1 Introduction

In the past decade, network traffic classification has been a heavily explored administrative means in enterprise networks to ensure critical e-business applications, cut off unwanted applications, enforce network security, and so on [1]. It is also proved to be helpful in Internet Studies to review the trend in social, technical, cultural, and other dimensions of the Internet.

Early network applications use well-known static port numbers assigned by the Internet Assigned Numbers Authority, e.g., most FTP (file transport protocol) clients use port 21 to communicate with a server. This fact indicates that an early traffic classifier that relies on transport layer port number of the communication channel could be largely successful. A port-based traffic classifier is easy to implement and incurs little calculation cost, however, as more and more new protocols using dynamic ports, the proportion of network traffic that can be identified by a simple port-based method keeps decreasing [2, 3]. Further advances lead to the introduction of network traffic classifiers based on more complicated characteristics. A popular scheme which makes use of signatures is

T. Huang et al. (Eds.): ICONIP 2012, Part IV, LNCS 7666, pp. 100–107, 2012.
© Springer-Verlag Berlin Heidelberg 2012

known as Deep Packet Inspection (DPI). DPI consists in searching for known string patterns of the application and perform classification on this basis. It is by far the most reliable way to classify network traffic and is widely used in com-mercial products. Another kind of methods first computes the statistical features that present the behavior of the application, and then feeds these features to classifiers or clustering methods to identify the application protocol [4–6].

Although there is a variety of methods proposed for traffic classification and most of them showed good performances in reported work, there are few tools that have been used to verify and compare the performance of different methods in a real-world high-performance backbone network. In response to the growing necessity for such a benchmark system, we introduce TrafficS, which is an network traffic classification benchmark system with good reusability and extensibility. The contributions of this paper could be summarized as follows. First, we analyze the present situation of traffic classification methods and work out five requirements that an evaluation system should satisfy in a real-world network environment. A special emphasis is placed on requirements in high-performance networks. Second, we present a real-time network traffic classification benchmark tool. The tool not only could figure out the importance of flow-features of certain application protocols, but also show the classification results to users in an easily perceptible fashion. In addition, it integrates packet and flow sampling techniques that help to design practical classification algorithms for high performance network environments.

2   Related Work

Recently, several traffic classification systems have been proposed. In[7] Dainotti et al. provided a novel community-oriented traffic classification system called Traffic Identification Engine (TIE). TIE is an open-source classification system which is able to combine multiple classification methods (implemented as separate plug-ins) and adopt different strategies of decision combination. Recently, another traffic classification system called NeTraMark is proposed by S. Lee [8]. It provides a benchmarking platform for eleven state-of-the-art traffic classifiers.

Compared with these proposed systems, our system features a traffic sampling method, which allows users to design high-performance-network-oriented classifiers. Furthermore, our system provides a way to track the change in different statistical features for inspected applications, which could help an operator to discover the most significant features that affect the classification performance of certain applications.

3   Requirements Analysis

We begin by discussing the basic criteria of a system that can be used to evaluate the different classification models. To formulate these requirements, we got inspiration from network management operators in our project and from related works especially when trying to reproduce their results. The basic aspects for a practical network traffic classification benchmark system are outlined as follows.

☐ Data and ground truth availability: The most obvious obstacle to progress on traffic classification is the lack of a variety of sharable traces to serve as test data for validation. Besides the problem of privacy, another reason is the difficulty to annotate a traffic flow with application label.

☐ Comparability: The system should support easy comparison between different traffic classification methods.

☐ Real-time testing: Real-world experiments is vital to evaluate the practicability of a method, therefore, the evaluation system should support efficient memory and data stream management mechanisms to treat with continuous one-pass network data.

☐ Sampling techniques: Network traffic sampling techniques could be an efficient means to alleviate the problem in managing the high-speed throughput of a modern network. However, different sample methods may have varying degrees of impact on the loss rate of key flows. Therefore, a system should provide multiple sampling methods for easy comparison.

☐ Reproducibility: To satisfy the requirement of real-time testing, a classifier and its results should be consistent for different datasets. And when different classifiers are chosen, reproduction of the same data trace should be possible.

4   TrafficS: Mechanisms and Functionalities

4.1   Performance Metrics and Statistical Features

Using appropriate measure metrics to evaluate the performance of traffic classification algorithms constitutes the basis of a benchmark system. In TrafficS, two kinds of metrics are adopted: coarse-grained and fine-grained.

Coarse-grained metrics include overall accuracy, false alarm rate, and missed alarm rate. They are used to measure the performance over all the application protocols. In the following, the overall accuracy is defined as the ratio of the sum of all True Positives (TP) to the sum of all TPs and False Positives (FP) for all classes.

Fine-grained measure metrics for certain protocols include precision, recall, and FPR (False Positive Ratio). All the metrics are defined as follows: For a certain protocol, precision is defined as $TP/(TP+FP)$, recall is defined as $TP/(TP+FN)$, and FPR is defined as $FP/(FP+TN)$. Here, TN and FN are the true negative and false negative values for the codified rule respectively [9].

4.2   Architecture and Implementation

This part describes the overall framework of our classification system. The framework of the system is shown in Fig.1. The system consists of two parts including a server and a client, which can run on different computers. The server receives packets from the monitored network, divide them into flows, and label the flows using a predefined mechanism. It also performs sampling and statistical-feature extraction for each flow. On the other hand, the client is responsible for receiving behavioral data from the server, choosing the classifier or clustering algorithm to predict the class label of input flows. It also provides visualization of the classification results.Some components in the figure can be extended to have new functionalities.

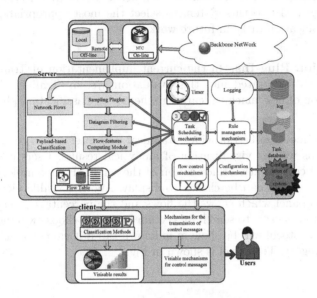

Fig. 1. The general framework of TrafficS

Ground Truth Plug-In. Performance comparison between different classifiers relies heavily on the ground truth label of the collected traces. TrafficS embraces a payload-based classification engine which applies signatures from L7-filter project [10], which is a popularized packet classifier based on DPI. This engine works more robust than the L7-filter on sampled network flows. Moreover, TrafficS allows users to choose any other method to establish the ground truth label and this flexibility helps to update and maintain TrafficS with the most accurate and complete ground truth information.

Sampling Plug-In. Traffic sampling is arguably the most widely accepted technique to cope with the high resource requirements imposed on high-performance networks. Sampling methods could be grouped into two categories: the packet-based sampling scheme [11] and the flow-based sampling scheme [12]. TrafficS has integrated the following sampling algorithms: periodic sampling, Poisson sampling, and random-add sampling. Users are allowed to choose the sampling methods to test the performance and robustness of the classification (clustering) methods.

Behavior Features Computation Component. Definition and selection of the statistical flow-behavior features plays a vital role in classifying the application protocols. In [4], Moore, et al. use 248 bidirectional statistical features to perform application protocol classification. In TrafficS, only 37 of the 248 bidirectional statistical features are adopted. On one hand, these statistical features have been proved expressive enough to distinguish among different applications [1]. On the other hand, TrafficS support an on-demand functionality to incorporate additional statistical features when needed. Feature selection algorithms will be integrated into this system to select the most appropriate features for certain classification tasks in future work.

Classification Plug-Ins. In the current implementation of TrafficS, WEKA [13] is directly integrated into this evaluation system. TrafficS also supports easy adoption of new machine learning schemes as classification plug-ins.

User Interface. TrafficS provides graphical user interface to configure system parameters as well as visualization of analytical results. Users can select and modify some configuration information of the server. For example, information such as the task list can be chosen to display. Another visible view is network flow analysis panel which shows the flows and the percentage of each protocol. This panel allows users to search for certain flows and choose whether to update the view. A protocol will be show in unknown tab if it cannot be classified into a known category. The panel is shown in Fig.2.

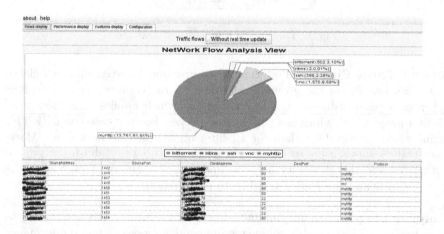

Fig. 2. Information of classified flows

Fig. 3 illustrates the flow statistical features. In this view, dynamic changing of chosen features could be perceived. An operator can learn which features likely have more effect on the performance of traffic classification methods.

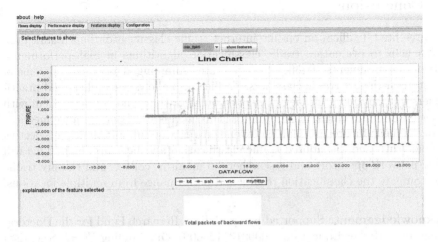

Fig. 3. Statistical feautre(mean-idle)

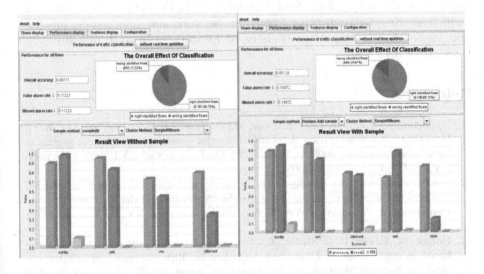

Fig. 4. Performance of the Simple-Kmeans algorithm before and after sampling

Fig. 4 shows the performance of a clustering algorism, named Simple-Kmeans, before and after a random sampling procedure. In this interface, users are allowed to select one of the classification/clustering methods and different sample methods to see the performance of the major applications. From Fig. 4, we can see that the random sampling method will deteriorate the performance of Simple-Kmeans in classifying the ssh protocol. In another word, such evaluation can help us to choose some sampling methods that are suitable for classification.

5   Conclusions

We presented TrafficS, an on-line internet traffic classification benchmark tool, which help to evaluation traffic classification algorithms in high-performance network environments. TrafficS has integrated sampling algorithms and some of the state-of-the-art traffic classifiers. TrafficS also allows researchers and practitioners to easily integrate new classification algorithms and compare them with other built-in classifiers, in terms of the two categories of performance metrics.

In the near future, we plan to extend TrafficS to support multiple selective ground truth and distributed traffic classification and benchmarking. This will allow network operators and researchers to monitor, collect, and classify traffic and compare the classification results and performance from multiple locations.

Acknowledgement. Supported by Specialized Research Fund for the Doctoral Program of Higher Education (20090131120009), Outstanding Young Scientists Foundation Grant of Shandong Province (BS2009DX018), the Key Science-Technology Project of Shandong Province(2010GGX10117), Open Research Fund from Key Laboratory of Computer Network and Information Integration In Southeast University, Ministry of Education,China (K93-9-2010-05), DNSLAB(K201206007).

References

1. Hyunchul, K., Claffy, K.C., Fomenkov, M., Barman, D., Faloutsos, M., Lee, K.Y.: Internet traffic classification demystified: myths, caveats, and the best practices. In: Proceedings of the 2008 ACM CoNEXT Conference, CoNEXT 2008, New York, NY, USA, pp. 11:1–11:12 (2008)
2. Sen, S., Spatscheck, O., Wang, D.M.: Accurate, scalable in-network identification of p2p traffic using application signatures. In: Proceedings of the 13th International Conference on World Wide Web, WWW 2004, New York, NY, USA, pp. 512–521 (2004)
3. Holger, B., Erwin, P.R., Stefan, Z.: Advanced p2p multiprotocol traffic analysis based on application level signature detection. In: 12th International Telecommunications Network Strategy and Planning Symposium, NETWORKS 2006, pp. 1–6 (2006)
4. Andrew, W.M., Zuev, D.: Internet traffic classification using bayesian analysis techniques. SIGMETRICS Perform. Eval. Rev. 33(1), 50–60 (2005)
5. Erman, J., Arlitt, M., Mahanti, A.: Traffic classification using clustering algorithms. In: Proceedings of the 2006 SIGCOMM Workshop on Mining Network Data, MineNet 2006, New York, NY, USA, pp. 281–286 (2006)
6. McGregor, A., Hall, M., Lorier, P., Brunskill, J.: Flow Clustering Using Machine Learning Techniques. In: Barakat, C., Pratt, I. (eds.) PAM 2004. LNCS, vol. 3015, pp. 205–214. Springer, Heidelberg (2004),
http://dblp.uni-trier.de/db/conf/pam/pam2004.html#McGregorHLB04
7. Dainotti, A., de Donato, W., Pescapé, A.: TIE: A Community-Oriented Traffic Classification Platform. In: Papadopouli, M., Owezarski, P., Pras, A. (eds.) TMA 2009. LNCS, vol. 5537, pp. 64–74. Springer, Heidelberg (2009)

8. Lee, S., Kim, H., Barman, D., Lee, S., Kim, C.K., Kwon, T., Choi, Y.: Netramark: a network traffic classification benchmark. SIGCOMM Comput. Commun. Rev. 41(1), 22–30 (2011)

9. Ghorbani, A.A., Lu, W., Tavallaee, M.: Network Intrusion Detection and Prevention - Concepts and Techniques. Advances in Information Security, vol. 47. Springer (2010)

10. http://l7-filter.sourceforge.net/

11. B. RFC 3954 - Cisco Systems NetFlow Services Export Version 9 (2004)

12. Hohn, N., Veitch, D.: Inverting sampled traffic. IEEE/ACM Trans. Netw. 14(1), 68–80 (2006)

13. Witten, I.H., Frank, E.: Data Mining: Practical Machine Learning Tools and Techniques with Java Implementations. Morgan Kaufmann, San Francisco (2000)

Harmony Search with Multi-Parent Crossover for Solving IEEE-CEC2011 Competition Problems

Iyad Abu Doush

Computer Science Department, Yarmouk University, Irbid, Jordan
iyad.doush@yu.edu.jo

Abstract. Harmony search algorithm (HSA) is a recent evolutionary algorithm used to solve several optimization problems. The algorithm mimic the improvisation behaviour of a group of musicians to find a good harmony. Several variations of HSA has been proposed to enhance its performance. In this paper, a new variation of HSA that uses multi-parent crossover is proposed (HSA-MPC). In this technique three harmonies are used to generate three new harmonies that will replace the worst three solution vectors in the harmony memory (HM). The algorithm has been applied to solve a set of eight real world numerical optimization problems (1-8) introduced for IEEE-CEC2011 evolutionary algorithm competition. The experiemental results of the proposed algorithm is compared with the original HSA, and two variations of HSA: global best HSA and tournament HSA. The HSA-MPC almost always shows superiority on all test problems.

Keywords: Harmony Search, Evolutionary Algorithms, Numerical Optimization.

1 Introduction

Evolutionary algorithms (EA) have been used to solve several kinds of real world optimization problems. Harmony Search Algorithm (HSA) [1] is a recent evolutionary algorithm successfully used to solve many practical optimization problems such as: structural optimization, multi-buyer multi-vendor supply chain problem, timetabling, flow shop scheduling [2–6]. HSA has the ability to deal with continuous and discrete variables.

HSA begins with a set of provisional solutions stored in Harmony Memory (HM). At each evolution, a new solution called new harmony is generated based on three operators: (i) Memory Consideration, which *selects* the variables of new harmony from HM solutions; (ii) Random Consideration, used to diversify the new harmony, and (iii) Pitch Adjustment which is responsible for local improvement. The new harmony is then evaluated and replaces the worst solution in HM, if it is better. The solutions in HM will evolve iteratively in the hope of obtaining a better solutions in the next evolutions. This process is looped until a stop criterion is satisfied.

T. Huang et al. (Eds.): ICONIP 2012, Part IV, LNCS 7666, pp. 108–114, 2012.

The HSA is initialized with different parameters as follows:

1. The Harmony Memory Consideration Rate (HMCR), used in the improving process to determine if the value of a decision variable is to be selected from the solutions stored in the Harmony Memory (HM).
2. The Harmony Memory Size (HMS) is an n-dimension vector similar to the population size in Genetic Algorithm.
3. The Pitch Adjustment Rate (PAR), decides whether the decision variables are to be modified to a neighboring value.
4. The distance bandwidth (bw) determines the adjustment value in the pitch adjustment operator.

Genetic algorithm (GA) is a population based search technique [7]. GA starts with initial population with a randomly generated set of solutions. Each population individual is called 'chromosome'. Crossover is the operation in which two randomly selected chromosomes are mixed to generate a pair of new chromosomes. This operation is applied with a certain probability in GA. The performance of GA is related with the use of crossover [8].

In the original HS algorithm, no crossover operation is applied. Generally speaking, the crossover operation directs the search into considering better individuals and enforce having a high diversity of population [9],[10]. The crossover operation has to maintain the diversity of the population and not fall into a premature convergence [10]. The number of solutions considered in the crossover determine the convergence level. Selecting many solutions from the memory will lead to a premature convergence, and selecting small number of solutions will make the algorithm progress slower [7],[10].

The HSA cannot converge sometime on the global optimal [11]. Several variations of HSA have been introduced to solve optimization problems [12–14]. The performance of these variations varies when they are considered on a wide range of problems. In this research, the objective is to improve the performance of HSA by introducing Multi-Parent Crossover (HSA-MPC) on a randomly selected harmony from a set of best solutions. This idea is adopted from [15], which is applied succesfully to genetic algorithms. This method applies the concept of survival of the fittest, and this can enhance the algorithm exploration for the optimal solution. This update on the original HSA is placed after updating the harmony memory (HM) with the new generated harmony, in case it is better than the worst harmony currently in HM.

The algorithm was applied to a group of real world optimization problems that have been proposed for the IEEE-CEC2011 evolutionary algorithm competition [16]. The results are then compared with the original HSA and two recent HSA variants.

This paper is organized as follows: after the introduction, section 2 presents hamony search algorithm with multi-parent crossover. The experimental results, and the analysis of those results, are presented in section 3. Finally, the conclusions are given in section 4.

2 Hamony Search Algorithm with Multi-Parent Crossover (HSA-MPC)

The proposed HSA-MPC works as follows, initial harmony with size HMS is generated randomly, the best m harmonies are stored in archieve pool. After that, a tournament selection with size three is applied to choose solutions randomly from the pool. This is performed according to the Multi-Parent Consideration Rate (MPCR), which specify the number of times this operation performed. Crosover is then applied to the selected three harmonies to generate three new harmonies. The genrated new harmonies are then merged with the HM by replacing them with the three worst individuals in the HM. The details of the algorithm are presnted in the end of this section.

If the new harmony generated make the harmony memory more narrow, then it will lose diversity and reach a premature convergence. On the other hand, if the generated new harmony make the HM widely distrbuted , it will have high diversity and will take longer time to converge.

The steps of the MPC in HSA is as follows (note that $\beta \in U(0, 1)$):

1. Select the harmony vector from the archive pool.
2. Order the harmonies according to thier fitness, the best (x_1) and the worst (x_3).
3. Generate three new harmonies (h_i) as follows:
   - $h_1 = x_1 + \beta \times (x_2 - x_3)$
   - $h_2 = x_2 + \beta \times (x_3 - x_1)$
   - $h_3 = x_3 + \beta \times (x_1 - x_2)$

Example of a Computer Program

```
1: Set HMCR, PAR, NI, HMS, BW, MPCR.
2: x_i^j = LB_i + (UB_i − LB_i) × U(0, 1), ∀i = 1, 2, . . . , N and ∀j = 1, 2, . . . , HMS
   {generate HM solutions}
3: Calculate(f(x^j)), ∀j = (1, 2, . . . , HMS)
4: Sort(HM)
5: itr = 0
6: while (itr ≤ NI) do
7:    x' = φ
8:    for i = 1, · · · , N do
9:       if (U(0, 1) ≤ HMCR) then
10:          x'_i ∈ {x_i^1, x_i^2, . . . , x_i^HMS} {memory consideration}
11:          if (U(0, 1) ≤ PAR) then
12:             x'_i = x'_i ± U(0, 1) × BW { pitch adjustment }
13:          end if
14:       else
15:          x'_i = LB_i + (UB_i − LB_i) × U(0, 1) { random consideration }
16:       end if
17:    end for
```

18:    **if** $(f(x') < f(x^{\text{worst}}))$ **then**
19:       Include x' to the **HM**.
20:       Exclude x^{worst} from **HM**.
21:    **end if**
22:    Sort(HM), and save the best $HMS/2$ harmonies in the archive pool (A).
23:    Apply tournament selection with size three on the archive pool, to select randomly three harmonies (x_1, x_2, x_3).
24:    **if** $(U(0,1) \leq \text{MPCR})$ **then**
25:       Rank the three harmonies $f(x_1) \leq f(x_2) \leq f(x_3)$
26:       calculate $\beta = U(0,1)$
27:       generate three new harmonies (h_i):
28:       $h_1 = x_1 + U(0,1) \times (x_2 - x_3)$
29:       $h_2 = x_2 + U(0,1) \times (x_3 - x_1)$
30:       $h_3 = x_3 + U(0,1) \times (x_1 - x_2)$
31:    **end if**
32:    Replace the three new harmonies (h_1, h_2, h_3) with the three worst harmonies in the harmony memory.
33:    $itr = itr + 1$
34: **end while**

3   Experiemental Results and Analysis

The performance of the proposed algorithm (HSA-MPC) is compared with the original HSA and other two variations of HSA: Global best HSA (HSA-GB) [12] and Tournament HSA (HSA-T) [17]. The algorithm has been coded using MATLAB, and performed on a Windows laptop with intel core 2 duo CPU at 2.10 GHz. As mentioned before, the algorithms behaviour is tested using a set of real world optimization problems (1-8) presented in CEC2011 [16].

The parameter setting is as follows: HMCR = 0.9, bw=0.01, HMS = 50, and PAR=0.3 these values are considered as they are the recommended values in the literature [18],[12]. Note that these parameters are the same for all the evaluated HSA. $\beta = N(0.7, 0.1)$ as this range gives better results according to [15]. The tournament size for HSA-T is $t = 2$, as this value is suggested by [17].

In the initial experiments different values for Multiparent Consideration Rate (MPCR) were used, the value of MPCR=0.3 gives better results. The tournament pool size is 3 and the archive pool size is $\frac{HMS}{2}$ these values are suggested by [15].

The best, median, average , worst, and standard deviation is calculated over 25 simulations, each is allowed to run for 50,000 evaluations of the objective function. These results are presented in Appendix A.

Table 1 report and compare, with respect to the eight real world IEEE-CEC2011 optimization problems (1-8). The optimization performance of HSA,

HSA-MPC, HSA-GB, and HSA-T is compared in terms of best, mean, median, worst, and standard deviation over 25 runs. It is observed that the proposed HSA-MPC shows a superior performance compared to HSA, HSA-GB, and HSA-T on all the optimization problems except for F4 and F6. In these two functions HSA-T gives a better best value The explatory nature of this algorithm gives it a better exploration on the solution space.

4 Conclusion

Harmony search algorithm is a recent evolutionary algorithm used to solve several optimization problems. In this paper, the efficiency of using a new proposed algorithm HSA-MPC is shown to improve the performance of HSA. The introduction of Multi-Parent Crossover to the HSA helps in better exploring the search space for different kinds of optimization problems. The proposed algorithm is compared with the original HSA and two other recent variations of HSA: global best (HSA-GB) and tournament (HSA-T), to solve problems (1-8) presented in the IEEE-CEC2011. HSA-MPC almost always outperforms its competitors on most test problems.

In the future, more detailed analysis on the effect of different parameters of the HSA-MPC algorithm will be measured (i.e., MPCR, tournament size, the pool proportionate of the HM size).

References

1. Geem, Z.W., Kim, J.H., Loganathan, G.V.: A New Heuristic Optimization Algorithm: Harmony Search. Simul. 76, 60–68 (2001)
2. Al-Betar, M.A., Khader, A.T.: A Harmony Search Algorithm for University Course Timetabling. Ann. Oper. Res., 1–29 (2010)
3. Geem, Z.W.: Harmony Search Applications in Industry. Soft Comput. Appl. Ind. 226, 117–134 (2008)
4. Ingram, G., Zhang, T.: Overview of Applications and Developments in the Harmony Search Algorithm. In: Geem, Z.W. (ed.) Music-Inspired Harmony Search Algorithm. SCI, vol. 191, pp. 15–37. Springer, Heidelberg (2009)
5. Taleizadeh, A.A., Niaki, S.T.A., Barzinpour, F.: Multiple-Buyer Multiple-Vendor Multi-product Multi-constraint Supply Chain Problem with Stochastic Demand and Variable Lead-Time: A Harmony Search Algorithm. Appl. Math. Comput. 217, 9234–9253 (2011)
6. Wang, L., Pan, Q.K., Tasgetiren, M.F.: A Hybrid Harmony Search Algorithm for the Blocking Permutation Flow Shop Scheduling Problem. Comput. Ind. Eng. 61, 76–83 (2011)
7. Goldberg, D.E.: Genetic Algorithms in Search, Optimization and Machine Learning. Addison Wesley (1989)
8. Durand, N., Alliot, J.M.: Genetic Crossover Operator for Partially Separable Functions. In: Proceedings of the Third Annual Genetic Programming Conference (1998)
9. Goldberg, D., Deb, K., Korb, B.: Messy Genetic Algorithms: Motivation, Analysis, and First Results. Complex Syst. 3, 493–530 (1989)

10. Mitchell, M.: An Introduction to Genetic Algorithms. MIT Press, Cambridge (1996)
11. Al-Betar, M.A., Khader, A.T., Liao, I.Y.: A Harmony Search Algorithm with Multi-pitch Adjusting Rate for University Course Timetabling. In: Geem, Z.W. (ed.) Recent Advances In Harmony Search Algorithm, vol. 270, pp. 147–162. Springer, Heidelberg (2010)
12. Omran, M.G.H., Mahdavi, M.: Global-Best Harmony Search. Appl. Math. Comput. 198, 643–656 (2008)
13. Pan, Q.K., Suganthan, P.N., Tasgetiren, M.T., Liang, J.J.: A Self-adaptive Global Best Harmony Search Algorithm for Continuous Optimization Problems. Appl. Math. Comput. 216, 830–848 (2010)
14. Wang, C.M., Huang, Y.F.: Self-adaptive Harmony Search Algorithm for Optimization. Expert Syst. Appl. 37, 2826–2837 (2010)
15. Elsayed, S.M., Sarker, R.A., Essam, D.L.: Ga with A New Multi-parent Crossover for Solving IEEE-CEC2011 Competition Problems. In: 2011 IEEE Congress on Evolutionary Computation (CEC), pp. 1034–1040 (2011)
16. Das, S., Suganthan, P.N.: Problem Definitions and Evaluation Criteria for the CEC 2011 Competition on Testing Evolutionary Algorithms on Real World Optimization Problems. Technical Report, Nanyang Technological University, Singapore (2011)
17. Al-Betar, M.A., Doush, I.A., Khader, A.T., Awadallah, M.A.: Novel Selection Schemes for Harmony Search. Appl. Math. Comput. (2011)
18. Mahdavi, M., Fesanghary, M., Damangir, E.: An Improved Harmony Search Algorithm for Solving Optimization Problems. Appl. Math. Comput. 188, 1567–1579 (2007)

Appendix

Table 1. Performance of HSA, HSA-MPC, HSA-GB, and HSA-T in terms of the best, mean, median, worst, and standard deviation over 25 runs (50,000 evaluations each) with respect to each of the 1-8 CEC-2011 problems

		HSA	HSA-MPC	HSA-GB	HSA-T
	Best	0.002825505	**0.001169411**	10.18959362	0.00453789
	Median	11.76850269	11.2518662	20.39105811	12.60136987
F1	Worst	21.14449973	22.81087549	25.87186641	24.60875994
	Mean	9.320810595	9.300249016	18.74778791	13.26065761
	St. d.	8.77917895	8.123201007	4.785734907	8.042664924
	Best	-11.92464615	**-12.3547751**	-12.01816406	-22.33112525
	Median	-21.86011923	-14.22405159	-17.98264158	-24.31308441
F2	Worst	-26.07972549	-25.46724021	-26.43566858	-26.75134593
	Mean	-20.93660167	-15.1047745	-18.05498781	-24.37721521
	St. d.	3.866023763	4.826485404	3.815616303	1.233446763
	Best	1.1514890584E-05	**1.1514890584E-05**	1.1514890595E-05	1.1514890587E-05
	Median	1.1514891362E-05	1.1514890643E-05	1.1514892242E-05	1.1514891275E-05
F3	Worst	0.000011514925	0.000011514900	0.000011514929	0.000011514907
	Mean	0.000011514895	0.000011514892	0.000011514896	0.000011514893
	St. d.	0.000000000009	0.000000000002	0.000000000009	0.000000000004
	Best	14.38019804	14.37670131	14.33941333	**14.35053211**
	Median	18.83940575	18.8193972	15.94638713	15.79395647
F4	Worst	20.96553687	21.08224015	21.01588349	20.96455807
	Mean	17.76917919	18.01447143	17.46044929	17.08416968
	St. d.	2.636306404	2.793641123	2.799854145	2.45993487
	Best	-19.64087474	**-21.42298826**	-31.48316383	-20.97577401
	Median	-17.48218828	-17.03697577	-34.10715195	-17.76295549
F5	Worst	-15.34288531	-15.76809718	-36.84320486	-15.5246342
	Mean	-17.5716699	-17.49035878	-33.9455887	-18.00415639
	St. d.	1.224392317	1.275052976	1.168303224	1.666554768
	Best	-13.36725857	-13.84622043	-27.42897903	**-14.51137227**
	Median	-11.09152916	-10.45893757	-29.16482696	-11.79301466
F6	Worst	-9.521013036	-9.990407624	-29.16558959	-10.58472177
	Mean	-11.28669543	-11.00292482	-28.58617689	-11.91304653
	St. d.	1.358840188	1.223270682	0.86773997	1.296554658
	Best	1.638863341	**1.591549478**	1.63812933	1.625272746
	Median	1.848944665	1.867508329	1.324447293	1.776957031
F7	Worst	2.050761766	1.977414949	1.034493172	2.071230767
	Mean	1.857213988	1.84056682	1.306082435	1.798751976
	St. d.	0.113937362	0.092722969	0.174947997	0.123784513
	Best	**220**	**220**	**220**	**220**
	Median	220	220	220	220
F8	Worst	220	220	220	220
	Mean	220	220	220	220
	St. d.	0	0	0	0

New Intelligent Interactive Automated Systems for Design of Machine Elements and Assemblies

Wojciech Kacalak and Maciej Majewski

Koszalin University of Technology, Department of Mechanical Engineering
Raclawicka 15-17, 75-620 Koszalin, Poland
{wojciech.kacalak,maciej.majewski}@tu.koszalin.pl
http://kmp.wm.tu.koszalin.pl

Abstract. This paper presents a new concept of intelligent interactive automated systems for design of machine elements and assemblies on the basis of its features described in a natural language. In the proposed system, computational intelligence methods allow for communication by speech and handwriting, meaning analyses of design engineer's messages, analyses of constructions, encoding and assessments of constructions, CAD system controlling and visualizations. The system uses an intelligent subsystem for assessment of engineer's ability for efficient designing. It is capable of control, supervision and optimization of the designing process. The system consists of spoken natural language and handwriting interfaces between the designing system and design engineers. They are equipped with several adaptive intelligent layers for human biometric identification, recognition of speech and handwriting, recognition of words, analyses and recognition of messages, meaning analyses of messages, and assessments of human reactions. The paper also makes a comparison of the proposed new automated designing system with the present system of realization of designing tasks. In the system also proposed are new concepts of a system of symbolic notation of construction features and language for notation, archiving and processing of construction description data (object oriented language for construction).

Keywords: Artificial Intelligence, User-Computer Interaction, Intelligent Designing System, Intelligent Interface, Natural Language Processing.

1 Introduction

The presented research involves the development of complex fundamentals of building new intelligent interactive systems for design of machine elements and assemblies on the basis of its features described in a natural language. The scientific aim of the research is to develop the bases of new design processes featuring the higher level of automation, objectual approach to problems and application of voice communication between design engineers and the data processing system. The comparison of the proposed new automated designing system with the present system of realization of designing tasks presents (fig. 1).

T. Huang et al. (Eds.): ICONIP 2012, Part IV, LNCS 7666, pp. 115–122, 2012.

The design and implementation of intelligent interactive automated systems for design is an important field of research. In these systems, a natural language interface using speech and handwriting is ideal because it is the most natural, flexible, efficient, and economical form of human communication [1-3]. This concept proposes a novel approach to intelligent interactive automated systems for design of machine elements and assemblies, with particular emphasis on their ability to be truly flexible, adaptive, human error-tolerant, and supportive both of design engineers and intelligent agents.

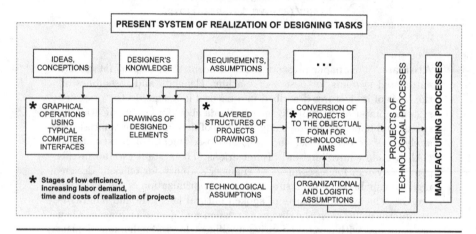

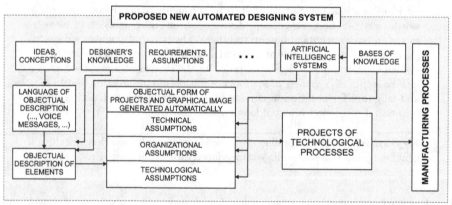

Fig. 1. The comparison of the proposed new automated designing system with the present system of realization of designing tasks

Application of intelligent interactive systems for design machine elements and assemblies using a natural language offers many advantages. It ensures robustness against design engineer errors and efficient supervision of machine design processes with adjustable level of automated supervision. Natural language interfaces also improve the cooperation between a design engineer and a design system in respect to the richness of communication. Further, intelligent inter-action allows for higher organization level of complex design processes, which is significant for their creativeness and efficiency. Design process decision and optimization systems can be remote elements of design processes.

The design of that intelligent system can be considered as an attempt to create a standard intelligent interactive automated system for design processes using natural language communication. It is very significant for the development of new effective and flexible designing methods. It can also contribute for increase of efficiency and decrease of costs of designing processes. This designing system provides an innovative solution allowing for more complete advantages of modern manufacturing processes nowadays.

At the Koszalin University of Technology, taking advantage of the own devel-oped solutions in the range of voice communication between users and technical devices, with the use of artificial intelligence, the research went on to be carried out concerning the development of complex fundamentals of building new intel-ligent interactive systems for design of machine elements and assemblies on the basis of its features described in a natural language.

2 The State of the Art

The most important disadvantages of the present systems for creating construc-tion notation can include:

1. Creation of constructions through executions of graphical operations, with the use of slow communication interfaces, in the form of a keyboard, tablet and mouse, on elementary components of the types of lines and graphic symbols.
2. Drawing is still excessively taking part in imposing the engineer designer's thinking processes.
3. Completing and processing of data occurs in layers, which contain graphical symbols of particular types. Because of that fact, it is difficult to take ad-vantages of objectual treatment of geometrical components of a particular object, e.g. particular grade of designed shaft or even particular cutting.
4. Completing and processing of data in layers containing graphical symbols of particular types (lines, circles,) causes that the software for technolog-ical process design has to perform operations of recognition of elementary graphical objects basing on analyses of graphical notation of these elements (reconstruction of drawings in the technological aspect).
5. Storing information of graphical image instead of storing information in the objectual form of elementary object components and relations between them.
6. Storing construction description data in typical formats for older vector graphic systems using elementary drawing components (lines) instead of using objects which will draw themselves as the result of code interpreter operations.

7. The last disadvantage can be clarified through a comparison of different description methods: construction drawings and description methods of document structures (XML) and internet pages (HTML). From this comparison we can find out that instead of storing a vector drawing, object features can be recorded and the interpreter recreates the drawing on any operating system with the use of universal software. After changing the object features, this drawing will be recorded as a set of features again.

It is worth to notice that the proposed solutions of object feature notation allow for arbitrary advancement of notation integration of construction features and technological process features, and also organizational instructions. It ought to be admitted that remain such cases for which the verbal or symbolic description would not be unequivocal enough [1]. Then data in the graphical form will play an important role. The supplementary information will also provide data from reconstruction processes of shapes and dimensions in the graphical and numerical form.

In the complex design tasks, the release of designer engineers from manual usage of slow interfaces, will allow for elimination of an indirect phase (composing of drawings from graphical symbols). The phase degrades objectual perception of designed object elements to the layered and fault level for further project usage. The application of intelligent interaction systems aims at increase of the designers' efficiency and convenience, and rapidity of creation of new constructions.

The current research has focused on the addition of a supplement to CAD systems [4,5], which consists of simple mechanisms of providing information in the vocal form (in a form of a simple interface for recognition of selected elementary shapes). In that work the objective was to simply support the tasks in traditional systems (simple interface for selective control of a CAD system) [5].

3   Description of the System

The new concept of intelligent interactive automated systems for design of machine elements and assemblies is presented in abbreviated form on Fig. 2. The intelligent designing system is equipped with a subsystem for intelligent assessment of design engineer's ability for efficient designing. The numbers in the cycle represent the successive phases of information processing. The system performs biometric identification of the design engineer whose spoken messages in a natural language are converted to text and numeric values. The recognized text is processed by the meaning analysis subsystem performing recognition of words and messages. The results from that natural language interface are recognized meaningful messages with essential information, which are sent to the subsystem of construction analyses. The analyzed constructions are processed by the subsystem of construction encoding. The novel language for construction notation is used for encoding of the constructions. The next phase of the processing is in the subsystem of construction assessment.

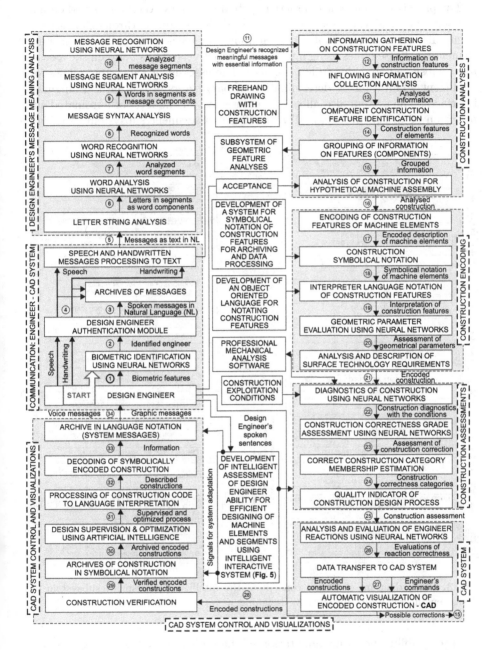

Fig. 2. Block diagram of a new concept of intelligent interactive automated systems for design of machine elements and assemblies on the basis of its features described in a natural language

The proposed intelligent interactive system between CAD systems and design engineers is capable of adaptation to the design engineer through an assessment subsystem that evaluates human ability for efficient designing of machine elements and assemblies using intelligent interactive systems between the assessment subsystem and the design engineer. The assessment subsystem allows for intelligent adaptation by determination of parameters of the natural language interfaces. The system also allows for adjustment of the level of automated supervision of design processes. The intelligent assessment subsystem of design

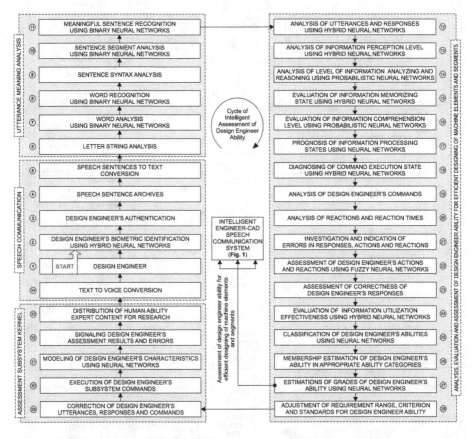

Fig. 3. Block diagram of a new concept of the intelligent assessment subsystem of design engineer's ability for efficient designing

engineer's ability performs adaptation of the designing system in a cycle presented on Fig. 3. After the sentence meaning analysis of the design engineer's utterance, response or message, the recognized meaningful sentences are subject to analysis, evaluation and assessment of the design engineer's ability for efficient designing. The assessment subsystem analyses the user's utterances or responses, and the level of information perception. It also evaluates the level of analyzing and reasoning of information. Then the constructions after the complex processes of analyzing, encoding and assessments are further processed with the CAD system control and visualizations, which is also composed of several specialized modules.

The intelligent designing system allows for optimal control of CAD systems using natural language communication by speech, handwriting and freehand drawing. The messages are processed by the intelligent interface using artificial intelligence methods. The processing involves meaning analysis of words, messages and sentences in a natural language. Therefore the system is capable of designing correct and optimal constructions of machine elements and assemblies. It is also capable of determination of optimal design process parameters and progress decisions with the aim of supporting the design engineer.

The novelty of the system also consists of inclusion of several layers for symbolic notation of construction features and archiving and processing of construction description data using a new object oriented language for construction.

4   Experimental Results

The experimental research of the developed new intelligent designing system allowed to achieve the following research work:

1. Development of a system of construction symbolic notation of element features.
2. Development of bases of a new language for notation, archiving and processing of data concerning construction description (hypertext object oriented language for construction).
3. Development of a specialized interface of voice communication between design engineers and the system of automatic feature recording and creating drawings of the designed elements.
4. Development of improved methods for recognition and processing of voice messages.
5. Development of new algorithms for recognition of handwriting and freehand drawing.
6. Development of elementary procedures of creating of a symbolic notation of construction basing on its description in a natural language using artificial intelligence.
7. Development of elementary procedures of creating of the notation basing on the symbolic notation.
8. Verification of the developed methods of creating of construction notation basing on its description in a natural language, for the following classes of machine elements: shafts, axles, spindles, gears, discs and others.
9. Assessment of quality of generated projects in the aspects of:
   (a) Conformity with the features of standard constructions,
   (b) Correctness of selection of tolerance, definition of dimensions, and dimension chains,
   (c) Correctness of construction of untypical element features.

10. Development of improvements and modifications of algorithms for design procedures and data processing. Application of constructional probability theory in the creation process of ordered variants of machine element constructions. Aggregation of construction standards for verification and evaluation of the system.
11. Development of modifications concerning the new language for construction notation and object language of construction description (with working name KM-XML, and KM-HTML).
12. Development of new directions for further research. Elaboration of foundations for realization of a development project and implementation of the work results.

5   Conclusions and Perspectives

The main effect of the realization of the research was the following:

1. Higher level of designing through complete advantage of designers' creativity, relieving designers from doing tasks involving creation of graphical image of elements.
2. Increase of rapidity of design process, particularly of complex elements.
3. Convenience of modifications and evaluation of many solution variants.
4. Automation of the most laborious tasks in the design of machine elements.
5. Development of an object oriented language for construction notation and methods for symbolic notation.
6. Improvement of operations of data processing and archiving.
7. Development of an artificial intelligence system aiding design processes.

The main results of the research are brand new effective systems for designing machine elements without the use of standard interfaces for inputting data. Directing of designer's creative potential to the conceptual tasks with relieve from performing graphical tasks with use of simple systems for communication with computer applications. The technological effect of the research is appreciable reduction of time for implementation of new and modern products.

References

1. Kacalak, W., Majewski, M.: Intelligent System for Automatic Recognition and Evaluation of Speech Commands. In: King, I., Wang, J., Chan, L.-W., Wang, D. (eds.) ICONIP 2006. LNCS, vol. 4232, pp. 298–305. Springer, Heidelberg (2006)
2. Kacalak, W., Majewski, M.: E-learning Systems with Artificial Intelligence in Engineering. Emerging Intell. Comput. Technol. Appl. 918–927 (2009)
3. Majewski, M., Kacalak, W.: Intelligent system for natural language processing. Comput. Intell. 742–747 (2006)
4. Piegl, L.A.: Ten Challenges in Computer-Aided Design. Comput. Aided Design 37, 461–470 (2005)
5. Kou, X.Y., Xue, S.K., Tan, S.T.: Knowledge-Guided Inference for Voice-Enabled CAD. Comput. Aided Design 42, 545–557 (2010)

Rough Sets and Neural Networks Based Aerial Images Segmentation Method

Xiao Fu*, Jin Liu, Haopeng Wang, Bin Zhang, and Rui Gao

Department of Fundamental Courses, Air Force Aviation University, Changchun, China
{fuxiao_cq}@163.com,
{liujin_aau,Wingroc,Xzhang_0814}@yahoo.com.cn

Abstract. The problem of aerial image segmentation using Rough sets and neural networks has been considered. Integrating the advantages of two approaches, this paper presents a hybrid system different from those previous works where rough sets were used only for accelerating or simplifying the process of using neural networks for aerial image segmentation. The hybrid system have been advanced to improve its performance or to explore new structures. These new segmentation algorithms avoids the difficulty of extracting rules from a trained neural network and possesses the robustness which are lacking for rough set based approaches. The proposed schemes are tested comparatively on a bank of test images as well as real world images.

Keywords: Aerial image segmentation, Rough sets, Neural networks.

1   Introduction

Image segmentation is a fundamental process in many image, video, and computer vision applications. It is often used to partition an image into separate regions, which ideally correspond to different real-world objects. It is a critical step towards content analysis and image understanding.

Many segmentation methods have been developed, but there is still no satisfactory performance measure, which makes it hard to compare different segmentation methods, or even different parameterizations of a single method. However, the ability to compare two segmentations (generally obtained via two different methods/parameterizations) in an application-independent way is important: (1) to autonomously select among two possible segmentations within a segmentation algorithm or a broader application; (2) to place a new or existing segmentation algorithm on a solid experimental and scientific ground [1]; and (3) to monitor segmentation results on the fly, so that segmentation performance can be guaranteed and consistency can be aintained [2].

Designing a good measure for segmentation quality is a known hard problemsome researchers even feel it is impossible. Each person has his/her distinct standard for a good segmentation and different applications may function better using different

---

* Corresponding author.

T. Huang et al. (Eds.): ICONIP 2012, Part IV, LNCS 7666, pp. 123–131, 2012.

segmentations. While the criteria of a good segmentation are often application-dependent and hard to explicitly define, for many applications the difference between a favorable segmentation and an inferior one is noticeable. It is possible to design performance measures to capture such differences.

Although development of image segmentation algorithms has drawn extensive and consistent attention, relatively little research has been done on segmentation evaluation. Most evaluation methods are either subjective, or tied to specific applications. Some objective evaluation methods have been proposed, but the majority of these have been in the area of supervised objective evaluation, which are objective methods that require access to a ground truth reference, i.e. a manually-segmented reference image. Conversely, the area of unsupervised objective evaluation, in which a quality score is based solely on the segmented image, i.e. it does not require comparison with a manually-segmented reference image, has received little attention.

The key advantage of unsupervised segmentation evaluation is that it does not require segmentations to be compared against a manually-segmented reference image. This advantage is indispensable to general-purpose segmentation applications, such as those embedded in real-time systems, where a large variety of images with unknown content and no ground truth need to be segmented. The ability to evaluate segmentations independently of a manually-segmented reference image not only enables evaluation of any segmented image, but also enables the unique potential for self-tuning.

The class of unsupervised objective evaluation methods is the only class of evaluation methods to offer segmentations algorithms the ability to perform selftuning. Most segmentation methods are manually tuned; the parameters for the segmentation algorithm are determined during system development, prior to system deployment, based on the set of parameters that generate the best overall segmentation results over a predetermined set of test images. However, these parameters might not be appropriate for the segmentation of later images. It would be preferable to have a self-tunable segmentation method that could dynamically adjust the segmentation algorithm's parameters in order to automatically determine the parameter options that generate better results. [3] recently proposed one such system, which uses unsupervised evaluation methods to evaluate and merge sub-optimal segmentation results in order to generate the final segmentation. Supervised segmentation evaluation methods only enable this capability on images for which a manually-segmented reference image already exists. Only unsupervised objective evaluation methods, which do not require a reference image for generating a segmentation evaluation metric, offer this ability for any generic image.

This paper provides a survey of the unsupervised evaluation methods proposed in the research literature. It presents a thorough analysis of these methods, categorizing the existing methods based on their similarities, and then discusses their specific differences. A number of empirical evaluations are performed, comparing the relative performance of nine of these unsupervised evaluation methods. Finally, based on the analysis and experimental results, we propose possible future directions for research in unsupervised segmentation evaluation.

2   Rough Sets and Neural Network

2.1   Rough Sets

Rough sets theory (RST) is a machine-learning method, which is introduced by [4] in the early 1980s, has proved to be a powerful tool for uncertainty and has been applied to data reduction, rule extraction, data mining and granularity computation [5].

The basic concept in rough set theory is an information system which can be expressed by a 4-tuple $S = (U, A, V, f)$, where $U = \{x_1, x_2,..., x_n\}$ is a finite set of objects, called the universe; A = C∪D is a finite set of attributes, which is a union of the condition attributes set C and decision attributes set D with C∩D = ∅; V =$\cup_{a \in A}Va$ is a domain of attribute a, and $f : U \times A \rightarrow V$ is an information function to determine each object x_i's attribute value in set U that is: f(x_i, a) $\in V_a$, for $\forall x_i \in U$, a∈A.

In rough set theory, the objects in universe U can be described by various attributes in attributes set A. When two different objects are described by the same attributes, then these two objects are classified as one kind in the information system S, thus we call their relationship is indiscernibility relation. In mathematical word, an indiscernibility relation $IND(\mathbf{B})$ generated by attribute subset $\mathbf{B} \subseteq A$ on U, is defined as follows:

$$IND(B) = \{(x_i, x_j) \in U \times U \mid f(x_i, a) = f(x_j, a), \forall a \in B\} \tag{1}$$

The partition of U generated by $IND(B)$ is denoted by $U/IND(B) = \{C_1, C_2, ..., C_k\}$ for every C_i is an equivalence class. For $\forall$ x $\in$ U the equivalence class of x in relation to $U/IND(B)$ is defined as follows:

$$[x]_{U/IND(B)} = \{y \in U \mid f(x, a), \forall a \in B\} \tag{2}$$

Let X $\in$ U be a target set and $P \subseteq A$ be a attribute subset, that we wish to represent X using attribute subset P. In general, X cannot be expressed exactly, because the set may include and exclude objects which are indistinguishable on the basis of attributes P. However, [6] present a method to approximating the target set P only by the information contained within P by constructing the P-lower and P-upper approximations of X, which is respectively defined as:

$$P - lower approaximations\, of\ X :$$
$$P * X = \{x \mid [x]_{U/IND(B)} \in X\}$$
$$P - upper approximations\, of\ X : \tag{3}$$
$$P * X = \{x \mid [x]_{U/IND(B)} \cap X \neq \varnothing\}$$

The P-lower approximation, also called the positive region, is the union of all equivalence classes in $[x]_{U/IND(P)}$ which are contained by (i.e., are subsets of) the target set X. In another word, the lower approximation is the complete set of objects in $U/IND(P)$ that can be positively classified as belonging to target set X.

The P-upper approximation is the union of all equivalence classes in $[x]_{U/IND(P)}$ which have non-empty intersection with the target set, that is the complete set of objects that in $U/IND(P)$ that cannot be unambiguously classified as belonging to the complement of the target set X. In other words, the upper approximation is the complete set of objects that are possibly members of the target set X.

One of the most important aspects in rough set theory is the discovery of attribute dependencies, that is, we wish to discover which variables are strongly related to which other variables. For this purpose, given two attribute subset $P, Q \subseteq A$, Then, the dependency of attribute set Q on attribute set P, $\gamma_P(Q)$, is given by

$$\gamma_P(Q) = \frac{card(\bigcup_{X \in U/IND(Q)} P * X)}{card(U)} \tag{4}$$

where $\bigcup_{X \in U/IND(Q)} P*X$ can be denoted as $POS_P(Q)$, which means that the objects in it can be classified to one class of the classification $U/IND(P)$ by attribute P.

An attribute a is said to be dispensable in P with respect to Q, if $\gamma_P(Q) = \gamma_{P-\{a\}}(Q)$; otherwise a is an indispensable attribute in P with respect to Q. Let $S = (U, A, V, f)$ be a decision table, the set of attributes $P(P \subseteq C)$ is a reduce of attribute, C if it satisfied the following conditions:

$$\gamma_P(D) = \gamma_P(D), \ \gamma_{P'}(D) \neq \gamma_C(D) \ \forall P' \subset P \tag{5}$$

A reduction of condition attributes C is a subset that can discern decision classes with the same accuracy as C, and none of the attributes in the reduced can be eliminated without decreasing its distrainable capability [7].

Though it is a kernel concept in rough set, it is difficult to calculate the reduction if the size of information system is large. Many scholars proposed a variety of attribute reduction algorithm, such as: consistency of data, dependency of attributes, mutual information, discernibility matrix and genetic algorithm which are employed to find reduction of an information system.

2.2   Neural Network

The BP neural network, which was first described by Paul Werbos in 1974, and gained recognition until 1986 through the work of David E. Rumelhart, Geoffrey E. Hinton and Ronald J. Williams, led to a "renaissance" in the field of artificial neural network research. The BP neural networks are the most widely used networks and are considered the workhorse of ANNs [8]. Thanks to its simplicity and excellent performance in extract useful information from samples, the BP neural network is widely applied recently. Commonly the BP neural network is used to solve the problems of classification and function approximation, which arise frequently in loan risk warning, stock market returns and price index prediction, the power system' short term load forecasting [9], box office revenue of movies forecasting, bank's efficiency evaluation and areas of decision support systems and management science.

An elementary neuron with R inputs of BP is shown in Fig. 1. Each input is weighted with an appropriate wi. The sum of the weighted inputs and the bias forms the input to the transfer function f(•), and f(•) transforms the sum of input values into output values of the node. Typical choices of the transform function consist of the logistic, the tangent, the sign, and the linear.

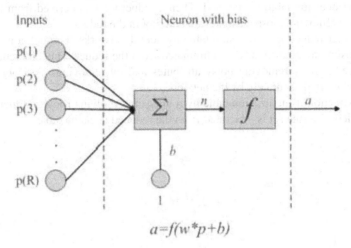

$$a=f(w*p+b)$$

Fig. 1. A neuron of BP

In this paper we apply a BP neural network with two hidden layer in which the neural neurons take tan-sigmoid function for transform, and purelin, a linear function, is used in output layer for transform to get a broad range of output values. The whole structure of our network is shown in Fig. 2, where a1 = tan − sig(IW11 * p1 + b1), a2 = tan − sig(LW21 * a1 + b2), and a3 =purelin(LW32 * a2 + b3), besides the number of neural cells in hidden layers is determined by the training process.

3   System Con_guration

In this paper, we proposed a model which combined the rough set theory and neural networks in aerial images segmentation.

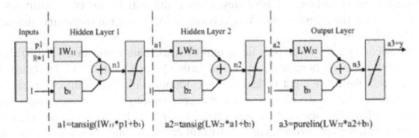

Fig. 2. The structure of BP

3.1   The Structure of System

Our hybrid approach of rough sets and neural networks for aerial images segmentation consists of three major phases:

(1) Attribute reduction by rough sets. Using rough set approach, a reduct of condition attributes of decision table is obtained. Then a reduct table is derived from the decision table by removing those attributes that are not in the reduct.
(2) The further reduction of decision table by neural networks. Through a neural network approach, noisy attributes are eliminated from the reduct. Thus the reduct table is further reduced by removing noisy attributes and by removing those objects that cannot be classified accurately by the network.
(3) Rule extraction from decision table by rough sets. Applying rough set method, the final knowledge–a rule set is generated from the reduced decision table.

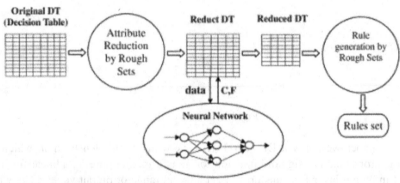

C: attribute subset reserved through neural network
F: object set that cannot be classified accurately by the network

Fig. 3. The procedures of system

3.2   The Algorithm

We develop our algorithms of attribute reduction and rule extraction based on a binary discernibility matrix, which replaces complex set operations by simple bit-wise operations in the process of finding reduct and provides a more simple and intelligible measure for the importance of attributes. Even if the initial number of attributes is very large, using the measure can effectively delete irrelevant and redundant attributes in a relatively short time.

In the second phase, we employ the neural-network feature selection (NNFS) algorithm introduced by Setiono and Liu [10] to further reduce attributes in the reduct. In this approach, the noisy input nodes (attributes) along with their connections are removed iteratively from the network without decreasing obviously the network's classification ability. The approach is very effective for a wide variety of classification problems including both artificial and real-world datasets, which was verified by a lot of experiments. Making use of the robustness to noise and generalization ability of the neural network method, these attributes and objects polluted by noise can be reduced from decision table.

Let $T = <U, C \cup D, V, f>$ be a decision table, $U = \{x_1, x_2, ..., x_m\}$, $C = \{c_1, c_2, ..., c_n\}$. In general, D can be transformed into a set that has only one element without changing the classification for U, that is, $D = \{d\}$. Every value of d corresponds to one equivalence class of $U/IND(D)$, which is also called the class label of object.

A binary discernibility matrix represents the discernibility between pairs of objects in a decision table.Let M be the binary discernibility matrix of S, its element $M((s, t), i)$ indicates the discernibility between two objects x_s and x_t with different class labels by a single condition attribute c_i, which is defined as follows:

$$M((s,t),i) = \begin{cases} 1 & c_i(x_s) \neq c_i(x_i) \\ 0 & otherwise \end{cases} \tag{6}$$

$$where\ 1 \leq s < t \leq m\ and\ d(x_s) \neq d(x_t),\ i \in \{1, 2, ..., n\}.$$

It can be seen that M has n columns and its maximal number of rows is $m(m-1)=2$. Each column of M represents a single condition attribute and each row of M represents an object pair having different d values.

Let M be a binary discernibility matrix having R rows and L columns, and its element value in the ith row jth column is a_{ij}. The discernibility degree of an attribute c_k for classification is defined as

$$Deg(c_k) = \frac{1}{2}\sum_{i=1}^{R} a_{ik} \tag{7}$$

$$where\ k \in \{1, 2, ..., L\}$$

The Deg of an attribute ck is in fact the rate of "1"s in the c_k column of M and can be used as a measure of classification capability of attributes.

4   Results

In this chapter, the performance of the proposed system will be tested in two aspects: Comparison of their convergence rates, quantitative scores obtained from synthetic test images.

4.1   Performance Rates

The rates of all the algorithms are shown in Table.1. It can be observed that, our method keeps the best convergence rate and it converges more rapidly.

Table 1. Comparison of three segmentation algorithms

Name	MRF-CSNN	CSNN	Our method
Speed(ms)	152	296	461

4.2   Qualitative Results

Results obtained from real world images by using CSNN, MRF-CSNN and our methods are shown in Fig.4. It can be noticed that segmentation boundary noise and absorption of the small segments by their larger neighbors are avoided in the CSNN and the MRF-CSNN algorithms. This is in contrast to our method that produces under-segmentation results as small segments are captured by the big ones.

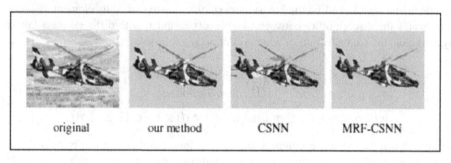

original   our method   CSNN   MRF-CSNN

Fig. 4. Segmentation results of three methods

5   Conclusion

In this study, various innovations of the image segmentation algorithm have been described and tested. The new algorithms are based on rough sets and neural networks. We have combined the rough set and BP neural networks to construct a model for aerial images segmentation. Our method gives a handle to the user to adjust the desired local detail or global morphology of the segmentation. If edge information is available or if the segmentation accuracy along the boundaries is paramount. It gives the overall best performance.

References

1. Bowyer, K.W., Phillips, P.J.: Empirical Evaluation Techniques in Computer Vision. Wiley-IEEE Computer Society Press (1998)
2. Erdem, C.E., Sanker, B., Tekalp, A.M.: Performance Measures for Video Object Segmentation and Tracking. IEEE Transactions on Image Processing 13, 937–951 (2004)
3. Gelasca, E.D., Ebrahimi, T., Farias, M., Carli, M., Mitra, S.: Towards Perceptually Driven Segmentation Evaluation Metrics. In: Proceedings of Conference on Computer Vision and Pattern Recognition Workshop (CVPRW 2004), vol. 4 (2004)
4. Pichel, J.C., Singh, D.E., Rivera, F.F.: Image Segmentation Based on Merging of Suboptimal Segmentations. Pattern Recognition Letters 10 (2006)
5. Pawlak, Z.: Rough sets: Theoretical Aspects of Reasoning about Data. Kluwer Academic Publishing, Dordrecht (1991)
6. Yeh, C.C., Chi, D.J., Hsu, M.F.: A Hybrid Approach of DEA, Rough set and Support Vector Machines for Business Failure Prediction. Expert Systems with Applications (2009)
7. Pawlak, Z.: Rough Set and Intelligent Data Analysis. Information Science 11, 1–12 (2002)

8. Basheer, I.A., Hajmeer, M.: Artificial Neural Networks: Fundamentals, Computing, Design, and Application. Journal of Microbiological Methods 43, 3–31 (2002)
9. Van Droogenbroeck, M., Barnich, O.: Design of Statistical Measures for the Assessment of Image Segmentation Schemes. In: Proceedings of International Conference on Computer Analysis of Images and Patterns (2005)
10. Ge, F., Wang, S., Liu, T.: Image-Segmentation Evaluation from the Perspective of Salient Object Extraction. In: Proceedings of IEEE Internatioanl Conference on Computer Vision and Pattern Recognition, vol. I, pp. 1146–1153 (2006)

Frontal Cortex Neural Activities
Shift Cognitive Resources Away from Facial Activities
in Real-Time Problem Solving

Shen Ren[*], Michael Barlow, and Hussein A. Abbass

School of Engineering and Information Technology
University of New South Wales (Canberra campus), Australia
{s.ren,m.barlow,h.abbass}@adfa.edu.au

Abstract. Extracting mental and task performance state-information from a human in real time is a challenging scientific endeavour. In this paper, we attempt to understand if there is a relationship between the frontal cortex activities in the F3 and F4 positions according to the 10-20 international system of electrode placement, which are known to correlate with executive control functions and working memory, and facial muscle activities. We demonstrate that in a highly demanding control, planning and problem solving task, as the human gets more engaged in the task, there is a consistent increase of correlation between the frontal cortex activities, an anti correlation between the cheeks and forehead muscles, and that the two correlations are perfectly anti-correlated with each other. The results suggest a resource shifting occurring during the task as the task progresses and the complexity of the task increases.

Keywords: Brain-computer interface, Cognitive science, EEG, EMG, Resource.

1   Introduction

Brain-computer interfaces and human-computer interfaces are two topics that can be linked together if we can understand the relationship between what is easily observable – such as human facial activities - and what is not – such as human neural activities. The fusion of these two fields paves the way towards the embodiment and situatedness of a human in virtual and/or synthetic environments. For example, the flight management system on an aircraft will be able to sense the pilot's neural activities, detecting disengagement, engagement, hyper-excitement, fatigue, etc. and adapting the aircraft performance accordingly.

Extracting mental and task performance state-information from a human in real time is a challenging scientific endeavour. Multiple metrics, indicators and a myriad of studies are needed to understand the interactions of different metrics.

In this paper, we attempt to understand if there is a relationship between the frontal cortex activities in the F3 and F4 positions according to the 10-20 international system [1], which are known to correlate with executive control functions, and facial

T. Huang et al. (Eds.): ICONIP 2012, Part IV, LNCS 7666, pp. 132–139, 2012.

expressions. Our hypothesis is that, as a human becomes more engaged in problem solving and planning-oriented tasks, more resources are needed to perform executive control functions. The demands on these resources require a shift of resources from synchronisation of facial activities to achieve synchronisation of executive control.

We use a simple game environment to study this hypothesis. The rest of the paper is structured as follows. Some key background information from the literature is discussed in Section 2. This is followed with the design of the procedure for conducting the experiments in Section 3, a discussion of the results in Section 4, and then conclusions are drawn.

2   Literature Review

The rapid development of neuroscience, Electroencephalography (EEG) and neuro-imaging has made the kind of interfaces that can directly receive inputs from the human brain possible. As brain-computer interfaces are becoming more mature, it is becoming possible to include them as an important sensory channel of Human-computer Interaction [2].

Resource theory conjectures that with finite processing resources, human information processing capabilities have an upper bound and are limited. If two processes use the same resources concurrently, the two processes interfere with each others. This interference is a two-way development, where each process interferes with the other process. These several active cognitive and physical processes competing for limited processing resources cause the performance of a human on a given task to diminish [3]. The primary resource time was originally conceived as non-sharable among tasks [4]. The theory has since been evolved into a limited but sharable "capacity-limited processor" [5], [6]. Different resource models have been studied and proposed, ranging from single channel bottle neck theory [6] to multiple resource models [7] in the last 50 years.

Internal resources are one factor that can influence human task performance. Other factors include goal setting, self-efficacy, ability, strategies, engagement and attention. The effect of goal settings on the enhancement of task performance as a motivation mechanism has long been established [8]. Self-efficacy captures the human ability to judge on his/her own ability to select and execute a course of action within a given context. It is normally associated with a number of factors including one's past experience, experience gained in watching others doing similar tasks, ability to persuade others and self, arousal and other mental activities [9]. Goal commitment, including goal commitment to task performance, is strongly affected by self-efficacy as a major predictor of future performances [10]. Attention has multiple components that can positively influence the ability of a human to achieve a task including strategies selected, self-regulation and efforts put in the task[11].

To study information processing in the human cortex, many techniques are employed including neuro-imaging, EEG, invasive or lesion studies. Among these, neuro-imaging techniques like Functional Magnetic Resonance Imaging (fMRI) and

Positron Emission Tomography (PET) are more suitable for collecting spatial information about the brain. Lesion studies of humans require particular subjects and the damage may show plasticity changes in other brain functions [12].

To study human information processing in real-time tasks, EEGs are chosen to measure electrical functions of the brain in our study. The scalp EEG recorded by a single electrode is a smoothed version of the local field potential (LFP). The spatio-temporally resolved wide-band LFP is probably the most important source of information in neural computations [13].The 10-20 international system proposed by Jasper [1] is usually used for electrode placement and the mapping of external scalp positions and underlying cortex perpendicular to the surface[14]. The electrical responses on F3 and F4 position are conventionally known to reach the dorsolateral prefrontal cortex (DLPFC) [14]. DLPFC is crucial in short-time processing of information [15]. Also, Gevins et al. (1997) found that the general non-specific enhancement of the frontal theta rhythm probably indicates the overall mental effort required for task performance and is also associated with working memory [16]. Note that theta rhythm is the low frequency component of EEG signal from 4-7 Hz.

We also rely in this study on Electromyography (EMG) techniques, which were used to measure facial muscle contraction. Electrodes were placed on forehead to measure the corrugators muscle activity, and the left cheek to measure the zygomatic muscle activity. The facial EMG activities are reacted to facial expressions known as responses to positive and negative stimulus [17].

3   Experimental Design

The experiment employs a version of the Greedy Snake games. The Greedy Snake game is a classic computer game which has been played by millions of people. The aim of the game is to control a continually moving snake using the arrow keys on a keyboard so as to reach apples on the game board, while trying to avoid crashing into the controlled snake's own body, and the walls around the edges of the game board. Once the snake eats an apple, the length of the snake automatically increases and another apple randomly appears. If the snake crashes into a wall or hit its own body, the game ends and the total score that the player gained during the game is calculated according to the number of apples that were eaten. Navigating the snake and eating more apples is the task to be performed by each participant. Once a game ends, a new one starts after filling in the questionnaires. The complexity of the game changes from one game to another due to different game configurations.

Our objective is to compare the player's game performance and the collected psycho-physiological data under different circumstances. Four variables are used as control parameters for game complexity; these are: the moving speed of the controlled snake (the snake moves ahead at low speed 100ms/move or high speed 70ms/move), the increase in the length of the snake after eating an apple (increase by 1 unit or 3 units), whether or not to add an extra apple in the environment in a random location after 7000ms, and whether or not to add a poisoned apple in a random location after

7000ms. The game ends if the snake eats a poisoned apple, it crashes into the wall, or it hits its own body. The different values for each of these four variables define 16 different games. The sequence of game presentation is shuffled for each player. Each player plays each configuration twice. Each player starts with two classic games to establish a baseline, before they then play the 32 different configurations. In total, a player plays 34 games in a session.

Experiment participants were 3 right-handed adults aging from 24 to 27. All had played the classic Greedy Snake Game before, and clearly understood the rules of the game as well as the scoring function. They were briefed about the test procedure before accepting to participate. Before the start of games, the participants were requested to complete questionnaires asking about their level of familiarity with the Greedy Snake Game. At the completion of each game, a questionnaire asking about the participant's self-reported frustration level during the games pops up for them to fill in.

Human response data was collected using EEG and EMG sensors. All these measurements (including self-reported subjective rating, psycho-physiological metrics, and game characteristics) provide both subjective and objective information of human performance and human responses to contribute to our analysis.

The EEG and EMG sensors were attached to the participant before the start of the session and actively collected data throughout the entire session. EEG sensors were attached to sites F3 and F4 on the participants' scalp, with a clip on both earlobes to stabilize the wires. EMG sensors were attached on the participant's forehead and left side of the cheek to measure electrical responses of facial muscle activities.

EEG sensors record small electrical signals on the scalp, said signals being generated by neurons in the brain. The typically used frequency band of EEG is between 1 to 40Hz. It has 3 electrodes: a positive site to measure the raw signal, a negative site as reference, and a ground site. Electrode placement on F3 and F4 followed the international 10-20 system with electrode caps filled with conductive paste to attach to the participant's scalp (shown in Fig.1). F3 and F4 are associated with executive control functions and working memory [16]. Ear references were used for all EEG sensors.

EMG sensors captured the muscle activities by measuring small electrical impulses when facial muscle fiber contract, with the active range of frequency of the raw signal between 20 and 500 Hz. The EMG sensors have 3 electrodes which are positive, negative and ground ones. During the experiment, the positive and negative electrodes were attached to the facial muscles and the ground electrode was placed at neural sites (cheekbones and brow ridge).

Fig. 1. EEG Placement, EEG Sensors and EMG Sensors

Both of the EEG and EMG signals were collected at a sample rate of 256Hz. The self-reported skill level, frustration level and game characteristics including game configurations, starting/ending time for each game, and final scores were collected by automated questionnaires and the game application itself.

4   Analysis

EEG correlation represents the degrees of functional cooperation between underlying neuron substrates and connections between different brain regions [18-19]. Therefore, inter-hemisphere correlation values could also show the degree of cooperation of both hemispheres required in given tasks.

The collected data was processed for each participant separately. After data cleaning, which removed the artifacts caused by sensor setup and the irrelevant parts when players were taking breaks or filling out the questionnaire between games, the time domain signals were divided into 34 different parts according to the time stamps of each game been played. The correlation between two different signal sequences is computed according to the convolution theorem with a 1 second window within each game. The correlation sequence is a function of time t which is known as time lag. The correlation between g and h is shown as Equation 1.

$$Corr(g,h) = \int_{-\infty}^{+\infty} g(t+\tau)h(t)d\tau \tag{1}$$

To compute the correlation between discrete sequences, the function is shown as

$$Corr_{xy}(m) = \begin{cases} \sum_{n=0}^{N-m-1} x_{n+m}y_n^* & m > 0 \\ Corr_{yx}^*(-m) & m < 0 \end{cases} \tag{2}$$

The correlation result is normalized so that the autocorrelations of the sequences themselves at time lag 0 are identical and equal to 1.0. The corresponding means and standard deviation are computed at the level of all 34 games played by each player.

In the first questionnaire, before game play, each participant was requested to identify him/herself into one level of Greedy Snake Game Player: Never played before, beginner, intermediate player, advanced player, and expert. In the inter-game questionnaires, he/she was asked to report his/her frustration level (low to high, from 0 to 15) on the most recently played game. Scores that the player gained were calculated based on the number of apples been eaten. The final score of each game was calculated as the total apples eaten multiplied by 100. The results on subjective and objective game performances were summarized in the first section of Table. 1.

The correlation between the two EEG signals collected from F3 and F4, and the two EMG signals collected from the forehead and left cheek muscle activities were then computed using a 1 second window. The average value reported is the mean of the correlation coefficient sequences in all games being played. The cross-correlation shown in the last row of Table 1 is the average value of the correlation coefficients computed from the correlation sequences of the 34 games.

Table 1. Analysis on Players' Performance, EEG and EMG Data

Players / Analysis	1	2	3
Self-reported skill level	Beginner	Intermediate	Intermediate
Mean frustration level (0-15)	7.1±1.5	7.0±1.5	8.0±2.3
Average Play time/game (in second)	76±47	71±64	51±44
Average Score/game	2140±1520	1280±1020	1920±1830
Correlation between EEG F3 and F4	0.87±0.11	0.50±0.08	0.51±0.08
Correlation between EMG on forehead and on left cheek	-0.79±0.09	-0.86±0.06	-0.84±0.06
Cross-correlation of correlation between EEG and EMG data	0.99±0.01	0.99±0.01	0.99±0.01

The game performance for each player is shown in the first section of Table 1. The total score shows the "level of goal achievement". It is a positive factor influencing the game performance – the higher score a player obtains, the better their performance will be rated. The play time shows "the duration of survival". Besides the objective of chasing apples, the player should try to navigate the snake on the game board and stay alive, as the snake continues to move.

The results in Table 1 show the subjective rating in the first two rows and objective indicators of game performance in the next two rows. Interestingly, player 1 who underrated themselves as a beginner had achieved the best performance among all 3 players, having both longest play time and highest scores. He/she also reported a stable and relatively low level of frustration during games.

For players 2 and 3, who both rated themselves as intermediate players, the results show that player 2 played longer while player 3 obtained higher scores during the entire session. This variance may indicate the different objectives of the two players, player 2 aims at staying alive, while chasing apples is a higher priority objective for player 3. The different goal setting affected performance in different ways. The overall trend shows that the best player had shown more stable performance while player 2 had higher deviation on play time and player 3 had higher deviation on final scores.

The best performing player (player 1) had the highest correlation between frontal F3 and F4 signals, indicating cooperation of both hemispheres during play which may indirectly contribute to better game performance. Players 2 and 3 had similar (lower) levels of mean EEG correlations. The objective performance indicators show that one of them obtained higher scores while the other played for longer. The results support previous research that self-efficacy, defined as the self-perception of ability, is related to the cognitive engagement in games [9]. The first player who identified him/herself as a beginner had higher levels of cooperation of both frontal hemispheres during games indicated by high correlation between EEG F3 and F4; than player 2 and 3 who were more self-confident in their skills. Frontal brain activity is associated with cognitive workload; hence these results regarding correlation appear to indicate the level of engagement in game play.

The correlation between the EMG on forehead and that on the left cheek shows almost the opposite trend as compared that of the EEG. That is, an inverse correlation. By further analyzing this trend, we looked into each game for each player and plotted the correlation sequences. Figure 2 shows the correlation coefficient sequences within the first game played by player 3. During the game, while the difficulty increases due to the increasing length of the snake, the correlation between EEG F3 and F4 increases while the correlation between the EMG measuring the forehead and cheek muscle activities decreases. This trend is shown in almost all other games. There appears to be an anti-correlation between these two processes. This can be interpreted as a resource shift from facial activities to brain functions across the duration of the game: a prioritizing of resources to planning and decision making as difficulty increases.

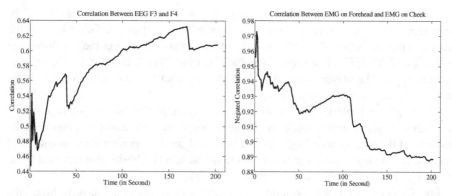

Fig. 2. Correlation between EEG F3 and F4 (on Left) and Correlation between EMG on Forehead and EMG on Cheek (on right) for a Single Game

5   Conclusion

The analysis of the subjective and objective task performance, as well as the factors contributing to the performance suggests that 1) the self-assessment of the player's own ability do not necessarily match the objective performance results - this accords with previous research results [20-21]; 2) the different goals setting by different players contribute to different performance results; 3) the correlation between EEG F3 and F4 could be an indicator of attention and engagement which influences the performance; 4) there appears to be a resource shift occurring during tasks as difficulty of the game increases, in that case the cognitive resource is concentrated on the higher-requirements and higher-priority parts of the brain functions to process game information.

References

1. Jasper, H.H.: The Ten Twenty Electrode System of the International Federation. Electroencephalogr. Clin. Neurophysiol. 10, 371–375 (1958)
2. Tan, D., Nijholt, A.: Brain-Computer Interfaces and Human-Computer Interaction. In: Tan, D.S., Nijholt, A. (eds.) Brain-Computer Interfaces, pp. 3–19. Springer, London (2010)
3. Norman, D.A., Bobrow, D.G.: On Data-Limited and Resource-Limited Processes. Cognitive Psychol. 7, 44–64 (1975)
4. Craik, K.J.W.: Theory of the Human Operator in Control System1. The Operator as an Engineering System. Brit. J. Psychol. General Section 38, 56–61 (1947)
5. Moray, N.: Where is Capacity Limited? A Survey and a Model. Acta Psychologica 27, 84–92 (1967)
6. Welford, A.T.: Single-Channel Operation in the Brain. Acta Psychologica 27, 5–22 (1967)
7. Wickens, C.D.: Multiple Resources and Performance Prediction. Theor. Issues Ergon. Sci. 3, 159–177 (2002)
8. Locke, E.A., Shaw, K.N., Saari, L.M., Latham, G.P.: Goal Setting and Task Performance: 1969–1980. Psychol. Bull. 90, 125–152 (1981)
9. Bandura, A.: Self-efficacy Mechanism in Human Agency. American Psychologist 37, 122–147 (1982)
10. Locke, E.A., Frederick, E., Lee, C., Bobko, P.: Effect of Self-efficacy, Goals, and Task Strategies on Task Performance. J. Appl. Psychol. 69, 241–251 (1984)
11. Greene, B.A., Miller, R.B.: Influences on Achievement: Goals, Perceived Ability, and Cognitive Engagement. Contemporary Educational Psychol. 21, 181–192 (1996)
12. Gerloff, C., Corwell, B., Chen, R., Hallett, M., Cohen, L.G.: Stimulation over the Human Supplementary Motor Area Interferes with the Organization of Future Elements in Complex Motor Sequences. Brain 120, 1587–1602 (1997)
13. Buzsáki, G., Anastassiou, C.A., Koch, C.: The Origin of Extracellular Fields and Currents — EEG, ECoG, LFP and Spikes. Nat. Rev. Neurosci. 13, 407–420 (2012)
14. Herwig, U., Satrapi, P., Schönfeldt-Lecuona, C.: Using the International 10-20 EEG System for Positioning of Transcranial Magnetic Stimulation. Brain Topogr. 16, 95–99 (2003)
15. Rossi, S., Cappa, S.F., Babiloni, C., Pasqualetti, P., Miniussi, C., Carducci, F., Babiloni, F., Rossini, P.M.: Prefontal Cortex in Long-Term Memory: an "Interference" Approach Using Magnetic Stimulation. Nat. Neurosci. 4, 948–952 (2001)
16. Gevins, A., Smith, M.E., McEvoy, L., Yu, D.: High-Resolution EEG Mapping of Cortical Activation Related to Working Memory: Effects of Task Difficulty, Type of Processing, and Practice. Cereb. Cortex 7, 374–385 (1997)
17. Dimberg, U.: Facial Reactions to Facial Expressions. Psychophysiology 19, 643–647 (1982)
18. Weiss, S., Mueller, H.M.: The Contribution of EEG Coherence to the Investigation of language. Brain Lang. 85, 325–343 (2003)
19. Clarke, A.R., Barry, R.J., Heaven, P.C.L., McCarthy, R., Selikowitz, M., Byrne, M.K.: EEG Coherence in Adults with Attention-Deficit/Hyperactivity Disorder. Int. J. Psychophysiol. 67, 35–40 (2008)
20. Dunning, D., Heath, C., Suls, J.M.: Flawed Self-Assessment. Psychol. Sci. Public Interest 5, 69–106 (2004)
21. Langendyk, V.: Not Knowing that They Do not Know: Self-assessment Accuracy of Third-Year Medical Students. Med. Educ. 40, 173–179 (2006)

Implement Real-Time Polyphonic Pitch Detection and Feedback System for the Melodic Instrument Player

Geon-min Kim[1], Chang-hyun Kim[2,*], and Soo-young Lee[1]

[1] Department of Electrical Engineering,
CNSL, Korea Advanced Institute of Science and Technology, Daejeon, Korea
[2] Department of Bio and Brain Engineering, CNSL,
Korea Advanced Institute of Science and Technology , Daejeon, Korea
{gmkim90,sy-lee}@kaist.ac.kr, flipflop98@gmail.com

Abstract. This research proposes an automatic transcription-feedback system of music which help people to learn musical instruments by themselves. The focus of this research is piano. We develop real-time polyphonic pitch detection-feedback system. For 'polyphonic pitch detection', we use inner product based similarity measure with discriminant note detection threshold and top down attention. Also, we develop two parallel processes on simulink and matlab separately for real-time system. On simulink workspace, real-time recording and signal flow management is implemented. This system takes 2mins. 12secs. for analyzing 1min. piece and have accuracy of pitch detection as 79.33% for test case (Chopin Nocturne Op.9 N.2).

Keywords: Real-time Polyphonic Pitch Detection, Feedback System, Note-scale filterbank, Multi-threshold, Top-down attention.

1 Introduction

More and more people want to learn new musical instruments, but there are not many possible ways for someone to study musical instruments by themselves. It is not easy for beginners to get self-feedback from playing the instrument alone. Thus, this research proposes an automatic transcription-feedback system which will help people to learn musical instruments by themselves.

Fundamental algorithms of pitch detection in time-frequency domain have been researched so far [1], [6]. Also Autotune[2] and Melodyne[7] are well known commercialized programs for monophony and polyphony pitch detection. However, those programs do not transcribe well on commercial CDs and real-time performances.

Therefore through this research, a system will be constructed which will provide real-time pitch detection for polyphonic music, express the music as sheet music, and give feedback to instrument player by comparing with the correct reference of the music. Also we want to compare the polyphonic pitch detection performance with

---

* Corresponding author.

T. Huang et al. (Eds.): ICONIP 2012, Part IV, LNCS 7666, pp. 140–147, 2012.

competitive algorithms [6], [8]. Emmanouil B. et al. [6] announced that multi pitch analyzer[8] algorithms have the best polyphonic pitch detection rate as 70.9% on Chopin Nocturne Op.9 N.2.

The focus of this research is the piano music. First, piano is a polyphonic musical instrument unlike other monophony instruments such as woodwinds or brasses. Also, piano is composed of 88 different sounds from A0 to C8, so it contains a long range of pitches which includes all the pitch ranges for many different musical instruments.

1.1   Note Scale Filterbank Output

In this system, we utilize note-scale filter bank output as feature of the music signal. It has 103 coefficients, which extract information from spectrogram of music. 103 coefficients are calculated by filtering spectrogram with the 103 filters. Each filters are a form of triangle, where their center frequencies are located at $27.5 \times 2^{[0 \cdot 102]/12}$. Note that first 88 coefficients are located at fundamental frequencies of 88 notes of piano according to the previous research.[3] Remaining 15 coefficients are for extracting higher harmonics of note frequencies. By using note-scale filter bank spectrogram, we can selectively emphasize fundamental frequency information from the spectrogram, which makes pitch detection more easy task.

2   System

2.1   Real-Time System

Figure 1 shows the system of real-time polyphonic pitch detection & feedback system that we developed in this research. There are 3 main parts.

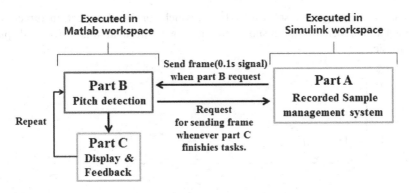

Fig. 1. Simplified real-time system diagram

- In part A, music signal is recorded from microphone and stored in queue with 0.1sec per frame length. Then it is sent to each frame one by one repeatedly to part B whenever part B requests for sending frame.

- In part B, the real-time pitch detection system gets a frame from queue in part A as first-in-first-out (FIFO) sense. Algorithms about poly-phonic pitch detection is shown in section 2.2

- In part C, we display the pitch detection result on time vs. pitch number axis, giving the player the feedback note correction information with ground truth music score.

The complete system diagram is shown in Figure 2. And the detail function of each subdiagrams are explained in section 3.2

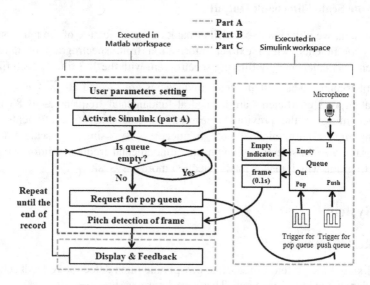

Fig. 2. Complete real-time system diagram

In part C, the program display a feedback, which consist of correct, incorrect, and missing note numbers. Figure 3 shows how feedback information is shown in display for 9s music.

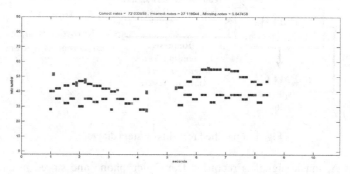

Fig. 3. Example of the feedback display (until 9s)

For given frame(0.1s), we filled each detected note with blue, red and green color on correct, incorrect and missing notes separately. Player can know whether they play correct pitch with correct beat or not. Program also shows the real-time accuracy of the performer's music by comparing with the score, which has the note sample index versus note pitch number axis.

2.2   Pitch Detection Algorithm

Figure 4 shows the algorithm of pitch detection. Firstly, standardized (Mean zero, Variance one) music signal is converted into note-scale filterbank output (=feature). Then this feature is normalized on every time frame. Next, inner product with the references(pitch templates), which are the average values of feature of 88 individual notes. Note that we can measure similarity of two different normalized vectors by inner product. The references are made by following sequences; 1) recording each individual notes of piano, 2) getting feature, 3) time-averaging and 4) normalize them. Since there are many overlapping harmonics between 88 notes, 88 features are not orthonormal with each other, which we can see the simulation result in section 2.3. We use two methods for supporting deficient parts of inner product as pattern recognition; 1) Different detection thresholds for each notes. 2) Top down attention. Details of these methods are explained in section 2.3 and 2.4 respectively.

After inner product process, if the value exceed detection threshold for each note, that note is regarded as played note candidates, otherwise regarded as silence. By using top down attention for these candidates, algorithm gives final detected notes.

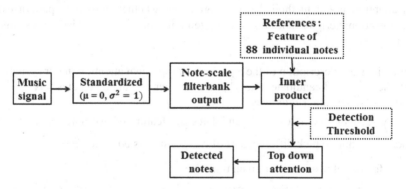

Fig. 4. System diagram for pitch detection algorithm

2.3   Method to Set Different Detection Thresholds for Each Notes

Before setting detection thresholds, we made an experiment by measuring similarity between 88 references. Fig 5 shows the inner product between 88 notes features. Red,yellow,and green parts except main diagonal shows references are not orthonormal with each other. Especially the notes lower than number 25(A2) and notes higher than number 80(E7) have high value of similarity with other notes pattern. Based on this observation, for each note, we set high detection threshold when the average value of similarity is high, and set low detection threshold when the average value of similarity is low.

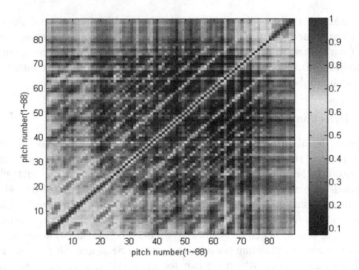

Fig. 5. Similarity between feature of 88 piano notes

2.4   Top Down Attention

Before top down attention step (see Fig 4), system already obtain the note candidates; some of them are 'really played notes', some of them are 'not played notes'. Top down attention is the method that can figure out whether individual pattern exists inside mixed pattern or not. For our system, it finds 'really exist' notes among candidate notes.

Followings are description of the algorithm of top down attention in our system;
1) Load feature of candidate notes .Let the number of candidate notes = N
2) If   N=1, The algorithm ends

3) If N≥2, pick two notes from candidates (let feature of two notes as $\vec{x}, \vec{y}$). The number of method to pick different sets of two features is equal to $\frac{N(N-1)}{2}$.

4) Let feature of test music at given time as $\vec{M}$.

5) Find a,b which minimizes error $e = \left\| \vec{M} - (a\vec{x} + b\vec{y}) \right\|$ by finding pseudo inverse of the linear system$(\vec{x} \quad \vec{y})\binom{a}{b} = \vec{M}$.

6) If one of $\binom{a}{b}$ makes 'e' smaller than given threshold, accept corresponds two notes.

7) Repeat from 3) until iterations run $\frac{N(N-1)}{2}$ times.

On each figure 6 and 7 are two test results with and without top-down attention.

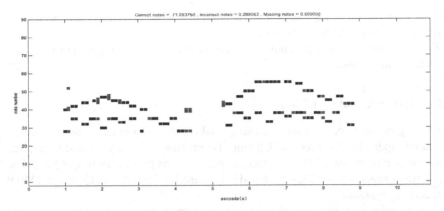

Fig. 6. Results without top-down attention (Accuracy : 71.09%)

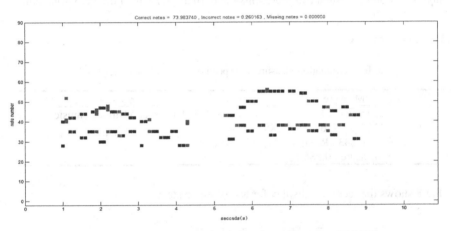

Fig. 7. Results without top-down attention (Accuracy : 73.98%)

We can see that top-down attention can reduce detecting false-positive note, which is note that is not played but detected as note alive by system.

3   Test and Performance

3.1   Performance Evaluation Criteria

We develop 'real-time pitch detection system' for this project. Thus performance can be evaluated by two criteria : Accuracy &Speed.

For accuracy,we usefollowingevaluation metrics : $\text{Accuracy} = \frac{Tp}{Tp+Fp+Fn} \times 100 \ [\%]$,

which is the simplest metric for evaluation of accuracy. Some of researcher [4], [5] uses 'Precision', 'Recall', and 'F-measure' as their evaluation metrics. (Where, Tp = "true positive": number of correct notes among played notes; Fp="false positive":

number of incorrect notes among played notes; Fn:"false negative": number of not played notes among reference notes).

For the speed of the algorithm, we measure the average computation times for analyzing whole music.

3.2   Test Data and Condition

For testing our system, in terms of accuracy and computation time, we use the piece ' Nocturne Op.9 N.2 'of Frederic Chopin. To measure accuracy of real-time pitch detection itself, we use MIDI reference, which contains correct answer of the piece, MIDI was created by the Prokeys 88(MIDI controller device) and Cubase 6 (MIDI sequencing program).

We are doing test with normal room (i.e. no silent condition), normal speaker and normal microphone, which can represents the normal user's recording environment.

3.3   Performance

Table 1. Evaluation measure and speed of our system(frame-based)

Measure	Speed
Accuracy = 79.33%	532.4s process/4min 2s music
(Tp = 3843 , Fp = 716 , Fn = 285)	≅ 132s/1min music
Precision = 0.843 , Recall 0.931,	
F-measure : 0.885)	

Fig 8 shows the feedback display for test of our system.

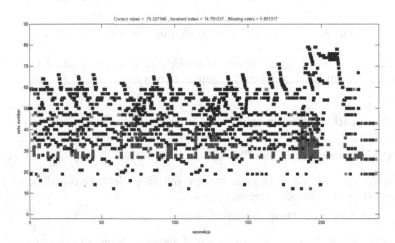

Fig. 8. Feedback display of the test

4   Conclusion and Future Work

In summary, we develop real-time polyphonic pitch detection-feedback system. For 'polyphonic pitch detection', we use inner product based similarity measure with discriminant detection threshold and top down attention (See Section 2.4). And for real-time system, we develop two simultaneously running process system in simulink and matlab. One is for real-time recording and signal flow management, and the other is for real-time pitch detection and displaying feedback to users.

As a final result,   accuracy of pitch detection of our system is 79.33% for 4min music and takes 132s to analyze 1min piece, which is over 8% improvement to the state of art system[8].

For the future work, we can add the system for reduce the effects of room acoustics for considering different user's environment. And the accuracy should be improved by considering musiccal knowledge such as key, beat, harmonic science etc.

Acknowledgement. This research was supported by Qualcomm incorporated. Chang-hyun and Geon-min are co-first authors, who equally contributed to this paper. We thank professor Sang-Hoon Oh and Dr. Wonil Chang for their helpful advices.

References

1. David, G.: Pitch Extraction and Fundamental Frequency: History and Current Techniques. Technical report. University of Regina, pp. 2–12 (2003)
2. Worldwide Standard in Vocal Processing Tools: Auto-Tune, http://www.antarestech.com/products/auto-tune-evo.shtml
3. Edward, M.B.: Intervals, Scales, and Tuning. In: Deutsch, D. (ed.) The Psychology of Music. Academic Press, San Diego (1999)
4. Mert, B., Andreas, F.E., Stephen, D.: Evaluation of Multiple-F0 Estimation and Tracking Systems. International Society for Music Information Retrieval 2009 (2009)
5. Graham, E.P., Daniel, P.W.E.: A Discriminative Model for Polyphonic Piano Transcription. EURASIP J. Appl. Sig. P. 8, 1–9 (2007)
6. Emmanouil, B., Simon, D.: Joint Multi-Pitch Detection Using Harmonic Enve-lope Estimation for Polyphonic Music Transcription. IEEE J-STSP 5, 1111–1123 (2011)
7. Melodyne, http://www.celemony.com/cms/
8. Kameoka, H., Nishimoto, T., Sagayama, S.: A Multipitch Analyzer Based on Harmonic Temporal Structured Clustering. IEEE Trans. Audio, Speech, Lang. Process. 15, 982–994 (2007)

Classification of Interview Sheets Using Self-Organizing Maps for Determination of Ophthalmic Examinations

Naotake Kamiura[1,*], Ayumu Saitoh[1], Teijiro Isokawa[1], Nobuyuki Matsui[1], and Hitoshi Tabuchi[2]

[1] Graduate School of Engineering, University of Hyogo, Himeji, Japan
{kamiura,saitoh,isokawa,matsui}@eng.u-hyogo.ac.jp
[2] Tsukazaki Hospital, Himeji, Japan

Abstract. In this paper, a method of determining examinations is presented for outpatients visiting the department of ophthalmology. It assumes that each of the interview sheets belongs to one of the four classes, and copes with the examination determination as the classification of the sheets using self-organizing maps. Training data presented to the maps are generated from handwriting sentences in the sheets. Some nouns, adjectives and adverbs that ophthalmologists consider to be of comparative importance are chosen as elements of the training data. The element values basically depend on frequencies of the chosen words appearing in the sentences. After map learning is complete, neurons in the map are labeled. The data class associated with the sheet to be checked is given as the label of the winner neuron for the presented data. It is established that the proposed method achieves as favorable classification accuracy as initial determination made by ophthalmologists.

Keywords: Data classification, Interview sheets, Self-organizing maps, Waiting time problems.

1 Introduction

Recently, a number of problems have surfaced in the Japanese medical arena. Long waiting time is one of such problems. It has been considered to be a serious reason that prevents sick persons from going to hospitals. The condition of such persons keeps on worsening while they hesitate to go to hospitals, and the fatal damage tends to suddenly befall to them. The straightforward approach to overcome this problem is to save even short amount time in the waiting room. In [1], an approach using event-driven network based on queuing theory is presented to reduce the waiting time of patients. The dispatching rules are suggested based on patients' expected visitation time and expected service time, and they are used to schedule the patients.

The effective utilization of the waiting time is also a promising approach. Some hospitals in Japan manage waiting time as part of examination time. Before seeing a new outpatient, a medical doctor generally reads an interview sheet filled out by the outpatient, and determines a set of examinations for the outpatient. In this medical protocol, the outpatient must wait for the determination made by the doctor. In

T. Huang et al. (Eds.): ICONIP 2012, Part IV, LNCS 7666, pp. 148–155, 2012.

addition, the protocol makes it difficult for the doctor to allot enough consultation time after examination results are available. The doctors are therefore anxious for the system automatically determining a set of examinations for new outpatients.

On the other hand, self-organizing maps (SOM's) have attracted much attention. A map consists of neurons with reference vectors. Map learning projects attributes of training data onto the reference vectors, and hence the attributes are visualized in the form of neuron clusters. This feature motivates researchers to apply SOM's as a means of expressing and/or processing clinical examination results [2]-[7].

In this paper, examination determination is proposed for new outpatients visiting the department of ophthalmology, using SOM's. It assumes that each of the interview sheets filled out by the outpatients belongs to one of the four classes. The proposed method therefore addresses the determination as the classification of the sheets. An open source engine developed for Japanese language morphological analysis is applied to handwriting sentences in the sheets, and some nouns and adjectives in them are picked up as elements of the training data for map learning. In addition, some words on which the ophthalmologists especially place great importance are also picked up. Frequencies of the above chosen words appearing in the sentences are basically assigned as element values. After general SOM learning is complete, labels corresponding to classes of training data are given to neurons in the map. The class of the sheet associated with the data presented to the map is therefore specified by the label of the winner neuron for the presented data. It is revealed that the proposed method is approximately close to ophthalmologists in classifying interview sheets.

2   Preliminaries

SOM learning constructs a map with neurons. The neuron has a reference vector with M element values if the M-dimensional training data is presented to the map. General SOM learning is conducted, based on the following formulas.

$$NF(t)=r_0(1-t/T), \tag{1}$$

$$\tau_i(t)= \tau_0(1-t/T), \tag{2}$$

$$W_i(t)\leftarrow W_i(t)+\tau_i(t)(X^l(t)-W_i(t)). \tag{3}$$

Each time the l-th training data, $X^l(t)$, is presented to the map, a winner neuron is determined. Eq. (1) is the neighborhood function defined around the winner at time t. Let C_i denote the i-th neuron with the reference vector $W_i(t)$. If C_i is located inside the area specified by $NF(t)$, $W_i(t)$ is modified according to Eqs. (2) and (3). $\tau_i(t)$ is the learning rate. Note that r_0 in Eq. (1) and τ_0 in Eq. (2) are initial values, and that T is the maximum epoch number employed as the learning-termination condition.

The proposed method determines examinations that new outpatients visiting the department of ophthalmology should undergo, based on sentences, which are handwritten in Japanese by them, in the interview sheets. It is probable that outpatients contracting several diseases undergo same examinations. In this context, according to common characteristics in terms of the examinations for the diseases, the examinations are divided into the following four classes: Class 1 associated with fundus examinations, Class 2 associated with glaucoma tests, Class 3 associated with slit lamp tests, and Class 4 associated with oculomotor tests. The relationships between these classes and concrete diseases are tabulated in Table 1. Since a new outpatient undergoes the examinations belonging to one of the four classes, the interview sheets can also be divided into such four classes. The proposed method therefore copes with determination of examinations as classification of interview sheets.

3   Examination Determination Using Self-Organizing Maps

3.1   Data Set Generated from Interview Sheets

Handwriting sentences in interview sheets are typed from the keyboard. The open source engine developed for Japanese language morphological analysis, MeCab [8], is applied to such electronically registered sentences to divide the words into some parts of speech. Several of the nouns and adjectives are next picked up per sentence in the sheet by consulting the list of prohibited words. Table 2 shows examples of the prohibited words. The words picked up are considered to be promising to describe the characteristic of the condition of the outpatient that fills out the sheet. In addition, some nouns, adjectives, and adverbs on which the ophthalmologists place great importance are also picked up. They are referred to as MD-designated words.

After all the promising words are picked up, the proposed method generates a matrix. If a set of N interview sheets whose classes are perfectly known is available and d words are chosen from the N sheets in the above-mentioned manner, the numbers of rows and columns are N and d, respectively. In other words, the words picked up (or sheets) are assigned to the columns (or rows). Frequencies of appearance are first given to element values as follows: if the p-th word appears in the l-th sentence m_{lp} times, the element specified by the l-th row and p-th column is set to m_{lp}, where $1 \leq l \leq N$ and $1 \leq p \leq d$. Fig. 1 depicts an example of the first matrix.

To emphasize the significance of the words, the proposed method employs the following weighting. Element values on the columns corresponding to the MD-designated words are usually weighted. Let us assume that each of such values is multiplied by α_{MDW}. The other targets for weighting are determined, based on the probability of words appearing. Let us assume that the p-th word, which is not the MD-designated word, appears $NA_p{}^q$ times in the set of registered sentences (i.e., the interview sheets) belonging to Class q, where $1 \leq p \leq d$ and $1 \leq q \leq 4$. In addition, let $R_p{}^q$ denote the ratio of the number of the p-th word appearing in the sentences belonging to Class q, compared to the total number of the sentences belonging to Class q. If the latter number is denoted by NS^q, $R_p{}^q$ is as follows.

$$R_p{}^q = NA_p{}^q / NS^q. \tag{4}$$

Table 1. Relationships between classes and diseases

Classes	Diseases that patients contract
Class 1	Cataract, Diabetes, Retinal disease, Uveitis, Pediatric ophthalmology
Class 2	Glaucoma
Class 3	Corneal and/or conjunctival disease, Lacrimal apparatus, Ametropia, The others
Class 4	Strabismus, Neuro-ophthalmologic disease, Trauma

Table 2. Examples of prohibited words

Types of prohibited words	Examples
Numbers	1, 2, 0, 100, 36.5
Symbols associated with SI base units	m, mm, cm, kg
Geographical names	姫路/Himeji, 佐用/ Sayoh
Words associated with time	週/ Week, 金曜日/ Friday, 去年/ last year, 先月/ last month

	Word 1	Word 2	Word 3	...	Word $d-1$	Word d
	涙 "Tears"	充血 "Hyperemia"	手術 "Surgery"	...	重い "Serious"	白内障 "Cataract"
Sheet 1	1	2	1	...	0	0
Sheet 2	0	0	0	...	2	1
Sheet 3	0	1	0	...	1	0
⋮	⋮	⋮	⋮	⋱	⋮	⋮
Sheet N	0	1	1	...	1	0

Fig. 1. Example of first matrix

All the possible R_p^q's are calculated. The words with comparatively high R_p^q's are probably of importance in specifying the attributes of Class q. The proposed method sets the threshold value to 0.055 when some word is checked whether its element values are weighted. A word determined as a target is referred to either as a powerful word or as a special word. The p-th word is considered to be powerful if $R_p^k \geq 0.055$ and $R_p^q < 0.055$ for some k and any q, where $k, q \in \{1, 2, 3, 4\}$ and $k \neq q$. Let α_{PW} denote the coefficient by which each of the element values on the column corresponding to the p-th word is multiplied.

If we have $R_p^q < 0.055$, $R_p^{k1} \geq 0.055$ and $R_p^{k2} \geq 0.055$ for any q, and some pair of $k1$ and $k2$, where $q, k1, k2 \in \{1, 2, 3, 4\}$ and $k1 \neq k2$, $k1 \neq q$, and $k2 \neq q$, the p-th word is considered to be special. The element values on the column corresponding to the p-th word are then multiplied by α_{SW}. It is difficult to specify the attribute of a unique class, using a word that equally appears in the numerous sentences belonging to arbitrary classes. This is why targets of weighting are restricted to the words, each of which has the value calculated by Eq. (4) exceeding the threshold (0.055) for at most two classes.

A simple example of weighting is as follows. If Word 2 in Fig. 1 is the powerful word, element values on the second column change from $(m_{12}, m_{22}, m_{32}, ..., m_{N2}) = (2, 0, 1, ..., 1)$ to $(2\alpha_{PW}, 0, \alpha_{PW}, ..., \alpha_{PW})$. If Word d is the MD-designated word, we have $(m_{1d}, m_{2d}, m_{3d}, ..., m_{Nd}) = (0, \alpha_{MDW}, 0, ..., 0)$ as weighted values.

3.2   Map-Based Classification for Interview Sheets

Maps are constructed in the general manner [3], [4] using Eqs. (1)-(3). A row in a matrix generated by the method in Subsect. 3.1 is presented to a map as a member of the training data set. The proposed method then expands the row by adding an element value associated with the age of the outpatient. This means that the matrix is enlarged by preparing the $(d+1)$-th column. One of the five-level values is given to each element on the rightmost $(d+1)$-th column. If the outpatient filling out the l-th sheet is younger than 20 years old, the value of 0 is given to that element specified by the l-th row and the $(d+1)$-th column, m_{ld+1}. If the age is more than 19 and less than 40, we have $m_{ld+1}=5$. If it is more than 39 and less than 60, m_{ld+1} is set to 10. If it is more than 59 and less than 80, $m_{ld+1}=15$ holds. For other outpatients, m_{ld+1} equals 20.

Once general SOM learning is complete, neurons are labeled as follows.

<Neuron labeling>

[Step 1] Let F_q^i denote the frequency of the i-th neuron (C_i) firing for the training data belonging to Class q, where $1 \leq q \leq 4$. Set four F_q^i's to 0, and set l to 1.
[Step 2] The l-th training data is presented, and F_q^i's of the winner are updated.
[Step 3] The value of l is incremented by 1. If $l \leq N$, go to Step 2; otherwise, go to Step 4. Note that N is the total number of training data.
[Step 4] Let LN^i denote the label of C_i. It is as follows.

$$LN^i = \arg\left\{ \max_q \left(F_q^i \right) \right\}. \tag{5}$$

Labels are assigned to the other neurons, using Eq. (5).

In this paper, each of the data unused for learning is referred to as pilot data. The set of pilot data is also generated in the manner described in Sect. 3.1. Note that the rightmost element value in each of the pilot data is also determined from the outpatient age as described above. When some pilot data is classified, it is presented to the map with labeled neurons. The class of the presented data is considered to be the label of the winner for it.

4   Experimental Results

The proposed method was applied to interview sheets provided from Tsukazaki hospital in Japan. A map with ten rows and ten columns is prepared. It is trained, subject to the learning termination condition $T=1000$. Besides, initial values of r_0 in Eq. (1) and τ_0 in Eq. (2) are set to 20 and 1.0, respectively.

Let us first discuss evaluation metrics. Let IS_q^k denote the number of pilot data, each of which is judged as Class q while its actual class is Class k, where $k, q \in \{1, 2, 3, 4\}$. The percentage of the number of pilot data whose classes are judged as Class q compared to the total number of pilot data actually belonging to Class k is calculated. Especially, the following value is referred to as the percentage of concordance associated with Class k, PC_k.

$$PC_k = IS_k{}^k \times 100 / \sum_{q=1}^{4} IS_q{}^k \qquad (6)$$

A set of data generated from interview sheets for 580 patients is used for experiments. The sheets were filled out from May through November 2010. The correspondence relationship between a sheet and its class is perfectly known in advance. All of the data are divided into four combinations by means of the four-fold cross-validation. A combination consists of 435 training data and 145 pilot data. Element values in each of the data are weighted in the manner described in Subsect.3.1, subject to (α_{PW}, α_{SW}, α_{MDW})=(10, 5, 12). Recall that one of the five-level values is added to each of the data as the final element value according to the age of the corresponding outpatient. The proposed method is evaluated ten times a combination of the training data and pilot data. The evaluation is made for all combinations. The averaged results are tabulated in Table 3. We have (PC_1, PC_2, PC_3, PC_4)=(76.5, 41.0, 41.3, 66.5). For Class 2, although the number of correctly judged data is larger than that of data wrongly judged as any other class, the resultant value is somewhat disappointing. This also applies to Class 3. The proposed method, however, copes well with the classification of data belonging to Classes 1 and 4.

Let us next discuss the simple comparison of the proposed method and ophthalmologists in terms of classification capability. In this paper, classes of data mean final results that ophthalmologists confirm by diagnoses after some medical treatments are applied to corresponding outpatients. It is therefore possible that the first impressions (i.e., the first data classes) that ophthalmologists get by reading handwriting sentences in the interview sheets once do not always accord with the final results employed as Classes 1-4 in this paper. The number of interview sheets actually belonging to each class is 145. The sheets in each class are divided into a set with 120 sheets to generate training data and that with 25 sheets to generate pilot data. The total number of handwriting sentences (i.e., sheets) prepared for pilot data generation is 100, and the classification of them is imposed on each of the seven ophthalmologists working at Tsukazaki hospital as a quiz. PC_k, the percentage of concordance associated with Class k, is then calculated using Eq. (6) for every ophthalmologist. Let $PC_k{}^{MED}$ denote the average of PC_k's. As a result, we have ($PC_1{}^{MED}$, $PC_2{}^{MED}$, $PC_3{}^{MED}$, $PC_4{}^{MED}$)=(65.7, 42.3, 77.7, 37.1).

The proposed method requires training data and pilot data associated with the above 120 sheets and 25 sheets a class, respectively. Three couples of training data set and pilot data set are then generated in the following cases: Case 1 specified by (α_{PW}, α_{SW}, α_{MDW})=(8, 4, 12), Case 2 specified by (α_{PW}, α_{SW}, α_{MDW})=(10, 5, 12), and Case 3 specified by (α_{PW}, α_{SW}, α_{MDW})=(12, 6, 12). The proposed method determines a starting point (i.e., neuron reference vectors randomly initialized), constructs a map and evaluates its classification ability, using a couple of training data set and pilot data set generated in each case. The above is repeated ten times, while changing starting points. We have averaged PC_1, PC_2, PC_3, and PC_4 in each of the three cases. The results are tabulated in Table 4. Note that an entry in the table equals averaged PC_k divided by $PC_k{}^{MED}$. $PC_k / PC_k{}^{MED} \geq 1$ implies that the proposed method is equivalent to or greater than ophthalmologists in classification capability.

Table 3. Classification results according to four-fold cross-validation

		Classification results (%)			
		Class 1	Class 2	Class 3	Class 4
	Class 1	76.5	3.1	5.6	14.8
Actualities	Class 2	36.8	41.0	5.9	16.3
	Class 3	20.7	12.2	41.3	25.9
	Class 4	19.1	2.1	12.4	66.5

As mentioned above, the classification of hundred complete sentences to which MeCab [8] has not been applied yet to divide words is imposed on each ophthalmologist. Since the data presented to the map have no elements associated with all of the verbs and most of the adverbs, the information quality given to the map is comparatively lower than that to the ophthalmologist. The proposed method however achieves favorable PC_1's and PC_4's as shown in Table 4. While somewhat inferior PC_2 appears in Table 3, each of PC_2/PC_2^{MED}'s in Table 4 exceeds the value of 1. In other words, the ophthalmologists tend to have difficulty of determining appropriate examinations for outpatients writing sentences belonging to Class 2 by hands. The proposed method unfortunately demonstrates the weakness in distinguishing Class 3 data. As a result, it is considered that the proposed method is almost about to approach ophthalmologists in classifying sentences handwritten in interview sheets.

5   Conclusions

In this paper, the SOM-based method of determining examination groups for outpatients was proposed, using handwriting sentences in their interview sheets. The proposed method chooses several of nouns, adjectives, and adverbs as powerful words, special words, and MD-designated words from the sentences to characterize conditions of outpatients. A matrix in which the sheets (or the chosen words) are related with the rows (or columns) is then generated. Frequencies of the chosen words appearing in the sentences are first given as element values in the matrix. The words with comparatively high frequencies for sentences belonging to two classes at most are referred to either as powerful words or as special words. A training data set is completed by weighting the element values corresponding to powerful words, special words and MD-designated words. After general SOM learning finishes, neurons are labeled. The proposed method determines the examination group for some outpatient by classifying the data generated from the interview sheet filled out by the outpatient. The label of the winner neuron for the presented data indicates its class. Experimental results have revealed that the proposed method achieves as high accuracy for data belonging to Classes 1, 2 and 4 as ophthalmologists.

Table 4. Simple comparison of proposed method and ophthalmologists

Class k	PC_k/PC_k^{MED}		
	Case 1	Case 2	Case 3
Class 1 (k=1)	1.30	1.28	1.30
Class 2 (k=2)	1.23	1.18	1.25
Class 3 (k=3)	0.62	0.62	0.57
Class 4 (k=4)	1.49	1.74	1.81

In future studies, the proposed method will be modified to improve the classification accuracy especially for data belonging to Class 3.

References

1. Ji, Y., Yanagawa, Y., Miyazaki, S.: Reducing Outpatient Waiting Times for Hospital Using Queuing Theory. Journal of Japan Industrial Management Association 60, 297–305 (2010) (in Japanese)
2. Kurosawa, H., Maniwa, Y., Fujimura, K., Tokutaka, H., Ohkita, M.: Construction of Checkup System by Self-Organizing Maps. In: Proc. of Workshop on Self-Organizing Maps, pp. 144–149 (2003)
3. Ohtsuka, A., Kamiura, N., Isokawa, T., Okamoto, M., Koeda, N., Matsui, N.: A Self-Organizing Map Approach for Detecting Confusion between Blood Samples. SICE Trans. 41, 587–595 (2005)
4. Ohtsuka, A., Tanii, H., Kamiura, N., Isokawa, T., Matsui, N.: Self-Organizing Map Based Data Detection of Hematopoietic Tumors. IEICE Transaction on Fundamentals of Electronics, Communications and Computer Sciences E90-A, 1170–1179 (2007)
5. Kamiura, N., Saitoh, A., Isokawa, T., Matsui, N.: Accuracy Improvement of SOM-based Data Classification for Hematopoietic Tumor Patients. In: Proc. of 2009 Ninth International Conference on Intelligent Systems Design and Applications, pp. 373–378. IEEE Press, New York (2009)
6. Kamiura, N., Takehara, N., Saitoh, A., Isokawa, T., Matsui, N., Tabuchi, H.: On Selection of Intraocular Power Formula Based on Data Classification Using Self-organizing Maps. In: Proceedings of 2010 IEEE International Conference on Systems, Man and Cybernetics, pp. 1147–1152. IEEE Press, New York (2010)
7. Fukuda, T., Kamiura, N., Saitoh, A., Isokawa, T., Matsui, N., Tabuchi, H.: Formula Selection for Intraocular Power Calculation Using Support Vector Machines and Self-organizing Maps. In: Proceedings of 2011 IEEE International Conference on Systems, Man and Cybernetics, pp. 1111–1116. IEEE Press, New York (2011)
8. Kudo, T., Yamamoto, K., Matsumoto, Y.: Applying Conditional Random Fields to Japanese Morphological Analysis. In: Proc. of Conference on Empirical Methods in Natural Language Processing, pp. 230–237 (2004)

Sound-Based Ranging System in Greenhouse Environment with Multipath Effect Compensation Using Artificial Neural Network

Slamet Widodo[*], Tomoo Shiigi, Naing Min Than, Yuichi Ogawa, and Naoshi Kondo

Division of Environmental Science and Technology
Graduate School of Agriculture, Kyoto University
Kitashirakawa, Oiwakecho, Sakyoku, Kyoto 606-8502, Japan
widodo@kais.kyoto-u.ac.jp

Abstract. In this study, sound-based ranging system in greenhouse environment with compensation of measurement error caused by multipath effect using Artificial Neural Network (ANN) was proposed. Greenhouse environment has special characteristic which is different with condition where the similar system that previously developed were applied. There are challenges need to be handled in developing accurate ranging system in greenhouse such as high humidity, temperature gradient, wind, and obstacles. In this study, error compensation was performed by first estimating the measurement error using some features extracted from the cross-correlation wave of the received signal by using ANN. Then, the estimated value was used to compensate the measurement error. The experiment result showed the feasibility to apply the proposed method to improve accuracy of sound-based ranging system in greenhouse environment.

Keywords: Sound-based ranging system, multipath effect, greenhouse, ANN.

1 Introduction

In recent work our research group has been developing local positioning system (LPS) using spread spectrum sound. There are some advantages offered by this positioning system such as high accuracy, robust to the presence of obstacle, and also inexpensive. This kind of system is actually already well developed and applied especially for office environment. *Cricket, Active Badge*, and *Dolphin* are some examples of famous project that successfully applied sound-based indoor positioning system [1-3]. However, as our knowledge, there is no report about development of sound-based LPS in greenhouse environment. This system is promising to be used as platform to develop many location aware applications for supporting greenhouse operations.

Developing sound-based positioning system in greenhouse is a challenging task due to the fact that its condition is much different from the environment where previously developed systems were applied. Inside the greenhouse, humidity and temperature are typically high. There is also influence of wind which usually fluctuates

---

[*] Corresponding author.

T. Huang et al. (Eds.): ICONIP 2012, Part IV, LNCS 7666, pp. 156–163, 2012.

all the time. This condition is challenging because sound velocity is strongly affected by humidity, temperature, and wind [4]. Hence, it may generate some error. Another issue is related to the presence of complex obstacles such as plant and greenhouse structure that may block or disturb the propagation of sound wave. Due to reflection and diffraction, there are many possible paths for the sound wave to propagate from transmitter to receiver. It is known as *multipath effect*. For Time-of-arrival (ToA)-based positioning systems, same as the developed system, the distance between a transmitter and a receiver is calculated based on the signal propagation delay. Therefore, accuracy of the estimated range is depends on the accuracy of ToA estimation. In the presence of multipath effect, it is difficult to accurately estimate the ToA.

This study would like to propose compensation method of measurement error caused by multipath effect by using Artificial Neural Network (ANN)-based error prediction model. The similar approach was also used for radio frequency (RF)-based geolocation in mining area [5]. Instead of trying to detect the direct ToA as accurate as possible to minimize the error; compensation was performed by estimating the error from the auto-correlation wave. Firstly, based on experiment data, ANN was used to find correlation between some features extracted from the received signal and to develop error prediction model. After getting valid model, then it is used to estimate and compensate the measurement error. Result of this study showed feasibility to use the proposed method to reduce the ranging error.

2   System Configuration

Schematic diagram of the sound-based ranging system is shown in Fig. 1. Two spread spectrum sound signals are created in PC1. One of the sounds is used as trigger signal and is emitted through wireless communication device. This signal is then received by PC2, also by using wireless communication device. Another signal is emitted through speaker and received by microphone connected to PC2. Then, cross-correlation processing was applied to both received signal (trigger and the second sound signal) to get Time of Arrival (ToA) of each signal. Signal propagation delay of sound wave can be calculated as difference of these two ToA values. From signal propagation delay (Δt) and sound wave velocity (c), then distance between speaker and microphone can be calculated. For detail please refer to our previous work in [6].

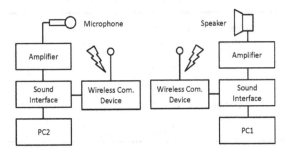

Fig. 1. Schematic of distance measurement system

Determination of ToA from cross-correlation wave is the critical point in order to obtain accurate measurement. In previous developed system, the ToA was estimated from the maximum peak of cross-correlation value of received signal. This method works well for line of sight (LOS) condition, however the performance is greatly reduce with the presence of multipath effect. Therefore we try to change it with simple first peak detection method combined with ANN-based error compensation.

In addition, the following are some properties of spread spectrum sound and trigger signal used in this study. M-sequence length = 511; Chip-rate = 6 kcps; frequency of carrier wave = 6 kHz, 18 kHz, 30 kHz; modulation = Binary Phase-Shift Keying (BPSK); sampling frequency = 192 kHz; and sampling bit = 16 bit. Trigger Signal's properties: M-sequence length = 3; chip-rate = 6 kcps; frequency of carrier wave = 6 kHz; modulation = Binary Phase-Shift Keying (BPSK).

3   ANN-Based Prediction Model of Ranging Error

In this work, a multilayer perceptron (MLP) neural network consisting three layers (input, hidden, and output layer) was used to develop prediction model of distance measurement error (Fig. 2). There are parameters which are suspected to have correlation with measurement error and can be used as input of the ANN model to estimate measurement error (ε). Parameters namely signal propagation delay (Δt), maximum peak value (P_{max}), average and standard deviation of cross-correlation value (P_{ave} and P_{std}), and number of detected peak (NP) were extracted from the cross-correlation value of the received signal and used for this purpose. For number of peak (NP), it is calculated from the peak value which is higher than certain threshold value.

In order to determine which variable that significantly correlated with measurement error and also as a benchmark for the developed ANN-based error prediction model, statistical method i.e. regression analysis was done by using Minitab software. The result is shown in Table 2. There are only three variables which have significant influence to the error: Δt, P_{std}, and NP. Prediction model developed using this statistical method has low R^2 value (0.574). It indicates that this model may not be able to predict the measurement error accurately. Those variables are then selected as input of the ANN model as shown in Fig. 2.

Table 1. Regression analysis result

Predictor	Coef.	SE-Coef.	T	P	VIF
Constant	-317.2	326.2	-0.97	0.337	
Δt	0.14288	0.03827	3.73	0.001	3.698
P_{std}	1.3E-07	6E-08	2.14	0.038	3.384
NP	-1.5666	0.8225	-1.9	0.064	3.234
S = 224.156		R-Sq = 57.40%			

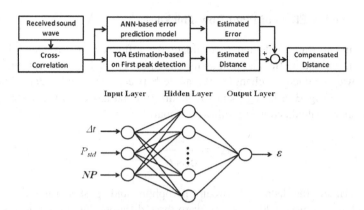

Fig. 2. Schematic of ANN-based error compensation and architecture of ANN

For ANN, each node on the network can be considered as single processing unit which performs summing process of incoming input and bias, then it generates output based on these input using certain transfer function. For every input (x), it will be processed/ summed into a net input (S) as follow:

$$S = \sum_{i=1}^{n} x_i w_i + b_i \tag{1}$$

where n, w, and b are number of inputs, weights, and biases. This value is then processed to generate output (y) using specified transfer function. In this work, *tansig* function was used to generate output of each node at hidden layer and simple *purelin* ($y = x$) at output layer. For *tansig* function, it can be written as:

$$y = \frac{2}{(1 + e^{-2S})} - 1 \tag{2}$$

The ANN model was developed by using MATLAB and back-propagation (BP) based on Levenberg-Marquardt algorithm (LMA) was chosen for training process. The goodness of the developed model then can be evaluated by comparing predicted and actual output (error) for both training and testing. One of measure that usually used to indicates the goodness of the model is coefficient of determination (R^2). It measures how well the developed model could represent the proportion of variability in a data set. The value is ranged from 0 to 1 and higher value indicates better model.

4   Experimental Setup

Two experiments were conducted to get typical case of sound wave propagation in greenhouse. The following are details of each experiment set up.

(a) Evaluation the Effect of Plant Row Thickness

This experiment was conducted to evaluate the influence of obstacle (plant row) thickness on multipath effect as well as measurement error. The different thickness of plant row was obtained by changing the angle between measured distance and horizontal line (see Fig. 3a). Using plant row width (w) and angle (θ) the obstacle thickness (Δx) can be calculated using following equation:

$$\Delta x = \frac{w}{\cos \theta} \tag{3}$$

In this experiment the distance between microphone and speaker was set as 4 m and width of plant row was 0.90 m. There were three different angle used in this experiment 0, 20, 40, and 60 degree. Therefore by using (3), the obstacle thicknesses can be calculated as 0.90 m, 0.96 m, 1.20 m, and 1.78 m.

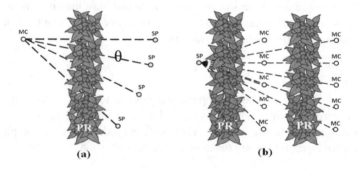

Fig. 3. Setup: a). Experiment 1 (plant row thickness); b). Experiment 2 (no. of plant row); [remark → MC : microphone; SP : speaker; PR : plant row]

(b) Evaluation the Effect of Number of Plant Row

The main objective of this experiment is to observe the multipath effect and measurement error generated when sound wave passing different number of plant row. For small-medium greenhouse, as one used in this experiment, typically there are 4-6 plant rows and for large greenhouse, it may have more plant row. In this work, the experiment was done in two conditions: 1 and 2 plant rows (Fig. 3b).

For all experiments, sound waves with frequency of 6 kHz, 18 kHz, and 30 kHz were used. This set up was chosen to understand the behavior of sound wave with different frequency in greenhouse environment, especially related to the multipath effect. It also provides basic information about appropriate frequency for the proposed system. All data then were used to develop error prediction model by using ANN.

5   Result and Discussion

5.1   Multipath Problem in Greenhouse Environment

As there are many obstacles such as plant, plastic wall, or other greenhouse structure, transmitted sound wave might be reflected and/or diffracted. This condition will generate multipath effect. Figure 4 shows the cross-correlation of received signal from measurement of 4 m distance with different obstacle thickness (Experiment 1). It can be seen that for thicker obstacle there are more multipath effect that indicated by increasing the number of detected peak. It also can be observed that the first peak is not always the maximum peak. In the presence of multipath effect, therefore, applying simple maximum peak algorithm is likely leads to false detection of true direct ToA and will result in measurement error.

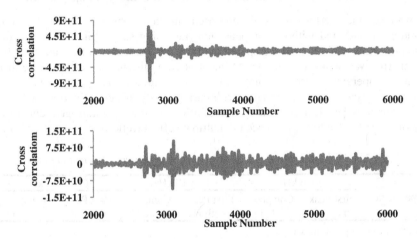

Fig. 4. Cross-correlation wave for distance measurement with different obstacle thickness

5.2   Performance of the Developed Predictive Model

After conducting the experiments, all of obtained data are then randomly divided into three data sets namely training, validation, and test data set. Training and validation data set were used during training process of ANN model while test data set is used to test the generalization capability of the developed model. The main advantage of this training and testing scenario (i.e. divide data into three groups) is it will guarantee there is no over-fitting or over-training during training process.

Comparison of actual and predicted measurement error from training process is shown in Fig.5. The result indicated that the network has been well trained. The test also showed that the developed model also has good performance in predicting new data which were not used for training. It indicates that the developed model has good generalization capability. This model also better than regression model obtained using statistical method that indicated by higher R^2 value (training: 0.996 and test: 0.949).

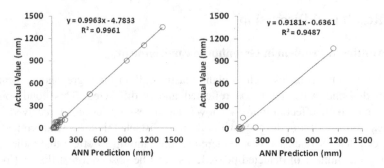

Fig. 5. Scatter plot of training and test of ANN-based error prediction model

5.3   Comparison of Measurement Error with and without Compensation

To evaluate the performance of proposed method, absolute error for both measurement with and without compensation (i.e. using simple first peak detection method to estimate ToA) were compared as shown in Table 2. In previous study [6], 3.4 mm error was obtained from measurement of 10 m distance in indoor where there was no temperature gradient, no wind, and minimum level of noise. This measurement range is about the same with current study however the obtained error value is much higher. It may come from the influence of temperature gradient, wind, and more specific due to the presence od multipath effect (reflection and diffraction).

Table 2. Measurement error with and without compensation (abs. error in mm)

Experiment	f1 = 6 kHz		f2 = 18 kHz		f3 = 30 kHz	
	First Peak method	Compen-sated	First Peak method	Compen-sated	First Peak method	Compen-sated
Exp. #1 (different plant thickness)						
0.90 m	**1.722**	2.258	8.861	**4.699**	23.688	**0.806**
0.96 m	**17.392**	20.368	29.028	**9.654**	34.077	**24.238**
1.20 m	19.377	**3.380**	4.457	44.173	46.216	**6.672**
1.78 m	**23.327**	187.214	72.217	**12.934**	151.360	**91.703**
Exp. #2a (one plant row)						
Point #1	12.942	**2.096**	**2.925**	30.639	83.231	**46.723**
Point #2	6.266	**5.643**	10.787	**5.386**	**14.725**	33.570
Point #3	**5.267**	17.176	**4.895**	21.146	**13.092**	36.147
Point #4	88.742	**15.989**	**78.255**	89.246	79.532	**12.878**
Point #5	114.085	**15.162**	186.504	**30.479**	115.605	**29.817**
Exp. #2b (two plant rows)						
Point #1	114.085	**4.477**	186.504	**93.770**	115.605	**42.008**
Point #2	90.294	**35.215**	51.388	**9.403**	**38.536**	39.507
Point #3	**4.870**	205.002	84.800	**14.782**	34.926	211.440
Point #4	909.267	**10.630**	558.614	**60.238**	1361.091	**1.003**
Point #5	1115.060	**19.836**	1073.872	**65.012**	458.609	**6.963**

*Note: number in bold means better result (smaller abs. error)

The result, especially from experiment 1, also shows the significant effect of frequency. The measurement error increases with the increasing of frequency. It indicates that the low frequency of sound wave is less suffering from multipath effect. This result is in a good agreement with characteristic of sound wave. Because of its high diffraction ability, when sound wave is approaching an obstacle, low frequency wave tends to be diffracted while high frequency wave tends to be reflected [7]. Therefore, it can handle the obstacle better than high frequency wave.

The result shows that for more cases (29 out of 42) error value of measurement with compensation was smaller than those without compensation. Also for almost all cases (39 ot of 42), the error value was less than 100 mm (the maximum value set as target for this research project). It indicates that the proposed method can be used effectively to compensate measurement error caused by mutipath effect.

6   Conclusion and Future Work

In this study, ANN-based error predictive model was used to estimate and compensate measurement error caused by multipath effect in greenhouse environment. In this model, several features extracted from the received signal and were used as inputs of ANN to predict the measurement error. The result showed the effectiveness of the proposed method on reducing measurement error. Beside error compensation, this study also provides basic information about behavior of sound-wave propagation in greenhouse, especially related the different multipath effect and measurement error experienced by sound wave with different frequency.

References

1. Priyantha, N.B.: The Cricket Indoor Location System. PhD Thesis, Massachusetts Institute of Technology (2005)
2. Want, R., et al.: The Active Badge Location System. ACM Trans. Information Systems 10(1), 91–102 (1992)
3. Fukuju, Y., et al.: Dolphin: An Autonomous Indoor Positioning System in Ubiquitous Computing Environment. In: Proceedings of the IEEE Workshop on Software Technologies for Future Embedded Systems, Hakodate, Japan (2003)
4. Truax, B. (eds.): Handbook for Acoustic Ecology (2012), http://www.sfu.ca/sonic-studio/handbook/Sound_Propagation.html (accessed May 13, 2012)
5. Nerguizian, C., Despins, C., Affers, S.: Geolocation in Mines with an Impulse Response Fingerprinting Technique and Neural Networks. IEEE Transactions on Wireless Communications 5(3), 603–611 (2006)
6. Shiigi, T., et al.: Position Detecting Method Using Spread Spectrum Sound: Correc-tion Method of Measurement Error by Compensating Wind and Temperature. In: Proceeding of IFAC Agricontrol Conference, Kyoto, Japan, December 6-8 (2010)
7. Mak, L.C., Furukawa, T.: A Time-of-Arrival-Based Positioning Technique with Non-Line-of-Sight Mitigation Using Low-Frequency Sound. Advanced Robotics 22(5), 507–526 (2008)

Multimedia Educational Content for Saudi Deaf

Yahya O. Mohamed Elhadj[*]

Center for Arabic and Islamic Computing,
Al-Imam Muhammad Ibn Saud Islamic University
P.O. Box 5701, Riyadh 11432, Kingdom of Saudi Arabia
yelhadj@ariscom.org

Abstract. Research demonstrates that deaf individuals are undereducated and most of them are illiterate or at least semi- illiterate. Educating individuals with disabilities, in general, is a good investment. It doesn't only reduce welfare costs and future dependence; it reduces current dependence and frees other household members from caring responsibilities, as well as allowing them to increase employment or other productive activities. In this scope, a national funded project is launched to develop an environment for teaching and learning for Saudi deaf, using both automatic translation from Arabic to Saudi Sign Language and 3D animation techniques called Avatars. As part of the project, this paper presents the development of educational material to allow access to vital information for deaf people by presenting essential knowledge needed in their daily lives in an easy manner to grasp and comprehend. Resources for the subject of Islamic Education is collected and indexed based on levels and depths of information to accommodate needs of various types of users targeted by our works.

Keywords: multimedia, educational material, 3D animation, Sign Language, Saudi Sign Language, Avatar technologies.

1   Introduction

United Nations (UN) estimates that around 10% of world population, or about 650 million people live with a disability [1]. Hearing loss is the most prevalent sensory disability globally. In 2004, over 275 million people globally had moderate-to-profound hearing impairment, 80% of them in developing countries [1-3]. In Saudi Arabia, deaf & hearing-impaired people represent 10.7% of disabled persons in the Kingdom according to a 2002 survey conducted by the World Bank [4].

Many studies indicate that deaf people around the world are undereducated, and most of them are illiterate or at least semi- illiterate. The World Federation of Deaf indicates that 80% of deaf people lack education [5]. Moreover, the global literacy rate for adults with disabilities is as low as 3% and 1% for women with disabilities according to a 1998 United Nations Development Program (UNDP) study [1], [6].

---

[*] Corresponding author.

T. Huang et al. (Eds.): ICONIP 2012, Part IV, LNCS 7666, pp. 164–171, 2012.
© Springer-Verlag Berlin Heidelberg 2012

For the above reasons, a lot of international bodies start developing specific educational curricula for the Deaf people to help them improve their living conditions and their integration in the society. New technologies have been used to ease multi-take advantage of the educational curricula, such as 3D animations and avatar technologies (e.g. [7-9]). In fact, enabling easy access for educational material to deaf people will be an empowerment for them toward more independence and self-reliance. This will engage them to be more independent, confident and participate proactively in their community.

Unfortunately, deaf Arabs are still encountering many difficulties related to their education due to several reasons: absence of official specification of many Arabic Sign Languages (e.g. unified dictionary, linguistic structure, etc.), which prevent their use as a medium of education, absence of educational signed content easily accessible and appropriate for deaf, and so on. In deed, we are aware of only few works related to the preparation of educational materials using appropriate technologies for deaf Arabs. The most relevant is the national Tunisian project[1], which represent great efforts aiming to develop learning materials and web-based environment for teaching deaf-pupils [10], [11].

In this paper, we present our efforts to collect and build a learning material for deaf people allowing them an easy access to essential knowledge needed in their daily lives in an easy manner to grasp and comprehend. These efforts represent a part of our work in a project, funded by the National Plan of Sciences, Technology, and Innovation, to build an environment for translation from Arabic texts to Saudi Sign Language (A2SaSL project) [12].

2   Development of the Educational Material

According to an in-depth survey, we conducted as a first phase of our ongoing project (A2SaSL), we noticed the scarcity of signed contents not only for Saudi Sign Language (SaSL) but also for almost all other Arabic Sign Languages (ArSLs) [13]. Abdel-Fattah indicated in [14] that not only very few Arabic signed-contents exit for just some ArSLs, but also the existing contents are available only in specific forms such as movies, TV series, and news bulletins. To overcome this deficit, we aim to help the Saudi deaf people community to improve their access to educational resources by providing them with a web-based environment for teaching and learning, based on automatic translation from Arabic to SaSL and 3D animation techniques called Avatars. This needs, firstly to prepare a convenient content that can be structured in an appropriate format easily accessible and comprehensible by deaf people.

2.1   Choosing the Domain

As a first work of its kind, we preferred considering the basic Islamic topics as they are highly required in Saudi Arabia, Arab world, and also all Islamic countries. In societies where Islamic education is primordial, providing deaf people community with access to such knowledge and principles will help them progress toward more

---

[1] www.utic.rnu.tn/websign

normal life and become well religiously educated. So, five basic topics known as pillars of Islam are chosen to be the core of our educational content. They are Prayer, Pilgrimage, Fasting, Zakat, in addition to the topic of purity (cleanliness) as it is considered a prerequisite to perform legitimacy obligations. We hope that other Islamic topics be included in future upcoming works to expand the coverage of the material.

2.2   Selecting Material

Arabic contents covering everything related to the forth mentioned topics were collected in terms of elements, functioning and provisions. These texts have been gathered from authentic Islamic references used in the actual Islamic teaching. They have been further analyzed, reviewed, and simplified to comply with the needs and cognitive capabilities of deaf people. Their contents are also linguistically and legitimately revised to be sure of its correctness after such modifications.

Table 1. Extracts from collected texts

Topic	ID	Content	Topic	ID	Content
Fasting (الصيام)	1	تعريف الصيام	Purity (الطهارة)	1	باب الوضوء
	2	الصيام لغة الإمساك		2	تعريف الوضوء
	3	أقسام الصيام		3	الوضوء شرعاً يعني استعمال الماء في أعضاء مخصوصة بكيفية مخصوصة
Pilgrimage (الحج)	1	تعريف الحج	Zakat (الزكاة)	1	تعريف الزكاة
	2	الحج لغة القصد إلى معظم		2	الزكاة لغة التطهير والنماء
	3	حكم الحج		3	حكم الزكاة
	4	فرض في العمر مرة على الفور		4	الزكاة ركن من أركان الإسلام الخمسة
Prayer (الصلاة)	1	حكم الصلاة	Prayer (الصلاة)	1	تعريف الصلاة
	2	الصلاة فرضت بالكتاب والسنة والاجماع		2	الصلاة لغة الدعاء
	3	فمن أنكر ذلك فهو مرتد عن دين الإسلام بلا خلاف		3	وشرعاً قربة فعلية ذات أقوال وأفعال مخصوصة

Texts are formatted and stored in an appropriate structure to ease their access and manipulation. Each document is split into separate sentences and/or short paragraphs conserving the meaning. This helps to be understood by deaf people and also to produce accurate translation to Sign Language. Obviously, all segments (sentences and/or paragraphs) are internally stored in relation with their original documents. In the following table (table 1), we report an extracted texts of the five considered topics to have a clear idea about their contents.

3   Preparations of Videao-Based Multimedia Content

It is known that the use of multimedia technologies in education increases the efficiency of learning. The use of images and videos in educational materials are generally highly required to support students understanding and grasping the material.

For the deaf people, they need to receive all the information in visual form as they cannot access the acoustic channel. So, it is very important that the material is completely presented for deaf people as visual signed-content using their own sign languages.

In our work, we are transforming the written content we collected to visual signed content following high quality standards. Firstly, we produced an accurate textual translation of all segments of texts. Second, high resolution video recording of all the written contents described above are performed. Third, a sign dictionary is deduced and being used as support for the educational material as we will see later.

3.1   Producing Video Recording of the Written Content

All the collected Arabic content we described above have been transformed to signed visual content by recording human signers performing sentences. This has been done as follows: 1) a team of deaf people and interpreters were selected to help executing this task; 2) working groups are formed to work in parallel on the texts collected for the five topics; 3) each group is asked to write for each segment of the text, the best translation they agreed on; 4) after cross-validation between groups, they proceed to their video recording; 5) video recording are also mutually revised and validated for accuracy not only for the quality of videos but also for the homogeneity of used-signs over all the contents. We notice, that this double translation (written and visual) of the content will allow many benefits: 1) easy learning for both deaf and hearing, as written concepts can be aligned with their visual realizations; 2) short video segments can be used separately to prepare specific instructional material; 3) transformational rules between Arabic and SaSL can be extracted; 4) some lexical/linguistic features can also be deduced.

3.2   Building Sign Dictionary

One of the main difficulties still encountered in Saudi Sign Language is related to the absence of official dictionaries [13]. Saudi Arabia is one of the Arabic countries that still have no unified sign dictionary despite an increasing official importance given to deaf. In fact, a large project for documenting Saudi Signal Language seems to be launched or will be in the near future by the Prince Salman Center for Disability Researches[2]. For the time being, no official dictionary is available at the best of our knowledge.

For the religious domain as it is our field of work, we notice an unofficial attempt recently initiated by some individuals at the Saudi Federation Sports for the deaf[3] to collect some words from the Holy Quran and then to create signs for them. It seems this work is still running and did not finish yet. Moreover it is very limited and concerns

---

[2] www.pscdr.org.sa
[3] www.deafsp-sa.com

only some Quranic words. We hear about another initiative launched recently in Qatar within the ongoing efforts to develop the unified Arab sign dictionary [15-17]. It consists of collecting religious terms that are mostly used in Islamic world and to create a unified signs for them. Unfortunately, we did not obtain enough information about this work. Indeed, we were interested in this work for just comparison purposes since we are working on Saudi sign language and not the unified sign language.

Our approach consists in building a large dictionary to be an infrastructure of the educational material. An animated version of this dictionary has also to be created using avatar technologies (see the next section). These versions (textual and animated) will allow automatic generation of multimedia learning contents. For this purpose, we resorted to building our sign dictionary from the content we collected. We extracted a list of unique words from the Arabic texts along with their meanings. This approach is better than the other attempts wishing to collect a list of separate words as in Arabic a same word may indicate completely different things depending on its meaning in a specific context.

Some relevant terms have been added to the dictionary to ensure a good coverage, like the Arabic sign alphabet and numbers. Islamic Education requires knowledge of numbers, counting and finger spelling. Finger spelling is a concept used by deaf people to sign unknown words using the sign alphabet instead of creating new signs. For more details on the creation of the sign dictionary, you can refer to [13].

4   Preparations of Avatar-Based Multimedia Content

Most adequate representation of signs is highly needed to allow deaf people memorizing concepts by visualizing them many times and from different angles. These circumstances motivate the use of 3D animated avatar technologies as a new medium of multimedia contents. In the next section, we will highlight the advantage of using avatar technology over video recording.

4.1   Advantage of Using Avatar for Building Educational Content

Methods for representing signs using computers have been evolved from images, to video-clips and finally to 3D technology called avatars. Avatars are virtual human signing as 3D animated images to simulate natural movements of people [18]. In an avatar-based approach, signs are created as text-files by avatar software (sign-editor) then they can be visualized using a component in the avatar software (sign-player) as a 3D animation. The created files for animated signs are very much smaller than videos and images of these signs. Thus the storage space required is minimal, and the download / visualization is very fast.

A report from the eSIGN (Essential Sign Language Information on Government Networks)[4], a well known European funded project, indicated many advantages of using avatar over video, such as: 1) browsing more quickly through information, 2) controlling the speed of signing, 3) changing the view angle of the virtual signer, 4) etc. [19]. Avatar has also many advantages on video when speaking about the creation of the signed content and its maintenance.

---

[4] http://www.visicast.cmp.uea.ac.uk/eSIGN

In addition to the above, important studies were interested in the comprehensibility of signed contents based on avatars compared to video-based contents. Hurdich from Vcom3D[5] conducted a research study at the Florida School for the Deaf and Blind and showed an increase of comprehension of a story from 17% to 67% after seeing it signed vs. being read [20]. Parton evokes in his paper [18] that another research study conducted by Seamless Solutions[6] reported that none of the students in their study encountered difficulty understanding the avatars. Kipp and al. [21] compare avatar performance with human signers and present a measure of real comprehensibility of the avatar (delta testing). They indicate that non-manual components and prosody are the most issues for a possible increase of comprehensibility beyond 60%. Naqvi and al. [22] conducted a study to test the effectiveness of digital representations of sign language content; they asked two groups of non-signers to watch a sign language lesson with either a digital video or a digital avatar. It was found that participants learning sign language with the Avatar had a higher learning rate than video.

4.2 Creating 3D Animation: Animated Sign Dictionary

One of the important features of 3D animation systems is that they try to be independent of SLs in terms of the possibility of creating and animating sequence of movements. So, we surveyed the avatar-based signing systems that are available and compared them in order to select an appropriate one to be used for our project.

Comparison of avatar-based signing software developed for animating SLs has been limited to parameters that respond to our needs: presence of sign-editor and plug-in sign-player, quality of graphics, reality of animation, possibility of controlling body parts and the motion speeds, etc. This means that we surveyed only systems that have such parameters. List of these software can be found in this paper [24] lastly published in Arabic.

After a convenient animation system was identified, we focused on the creation of our animated signs and how they can be visualized later-on from inside our system (plug-in player). The eSIGN software was selected to be used in our work. It has been chosen based on the features he offers, which are superior of those given by the other available software. For example, he has two separate components, one for creating signs (offline) and another for playing them (on fly) and can be plugged in our application. Also, he has other important parameters, such as quality of graphics, reality of animation, full control of body parts, facial expression, etc.

The sign dictionary we built from the collected Islamic content is being transformed progressively to an animated sign dictionary [25]. For each sign, we have to simulate its high quality video by transforming it to an animated format. This is done by manipulating hand-shapes, body parts, and many other parameters of the avatar through the features provided by the eSign software. We have to notice the difficulty of this task, which needs long time and big efforts. Every time an animated sign is created, it has to be validated by the deaf people team to ensure its correctness. This manner of working guarantees a good quality, but it is very time consuming. To help accelerating this work and to be able to build a large animated dictionary, we called for a collaborative strategy (see the next section).

---

[5] http://www.vcom3d.com
[6] http://www.signingbooks.org/doku.php

4.3   Strategy for Expansion: Collaborative Strategy

To help building a large animated dictionary, we proposed a collaborative strategy to allow contribution from SaSL experts and interested people. Two stages of contribution are allowed to external collaborators: 1) creating animation corresponding to video-clips we provided; 2) proposing new contents by providing, and uploading on our system, high quality compressed video files with their corresponding animated files created using our integrated sign-editor avatar. Based on the opinion of an official focus group, signs are accepted and inserted in the dictionary or refused and deleted from the system. Video files are requested to verify that the corresponding animation was correctly and accurately created. It may happen also that the person proposing the sign may not be so familiar with the creation of animation; so, the video will be used by our team to create the animation. Video files are removed once animations are verified / created.

5   Conclusions

In this paper we presented the development of educational material for deaf people at different stages: 1) Islamic resources in well chosen topics are collected and indexed; the choice of this subject was driven by two essential factors: the centrality of learning and disseminating teachings of Islam to all segments of our society, as well as the potential for expansion of the project, knowing that the lexicon and keywords for this subject are not just common to Arabic only, but it could easily be expanded to all Muslims around the world; 2) the written content has been transformed to visual signed content by recording human signers performing textual segments following high quality standards; 3) An avatar-based 3D animations are created for religious terms; they are used to generate multimedia sequences to build a free context materials for educational purposes; 4) a strategy of expansion for creating large animated contents is sketched. We expect that this material will greatly help deaf centers to better educating deaf children and adults.

Acknowlegments. This work is part of the A2SaSL project funded by Al-Imam University Unit of Sciences and Technology under the framework of the National Plan for Sciences and Technology (grant number 08-INF432-8). We thank all the other project team members, especially our deaf groups.

References

1. Convention on the Rights of Persons with Disabilities: Some Facts about Persons with Disabilities. United Nations (2011)
2. Deafness and hearing impairment. World Health Organization (2012)
3. Situation Review and Update on Deafness, Hearing Loss and Intervention Programs. World Health Organization. Regional Office of South-East Asia (2007)
4. Country Profile on Disability KINGDOM OF SAUDI ARABIA. Japan International Cooperation Agency Planning and Evaluation Department. World Bank (2002)

5. Convention on the Rights of Persons with Disabilities. WFD (2003)
6. Human Development Report 1998. Published for the United Nations Development Programme (UNDP). Oxford University Press, New York, Oxford (1998)
7. Hilzensauer, M., Dotter, F.: "SignOn", a Model for teaching Written Language to Deaf People. In: Proc. of IST-Africa 2011 Conference (2011)
8. Official Website of SignOnOne and SignOn, http://www.signonone.eu
9. Adamo-Villani, N., Doublestein, J., Martin, Z.: The MathSigner: an Interactive Learning Tool for American Sign Language K-3 Mathematics. In: Prof. of the 8th IEEE/International Conference on Information Visualization (2004)
10. El Ghoul, O., Jemni, J.: Multimedia Courses Generator for Deaf Children. The International Arab Journal for Information Technology 6 (2009)
11. Jemni, M., El Ghoul, O.: Using ICT to Teach Sign Language. In: The 8th IEEE International Conference on Advanced Learning Technologies, Santander, Cantabria, Spain (2008)
12. Elhadj, Y.O.M., Zemirli, Z.A.: Virtual Translator from Arabic text to Saudi Sign-Language (A2SaSL). Technical Annual Report, Riyadh, Saudi Arabia (2011)
13. Elhadj, Y.O., Ayadi, K.: A Religious Sign-Dictionary for Saudi Sign Language. Communications of Arab Computer Society Journal (2012)
14. Abdel-Fattah, M.: Arabic Sign Language: A Perspective. Deaf Studies and Deaf Education 10, 212–221 (2005)
15. LAS.: First part of the Unified Arabic Sign Dictionary. The League of Arab States & the Arab League Educational, Cultural and Scientific Organization, Tunisia (2000) (in Arabic)
16. LAS.: Second part of the Unified Arabic Sign Dictionary. The League of Arab States & the Supreme Council for Family Affairs, Qatar (2006) (in Arabic)
17. SCFA.: The Unified Arabic Sign Dictionary (DVD). The Supreme Council for Family Affairs, Qatar (2007) (in Arabic)
18. Parton, B.: Sign Language Recognition and Translation: A Multidisciplined Approach From the Field of Artificial Intelligence. Deaf Studies and Deaf Education 11 (2006)
19. eSIGN Team. Animating Sign Language: The eSIGN Approach (2005)
20. Hurdich, J.: Utilizing Lifelike, 3D Animated Signing Avatar Characters for the Instruction of K-12 Deaf Learners. In: International Tech Symposium on Instructional Technology and Education of the Deaf, National Technical Institute of the Deaf, NTID (2008)
21. Kipp, M., Heloir, A., Nguyen, Q.: Sign Language Avatars: Animation and Comprehensibility. In: Vilhjálmsson, H.H., Kopp, S., Marsella, S., Thórisson, K.R. (eds.) IVA 2011. LNCS, vol. 6895, pp. 113–126. Springer, Heidelberg (2011)
22. Naqvi, S., Ohene-Djan, J., Spiegel, R.: Testing the effectiveness of digital representations of sign language content. Paper presented at the Instructional Technology and Education of the Deaf Symposium, Rochester, NY (2005)
23. Villani, N.: 3D Rendering of American Sign Language Finger-Spelling: A Comparative Study of Two Animation Techniques. PWASET 34 (2008) ISSN: 2070-3740
24. Elhadj, Y.O.M., Zemirli, Z.A.: Using 3-D Animation to build an Animated Religious Dictionary for Saudi Sign Language: Version 1. International Journal of Computer Science and Engineering in Arabic (2011)
25. Elhadj, Y.O.M., Zemirli, Z.A., Alfaraj, B.: Towards a Unified 3D Animated Dictionary for Saudi Sign Language. Accepted for Publication in the ACM ICACCI 2012 Conference, August 3-5, 2012, Chennai, India. It will appear in the ACM Library (2012)

Grasping Region Identification in Novel Objects Using Microsoft Kinect

Akshara Rai, Prem Kumar Patchaikani, Mridul Agarwal, Rohit Gupta,
and Laxmidhar Behera

Department of Electrical Engineering, Indian Institute of Technology Kanpur, India
{akshara,premkani,mridagar,grohit,lbehera}@iitk.ac.in

Abstract. We present a novel solution to the problem of robotic grasping of unknown objects using a machine learning framework and a Microsoft Kinect sensor. Using only image features, without the aid of a 3D model of the object, we implement a learning algorithm that identifies grasping regions in 2D images, and generalizes well to objects not encountered previously. Thereafter, we demonstrate the algorithm on the RGB images taken by a Kinect sensor of real life objects. We obtain the 3D world coordinates utilizing the depth sensor of the Kinect. The robot manipulator is then used to grasp the object at the grasping point.

1 Introduction

We consider the problem of grasping of novel objects by the robot. If we are aiming at grasping a previously known object, with a known 3D model, there are methods available, such as those described in Miller et al., 2003 based on pre-stored primitives. However, obtaining a full and accurate 3D reconstruction of new objects in a practical scenario is infeasible, more so with only two images available. In other works, an estimate of the 3D model of the object is created by manipulating it using a robotic hand, which is typically time consuming and not robust.

In contrast to these approaches, we employ a learning algorithm that neither requires nor tries to build a 3D model of the object. Instead it directly identifies, as a function of the image features and properties, a point at which to grasp the object. Informally, the algorithm takes a picture of the object, and then tries to identify a point within the 2D image that corresponds to a good point at which to grasp the object. (For example, if trying to grasp a coffee mug, it might try to identify the midpoint of the handle.) The learning is based solely on image features and no 3D information is required. The real world 3D coordinates were determined from depth stream of Kinect sensor. This eliminates computationally expensive steps required for stereo vision (as done by Saxena et al.).

In the experiments conducted, a grasping region is identified on the RGB image of the scene from the RGB camera of the Kinect. The depth and image sensors are calibrated intrinsically as well as extrinsically to a robot base frame. Using the above identified grasp region, a grasping point in 3D is isolated with respect to the robot base frame and the robot is programmed to grasp it at that location.

T. Huang et al. (Eds.): ICONIP 2012, Part IV, LNCS 7666, pp. 172–179, 2012.

2   Related Work

Most work in robotic grasping deals with control and planning methods to do the grasping. Fewer focus on identifying the grasp point.

Some work has been done on grasping planar objects based on object features in 2D images. For example, contours of multi-coloured flat objects lying on a uniformly coloured background can be obtained quite reliably. Using local visual features (like the 2D contour) and other properties, 2D locations of points where to put the finger for grasping can be obtained. For example, Morales et al. (2002) calculated 2D positions of 3 fingered grasps based on feasibility and closure criteria.

Other recent works on grasping include probabilistic models of the object geometry. These aim at obtaining a probabilistic 3D model of a previously learnt object using 2D features like object contours. For example, Glover et al. (2008) found grasping locations of non-rigid objects using a maximum-likelihood correspondence between the observed contours and known models. This was followed by grasping based on a previously planned grasp strategy.

Other works involve 3D shape estimation based on Box-based volume decomposition. For example, Huebner et al. (2008) used a minimum bounding box based best-split algorithm for determining the shape of the object.

3   Learning the Grasping Point

Several very different objects may have similar grasping points, such as a book, eraser or boxes may be picked up similarly while pencil, tubes, bottles need to be picked up in a similar fashion. We use simulated images of different types of objects with grasping points highlighted to train a network and hope that it generalizes to new real world objects.

In this work, we predict the 2D location of the grasping point in each image. More formally, we try to identify the projection of a good grasping point onto the image plane.

For most objects, there is typically a small region that a human (using a two or three finger pinch grasp) or a three-fingered robot would choose to grasp it. We refer to this region as the "grasping point" (similar to Saxena et al.) and our training set will contain labelled examples of this region. Examples of grasping points include the centre region of the neck for a martini glass, the centre region of the handle for a coffee mug, etc.

3.1   Features

A high-dimensional feature vector is created by dividing the image into small rectangular patches, and analysing each patch and its surroundings to predict whether or not it contains a projection of a grasping point onto the image plane. Figure 1 shows the size of a patch in an image.

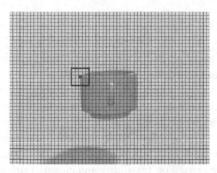

Fig. 1. Image of Coffee Mug showing grid lines and a patch whose features are to be calculated

By using a large number of different visual features, edge, colour and texture and training on a huge training set with varying object colours, shapes and sizes, method is obtained for predicting grasping points that is robust to changes in the appearance of the objects and is also able to generalize well to new objects not included in the training set.

We start by computing features on three types of local cues: edges, texture and colour. The image is transformed to YCbCr colour space, where Y is the intensity and Cb and Cr are the colour channels. We then calculate the edge features of the image by convolving the intensity image with six oriented edge filters. These are the Nevatia-Babu filters for edge detection. These are six different masks oriented in six directions- $0°, 30°, 60°, 90°, 120°$ and $150°$.

Texture information is mostly contained within the image intensity channel, so we apply the nine Laws' masks to this channel to compute the texture energy. Colour features are computed by applying a local averaging filter (the first Laws' mask) to the two colour channels. The sum squared energy of each of these filters' outputs is then computed. This gives us a $(9 + 6 + 2 = 17)$ dimensional feature vector.

However, local image features centred on the patch may be insufficient to predict whether a patch contains a grasping point, and one has to use more global properties of the object. This information is captured by using image features extracted at multiple spatial scales (three in our experiments- 1, 2, 5) for the patch. This gives us 17×3 dimensional feature vector. Objects exhibit different behaviours across different scales, and using multi-scale features allows us to capture these variations.

Also, the 17 features described above are computed from that patch as well as the 24 neighbouring patches (in a 5×5 window centred on the patch of interest) to take into account neighbourhood cues.

This results in a feature vector x of dimension $1 \times 17 \times 3 + 24 \times 17 = 459$.

3.2   Synthetic Data for Training

Supervised learning was applied to identify regions in the image that contain the grasping points. Such training needed images that had the grasping region labelled. The features of these selected points were used to learn the weight vector for the network. Instead of real world images, a synthetic data set was used for training which was taken from the Cornell University webpage for Personal Robotics http://pr.cs.cornell.edu/grasping/point_data/data.php

Synthetic training images were used for a book, thick pencil, coffee mug and martini glass. A sample image is shown in Figure 2 with grasping region marked with red colour.

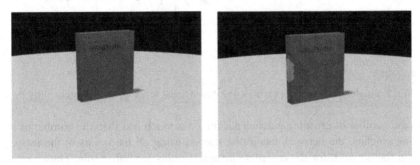

Fig. 2. Synthetic Training set without and with the grasping region labelled

3.3   Training the Network

A single neuron network is used to model the probability that a particular point (u, v) in image C is a grasping point.

A label $z(u, v)$ is defined for each location (u, v) on the image C,

$$z(u, v) = 1, \quad if\ (u, v)\ is\ a\ grasping\ point$$
$$= 0, \quad otherwise$$

As explained before, by (u, v) being a grasping point, it is meant that a projection of a grasping point onto the image plane. Since a binary classification is performed, probability that (u, v) is a grasping point is modelled using binomial logistic regression and is given as:

$$P(z(u, v) = 1|C) = \frac{1}{1+e^{-S}} \tag{1}$$

where $S = \sum_{i=1}^{459} X_i \times \theta_i$ in which $X \in \mathbb{R}^{459}$ are the features for the rectangular patch centred on (u, v) in image C and $\theta \in \mathbb{R}^{459}$ is weight vector, shown in Figure 3.

The goal of the logistic regression is to estimate the 459 parameters in the weight matrix of the above network. Since an analytical solution does not exist in logistic regression, maximum likelihood is used to determine the parameter $\theta^* \in \mathbb{R}^{459}$ of this model.

$$\theta^* = arg\ max_\theta \prod_i P(z_i|X_i; \theta) \tag{2}$$

where (X_i, z_i) are the synthetic training examples (features of image patches and labels). The initial guess weight matrix consists of all ones and is updated iteratively until the solution converges. This operation ensures that the probability is maximum for a grasping patch and minimum for a non-grasping patch. These learned parameters are used later in identifying grasp points in test images.

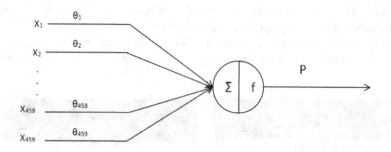

Fig. 3. Single neuron network with a 459 dimensional input, giving the probability P

As the number of grasping patches per image is much less than the number of non-grasping patches, the network cannot be trained using all the points in the image. If we train the network for the complete image, it trains for only zeros. Thus a ratio of 1:2 is taken, with 1 part of grasping patches and 2 parts non-grasping patches, chosen randomly over the object excluding the graspable patches. This is done for approximately 1000 images per object and the parameter θ^* is learnt.

3.4   Identifying the Grasp Point

Given a new image, the image in divided into a grid of 10 pixels × 10 pixels. Thereby, the features are calculated for each patch as described in the previous section.

Using the learnt model and parameter θ, the probability of each patch being a grasping region is calculated as:

$$P(z(u,v) = 1|C) = \frac{1}{1+e^{-S}} \tag{3}$$

where $S = \sum_{i=1}^{459} X_i \times \theta_i^*$. Here $X_i s$ are patch features, $\theta_i^* s$ are the learnt weights learnt from training.

Once the patch with the maximum probability is identified, it is labelled as 1 (grasping point). Now, neighbouring patches are tested and if their evaluated probability is greater than half the maximum probability, it is labelled as grasping too. This is done to cluster the grasping patches together. If the neighbouring patches' probabilities are less than the above value, we move on to the next highest probability. 5 such points are displayed.

4   Integrating with Microsoft Kinect and Power Cube Robotic Arm

4.1   Kinect Calibration

Microsoft Kinect is used to obtain, simultaneously at 30 Hz, a 640 × 480 pixel monochrome intensity coded depth map and a 640 × 480 RGB video stream. The

RGB stream of Kinect is calibrated with the depth stream to obtain 3D world coordinates of all the points captured in the depth image. To operate a manipulator robot, this needs to be calibrated with the robot base frame. This necessitates a camera model and Tsai algorithm is used to calibrate the camera.

Figure 4 shows the output of RGB stream and depth image from Kinect sensor, and the calibrated image superimposing the depth information on the RGB image.

Fig. 4. From Left to Right: RGB Image, Depth Image, Superimposed RGB and Depth Image

The 2D grasp point location in the RGB stream is obtained using the algorithm described in section 3.4 and the corresponding 3D point location of the grasping point is calculated form the superimposed RGB and depth image,

4.2   Guardian WAM, PowerCube and Robotic Grasping

We use a Guardian WAM and a 7DOF PowerCube Robotic manipulator arm for the actual grasping. The Jacobian pseudo-inverse method is used to determine the manipulator link angles from the Cartesian coordinates of the grasping point identified earlier. The joint angles are used to achieve the grasping point with lazy-arm like movement. Figure 6 shows the powercube manipulator arm in home and grasping a jug and a guardian wam grasping a spray can.

5   Results and Discussion

For synthetic data: The algorithm was first tested on the synthetic data set (described in section 3.2). As noted there the data consists of labelled grasping patches. The task was to identify whether a 2D image patch is a projection of a grasping point or not. The average success was 80.1% for a variety of objects. Table 1 shows the aggregated results of tests done on the synthetic data set.

Table 1. Results of algorithm tested for synthetic data set

Object	Test Points	Error Points	Success Rate
Book	12064	2522	79%
Martini Glass	4268	568	86.6%
Mug	6686	1206	82%
Pencil	10925	2450	77.6%
Total	33943	6746	80.1%

For real life novel objects: The algorithm was tested using Microsoft Kinect, which was stationary with respect to the base of the 7DOF robotic manipulator and a Guardian WAM which has kinect camera mounted in its front (The white box in Figure 6 shows the location of kinect camera on Guardian WAM). The task was to identify the grasping point of the object and move the tip of the manipulator arm to it and an attempt is considered to succeed if the manipulator arm grasps the object. The objects used were a water bottle, a spray can, a jug, and a Rubik's cube. For each object 5 trials were made, varying orientation and position. On an average the manipulator managed to grasp the object 70% of times. Table 2 shows the results of tests done for novel objects. The results show that algorithm generalizes well for real life images. Most of the failures were due to the presence of graspable points in the environment around the object. It can be concluded that the environment should not have such objects which resembles the grasping point. Figure 5 shows identified grasping points on some of the objects which are being grasped by the manipulator arm as shown is Figure 6.

Table 2. Results of algorithm tested on novel objects

Objects	Success Rate
Water bottle	60%
Jug	80%
Spray can	100%
Rubik's Cube	40%
Overall	70%

Fig. 5. Grasping Points Highlighted in a Jug

Fig. 6. From Left to Right: Power Cube in home position, reaching grasping position of jug and a Guardian WAM grasping a spray can

6   Conclusion

The algorithm developed enables a robot to grasp novel object. The learning algorithm identifies, directly as a function of the image features, a point at which to grasp the object. In experiments, the algorithm generalizes fairly well to novel objects and environments, and identifies good grasp points in most of the cases, in real-time.

This algorithm implemented on RGB images taken by the Kinect sensor, returned the grasping point in the image. Using a calibrated depth map, the exact 3D world coordinate of the object were obtained.

This information can now be used by a robot to reach and grasp the novel object. The validity of the algorithm is demonstrated using a Power Cube Robot Arm based implementation.

References

1. Saxena, A., Driemeyer, J., Kearns, J., Ng, A.Y.: Robotic Grasping of Novel Objects Using Vision. International Journal of Robotics Research (2008)
2. Milller, A.T., Knoop, S., Allen, P.K., Christensen, H.I.: Automatic grasp planning using shape primitives. In: Proceedings of the International Conference on Robotics and Automation (2003)
3. Morales, A., Sanz, P.J., del Pobil, A.P.: Vision based computation of the three finger grasps on unknown planar objects. In: Proceedings of the IEEE/RSJ International Robots and System Conference (2002a)
4. Morales, A., Sanz, P.J., del Pobil, A.P., Fagg, A.H.: An experiment in constraining vision-based finger contact selection with gripper geometry. In: Proceedings of the IEEE/RSJ Intelligent Robots and Systems Conference (2002b)
5. Glover, J., Rus, D., Roy, N.: Probabilistic Models of Object Geometry for Grasp Planning. Robotics: Science and Systems IV (2008)
6. Geidenstam, S., Huebner, K., Banksell, D., Kragic, D.: Learning of 2D Grasping Strategies from Box-Based 3D Object Approximations. Robotics: Science and Systems V (2008)
7. Nevatia, R., Babu, K.R.: Linear Feature Extraction and Description. Computer Graphics and Image Processing 13, 257–269 (1980)
8. Laws, K.I.: Textured image segmentation. Ph.D. Thesis, University of Southern California (1980)

Annotating Words Using WordNet Semantic Glosses

Julian Szymański[1,*] and Włodzisław Duch[2,3]

[1] Department of Computer Systems Architecture, Gdańsk University of Technology, Poland
julian.szymanski@eti.pg.gda.pl
[2] Department of Informatics, Nicolaus Copernicus University, Toruń, Poland
[3] School of Computer Engineering, Nanyang Technological University, Singapore
Google: W. Duch

Abstract. An approach to the word sense disambiguation (WSD) relaying on the WordNet synsets is proposed. The method uses semantically tagged glosses to perform a process similar to the spreading activation in semantic network, creating ranking of the most probable meanings for word annotation. Preliminary evaluation shows quite promising results. Comparison with the state-of-the-art WSD methods indicates that the use of WordNet relations and semantically tagged glosses should enhance accuracy of word disambiguation methods.

Keywords: Word Sense Disambiguation, WSD, WordNet, Wikipedia, NLP.

1 Introduction

Ambiguity of natural language is the source of many problems in automatic text processing. It is quite evident for example in classification or clustering of documents represented by features derived from word frequencies. Automatic semantic annotation is still a great challenge, requiring solution to the word sense disambiguation (WSD) problem. To realize the promise of semantic Internet, with machine-enabled interpretation, words in the text should be annotated with specific senses. Adding elementary semantic information during the initial processing phase greatly facilitates text annotation and interpretation. It can be achieved by creating text representation based on word senses instead of string tokens extracted from the content of the text [5]. Other types of annotations, not discussed here, include syntactic parts of speech tagging, grammatical, anaphoric, prosodic and affective annotations.

In analogy to the NP-complete problems in computational complexity theory [7] the word disambiguation and many other Natural Language Processing (NLP) problems are defined as AI-complete [14], meaning that their solution is as hard as passing the Turing Test [17]. The problem of words disambiguation implies several issues that should be taken into consideration.

First, how to distinguish and represent word meanings? The level of granularity of senses and how they relate to each other needs to be defined. The use of synonyms and/or homonyms must be considered. Many approaches have been developed to acquire word senses in an automatic way [18] eg. using Latent Semantic Indexing based

---

* Corresponding author.

T. Huang et al. (Eds.): ICONIP 2012, Part IV, LNCS 7666, pp. 180–187, 2012.

on Singular Value Decomposition applied to statistics of words occurrences. Still the most successful results are achieved creating word senses by a hand.

Second, how to encode lexical knowledge to enable effective interpretation similar to the way people use natural language? Many approaches have been devised here, using for example manually crafted ontologies, ex. Sumo/Milo [12], or semantic networks [16], such as manually created WordNet [11], or semi-automatically created Concept-Net [8] and MindNet [13]. Other approaches try to acquire linguistic knowledge using statistical methods applied to a large collections of data eg. HAL [9] or ADIOS [15].

One fruitful approach to semantic annotation of texts is to move beyond bag-of-words representation, using atoms of lexical knowledge to represent the elementary word meanings (senses), and converting the text into a graph linking senses rather than words. We shall focus here only on the word sense disambiguation during initial text processing phase, mapping words form texts to the structures that carry elementary meanings that may be treated as semantic atoms (senses). WordNet synsets are well-suited for that purpose, grouping words into sets of synonyms related to word definitions, providing sense identifiers and recording semantic relations between synsets. Different people rarely use the same words describing the same object, scene or situation. The use of synsets helps to capture similarities of texts that contain different words but have similar meaning. Employing synsets allows for using WordNet semantic network formed by relations between synsets. Text annotated at a higher abstraction level is can be clustered in a better way because similarities between texts are more clear.

Enhancing document representation with superordinate categories works even better for clustering [4], simulating spreading neural activation responsible for simple inference processes in the reader's mind. New features expose content of the text in more obvious way, simplifying conceptual processing. This elementary representations of words meanings has been already used in our projects aimed at constructing algorithms for automatic analysis of texts[1]. The approach to the word sense disambiguation introduced here is based on contextual information obtained from synsets related to a given synset by exploiting its definition. Description of the used algorithm, examples of the disambiguations and evaluation on the set of several polysemous test words is given below.

2   Disambiguation with WordNet Semantic Glossses

Semantic Glosses (SG) approach employs relations between synsets, or more precisely relations obtained from references between synsets that are related to their definitions (gloss tags)[2]. This idea differs from one of the most popular approaches – adapted Lesk algorithm [1] that uses structural information and traverses the hypernym hierarchy formed as a tree of senses. A graph of related synsets is used here to strengthen mutual associations of synsets. It resembles the spreading activation process [2], automatic activation of related concepts during sentence comprehension, formation of patterns of activations related to word meanings. Activations of the network of synsets may serve

---

[1] http://kask.eti.pg.gda.pl/CompWiki/
[2] http://wordnet.princeton.edu/glosstag.shtml

Algorithm 1. Pseudocode of Semantic Glosses algorithm

```
 1: function SEMANTICGLOSS(text)
 2:     wordSenses ← empty set
 3:     Nwords ← number of adjacent words included
 4:     for word in text do
 5:         for sense in word.def.senses do
 6:             wordSenses[sense] ← 0;
 7:         end for
 8:     end for
 9:     for word in text do
10:         Refs ← Nwords words adjacent to the word
11:         for reference in Refs do
12:             for wordSense in word.def.senses do
13:                 if sense ∈ reference.def.senses then
14:                     scores[sense] + +                    ▷ Add a point to the score
15:                 end if
16:             end for
17:         end for
18:     end for
19:     result ← an empty set
20:     for word in wordSenses do
21:         result ← a sense for word with the highest score
22:     end for
23:     return result
24: end function
```

---

as an approximation of semantic representation [3], where the already active network constrains selection of the next synset.

SG approach for disambiguation of word meanings employs relations between synsets. WordNet not only lists various word meanings, representing them by synsets, but also provides several types of relations between them. Many structural relations, such as hypernyms, troponyms, or meronyms, are defined, usually for a particular part of speech. Starting with the version 3.0 WordNet also provides *semantically annotated disambiguated gloss corpus*. Glosses are short definitions providing proper meanings of words and thus whole synsets. The gloss annotations cover also concepts, collocations (multi-word forms), tagging discontiguous spans of text, for example converting "personal or business relationship" to "personal_relationship", "business_relationship". Glosses have been linked manually to the context-appropriate sense in WordNet, disambiguating the corpus. Tagging includes part of speech, potential lemma forms, a few semantic classes (acronym, number, year, currency, etc). This information creates many new opportunities, but in this paper only associations between synset definitions are used. With semantically annotated gloss corpus each synsets is loosely coupled with several others, related to its definition. In this way additional contextual relations are provided and these relations are not restricted to the one part of speech, as is the case with most structural relations.

The main steps in the disambiguation process are as follows:

- Disambiguated word W is mapped on its possible meanings (synsets) $\{Ts(W)\}$.
- For each synset from $\{Ts(W)\}$ set retrieve all synsets Tgs that may be derived from its glosses.
- Rank all Ts synset according to the number of relations with glosses in Tgs.

The details of this algorithm are described in the pseudocode Algorithm 1. Wordnet glosses may be extended by Wikipedia articles that may be linked directly to the appropriate synsets.

3 Results

Examples of disambiguations performed using relations between synset glosses are presented below. For illustration different texts using different meanings of a test word *horse* have been selected. Disambiguation results for different meanings of the word are presented in Tables 2–6. To compare the results achieved with this approach, denoted as SG (*Semantic Glosses*), results obtained by the Stanford Parser[3] (SP) are also given. Probabilities in percentages indicating the most suitable sense are presented. Results are presented in the form $\boxed{word_x^y}$, where x is the sense number and y is a Greek letter used to enumerate consecutive words that appear in the text.

WordNet offers following senses of the word *horse*:

1. **horse, Equus caballus** – solid-hoofed herbivorous quadruped domesticated since prehistoric times.
2. **horse, gymnastic horse** – a padded gymnastic apparatus on legs.
3. **cavalry, horse cavalry, horse** – troops trained to fight on horseback; 500 horse led the attack.
4. **sawhorse, horse, sawbuck, buck** – a framework for holding wood that is being sawed.
5. **knight, horse** – a chessman shaped to resemble the head of a horse; can move two squares horizontally and one vertically (or vice versa).

The Wikipedia articles about these different meaning of *horse*, with each appearance labeled, are shown below, followed by tables showing results of the Stanford Parser (SP) and our Semantic Glosses (SG) method.

1) The $\boxed{horse_1^\alpha}$ is a hooved (ungulate) mammal, a subspecies of the family Equidae. $\boxed{Horses_1^\beta}$ and humans interact in a wide variety of sport competitions and non-competitive recreational pursuits, as well as in working activities such as police work, agriculture, entertainment, and therapy. $\boxed{Horses_1^\gamma}$ were historically used in warfare, from which a wide variety of riding and driving techniques developed, using many different styles of equipment and methods of control. Humans provide domesticated

---

[3] http://nlp.stanford.edu/software/lex-parser.shtml

$\boxed{\text{horses}_1^\delta}$ with food, water and shelter, as well as attention from specialists such as veterinarians and farriers. The results of disambiguation of the word "horse" in the first sense are shown in Table 1. Scores are given in percents, and bold face shows the highest score for a given method and a given word position. SG has slight preference for wrong sense 3 and 5 over correct 1.

Table 1. Results of disambiguating four occurrences of word "horse" in sense 1 (horse_1)

	horse_1^α		horses_1^β		horses_1^γ		horses_1^δ	
	SP	SG	SP	SG	SP	SG	SP	SG
#1	**48%**	22%	**40%**	22%	**41%**	**22%**	**62%**	22%
#2	26%	18%	17%	19%	11%	19%	7%	19%
#3	18%	**24%**	18%	**23%**	10%	21%	17%	21%
#4	8%	15%	24%	15%	0%	16%	9%	16%
#5	1%	21%	1%	21%	38%	**22%**	4%	**22%**

2) The $\boxed{\text{horse}_2^\alpha}$ is an artistic gymnastics apparatus. It is used by only male gymnasts, due to intense strength requirements. Originally made of a metal frame with a wooden body and a leather cover, modern pommel $\boxed{\text{horses}_2^\beta}$.

Both parsers found for the second occurrence collocations 'pommel horse', defined as "a metal body covered with foam rubber and leather, with plastic handles (or pommels)", therefore the second occurrence of the word has not been tagged. The results of disambiguation of the word "horse" in that sense are shown in Table 2. SG shows very stron preference for correct meaning, while SP fails here.

Table 2. Results of disambiguating sense horse_2

	horse_2^α		horses_2^β	
	SP	SG	SP	SG
#1	**57%**	0%	n/a	n/a
#2	42%	**98%**	n/a	n/a
#3	1%	2%	n/a	n/a
#4	0%	0%	n/a	n/a
#5	0%	0%	n/a	n/a

Table 3. Results of disambiguating sense horse_3

	horse_3^α	
	SP	SG
#1	31%	n/d
#2	6%	n/d
#3	**63%**	n/d
#4	0%	n/d
#5	0%	n/d

3) $\boxed{\text{Horse}_3^\alpha}$ cavalry were soldiers or warriors who fought mounted on horseback. Cavalry were historically the third oldest (after infantry and chariotry) and the most mobile of the combat arms. A soldier in the cavalry is known by a number of designations such as cavalryman or trooper.

In this case SG approach found collocation 'horse cavalry' with score 69% and annotated it as 'an army unit mounted on horseback', and with score 31% 'troops trained to fight on horseback'. Although this is essentially correct single word has not been annotated, hence n/d in Table 3.

4) A $\boxed{\text{horse}_4^\alpha}$ is a beam with four legs used to support a board or plank for sawing. The sawhorse may be designed to fold for storage. A sawhorse with a wide top is particularly useful to support a board for sawing or as a field workbench, and is more useful as a single, but also more difficult to store. The results of disambiguation of word horse in that sense have been shown in Table 4. Here SG has very stron correct preference while SP fails.

Table 4. Results of disambiguating sense horse$_4$

	SP	SG
#1	32%	0%
#2	19%	10%
#3	10%	0%
#4	2%	**90%**
#5	**37%**	0%

horse$_4^\alpha$

Table 5. Results of disambiguating sense horse$_5$

	horse$_5^\alpha$		horse$_5^\beta$	
	SP	SG	SP	SG
#1	28%	7%	26%	11%
#2	8%	6%	19%	9%
#3	27%	36%	**36%**	33%
#4	0%	5%	3%	8%
#5	**36%**	**47%**	15%	**38%**

5) The $\boxed{\text{horse}_5^\alpha}$ is a piece in the game of chess, representing a knight (armored cavalry). It is normally represented by a $\boxed{\text{horse}_5^\beta}$'s head and neck. Each player starts with two knights, which start on the rank closest to the player, one square from the corner. The results of disambiguation of the word horse in that sense are shown in Table 5. SG is correct by a wide margin over other senses, SP fails for the second position in favor of sense 3.

The evaluation of the SG approach has been performed on a test set of eight multi-sense words. For different senses of these words 51 test texts have been prepared, and evaluation of disambiguation performed in the same way as in the case of a word *horse*.

The results of disambiguation for 8 test words are shown in Table 6. The percentage values describes the fraction of proper disambiguations achieved for each word aggregated for all senses.

Table 6. Accuracy of disambiguating 8 test words

	SP	SG
Horse	66%	75%
King	30%	53%
Road	100%	66%
Computer	100%	50%
Grass	60%	100%
Kernel	33%	100%
Shell	20%	55%
Root	50%	80%

Table 7. Aggregated graded scores of disambiguating 8 test words (see text)

	SP	SG
Horse	63%	81%
King	38%	50%
Road	83%	66%
Computer	100%	50%
Grass	30%	100%
Kernel	33%	100%
Shell	20%	50%
Root	41%	100%

Results presented in Table 6 describe only the precision of disambiguation in terms of binary correct-incorrect decisions. However, sometimes the difference between two top-scored synsets may be very small, or even zero, as for example in Table 1 where the horse$_1^\gamma$ Semantic Glosses algorithm equally score sense #1 and #5. A graded evaluation that values bigger differences in scores gives more useful information. If the proper sense of the word has been determined with high confidence (more than 10% difference from the next sense) the graded score is 1. If this difference is in the range $3 - 10\%$ it obtains 0.75, and for differences within $\pm 3\%$ the graded score is 0.5 (even if the correct synset is cored below the winner). Finally, if the proper meaning fells below the 3% of the winner, or for some reason could not be evaluated the score is 0.

Summing all graded scores and dividing this sum by the number of performed disambiguations expressed in percentages allows for measuring results including some estimation of disambiguation confidence. These results are presented in Table 7.

4   Discussion and Future Directions

The algorithm that employs semantically annotated glosses provides quite promising results. So far it has been evaluated only on a small test set of 8 multi sense words (51 different meanings). As the preliminary results are promising the method is now being tested on a larger scale, and some improvements will be introduced.

The approach can run into problems while disambiguating different meanings of the same word in one sentence eg. „*Turtle's shells provide protection to parts of the animal body, like egg shell protects birds' embryo.*" The first 'shell' is related to the turtle shell the second to the egg shell. The task for disambiguating such cases is relatively easy for humans because using semantic memory collocations are easily discovered and require much smaller context for proper sense classification. Experiments with variable context length dependent on the number of identical words with different meanings in one sentence will be performed to check how to deal with such difficulties.

Some WordNet synsets are larger and have more relations than others, the distribution is very uneven. This causes preference for larger synsets that may confuse many algorithms degrading results for meanings that correspond to synsets with small number of relations. To simulate effects of spreading activation weighed relations between synsets may be introduced, describing patterns of more and less important activations. One should also explore the use of WordNet structural information given in predefined relations that extends the network of relations between synsets. Also it is possible to use references between glosses obtained from higher order relations that should have smaller weights.

It should be fruitful to employ additional relations from mining Wikipedia hyper-references [10] to introduce more relations between synsets. This task requires first a mapping between WordNet synsets and Wikipedia articles. Results of the semi-automatic approach [6] to perform such mapping are quite good. Another aspect is the use of negative knowledge about the words present in glosses that do not appear in the wider context.

To perform experiments presented in this article the application for testing different methods of word sense disambiguation has been created. Using it one can enter the sentence and obtain the text with semantic annotations. This application integrates selected parsers and allows for experimentation displaying results in the user-friendly form. The source code of this application is freely available for academic use at the address: http://kask.eti.pg.gda.pl/semagloss/annotations.zip. This project resulted also in development of API in C# and Java for WordNet semantically annotated gloss corpus. The API is available for download at the address http://kask.eti.pg.gda.pl/semagloss/index.html.

Acknowledgments. The work has been supported by the Polish Ministry of Science and Higher Education under research grant N N 516 432338. Authors would like also to thank Adam Kuśmierz for his contribution to the application development.

References

1. Banerjee, S., Pedersen, T.: An Adapted Lesk Algorithm for Word Sense Disambiguation Using WordNet. In: Gelbukh, A. (ed.) CICLing 2002. LNCS, vol. 2276, pp. 136–145. Springer, Heidelberg (2002)
2. Collins, A., Loftus, E.: A Spreading-Activation Theory of Semantic Processing. Psychol. Rev. 82, 407 (1975)
3. Duch, W., Matykiewicz, P., Pestian, J.: Towards Understanding of Natural Language: Neurocognitive Inspirations. In: de Sá, J.M., Alexandre, L.A., Duch, W., Mandic, D.P. (eds.) ICANN 2007. LNCS, vol. 4669, pp. 953–962. Springer, Heidelberg (2007)
4. Duch, W., Matykiewicz, P., Pestian, J.: Neurolinguistic Approach to Natural Language Processing with Applications to Medical Text Analysis. Neural Netw. 21, 1500–1510 (2008)
5. Kehagias, A., Petridis, V., Kaburlasos, V.G., Fragkou, P.: A Comparison of Word and Sense-based Text Categorization Using Several Classification Algorithms. J. Intell. Inf. Syst. 21, 227–247 (2003)
6. Korytkowski, R., Szymański, J.: Collaborative Approach to WordNet and Wikipedia Integration. In: The Second International Conference on Advanced Collaborative Networks, Systems and Applications, COLLA 2012, pp. 23–28 (2012)
7. Kubale, M.: Introduction to Computational Complexity and Algorithmic Graph Coloring, Gdańskie Towarzystwo Naukowe, Poland (1998)
8. Liu, H., Singh, P.: ConceptNet: A Practical Commonsense Reasoning Tool-Kit. BT Technol. J. 22, 211–226 (2004)
9. Lund, K., Burgess, C.: Producing High-Dimensional Semantic Spaces from Lexical Cooccurrence. Behav. Res. Methods 28, 203–208 (1996)
10. Medelyan, O., Milne, D., Legg, C., Witten, I.: Mining Meaning from Wikipedia. Int. J. Hum-Comput. St. 67, 716–754 (2009)
11. Miller, G.A., Beckitch, R., Fellbaum, C., Gross, D., Miller, K.: Introduction toWordNet: An On-line Lexical Database. Princeton University Press, New Jersey (1993)
12. Niles, I., Pease, A.: Towards a Standard Upper Ontology. In: Proceedings of the International Conference on Formal Ontology in Information Systems, pp. 2–9. ACM (2001)
13. Richardson, S., Dolan, W., Vanderwende, L.: MindNet: Acquiring and Structuring Semantic Information from Text. In: Proceedings of the 17th International Conference on Computational Linguistics, vol. 2, pp. 1098–1102. Association for Computational Linguistics (1998)
14. Shahaf, D., Amir, E.: Towards a Theory of AI Completeness. In: AAAI Spring Symposium: Logical Formalizations of Commonsense Reasoning, pp. 150–155 (2007)
15. Solan, Z., Horn, D., Ruppin, E., Edelman, S.: Unsupervised Learning of Natural Languages. Proceedings of the National Academy of Sciences of the USA 102(33), 11629 (2005)
16. Sowa, J.: Principles of Semantic Networks. Morgan Kaufmann Series in Representation and Reasoning. Morgan Kaufmann, San Mateo (1991)
17. Turing, A.: Computing Machinery and Intelligence. Mind 59, 433–460 (1950)
18. Turney, P., Pantel, P.: From Frequency to Meaning: Vector Space Models of Semantics. J. Artif. Intell. Res. 37, 141–188 (2010)

Identification of Moving Vehicle Trajectory
Using Manifold Learning

Giyoung Lee, Rammohan Mallipeddi, and Minho Lee*

School of Electrical Engineering and Computer Science, Kyungpook National University,
1370 Sankyuk-Dong, Puk-Gu, Taegu 702-701, South Korea
gylee@ee.knu.ac.kr, mallipeddi.ram@gmail.com, mholee@knu.ac.kr

Abstract. We present a method to identify the trajectories of moving vehicles from various viewpoints using manifold learning to be implemented on an embedded platform for traffic surveillance. We use a robust kernel Isomap to estimate the intrinsic low-dimensional manifold of input space. During training, the extracted features of the training data are projected on to a 2D manifold and features corresponding to each trajectory are clustered in to k clusters, each represented as a Gaussian model. During identification, features of test data are projected on to the 2D manifold constructed during training and the Mahalanobis distance between test data and Gaussian models of each trajectory is evaluated to identify the trajectory. Experimental results demonstrate the effectiveness of the proposed method in estimating the trajectories of the moving vehicles, even though shapes and sizes of vehicles change rapidly.

Keywords: trajectory identification, manifold learning, robust kernel Isomap, Mahalanobis distance, traffic surveillance system.

1 Introduction

The robust tracking and trajectory identification of multiple moving objects from various viewpoints is a challenging and an important task in the field of computer vision and artificial intelligence with many practical applications, such as video surveillance and traffic control. The vehicle tracking in a real traffic scene involves many challenging vision problems like rotations, scaling, lighting conditions and occlusions [1]. In an embedded surveillance system with limited memory and computing resources, the throughput rate of video sampling dramatically decreases even at the low speed of the moving vehicles. Due to the low sampling video rate the shapes and sizes of the moving vehicles vary rapidly. Therefore, it is difficult to implement a low-cost intelligent camera on an embedded platform for tracking and identification of moving vehicle trajectories.

In literature, various methods for tracking and trajectory identification have been proposed. The approaches can be categorized as Blob-based, Contour-based, model-based, region-based and feature-based [2-4]. Among them, the feature-based

---

* Corresponding author.

T. Huang et al. (Eds.): ICONIP 2012, Part IV, LNCS 7666, pp. 188–195, 2012.

approaches are robust and perform well even in the presence of partial occlusions as some of the features of the moving vehicles are visible [4]. As well, the feature-based approaches do not employ computationally expensive operations such as background subtraction [3], Kalman filtering [4], etc.

In feature-based trajectory identification, though, the dimensionality of the input images is high, the perceptually meaningful structure of these images has many fewer independent degrees of freedom [5]. Therefore, for robust tracking and trajectory identification of the moving vehicles, it is crucial to reduce the dimensionality of the input to capture the meaningful axes in data space. Also, the feature-based trajectory identification uses the relationship between the object information in previous and current frames, which is difficult to be implemented on an embedded platform due to the abrupt changes in the shape and size. To address this problem, the relationship among the previous, current and next positions of a moving vehicle is necessary. Actually, the change in the shape of a moving vehicle is sequential even though the change is abrupt. Our idea is to learn the complete structure of this sequential change in the shape of the vehicles in an intrinsic low-dimensional subspace and embed every shape instance from the high-dimensional image space into this structure.

To find the independent degrees of freedom, there are different dimensionality reduction techniques like principal component analysis (PCA) and multidimensional scaling (MDS) [6], locally linear embedding (LLE) [7] and isometric feature mapping (Isomap) [5], [8]. In this paper, the manifold learning using the robust kernel Isomap is employed to address the problems of dimensionality reduction and the low-dimensionality embedding of the sequential shape change of the moving vehicles. After training the possible trajectories of the moving vehicles in advance by using the manifold learning, the proposed method can recognize the trajectory of a new moving vehicle with large variation of shapes and sizes in successive sampled images in the identification phase, which is implemented in an embedded platform. If we know the trajectory of a moving vehicle, we can predict the next position of the moving vehicle and we can track the same vehicle using this trajectory information.

The rest of the paper is organized as follows: Section 2 presents a brief literature review on the robust kernel Isomap. Section 3 presents the implementation of the proposed algorithm. Section 4 presents the experimental results and discussions, while Section 5 concludes the paper with some future directions.

2   Robust Kernel Isomap

In video surveillance, each image can be viewed as a point in the high dimensional vector space whose dimensionality is equal to the number of pixels in the image [12]. The images can be characterized by far fewer degrees of freedom than the actual number of pixels per image. Dimensionality reduction is the transformation of high-dimensional data into a meaningful representation in the reduced dimensionality. Ideally, the reduced representation should have a dimensionality that corresponds to the intrinsic dimensionality of the data, which is the minimum number of parameters needed to account for the observed properties of the data.

PCA and MDS, guaranteed to discover the true structure of data lying on or near a linear subspace of the high dimensional input space [5], but fail in datasets which contain non-linear structures [1]. LLE [7] is a nonlinear dimensionality reduction that uses locally linear embedding to compute low-dimensional neighborhood and hence results in poor performance. Manifold learning is a dimensionality reduction technique that extracts a smooth non-linear low dimensional manifold form high dimensional data points. Isomap [5], [9] is a manifold learning algorithm, which extends the classical MDS by considering approximate geodesic distance instead of Euclidean distance [10]. Unlike LLE, Isomap has no local assumptions, and only needs a sparse set of distance measures [11], thus facilitating the implementation on an embedded platform. Isomap algorithm also suffers from topological instability and weak generalization ability [8], [12]. Robust kernel Isomap [12-14] is an improved version of Isomap, with more topological stability and generalization ability.

Given N objects, each represented as a n-dimensional vector x_i, $i = 1,... N$, the kernel Isomap algorithm finds a mapping which places the N points in a low-dimensional space. The robust kernel Isomap algorithm can be summarized as [10]:

1. Identify k nearest neighbors to each input data point and construct a neighborhood graph. The edge lengths between points in a neighborhood are the Euclidean distances.
2. Construct a matrix $K(D^2)$ by computing the geodesic distances, D_{ij}, that are associated with the sum of the edge weights along the shortest paths between the N pair of points as shown by Eq. 1 & 2, where H is the centering matrix.

$$K(D^2) = -\frac{1}{2}HD^2H \tag{1}$$

$$D^2 = \left[D_{ij}^2\right] \in \Re^{N\times N}, \; H = I - \frac{1}{N}ee^T, \; e = [1, ... ,1]^T \in \Re^N \tag{2}$$

3. Compute the largest eigenvalue (c^*) of the matrix

$$\begin{bmatrix} 0 & 2K(D^2) \\ -I & -4K(D) \end{bmatrix} \tag{3}$$

and construct a Mercer kernel matrix $\tilde{K}$, which is guaranteed to be positive semi-definite for $c \geq c^*$ using the Eq. (4)

$$\tilde{K} = K(D^2) + 2cK(D) + \frac{1}{2}c^2H \tag{4}$$

4. Compute top d eigenvectors of $\tilde{K}$, which leads to the eigenvector matrix $V \in \Re^{N\times d}$ and the eigenvalue matrix $\Lambda \in \Re^{d\times d}$.
5. The coordinates of the N points in the d-dimensional Euclidean space are given by the column vectors of $Y = \Lambda^{\frac{1}{2}}V^T$.

3   Proposed Method for Trajectory Identification

The proposed system consists of two phases, training and identification, which are performed offline and online, respectively as shown in Fig. 1.

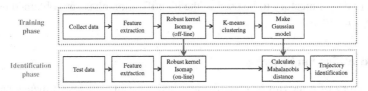

Fig. 1. Flow chart

3.1   Feature Extraction

As a vehicle moves away or towards the camera mounted on a streetlight, the size and/or shape of the vehicle vary from frame to frame. Because of the variations in the size and shape of a detected vehicle, the dimensionality of input images varies. However, it is necessary to extract features with same dimensionality for the manifold learning. In this paper, we use the down sampled edge features to classify the trajectory of moving vehicles as shown in Fig. 2 [13-15], because the edge features are sensitive to vehicle shapes depending on the viewpoints but insensitive to the different kinds of vehicles. As shown in Fig. 2, an edge image is obtained from the original image by applying the Sobel filter [9]. The extracted edge image is resized into a 16 by 16 image, so that the extract features have the same input dimension for all the input images with different sizes of vehicles. From the extracted 16 by 16 edge image, a 256 dimensional feature vector can be prepared for manifold learning.

Fig. 2. Feature extraction for manifold learning

3.2   Training Phase

The possible trajectories of moving vehicles are defined in advance and Gaussian models that can better represent the data pertaining to each trajectory are created.

To achieve efficient performance, the collected training data should well represent the various viewing conditions of the moving vehicles to be tracked and hence the training data should include the images of the moving vehicles with various shapes and sizes. Also, the training data should be dense enough to have a smooth variation in the shape and/or size of the vehicles. Therefore, we collect the training data in high video frame rate (10 FPS) using a server PC (Intel Quad Core 2.8 Ghz, 4G RAM).

After building a training set, we create a training data matrix in which each row represents a training sample and the number of columns represents the dimensionality of features described in section 3.1. After obtaining the training data matrix, robust kernel Isomap is computed off-line as described in section 2. Hence, the dimensionality of the training data matrix is reduced to a low-dimensionality (usually 2 or 3).

In low-dimensional manifolds, the training data corresponding to each trajectory is clustered into k (= 5) clusters and the Gaussian mean vectors and the covariance matrices of the Gaussian models are computed. In other words, Gaussian models corresponding to each trajectory to be used in the identification phase are obtained using the K-means clustering.

3.3   Identification Phase

The trajectory of a moving vehicle is identified by using the low-dimensional manifolds which are obtained in the training phase. However, due to the limited throughput rate of video sampling in an embedded platform (Davinci Board: TMS320DM6437, 2 FPS), the test samples may not include the smooth variation in the shape and size of the moving vehicles.

At first, the feature vector of the detected moving vehicle is extracted and projected on-line onto the low-dimensional manifolds which are built by using the robust kernel Isomap in the training phase.

If χ_t is the projected test data and t is the time index, the Mahalanobis distance between the Gaussian of cluster i (mean : μ_i, covariance : Σ_i) obtained in the training phase and the projected test data are evaluated using Eq. (5).

$$D(\chi_t, \mu_i) = (\chi_t - \mu_i)\Sigma_i^{-1}(\chi_t - \mu_i)^T \qquad (5)$$

After obtaining the Mahalanobis distance between the test data and all the Gaussian models in the training data, we select the Gaussian models that are close to the test sample and within a threshold value (=10) in the 2D manifold using Eq. (6).

$$S(t) = \{ i \mid D(\chi_t, \mu_i) < Threshold \} \qquad (6)$$

For the test data χ_t, the possible trajectories $TRJ(t)$ can be obtained by selecting the trajectories corresponding to the Gaussian models $S(t)$ obtained by the Eq. (6).

Finally, we estimate the trajectory of test sample by using a simple majority voting on the possible trajectories $TRJ(t)$, $TRJ(t-1)$, $TRJ(t-2)$, ..., $TRJ(1)$. In other words, a trajectory that appears majority of times as a "possible trajectory" up to that particular point of time, where the test sample is taken, is considered as the trajectory to which the test sample belongs.

4   Experimental Results

To evaluate the proposed system, we identify the trajectory of vehicles moving on four different trajectories on a video clip, which is captured by a fixed camera as shown in Fig. 3 (a).

The data corresponding to 7 different vehicles, when moving along each of the four different trajectories is gathered to form a training data set. The vehicles have various colors such as white, silver, black and so on.

Training data matrix is constructed using the 256-dimensional edge feature vectors of training data. The dimensionality of the training data matrix is reduced to 2 by using the robust kernel Isomap as shown in Fig. 3 (b). The dotted colors in Fig. 3 (b) correspond to the respective colored trajectories in Fig. 3 (a). The robust kernel Isomap obtained uses a nearest neighborhood size l (=16) to construct the neighborhood graph. Fig. 4 shows the clustering of the points corresponding to each trajectory in the 2D manifolds into k (= 5) clusters using the K-means clustering algorithm. Each cluster is represented by a Gaussian model and the labeling of the Gaussian models corresponding to each trajectory is presented in Fig. 4.

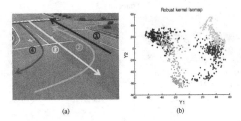

Fig. 3. (a) Trajectory of moving vehicles (b) 2D plots of low-dimensional training data

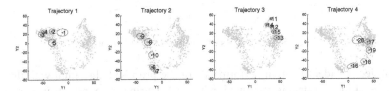

Fig. 4. The Gaussian model of each trajectory using K-means clustering

For testing, the data corresponding to 10 different vehicles moving along each of the 4 different trajectories are gathered. To demonstrate the effectiveness of the proposed method, we have done the experiment by collecting the test data not only at the same frame rate as the training data (10 FPS), but also at the low video frame rates (5 FPS and 2 FPS) compared to the training data.

Fig. 5 shows the images of the moving vehicles and their corresponding projections on the 2D manifold at different points in time (t-3, t-2, t-1, t, from right to left) while moving along the trajectories 3 and 4 as shown in Fig. 3 (a). From the Figs. 3, 4 and 5, it can be observed that the trajectories 3 and 4 have a common area to be traversed by the moving vehicles and hence occupy similar positions on the 2D manifold. In Fig. 5, it can be observed that during the time instances t-3, t-2 and t-1 both the vehicles move on the same path and hence occupy the similar positions on the 2D manifold. However, at the time instance t, when the two vehicles move on to different trajectories, the projections on the 2D manifold vary significantly. Therefore, it can be concluded that data which belongs to the same trajectory have similar coordinates in the 2D kernel Isomap and the trajectory corresponding to the test data projected on the 2D manifold can be identified by comparing with the nearest training data.

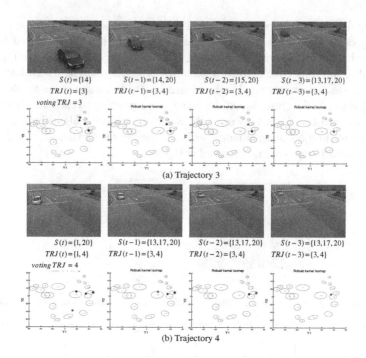

Fig. 5. The process of trajectory identification

The trajectory identification performance of the proposed algorithm with different frame rates of the test data is presented in Table 1. During the experimentation, it was observed that, three times the test data of trajectory 1 is incorrectly identified as trajectory 2 due to the similarity in the trajectory shapes and the common area between the two trajectories. A similar kind of misclassification has been observed during the identification of trajectories 3 and 4.

Table 1. Trajectory identification result

	# of total test data	Identification accuracy
10 FPS	40	87.5 %
5 FPS	40	87.5 %
2 FPS	40	87.5 %

5   Conclusion and Future work

In this paper, we proposed a vehicle trajectory identification method using manifold learning. Experimental results show that the proposed method can efficiently estimate the trajectories of the moving vehicles whose shape and/or size change rapidly.

As a future work, we would like to track moving vehicles using the pre-trained trajectory information based on the manifold learning. After identification of the vehicle trajectory, we can link the vehicle having the same trajectory, and then the linked results can be regarded as the tracking results. So we can track moving vehicles in real time even though their shapes and sizes rapidly change in an embedded platform with low video frame rate such as embedded platform.

Acknowledgments. This research was supported by the MKE(The Ministry of Knowledge Economy), Korea, under the CITRC(Convergence Information Technology Research Center) support program (NIPA-2012-H0401-12-1006) supervised by the NIPA(National IT Industry Promotion Agency).

References

1. Rane, N., Birchfield, S.: Isomap Tracking with Particle Filtering. In: International Conference on Image Processing, pp. 513–516 (2007)
2. ZuWhan, K., Meng, C.: Evaluation of Feature-Based Vehicle Trajectory Extraction Algorithms. In: International Conference on Intelligent Transportation Systems (ITSC), pp. 99–104 (2010)
3. Qiu, T., Yiping, X., Manli, Z.: Robust Vehicle Tracking Based on Scale Invariant Feature Transform. In: International Conference on Information and Automation, pp. 86–90 (2008)
4. Saunier, N., Sayed, T.: A Feature-Based Tracking Algorithm for Vehicles in Intersections. In: Canadian Conference on Computer and Robot Vision, pp. 59–59 (2006)
5. Tenenbaum, J.B., Silva, V.D., Langford, J.C.: A Global Geometric Framework for Nonlinear Dimensionality Reduction. Science 290, 2319–2323 (2000)
6. Cox, T.F., Cox, M.A.A.: Multidimensional Scaling. Chapman and Hall, London (1994)
7. Roweis, S.T., Saul, L.K.: Nonlinear Dimensionality Reduction by Locally Linear Embedding. Science 290, 2323–2326 (2000)
8. Van der Maaten, L.J.P., Postma, E.O., Van den Herik, H.J.: Dimensionality Reduction: A Comparative Review. Review Literature and Arts of the Americas 10 (2007)
9. Choi, H., Choi, S.: Kernel Isomap. Electronics Letters 40, 1612–1613 (2004)
10. Pless, R.: Image Spaces and Video Trajectories: Using Isomap to Explore Video Sequences. In: International Conference on Computer Vision, vol. 2, pp. 1433–1440 (2003)
11. Choi, H., Choi, S.: Robust Kernel Isomap. Pattern Recognition 40, 853–862 (2007)
12. Nagao, K., Sohma, M., Kawakami, K., Ando, S.: Detecting Contours in Image Sequences. IEICE Transactions on Information and System 76, 1162–1173 (1993)
13. Black, M.J.: Combining Intensity and Motion for Incremental Segmentation and Tracking Over Long Image Sequences. In: Sandini, G. (ed.) ECCV 1992. LNCS, vol. 588, pp. 485–493. Springer, Heidelberg (1992)
14. Boumediene, M., Ouamri, A., Keche, M.: Vehicle Detection Algorithm Based on Horizontal/Vertical Edges. In: International Workshop on Systems, Signal Processing and their Applications (WOSSPA), pp. 396–399 (2011)
15. Mae, Y., Shirai, Y., Miura, J., Kuno, Y.: Object Tracking in Cluttered Background Based on Optical Flow and Edges. In: International Conference on Pattern Recognition, vol. 1, pp. 196–200 (1996)

Approaches for the Detection of the Keywords in Spoken Documents Application for the Field of E-Libraries

Bendib Issam[*] and Laouar Mohamed Ridda

LAMIS Laboratory, University of Tebessa,
Route de Canstantine, 12002, Tebessa, Algeria
bendib2012@gmail.com, ridda_laouar@yahoo.fr

Abstract. Automatic indexing of multimedia documents across several different application tasks, including searching for words spoken, the detection of keywords and audio information retrieval. Thus, despite the changes made in the field of indexing speech, much remains to be done particularly for the key word search in spontaneous speech. Although the research areas of spoken words and audio retrieval has been well addressed, but still significant limitations to achieve, especially in terms of resource available today on the web.

The goal of this paper is to propose an approach for document management based multimedia indexing techniques to detect speech and keywords. We present in this article the various methods of indexing with the techniques of detection of key words. These methods derive three principal approaches from vocal indexing: the detection of key word, the detection of key words on phonetic flow (PSPL, CN,...) and the indexing containing the recognition with great vocabulary (LVR). We present, thereafter the step suggested for an approach based on the combinations of two techniques (PSPL, S-PSPL and CN, like on technique LVR.

A validation of this approach of indexing and information retrieval is in the course of validation for the field of the E-libraries.

Keywords: Multi-media information, vocal Indexing, spoken Documents, Keyword Spotting, PSPL, S-PSPL, CN, LVR.

1   Introduction

The automatic speech recognition (ASR) covers quite large areas of application, going from the isolated word recognition, for speech-based systems for example: recognition systems for continuous speech of automatic transcription. Thus, the indexing of multimedia documents, in particular, the detection of keywords in audio files currently was attracting great interest both in terms experimentally and theoretically. However, despite the progress made in the field of indexing speech, much remains to be done notably for the key word search in spontaneous speech. Further, this evolution has allowed to pass from indexing structured data (Radio broadcasts, the

---

[*] Corresponding author.

T. Huang et al. (Eds.): ICONIP 2012, Part IV, LNCS 7666, pp. 196–203, 2012.

documentary), the indexing of unstructured data (such as spontaneous speech). In literature, among the indexing systems of unstructured data, we find the indexing systems of vocal emails, video conferences and voice annotations of images [7], [11]. These systems have introduced new problems such as hesitations, false starts, poor structuring of sentences...etc. This has encouraged researchers to suggest implementing new techniques to deal with these problems.

Several retrieval techniques dealing with multiple hypotheses from an ASR system have been proposed, in which word and/or sub-word lattices are used to index each utterance. In this context, a very successful new approach of indexing speech information with a very compact structure, PSPL, has been recently proposed [8]. This approach efficiently considers all possible paths in the recognized lattice, as well as word proximity information within the lattice. However, the OOV problem is still left unaddressed in PSPL; that is, OOV words generally do not appear in the recognized lattice. However, sub-word based representations such as phone lattices are crucial especially for OOV words. It is also effective to combine word and subword lattices to achieve high retrieval performance for both IV and OOV queries since subword-based indices generally yield a lower precision for IV queries compared with word-based ones[6].

In this paper, we propose an approach for detecting keywords in multimedia documents based on a combination for these two methods. In addition, to cover the OOV/rare word problems, we incorporate subword posterior probabilities in both PSPL and CN to produce subword-based PSPL (S-PSPL) and CN (S-CN).

2   Problems of Indexing Multimedia Documents

The Navigation technologies and information retrieval are rapidly evolving to meet the diverse needs of multimedia applications. However, the Masses of data which are absolutely enormous what making difficult for experts to automate their indexing, from where new challenges are then defined to catch up with the rapid development of the volume of data and appropriate applications.

In general spoken utterance retrieval (SUR), the process is separated into two parts. The first one is indexing and the second one is search. Indexing is an offline process in which linguistic information is extracted from all speech data in the archive, and an index table is built. A speech recognizer is used to extract such linguistic information. The index table maps each extracted linguistic symbol (word, subword or phrase) to a set of utterances that the symbol matches. Search is an online process in which users' queries are accepted and utterances that each query matches are found [5]. The system finds the target utterances efficiently using the previously built table. The index table helps to search with a desirable speed that is almost independent of the size of the archive.

It is in this perspective that we targeted through this work to contribute by proposing an approach for indexing of multimedia documents. This will be applied later in the field of digital libraries (E-Library).

3   Lattices for Spoken Document Retrieval

The phone Lattice Scanning (PLS) is a method of indexing audio files from phonetic transcriptions from an acoustic-phonetic decoding (DAP) [3]. This method does not require a priori definition of a vocabulary of keywords as in keyword spotting method. This method is able to detect any term based on its phonetic representation.

A new method has been proposed for speech file indexing for an unlimited vocabulary. This approach consists in generating in differed time a phoneme lattice for each audio file using a modified version of the Viterbi algorithm [12]. The detection of keywords is performed by dynamic comparison between the search word and phoneme sequences in the lattice.

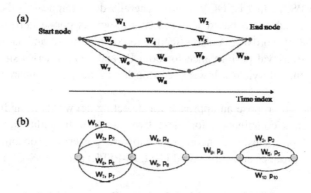

Fig. 1. Structure of indexing containing lattice [4], [1]

Nevertheless, the calculations can become excessively enormous, if one considers lattices for the files of the spoken documents. Great efforts were therefore made to reduce such lattices in simplified forms structures called indexing. For example in Fig. 1, the indexing structure is simplified linear sequence of a segment, which includes a number of assumptions with word posterior probabilities. In this way, we reduce dramatically the memory space and time calculations. However, one can create new non-existent paths in the initial network. In Fig 1. (a) the word W_3 cannot be followed by the word W_8, but this becomes possible in (b).

4   Approaches of Indexing

4.1   Position-Specific Posterior Lattices (PSPL)

The basic idea of PSPL is to calculate the posterior probability *prob* of a word *W* at a specific position *pos* in a lattice for a spoken segment *d* as a tuple *(W, d, pos, prob)*. Such information is actually hidden in the lattice L of d since in each path of L we clearly know each word's position. Since it is very likely that more than one path includes the same word in the same position, we need to aggregate over all possible paths in a lattice that include a given word at a given position.

A variation of the standard forward-backward algorithm can be employed for this computation. The forward probability mass $\alpha(W, t)$ accumulated up to a given time t at the last word W needs to be split according to the length l measured in the number of words:

$$\alpha(w,t,l) \doteq \sum_{\substack{\pi:a \ partial \ path \ ends \ at \ time \ t, \\ has \ last \ word \ W, and \ inludes \\ l \ ssubword \ units}} P(\pi) \tag{1}$$

Where π is a partial path in the lattice. The backward probability $\beta(W, t)$ retains the original definition [9].

The elementary forward step in the forward pass can now be carried out as follows:

$$\alpha(W,t,l) = \sum_{w'} \sum_{\substack{t':\exists \ edge \ e \\ starting \ at \\ time \ t', \ end- \\ ing \ at \ time \ t, \\ and \ with \ word(e)=W}} [\alpha(W',t',l') \cdot P_{AM}(w) \cdot P_{LM}(w)] \tag{2}$$

Where $P_{AM}(W)$ and $P_{LM}(W)$ denote the acoustic and language model scores of W respectively; e is a word arc in the lattice and $word(e)$ means the word entity of arc e.

The position specific posterior probability for the word W being the lth word in the lattice is then:

$$P(W,b,b+Sub(W)-1|L) = \sum_t \frac{\alpha(W,t,b+Sub(W)-1) \cdot \beta(W,t)}{\beta_{start}} \cdot Adj(W,t) \tag{3}$$

Where $\beta start$ is the sum of all path scores in the lattice, and $Adj(W, t)$ consists of some necessary terms for probability adjustment, such as the removal of the duplicated acoustic model scores on W and the addition of missing language model scores around W [9]. In this context, we regard the tuples $(W, d, pos, prob)$ for a specific spoken segment d and position pos as a *cluster*, which in turn includes several words along with their posterior probabilities.

4.2   Confusion Network (CN)

The confusion network is the most compact structure representing multiple hypotheses while keeping the order of symbols (phones/words) along the time axis, the space for the index table can be reduced. In addition, the confusion network essentially has more paths than the original lattice that has only paths allowed by the recognizer. The confusion network has many additional paths on which any connection of hypothesized symbols is allowed unless breaking their original order. Fig.2 shows an example of a lattice (a) and the confusion network (b) that is converted from the lattice. Thus it is compact and potentially achieves more robust keyword matching for OOV words and errorful recognition hypotheses [5].

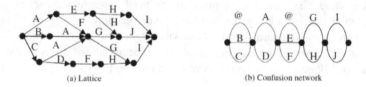

(a) Lattice   (b) Confusion network

Fig. 2. Lattice vs. confusion network. A, B, ..., J indicate hypothesized symbols. @ is a label that stands for allowing null transitions.

A confusion network is a finite state network, and therefore the same approach as the lattice-based method can be applied. However, in a confusion network, arcs are aligned to columns as in Fig. 2 (b) and the weight of each arc has already been normalized so that the sum of weights in each column becomes 1. As the result, $f(p[a])$ is always 1, so this element can be eliminated. In addition, $n[a]$ can also be eliminated since $n[a]$ is necessarily located right next to $p[a]$.

5   Proposed Approach

5.1   Problematic of Digital Libraries

Digital libraries have now become the most popular area for users of documentary resources. Thus, today, and with the great advances of web technologies, structuring, and access to information and multimedia resources by the speech recognition has become an important necessity for managers and users of these libraries.

In summary, the definitions of technical PSPL and CN, we see that the posterior probability of cluster K with CN technique is more reliable because it also incorporates the posterior probability of clusters (k +1) and (k-1).

Moreover, we also find that the computations with the CN are faster because the number of clusters produced by the network is optimal in accordance with search terms. However, the large volume of documents available for the E-library we insist to use all possibility of accelerating this process.

It is this necessity that encouraged us to propose an approach for a fast and efficient access to multimedia resources. This approach will then be the basis for developing an indexing system for multimedia digital libraries (E-Library).

5.2   System Architecture

This section aims to explain our methodological step for the proposal for an approach of indexing of the documentary resources in order to exploitation by the numeric libraries shown in Fig 4.

In this context, we try to use the multi-level information (word and subword) to improve the performance for both IV and OOV words. It is verified that the phone-based indexing method is effective especially for OOV keywords. However it yields generally a lower precision for IV queries than word-based indexing. The benefit with combined word and phone hypotheses has been shown in recent works [9]. It is reported that the combination of word and phone confusion networks is effective to achieve high retrieval performance for both IV and OOV queries.

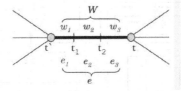

Fig. 3. An edge of W with the word of $w1w2w3$ Subword units starting at time t' and ending at time t

Against this background, we compute the subword Posterior Probability and subword Confusion Network. Consider a word W as shown in Fig 3 with characters $\{w_1w_2w_3\}$ corresponding to the edge e starting at time t' and ending at time t in a word lattice. During decoding the boundaries between w_1 and w_2, and w_2 and w_3 are recorded respectively as t_1 and t_2. The posterior probability (PP) of the edge e given the lattice A, $P(e|L)$, is:

$$P(e|L) = \frac{\alpha(t') \cdot P\left(x_{t'}^t|W\right) \cdot P_{LM}(W) \cdot \beta(t)}{\beta_{start}} \tag{4}$$

where $\alpha(t')$ and $\beta(t)$ denote the forward and backward probability masses accumulated up to time t' and t obtained by the standard forward-backward algorithm, $P\left(x_{t'}^t|W\right)$ is the acoustic likelihood function, $P_{LM}(W)$ the language model score, and β_{start} the sum of all path scores in the lattice. Equation 4 can be extended to the PP of a subword of W, say w_1 with edge e_1:

$$P(e_1|L) = \frac{\alpha(t_1) \cdot P\left(x_{t'}^{t_1}|w_1\right) \cdot P_{LM}(w_1) \cdot \beta(t_1)}{\beta_{start}}$$

After we obtain the PPs for each subword arc in the lattice, such as $P(ei|L)$ as mentioned above, we can perform the same clustering method proposed in related work [10] to convert the word lattice to a strict linear sequence of clusters, each consisting of a set of alternatives of subword hypotheses, or a subword confusion network (CN) [2]. In CN we collect the PPs for all character arc w with beginning time t' and end time t as $P([w; t', t]|L)$:

$$P([w; t', t]|L) = \frac{\sum_{\substack{H=W_1...W_N \in \ lattice:\ P(H)P(L|H) \\ \exists i \in \{1...N\}: \\ W_i \ contains \ [w;t',t]}}}{\sum_{path \ H' \in \ lattice} P(H')P(L|H')} \tag{5}$$

Where H stands for a path in the word lattice. P(H) is the language model score of H (after proper scaling) and $P(L|H)$ is the acoustic model score. CN was known to be very helpful in reducing subword error rate (SER) since it minimizes the expected SER [2]. Given a CN, we simply choose the subword with the highest PP from each cluster as the recognition results.

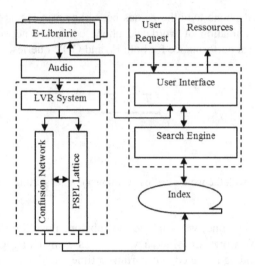

Fig. 4. Approach to indexing information resources multimedia

6   Conclusion

In the field of automatic speech recognition (ASR), existing techniques are mostly based on the recognition of large vocabulary, they offer very good results on structured data, but are still far from being able to handle so successfully for spontaneous speeches and interviews.

In this context, we attempted to propose a hybrid approach combining speaker indexing techniques for automatic speaker recognition. The objective is to propose a technique of hierarchical models of speakers whose purpose is to:

− Build an index structure for easier navigation and updated databases
− To adapt the structure to the problem incrementally.
− Reduce the time and complexity of query search of speaker model (document spoken).

Many other objectives of the indexing of multimedia documents are the subject of our research for things like: (video, audio, integrated audio information in an image, ...). As perspectives to our work, we are interested in defining methods to evaluate the performance of indexing approaches and hierarchical classification.

Thus, to have classification and indexing techniques for applications in the field of digital libraries (E-Library).

References

1. Lee, L.S., Pan, Y.C.: Voice-Based Information Retrieval — How Far are We from the Text-Based Information Retrieval? In: Automatic Speech Recognition & Understanding, pp. 26–43 (2009)

2. Yao, Q., Soong, F.K., Lee, T.: Tone Enhanced Generalized Character Posterior Probability (GCPP) for Cantonese LVCSR. Comp. Speech Lang. 22, 360–373 (2008)
3. Chelba, C., Silva, J., Acero, A.: Soft Indexing of Speech Content for Search in Spoken Documents. Computer Speech and Language 21, 458–478 (2007)
4. Seide, F., Yu, P., Shi, Y.: Towards Spoken Document Retrieval for the Enterprise: Approximate Word-Lattice Indexing with Text Indexers. In: Automatic Speech Recognition & Understanding, pp. 629–634 (2007)
5. Hori, T., Hetherington, I.L., Hazen, T.J., Glass, J.R.: Open Vocabulary Spoken Utterance Retrieval Using Confusion Networks. In: ICASSP, pp. 73–76 (2007)
6. Pan, Y.C., Chang, H.L., Lee, L.S.: Subword-Based Position Specific Posterior Lattices (S-PSPL) for Indexing Speech Information. In: Interspeech, pp. 318–321 (2007)
7. Park, A., Hazen, T., Glass, J.: Automatic Processing of Audio Lectures for Information Retrieval: Vocabulary Selection and Language Modeling. In: Proc. ICASSP, Philadelphia, PA (2005)
8. Chelba, C., Acero, A.: Position Specific Posterior Lattices for Indexing Speech. In: Proceedings of the 43rd Annual Meeting of the Association for Computational Linguistics (ACL 2005), pp. 443–450. Association for Computational Linguistics, Michigan, Ann Arbor (2005)
9. Wessel, F., Schluter, R., Macherey, K., Ney, H.: Confidence Measures for Large Vocabulary Continuous Speech Recognition. In: SAP, vol. 9, pp. 288–298 (2001)
10. Mangu, L., Brill, E., Stolcke, A.: Finding Consensus in Speech Recognition: Word Error Minimization and other Applications of Confusion Networks. Computer Speech and Language 14, 373–400 (2000)
11. Mills. T., Pye, D., Sinclair, D., Wood, K., A Digital Photo Management System. Technical, AT&T Laboratories Cambridge, Cambridge (2000)
12. El Meliani, R., O'Shaughnessy, D.: Lexical Fillers for Task-Independent-Training Based Keyword Spotting and Detection of New Words. In: Proc. EUROSPEECH, pp. 2129–2133 (1995)

Fast Affine Invariant Shape Matching from 3D Images Based on the Distance Association Map and the Genetic Algorithm

Peter Wai-Ming Tsang, W.C. Situ,
Chi Sing Leung, and Kai-Tat Ng

Dept. of Electronic Engineering,
City University of Hong Kong, Hong Kong
eeleungc@cityu.edu.hk

Abstract. The decision on whether a pair of closed contours is derived from different views of the same object, a task commonly known as affine invariant matching, can be encapsulated as the search for the existence of an affine transform between them. Past research has demonstrated that such search process can be effectively and swiftly accomplished with the use of genetic algorithms. On this basis, a successful attempt was developed for the heavily broken contour situation. In essence, a distance image and a correspondence map are utilized to recover a closed boundary from a fragmented scene contour. However, the pre-processing task involved in generating the distance image and the correspondence map consumes large amount of computation. This paper proposes a solution to overcome this problem with a fast algorithm, namely labelled chamfer distance transform. In our method, the generation of the distance image and the correspondence map is integrated into a single process which only involves small amount of arithmetic operations. Evaluation reveals that the time taken to match a pair of object shapes is about 10 to 30 times faster than the parent method.

Keywords: Affine invariant matching, chamfer distance transform.

1 Introduction

Numerous works have been conducted to remove the effect of the viewpoint dependent affine transform in the matching process [1,2]. One popular approach for the matching process is based on the assumption that if a pair of contours is emerged from the same object, there should exist an affine transform between them. This rationale encapsulates the matching process as a search problem which can be effectively solved with the use of genetic algorithms (GA). Early attempts of such approach were reported in [3–5]. Lately, enhanced schemes for increasing the success rate in identifying matched contours with the incorporation of the migrant principle [6], wavelet Transform [7], and latin square [8]. Amongst the above-mentioned investigations, the work in [4] proposed a reliable means of identifying matched contours. It reduces the dimension of the search space to 3, hence enabling the use of GAs to local the optimal solution within small number of generations. However, despite the success of the above methods, they are not

T. Huang et al. (Eds.): ICONIP 2012, Part IV, LNCS 7666, pp. 204–211, 2012.

applicable to fragmented contours. Although different solutions [9,10], the computation load is high which jeopardizes the practical value of these methods.

Recently, this shortcoming has been overcome by a scheme reported in [11] which enables a closed boundary to be constructed from the distance image and correspondence map of a fragmented contour. However, a drawback of this approach is that the computation involved in generating the Euclidean distance image and the correspondence map is heavy, and much higher as compared with the part spent on the GA. This paper overcomes this problem with a method, namely the labelled chamfer distance transform (LCDT), which enables fast generation of the distance image and the correspondence map. The adoption of the proposed method in the parent scheme in [11] could lead to at least one order of magnitude reduction in the time taken to match a pair of fragmented contours.

The outline of the parent method in [11] is provided in section 2. The proposed LCDT method, and the subsequent use of SGA in matching the reconstructed contours, is reported in section 3. Experimental evaluation on the speed of contour matching, as compared with the parent scheme, is provided in section 4. This is followed by a conclusion summarizing the essential findings.

2   Affine Invariant Matching Based on Simple Genetic Algorithm and Contour Reconstruction

The parent scheme in [11] consists of of two stages. First, given a reference and a scene contour to be matched, a Euclidean distance image is generated for each of them. Subsequently a correspondence map is derived, with which a closed outermost boundary is extracted from the distance image. In the second stage, a simple GA is applied to match the closed outermost boundaries of the scene and the reference contours. The matching process is realized by searching the existence of three pairs of corresponding points on the pair of contours. For the sake of clarity, the contour reconstruction stage in the parent scheme will be summarized in this section. Whenever possible, we shall stick to the terminology and equations in [11].

Consider an arbitrary contour O_B (a reference or a scene contour) represented by a set of edge points given by $O_B = \{(o^x_{B;0}, o^y_{B;0}), \cdots, (o^x_{B;m-1}, o^y_{B;m-1})\}$, where m is the length of the contour, and $(o^x_{B;0}, o^y_{B;0})$'s are the rectangular coordinates of the ith point in O_B. A distance image and a correspondence map are computed from O_B. The distance image is the weighted distance transform (WDT) [12] of the contour with each pixel representing its distance to the nearest point on O_B as

$$d_B(x,y) = \min_{0 \le j < m} \left\{ \sqrt{(x - o^x_{B;j})^2 + (x - o^y_{B;j})^2} \right\} \tag{1}$$

where x and y denote the horizontal and vertical positions of a pixel. The correspondence map $c_B(x,y)$ is a two-dimensional image in which each point is an ordered pair representing the position of the contour point in O_B nearest to it, i.e.,

$$c_B(x, y) = (o^x_{B;\hat{j}}, o^y_{B;\hat{j}}), \tag{2}$$

where $\hat{j} = \arg \min_{0 \leq j < m} \left\{ \sqrt{(x - o^x_{B;j})^2 + (x - o^y_{B;j})^2} \right\}$.

Next, a closed iso-contour sequence IS_B is extracted from the distance image by tracing the pixels with the same distance value T on the distance image along the clockwise direction as

$$IS_B = \left[(is^x_{B;0}, is^y_{B;0}), \cdots, (is^x_{B;N-1}, is^y_{B;N-1}) \right], \tag{3}$$

such that $d_B(is^x_{B;k}, is^y_{B;k})|_{(0 \leq k < N)} = T$. T is a constant defining the distance of IS_B from O_B, and $(is^x_{B;0}, is^y_{B;0})$ is an arbitrary start point on the iso-contour. N is the number of points in IS_B. Finally, the outermost boundary sequence

$$OM_B = \left[(om^x_{B;0}, om^y_{B;0}), \cdots, (om^x_{B;N-1}, om^y_{B;N-1}) \right] \tag{4}$$

is obtained by associating each member in IS_B to a point in the boundary O_B, where $(om^x_{B;0}, om^y_{B;0}) = c_B(is^x_{B;i}, is^y_{B;i})|_{(0 \leq i < N)}$.

From equations (1) and (2), the computation of the distance image and the correspondence map involve enormous amount of computation. In view of this, we proposed a new method for the contour reconstruction process. The latter, together with the subsequent use of SGA in contour matching, are described in the next section.

3   Proposed Method for Matching of Fragmented Contours

Our proposed method for fast derivation of the distance image and the correspondence map is known as the labeled chamfer distance transform (LCDT). It is a variation of the classical chamfer distance transform (CDT). In the following subsections we describe the CDT, and how it can be modified, with the incorporation of a simple labelling mechanism, to derive the correspondence map. Next we shall describe the use of SGA in matching the contours that are reconstructed with the correspondence map.

Chamfer Distance Transform

In CDT, given an image $I(x, y)$ that contains an object boundary O_B, its chamfer distance image $d_s(x, y)$ can be deduced with a forward and a backward processes which are outlined as follow. We can assumed that $I(x, y)$ and $d_s(x, y)$ have the same resolution. Initially, all the pixels in $d_s(x, y)$ that are at the same position as O_B are set to zero. The remaining pixels of $d_s(x, y)$ are set to the maximum distance value D_{max}. In the forward pass, the values of the pixels in $d_s(x, y)$ are updated recursively according to a raster order manner. Mathematically, we have

$$d_s(x, y) = \min_{0 \leq i < 4} \{p_i\}, \tag{5}$$

where $p_0 = d_s(x+1, y-1) + 4$, $p_1 = d_s(x, y-1) + 3$, $p_2 = d_s(x-1, y-1) + 4$, and $p_2 = d_s(x-1, y) + 3$. In the backward pass, the values of the pixels in $d_s(x, y)$ are updated recursively according to an anti-raster order manner as given by

$$d_s(x, y) = \min_{4 \leq i < 8} \{p_i\},\tag{6}$$

where $p_4 = d_s(x-1, y+1) + 4$, $p_5 = d_s(x, y+1) + 3$, $p_6 = d_s(x+1, y+1) + 4$, and $p_7 = d_s(x+1, y) + 3$.

Fast Generation of the Correspondence Map

In the parent scheme, each entry $c_B(x, y)$ in the correspondence map is derived by blind searching, based on Eq.(2), the boundary point with the nearest Euclidean distance. While this approach is effective, the computation load is intensive and escalates with the number of boundary points. This leads to a substantial increase in the time required to match a pair of object contours. To overcome this problem, we propose to generate the correspondence map and the distance image in a near concurrent manner.

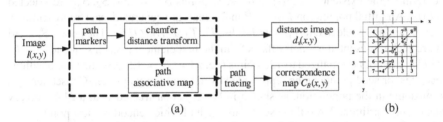

(a)   (b)

Fig. 1. (a) The flow chart for the fast generation of correspondence map. (b) A small 5×5 chamfer distance map overlay with the path association map.

The proposed method is shown in Fig.1(a). From the generation process of the chamfer distance map (based on (5) and (6)), a path marker can be deduced at the same time linking a pixel to one of the eight neighbors that is closest to its current position. Collection of all the path markers form a path association map recording the linkage of each pixel with its nearest neighbor. As an example, a small 5x5 distance image overlay with the path association map is shown in Fig.1(b). Pixels with value of zero are the boundary points and the linkage between pair of pixels is symbolized with directional arrows. Consequently, each point on the correspondence map can be easily derived by simply following the path (guided by the sequence of arrows) that lead to the object boundary. Referring back to the example in Fig.1(b), we observe that pixels 'A', 'B', and 'C' are corresponding to the boundary pixel 'D'.

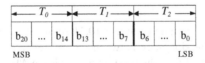

Fig. 2. Structure of the chromosome

GA Based Matching of the Reconstructed Contours

Based on our proposed LCDT method, a pair of complete contours are resulted as given by

$$OM_R = \left[(om_{R;0}^x, om_{R;0}^y), \cdots, (om_{B;M-1}^x, om_{B;M-1}^y)\right] \qquad (7)$$

$$OM_S = \left[(om_{S;0}^x, om_{S;0}^y), \cdots, (om_{S;M-1}^x, om_{S;M-1}^y)\right] \qquad (8)$$

where m, n, M, and N are the number of points in O_R, O_S, OM_R, and OM_S, respectively. If both contours are rigid, near-planar and projections of the same object, their edge points are approximately related by the affine transform. The forward affine transform $A(\cdot)$ is given by

$$A(O_R) = \begin{bmatrix} a & b \\ c & d \end{bmatrix} \begin{bmatrix} o_{R;i}^x \\ o_{R;i}^y \end{bmatrix} + \begin{bmatrix} e \\ f \end{bmatrix}_{i=0,\cdots,m-1}. \qquad (9)$$

To determine the existence of $A(\cdot)$, three seed points $SP = [S_1, S_2, S_3]$ are selected on OM_S. An initial population Pop of P individuals are generated, each representing a triplet of randomly selected test points $TP = [T_1, T_2, T_3]$ on OM_R. The points S_i's and T_i's are each represented by their offsets from the first element in each sequence. OM_R and OM_S are both normalized to 128 points (i.e. $M = N = 128$), so that the positions of each test or seed point can be represented by a 7-bit binary number. Consequently a chromosome in the population, as shown in Fig.2, is constructed with a 21-bit binary string evenly partitioned into three segments, each of which encodes a test point.

From the triplets defined in TP and SP, the affine transform and its inverse can be calculated [4]. For each chromosome, a fitness value is deduced from the matching score MS between the pair of contours based on the given set of test points is determined. The fitness value MS is given by

$$MS = AS_1 \times AS_2. \qquad (10)$$

The component AS_1 is called forward score. It is obtained by overlaying the affine transformed reference contour $(A(O_R))$ onto the distance image of the scene contour as

$$AS_1 = [1 + \frac{1}{2} \sum_{i=1}^{m} d_S(x_{R:i}, y_{R;i})]^{-1}|_{(x_{R:i}, y_{R;i}) \in A(O_R)}. \qquad (11)$$

Another component AS_2 is called backward score. It is defined as

$$AS_2 = [1 + \frac{1}{2} \sum_{i=1}^{m} d_R(x_{S:i}, y_{S;i})]^{-1}|_{(x_{S:i}, y_{S;i}) \in A^{-1}(O_S)}. \qquad (12)$$

The matching score MS is bounded between [0,1], representing the fitness of the chromosome and reflecting the similarity between the reference and scene contours. The genetic algorithm (integrated with the migrant principle) in Algorithm 1 is employed to determine the existence of an affine transform between OM_R and OM_S. An affirmative result reflects the matching between the reference and the scene contours, and vice versa.

Algorithm 1. Genetic algorithm for Shape Matching.

1. Generate an initial random population with N_c chromosomes. Set generation count 't' to 0.
2. Increase t by 1.
3. Evaluate the fitness of all chromosomes in the population.
4. Select $L2$ pairs of parents into mating pool with probabilities according to their fitness.
5. For each pair of parents, perform either a mutation or a crossover operation to generate two child chromosomes, forming a local population of L chromosomes.
6. Inject Q randomly generated migrant chromosomes.
7. Evaluate fitness of all chromosomes using equation (10).
8. Apply the elite principle by finding the weakest individual and replacing it with the strongest one in the previous generation.
9. Find and record the strongest individual in the new generation. If the maximum fitness exceeds a predefined threshold, go to step 12.
10. If number of generation exceeds the upper limit, go to step 13.
11. Go to step 4.
12. The pair of object contours belongs to the same object. End of process.
13. The pair of object contours belongs to different objects. End of process.

4   Experimental Results

The reference model contours of six objects: open-end wrench (A), adjustable wrench (B), long-nose pliers (C), Lineman's pliers (D), hammer (E), and scissors (F) as

ref. model scene 1 scene 2 scene 3 ref. model scene 1 scene 2 scene 3

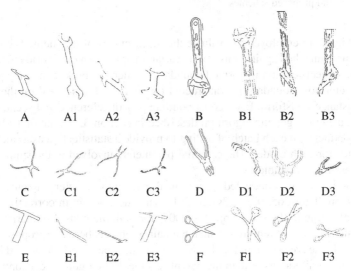

A   A1   A2   A3   B   B1   B2   B3

C   C1   C2   C3   D   D1   D2   D3

E   E1   E2   E3   F   F1   F2   F3

Fig. 3. Reference contours and scene contours

Table 1. Setting of coefficients in the simple genetic algorithm (SGA)

Coefficient	Properties
Population Size	100 Properties local population:70 migrant population:30
Chromosome Length	7 bits × 3 parameters= 21 bits
Encoding of parameters	Binary
Maximum generation	200
Crossover method/rate	Single point crossover/0.8
Mutation method/rate	Single point mutation/0.2

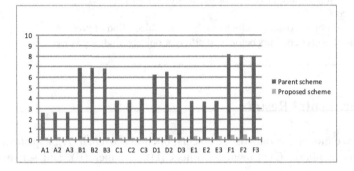

Fig. 4. Comparison of average time (in second) required to match a pair of objects between the parent scheme and proposed schemes

shown in Fig.3, are employed to evaluate the performance of the method. Each reference contour is matched against three of its variants captured in real world environment. All the test subjects exhibit ill-formed boundaries which are broken in many places. A complete outermost boundary is constructed for each contours based on the labelled chamfer distance transform. Each corresponding pair of reference and scene contours is matched using the genetic algorithm described in section 3. A total of 100 repeated trials are conducted in each batch of testing to provide a statistical measurement on the success rates for each model. The essential parameters involved in the simple genetic algorithm are listed in Table 1.

Our experiments are conducted on a typical personal computer implemented with the "Intel Core2Duo E6550", 2.33GHz CPU. The success rate in correctly matching contours belonging to the same object is 100% for both methods. Comparisons of the average time (in second) required to match a pair of objects between proposed scheme and parent scheme are shown in Fig.4, which shows that the proposed method is at least one order of magnitude faster than the parent scheme. For certain object contours such as B2, B3, and D3, improvement of over 30 times is noted.

5   Conclusions

In this paper, we propose a method to enhance the computation efficiency of an affine invariant contour matching scheme. The latter has been proven to be effective in identifying fragmented contours that are belonged to the same object. We note that in the parent scheme, the majority of computation load is concentrated on the generation of the Euclidean distance image and the correspondence map. To overcome this problem, we propose to replace the Euclidean distance image by the chamfer distance image which requires substantially less amount of arithmetic operations. In addition, by labelling the association of each pixel with its nearest neighbor in the course of generating the chamfer distance image, the correspondence map can also be deduced with a simple path tracing process. The new approach is referred to as the "labelled chamfer distance transform". Experimental results reveal that our method is significantly faster, and at the same time capable of attaining similar performance as the parent scheme.

Acknowledgment. The work was supported by a research grant (CityU 116511) from General Research Fund, Hong Kong.

References

1. Wang, Y., Teoh, E.K.: 2D Affine-Invariant Contour Matching Using B-Spline Model. IEEE Trans. PAMI 29(10), 1853–1858 (2007)
2. Tzimiropoulos, G., Mitianoudis, N., Stathaki, T.: Robust recognition of planar shapes under affine transforms using Principal Component Analysis. IEEE Signal Process. Lett. 14(10), 723–726 (2007)
3. Toet, A., Hajema, W.P.: Genetic contour matching. Patt. Recog. Lett. 16, 849–856 (1995)
4. Tsang, P.W.M.: Genetic Algorithm for Affine Invariant Object Shape Recognition. Proc. Instn. Mech. Engrs. 211, 385–392 (1997)
5. Ozcan, E., Mohan, C.K.: Partial Shape Matching using Genetic Algorithm. Patt. Recog. Lett. 18, 987–992 (1997)
6. Tsang, P.W.M.: Enhancement of a Genetic Algorithm for Affine Invariant Object Shape Matching using the Migrant Principle. IEE Proc. Vis. Img. Sig. Proc. 150(2), 107–113 (2003)
7. Rube, I.A.E., Ahmed, M., Kamel, M.: Coarse to Fine Affine Invariant Shape Matching and Classification. In: Proc. 17th ICPR 2004 (2004)
8. Wu, A., Tsang, P.W.M., Yuen, T.Y.F., Yeung, L.F.: Affine invariant object shape matching using genetic algorithm with multi-parent orthogonal recombination and migrant principle. Appl. Soft Comp. 9(1), 282–289 (2009)
9. Lim, H.S., Cheraghi, S.H.: An optimization approach to shape matching and recognition. Compt. Engg. 24, 183–200 (1998)
10. Tsang, P.W.M., Yuen, T.Y.F.: Affine invariant matching of broken boundaries based on an enhanced genetic algorithm and distance transform. IET Comp. Vis. 2(3), 142–149 (2008)
11. Tsang, P.W.M., Situ, W.C.: Affine invariant matching of broken boundaries based on simple genetic algorithm and contour reconstruction. Patt. Recog. Lett. 31(9), 771–780 (2010)
12. Borgefors, G.: Hierarchical chamfer matching: a parametric edge matching algorithm. IEEE Trans. Patt. Anal. Mach. Intell. 10(6), 849–865 (1998)

Sparse Gradient-Based Direct Policy Search

Nataliya Sokolovska

Department of Computing, Macquarie University, Sydney, Australia
nataliya.sokolovska@mq.edu.au

Abstract. Reinforcement learning is challenging if state and action spaces are continuous. The discretization of state and action spaces and real-time adaptation of the discretization are critical issues in reinforcement learning problems.

In our contribution we consider the adaptive discretization, and introduce a sparse gradient-based direct policy search method. We address the issue of efficient states/actions selection in the gradient-based direct policy search based on imposing sparsity through the L_1 penalty term. We propose to start learning with a fine discretization of state space and to induce sparsity via the L_1 norm.

We compare the proposed approach to state-of-the art methods, such as progressive widening Q-learning which updates the discretization of the states adaptively, and to classic as well as sparse Q-learning with linear function approximation. We demonstrate by our experiments on standard reinforcement learning challenges that the proposed approach is efficient.

Keywords: Direct policy search, Q-learning, model selection.

1 Introduction

In a large number of reinforcement learning tasks, states and actions are continuous. This holds for e.g., states in autonomous robot navigation and for both states and actions in energy or power-producing applications. Many methods to solve the reinforcement learning challenges have been already proposed (see e.g. [15] for an up-to-date panorama).

Although some reinforcement learning problems can be modeled as Markov decision processes, there are a lot of applications where the learning environment is unknown, and therefore the model of the environment cannot be built. Q-Learning [17] is a model-free reinforcement learning algorithm, where the goal is to estimate Q-values which are associated with the expected reward for taking a particular action in a given state.

Another branch of model-free methods is direct policy search [1]. Direct policy search is an approach which allows to search directly in the policy space. This method is particularly useful if state and action spaces are too large for analytic solutions. Gradient-based direct policy search approximates the gradient of the average reward. The gradient is used to adjust the parameters to maximize the cumulated average reward.

T. Huang et al. (Eds.): ICONIP 2012, Part IV, LNCS 7666, pp. 212–221, 2012.

When a learning method is applied to a task, it is usually assumed that the domain is discrete, or discretized. If the initial grid is rather coarse, and if all vertices of the grid are far enough from the goal, it is possible that an agent never reaches a goal. The important problem is to refine the discretization grid of states (and actions) adaptively, especially around the areas of interest, e.g., around the goal.

In this contribution we consider how to introduce sparsity on estimated values. An advantage of such approaches were the possibility to start a learning procedure with a very fine grid and to keep only relevant state/action dependencies.

The paper is organized as follows. Section 2 presents related work on state and action spaces discretization approaches as well as discusses methods which take continuous values into consideration directly. Section 3 provides an overview of the gradient-based direct policy search. Section 4 introduces sparse parametric direct policy search. We show the results of our experiments in Section 5. Concluding remarks and perspectives close the paper.

2   Related Work

In this section we provide some details on two state-of-the art methods, double progressive widening Q-Learning and Q-Learning with linear function approximation, since we compare our results to these approaches as well as to the classical Q-Learning in Section 5.

To solve a goal-planning task means to find an optimal policy, i.e. a policy $\pi^\star$ that is equal to or better (in terms of cumulated expected reward) than any other policy π. It is known [14] that optimal policies share the same optimal action-value function $Q^\star$. Given a set of states $\mathcal{S}$ and a set of actions $\mathcal{A}$, the optimal action-value function is defined as

$$Q^\star(s, a) = \max_\pi Q^\pi(s, a)$$
$$= \mathrm{E}\{r_{t+1} + \gamma \max_{a'} Q^\star(s_{t+1}, a')|s_t = s, a_t = a\}$$
$$= \sum_{s'} \mathcal{P}^a_{ss'} \left(\mathcal{R}^a_{ss'} + \gamma \max_{a'} Q^\star(s', a') \right), \tag{1}$$

where $\mathcal{P}^a_{ss'} = p(s'|s, a)$, $\mathcal{R}$ is the reward, $s, s' \in \mathcal{S}$, and $a, a' \in \mathcal{A}$. In other words, the action-value function $Q : \mathcal{S} \times \mathcal{A} \longrightarrow \mathbb{R}$ defines the quality of each (state, action) pair. In the following, s is the current state, a is the current action, s' is the next state, so that $(s, a) \to s'$.

Q-Learning is a general term for approaches which compute the expected reward given an action a in a given state s, and allow to choose an action maximizing the reward value. The strength of Q-Learning methods is that they do not require any knowledge of a model of environment $\mathcal{P}^a_{ss'}$, which is not available in a number of real-world applications.

An example of a policy is the greedy policy, given by

$$\pi(s) = \arg\max_a Q^\pi(s, a), \tag{2}$$

which we use in the following.

One-step Q-Learning is proposed by [17]. The approach is based on the following update rule:

$$Q(s, a) = Q(s, a) + \alpha\Big(r + \gamma\max_{a'} Q(s', a') - Q(s, a)\Big), \tag{3}$$

where α is usually called learning rate, $\alpha \in (0, 1]$, and γ – discount factor, $\gamma \in [0, 1)$.

If the state and/or action spaces are continuous, the application of Q-Learning is not straightforward. If the state/action domains are continuous (or very large), it becomes hardly possible to keep (and to update) a look-up table which contains Q-values for each state-action pair.

A number of approaches to work in continuous state and action spaces has been proposed [5,9,10]. The idea is either to work with continuous states and actions directly, or to discretize them. The discretization of state and action spaces is a challenge, since if a discretization is too rough, it will be impossible to find the optimal policy; if a grid is too fine, the generalization will be lost.

An adaptive approach to refine the initial grid has been recently proposed by [7]. The idea is to provide pseudo-goals which lie on the vertices of the initial grid. It has been shown that the method is efficient, however, its serious disadvantage is that the knowledge of a location of a goal is required. The initialization of the grid with the pseudo-goals, which are in a proximity to the true goal, is not obvious.

Recently an approach for adaptive discretization of the continuous setting, called double progressive widening Q-Learning [12] has been introduced. The method is inspired by techniques developed in Monte-Carlo Tree Search and used in bandit-based algorithms [3,4,11]. The double progressive widening means that the state and action spaces are explored progressively and the current discretization both of states and actions is adapted at each time step. The main idea of the algorithm is that if a particular state is visited often (compared to the number of times the previous action has been taken in the previous state), then the state is added to the pool of explored states. The discretization of states and actions in the double progressive widening is carried out based on the Euclidean distance. A newly observed state (action) gets the same discrete value as its closest state (action) in the already explored set of states $\mathcal{S}$ (set of actions $\mathcal{A}$).

Another state-of-the art approach is Q-Learning with linear function approximation [15]. Reinforcement learning with linear function approximation has been studied extensively in the last years, and it is a direct extension of Q-Learning. The Q is represented as a parametric function of $\theta \in \mathbb{R}^d$

$$Q_\theta = \exp(\theta^{\mathrm{T}}\phi(s, a)), \tag{4}$$

where $\phi(s, a)$ is a feature vector, e.g., a vector of binary values testing co-occurrence of s and a.

3   Gradient-Based Direct Policy Search

In this section we describe briefly the parametric direct policy search.

The performance of the greedy policy derived from the approximate value function is not guaranteed to improve on each iteration [1]. Parametric approaches, i.e., approaches based on stochastic policies are known to have better theoretical properties [2]. In a parametric approach, we consider a class of stochastic policies parameterized by $\theta \in \mathbb{R}^d$. The optimization involves a gradient-based method, where the gradient of the average reward is computed with respect to parameters θ.

In large-scale problems or in partially observable problems (i.e., in applications where states are not observed) the gradient cannot be computed in closed form. However, the gradient can be estimated via simulation. The gradient can even be approximated by a single simulation using the technique called the score function or likelihood ratio [1,2].

Let $r(X)$ be a reward value which depends on the simulation $X = X_1, \ldots, X_T$. Then the expected performance takes the following form:

$$\eta(\theta) = E\{r(X)\}. \tag{5}$$

Applying the likelihood ratio technique we get

$$\nabla \eta(\theta) = E\{r(X)\frac{\nabla q(\theta, X)}{q(\theta, X)}\}, \tag{6}$$

where $q(\theta, X)$ is the marginal probability of a particular sequence X, and an unbiased estimate of the gradient can be computed as follows:

$$\hat{\nabla} \eta(\theta) = \frac{1}{N} \sum_{i=1}^{N} r(X^{(i)}) \frac{\nabla q(\theta, X^{(i)})}{q(\theta, X^{(i)})}. \tag{7}$$

The expression can be rewritten as

$$\hat{\nabla} \eta(\theta) = \frac{1}{N} \sum_{i=1}^{N} r(X^{(i)}) \sum_{t=1}^{T} \frac{\nabla \mu_{y_t}(\theta, X_t^{(i)})}{\mu_{y_t}(\theta, X_t^{(i)})}, \tag{8}$$

where $\mu_{y_t}(\theta, X_t^{(i)})$ denotes the conditional probability of the label y at time step t given an observation $X_t^{(i)}$.

4   Sparse Parametric Direct Policy Search

The original idea to induce sparsity in parametric models via the L_1 regularization belongs to [16]. We propose to penalize the parameterized average reward function by the L_1 penalty term

$$\eta(\theta) = E\{r(X)\} - \rho\|\theta\|_1, \tag{9}$$

where $\|\theta\|_1 = \sum_{i=1}^{d} \theta_i$, and ρ is a parameter to be adjusted. Then the approximation of the gradient takes the following form:

$$\hat{\nabla}\eta(\theta) = \frac{1}{N}\sum_{i=1}^{N} r(X^{(i)}) \sum_{t=1}^{T} \frac{\nabla\mu_{y_t}(\theta, X_t^{(i)})}{\mu_{y_t}(\theta, X_t^{(i)})} - \rho\sum_i \text{sgn}(\theta_i), \quad (10)$$

where

$$\text{sgn}(a) = \begin{cases} -1, & \text{if } a < 0, \\ 0, & \text{if } a = 0, \\ 1, & \text{if } a > 0. \end{cases} \quad (11)$$

To compute the gradient efficiently, we use the eligibility traces, which are filtered versions of the sequence $\nabla\mu_{y_t}(\theta, X_t^{(i)})/\mu_{y_t}(\theta, X_t^{(i)})$.

Let $v_t = \nabla\mu_{y_t}(\theta, X_t^{(i)})/\mu_{y_t}(\theta, X_t^{(i)})$, then the discounted eligibility trace at time t is defined as

$$\begin{cases} z_0 = 0, \\ z_{t+1} = \beta z_t + v_t, \end{cases} \quad (12)$$

where $\beta \in [0, 1)$, and the gradient expression can be rewritten

$$\hat{\nabla}\eta(\theta) = \frac{1}{N}\sum_{i=1}^{N} r(X^{(i)})\frac{1}{T}\sum_{t}^{T} z_t - \rho\sum_i \text{sgn}(\theta_i). \quad (13)$$

Since the eligibility trace is discounted, the biased estimate of the performance gradient can be written

$$\hat{\nabla}_\beta\eta(\theta) = \frac{1}{T}\sum_{t=1}^{T} r(X_t)z_t(\beta) - \rho\sum_i \text{sgn}(\theta_i). \quad (14)$$

Note that [1,2] consider the same form of the gradient approximation as eq. (14), however, they do not apply any regularization term.

The update of θ takes the following form:

$$\theta_{t+1} = \theta_t + \gamma(r_{t+1}z_{t+1} - \rho\text{sgn}(\theta_t)), \quad (15)$$

or

$$\theta_{t+1} = S(\theta_t + \gamma r_{t+1}z_{t+1}, \rho'), \quad (16)$$

where $\rho' = \gamma\rho$, and S is the thresholding function

$$S(a, \lambda) = \begin{cases} a - \lambda, & \text{if } a \geq 0, \lambda \leq |a| \\ a + \lambda, & \text{if } a \leq 0, \lambda \leq |a| \\ 0, & \text{if } \lambda > |a|. \end{cases} \quad (17)$$

In our experiments μ is defined by the logistic regression function. The logistic regression models posterior probability of a class y, $y \in \{1, \ldots, K\}$ via a linear function of observations x, $x \in \mathbb{R}^d$, what results in

$$\frac{\nabla \mu_{y_t}(\theta, X_t)}{\mu_{y_t}(\theta, X_t)} = \begin{cases} 1 - p(y_t|X_t; \theta) \text{ , if } y_t = k, \\ -p(y_t|X_t; \theta) \text{ , if } y_t \neq k. \end{cases} \tag{18}$$

The learning procedure we propose and we use in our experiments is summarized in Algorithm 1.

Algorithm 1. Sparse Gradient-Based Direct Policy Search

---

  for $i = 1 \ldots N$ **do**
    for $t = 1 \ldots T$ **do**
      $z_{t+1} = \beta z_t + \frac{\nabla \mu_{y_t}(\theta, X_t^{(i)})}{\mu_{y_t}(\theta, X_t^{(i)})}$
      $\theta_{t+1} = S(\theta_t + \gamma r_{t+1} z_{t+1}, \rho')$ (see definition of thresholding function S presented
      as eq. (17))
    end for
  end for

---

5 Experiments

In this section we show our results on two standard reinforcement learning challenges, on Mountain Car and Puddle World tasks. Note that in the problems we consider, the actions are discrete but the states are continuous. In our experiments we compared the proposed sparse gradient-based direct policy search (SGBDPS) with several state-of-the art methods, as well as with some heuristic approaches.

The state-of-the-the art methods we implemented and tested are standard Q-Learning, Q-Learning with double progressive widening (PW Q-learn), and Q-Learning with linear function approximation, its dense (LFA) and sparse versions (LFAS). The heuristics we use are based on thresholding: we put to zero values of parameters if their absolute values are less than a threshold.

5.1 Mountain Car

Mountain Car is a standard realistic challenge. The task was introduced by [14]. The goal is to learn to drive a car to a steep mountain. The car is underpowered, and gravity is stronger than the engine of the vehicle. The state is a two-dimensional vector, and its first coordinate corresponds to the current position of the car, and the second coordinate is the car's current speed. There are only three discrete actions: "forward throttle", "no throttle", and "backward throttle". Figures 1 and 2 illustrate our results for the Mountain Car task.

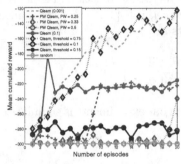 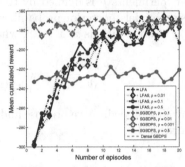

Fig. 1. Mountain Car task. Cumulated reward as a function of number of learning episodes. On the left: standard Q-Learning, progressive widening Q-learning, Q-learning with thresholding, and random. On the right: sparse and dense gradient-based direct-policy search, sparse and dense Q-learning with linear function approximation.

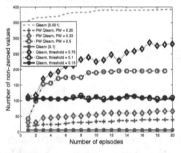 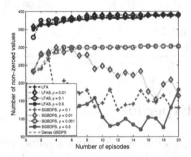

Fig. 2. Mountain Car task. Number of non-zeroed values. On the left: standard Q-Learning, progressive widening Q-learning, Q-learning with thresholding, and random. On the right: Sparse and dense gradient-based direct-policy search, sparse and dense Q-learning with linear function approximation.

Figure 1 shows the cumulated reward values for each episode number. Figure 2 illustrates the number of non-zeroed values in each method. As expected, the performance is very poor, if actions are taken completely randomly. To evaluate the performance of the classical Q-Learning we discretize the state space and impose a fine grid (with a step 0.001). Note that the Q-Learning cumulated reward values are coherent with results reported earlier [6] on Mountain Car with Q-Learning. However, the number of Q-values to be estimated is maximal, i.e., the look-up Q-table is dense. We carry out experiments with a much less fine discretization grid (discretization with a step 0.1), what results in a compact look-up table but leads to bad performance. The progressive widening Q-Learning approach is not efficient for the Mountain Car problem. The intuition behind its failure is that the progressive widening in this particular problem refines a grid too much, and loses its generalization abilities. It is easy to see that in case where $\lambda = 0.25$, i.e., the exploration is rather slow, the cumulated reward is higher than when $\lambda = 0.5$ and the number of explored states increases fast. The Q-Learning with thresholding is rather efficient, and reaches the state-of-the art performance. However, the thresholding parameter should be carefully chosen by cross validation.

The state-of-the art Q-Learning with linear approximation, both its dense and sparse versions, converge to the optimum. However sparsity induction is not obvious in the given task. If the value of ρ is quite small, the performance is very reasonable but the model is not sparse. Increasing the value of ρ leads to a bad performance. The gradient-based direct policy search without L_1 regularization term reaches the state-of-the art performance but it is not sparse. The newly introduced sparse parametric direct policy search achieves the state-of-the art performance using a rather modest number of parameters. An important remark is that the proposed sparse gradient-based method converges much faster than all other tested approaches.

5.2   Puddle World

The problem was originally introduced in [13]. The goal is to learn a robot to navigate in a two-dimensional space. The state is represented as a two-dimensional vector, where the two coordinates are the robot's current position. The actions are discrete, and there are only four actions: "left", "right", "up", and "down".

Figures 3 and 4 illustrate our experiments on Puddle World. In comparison to random decision making, all considered methods reach reasonable performances. However, there is a trade off between the sparsity, i.e., the number of non-zeroed values, and performance. The standard Q-Learning (with the discretization step 0.01) reaches the state-of-the art performance. As in the experiments with Mountain Car, the heuristic thresholding of estimated Q-values is an efficient approach, and it slightly outperforms the sparse Q-learning with linear function approximation. Note that the Q-Learning with thresholding not only achieves a higher cumulated reward but its sparsity is higher than for the Q-learning with lineard function approximation.

Figures 3 and 4 on the right compare the progressive widening Q-Learning and the dense and sparse versions of the gradient-based direct policy search. Note that the plots of Figure 3 are differently scaled, and the performance of the approaches of the right plot is very reasonable. The dense version of the gradient-based direct policy search performs slightly better than other algorithms. The progressive widening Q-learning seems to be very efficient, moreover, the number of values in the look-up Q-table is minimal (about 50 values are enough). The best performance is achieved by the progressive widening Q-Learning, which starts the learning procedure with a rather rough discretization and which explores new states rather aggressively with the parameter value $\lambda = 0.5$; however, the number of explored states is too high.

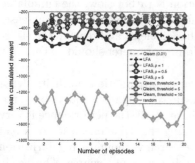 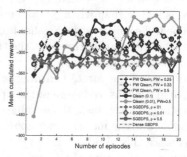

Fig. 3. Puddle World task. Cumulated reward. On the left: random action, Q-Learning, Q-Learning with thresholding, and dense and sparse Q-Learning with linear function approximation. On the right: progressive widening Q-Learning, dense and sparse gradient-based direct policy search.

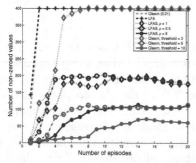 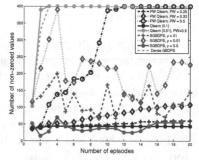

Fig. 4. Number of non-zeroed values. On the left: random action, Q-Learning, Q-Learning with thresholding, and dense and sparse Q-Learning with linear function approximation. On the right: progressive widening Q-Learning, dense and sparse gradient-based direct policy search.

6   Conclusion

In this contribution, we challenged the reinforcement learning problem in continuous state and action spaces. The technique which is widely used is the state (and action) space discretization, however, a static discretization usually leads to a bad performance, and some adaptation of the discretization is needed. Progressive widening Q-Learning is a method which implements the idea of discretization adaptation at each time step.

We proposed a method called sparse gradient-based direct policy search, which induces sparsity on a fine discretization of state (and action) space. Since the approach is parametric, the sparsity can be induced via the L_1 regularization term. The advantages of the proposed method are its scalability to problems of large dimensions and its simplicity of implementation. It requires storage of only twice the number of parameter values. Note that the logistic regression function used in the sparse gradient-based direct policy search is not the only possible choice. In particular, we are currently interested in extending the proposed approach to the case of structured output prediction with conditional random fields. We are particularly interested in applying the proposed method to large-scale applications.

References

1. Baxter, J., Bartlett, P.L., Weaver, L.: Experiments with infinite-horizon, policy-gradient estimation. Journal of Artificial Intelligence Research 5, 351–381 (2001)
2. Baxter, J., Bartlett, P.L.: Infinite-horizon policy-gradient estimation. Journal of Artificial Intelligence Research 15, 319–350 (2001)
3. Couëtoux, A., Hoock, J.-B., Sokolovska, N., Teytaud, O., Bonnard, N.: Continuous Upper Confidence Trees. In: Coello, C.A.C. (ed.) LION 2011. LNCS, vol. 6683, pp. 433–445. Springer, Heidelberg (2011)
4. Coulom, R.: Monte-carlo tree search in crazy stone. In: Game Programming Workshop (2007)
5. Davies, S.: Multidimensional triangulation and interpolation for reinforcement learning. In: Mozer, M.C., Jordan, M.I., Petsche, T. (eds.) NIPS (1997)
6. Dutech, A., et al.: Reinforcement learning benchmarks and bake-offs. In: Workshop at the 2005 NIPS Conference (2005)
7. Lampton, A., Valasek, J.: Multiresolution state-space discretization method for Q-learning. In: American Control Conference (2009)
8. Melo, F.S., Meyn, S.P., Ribeiro, M.I.: An analysis of reinforcement learning with function approximation. In: ICML (2008)
9. Munos, R., Moore, A.: Variable resolution discretization in optimal control. Technical report, Robotics Institute, CMU (1999)
10. Munos, R., Moore, A.W.: Variable resolution discretization for high-accuracy solutions of optimal control problems. In: IJCAI (1999)
11. Rolet, P., Sebag, M., Teytaud, O.: Boosting Active Learning to Optimality: A Tractable Monte-Carlo, Billiard-Based Algorithm. In: Buntine, W., Grobelnik, M., Mladenić, D., Shawe-Taylor, J. (eds.) ECML PKDD 2009, Part II. LNCS, vol. 5782, pp. 302–317. Springer, Heidelberg (2009)
12. Sokolovska, N., Teytaud, O., Milone, M.: Q-Learning with Double Progressive Widening: Application to Robotics. In: Lu, B.-L., Zhang, L., Kwok, J. (eds.) ICONIP 2011, Part III. LNCS, vol. 7064, pp. 103–112. Springer, Heidelberg (2011)
13. Sutton, R.: Generalization in reinforcement learning: successful examples using sparse coarse coding. In: NIPS (1996)
14. Sutton, R.S., Barto, A.G.: Reinforcement learning: an introduction. MIT Press (1998)
15. Szepesvári, C.: Algorithms for reinforcement learning. Morgan and Claypool (2010)
16. Tibshirani, R.: Regression shrinkage and selection via Lasso. Journal of the Royal Statistical Society. Series B 58(1), 267–288 (1996)
17. Watkings, C.J.C.H.: Learning from Delayed Rewards. PhD thesis. Cambridge University (1989)

Application of Sampling Theory to Forecast Ozone by Neural Network

Armando Pelliccioni[1] and Rossana Cotroneo[2]

[1] Inail ex-Ispesl, via F. Candida 1, 00040 Monteporzio Catone, Italy
a.pelliccioni@inail.it
[2] Istat, Viale Liegi, 13,00198 Roma Italy
cotroneo@istat

Abstract. In the present work, we analyzed environmental data by using neural net techniques for ozone prediction. The data concerns a period of two years (2006 and 2007) and comes from a monitoring station of air quality of Rome. The aim of this paper is to suggest a strategy for choosing an optimal set of input patterns to optimize the learning process during training and generalization phase, and to improve computation reliability of a Neural Net (NN). The selection of patterns combined with NN improves capability and accuracy of ozone prediction and goodness of models obtained. In particular, the approach considers two different methodologies for selecting an optimal set of input patterns: random patterns selection and cluster (K-means algorithm) ones. Results show significant differences between the methodologies: the NN's performance is always better when the patterns are obtained using our method based on cluster analysis than the conventional random pattern choice.

Keywords: Ozone, pattern selection, data mining, K-means clustering, neural networks, optimization.

1   Introduction

Ozone prediction is one of more important questions to be solved in urban area. In fact, the air quality problems, linked to the ozone (O_3) could produce effects on human health related to respiratory problems, damage to ecosystems, agricultural crops and materials (World Health Organisation, 2003). The impacts on human health due to ozone are critical especially in large metropolitan areas, like Rome, where emissions due to transport are relevant and they cause an increase of exposure on population with consequent health problems especially during summer season and/or under stable turbulence conditions, as in the winter season. The ozone can be classified as a secondary pollutant, and its levels are determined by complex photochemical reaction with primary pollutants (EPA, 2006). The concentrations are strongly dependent both from micro-meteorological conditions linked to the turbulence and the effects related to the seasons, to the long range transport, to the incoming solar radiation and to the atmospheric turbulence conditions   (Penkett et al., 2004), (Finlayson-Pitt et al.. 1986).

T. Huang et al. (Eds.): ICONIP 2012, Part IV, LNCS 7666, pp. 222–230, 2012.
© Springer-Verlag Berlin Heidelberg 2012

As described in different works, the prediction of ozone levels is very complex to obtain by mathematical models ((Carter, 1990), (Comrie, 1997), (Gardner et.al, 2000.), (Gardner et al., 1998), (Dutot et al., 2007), (Pelliccioni et. Al, 2010)). The analytical solution for ozone obtained by Eulerian solution is difficult to model especially in complex situations, such as an urban site or with a complex orography where is hard to obtain a valid overall analytical formulation.

As regard ozone's forecasting by Regression Models (RegMod), one of the most difficult problems to deal with is the simulation of the chemical reactions that occur in atmosphere (Penkett et al, 2004). In addition, the results obtained are affected by a significant margin of uncertainty connected to intrinsic variability of used variables and to the local orography complexity.

Unlike the deterministic models and regression's ones, the neural networks (Rojas, 1996) are easier to implement and more functional to the pollutant prediction especially in complex situations.

Given the importance of prediction of chemical reactions of O_3, the learning ability of the NN of working as universal *approximator* of non-linear functions can be considered as a mathematical crucial feature for improving the problem of ozone prediction. The NNs can be adequate to evaluate the dynamics of environmental systems and can capture the hidden relations between the input variables and the ozone's. For NN model, in general, people approach the training phase to learning through random patterns selection. In our approach, to improve the ability of NN to "capture" the true hidden relation inside the data set by of pre-choice of information, we applied cluster analysis techniques. In general, the optimization of the input patterns is always a critical asset and the selection of patterns improve the impact on the forecasting performance of the NN.

The aim of this study is to investigate the usefulness of neural networks to predict ozone levels by using the inner information coming from a classification algorithm.

2   Dataset Description

The environmental data set come from urban background monitoring station of the ARPAL (Environmental Protection Agency of Lazio Region) of Rome (Villa Ada monitoring station), and regard hourly data during the two calendar years 2006 and 2007.

At Villa Ada monitoring station is available a set of meteorological variables such as solar radiation, temperature and humidity, and the main primary air pollutant variables (O3, NO, NO2, CO) and data set regards about 14324 hourly patterns.

Table 1 shows the general statistics calculated for 2006 and 2007. The table examines for each year the main statistical parameters: mean, standard deviation, maximum and minimum, variation coefficient (CV) that represents the ratio of the standard deviation to the mean. Our dataset shows the maximum hourly value of ozone for 2006 around 227.6µg/m³ vs. around 189.1µg/m³ for 2007, verified during summer season (15/07/2007 h.15.00), whereas in 2007 the maximum hourly of CO is about 4.1 mg/m³ during winter season (11/01/2007 h.23.00) vs. the maximum hourly of CO is about 3.8 mg/m³ in 2006. For 2007, we also observed the maximum of the variability in the time series (CV=102.3%) for O_3. Further, we also examined the correlation matrix (not shown) and we found that ozone is anti-correlated with NO_2. The global solar radiation (GSR) is usually correlated with the ozone production and is main parameter linked to photochemical reactions. Other fundamental variable is temperature (keeping in mind that photochemical cross section for the ozone production increased with the temperature). At the same time, the NO_2 and the atmospheric pressure show low values. Finally, relative humidity appears as input variable of less relevance, but this does not mean that they do not influence the ozone values.

Other fundamental information is shown in the ozone's distributions for 2006 and 2007.

Usually, the pollutants distribution is skewness, because low values are more frequent than the higher values.

Table 1. General statistics

	CO (mg/m²)	NO (µg/m²)	NO₂ (µg/m²)	O₃ (µg/m²)	T (C°)	RH (%)	GSR (W/m²)
2006							
Mean	0.61	20.84	41.46	41.55	13.07	73.41	124.15
SD	0.35	42.87	24.7	41.21	7.09	19.23	219.73
CV (%)	58.03	205.7	59.58	99.18	54.25	26.2	176.98
Min	0.1	0	0	0	0	9	0
Max	3.8	1228.4	165.3	227.6	33	97	1005
N	8124	7957	7957	8056	8287	8692	8692
Missing	636	803	803	704	473	68	68
2007							
	CO (mg/m²)	NO (µg/m²)	NO₂ (µg/m²)	O₃ (µg/m²)	T (C°)	RH (%)	GSR (W/m²)
Mean	0.61	22.32	43.99	36.62	12.97	73.23	126.42
SD	0.38	41.8	26.25	37.46	7.09	19.62	221.95
CV (%)	61.98	187.33	59.67	102.32	54.64	26.8	175.57
Min	0	0	0.6	0	0	10	0
Max	4.1	398.7	156.8	189.1	37	97	1002
N	8260	8277	8277	8279	8738	8760	8760
Missing	500	483	483	481	22	//	//

Asymmetry of the distribution, and the identification of outlier situations (Hawkins, 1980) are very important problems. In our case, ozone's distribution is highly skewed (see Fig.1).

In fact, about 97% for 2007 and 96% for 2006 of patterns belong to the class 0-120µg/m³, whereas less than 0.1% for 2007 and 0.2% for 2006 is above the information threshold (180µg/m³).

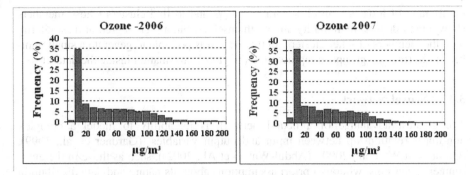

Fig. 1. Ozone distribution (2006 and 2007)

3   Methodologies

The main target of the study is to suggest a way to implement the learning of neural network models for the ozone forecasting. Generally, NN tends to train a large data set with a large amount of computation time that is necessary in order to classify unknown patterns. During the training, irrelevant or redundant patterns could conduct to a degradation of NN's performance, linked mainly to the inner NN weights that connect the patterns themselves. When neural network becomes oversensitive, it behaves with a poor generalization capacities to the never seen data (Bishop, 1995). The patterns selection is a very important task that should be solved in order to achieve good generalisation of the net, above all if the net is used to simulate chemical reactions in atmosphere. This task ensures quality of efficiency during elaboration phase and improves learning process of NN. We tested NN's performances through two ways of patterns selection. The first one is random pattern selection and the second one is cluster analysis pattern selections.

3.1   Random Patterns Selection

Random patterns selection chooses each unit of dataset that has an equal probability of being in the sample. The selection of each unit is independent from selection of every other unit. The selection of one unit does not affect the chances of any other pattern. The random pattern selection is the most utilized techniques and provides a sampling error easily measured, but it not always get the best representation above all when the information is not had the same probability.

3.2   Cluster Analysis Patterns Selection

The proposed method to select the environmental patterns consists on the use of the cluster analysis as discriminate tool for information (EUROSTAT, 2008). The way to select patterns by using cluster techniques has to be seen as an important unsupervised learning technique able to discover the inner structure present in data. Its purpose is the partitioning of a dataset into k separate clusters and to find clusters whose members show a high degree of similarity among themselves, but a high degree of dissimilarity with the members of other clusters. In this way, it is possible to generate a small number of discriminate groups to representing the global information inside the dataset.

3.3   Neural Network

The greatest advantage of a neural network is its ability to model a complex non-linear relationship between input and output variables (((Gardner et. al., 1999), (Gardner M.W et. al., 2000), (Abdul-Wahab et Al., 2002)), such as those in the environmental systems, without a priori assumptions about its nature and data distribution from which the modelling sample is drawn (BuHamra, 2003). The selection of appropriate network topology depends on the number of parameters, the weights, the selection of an appropriate training algorithm and the type of transfer functions used. For transfer function, we use the most common architecture, the Multi Layer Perceptron (MLP) (Fausett, 1994), (Ripley, 1996), which is a type of feed-forward of neural network and generally uses the back-propagation algorithm to develop a model to illustrate relationships between inputs and desired output for training data.

We tested different NN parameters in order to choice the best model. After several simulations, running for 3000 epochs and varying the number of hidden layer (9, 6 and 12), we selected 12 hidden layers with sigmoid activation function for the hidden unit as the best performance of perceptron network. The input layer contains the main and essential variables for ozone, such as the hourly CO, NO, NO_2, T, RH and SR, whereas the target neuron is ozone. Our final NN is constituted by 6-12-1 neurons, where the hidden is the best choice as coming from above test elaborations.

4   Results and Discussion

We applied NN to the results coming from the patterns selection process to forecast ozone concentrations using as input data, meteorology, as well as primary and secondary pollutants (CO, NO, NO_2). We execute 27 tests using different percentages of input patterns for the training. All results are referred to generalization's phase, where the patterns are never seen by the NN (N_G in tables). The results obtained by Cluster Analysis applied to NN (CANN) are compared to the Conventional Random Pattern Selection applied to NN (CRPSNN), our benchmark, with different percentages of input patterns from 0.01% to 100% (from S1 to S27). We perform also the analysis of negative O_3 concentrations (not shown here) that are verified during the generalizations phase. The performance of CANN is evaluated with CRPSNN through different statistical indicators. These indicators measure residual errors and give a global idea of the difference between observed and forecasted values. The main values of indicators obtained for the quality indexes are shown in the tables (see Table 3).

In general, we observed that the NN's performance shows different values for the ozone predictions. The results show that CANN is performed and predicted better than CRPSNN. In term of global fit, CANN has performance better (R^2 from 0.59 to

0.89) than CRPSNN (R^2 from 0.05 to 0.97). For example, when we consider S7 (0.40% of input pattern) the R^2 is 0.75 for CANN vs. 0.37 for CRPSNN. These results show a very meaningful difference. Moreover, we observed that in S10 we obtain R^2=0.56 utilizing CANN, whereas R^2=0.41 for CRPSNN (see Table 3). This first important result shows that CA sampling is more efficient than CRPSNN using small amount of patterns during the training and, consequently, could be adapt to simulate rare events.

In Fig. 2, the R^2 coefficients for the two choices in relation to different percentages of input patterns are given. By the figure, it is evident that the performances become similar after the use of 4% of total data set. As consequence, the cluster analysis selection is mostly efficient respect to the random patterns choice. However, it can be observed that CANN is always more performing starting from 7% up to 90%. The best performance (linked to the values of R^2=0.86) was obtained with 20%, 30% and 40% of centroids (S20, S21, S22) for CANN. While in the random selection by CRPSNN the performances decrease in meaningful way when we move from this percentage of input patterns (note the value of 0.51 if we use 2% of data), in the selection by cluster sampling the performance can be considered satisfactory in the generalization cases.

Figure 3 (a, b) show the full set of scatter diagrams illustrating observed versus predicted ozone by CANN and CRPSNN, including the best-fit lines. Scatter plots are referred to S22 for CRPSNN and S21 for CANN. The differences in statistical performance between the two procedures are easily discernible. In CANN, the high R^2 (0.86-0.87) indicates that a majority of the variability in the air pollutant outputs is more explained by this approach than CRPSNN.

Our results seem to indicate the right way to optimize the training by using NN model utilising the optimal input pattern choice as CANN.

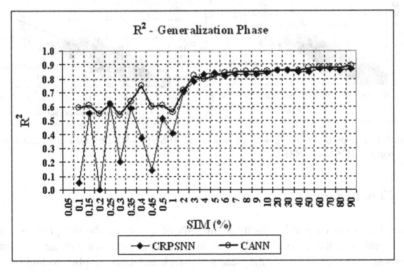

Fig. 2. R^2 performance

Table 2. Conventional Random Patterns Selection (CRPSNN) and Cluster Patterns Selection (CANN): Generalization phase (N=14324)

	Train (%)	Gen (%)	N$_G$	2006			2007		
				BIAS μg/m³	R²	RMSE μg/m³	BIAS μg/m³	R²	RMSE μg/m³
S1	0.1	99.9	14310	44,82	0,05	56,96	16,68	0,59	24,97
S2	0.15	99.85	14303	67,74	0,56	66,31	8,38	0,61	33,00
S3	0.2	99.8	14296	4,58	0,00	58,72	7,45	0,55	38,64
S4	0.25	99.75	14289	87,22	0,62	113,33	8,99	0,62	30,05
S5	0.3	99.7	14282	23,07	0,20	57,05	7,25	0,54	41,98
S6	0.35	99.65	14274	56,32	0,59	50,63	4,94	0,64	31,62
S7	0.4	99.6	14267	4,08	0,37	49,04	4,96	0,75	20,89
S8	0.45	99.55	14260	16,41	0,14	42,76	7,61	0,60	28,78
S9	0.5	99.5	14253	10.51	0,51	30.64	7.91	0,61	30.05
S10	1	99	14181	14.1	0,41	38.89	4.88	0,56	36.13
S11	2	98	14038	9.7	0,71	20.76	7.22	0,72	22.64
S12	3	97	13895	8.19	0,78	17.99	6.35	0,82	16.84
S13	4	96	13752	6.38	0,83	15.86	5.89	0,80	17.48
S14	5	95	13608	5.86	0,84	15.5	6.35	0,82	16.62
S15	6	94	13465	6.07	0,82	16.22	6.24	0,84	15.36
S16	7	93	13322	5.56	0,83	15.94	6.05	0,85	15.24
S17	8	92	13179	5.3	0,83	15.93	6.04	0,85	15.23
S18	9	91	13035	4.94	0,83	16.12	6.68	0,85	14.94
S19	10	90	12892	5.45	0,84	15.81	6.68	0,85	15.04
S20	20	80	11460	5.18	0,86	14.58	5.92	0,86	14.37
S21	30	70	10027	5.45	0,86	14.53	6.24	0,86	13.99
S22	40	60	8595	5.68	0,85	14.67	6.43	0,86	14.13
S23	50	50	7162	5.66	0,85	14.53	5.62	0,88	13.38
S24	60	40	5730	5.17	0,87	13.94	5.6	0,88	13.46
S25	70	30	4298	4.86	0,87	13.68	5.55	0,88	13.31
S26	80	20	2865	5.26	0,86	14.01	5.49	0,88	13.3
S27	90	10	1433	5.68	0,87	13.89	5.16	0,89	12.97

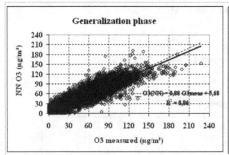
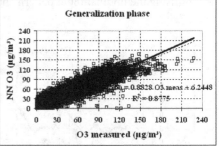

Fig. 3. a). Ozone predicted using 40% of patterns (CRPSNN) –S22; b). Ozone predicted using 30% of patterns (CANN) – S21.

5   Conclusions

Our research shows a good capacity of the NN of analyzing the complex and large data sets and modelling ozone levels using in pattern pre-processing phase the clustering approach. The capability of the Neural Network technique, applied to multivariate and non-linear problems, to capture the environmental information inside the data depended not only by the learning methods used, but also by the preliminary study of patterns, related to the quality of the data related to train the network. The problem of pre-processing and proper sampling plan of input data is essential to obtain a good forecasting performance of NN. The generalization capacity of NN to forecast ozone should be connected to the essential information inside the data set and this information is not necessarily regularly distributed inside all patterns. We observed that the neural classifier trained after random patterns choice, is able to distinguish only average/stable situations. On the contrary, NN after cluster pattern choice is able to distinguish also outlier situations.

In conclusion, clustering approach, adopted as patterns selection approach, obtains better predictions of pollutant phenomena. Results obtained are very encouraging and our simulations based on cluster analysis demonstrated that this method is feasible and effective, resulting in a substantial reduction of data input requirement and outperform respect to other techniques applied in this contest

References

1. World Health Organisation. Health aspects of air pollution with par-ticulate matter, ozone and nitrogen dioxide. Report on a WHO Working Group, Bonn, Germany, pp. 13–15 (2003)
2. EPA. Air Pollutants. Epa.gov. (June 28, 2006), http://www.epa.gov/ebtpages/airairpollutants.html (retrieved August 29, 2010)
3. Penkett, S.A., Evans, M.J., Reeves, C.E., Law, K.S., Monks, P.S., Bauguitte, S.J.B., Pyle, J.A., Green, T.J., Bandy, B.J., Mills, G., Cardenas, L.M., Barjat, H., Kley, D., Schmitgen, S., Kent, J.M., Dewey, K., Methven, J.: Long-range transport of ozone and related pollutants over the North Atlantic in spring and summer. Atmospheric Chemistry. Physic Discussion 4, 4407–4454 (2004)
4. Finlayson-Pitt, J.B., Pitts, W.J.: Fundamental and Experimental Techniques, pp. 108–136. John Wiley and Sons, Inc., New York (1986)
5. Carter, W.P.L.: A detailed mechanism for the gas-phase atmospheric reac-tions of organic compounds. Atmospheric Environment 24A, 481–518 (1990)
6. Comrie, R.S.: Comparing neural network and regression models for ozone forecasting. J. Air. Waste. Manage. 47, 653–663 (1997)
7. Gardner, M.W., Dorling, S.R.: Statistical surface ozone models: an improved methodology to account for non-linear behaviour. Atmos. Environ. 34, 21–34 (2000)
8. Gardner, M.W., Dorling, S.R.: Artificial Neural Networks (the Multilayer Per-ceptron)- E Review of applications in the atmospheric sciences. Atmos. Environ. 32(14/15), 2627–2636 (1998)
9. Dutot, A.L., Rynkiewicz, J., Steiner, F.E., Rude, J.: A 24-h forecast of ozone peaks and exceedance levels using neural classifiers and weather predictions. Environ. Modell. Softw. 22, 1261–1269 (2007)
10. Pelliccioni, A., Lucidi, S., La Torre, V., Pungì, F.: Optimization of Neural Network performances by means of exogenous input variables for the fore-cast of Ozone pollutant in Rome Urban Area. In: Eighth Conference on Artificial Intelligence and its Applications to the Environmental Sciences. AMS 90th Annual Meeting, pp. 17–21 (2010)
11. Rojas, R.: Neural Networks. Springer, Berlin (1996)
12. Hawkins, D.: Identification of Outliers. Chapman and Hall, London (1980)

13. Bishop, C.M.: Neural Networks for Pattern Recognition. Oxford Univ. Press, New York (1995)
14. EUROSTAT, Survey sampling reference guidelines-Introduction to sample design and estimation techniques (2008), http://ec.europa.eu/eurostat
15. Gardner, M.W., Dorling, S.R.: Neural network modelling and prediction of hourly NOx and NO2 concentrations in urban air in London. Atmos. Environ. 33, 709–719 (1999)
16. Abdul-Wahab, S.A., Al-Alawi, S.M.: Assessment and prediction of tropospheric ozone concentration levels using artificial neural networks. Environ. Modell. Softw. 17, 219–228 (2002)
17. BuHamra, S., Smaoui, N., Gabr, M.: The Box-Jenkins analysis and neural networks: prediction and time series modelling. Appl. Math. Model. 27(10), 805–815 (2003)
18. Fausett, L.: Fundamentals of Neural Networks. In: Architectures, Algorithms and Applications. Prentice Hall, Englewood Cliffs (1994)
19. Ripley, B.D.: Pattern Recognition and Neural Networks. Cambridge University Press (1996)

Application of Genetic Neural Networks for Modeling of Active Devices

Anwar Jarndal

Electrical and Computer Engineering Department
University of Nizwa
P.O. Box 33, PC 616
Nizwa, Sultanate of Oman
jarndal@ieee.org

Abstract. This paper presents detailed procedure of genetic neural networks modeling and application of this approach on GaN high electron mobility transistors (HEMTs). The developed model has been validated by RF large-signal measurements of the considered devices. The model shows very good capability for simulating the nonlinear behavior of the devices with higher rate of convergence.

Keywords: GaN HEMT, large-signal modeling, neural networks, genetic optimization, power amplifier design.

1 Introduction

Today, Neural-networks have gained considerable attention as a useful tool for RF and microwave modeling and design [1]. Neural networks can be trained to "learn" the behavior of active device. The trained neural network model can then be used for design purposes. This technique is an attractive alternative to other methods such as physical modeling method, which could be computationally expensive, or analytical method which requires assumption of particular analytical functions to be used and it could be difficult to obtain for new devices, or table-based modeling method, which has limited prediction capability and lower rate of convergence. In this paper, neural networks as a modeling technique for active devices will be investigated. The model prediction capability has been improved by using a knowledge-based approach to choice a suitable activation function. For the sake of simplicity in the construction and implementation of the proposed model, a single hidden layer topology has been used. To improve the model accuracy and finding the optimal values for weights, genetic algorithm optimization based has been adopted. The developed approach is applied for large-signal modeling of GaN devices. These transistors are becoming the most appropriate technology for high power amplifier (HPA) design [2]. This accordingly increases the need for rigorous modeling of the electrical and electro-thermal behavior of these devices. The developed models will be validated by DC and RF large-signal measurements.

T. Huang et al. (Eds.): ICONIP 2012, Part IV, LNCS 7666, pp. 231–239, 2012.

2   Genetic Neural Network Model

The main sources of nonlinearity in active devices are the drain and gate currents and capacitances. The nonlinear behavior of these elements at pinch-off and forward voltages enhances the AM-PM distortion of the output of the device amplifier [3]. Each one of elements can be described by neural network model. The model can then be embedded in the large-signal equivalent circuit of the considered device.

2.1   DC IV Model

The main nonlinear elements of the equivalent circuit model are the drain and gate currents. The DC IV characteristics can be used to describe these currents even under RF of operation by using a frequency dependent correction factor to simulate the DC-RF dispersion as it can be explained later. A neural network based model is used to represent the DC drain and gate currents. The model topology includes only a single hidden layer with unit biases, as illustrated in Fig. 1.

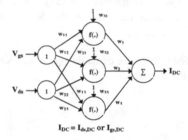

$$I_{DC} = I_{ds,DC} \text{ or } I_{gs,DC}$$

Fig. 1. Neural network model for dc drain or gate current

The activation function f(.) of the model (see Fig. 1) is hyperbolic tangent in case of the drain current, but it is exponentional for the gate current. It is found that using of these functions improves the prediction capability of the model. According to this model, the drain or gate current can be expressed as

$$I_{DC} = \sum_{i=1}^{3} w_i f\left(w_{1i}V_{gs} + w_{2i}V_{ds} + w_{3i}\right) \tag{1}$$

where I_{DC} represent the gate or the drain current, V_{gs} and V_{ds} are the intrinsic gate and drain voltages, w_{1i}, w_{2i} and w_{3i} are the input weights and w_i is the output weight. f(.) equal to $tanh(.)$ for the drain current and equal to $e^{(.)}$ for the gate current. An optimization technique can be used in training the neural-network model and finding the optimal values for the input and output weights. Here, the problem is nonlinear multi-dimensional optimization of 12 variables and it is likely to have multiple local minima. Therefore, to overcome this problem and to find the global minimum, the genetic algorithm, as a global optimization technique, has been used. The procedure of the implemented genetic optimization is presented in Fig. 2 and it can be summarized as follows:

1. Randomly, generation of initial population of individuals. Each individual consist of 12 variables (9 input weights and 3 output weights). The generated values of these weights are within -1 and 1. The current of generation No.1 will be the parents of the next generation individuals and the optimization will continue over N_{max} generation.
2. Calculating the corresponding error between simulated and measured DC IVs for each individual as follow.

- Computing $I_{ds,DC}$ or $I_{gs,DC}$ in (1) using the values of the input and output weight in the individual over the entire grid of the measured V_{gs} and V_{ds}.
- Determining the total error between the simulated $I_{ds,DC}$ or $I_{gs,DC}$ and the corresponding measured one as follow

$$Error = \frac{1}{N} \sum_{m=1}^{N} \left(I_{DC}^{meas} - I_{DC}^{sim} \right)^2 \qquad (2)$$

where N is the total number of measurements, I_{DC}^{meas} is the measured DC drain or gate current and I_{DC}^{sim} is the corresponding simulated one.

3. Ranking the individuals of the selected population and their errors to reject some of the maximum error individuals in the population.
4. Recombining the selected individuals to perform crossover reproduction by using double-point crossover routine [4]. The individuals are ordered such that individuals in odd numbered positions are crossed with the individuals in the adjacent even numbered positions.
5. Mutating (the values of each individual are altered randomly) the reproduced offspring from the crossover process using low probability mutation technique [5].
6. Repeating step no. 2 to calculate the error of each reproduced individual.
7. The next step is reinsertion. Reinsertion replaces the most error individuals in the old population (parents) with individuals in the new reproduced population (offspring) [5].
8. The generational counter is incremented, and the steps from 2 to 7 are repeated until generation No. = N_{max}.
9. When the number of generational counter equal to N_{max} or the minimum error is smaller than a fixed threshold value ε, the algorithm reaches the last generation and stops.
10. The values included in the minimum error individual will be chosen as optimal values for the network model weights.

This procedure has been applied to DC IV measurements of a packaged 4-W GaN HEMT device on Si Substrate from Nitronex corporation to determine the optimal weights of the neural network models of the drain and gate currents. The procedure is started by generating a uniformly distributed random initial population of 500 individuals. Each individual consists of 12 variables (input and output weights). The maximum number of generations is set to 50 and ε is defined to be equal to 0.001. The constructed Genetic Neural Network model (GNN) is embedded in the adopted large-signal equivalent circuit shown in Fig. 3 and reported in [6]. The model has been verified by DC IV measurements and it showed accurate simulation for these measurements even for the typical self-heating induced collapse of the drain current in the high power dissipation area. In general, this current reduction is significant under static and quasi-static operation. However, it is reduced by increasing the frequency of operation since the input signal is not slow enough to heat up the device.

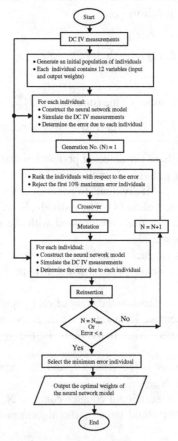

Fig. 2. Flowchart of the neural network weights optimization using genetic algorithm

To simulate this effect, the drain current is formulated as [6]

$$I_{ds}(V_{ds},V_{gs}) = $$
$$I_{ds,DC}(V_{ds},V_{gs})\left[1 + K_T H(\omega)\begin{pmatrix} I_{ds,DC}(V_{ds},V_{gs})V_{ds} \\ -I_{ds,DC}(V_{dso},V_{gso})V_{dso} \end{pmatrix}\right]. \tag{3}$$

$I_{ds,DC}$ is the DC drain current, which is represented by the GNN model and K_T is a thermal constant describing the dependency of the drain current on the device self-heating. V_{ds} and V_{gs} are the instantaneous intrinsic drain and gate voltages, which are dynamically changed around their average voltages or DC values V_{dso} and V_{gso}.

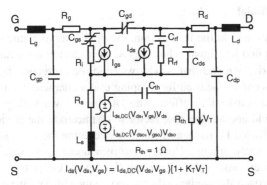

Fig. 3. Large-signal equivalent circuit model for GaN HEMTs including self-heating and output conductance dispersion effects [6]

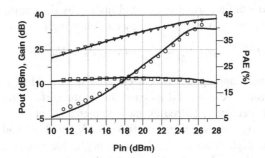

Fig. 4. Measured (symbols) and simulated (lines) output power, gain and efficiency at class-AB (V_{GS} = -1.6 V and V_{DS} = 28 V) for a 4-W GaN HEMT in a 50 Ω source and load environment at 2.35 GHz

$H(\omega)$ is a thermal frequency response function, which can be defined as [6]

$$H(\omega) = j\omega\tau / (1 + j\omega\tau) \qquad (4)$$

where ω is the operating frequency and τ is a thermal time constant. This function describes the smooth transition of the drain current from static or quasi-static to RF and it is implemented using the high-pass thermal sub-circuit in the model (see Fig. 3). The large-signal equivalent circuit model was implemented in a commercial RF circuit design tool [Agilent Advanced Design System (ADS)]. The implemented model has been used for simulating single-tone large-signal measurements carried out on the same considered device. As it can be seen in Fig. 4, for class-AB operated device, a very good simulation can be observed and also the model showed a higher rate of convergence with respect to the developed table-based model for the same device in [6].

2.2   Pulsed IV Model

As it has been mentioned in the last section, the device temperature does not clearly change with the applied RF signal. Thus the DC IV measurements cannot be used for RF device characterization and that is why we used the correction factor $H(\omega)$. Also these measurements cannot be used for trapping effect characterization because of its correlation with the self-heating effect. The DC breakdown voltage and maximum power dissipation add another limiting factor to characterize the device at high drain and gate voltages. Pulsed IV measurements can overcome the main limitations of the corresponding DC measurements. Under these measurements it is possible to extend the range of measurements, without harming the device under test. These measurements can also approach the isothermal condition, since all IV characteristics can be obtained at a constant device temperature defined by quiescent bias condition and ambient temperature (typically the pulse width is < 1 µs). Under very low quiescent voltages or currents, the drain current is affected mainly by the trapping effect. Thus the drain current in this case can be described by the following formula [7]

$$I_{ds} = I_{ds,iso}^{DC} + \alpha_G (V_{gs} - V_{gso}) + \alpha_D (V_{ds} - V_{dso}). \tag{5}$$

$I_{ds,iso}^{DC}$ is the isothermal trapping-free dc current after deembedding the self-heating effect. α_G and α_D measure the current dispersion due to the surface-trapping and buffer-trapping effects, respectively. Pulsed IV measurements I_{ds1} at (V_{gso}= 0 V, V_{dso}= 0 V) quiescent voltages, I_{ds2} at (V_{gso}= pinch-off V_p V, V_{dso}= 0 V) and I_{ds3} at (V_{gso}= pinch-off V_p V, V_{dso}= high voltage V_{dsm} V) have negligible power dissipation. By applying (5) to these three characteristics, the following three equations will be obtained

$$I_{ds1} = I_{ds,iso}^{DC} + \alpha_G V_{gs} + \alpha_D V_{ds}. \tag{6}$$

$$I_{ds2} = I_{ds,iso}^{DC} + \alpha_G (V_{gs} - V_p) + \alpha_D V_{ds}. \tag{7}$$

$$I_{ds3} = I_{ds,iso}^{DC} + \alpha_G (V_{gs} - V_p) + \alpha_D (V_{ds} - V_{dsm}). \tag{8}$$

By solving these three equations, the values of α_G, α_D, and $I_{ds,iso}^{DC}$ can be calculated as follow:

$$\alpha_G = \frac{I_{ds1} - I_{ds2}}{V_p}. \tag{9}$$

$$\alpha_D = \frac{I_{ds2} - I_{ds3}}{V_{dsm}}. \tag{10}$$

$$I_{ds,iso}^{DC} = I_{ds1} - (\frac{I_{ds1} - I_{ds2}}{V_p})V_{gs} - (\frac{I_{ds2} - I_{ds3}}{V_{dsm}})V_{ds}. \tag{11}$$

The general model of the drain current can be obtained by adding another term to describe the current deviation due to self-heating as follow:

$$I_{ds} = I_{ds,iso}^{DC} + \alpha_G(V_{gs} - V_{gso}) + \alpha_D(V_{ds} - V_{dso}) \\ + \alpha_T(I_{dso}V_{dso}). \tag{12}$$

Under DC operation $V_{gs} = V_{gso}$ and $V_{ds} = V_{dso}$ and $I_{ds} = I_{ds,iso}^{DC}$ thus α_T can be calculated from the determined $I_{ds,iso}^{DC}$ in (11) and dc IV measurements as follows:

$$\alpha_T = \begin{cases} \dfrac{I_{ds}^{DC} - I_{ds,iso}^{DC}}{I_{ds}^{DC}V_{dso}}, for \ I_{ds}^{DC}V_{dso} > 0 \\ 0 \qquad\qquad , for \ I_{ds}^{DC}V_{dso} = 0. \end{cases} \tag{13}$$

I_{ds1}, I_{ds2}, I_{ds3} and I_{ds}^{DC} characteristics can be fitted by three neural networks models of the same topology illustrated in Fig.1 with tangent activation function. The same genetic algorithm optimization procedure detailed in the last section can be used to find the weights of each neural networks model. The simulated values I_{ds1}, I_{ds2}, I_{ds3} and I_{ds}^{DC} are then used to calculate α_G, α_D, α_T and $I_{ds,iso}^{DC}$. These fitting parameters can then be inserted in (12) in order to reproduce the value of the drain current at any applied voltages. DC IV and Pulsed IV characteristics of on-wafer 1-mm GaN HEMT on SiC substrate device at quiescent bias conditions (V_{gso} =0 V,V_{dso} = 0 V), (V_{gso} = -7 V, V_{dso} = 0 V), and (V_{gso} = -7 V,V_{dso} = 25 V) are used for modeling the drain current. Fig. 5 shows the measured and simulated characteristics. As it can be seen the neural networks models can accurately reproduce the measurements.

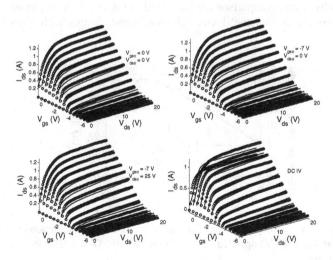

Fig. 5. Measured (symbols) and simulated (lines) IV characteristics for 1mm gate width on-wafer GaN on SiC substrate

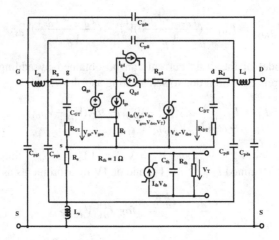

Fig. 6. Equivalent circuit large-signal model for GaN HEMT on SiC substrate [8]

The developed neural networks based model of the drain current has been embedded in the large-signal equivalent circuit shown in Fig. 6 instead of the previously implemented table-based one in [8]. The developed large-signal model was implemented in a commercial RF circuit design tool (Agilent Advanced Design System [ADS]). The extrinsic bias-independent passive elements are represented by lumped elements, whereas the intrinsic nonlinear part including the drain current is represented by a symbolically defined device component. Single-tone large-signal on-wafer measurements has been also performed for the considered 1-mm GaN HEMT in 50 Ω source and load terminations under different bias conditions (classes of operation) and different input drive levels. The corresponding simulations have been performed using the ADS implemented model and compared with the measurements. The results of this comparison are presented in Figures 7. As it can be seen, the model can accurately predict the device RF characteristics also with higher rate of convergence.

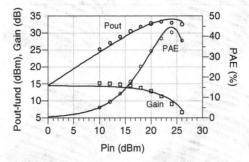

Fig. 7. Single-tone power sweep simulations (lines) and measurements (symbols) for class AB (V_{GS} = -1.5 V and V_{DS} = 15 V) operated 1mm GaN on SiC HEMT at 2 GHz in a 50-Ω source and load environment

3   Conclusion

A genetic neural networks modeling method for active devices is presented. The modeling approach has been used for representing the drain current based on DC and pulsed IV measurements of packaged and on-wafer GaN HEMTs. The developed models have been validated by RF large-signal measurements of the considered devices. The models show very good capability for describing the nonlinear behavior of the device with higher rate of convergence.

References

1. Change, Q.J., Gupta, K.C.: Neural networks for RF and microwave design. Artech House, Norwood (2000)
2. Mishra, U.K., Shen, L., Kazior, T.E., Wu, Y.-F.: GaN-based RF power devices and amplifiers. Proceeding IEEE 96, 287–305 (2008)
3. Aaen, P., Plá, J., Wood, J.: Modeling and Characterization of RF and Microwave Power FETs. Cambridge University Press (2011)
4. Chipperfield, A., Fleming, P., Pohlheim, H., Fonseca, C.: Genetic Algorithm Toolbox for Use with MATLAB. ver. 1.2, Department of Automatic Control Systems Engineering, University of Sheffield, UK (1993)
5. Melanie, M.: An introduction to genetic algorithms. MIT Press, Cambridge (1996)
6. Jarndal, A., Aflaki, P., Negra, R., Kouki, A., Ghannouchi, F.M.: Large-signal modeling methodology for GaN HEMTs for RF switching-mode power amplifiers design. International Journal of RF and Microwave Computer-Aided Engineering 21, 45–50 (2010)
7. Filicori, F., Vannini, G., Santarelli, A., Sanchez, A.M., Tazon, A., Newport, Y.: Empirical modeling of low-frequency dispersive effects due to traps and thermal phenomena in III-V FET's. IEEE Transaction Microwave Theory Technique 43, 2972–2981 (1995)
8. Jarndal, A., Kompa, G.: Large-Signal Model for AlGaN/GaN HEMT Accurately Predicts Trapping and Self-Heating Induced Dispersion and Intermodulation Distortion. IEEE Transaction on Electron Devices 54, 2830–2836 (2007)

Displacement Prediction Model of Landslide Based on Ensemble of Extreme Learning Machine

Cheng Lian[1], Zhigang Zeng[1,*], Wei Yao[2], and Huiming Tang[3]

[1] Department of Control Science and Engineering,
Huazhong University of Science and Technology,
and Key Laboratory of Image Processing and Intelligent Control of Education
Ministry of China, Wuhan, Hubei, China
[2] School of Computer Science, South-Central University for Nationalities,
Wuhan, Hubei, China
[3] Faculty of Engineering, China University of Geosciences,
Wuhan, Hubei, China
zgzeng@gmail.com

Abstract. Based on time series analysis, total accumulative displacement of landslide is divided into the trend component displacement and the periodic component displacement according to the response relation between dynamic changes of landslide displacement and inducing factors. In this paper, a novel neural network technique called the ensemble of extreme learning machine (E-ELM) is proposed to investigate the interactions of different inducing factors affecting the evolution of landslide. Trend component displacement and periodic component displacement are forecasted respectively, then total predictive displacement is obtained by adding the calculated predictive displacement value of each sub. A case study of Baishuihe landslide in the Three Gorges reservoir area is presented to illustrate the capability and merit of our model.

Keywords: Extreme learning machine, Artificial neural networks, Ensemble, Landslide, Displacement prediction.

1 Introduction

Landslides are a recurrent problem throughout the Three Gorges Reservoir area, which is located at the upper reaches of the Yangtze River in China. Frequent landslides often result in significant damage to people and property, hence, the prediction of landslide-prone regions is essential for carrying out quicker and safer mitigation programs, as well as future planning of the area. It is well known that landslide hazard is a complex nonlinear dynamical system with the uncertainty, in which tectonic, rainfall and reservoir level fluctuation and so on all influence the evolution of landslide. These factors are divided into the trigger and the

---

* Corresponding author.

T. Huang et al. (Eds.): ICONIP 2012, Part IV, LNCS 7666, pp. 240–247, 2012.

primary cause [1][2]. Studies on the interactions of the different factors affecting landslide occurrence are very important for the prediction of landslide. A time series decomposable model was proposed to establish the response relation between dynamic changes of landslide displacement and inducing factors[3][4]. Total displacement of landslide can be divided into the trend component displacement and the periodic component displacement. The trend component displacement is determined by the potential energy and constraint condition of the slope. The periodic component displacement is affected by the periodic dynamic functioning of inducing factors such as rainfall, reservoir level fluctuation and so on.

In recent years, artificial neural networks (ANNs) have been widely applied in the area of landslide forecasting[5][6]. Compare with logistic regression analysis, ANNs give a more optimistic evaluation of landslide susceptibility[6]. However most ANN based landslide forecasting methods used gradient-based learning algorithms such as back-propagation neural network (BPNN), which are relatively slow in learning and may easily get stuck in a local minimum[7][8]. Recently, a novel learning algorithm for single-hidden-layer feedforward neural networks (SLFNs) called extreme learning machine (ELM) has been proposed[9][10]. ELM not only learns much faster with higher generalization performance than the traditional gradient-based learning algorithms but also avoids many difficulties faced by gradient-based learning methods such as stoping criteria, learning rate, learning epochs and local minimum[11]-[14]. Although ELM has many advantages, a disadvantage is that its output is usually different from time to time because the input weights and biases are randomly chosen. So we don't know exactly on which time the initiation will give a good result. The idea of ELM ensembles has been proposed which can significantly improved the generalization ability of learning systems through training a finite number of ELMs and then combining theirs results [15][16]. The final output of E-ELM is the average of the outputs of each ELM network. A case study of Baishuihe landslide in the Three Gorges reservoir area is presented to illustrate the capability and merit of the model.

2 Methodology

The change of landslide displacement is determined by its own geological conditions and dynamic functioning of inducing factors. The displacement of landslide sequence is an instability time series. Based on time series analysis, total displacement of landslide can be broken down into different corresponding components according to the different influential factors. Based on the above analysis, the displacement of landslide sequence can be described in terms of 4 basic classes of components: trend component, periodic component, impulse component and random component. It can be expressed as follows [3][4]:

$$A_t = T_t + P_t + I_t + R_t, \quad t = 1, 2, \ldots, N \tag{1}$$

where A_t is the time series value (the accumulative change of landslide displacement) at time t. T_t is the trend component revealed the long-term trend of the sequence which is determined by the potential energy and constraint condition of the slope. P_t is the periodic component which is influenced by changes in the natural environment cycle like rainfall and difference in temperature of day and night. I_t is the impulse component which responds to the abrupt events like reservoir level fluctuation and R_t is the random component which reflects the impacts of random factors like earthquake. We can treat I_t as P_t when the impulse factors show characteristic of periodicity. In this paper, we present the prediction model without taking into account of influences of random component. Then the model can be simplified as follows:

$$A_t = T_t + P_t, \quad t = 1, 2, \ldots, N \tag{2}$$

It has been found that single hidden-layer feedforward neural networks (SLFNs) can approximate any continuous non-linear function with arbitrary precision [17]. Based on this, E-ELM model is used to forecast the trend component displacement and the periodic component displacement. Total predictive displacement is obtained by adding the calculated predictive displacement value of each sub. E-ELM consists of λ ELM networks with same structure, including the number of hidden nodes and same activation function. The final output of E-ELM is the average of the outputs of each ELM network. It is the same to repeat run the ELM for λ times with the same training data and calculate as follows:

$$\bar{y}_i = \frac{1}{\lambda} \sum_{j=1}^{\lambda} y_i^j, \quad j = 1, 2, \ldots, \lambda \tag{3}$$

where y_i is the output of each ELM network with the input of $\mathbf{x}_i$. Obviously, the outputs obtained by E-ELM will become more stable when the parameter λ becomes larger, but the computation time also increases. The scheme of the E-ELM is shown in Fig. 1.

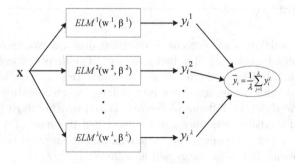

Fig. 1. The scheme of the E-ELM integration system

3   Simulation Studies

3.1   Date Collection

Baishuihe landslide is located on the south bank of Yantze River and its 56km away from the Three Gorges Dam. The bedrock geology of the study area consists mainly of sandstone and mudstone, which is an easy slip stratum. The slope is of the category of bedding slopes. Fig. 2 shows the monitoring data of landslide accumulative displacement at ZG118 monitoring point and Fig. 3 shows the monitoring data of rainfall and reservoir level elevation [3]. The total number of the data was 38 groups from June 2004 to July 2007. The data between June 2004 to December 2006 were selected as training data in order to construct the forecasting model and the rest of 7 groups of data from January 2007 to July 2007 were selected as predicting data.

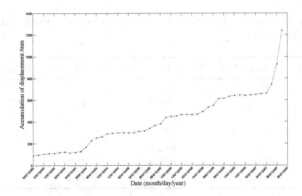

Fig. 2. Monitoring curves of landslide accumulative displacement

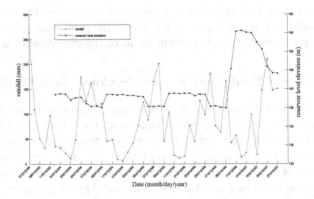

Fig. 3. Monitoring curves of rainfall and reservoir level elevation

Double moving average method is used to separate the trend component displacement from the total displacement of landslide. The period is chosen 12 months. Then, the periodic component displacement can be obtained by removing the trend component displacement from the total displacement of landslide.

3.2 Analysis and Prediction of Trend Component Displacement

Trend component displacement is determined by the potential energy and constraint condition of the slope, which nearly increases under large time scales. ELM can approximate any continuous function which is used to forecast the trend term displacement. The activation function of E-ELM is the sigmoidal function $g(x) = 1/(1 + e^{-x})$. The number of hidden nodes is 3. Considering both computation time and the stability of E-ELM, the parameter λ is selected 1000. The predicted values of trend component displacement is shown in Table 1.

Table 1. Trend component displacement comparison between predicted values(mm) and measured values(mm)

Time	Measurement Value	Predicted Value	Absolute Error	Relative Error(%)
01/ 01/ 2007	599.6	580.6	19.0	3.17
02/ 01/ 2007	615.5	596.2	19.3	3.14
03/ 01/ 2007	627.6	611.7	15.9	2.53
04/ 01/ 2007	636.3	627.3	9.0	1.41
05/ 01/ 2007	644.6	642.8	1.8	0.28
06/ 01/ 2007	658.6	658.4	0.2	0.03
07/ 01/ 2007	688.1	673.9	14.2	2.06

As shown in Table 1, E-ELM model shows a good extrapolation capability. The predictive values and measurement values are very close for every calculation, and the relative error all falls into 5 percent, the predicting precision is high enough which can revealed the long-term trend of the evolution of landslide.

3.3 Analysis and Prediction of Periodic Component Displacement

As we know, the periodic component displacement of landslide will be affected by many factors. Baishuihe landslide is located at the Three Gorges Reservoir Area, based on empirical knowledge, the periodic component is mainly affected by rainfall, and the impulse component is mainly affected by reservoir level fluctuation. Because of the reservoir level adjustment shows characteristic of periodicity with one year cycle in Three Gorges Reservoir Area. We can treat reservoir level fluctuation as the periodic component inducing factors. The reservoir level fluctuation is calculated as the difference between the water level at this month and last month. As a result of rainfall has the lagged effect on the evolution of landslide. Here are three inputs of ELM: cumulative of rainfall anterior two month, cumulative of rainfall of current month and anterior two month, and reservoir level fluctuation.

All the data sets should be normalized into the rang of [-1, 1]. The activation function of E-ELM is the sigmoidal function $g(x) = 1/(1 + e^{-x})$. The number of hidden nodes is 18. Considering both computation time and the stability of E-ELM, the parameter λ is selected 1000. The predicted values of periodic component displacement is shown in Table 2.

Table 2. Periodic component displacement comparison between predicted values (mm) and measured values (mm)

Time	Measurement Value	Predicted Value	Absolute Error	Relative Error(%)
01/ 01/ 2007	44.6	48.2	3.6	8.07
02/ 01/ 2007	32.8	55.9	23.1	70.43
03/ 01/ 2007	29.0	52.7	23.7	81.72
04/ 01/ 2007	24.5	30.0	5.5	22.45
05/ 01/ 2007	99.2	101.5	2.3	2.32
06/ 01/ 2007	272.3	216.0	56.3	20.68
07/ 01/ 2007	550.3	122.4	427.9	77.76

Rainfall and reservoir level fluctuation are the major factors that affect the stability of landslide, but there are many other factors such as temperature difference between day and night and other random factors such as human project activities. So the predictive values in February 2007 and March 2007 are not very precise, but that lack of precision may not matter in engineering. The predictive values successfully reflect the evolution tendency of landslide from January 2007 to June 2007, especially the model successfully predicts the obvious deformation from April 2007 to June 2007 which is able to provide early warnings. The landslide collapsed in July 2007. Once the collapse happens, the displacement will increase exponentially without constrain. Then, the landslide is at the unsteady state. Obviously, without any training samples belong to the same balancing system, neural networks model is not suited to forecast the evolution of displacement in this condition. Actually, the impact factors of neural network model are also changed in this condition.

3.4   Analysis and Prediction of the Total Displacement of Landslide

The total displacement prediction is obtained by adding the predictive values of trend component displacement and periodic component displacement. The predicted values are shown in Table 3.

As shown in Table 3, the predictive values and measurement values are very close for every calculation except in July 2007 and the model successfully predicts the obvious deformation from April 2007 to June 2007. On the basis of the mentioned analysis in Section 3.3, the total displacement of landslide are

Table 3. Total displacement of landslide comparison between predicted values (mm) and measured values (mm)

Time	Measurement Value	Predicted Value	Absolute Error	Relative Error(%)
01/ 01/ 2007	644.2	628.8	15.4	2.39
02/ 01/ 2007	648.3	652.1	3.8	0.59
03/ 01/ 2007	656.6	664.4	7.8	1.19
04/ 01/ 2007	660.8	657.3	3.5	0.53
05/ 01/ 2007	743.8	744.3	0.5	0.07
06/ 01/ 2007	930.9	874.4	56.5	6.07
07/ 01/ 2007	1238.4	796.3	442.1	35.70

unpredictable in July 2007 using our model. But the forecasting ability of our model is good enough to provide early warnings.

4   Conclusion

Landslides are a recurrent problem throughout the Three Gorges Reservoir area in China. It's very important for us to improve the technology of landslides forecasting to prevent and reduce the loss caused by landslides. Based on time series analysis, total accumulative displacement of landslide is divided into the trend component displacement and the periodic component displacement according to the response relation between dynamic changes of landslide displacement and inducing factors. Trend component displacement and periodic component displacement are forecasted respectively, then total predictive displacement is obtained by adding the calculated predictive displacement value of each sub. In this paper, we apply a relatively novel neural network technique, E-ELM, to study the interactions of the different factors affecting landslide occurrence.

The application shows that our method can achieve a good prediction result. Especially, the model successfully predicts the obvious deformation of landslide, which is able to provide early warnings. Therefore, this method has a good perspective in application and further development. It must be pointed out that landslide hazard has its own characteristics which varied with geological environment, landslide forecast model should be established according to the concrete types of landslides. Actually, expert judgment still should be taken into account in practical applications.

Acknowledgments. The work is supported by the Natural Science Foundation of China under Grant 60974021, the 973 Program of China under Grant 2011CB710606, the Fund for Distinguished Young Scholars of Hubei Province under Grant 2010CDA081, the Specialized Research Fund for the Doctoral Program of Higher Education of China under Grant 20100142110021.

References

1. Kawabata, D., Bandibas, J.: Landslide susceptibility mapping using geological data, a DEM from ASTER images and an Artificial Neural Network (ANN). Geomorphology 113, 97–109 (2009)
2. Cubito, A., Ferrara, V., Pappalardo, G.: Landslide hazard in the Nebrodi Mountains (Northeastern Sicily). Geomorphology 66, 359–372 (2005)
3. Du, J., Yin, K.L., Chai, B.: Study of displacement prediction model of landslide based on respones analysis of inducing factors. Chinese Journal of Rock Mechanics and Engineering 28, 1783–1789 (2009)
4. Wang, J.F.: Quantitative prediction of landslide using S-curve. Chinese Journal of Rock Mechanics and Engineering 14, 1–8 (2003)
5. Chen, H.Q., Zeng, Z.G.: Deformation prediction of landslide based on genetic-simulated annealing algorithm and BP neural network. In: 4th International Workshop on Advanced Computational Intelligence, pp. 675–679 (2011)
6. Nefeslioglu, H.A., Gokceoglu, C., Sonmez, H.: An assessment on the use of logistic regression and artificial neural networks with different sampling strategies for the preparation of landslide susceptibility maps. Eng. Geol. 97, 171–191 (2008)
7. Amro, E.J., John, M.: A new error criterion for posterior probability estimation with neural nets. In: Proceedings of Iteration Joint Conference on Neural Networks, pp. 185–192 (1990)
8. Harris, D., Yann, L.C.: Improving generalization performance using double back-propagation. IEEE Trans. Neural Networks 3, 991–997 (1992)
9. Huang, G.B., Zhu, Q.Y., Siew, C.K.: Extreme learning machine: Theory and applications. Neurocomput. 70, 489–501 (2006)
10. Zhu, Q.Y., Qin, A.K., Suganthan, P.N., Huang, G.B.: Evolutionary extreme learning machine. Pattern Recogn. 38, 1759–1763 (2005)
11. Huang, G.B.: Learning capability and storage capacity of two-hidden-layer feed-forward networks. IEEE Trans. Neural Networks 14, 274–281 (2003)
12. Huang, G.B., Haroon, A.B.: Upper bounds on the number of hidden neurons in feedforward networks with arbitrary bounded nonlinear activation functions. IEEE Trans. Neural Networks 9, 224–229 (1998)
13. Huang, G.B., Chen, L., Siew, C.K.: Universal approximation using incremental constructive feedforward networks with random hidden nodes. IEEE Trans. Neural Networks 17, 879–892 (2006)
14. Tamura, S., Tateishi, M.: Capabilities of a four-layered feedforward neural network: four layers versus three. IEEE Trans. Neural Networks 8, 251–255 (1997)
15. Sun, Z.L., Choi, T.M., Au, K.F., Yu, Y.: Sales forecasting using extreme learning machine with applications in fashion retailing. Decis. Support Syst. 46, 411–419 (2008)
16. Lan, Y., Chai, Y., Huang, G.B.: Ensemble of online sequential extreme learning machine. Neurocomput. 72, 3391–3395 (2009)
17. Hornik, K.: Approximation capabilities of multilayer feedforward networks. Neural Networks 4, 251–257 (1991)

Effective Handwriting Recognition System Using Geometrical Character Analysis Algorithms

Wojciech Kacalak and Maciej Majewski

Koszalin University of Technology, Department of Mechanical Engineering
Raclawicka 15-17, Koszalin, Poland
{wojciech.kacalak,maciej.majewski}@tu.koszalin.pl

Abstract. We propose a new method for natural writing recognition that utilizes geometric features of letters. The paper deals with recognition of isolated handwritten characters using an artificial neural network. As a result of the geometrical analysis realized, graphical representations of recognized characters are obtained in the form of pattern descriptions of isolated characters. The radius measurements of the characters obtained are inputs to the neural network for natural writing recognition which is font independent. In this paper, we present a new method for off-line natural writing recognition and also describe our research and tests performed on the neural network.

Keywords: handwriting recognition, artificial neural networks, artificial intelligence, human-computer interaction, natural writing processing.

1 Introduction

Natural writing recognition has been studied for nearly forty years and there have been many proposed approaches. The problem is quite complex, and even now there is no single approach that solves it both efficiently and completely in all contexts. In written language recognition processes, an image containing text must be appropriately supplied and preprocessed. Then the text must either undergo segmentation or feature extraction. Small processed pieces of the text will be the result, and these must undergo recognition by the system. Finally, contextual information should be applied to the recognized symbols to verify the result. Artificial neural networks, applied in handwriting recognition, allow for high generalization ability and do not require deep background knowledge and formalization to be able to solve the written language recognition problem.

Handwriting recognition can be divided by its input method into two categories: off-line handwriting recognition and on-line handwriting recognition. For off-line recognition, the writing is usually captured optically by a scanner. For on-line recognition, a digitizer samples the handwriting to time-sequenced pixels as it is being written. Hence, the on-line handwriting signal contains additional time information which is not present in the off-line signal.

T. Huang et al. (Eds.): ICONIP 2012, Part IV, LNCS 7666, pp. 248–255, 2012.

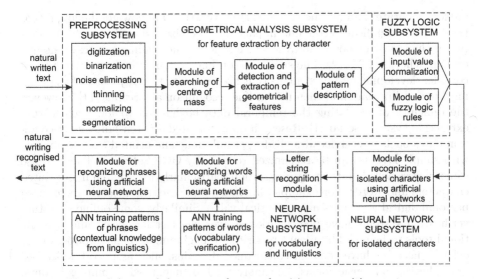

Fig. 1. Scheme of the proposed natural writing recognition system

In the proposed new method of natural writing recognition in Fig. 1, the handwritten text is produced subject to the following preprocessing: digitization, binarization, noise elimination, thinning, normalizing and segmentation. The next step is to find the center of mass of the character image. With the center of mass as a reference point, radiuses are drawn, creating a set of points describing the contour of the character so that its pattern description is made. In the proposed hybrid system, the pattern description of each isolated character, after the process of input value normalization and application of letter description rules using fuzzy logic, are the input signals for the neural networks for isolated character recognition. The recognized characters are grouped into more quantitative units with the letter string recognition module, which are coded as binary images of vectors and then become inputs of the module for recognizing words. The module uses a 3-layer Hamming neural network. The network of this module uses a training file containing patterns of words. The recognized vocabulary words represented by the output neurons are processed by the module for recognizing phrases which uses the Hamming Maxnet network equipped with a training file containing phrases built with contextual knowledge from linguistics.

2   The State of the Art

The state of the art of automatic recognition of handwriting at the beginning of the new millennium is that as a field it is no longer an esoteric topic on the fringes of information technology, but a mature discipline that has found many commercial uses. On-line systems for handwriting recognition are available in hand-held computers such as personal digital assistants. Their performance is acceptable for processing handprinted symbols, and when combined with keyboard entry, a powerful method for data entry has been created. Off-line systems are less accurate than on-line systems. However, they are now good enough that they have a significant economic impact on specialized domains such as interpreting handwritten postal addresses on envelopes and reading courtesy amounts on bank checks [1,2,3,12]. The success of on-line systems makes it attractive to consider developing off-line systems that first estimate the trajectory of the writing from off-line data and then use on-line recognition algorithms [11]. However, the difficulty of recreating the temporal data [4] has led to few such feature extraction systems so far [1]. Research on automated written language recognition dates back several decades. Today, cleanly machine-printed text documents with simple layouts can be recognized reliably by OCR software. There is also some success with handwriting recognition, particularly for isolated handprinted characters and words. For example, in the on-line case, the recently introduced personal digital assistants have practical value. Similarly, some online signature verification systems have been marketed over the last few years and instructional tools to help children learn to write are beginning to emerge. Most of the off-line successes have come in constrained domains, such as postal addresses, bank checks, and census forms. The analysis of documents with complex layouts, recognition of degraded printed text, and the recognition of running handwriting continue to remain largely in the research arena. Some of the major research challenges in on-line or off-line processing of handwriting are in word and line separation, segmentation of words into characters, recognition of words when lexicons are large, and the use of language models in aiding preprocessing and recognition. In most applications, machine performance is far from being acceptable, although potential users often forget that human subjects generally make reading mistakes [2,3]. The design of human-computer interfaces [5,6,7,8,9] based on handwriting is part of a tremendous research effort together with speech recognition, language processing and translation to facilitate communication of people with computers. From this perspective, any successes or failures in these fields will have an important impact on the evolution of languages [10].

3   Description of the Method

The proposed system attempts to combine two methods for natural writing recognition, neural networks and preprocessing for geometric features extraction. The system consists of the preprocessing subsystem, geometrical analysis subsystem, fuzzy logic subsystem, neural network subsystem for isolated characters as well as neural network subsystem for vocabulary and linguistics, as shown in Fig. 2. The motivation behind that preprocessor is to reduce the dimensionality of the neural network input. However, another benefit given by the preprocessor is immunity against image translation, because all the information is relative to the image's center of mass.

The extraction process of the selected geometrical features of letters is based on application of the center of mass of a letter with a method of data clustering. The selected Fuzzy C-Means algorithm (Fig. 3) is described with typical denotations of data clustering algorithms and can be aliased as unsupervised

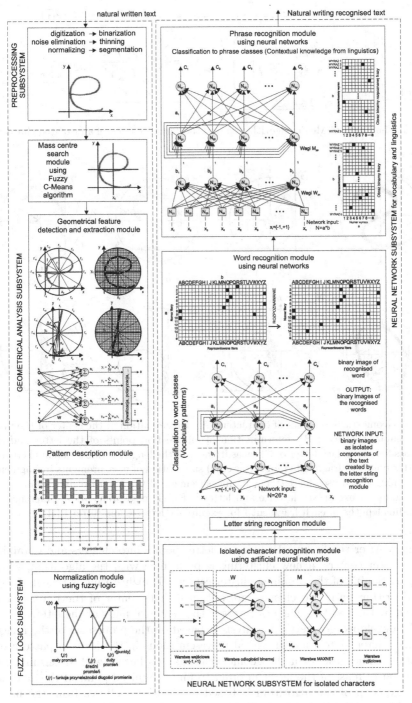

Fig. 2. Algorithm of the proposed system of effective handwriting recognition

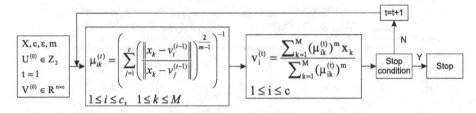

Fig. 3. The Fuzzy C-Means algorithm to find the center of mass of an isolated character

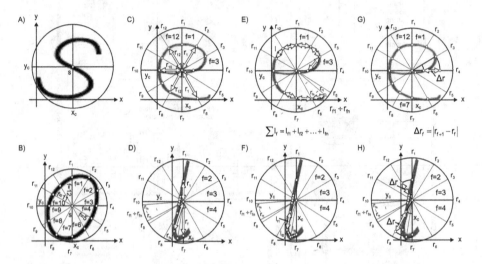

Fig. 4. Geometrical analyses of characters: A) determination of the center of the mass for exemplary letter s; B) determination of intersection points of the letter and the radiuses for exemplary letter o; C) measurement of the length of line segments l created by letter points in fragments f for letter e; D) summation of measurements in fragments f containing n radiuses for letter l; E) measurement of the length of line segments of each radius for exemplary letter e and letter l (F); G) measurements of differences of the radius lengths in each fragment f for exemplary letter e and letter l (H).

learning. After the first partitioning of letter points into clusters and obtaining their cluster centers, a new clustering is performed with the algorithm, which is a partitioning of the obtained cluster centers. The clustering is repeated with the algorithm until two clusters are obtained. The center of the line segment created by the last two cluster centers is the center of mass of the letter.

The developed geometrical analysis is based on the processing of the images of letter shapes into their graphical representations in the form of pattern descriptions. The process of the geometrical analysis begins with determining of the center of mass of the letter in order to find the initial point of the analysis. The next step of the algorithm is based on drawing radiuses from the initial point, the lengths of which are equal to the length of the line segment created by the initial point and the point on the letter furthest from this point. The

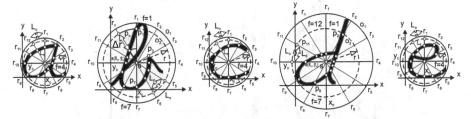

Fig. 5. Geometrical analyses of characters for exemplary letters

creation of a circle of that radius makes it visible that the analysis covers the whole letter. The precision of this geometrical analysis method is proportional to the number of radiuses. Where the radiuses intersect with the letter, points are obtained, which makes it possible to obtain the measures of the line segment created by the initial point and the letter intersection point. The lengths of the created line segments obtained are represented in the form of pattern descriptions of isolated characters which are inputs of the neural network. Geometrical analyses of characters for exemplary letters are shown in Fig. 4 and Fig. 5.

4   Experimental Results

The research on the developed method concerns the ability of the neural network to learn to recognize specific letters. The neural networks are trained with the model of isolated written language characters.

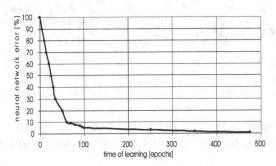

Fig. 6. The error rate of the neural network for recognition of isolated handwritten characters

The ability of the neural network to learn to recognize specific letters depends on the number of learning epochs. The specified time of learning enables the network to minimize the error so that it can work more efficiently. Based on the research, the error rate achieved is as shown in Fig. 6.

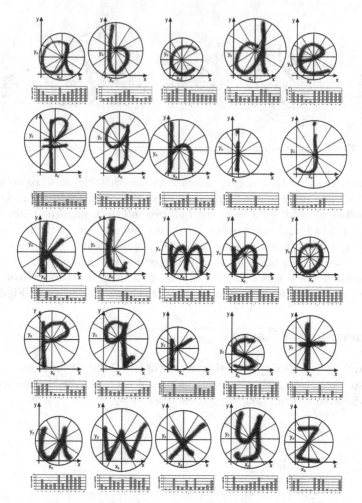

Fig. 7. Geometrical analysis and pattern description of isolated characters

Error rate is about 20% at learning time equals 50 epochs and 5% at 100 epochs. The error rate dropped by about 90% after training with 60 series of all patterns.

Several geometrical analyses of isolated characters and their pattern description were realized, which made it possible to draw significant conclusions (Fig. 7) and apply them in the proposed algorithms.

5   Conclusions and Perspectives

Many advances and changes have occurred in the field of automated written language recognition, over the last decade. The different sources of variability in psychophysical aspects of the generation and perception of written language make handwriting processing difficult.

Considerable progress has been made in natural writing recognition technology. Written language recognition systems have been limited to small and medium vocabulary applications, since most of them often rely on a lexicon during the recognition process. The capability of dealing with large lexicons, however, opens up many more applications.

The advantages of this new method of natural writing recognition are flexibility with regards to writing style, geometrical analysis enabling font independent character recognition, possibility of application of other types of neural networks, extension of the range of geometrical analysis and other possibilities for further development.

References

1. Artieres, T., Gauthier, N., Gallinari, P., Dorizzi, B.: A Hidden Markov Models combination framework for handwriting recognition. International Journal on Document Analysis and Recognition 5(4), 233–243 (2003)
2. Besner, D., Humphreys, G.W.: Basic Processes in Reading: Visual Word Recognition. Lawrence Erlbaum Associates, Hillsdale (1991)
3. Bishop, C.M.: Neural Networks for Pattern Recognition. Oxford University Press Inc., New York (2004)
4. Dori, D., Bruckstein, A.: Shape, Structure and Pattern Recognition. World Scientific Publishing Co., New Jersey (1995)
5. Kacalak, W., Majewski, M.: Intelligent System for Automatic Recognition and Evaluation of Speech Commands. In: King, I., Wang, J., Chan, L.-W., Wang, D. (eds.) ICONIP 2006. LNCS, vol. 4232, pp. 298–305. Springer, Heidelberg (2006)
6. Kacalak, W., Majewski, M.: E-Learning Systems with Artificial Intelligence in Engineering. In: Huang, D.-S., Jo, K.-H., Lee, H.-H., Kang, H.-J., Bevilacqua, V. (eds.) ICIC 2009. LNCS, vol. 5754, pp. 918–927. Springer, Heidelberg (2009)
7. Majewski, M., Kacalak, W.: Intelligent System for Natural Language Processing. In: Huang, D.-S., Li, K., Irwin, G.W. (eds.) ICIC 2006. LNCS (LNAI), vol. 4114, pp. 742–747. Springer, Heidelberg (2006)
8. Kacalak, W., Majewski, M., Zurada, J.M.: Intelligent E-Learning Systems for Evaluation of User's Knowledge and Skills with Efficient Information Processing. In: Rutkowski, L., Scherer, R., Tadeusiewicz, R., Zadeh, L.A., Zurada, J.M. (eds.) ICAISC 2010, Part II. LNCS, vol. 6114, pp. 508–515. Springer, Heidelberg (2010)
9. Majewski, M., Zurada, J.M.: Sentence recognition using artificial neural networks. Elsevier Knowledge-Based Systems 21(7), 629–635 (2008)
10. Mori, S., Nishida, H., Yamada, H.: Optical Character Recognition. John Wiley & Sons, Inc., New York (1999)
11. Nishida, H.: An Approach to Integration of Off-Line and On-Line Recognition of Handwriting. Pattern Recognition Letters 16(11), 1213–1219 (1995)
12. Zhou, J., Krzyzak, A., Suen, C.Y.: Verification-a Method of Enhancing the Recognizers of Isolated and Touching Handwritten Numerals. Pattern Recognition 35(5), 1179–1189 (2002)

The Use of ASM Feature Extraction and Machine Learning for the Discrimination of Members of the Fish Ectoparasite Genus Gyrodactylus

Rozniza Ali[1], Amir Hussain[1], James E. Bron[2], and Andrew P. Shinn[2]

[1] Institute of Computing Science and Mathematics, University of Stirling, UK
{ali,ahu}@cs.stir.ac.uk
[2] Institute of Aquaculture, University of Stirling, UK
{jeb1,aps1}@stir.ac.uk

Abstract. Active Shape Models (ASM) are applied to the attachment hooks of several species of *Gyrodactylus*, including the notifiable pathogen *G. salaris*, to classify each species to their true species type. ASM is used as a feature extraction tool to select information from hook images that can be used as input data into trained classifiers. Linear (*i.e.* LDA and KNN) and non-linear (*i.e.* MLP and SVM) models are used to classify *Gyrodactylus* species. Species of *Gyrodactylus*, ectoparasitic monogenetic flukes of fish, are difficult to discriminate and identify on morphology alone and their speciation currently requires taxonomic expertise. The current exercise sets out to confidently classify species, which in this example includes a species which is notifiable pathogen of Atlantic salmon, to their true class with a high degree of accuracy. The findings from the current exercise demonstrates that data subsequently imported into a K-NN classifier, outperforms several other methods of classification (*i.e.* LDA, MLP and SVM) that were assessed, with an average classification accuracy of 98.75%.

Keywords: Attachment hooks, image processing, SEM, parasite, machine learning classifier.

1 Introduction

There are over 440 described species of *Gyrodactylus* which are typically small (<1mm), ectoparasitic monogenetic flukes of fish (REF). Most species are imperfectly known, with many descriptions limited to an incomplete morphological description of their attachment hooks. Whilst molecular techniques have, in recent years, made a vast contribution to the discrimination of one species from another, species definitions often continue to rely on morphological characteristics (*i.e.* attachment hook morphology and in particular the shape of the sickle of the 16 small peripheral marginal hooks which are regarded as the key taxonomic feature) (REF). While most species of *Gyrodactylus* are non-pathogenic, causing little harm to their hosts, other species like *Gyrodactylus salaris* Malmberg, 1957, which is an OIE (Office International des Epizooties) - listed pathogen of Atlantic salmon, has led to a catastrophic decimation in the size of the juvenile salmon population in over 40 Norwegian rivers (REF). Uncontrolled increases in

T. Huang et al. (Eds.): ICONIP 2012, Part IV, LNCS 7666, pp. 256–263, 2012.

the size of the parasite population on resident salmon populations have necessitated extreme measures such as the use of the biocide rotenone to kill-out entire river systems, to remove the entire fish population within a river and the parasite (REF). Given the impact that *G. salaris* has had in Norway and elsewhere in Scandinavia (REF), many European states including the UK now have mandatory surveillance programmes screening wild salmonid populations (*i.e.* brown trout, charr, grayling, Atlantic salmon etc) for the presence of notifiable pathogens including *G. salaris*. Current OIE methodologies for the identification of *G. salaris* from other species of *Gyrodactylus* that occur on salmonids require confirmation from both morphological and molecular approaches, which can be time consuming. If *G. salaris* specimens, however, are overlooked in a diagnostic sample or misclassified (*i.e.* type I error - where *G. salaris* is misidentified as another species and *G. salaris* goes undetected resulting in the death of fish, or, type II error where another non-pathogenic species is misclassified as *G. salaris* with the result that a population of fish is treated unnecessarily), the environmental and economic implications can be severe (REF). For this reason and because of the widely varying pathogenicity seen between closely related species, accurate pathogen identification is of paramount importance. The discrimination of species from their congeners, however, is compounded by a limited number of morphological discrete characteristics which makes identification difficult. The task of morphological identification is, therefore, currently heavily reliant upon a limited number of domain experts available to analyse and determine species groups. This time can be dramatically reduced if the initial identification of *G. salaris* or *G. salaris*-like specimens by the morphology step can be improved and accelerated. In the event of a suspected outbreak, the demand for identification may significant exceed the available supply of suitable expertise and facilities. There is, therefore, a real need for the development of rapid, accurate, semi-automatic / automatic diagnostic tools that are able to confidently identify *G. salaris* in any population of specimens.

The aims of the current study were to explore the potential use of an Active Shape Model (ASM) to extract feature information from the attachment hooks of each species of *Gyrodactylus*. Given the small size of the marginal hook sickles (*i.e.* <7m), which are regarded as the most taxonomically informative morphological structure, this study will begin with an assessment of scanning electron microscope (SEM) images which give the best quality images. Given the subtle differences in the hook shape of each species, it is hoped that this approach moves towards the rapid automated classification of species with improved rates of correct classification over existing methods and negates the current laborious process of taking manual measurements which are used to assist experts in identifying species.

2   Attachment Hook Extraction

To improve the correct identification of *G. salaris*, a number of morphometric techniques based on statistical classification techniques [10], [14], [17] and molecular techniques [6], [7], [15], [9] have been developed to classify this pathogen from its close relatives on salmonid hosts. Whilst expert taxonomists may be able to classify *G. salaris* from other closely related species, morphometric speciation and molecular characterisation of the *Gyrodactylus* species is frustrating and difficult for the reasons described earlier. Computerised recognising species group from the hook features makes the species recognition process more accurate and effective. For this purpose, the feature extraction based on ASM is explored. Example of the potential features are length, width, shape, angle and etc. In the manual measurement of features, these tasks heavily depend on the concentration of the person who did the measurement, otherwise, incorrectly prediction will happened. And of course, the time taken to finish up the process are longer than expected. With the computer processing techniques, these process are possible to be done efficiently and effectively.

Basically, the extraction of object feature points plays an important role in many applications, such as face recognition [4], cancer detection [3], leaf species recognition [8] and segmentation protozoan parasites from microscope images [11]. The objective of *Gyrodactylus* image analysis is to extract the intended information from SEM hook images as a human would.

Feature extraction is the key to both face segmentation and recognition, as it is to any pattern classification task. For this purpose, the ASM have been explored and used to extract the feature information on *Gyrodactylus* hook species. This technique is selected because of the highest rate successful in medical images; which are using more or less similar type of images. Similar approach, which is also consider texture information is Active Appearance Model (AAM). This approach has been applied in fish species identification [16]. Inspiration from these worked, ASM is chosen to be applied in *Gyrodactylus* species image recognition.

ASM offer a lots of benefits, especially in shape recognition. This approach offers the users to landmark the image areas for every given images. In such way, it provides the patterns represent the varieties of shape images. More train images have been landmarked, more pattern available in fitting the testing images. This technique is chosen because it has ability in analysing the SEM images of *Gyrodactylus* species that accurately locate the contour of the specimen hook. The mistake in measurement of these contour, may result inaccuracy of species classification. Many applications have proven successful in applying this method, such as lung segmentation [12], [1]. The ASM approach is decided to apply due to the effectiveness in selection and extraction the features information from the SEM images.

2.1 ASM Construction

The ASM has been initially developed for medical image recognition by analysing the landmark points of the x-ray film. Such landmark points can be acquired by applying a sample template to the problem area. This strategy is better than the edge detection approach [13] because, any noises or unwanted object in an image can be ignored in selection of the shape contour Fig. 1.

The shape of each attachment hook images is presented by a vector consisting of the positions of the landmarks, $D = (d_1, e_1, ..., d_n, e_n)$, where $(d_i e_i)$ denotes the 2D image coordinate of the ith landmark points. All the shape vectors of hooks are normalised into a common coordinate system. The Procrustes Analysis is implemented in aligning the training set. This aligns each shape so that the sum of distances of each shape to the mean $F = \sum |D_i - \bar{D}|^2$ is minimised. For this purpose, choose one example as an initial estimate of the mean shape and scale so that $|\bar{D}| = 1$, which minimises the F.

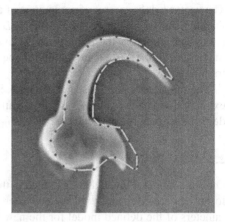

Fig. 1. The landmark point defined during the ASM construction

Fig. 2. To fit the ASM model to the new image

Assuming s sets of points D_i which are aligned into a common co-ordinate frame, if this distribution can be modelled, then new examples can be generated similar to those in the original training set s, and wand then these new shapes can be examined to decide whether they are reasonable example. In particular, a parameter model of the form $D = M(b)$, where b is a vector of parameters of the model. Such a model can be used to generate new vectors, D. If we can model the distribution parameters, $p(b)$, we can limit them so that the generated D's are similar to those in the training set. Similarly it should be possible to estimate $p(D)$ using the model.

To simply the problem, the Principle Component Analysis (PCA) is applied to the data. PCA is applied to reduce the dimensionality of the data to something more manageable. PCA computes the main axes of points, allowing one to approximate any of the original points using a model. The approach is as follows:

1. Compute the mean shape of the data,

$$\bar{D} = \frac{1}{s} \sum_{i=1}^{s} D_i \tag{1}$$

2. Compute the variance of the data,

$$S = \frac{1}{s-1} \sum_{i=1}^{s} (D_i - \bar{D}(D_i - \bar{D})^T) \tag{2}$$

3. Compute the eigenvectors, ϕ and corresponding eigenvalues λ_i of S. If ϕ contains the t eigenvectors corresponding to the largest eigenvalues, then it can approximate any of the training set, using:

$$D \approx \bar{D} + \phi b \tag{3}$$

where $\phi = (\phi_1|\phi_2|...|\phi_t)$ and b is a t dimensional vector given by:

$$b = \phi^T (D - \bar{D}) \tag{4}$$

The vector b defines a set of parameters of a different kind of shape model. By varying the elements of b, it can vary the shape D using Equation 3. The variance of the ith parameter b_i, across the training set is given by λ_i. By applying limits of $\pm 3\sqrt{\lambda i}$ to the parameter b_i, it is ensure that the shape generated is similar to those in the original training set.

The model was built based on 68 hook images, each with 45 points. While, for the extraction of shape features, only 22 parameters are identified the valuable features in describing about the particular *Gyrodactylus* hooks.

2.2   ASM Fitting

Once the model is created as described in the ASM construction section, it is important to fit the defined model with the new input images. It will find the most accurate parameters of the defined model for the new hook images. The ASM fitting try the best to locate the defined model parameter to a given new image. Cootes *et al.* [5] explained that an adjusting each model parameter from the correct values (defined model) will create the extraction pattern of the SEM hook images. During the model fitting, it measures the new coming images and uses this model to correct the values of current parameters, leading to a better fit. In ASM, such information indicates the shape of the model. Fig. 2 shows the iterative ASM fitting process between new input image with defined model. While the Fig. 3 shows the variation of shapes modelled train using ASM method.

Fig. 3. Variations in the some of principle components to generate shape variations

3   Experimental Results

Specimens of *Gyrodactylus* (*G. derjavinoides* n = 25, *G. salaris* n = 34, and *G. truttae* n = 9) were removed from their respective fish hosts and fixed in 80% ethanol until required. Individual specimens were subsequently rinsed in distilled water, transferred to a glass slide, had their posterior attachment organ excised with a scalpel and the attachment hooks released using a proteinase-K based digestion fluid (*i.e.* 100 μg/ml proteinase K (Cat. No. 4031-1, Clontech UK Ltd., Basingstoke, UK), 75 mM Tris-HCl, pH 8, 10 mM EDTA, 5% SDS). The digestion process was stopped through the addition of 3 μl of a 50:50 formaldehyde:glycerine solution. A coverslip was added to the preparation, which was sealed using a commercial nail varnish. An image of the attachment hooks from each specimen were captured using an AxioCam MRC (Zeiss) 1.5 megapixel camera fitted with a MicroCam Olympus LB Neoplan D-V C mount 0.75× interfacing lens attached to an Olympus BX51 compound microscope. The specimens were viewed under 100× oil immersion objective using MRGrab v. 1.0.0.0.4 (Carl Zeiss Vision GmbH, Munchen, Germany) software.

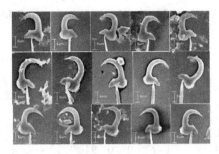

Fig. 4. Samples from a training set (SEM images of attachment hook)

Comparing to the manual measurement of hook, there are three parts of hook images were used, namely are hamuli, ventral bar and marginal hook. But, for this image processing algorithm, only marginal hook is used in extracting the feature vectors. Fig. 4 shows some sample training images used in this experiment. Due to the inconsistency and unstandardised format of images, the pre-processing images are needed to be applied. Scaling and rotation of images are applied to reduce the processing time and complexity during the training and construction the ASM model of *Gyrodactylus* image attachment hooks.

Classification accuracy is measured and compared among a number of different classifies. For this purpose, linear and non-linear models are applied and evaluated according to the accuracy in predicting the correct *Gyrodactylus* species. To avoid overfitting, for each of classifier, a 10-fold cross validation strategy is applied. Using the ASM approach, we manage to define 45 landmark points and 22 features have been extracted. From the results tabulated on the following Table 1, we find that the K-NN classifier outperform others compared classifiers, scoring an accuracy of 98.75%. Beside accuracy performance of classifiers, the confusion matrix is also considered, in order to provide a quantitative performance representation for each classifier in terms of species recognition.

Referring to the following Table 2, using the LDA classifier extracted 22 features, all specimens are correctly classified to their true species class except for *G. derjavinoides* (d). By using the K-NN classifier (Table 3), there is an improvement in the classification of *G. derjavinoides* (d) specimens over the method using LDA. Unfortunately, the *G. truttae* (t) had totally been misclassified as *G. salaris* (s). Two species are correctly allocated to the true class while only *G. truttae* (t) is misclassified. While for MLP classifier (Table 4), only *G. derjavinoides* (d) manage to get full classification of 25 specimens. Using SVM classifier (Table 5) has achieved good classification rate, even though not all the species achieved full classification. Only one for each species has been misclassified to other species group. Inspired from this result, work is ongoing to investigate the feasibility of the extraction method and to combine multiple classifier and feature selection technique to improve classification accuracy of each true species.

Now we compare the results of our proposed model with manual measurement of features. In [2], an accuracy of 92.59% using Linear Discriminant Analysis (LDA) was reported with 557 data points, but it relies on 25 points manual location of the three different part of *Gyrodactylus* attachment hooks. In our newly proposed extraction method,

Table 1. The average classification rate of *Gyrodactylus* species extracted features using ASM approach

Classifier	Accuracy (%)
LDA	85.71 ±7.59
KNN	**98.53** ±3.95
MLP	95.59 ±4.81
SVM	95.89 ±6.70

Table 2. A confusion matrix of the ASM extracted features implemented to the LDA classifier

	d	*s*	*t*	sum
d	**25**	0	0	25
s	10	**24**	0	34
t	0	0	**9**	9
sum	35	24	9	68

Table 3. A confusion matrix of the ASM extracted features implemented to the K-NN classifier

	d	*s*	*t*	sum
d	**25**	0	0	25
s	0	**34**	0	34
t	0	1	**8**	9
sum	25	35	8	68

Table 4. A confusion matrix of the ASM extracted features implemented to the MLP classifier

	d	*s*	*t*	sum
d	**25**	0	0	25
s	1	**32**	1	34
t	0	1	**8**	9
sum	26	33	9	68

Table 5. A confusion matrix of the ASM extracted features implemented to the SVM classifier

	d	*s*	*t*	sum
d	**24**	1	0	25
s	1	**33**	0	34
t	0	1	**8**	9
sum	25	35	8	68

the recognition procedure is more efficient because only one part of attachment hook was used and the results outperform other state-of-the-art methods in literature.

4 Conclusions

In this paper, we suggested novel effective approach of hook attachment features extraction for *Gyrodactylus* species recognition. ASM was tested on a dataset of 68 images and this technique have shown to overcome the limitation and difficulties in extracting the feature information. Different of classifiers were compared to find the best model for classification the *Gyrodactylus* image species. For future work, further large-scale investigations using larger dataset will be carried out and different modelling techniques will be applied (including model based on ensemble classifiers, which have shown promising results) with a view to providing a reliable model that could be deployed and re-used in classification of fish pathogen attachment hook, especially, *Gyrodactylus* species.

Acknowledgment. This research project was funded by the Ministry of Higher Education, Malaysia and the Universiti Malaysia Terengganu (UMT), which are greatly acknowledged.

References

1. Algabary, K.M., Omar, K.: A comparative study of face recognition using improved AAM, PCA and ICA via Feret Date Base. European Journal of Scientific Research (2010)
2. Ali, R., Hussain, A., Bron, J.E., Shinn, A.P.: Multi-stage classification of *Gyrodactylus* species using machine learning and feature selection techniques. In: Int. Conf. on Intelligent Systems Design and Applications (ISDA), pp. 457–462 (2011)
3. Blackledge, J.M., Dubovitskiy, A.: Object detection and classification with applications to skin cancer screening. ISAST Transactions on Intelligent Systems 1(1), 34–45 (2008)
4. Choi, J., Chung, Y., Kim, K., Yoo, J.: Face recognition using energy probability in DCT domain, pp. 1549–1552. IEEE (2006)
5. Cootes, T.F., Edwards, G.J., Taylor, C.J.: Active appearance models. IEEE Transactions on Pattern Analysis and Machine Intelligence 23(6), 681–685 (2001)
6. Cunningham, C.O., McGillivray, D.M., MacKenzie, K., Melvin, W.T.: Discrimination between *Gyrodactylus salaris*, *G. derjavini* and *G. truttae* (Platyhelminthes: Monogenea) using restriction fragment length polymorphisms and an oligonucleotide probe within the small subunit ribosomal RNA gene. Parasitology 111, 87–94 (1995a)
7. Cunningham, C., McGillivray, D., MacKenzie, K., Melvin, W.: Identification of *Gyrodactylus* (monogenea) species parasitizing salmonid fish using DNA probes. Journal of Fish Diseases 18, 539–544 (1995b)
8. Du, J.X., Wang, X.F., Zhang, G.J.: Leaf shape based plant species recognition. Applied Mathematics and Computation 185(2), 883–893 (2007)
9. Hansen, H., Bachmann, L., Bakke, T.A.: Mitochondrial DNA variation of *Gyrodactylus* spp (Monogenea, Gyrodactylidae) populations infecting Atlantic salmon, grayling, and rainbow trout in Norway and Sweden. International Journal for Parasitology 33, 1471–1478 (2003)
10. Kay, J.W., Shinn, A.P., Sommerville, C.: Towards an automated system for the identification of notifiable pathogens: using *Gyrodactylus salaris* as an example. Parasitology Today 15(5), 201–203 (1999)
11. Lai, C.H., Yu, S.S., Tseng, H.Y., Tsai, M.H.: A protozoan parasite extraction scheme for digital microscope images. Computerized Medical Imaging and Graphics 34, 122–130 (2010)
12. Lee, J.S., Wu, H.H., Yuan, M.Z.: Lung segmentation for chest radiograph by using adaptive active shape models. In: Int. Conf. on Information Assurance and Security, pp. 383–386 (2009)
13. Maini, R., Aggarwal, H.: Study and comparison of various image edge detection techniques. International Journal of Image Processing (IJIP) 3(1), 1–60 (2009)
14. McHugh, S.E., Shinn, A.P., Kay, J.W.: Discrimination of *G. salaris* and *G. thymalli* using statistical classifiers applied to morphometric data. Parasitology 121, 315–323 (2000)
15. Meinilä, M., Kuusela, J., Zietara, M., Lumme, J.: Brief report: Primers for amplifying 820 bp of highly polymorphic mitochondrial COI gene of *Gyrodactylus salaris*. Hereditas 137, 72–74 (2002)
16. Quivy, C.H., Kumazawa, I.: Normalization of active appearance model for fish species identification. International Scholarly Research Network, ISRN Signal Processing (2011)
17. Shinn, A.P., Kay, J.W., Sommerville, C.: The use of statistical classifier for the discrimination of species of the genus *Gyrodactylus* (Monogenea) parasitizing salmonids. Parasitology 120, 261–269 (2000)

DBNs-BLR (MCMC) -GAs-KNN: A Novel Framework of Hybrid System for Thalassemia Expert System

Patcharaporn Paokanta

Chiang Mai University, Thailand
Patcha535@gmail.com

Abstract. Genetic Algorithms (GAs) is one of the most effective technique applied to feature selection in medical diagnostic decisions. In particular, Thalassemia, which is one of the most common genetic disorders found around the world. The main problems of diagnosing this disease are the complex processes for identifying the several types of Thalassemia. Moreover, diagnostic methods are slow and rely on expert knowledge and experience as well as expensive equipment. For these reasons, in this study, a new framework of applied DBN and BLR (MCMC)-GAs-KNN for Thalassemia Expert System is proposed. The filter techniques called DBNs and the hybrid classification technique namely BLR (MCMC)-GAs-KNN will be used for classifying the types of β-Thalassemia. The obtained result will be compared to the results of other techniques such as BNs, BLR based on Classical (ML) and Bayesian (MCMC) approach, SVM, MLP, KNN, C5.0, and CART for selecting the best results to implement Thalassemia Expert System.

Keywords: Hybrid System, Diagnostic Bayesian Networks (DBNs), Binomial Logistic Regression (BLR), Genetic Algorithms (GAs), Bayesian approach (MCMC), KNN, β-Thalassemia, Thalassemia Expert System (TES).

1 Introduction

Genetic Algorithms (GAs) play an important role in classification task especially in the feature selection process which this process will be used before classifying [1-4]. These techniques well known as a method for selecting varibles stored in the multi-dimensional data sets. In the past decades, various techniques have been used for feature selection, among these algorithms; GAs is a popular method for obtimization purpose and Machine Learning algorithms based on biological evoulation process. It is a powerful tool in particular, when the dimentions of the original feature set are large. Reducing the dimentions of the feature space not only improves the quality of data by obtained satisfied classification performance (high accuracy percentage) but also reduces the computational complexity. In classification task, it can increase estimated performance of the classifiers with the selecting a number of appropriate features used. For the hybrid feature selection approach, DBNs-BLR (MCMC)-GAs-KNN is the combination of optimization and classification techniques which the goal of DBNs, BLR and GAs is to select the appropriate feature subset, and the purpose of

T. Huang et al. (Eds.): ICONIP 2012, Part IV, LNCS 7666, pp. 264–271, 2012.

using KNN is the classification task. One of the interesting research topics, the applied soft computing for discovering the medical knowledge to support the decision making has grown dramatically over the past decades, especially, the studying of cancer, diabetes, and heart disease. Diagnosing these diseases require the knowledge from some experts to diagnosis and give the appropriate treatment methods to patients.

Nowadays, one of the most common genetic disorders in the world is Thalassemia syndrome. In Thailand, there are several types of Thalassemia, including Thalassemia minor, Thalassemia intermedia and Thalassemia major. It is estimated 1.5% of the worldwide population is diagnosed with a minor Thalassemia called β-Thalassemia. This disorder is common in areas where malaria was once prevalent, such as Africa, the Mediterranean region, the Middle East, Southeast Asia (India, Thailand and Indonesia), and the Far East. Southeast Asia accounts for approximately 50% of worldwide carriers, while European and American countries account for 10–15%. Thalassemia has the highest prevalence in Southeast Asia, where approximately 55 million people are carriers. The gene frequency of alpha-Thalassemia reaches 30-40% in Northern Thailand and Lao PDR, while β-Thalassemia varies between 1- 9%, and HbE, which is one type of minor Thalassemia has a frequency of 50-60% at the junction between Thailand, and Lao PDR. This high magnitude in the border regions poses public health problems in Thailand. Approximately 40% of Thai people are heterozygous carriers of these genes [5]. In Thailand, Thalassemia is currently diagnosed via a two or three step processes, depending on the suspected variant of disease. These methods however, are slow and rely on expert knowledge and experience as well as expensive equipment. The number of genetic variations of Thalassemia in northern Thailand is large and therefore also poses particular challenges in diagnosis [6].

For these reasons, the purposes of this study are to select the appropriate feature for screening Thalassemia using DBN-BLR (MCMC)-GAs and to classify the types of Thalassemia by using KNN. The paper is organized as follows; after the introductory section, BNs, GAs and KNN will be demonstrated in theoretical terms, next the materials and methodology for selecting and classifying the types of Thalassemia will be explained. This includes: Procedure of using GAs and KNN. The results of each process and the discussions are illustrated in the fifth section. Finally, a novel framework of applied DBNs-BLR (MCMC)-GAs-KNN for Thalassemia Expert System (TES) is discussed.

2   Beyesian Networks (BNs)

Bayesian Networks (BNs), a directed acyclic graphs model (DAGs), is a well known adaptive idea from proability theory called Bayes's theorem proposed by Tomas Bayes. BNs are the graphical models that have a Joint Probability Distribution which it can identify the relationship between nodes. Moreover, each node represents the Conditional Probability Distributions (CPD) that belong to their parent nodes, and these prior probabilities can be used to calculate the posterior probability of each interested event. BNs are composed of

1) Nodes that are a set of random variables.
   e.g.

2) Directed links.
   e.g.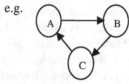

3) Directed acyclic graphs (DAG).
   e.g.

A table of CP contains a Joint Probability Distribution. In the case of a directed model, users must specify CPD for each node. If the variables are discrete, this can be represented as a table, which lists the probability that the child node takes on each of its different values for each combination of values of its parents. The number of possible events that can occur is 2^{nodes} [7].

3   Genetic Algorithms and K-Nearest Neighbor

GAs is one of optimization technique which is used to search an optimal binary vector where bit set represents the value of variable. This technique provides the linear and nonlinear transformation. The concept of GAs for filtering task is to select the related variable for classifying purpose by making the decision if the jth bit of this vector is 1, the jth record is appropriate for classifying. On the other hand, if the jth bit of this vector is 0, this record does not appropriate for classifying [8]. Moreover, it defines a set of weight vectors W^*, where the number of wight vectors equals to the number of the data pattern $\overline{X}$ for each data sample. Obtained W^* from using GAs is multiplied by every sample's data pattern vector $\overline{X}$, resulting a new feature vector   for the input data [9].

$$\overline{Y} = \overline{X} * \overline{W^*} \tag{1}$$

The GAs processes consist of three main steps, including selecting the population, chromosomal crossover and gene mutation process. These processes are the iterative processes which the random chromosomes were calculated and compared using fitness function to select the optimized chromosomes to the population side.

One of the several Neural Network Algorithms, K-Nearest Neighbor (KNN), is the popular classification technique and a non-parametric pattern recognition method to appraise the discriminative ability of mathematical model generated from the data set. The procedure of KNN are listed below [10],

- To calculate the value of the pair-wise sample using Euclidean distances or the other measures such as Euclidean Squared, Cityblock and Chebyshev, for all pairs of samples in the relation to each value of variables.

- To fine KNN (k=i) for each sample.
- To assign class membership ($I_1=C_2$, $I_2=C_2$, ..., $I_n=C_n$) based on Neighbors.

4   Materials and Methodology

Materials and methodology of applying DBN-BLR (MCMC)-GAs-KNN for implementing TES will be illustrated below.

4.1   Materials

Primary resources for implementing TES are Thalassemia knowledge including diagnostic processes, Thalassemia indicators, which these knowledges gathered from experts (medical practitioner, Biochemist) using the Diagnosis template of Common-KADS suite which is a technique of KE. These knowledges used to identify collected variables for classifying the genotypes of patients who have Thalassemia genes but are not usually expressed at the clinical level. In this paper, because of the prevalence of several types of Thalassemia in northern Thailand, the data of 351 Thalassemia patients were collected from the out-patient records in a hospital in northern Thailand. Among several types of Thalassemia, β-Thalassemia is a crucial type that can be found around the world. For this study, 127 β-Thalassemia used to classify subtypes of Thalassemia. The data set for this experiment are 4 related indicators including genotype of children (Nominal scale and output), F-cell of children (Ordinal scale and input), Hb A_2 of children (Interval scale and input) from 12 Thalassemia indicators which were filtered using Pearson's Chi-square.

4.2   Methodology

The procedure of using DBN-BLR(MCMC)-GAs-KNN to implement TES starts at the first step, Thalassemia experts were interviewed to elicit Thalassemia knowledge such as screening processes and types of Thalassemia using CommonKADs Templates. In addition, documents of Thalassemia screening processes and types of Thalassemia were reviewed to develop this system. In the second step, variables were defined and collected from some experts (biochemists and medical practitioners), documents (papers and articles of the Thalassemia foundation of Thailand), and Out-Patient Department records from a hospital in northern Thailand). These variables were elicited through a Diagnosis template of CommonKads model suite. In the third step, variables of obtained data from the previous step were filtered using feature selection technique called Pearson's Chi-square. According to the obtained results of using Pearson's Chi-square, the reasoning matrices and DBNs for screening β-thalassemia were constructed to represent Thalassemia knowledge (the related β-Thalassemia indicators) in form of graphical model. The fifth step, not only the functions of related factors were created but also integer data sets were transformed to binary data sets. These processes were prepared to construct BLR models which the obtained parameter estimation generated based on Classical and Bayesian approaches.

Next, the binary data sets were calculated using BLR model (fitness functions) of GAs to filter the appropriate variables and then these results were classified to find subtypes of Thalassemia by KNN. The seventh step, all obtained results were used to design the achitectural and detailed design such as user interface, and components for implementing the TES. Moreover, this developed system was tested and fixed in unit testing, integration testing, system testing, acceptance testing and usability testing by stakeholders (users, programers, etc.). Finally, this sytem was delivered to users for supporting the decision making on the Thalassemia diagnosis.

5   Results

The construction processes of TES start from the knowledge elicitation processes, data collection processes, filter processes, classification processes, requirement elicitation processes, requirement analysis processes, system and funtional design processes, testing processes, and delivery and mantanance processes respectively. Due to these processes, the results of the first step are the thalassemia knowledge such as thalassemia indicators which are 12 variables, including genotype, F-cell, HbA_2 and inclusion body obtained from children who have thalassemia genes and their parents [6]. Moreover, the knowledge of thalassemia diagnostic processes was elicited from experts (medical practitioner, biochemist and nurse) using CommonKADs model suite for implementing this system. Then Pearson Chi-square was used to find the relationships of 12 collected variables which the   result presents that there are 4 related variables of β-Thalassemia. These factors were represented as Reasoning

Table 1. Resoning Matrices of β-Thalassemia

Types of Thalassemia	Thalassemia indicators												
	f1	f2	f3	f4	f5	f6	f7	f8	f9	f10	f11	f12	f13
F1	0	0	0	0	0	0	1	0	0	0	1	0	0
F2	1	1	1	1	1	1	1	1	1	1	1	1	1
F3	0	0	1	0	0	1	0	1	0	0	0	0	0
F4	1	0	0	0	0	1	1	0	0	0	0	0	0
F5	0	0	0	1	0	1	1	0	0	1	1	1	0

Table 2. Functions for DBN-BLR(MCMC)-GAs-KNN

Factors	Functions	Binary data sets
F1	(f7,f11)	00000010000100
F2	(f1,f2,f3,f4,f5,f6,f7,f8,f9,f10,f11,f12,f13)	11111101011110
F3	(f3,f6,f8)	00100101000000
F4	(f1,f6,f7)	10000110000000

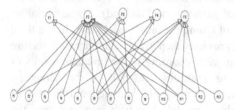

Fig. 1. DBNs for Thalassemia Expert System [7]

Table 3. Classification Results of DBN-BLR (MCMC)-GAs-KNN

Class	Correct items	Incorrect items	Accuracy Percentage
1	103	6	81.1000
2	6	5	4.7200
3	3	2	2.3600
4	1	0	0.7900

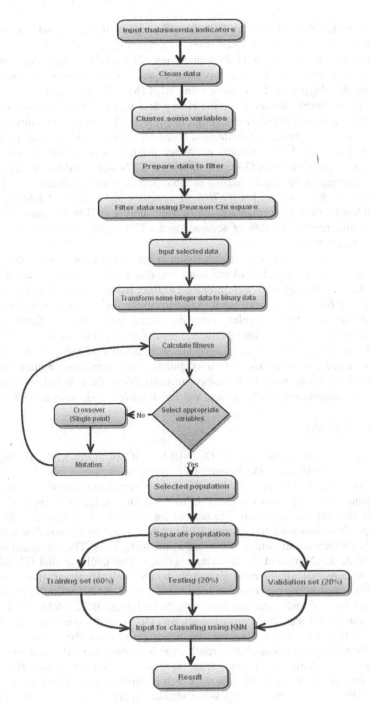

Fig. 2. A novel framework of applied DBN-BLR (MCMC)-GAs-KNN for constructing TES

matrices and DBNs for screening β-Thalassemia shown in Fig. 1. and table 1. DBNs presents a novel framework for screening β-thalassemia, including 12 related thalassemia sub-indicators and 4 β-Thalassemia types. These related variables were transformed from integer data set to binary data set and created in forms of functions for classifying β-subtypes of thalassemia using BLR (MCMC)-GAs-KNN that shown in table 2. These functions used to define binary data sets which are randomly selected to generate the initial population through using GAs. The fitness functions are BLR models obtained by parameter estimation based on classical and are used to Bayesian approach. There are 5 BLR models which are the fitness functions for selecting the appropriate population using GAs, and afterward KNN were used to classify subtypes of β-thalassemia as the results shown in table 3. The obtained results of using BLR (MCMC)-GAs-KNN show that 89.76% of accuracy and 0.1024 of RMSE. For the highest classification performance result of each suptype of β-thalassemia is type 1 of β-thalassemia, reaches 81.10% of accuracy and 4.72%, 2.36%, 0.79% 0.79% for type 2, 3, 4, 5, respectively.

Fig. 2. starts in the first step, the thalassemia indicators were input to this system then these variables were cleaned and some variable were clustered to identify groups of data. Next, these variables were prepared to filter using Pearson Chi-Square. In the fifth step, the filtered data were input to transform the integer data to binary data and the randomly selected variables were calculated using fitness functions BLR (MCMC) models afterward the obtained results were compared to select the high calculated values. For the remained variables were crossoverd using single point crossover and these crossover data sets were mutated and then these mutated variables were calculated again using BLR fitness functions. Next, the selected population was separated to training set, testing set and validation set for classifying using KNN.

6   Conclusion

According to a new framework of DBN-BLR (MCMC)-GAs-KNN, the reasoning matrices was constructed using Pearson-Chi Square before the graphical model called DBNs was developed to present the related thalassemia indicators for screening β-thalassemia. This graph can represent thalassemia knowledge in terms of the cuase-effect relationships of thalassemia indicators for screening β-thalassemia. Moreover, next step DBNs was used to generate BLR models based on Classical and Bayesian approaches which are the fitness functions in the using GAs. The classification result of using GAs-KNN presents a satisfying accuracy. This confirms that GA-KNN can be used to construct Thalassemia Expert System as well. However, if compared the GAs-KNN result to the previous studied of using PCA, KNN, MLP, BLR and MLR based on Classical and Bayesian approach, NaiveBayes, BNs, SVM, C5.0, CART, and Reasoning Matrices, Polychromatic set, and DBNs on the same data set [5, 7, 11-14] this hybrid framework obtained the better result than the other techniques. Among these obtained classification results, the best result is BLR based on Classical and Bayesian statistics with more than 90% of accuracy in almost types of β-Thalassemia due to the types of data set which is appropriate to using this technique. On the law of large number of Classical statistic approach, there are the limitation points about the source of parameter estimation and the population size which affects the obtained error. On the other hand Bayesian statistic approach can be applied as well for a small population size due to this approach allows user to use their experience about the characteristics of the distribution for estimating parameter through the prior distribution. In the future, Fuzzy approach and GAs [15] will be used to construct the TES to compare the results.

References

1. Elami, M.E.: A Filter Model for Feature Subset Selection Based on Genetic Algorithm. Journal of Knowledge-Based Systems 22, 356–362 (2009)
2. Yannis, M., Georgios, D., Jan, J.: Pap Smear Diagnosis Using a Hybrid Intelligent Scheme Focusing on Genetic Algorithm Based Feature Selection and Nearest Neighbor Classification. Journal of Computers in Biology and Medicine 39, 69–78 (2009)
3. Jin-Hyuk, H., Sung-Bae, C.: Efficient Huge-scale Feature Selection with Speciated Genetic Algorithm. Journal of Pattern Recognition Letters 27, 143–150 (2006)
4. Zexuan, Z., Yew, S., Manoranjan, D.: Markov Blanket-embedded Genetic Algorithm for Gene Selection. Pattern Recognition 40, 3236–3248 (2007)
5. Patcharaporn, P., Michele, C., Somdet, S.: The Effeciency of Data Types for Classification Performance of Machine Learning Techniques for Screening β-Thalassemia. In: 3rd International Symposium on Applied Sciences in Biomedical and Communication Technologies, pp. 1–4. IEEE Press, New York (2010)
6. Patcharaporn, P., Napat, H., Nopasit, C., Michelle, C., Somdet, S.: Parameter Estimation of Binomial Logistic Regression Based on Classical (Maximum Likelihood) and Bayesian (MCMC) Approach for Screening β-Thalassemia. International Journal of Intelligent Information Processing 3, 90–100 (2012)
7. Patcharaporn, P., Napat, H.: Risk Analysis of Thalassemia Using Knowledge Representation Model: Diagnostic Bayesian networks. In: International Conference on Biomedical and Health Informatics, pp. 155–158. IEEE Press, New York (2012)
8. Te-Sheng, L.: Feature Selection for Classification By Using A GA-Based Neural Network Approach. Journal of the Chinese Institute of Industrial Engineers 23, 55–64 (2006)
9. Sihua, P., Qianghua, X., Xuefeng, B.L., Xiaoning, P., Wei, D., Liangbiao, C.: Molecular Classification of Cancer Types from Microarray Data Using the Combination of Genetic Algorithms and Support Vector Machines. FEBS Letters 555, 358–362 (2003)
10. Wu, X., Kumar, V., Quinlan, J.R., Ghosh, J., Yang, Q., Motoda, H.: Top 10 Algorithms in Data Mining. Journal of Knowledge Information System 14, 1–37 (2008)
11. Patcharaporn, P., Napat, H., Nopasit, C.: The Classification Performance of Binomial Logistic Regression Based on Classical and Bayesian Statistics for Screening β-Thalassemia. In: 3rd International Conference on Data Mining and Intelligent Information Technology Applications, pp. 427–432 (2011)
12. Patcharaporn, P., Michele, C., Napat, H., Nopasit, C., Somdet, S.: Rule Induction for Screening Thalassemia Using Machine Learning Techniques: C5. 0 and CART. ICIC Express Letter 6, 301–306 (2012)
13. Patcharaporn, P., Michele, C., Napat, H., Somdet, S.: The Comparison of Classification Performance of Machine Learning Techniques Using Principal Components Analysis: PCA for Screening β-Thalassemia. In: 4th International Conference on Computer Science and Information Technology, pp. 316–319 (2011)
14. Patcharaporn, P.: Reasoning Matrices and Polychomatic Set for Screening Thalassemia. Journal of Medical Research and Science 2, 144–152 (2012)
15. Patcharaporn, P.: Knowledge and Data Engineering: Fuzzy C-Mean and Genetic Algorithms for Clustering?-Thalassemia of Knowledge Based Diagnosis Decision Support System. ICIC Express Letter (in Press, 2012)

Hybrid Approach for Diagnosing Thyroid, Hepatitis, and Breast Cancer Based on Correlation Based Feature Selection and Naïve Bayes

Mohammad Ashraf, Girija Chetty, Dat Tran, and Dharmendra Sharma

Faculty of Information Sciences and Engineering University of Canberra, ACT 2601, Australia
{mohammad.baniahmad,girija.chetty,
dat.tran,dharmendra.sharma}@canberra.edu.au

Abstract. Feature selection techniques have become an obvious need for researchers in computer science and many other fields of science. Whether the target research is in medicine, agriculture, business, or industry; the necessity for analysing large amount of data is needed. In Addition to that, finding the most excellent feature selection technique that best satisfies a certain learning algorithm could bring the benefit for researchers. Therefore, we proposed a new method for diagnosing some diseases based on a combination of learning algorithm tools and feature selection techniques. The idea is to obtain a hybrid approach that combines the best performing learning algorithms and the best performing feature selection techniques in regards to three well-known datasets. Experimental result shows that co-ordination between correlation based feature selection method along with Naive Bayse learning algorithm can produce promising results.

Keywords: Feature selection methods, Learning algorithms, Hybrid systems, Data mining, Breast cancer dataset, Thyroid, Hepatitis.

1 Introduction

Nowadays, we are capable to collect and generate data more than before. Contributing factors include the steady progress of computer hardware technology for storing data and the computerization of business, scientific, and government transactions. In addition, the use of the internet as a wide information system has flooded us with incredible amount of data and information. Data mining has attracted a big attention to information systems researchers in the recent years due to the wide availability of big amount of data and the need for tuning such data into knowledge and useful patterns. The gained knowledge and patterns can be used in many fields such as marketing, business analysis, and health information systems [1]. However, the quality of data of paramount importance, for this large amount of data, and implies the existence of low quality, unreliable, redundant and noisy artifacts and outliers, which affect the process of extracting knowledge and useful patterns, and then knowledge discovery during training is more difficult. Therefore, researchers have felt the necessity for producing more reliable data from large amount of records such as using feature selection

T. Huang et al. (Eds.): ICONIP 2012, Part IV, LNCS 7666, pp. 272–280, 2012.

methods. Feature selection or attribute subset combination is the process of identifying and utilizing the most relevant attributes and removing as many redundant and irrelevant attributes as possible [2] [3]. In addition, features selections mechanisms do not alter the original representation of data in any way. It just selects an optimal useful subset. Recently, the inspiration for applying features selection techniques in machine learning has shifted from theoretical approach to one of steps in model building. Many attribute selection methods use the task as a search problem, where each result in the search space groups a distinct subset of the possible attributes [4]. Since the space is exponential in the number of attributes which produce lots of possible subsets, this requires the use of a heuristic search procedure for all data sets. The search procedure is combined with an attribute utility estimator in order to evaluate the relative merit of alternative subsets of attributes [2]. This large number of possible subsets and the computation cost involved necessitate researchers to conduct a benchmark feature selection methods that produce the best possible subset in regards to more accurate results as well as low computation overhead.

Feature selection techniques could perform better if the researcher chooses the right learning algorithm. Therefore, we propose a new approach that combine a promising feature selection technique and one of well-knows learning algorithm. In the current work, we have focused on publicly available diseases datasets (Thyroid, Hepatitis, and Breast cancer) to evaluate the proposed approach. Yearly around the world, millions of ladies suffer from breast cancer, making it the second common non-skin cancer after lung cancer, and the fifth cause of death among cancer diseases in the world [5]. Thyroid disorder in women is much more common than thyroid problems in men and may lead to thyroid cancer [6]. Hepatitis can be caused by chemicals, drugs, drinking too much alcohol, or by different kinds of viruses and may lead to liver problems [7]. This paper begins with brief related work, then a description of benchmark feature selection methods, a description of our methodology in the current paper, and the results obtained by using the three datasets. Finally, a brief discussion and future work are presented.

2   Feature Selection Methods

2.1   Information Gain

The information gain method was proposed to approximate quality of each attribute using the entropy by estimating the difference between the prior entropy and the post entropy [8]. This is one of the simplest attribute ranking methods and is often used in text categorization. If x is an attribute and c is the class, the following equation gives the entropy of the class before observing the attribute:

$$H(x) = -\sum_x P(x)\log_2 P(x) \tag{1}$$

where $P(c)$ is the probability function of variable c. The conditional entropy of c given x (post entropy) is given by:

$$H(c \mid x) = -\sum_x P(x)\sum_C P(c \mid x)\log_2 P(c \mid x) \tag{2}$$

The information gain (the difference between prior entropy and postal entropy) is given by the following equations:

$$H(c,x) = H(c) - H(c \mid x) \tag{3}$$

$$H(c,x) = -\sum_c P(c) \log_2 P(c) - \sum_x \left(-P(x) \sum_c P(c \mid x) \log_2 P(c \mid x) \right) \tag{4}$$

2.2   Correlation Based Feature Selection (CFS)

CFS is a simple filter algorithm that ranks feature subsets and discover the merit of feature or subset of features according to a correlation based heuristic evaluation function. The purpose of CFS is to find subsets that contain features that are highly correlated with the class and uncorrelated with each other. The rest of features should be ignored. Redundant features should be excluded as they will be highly correlated with one or more of the remaining features. The acceptance of a feature will depend on the extent to which it predicts classes in areas of the instance space not already predicted by other features. CFS's feature subset evaluation function is shown as follows [9]

$$Merit_s = \frac{kr_{cf}}{\sqrt{K + (K+1)r_{ff}}} \tag{5}$$

where $Merit_s$ is the worth of feature subset s that contain k features, r_{cf} is the average feature correlation to the class, and r_{ff} is the average feature to feature correlation. In order to apply this equation to calculate approximately the correlation between features, CFS uses a modified information gain method called symmetrical uncertainty to compensate the information gain bias for attributes with more values as follows [10]

$$SU = \frac{H(x_i) + H(x_j) - H(x_i, x_j)}{H(x_i) + H(x_j)} \tag{6}$$

2.3   Relief

One of the most known feature selection techniques is Relief. Its aim is to measure the quality of attributes according to how their values distinguish instances of different classes. Relief uses instance based learning (lazy learning such as k Nearest Neighbour) to assign a grade to each feature. Each feature's grade reflects its ability to distinguish among the class values. Features are ranked by weight and those that exceed a threshold -determined by the user- are selected to form the promising subset. For each instance, the closest neighbour instance of the same class and the closest instance of a different class are selected. The score W_x of the x variable is computed as the average over all examples of magnitude of the difference between the distance to the nearest hit and the distance to the nearest miss as follows [11].

$$W_x = W_x - \frac{diff(x,r,h)^2}{m} + \frac{diff(x,r,h')^2}{m} \tag{7}$$

where W_x is the grade for the attribute x, r is a random sample instance, h is the nearest hit, h' is the nearest miss, and m is the number of samples.

2.4 Principle Components Analysis (PC)

The purpose of PC is to reduce the dimensionality of data set that contains a large number of correlated attributes by transforming the original attributes space to a new space in which attributes are uncorrelated. The algorithm then ranks the variation between the original dataset and the new one. Transformed attributes with most variations are kept; meanwhile discard the rest of attributes. It is also important to mention that PC is valid for unsupervised data sets because it does not take into account the class label [12].

2.5 Consistency Based Subset Evaluation (CB)

CB adopts the class consistency rate as the evaluation measure. The idea is to obtain a set of attributes that divide the original dataset into subsets that contain one class majority [2]. One of well-known consistency based feature selection is consistency metric [13] proposed by Liu and Setiono:

$$Consistency_s = 1 - \frac{\sum_{j=0}^{k} |D_j| - |M_j|}{N} \tag{8}$$

where s is feature subset, k is the number of features in s, $|D_j|$ is the number of occurrences of the jth attribute value combination, $|M_j|$ is the cardinality of the majority class for the jth attribute value, and N is the number of features in the original data set. For continuous values, we may use Chi2 [14]. Chi2 automatically discretises the continuous features values and removes irrelevant continuous attributes.

3 Experiment Methodology

Different sets of experiments were performed to evaluate the above mentioned attribute selection methods on well-known publicly available datasets from UCI machine learning repository [15]. We have chosen three datasets that of different size. The smallest dataset contains 155 attributes and the largest dataset contains 3772 attributers. Number of attributes also ranges from 9 to 30 attributes while all the datasets contain 2 classes. We used k-fold cross validation technique to separate the training set from test set with k=10. Table 1 shows a full description of datasets.

Table 1. Datasets characteristics

Dataset	Traiing	Testing	Attributes	class
BreastCancer	699	Cvalidation	9	2
Hepatitis	155	Cvalidation	20	2
Thyroid	3772	Cvalidation	30	2

For obtaining a fair judgment, as possible, between feature selection methods, we have considered three machine learning algorithms from three categories of learning methods. The first algorithm is k nearest neighbours (kNN) from lazy learning category. kNN is an instance-based classifier where the class of a test instance is based upon the class of those training instances alike to it. Distance functions are common to find the similarity between instances. Examples of distance functions are Euclidean and Manhattan distance functions [16].

The second algorithm is Naïve Bayes classifier (NB) form Bayes category. NB is a simple probabilistic classifier based on applying Bayes' theorem. NB is one of the most efficient and effective learning algorithms for machine learning and data mining because the condition of independency (no attributes depend on each other) [17]. The last machine learning algorithm is Random Tree (RT) or classification tree. RT is used to classify an instance to a predefined set of classes based on their attributes values. RT is frequently used in many fields such as engineering, marketing, and medicine [18].

After applying the feature selection techniques and the learning algorithms on the datasets and obtaining classification accuracy results, we shall construct a hybrid method that combine the advantages of best performed feature selection technique and the advantages of best perform learning algorithm as shown in Figure 1.

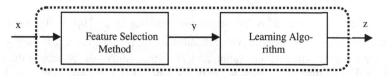

Fig. 1. Hybrid method of feature selection technique and a learning algorithm

The software package used in the present paper is Waikato Environment for Knowledge Analysis (Weka). Weka provides the environment to perform many machine learning algorithm and feature selection methods. Weka is an open source machine learning software written in JAVA language. WEKA contains some data mining and machine learning methods for data pre-processing, classification, regression, clustering, association rules, and visualization [19].

4   Experimental Results

We use the notations "+","-", and "=" to show the feature selection methods classification performance in compared with the original dataset (before performing feature selection methods); where "+" denotes to improvement, "-" denotes to degradation, and "="denotes unchanged. The experimental results of using Naïve Bayes (NB) as a machine learning algorithm on three datasets (Thyroid, Hepatitis, and BreastCancer) are shown in Table 2.

Table 2. Results for Attributes Selection Methods with Naïve Bayes

Method	Thyroid	Hepatitis	BreastCancer
NB	92.60%	84.52%	95.99%
CFS	96.53%+	87.74%+	95.99%=
IG	93.88%+	85.16%+	95.99%=
RF	92.60%=	84.52%=	95.99%=
PC	94.30%+	84.52%=	96.14%+
CB	94.59%+	84.52%=	96.28%+
SU	93.88%+	85.16%+	95.99%=

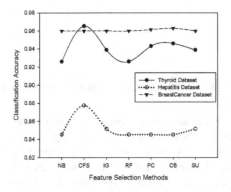

Fig. 2. Results for attributes selection methods with Naïve Bayes

Table 2 shows that the classification accuracy of using NB on original Thyroid dataset is 92.60%, where it shows improvement by applying the feature selection methods CFS, IG, PC, CB, and SU with best result performed by CFS (96.53%). On the original Hepatitis dataset, the classification accuracy is 84.52%, where it shows improvement by applying the feature selection methods CFS, IG, and SU with best results performed by CFS (87.74%). By using BreastCancer original dataset, the classification accuracy is 95.99% with rooms of performance in classification accuracy using the feature selection methods PC and CB with best result performed by CB (96.28%). Figure 2 illustrates the results on Table 2.

The second machine learning classifier for testing the feature selection methods is *k*NN. The experimental results of using *k*NN as a machine learning algorithm on three datasets (Thyroid, Hepatitis, and BreastCancer) are shown in Table 3.

Table 3. Results for Attributes Selection Methods with kNN

Method	Thyroid	Hepatitis	BreastCancer
KNN	95.92%	81.94%	95.42%
CFS	96.10%+	84.52%+	95.42%=
IG	96.50%+	81.29%-	95.42%=
RF	95.92%+	81.94%=	95.42%=
PC	95.78%-	81.29%-	95.42%=
CB	96.37%+	81.94%=	95.85%+
SU	96.50%+	81.29%-	95.42%=

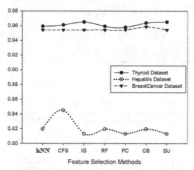

Fig. 3. Results for attributes selection methods with *kNN*

Table 3 shows that the classification accuracy of using *k*NN on the original Thyroid dataset is 95.92%, where it shows improvement by applying the feature selection methods CFS, IG, RF, CB, and SU with best result performed by IG and SU (96.53%). On the original Hepatitis dataset, the classification accuracy is 81.94%, where it shows improvement by applying feature selection methods CFS. However, it shows degradation by using IG, PC, and SU. The best results performed by CFS (84.52%) and the worst results obtained by IG, PC, and SU (81.29%). By using BreastCancer original dataset, the classification accuracy is 95.42%. The classification accuracy has not changed by applying CFS, IG, RF, PC, and SU. However, the feature selection CB obtained the finest result (95.85%). Figure 3 illustrates the results on Table 3.

The last machine learning classifier in our experiment is RT. The experimental results of using RT as a machine learning algorithm on three datasets (Thyroid, Hepatitis, and BreastCancer) are shown in Table 4.

On Table 4, we can observe improvement in classification accuracy by applying the feature selection RF, PC, and CB on Thyroid dataset, while there is degradation by using CFS, IG, and SU on the same dataset. The best performed feature selection method is RF (97.22%), while the worst is CFS (96.29% which is still same classification accuracy on the original dataset). By testing on Hepatitis dataset, classification accuracy has been increased by using the feature selection methods CFS and CB while decreased on IG, PC, and SU. On the last dataset (BreastCancer), there is improvement on feature selection method PC, degradation on CB while unchanged by using CFS, IG, RF, and SU. Figure 4 illustrates the results on Table 4.

Table 4. Results for Attributes Selection Methods with Decision Tree

Method	Thyroid	Hepatitis	BreastCancer
RT	96.92%	76.77%	94.56%
CFS	96.29%-	77.42%+	94.56%=
IG	96.63%-	74.19%-	94.56%=
RF	97.22%+	76.77%=	94.56%=
PC	97.03%+	76.13%-	94.85%+
CB	97.16%+	80.65%+	93.56%-
SU	96.63%-	74.19%-	94.56%=

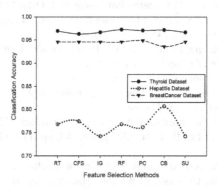

Fig. 4. Results for attributes selection methods with Decision Tree

5   Discussion and Conclusion

The experiment results showed that CFS is one of best performed feature selection method. It also showed that Naïve Bayes learning algorithm has performed the best according to classification accuracy on three datasets. According to the results obtained by the current work on three different sizes datasets, Naïve Bayes has performed the supreme in regard to classification accuracy. kNN and DT have performed just better on datasets after applying feature selection methods. In general, attribute feature selection methods can improve the performance of learning algorithms. However, no single feature selection method that best satisfy all datasets and learning algorithm. Therefore, machine learning researcher should understand the nature of datasets and learning algorithm characteristics in order to obtain better outcome as possible.   Overall, CFS and CB feature selection methods performed the better than IG, SU, RF, and PC. We have also found that IG and SU performed typically because SU is a modified version of IG. Future work should compare between feature selection methods and the associated learning algorithms in regard to speed, tolerance to noise, as well as applying feature selection methods on more datasets.

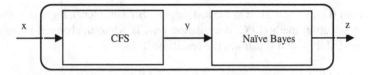

Fig. 5. Hybrid method of feature selection technique and a learning algorithm

References

1. Han, J., Kambler, M.: Data Mining Concepts and Techniques, vol. 3. Morgan Kaufmann, San Franscisco (2011)
2. Hall, M.A., Holmes, G.: Benchmarking Attribute Selection Techniques for Discrete Class Data Mining. IEEE Transactions on Knowledge and Data Engineering 15(3) (2003)
3. Ashraf, M., et al.: A New Approach for Constructing Missing Features Values. International Journal of Intelligent Information Processing 3(1), 110–118 (2012)
4. Blum, A.L., Langley, P.: Selection of relevant features and examples in machine learning. Artificial Intelligence 97(1-2), 245–271 (1997)
5. Cancer, I.A.R. Mammography Screening can Reduce Deaths from Breast Cancer, http://www.iarc.fr/en/media-centre/pr/2002/pr139.html
6. Lee, S.L.: Thyroid Problems (2012), http://www.emedicinehealth.com/thyroid_problems/article_em.htm
7. Introducing Hepatitis C, http://www.hep.org.au/.
8. Kononenko, I.: Estimating Attributes: Analysis and Extensions of RELIEF. In: Bergadano, F., De Raedt, L. (eds.) ECML 1994. LNCS, vol. 784, pp. 172–182. Springer, Heidelberg (1994)
9. Hall, M.A.: Correlation-based Feature Selection for Machine Learning in Department of Computer Science. The University of Waikato, Hamilton (1999)
10. Rutkowski, L., et al. (eds.): Artificial Intelligence and Soft Computing, Part I. ed. L.N.i.C.S. 6113, vol. 1, pp. 487–498 Springer, Poland (2010)
11. Guyon, I., Elisseeff, A.: An Introduction to Variable and Feature Selection. Journal of Machine Learning Research 3, 1157–1182 (2003)
12. Jolliffe, I.T.: Principal Component Analysis. Springer, NY (2002)
13. Liu, H., Setiono, R.: A probabilistic approach to feature selection: A fiter solution. In: Proc. of the 13th International Conference on Machine Learning (1996)
14. Liu, H., Setiono, R.: Chi2:Feature selection and discretization of numeric attributes. In: Proc. of the 7thIEEE International Conference on Tools with Articial Intelligence (1995)
15. Wolberg, W., Mangasarian, L.: Multisurface method of pattern separation for medical diagnosis applied to breast cytology. Proceedings of the National Academy of Sciences 87, 9193–9196 (1990)
16. Pevsner, J.: Bioinformatics and Functional Genomics, 2nd edn. Wiley-Blackwell (2009)
17. Zhang, H., Su, J.: Naive Bayes for optimal ranking. Journal of Experimental & Theoretical Artificial Intelligence 20(2), 79–93 (2008)
18. Rokach, L., Maimon, O.: Data Mining With Decision Trees. World Scientific Publishing (2008)
19. Ashraf, M., Le, K., Huang, X.: Information Gain and Adaptive Neuro-Fuzzy Inference System for Breast Cancer Diagnoses. In: Proc. of ICCIT 2010, pp. 911–915. IEEE Press, Seoul (2010)

Data Discretization Using the Extreme Learning Machine Neural Network

Juan Jesús Carneros, José M. Jerez, Iván Gómez, and Leonardo Franco

Universidad de Málaga, Department of Computer Science, ETSI Informática, Spain
jcarnego@hotmail.com, {jja,ivan,lfranco}@lcc.uma.es

Abstract. Data discretization is an important processing step for several computational methods that work only with binary input data. In this work a method for discretize continuous data based on the use of the Extreme Learning Machine neural network architecture is developed and tested. The new method does not use data labels for performing the discretization process and thus is suitable for supervised and supervised data and also, as it is based on the Extreme Learning Machine, is very fast even for large input data sets. The efficiency of the new method is analyzed on several benchmark functions, testing the classification accuracy obtained with raw and discretized data, and also in comparison to results from the application of a state-of-the-art supervised discretization algorithm. The results indicate the suitability of the developed approach.

Keywords: Neural networks, Supervised learning, Extreme learning machine, Generalization, Discretization.

1 Introduction

Discretization techniques play an important role in the areas of data mining and knowledge discovery. Not only they help to produce a more concise data representation but also are a necessary step for the application of several classification algorithms, like Decision Trees, Logical Analysis of Data, DASG algorithm, etc. [1,8]. Discretization methods can be characterized according to at least five different features: supervised or unsupervised, static or dynamic, global or local, bottom-up or top-down, and direct or indirect methods. We will not do an in-depth analysis of these characteristics (see [6] for a more detailed analysis) but will mention the relevant ones in relationship to the method to be proposed below. The use or not of the label (or class) of the data for the discretization process distinguish a supervised from an unsupervised method; global methods use all set of features and data for the processing while local methods use only a portion of them; direct methods requires the user to decide on the characteristics of the output while, in general, incremental methods works by applying some predefined conditions towards the fulfillment of a goal. We present in this work a discretization algorithm constructed by using the Extreme Learning Machine (ELM) model that will be unsupervised, local and incremental. The ELM algorithm proposed by Huang et al. [5,4], as a fast and accurate neural network

T. Huang et al. (Eds.): ICONIP 2012, Part IV, LNCS 7666, pp. 281–288, 2012.

based classification system, construct feedforward neural network architectures containing a single hidden layer of neurons with the peculiarity that only the set of synaptic weights connecting the hidden layer to the output need to be learnt, as the connections between the input and the middle layer have randomly chosen values. In this way, not only the process is much faster but also can be performed deterministically by solving a system of equations, without depending on gradient descent based algorithms. Despite its simplicity, and the use of random weights in the first layer, the results obtained when using the ELM are surprisingly good, in most cases reaching a similar level of performance than more complex algorithms. The ELM normally uses continuos valued neurons in the middle layer but a version operating with discrete neurons has been proposed [4], and it is this version that we used to build a discretization algorithm.

2   Methods

2.1   A Discretizer Based on the Extreme Learning Machine

The Extreme Learning Machine is a neural based architecture for classification problems that can be trained in a supervised way. The network architecture contains a single layer of neurons that receive the information from the input neurons through random valued weights, and thus the training of the network only involves finding the value of the synaptic weights connecting the middle layer to the output neurons. This training process can be performed very fastly and straightforward as essentially consists in a matrix inversion.

For N input patterns $(\mathbf{x}_i, \mathbf{t}_i)$, where $\mathbf{x}_i = [x_{i1}, x_{i2}, \ldots, x_{in}]^T \in \mathbf{R}^n$ and $\mathbf{t}_i = [t_{i1}, t_{i2}, \ldots, t_{im}]^T \in \mathbf{R}^m$, the output vector $\mathbf{o}_i$ of an ELM with L neurons in the hidden layer and activation function $g(x)$ can be modelled mathematically as:

$$\sum_{i=1}^{L} \beta_i g\left(\mathbf{w}_i \cdot \mathbf{x}_j + b_i\right) = \mathbf{o}_j, \quad j = 1, \ldots, N, \tag{1}$$

where $\mathbf{w}_i = [w_{i1}, w_{i2}, \ldots, w_{in}]^T$ is the vector of synaptic weights connecting the ith hidden neuron to the input neurons, $\beta_i = [\beta_{i1}, \beta_{i2}, \ldots, \beta_{im}]^T$ is the vector connecting the hidden neurons to the output ones, and b_i indicates the bias of the ith hidden neuron. An ELM with m output neurons that can approximate N patterns with zero error implies that the $\sum_{i=1}^{N} \|\mathbf{o}_i - \mathbf{t}_i\| = 0$, expression that can be written in matrix form as:

$$\mathbf{H}\beta = \mathbf{T}, \tag{2}$$

where $\mathbf{H}$ represents the activity of the hidden neurons when an input pattern is presented, $\mathbf{T}$ represents the target outputs, and β are the weights to be find in order to verify the learning of the input-output relationship. The value of β with smallest norm that verifies in the minimum square sense the previous lineal system is:

$$\hat{\beta} = \mathbf{H}^\dagger \mathbf{T}, \tag{3}$$

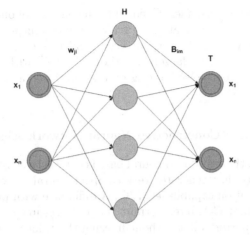

Fig. 1. Structure of an ELM used for data discretization. The number of output neurons is equal to the number of inputs and for each pattern of the data set the desired outputs (targets) are equal to the input values. Using discrete units in the hidden layer permits to obtain after training the network a discrete representation of the inputs.

where $\mathbf{H}^{\dagger}$ is the Moore-Penrose generalized inverse of matrix $\mathbf{H}$ [5,4]. In the original ELM architecture [5] the activation function of the neurons in the middle layer is continuous valued but binary units can be used, and it has been shown that its functioning and the results that can be obtained are quite similar [4].

To build a discretizer based on the ELM, it is only necessary to use discrete valued hidden units, select the number of neurons to include in this hidden layer, and train the network as described above, with patterns that have the same output as the input. A network architecture of a discretizer constructed for the case of input vectors of length 2 is depicted in Figure 1, where the hidden layer, from which the discretized representation will be obtained, contains 3 neurons in this example. If the total error of the network, measured as the difference between input and output across all training patterns, is zero then the discretization obtained in the hidden layer contains all the information of the original input; in other cases the error quantifies the information loss during the discretization process.

2.2 The CAIM Discretization Algorithm

We briefly explain some characteristics of the CAIM discretization algorithm because it will be used as a reference for comparing the efficiency of the new developed algorithm. The CAIM algorithm is a supervised discretization method that tries to maximize the class-attribute interdependency by exploring a series of discretization points, to choose those that optimize an heuristic measure related to the dominancy of a given class in the created intervals, including also a factor so to minimize the number of intervals [6]. It is a local algorithm that consider the input attributes independently, usually generating for each input variable a discrete representation with a length equal to the number of output classes (e.g., twice the original number of attributes for a binary output). The algorithm has been extensively tested leading to very good results and in several cases leading to classification rates larger than those obtained with the original data, fact that can be due to a noise-filtering process produced as a side-effect in the discretization process.

2.3   The C-Mantec Constructive Neural Network Algorithm

C-Mantec is a recently introduced neural based constructive algorithm for supervised learning tasks that generates very compact neural network architectures with high generalization capabilities [7,9] . A difference with previous constructive algorithms is that C-Mantec permits the architectures to grow as required by the learning of patterns but without freezing the values of existing synaptic weights, in a scheme where the neurons compete for the incoming knowledge and in which the learning at the level of individual neurons is performed by the thermal perceptron. The algorithm is used in this work to test the generalization ability of the several representation of the data obtained by applying the ELM discretizer because of its observed good generalization capabilities but also because training computational are much reduced in comparison to traditional MLPs.

2.4   The Generalization Complexity Measure

The GC measure was introduced by Franco and colleagues [2] and was derived from evidence that pairs of bordering examples, those lying closely at both sides of the separating hyperplanes that classify different regions, play an important role on the generalization ability obtained in classification problems when neural networks are used as predictive methods. The idea behind the GC measure is to build an estimate of the generalization ability that can be obtained beforehand but also to be used to estimate the size of an adequate neural architecture [3]. As the GC measure has been defined only for binary input values, we used it in this work as an additional test of the quality of the representations obtained using the ELM discretizer developed.

3   Results

We have applied the ELM discretizer using several size architectures to a set of 20 benchmark data sets. To choose the number of neurons in the hidden layer of the ELM, i.e., the size of the representation, we took the number of neurons generated by CAIM and consider this value as a reference, and thus we try representations with the same number of units used by CAIM, twice and five times this number, and also consider a number of hidden units necessary to achieve a 0.01 level of error.

Table 1. Characteristics of the 20 data sets selected from the PROBEN1 benchmark set to carry out the tests (see the text for more details)

Id	Set	Patterns	Class distribution % (0/1)	Inputs	Neurons CAIM	Neurons ELM (Error < 0.01)
f_1	cancer1	699	65 - 35	9	18	527
f_2	card1	690	45 - 55	51	100	641
f_3	diabetes1	768	35 - 65	8	16	818
f_4	gene1a	3175	76 - 24	120	240	2977
f_5	gene1b	3175	76 - 24	120	240	2600
f_6	gene1c	3175	48 - 52	120	240	2600
f_7	glass1a	214	67 - 33	9	18	215
f_8	glass1a	214	64 - 36	9	18	243
f_9	glass1b	214	92 - 8	9	18	251
f_{10}	glass1c	214	94 - 6	9	18	235
f_{11}	glass1d	214	96 - 4	9	18	243
f_{12}	glass1e	214	86 - 14	9	18	251
f_{13}	heart1a	920	44 - 56	35	67	824
f_{14}	heartc1b	303	54 - 46	35	59	300
f_{15}	horse1a	364	38 - 62	58	116	347
f_{16}	horse1b	364	75 - 25	58	116	347
f_{17}	horse1c	364	85 - 15	58	116	347
f_{18}	thyroid1a	7200	97 - 13	21	42	538
f_{19}	thyroid1b	7200	95 - 5	21	42	529
f_{20}	thyroid1c	7200	7 - 93	21	42	520

Table 1 shows the characteristics of the 20 binary output data sets taken from the Proben benchmark set used for testing the discretization performance of the proposed algorithm. The first column of the table indicates the identification reference of the data set, and the rest of columns indicate its the name, the number of patterns available, the class distribution, the number of inputs, the number of discretization intervals (or equivalently the number of neurons) needed by the CAIM algorithm, and the number of neurons used in the developed discretizer needed to achieve a codification error below 0.01.

Table 2 shows the generalization ability results obtained in a ten fold cross-validation approach. The first column indicates the function used and the rest of columns the generalization ability obtained using the several representations of the function: the original continuous input data, the CAIM discretized inputs, the ELM based discretization with the same number of neurons as used with CAIM, twice and five times the previous value and finally with a number of attributes needed to achieve a 0.01 or less of training error (value indicated in the last column of Table 1).

Regarding the computational costs involved in the tests carried out, we report the average CPU times needed on an Intel Core 2 Duo E7300 PC running at 2.66GHz with Windows Vista OS for discretizing the 20 data sets analyzed. On average the CAIM algorithm needed 24.03 seconds per function, while the ELM based discretizer needed 0.05, 0.30 and 1.91 seconds respectively for the case of

Table 2. Generalization ability obtained for the 20 benchmark functions extracted from the Proben benchmark set. The function identifier is shown in the first column, and the rest of values shown indicate the generalization for the original continuous input data (no discretization applied), for the discretization obtained from CAIM, and for the ELM based discretizer with the different representations sizes considered. (See the text for details).

Function id	Continuos input	CAIM	ELM nh	ELM 2nh	ELM 5nh	ELM (Error < 0.01)
f_1	96.19	96.57	93.05	95.24	94.57	94.76
f_2	74.20	75.56	70.05	74.88	72.46	72.85
f_3	54.96	75.13	67.57	68.78	60.17	54.26
f_4	85.15	86.09	80.55	82.00	82.92	83.34
f_5	79.24	81.37	75.84	77.21	78.28	77.65
f_6	78.38	78.32	71.66	73.93	75.82	75.80
f_7	69.69	80.31	67.19	75.31	74.06	73.44
f_8	66.88	78.75	66.56	68.13	71.25	70.94
f_9	84.06	88.44	93.13	90.94	90.00	90.63
f_{10}	94.69	91.88	94.38	93.44	93.75	96.88
f_{11}	97.50	99.38	95.63	97.19	97.50	96.56
f_{12}	94.06	95.31	95.31	95.31	95.00	95.63
f_{13}	67.68	72.68	63.12	67.03	64.35	64.49
f_{14}	74.73	80.00	73.85	70.77	76.48	73.63
f_{15}	68.07	70.28	68.26	67.34	68.81	68.62
f_{16}	74.68	76.51	74.31	72.84	76.33	74.50
f_{17}	85.14	85.32	81.28	82.39	82.02	84.22
f_{18}	98.59	99.03	97.90	94.72	98.05	98.30
f_{19}	93.85	97.35	94.63	91.29	94.69	94.48
f_{20}	93.10	98.95	92.61	90.82	92.73	92.87
Mean	81.54	85.36	80.84	81.48	81.96	81.69

networks having the same number of outputs than the CAIM algorithm, twice and five times this value respectively.

Further, we have computed the Generalization Complexity (GC) of the several data sets used. This measure is related to the generalization ability that can be expected when a function is implemented in a neural network or other prediction system. The GC measure has been defined only for Boolean inputs and thus it is useful to analyze the values obtained with the different discretized representations considered previously. Table 3 shows the value obtained for the Generalization complexity for the 20 functions indicated in the first column. Columns 2 to 5 show the values of GC using the 5 considered discretization schemes: CAIM, ELM with same number of attributes than CAIM, twice and five times the previous value, and with a number of neurons enough to reduce the representation error below 0.01. The last row of the table shows the average of column values.

As the GC measure can only be computed for binary data, given a continuous input data it is not clear its true value as the data needs to be transformed.

Table 3. The Generalization Complexity values computed the 20 test data sets using 5 different discretization representation: CAIM, ELM nh, ELM 2nh, ELM 5nh, ELM with an error less than 0.01. Average column values are indicated in the last row.

Function id	CAIM	ELM nh	ELM 2nh	ELM 5nh	ELM (Error < 0.01)
f_1	0.0525	0.1440	0.0700	0.0473	0.0401
f_2	0.2033	0.2191	0.1969	0.2053	0.2109
f_3	0.3761	0.4481	0.4219	0.3356	0.3027
f_4	0.1617	0.2139	0.1860	0.1854	0.1761
f_5	0.1764	0.2334	0.2059	0.1792	0.1762
f_6	0.2417	0.2935	0.2763	0.2744	0.2601
f_7	0.3105	0.3860	0.3397	0.2823	0.2103
f_8	0.4048	0.4263	0.3765	0.2614	0.2240
f_9	0.1570	0.1543	0.1262	0.1262	0.1090
f_{10}	0.0591	0.1210	0.0944	0.0380	0.0506
f_{11}	0.0566	0.0768	0.0613	0.0389	0.0257
f_{12}	0.0573	0.1801	0.0491	0.0254	0.0288
f_{13}	0.2412	0.2562	0.2381	0.2579	0.2458
f_{14}	0.2333	0.2693	0.2294	0.2393	0.2525
f_{15}	0.3097	0.3300	0.3574	0.3810	0.3537
f_{16}	0.2454	0.2689	0.2889	0.2540	0.2807
f_{17}	0.1861	0.1832	0.1964	0.1882	0.1850
f_{18}	0.0120	0.0503	0.0271	0.0172	0.0130
f_{19}	0.0553	0.0915	0.0917	0.0919	0.0819
f_{20}	0.0543	0.1266	0.1269	0.1052	0.0878
$\bar{f}$	0.1797	0.2236	0.1980	0.1767	0.1657

Nevertheless, as the GC measure has been validated before [2] as a good predictor for the generalization ability, we computed the correlation between the generalization ability obtained using the original continuos data and the GC obtained from the several discretization schemes. The results show a strong correlation value in all cases (measured by the absolute value of the Pearson correlation coefficient), with largest values observed for the case of the CAIM discretized data (r=0.89) and also for the ELM based discretizer with twice as much bits as those generated by CAIM (r=0.86).

4 Conclusions

We have tested in this work the implementation of a non-supervised discretization algorithm based on the ELM neural network architecture and compared the results with those obtained using a supervised discretization method, named CAIM, known for its good performance. To compare the quality of the representations obtained we have analyzed the generalization ability using 20 benchmark data sets. The results show that the same of level of generalization than for the case of using the original continuous data can be obtained by using approximately five times more neurons than those selected by the CAIM algorithm, that normally chooses a representation two times larger than the number of input features for binary classification problems, as it is the case of the analyzed data sets. Considering that the CAIM algorithm is a supervised method, the previous result can be considered useful, also because of the speed of the ELM based discretizer, several times faster in comparison to the CAIM algorithm. A further test of the quality of the discretization representation obtained with the ELM-based discretizer comes from an experiment carried to measure the value of the Generalization Complexity of the data sets, finding results quite close to those obtained when using the CAIM algorithm. As an overall conclusion we can state that the experiments carried out in this work shows the suitability of using the ELM as a discretization algorithm, highlighting that even if it generates larger data representations in comparison to a performant supervised algorithm like CAIM, the process is much faster and permits to work both with supervised and unsupervised data.

Acknowledgements. The authors acknowledge support from CICYT (Spain) through grants TIN2008-04985 and TIN2010-16556 (including FEDER funds) and from Junta de Andalucía through grant P08-TIC-04026.

References

1. Boros, E., Hammer, P.L., Ibaraki, T., Kogan, A., Mayoraz, E., Muchnik, I.B.: An Implementation of Logical Analysis of Data. IEEE Trans. Knowl. Data Eng. 12, 292–306 (2000)
2. Franco, L., Anthony, M.: The Influence of Oppositely Classified Examples on the Generalization Complexity of boolean functions. IEEE Trans. Neural Netw. 17(3), 578–590 (2006)
3. Gómez, I., Franco, L., Jerez, J.M.: Neural Network Architecture Selection: Can Function Complexity Help? Neural Proc. Lett. 30, 71–87 (2009)
4. Huang, G.B., Zhu, Q.Y., Mao, K.Z., Siew, C.K., Saratch, P., Sundararajan, N.: Can Threshold Networks be Trained Directly. IEEE Trans. Circuits Syst. II. 53, 187–191 (2006)
5. Huang, G.B., Zhu, Q.Y., Siew, C.K.: Extreme Learning Machine: A New Learning Scheme of Feedforward Neural Networks. In: Proc. Int. Joint Conf. Neural Networks, pp. 985–990 (2004)
6. Kurgan, L.A., Cios, K.J.: Caim Discretization Algorithm. IEEE Trans. on Knowl. and Data Eng. 16(2), 145–153 (2004)
7. Subirats, J.L., Franco, L., Jerez, J.M.: C-mantec: a Novel Constructive Neural Network Algorithm Incorporating Competition between Neurons. Neural Networks 26, 130–140 (2012)
8. Subirats, J.L., Jerez, J.M., Franco, L.: A New Decomposition Algorithm for Threshold Synthesis and Generalization of Boolean Functions. IEEE Trans. on Circuits and Systems 55-I(10), 3188–3196 (2008)
9. Urda, D., Cañete, E., Subirats, J.L., Franco, L., Llopis, L., Jerez, J.M.: Energy Efficient Reprogramming in wsn Using Constructive Neural Networks. International Journal of Innovative Computing, Information and Control 8 (2012)

Bayesian Variable Selection in Neural Networks for Short-Term Meteorological Prediction

Pierrick Bruneau and Laurence Boudet

CEA, LIST, Information, Models & Learning Laboratory,
91191 Gif sur Yvette CEDEX FRANCE
firstname.lastname@cea.fr

Abstract. This work examines the influence of Bayesian variable selection on neural architectures for global solar irradiation and air temperature time series prediction. These models, 3 neural architectures with differing input and output processing strategies [2], predict all time slots in the 24 hours ahead period, with inputs solely taken from local measurements of the 24 last hours. Qualitative and computational points of view are considered for the comparison of Bayesian and non-Bayesian learning, with a specific care for salient variable sets analysis. For generalization purpose, models are assessed and compared on data from two contrasted sites in France. The input space appeared to be reduced by at least 34%, and up to 73%, with a small prediction quality loss (1.3% on average), and a good repeatability of selected salient variables across sites.

Keywords: Meteorological time series prediction, Neural architectures, Bayesian variable selection, Solar irradiation, Air Temperature.

1 Introduction

Accurate predictions of global solar irradiation and air temperature are useful for anticipative control in autonomous energy management systems that use solar energy. For example, the system could then anticipate future production and needs, and adapt its behavior accordingly. This information may be provided by an external numerical weather prediction system, but to prevent any service versatility, or its excessive cost relatively to the expected gains, some system designers require autonomous predictions.

Naive models such as persistence or monthly means can then be effective [4]. Alternatively, Multi-Layer Perceptrons (MLP's) have been used to learn predictors from meteorological time series' data [3], [7], [8]. These approaches have recently been unified and extended to handle the prediction of every time slot in the upcoming 24 hours, using only local measurements from the 24 latest hours [2]. In short, absolute versus relative and daily versus tri-hourly strategies have been explored for neural predictions. However, all available variables were used for prediction. In this paper, a Bayesian variable selection procedure is applied to these predictors.

T. Huang et al. (Eds.): ICONIP 2012, Part IV, LNCS 7666, pp. 289–296, 2012.

First, we define some basics about the MLP and introduce the evidence procedure [5] for the selection among its inputs. Then, we give a short reminder of the neural architectures introduced in [2] for this problem. An experimental protocol, including available data sets, is described in section 3. Experimental results are given in the next section, that includes an analysis of the selected variables and its potential generality across different sites. The interest of this approach is finally outlined, and perspectives to this work are drawn.

2 Neural Architectures Description

2.1 Multi-layer Perceptron Basics

Let us consider a regression problem with D-dimensional inputs $\mathbf{x} = \{x_1, \ldots, x_D\}$, and S-dimensional outputs $\mathbf{y} = \{y_1, \ldots, y_S\}$. The regression MLP with S outputs and K hidden neurons is determined by the following formula:

$$f_s(\mathbf{x}) = \sum_{s,k} w_{sk}\sigma\left(\sum_{k',d} w_{k'd}x_d\right) \tag{1}$$

where w_{kd} (respectively w_{sk}) denotes the weight from the dth input node (respectively k^{th} hidden node) to the kth hidden node (respectively s^{th} output node), and $\mathbf{f} = \{f_s\}$ is the model output. The full set of weights is generally summarized as $\mathbf{w} = \{\{w_{kd}\}, \{w_{sk}\}\}$. Hidden and output layer nodes also have bias weights, implicitly associated to an additional input dimension, clamped at 1 in formula (1). $\sigma(.)$ is a non-linear function, chosen as the logistic function in this paper. Learning a MLP then amounts to set $\mathbf{w}$ such as $\mathbf{f}(\mathbf{x})$ approximates correctly $\mathbf{y}$.

Values for $\mathbf{w}$ are fitted to a set $\{\mathbf{x}_n, \mathbf{y}_n\}_{n=1\ldots N}$ using the back-propagation algorithm [5]. This algorithm optimizes the quadratic loss of the model output with respect to target $\mathbf{y}_n$ vectors. The estimated model is then able to predict $\mathbf{y}$ for an unknown $\mathbf{x}$. Let us note that the cardinality of the weights $|\mathbf{w}|$ in the MLP roughly equals $K(D + S)$.

2.2 Bayesian Variable Selection

For a given problem, some input dimensions may be useless or redundant. Bayesian variable selection aims at detecting the relevant inputs, and deciding which inputs should be removed. This selection is not trivial for MLP's: inputs associated to null weights play no role, but the reciprocal generally does not hold [1]. The Bayesian view of MLP's aims at enforcing this equivalence, thus providing an indicator for non-relevant inputs.

Formally, each input weight is associated to a prior centered on 0:

$$p(w_{kd}|\alpha_d) = \mathcal{N}(w_{kd}|0, \alpha_d^{-1}) \tag{2}$$

where $\mathcal{N}(.)$ is the Gaussian law, and α_d is the precision parameter (i.e. inverse variance) for the dth input. Posterior α_d values are obtained from the *evidence procedure* [5], along with the associated estimation of $\mathbf{w}$. This method combines the back-propagation algorithm with an iterative re-estimation of α_d coefficients. A high posterior value for α_d is then equivalent to the redundancy of the dth input.

2.3 Meteorological Time Series Prediction

The prediction of solar irradiation and air temperature is made for the 24 hours ahead with data from the latest 24 hours. Thus, the timestamps t can be identified relatively to the current time slot h_c and prediction time slot h_p without any ambiguity. The predicting *horizon* is defined as the difference between h_p and h_c with a 24h modulo function. In this work, we restrict to tri-hourly time series, i.e. time slots range in $\{0, 3, \ldots, 21\}$. Computational complexity issues taken apart, the extension to hourly data would be straightforward.

New architectures for meteorological time series prediction were proposed in [2]. They incorporate strategies specially adapted to cyclic inputs and outputs. In short, the absolute (respectively relative) strategy amounts to specialize MLP's to one admissible h_c value (respectively to use only one MLP for all h_c values). The tri-hourly (respectively daily) strategy specializes MLP's to one admissible h_p value (respectively defines only one MLP for all h_p values). The resulting predictors are composed of a different number of MLP's according to the chosen strategies, that can range from 1 (relative-daily) to 64 (absolute-tri-hourly) with 1 or 8 outputs.

3 Experimental Protocol

3.1 Data Sets Description and Pre-processing

Four meteorological time series are available for prediction: solar irradiation (W.m^{-2}), air temperature $(^{\circ}C)$, air pressure (hPa) and hygrometry $(\%)$. Two 5-years long data sets were recorded in Saclay (near Paris, France) and in Marcoule (South-East of France). Let us note that solar irradiation time series is null at 21h, 0h and 3h at French latitude. In the remainder, these time slots were removed when needed (pre-processing, learning and evaluation stages). Therefore, the number of inputs D considered for prediction equals 29.

Meteorological time series have strong yearly and daily cycles, they should be converted to an acceptable stationary process before being used in a learning procedure [6], [7], [8]. To this aim, let us define the monthly-hourly sets of a time series data set:

$$\mathbf{x}_{m,h} = \{x_t | \text{month}(t) = m \wedge \text{time slot}(t) = h\}. \tag{3}$$

This results in 96 sets (12 months and 8 time slots). Then the time series are normalized in order to make the time series' mean and variance approximately constant $\forall t$, which is the weak stationarity definition:

$$x_t^{\text{norm}} = \frac{x_t - \text{mean}(\mathbf{x}_{m,h})}{\text{standard deviation}(\mathbf{x}_{m,h})} \forall t, \text{s.t. } x_t \in \mathbf{x}_{m,h}. \tag{4}$$

3.2 MLP Learning Protocol

Training and testing sets are defined by their chronological bounds. For each admissible time t in the training set (i.e. that matches admissible h_c values for the MLP), an input vector is made with the time series restricted to $[t - 21h; t]$. Matching output data entries are either scalars or vectors, depending on admissible h_p values for this MLP. Training entries with missing values in either input or output vectors are ignored.

The determination of K, the number of neurons in the hidden layer, is left open by the back-propagation algorithm and the evidence procedure. Thus, a cross-validation scheme is used to learn a model with a good generalization ability. More specifically:

- the training set is randomly divided in 5 equal-sized parts,
- each of these 5 parts is used for the validation of a MLP. Each MLP is first trained using remaining parts.
- observed validation errors are averaged, and serve as a criterion for selecting K: initialized to 1, K increases until validation error stops decreasing.

For each site, 3 years of consecutive data are used for training, and the 2 following years are used for testing. Transformed data (see equ. (4)) is used for learning and validation, but testing errors are measured with back-transformed outputs and the following *normalized Root Mean Squared Error* quality metric:

$$nRMSE = \sqrt{\frac{1}{N} \sum_{n=1}^{N} \left(\frac{y_n - \hat{y}_n}{\max(\mathbf{y}_{m,h}) - \min(\mathbf{y}_{m,h})} \right)^2}$$

where y_n is the time series to predict, $\hat{y}_n$ the prediction, $\mathbf{y}_{m,h}$ is the monthly-hourly set associated to time series data y_n (see equ. (3)). Errors are thus scaled to the domain they relate to, and can be seen as the percentage to maximal possible error on the original scale (indeed, maximal daytime solar irradiation ranges from 100 to 1000 $W.m^{-2}$). Thus nRMSE estimates will be further denoted as percentages.

3.3 Variable Selection Protocol

Non-Bayesian MLP's may be learnt using the back-propagation algorithm, and all available input variables. Bayesian MLP's are obtained by the evidence procedure, that combines a variable selection procedure to the back-propagation approach. This model outputs a set of α_d coefficients. In this paper, the saliency of a variable d is defined as the inverse of α_d (i.e. the tendency of its weights to deviate from 0). The smallest variable set that gathers at least 95% of the saliency is selected. This set of relevant variables is then used to learn Bayesian MLP's afresh. Depending on their position in this sequence, Bayesian predictors are said to occur *before* or *after* variable selection.

4 Experimental Results

In [2], the non-Bayesian versions of the architectures described in this paper were evaluated for the prediction of solar irradiation and air temperature time series. Comparisons to a set of naive models (persistence, monthly means) showed that neural predictors were more performant. Among them, the absolute-daily architecture emerged as the best compromise between predictive quality and computational cost. In this section, the non-Bayesian versions are compared to their Bayesian counterparts, before and after variable selection. All experiments were performed 5 times in order to improve the robustness of given results.

Table 1. Comparative performance results of neural architectures. Associated standard deviations range from 0.1% to 0.3%.

	$h_p - h_c=$	3h	6h	9h	12h	15h	18h	21h	24h
Irradiation - Saclay									
Relative-daily	Non-Bayesian	17.0%	19.4%	20.3%	20.6%	20.9%	21.2%	21.3%	21.5%
	Bayesian (before sel.)	17.0%	19.4%	20.3%	20.6%	20.9%	21.2%	21.3%	21.5%
	Bayesian (after sel.)	17.2%	19.7%	20.6%	20.8%	21.1%	21.4%	21.4%	21.6%
Absolute-daily	Non-Bayesian	15.8%	18.8%	19.9%	20.5%	20.9%	21.1%	21.3%	21.4%
	Bayesian (before sel.)	15.7%	18.7%	19.8%	20.4%	20.9%	21.2%	21.3%	21.5%
	Bayesian (after sel.)	16.0%	19.2%	20.2%	21.2%	21.5%	21.7%	21.8%	21.8%
Absolute-tri-hourly	Non-Bayesian	15.5%	18.8%	19.9%	20.4%	21.5%	21.1%	21.3%	21.5%
	Bayesian (before sel.)	15.4%	18.8%	19.9%	20.3%	20.9%	21.1%	21.2%	21.4%
	Bayesian (after sel.)	15.7%	19.4%	20.2%	20.9%	21.1%	21.8%	21.7%	21.4%
Irradiation - Marcoule									
Relative-daily	Non-Bayesian	15.6%	18.0%	18.9%	19.2%	19.7%	20.0%	20.2%	20.2%
	Bayesian (before sel.)	15.4%	18.0%	18.7%	19.1%	19.6%	19.9%	20.1%	20.1%
	Bayesian (after sel.)	15.7%	18.0%	18.8%	19.2%	19.7%	20.0%	20.3%	20.2%
Absolute-daily	Non-Bayesian	14.4%	17.2%	18.2%	18.6%	19.2%	19.4%	19.6%	19.7%
	Bayesian (before sel.)	14.5%	17.1%	18.2%	18.5%	19.1%	19.4%	19.7%	19.8%
	Bayesian (after sel.)	14.9%	17.7%	18.8%	19.2%	19.6%	19.7%	19.7%	19.9%
Absolute-tri-hourly	Non-Bayesian	14.0%	17.0%	18.1%	18.6%	19.1%	19.5%	19.6%	20.0%
	Bayesian (before sel.)	14.0%	16.9%	18.0%	18.5%	19.0%	19.3%	19.4%	19.7%
	Bayesian (after sel.)	14.4%	17.3%	19.0%	19.7%	19.3%	19.6%	20.0%	19.8%
Temperature - Saclay									
Relative-daily	Non-Bayesian	6.0%	8.9%	10.7%	11.8%	12.4%	12.8%	13.1%	13.3%
	Bayesian (before sel.)	6.0%	8.9%	10.6%	11.7%	12.3%	12.7%	13.0%	13.2%
	Bayesian (after sel.)	6.0%	8.9%	10.6%	11.6%	12.2%	12.6%	12.9%	13.2%
Absolute-daily	Non-Bayesian	5.9%	8.2%	9.8%	10.9%	11.7%	12.3%	12.7%	13.1%
	Bayesian (before sel.)	5.9%	8.3%	9.9%	10.9%	11.7%	12.3%	12.8%	13.1%
	Bayesian (after sel.)	6.0%	8.6%	10.1%	11.0%	11.7%	12.2%	12.7%	13.1%
Absolute-tri-hourly	Non-Bayesian	5.4%	8.2%	9.8%	10.9%	11.6%	12.2%	12.7%	13.1%
	Bayesian (before sel.)	5.4%	8.2%	9.9%	10.9%	11.7%	12.2%	12.6%	13.0%
	Bayesian (after sel.)	5.7%	8.4%	10.0%	11.0%	11.8%	12.3%	12.7%	13.1%
Temperature - Marcoule									
Relative-daily	Non-Bayesian	6.4%	9.6%	11.1%	11.7%	12.1%	12.3%	12.5%	12.8%
	Bayesian (before sel.)	6.3%	9.5%	11.1%	11.7%	12.1%	12.3%	12.5%	12.8%
	Bayesian (after sel.)	6.3%	9.6%	11.1%	11.7%	12.0%	12.2%	12.4%	12.7%
Absolute-daily	Non-Bayesian	6.4%	9.1%	10.8%	11.6%	12.0%	12.3%	12.5%	12.9%
	Bayesian (before sel.)	6.5%	9.2%	10.8%	11.6%	12.1%	12.4%	12.6%	12.9%
	Bayesian (after sel.)	6.4%	9.3%	10.9%	11.6%	12.0%	12.2%	12.5%	12.9%
Absolute-tri-hourly	Non-Bayesian	5.9%	9.0%	10.6%	11.5%	12.0%	12.3%	12.5%	12.8%
	Bayesian (before sel.)	5.8%	8.9%	10.6%	11.5%	11.9%	12.3%	12.5%	12.8%
	Bayesian (after sel.)	6.1%	9.1%	10.8%	11.5%	12.0%	12.3%	12.6%	12.8%

4.1   Performance Evaluation

For predicting horizons greater than 9 hours, the non-Bayesian versions of the 3 evaluated architectures perform equally well (table 1). For shortest termed predictions, as reported in [2], the absolute-tri-hourly architecture generally performs slightly better than the absolute-daily architecture, which itself performs better than the relative-daily architecture. Let us note that the absolute-tri-hourly architecture reduces the error made by the relative-daily approach at most by 10.3%.

Bayesian learning (i.e. also involving the evidence procedure) produces models that perform similarly to non-Bayesian ones. Indeed, the cross-validation procedure alone already prevents over-fitting. As suggested by the end of section 3.2, the interest of the Bayesian approach lies in encouraging the minimal amount of input variables to be effectively used. Bayesian learning that occurs after variable selection performs at best similarly to learning without variable selection. At worst, the associated error is increased by 5.6%. Such a degeneracy is rare though, the average error increase being 1.3%. Specifically, this means that an absolute architecture learnt after variable selection is generally more performant than the simplest one (relative-daily) before variable selection.

4.2   Variables Selection Analysis

In table 2, the nature of the inputs selected by the Bayesian method is highlighted. This gives an overview of the relative importance of each meteorological input for prediction. For the relative-daily architecture, 17 to 19 variables (over 29) are selected after Bayesian learning. This number drops down to [8,13] for the absolute-daily architecture, and down to [7,9] for the absolute-tri-hourly architecture. Air pressure seems to play a dominant role in this selection: indeed, on average it represents 47.8% of selected inputs, and its slots are almost always selected more frequently than any other meteorological variable. Air temperature is the second most represented variable, with an average of 30.2% of selected inputs. Past irradiation measurements are also important for irradiation prediction, but are almost insignificant for temperature prediction. Finally, past hygrometry measures play a minor role in both cases.

To evaluate to which extent this selection may be site-specific, the saliency of overlapping variables among those selected for each site is given in the table 2 (95% being the threshold for selection). The overlap is high for the relative-daily architecture (from 85% to 94%), but weakens for absolute architectures (from 69.5% to 85%). Yet 70% is a satisfactory part of the total saliency: thus input selections may be generalized for all architectures. However, there could be an important variability of this estimate among the MLP's that constitute an absolute architecture: site-specific variable sets for the MLP's with the lowest overlap rates might then be used for deployment.

Table 2. Total and average number of selected inputs for each meteorological variable, with the saliency statistics of overlapping selected inputs between Saclay and Marcoule

	Hygro. (8 slots)	Irradiation (5 slots)	Temp. (8 slots)	Pressure (8 slots)	Total (29 slots)	% saliency of overlapping inputs
Irradiation - Saclay						
Relative-daily	2.0	4.0	3.0	8.0	17.0	85.2%
Absolute-daily	0.9	1.6	2.1	5.1	9.7	74.4 [± 22.1]%
Absolute-tri-hourly	0.5	1.0	1.7	4.7	7.9	70.7[± 25.8]%
Irradiation - Marcoule						
Relative-daily	3.0	5.0	4.0	7.0	19.0	86.7%
Absolute-daily	1.0	1.6	1.5	4.3	8.4	84.2[± 12.3]%
Absolute-tri-hourly	0.8	1.2	1.5	4.2	7.7	79.0 [± 18.0]%
Temperature - Saclay						
Relative-daily	2.0	1.0	8.0	8.0	19.0	94.3%
Absolute-daily	1.2	0.6	4.8	6.4	13.0	69.5[± 24.4]%
Absolute-tri-hourly	0.8	0.1	3.2	4.4	8.5	70.8 [± 18.8]%
Temperature - Marcoule						
Relative-daily	2.0	1.0	8.0	8.0	19.0	93.8%
Absolute-daily	1.2	0.3	5.0	5.6	12.1	85.0[± 3.0]%
Absolute-tri-hourly	0.9	0.4	3.4	3.2	7.9	76.9[± 20.0]%

Table 3. Comparative analysis of architecture complexities, after variable selection. Respective results for non-Bayesian architectures are given in [2]: average K values are reported from this reference to the second column of the table, for discussion clarity.

		# of MLP	avg(K) / MLP	After Bayesian variable selection				
				avg(K) / MLP	avg(|**w**|) / MLP	total # of ops. ∝ (prediction)	# of ops. / MLP ∝ (learning)	total # of ops. ∝ (learning)
Irradiation								
Relative daily	Saclay	1	12.0	16.2	356	356	127021	127021
	Marcoule	1	10.8	13.8	331	331	109693	109693
Absolute daily	Saclay	8	2.1	3.5	51	51	2647	21177
	Marcoule	8	1.4	3.1	42	42	1726	13805
Absolute tri-hourly	Saclay	40	1.1	2.7	24	120	577	23098
	Marcoule	40	1.0	2.0	17	87	303	12110
Temperature								
Relative daily	Saclay	1	17.4	12.6	340	340	115736	115736
	Marcoule	1	10.8	12.6	340	340	115736	115736
Absolute daily	Saclay	8	2.7	4.0	84	84	7056	56448
	Marcoule	8	2.3	3.7	74	74	5531	44247
Absolute tri-hourly	Saclay	64	1.1	3.2	30	243	924	59146
	Marcoule	64	1.1	3.2	28	228	811	51911

4.3   Complexity Analysis

The counterpart of a reduced input space is generally a higher number of hidden neurons K, as reported in table 3. In average, with variable selection this value is 83% higher than for the respective non-Bayesian model, and up to 3 times higher when considering absolute architectures. Yet the influence of the input space reduction remains dominant: learning and prediction computational costs are always reduced by the Bayesian approach with variable selection, by 30.1% in average, and up to 72.1%. The average prediction and learning cost reduction amounts respectively to 20.1% and 36.1% for the absolute-daily architecture in this context. This architecture also compares favorably to alternatives such as the relative-daily architecture, and divides associated learning and prediction costs by up to 10. Following the conclusions in [2], now under a Bayesian setting, and the remarks in section 4.1 regarding predictive accuracy, this architecture remains the best compromise between quality and cost.

5   Conclusion

In this paper, the qualitative and computational consequences of Bayesian variable selection for solar irradiation and air temperature prediction with neural architectures were detailed. Neural architectures learnt with relevant subsets of variables were compared to their Bayesian and non-Bayesian counterparts.

The usage of a subset of variables implies a small performance loss and a gain of computational resources. The best compromise is achieved by the absolute-daily architecture after variable selection. This model uses one distinct MLP for each admissible current time slot that predicts a variable for the upcoming 24 hours. This study also outlines the preponderant role of the latest air pressure measurements, which highly populate the selected inputs of all evaluated models. This selection appeared to generalize well across the 2 tested sites.

Extensions to this work may include the combination of neural architectures, such as presented in this paper, with a mixture of experts. This combination would aim at adapting existing models to a new site with little or no prior data.

References

1. Bishop, C.M.: Pattern Recognition and Machine Learning. Springer (2006)
2. Bruneau, P., Boudet, L., Damon, C.: Neural Architectures for Global Solar Irradiation and Air Temperature Prediction. In: International Conference on Artificial Neural Networks (2012)
3. Cao, J., Lin, X.: Study of Hourly and Daily Solar Irradiation Forecast Using Diagonal Recurrent Wavelet Neural Networks. Energy Conversion and Management 49, 1396–1406 (2008)
4. Giebel, G., Kariniotakis, G.: Best Practice in Short-term Forecasting. A Users Guide. In: European Wind Energy Conference & Exhibition (2007)
5. Nabney, I.T.: Netlab Algorithms for Pattern Recognition. Springer (2002)
6. Qi, M., Zhang, G.P.: Trend Time series Modeling and Forecasting with Neural Networks. IEEE Transactions on Neural Networks 19(5), 808–816 (2008)
7. Voyant, C., Muselli, M., Paoli, C., Nivet, M.L.: Numerical Weather Prediction (nwp) and Hybrid Arma/ann Model to Predict Global Radiation. Energy 39, 341–355 (2012)
8. Wu, J., Chan, C.K.: Prediction of Hourly Solar Radiation Using a Novel Hybrid Model of Arma and Tdnn. Solar Energy 85, 808–817 (2011)

Integrated Problem Solving Steering Framework on Clash Reconciliation Strategies for University Examination Timetabling Problem

J. Joshua Thomas[1,2], Ahamad Tajudin Khader[1],
Bahari Belaton[1], and Choy Chee Ken[2]

[1] School of Computer Sciences, University Sains Malaysia &
[2] KDU College Penang, Malaysia
joshopever@yahoo.com, {tajudin,bahari}@cs.usm.my,
kennychoy27@hotmail.com

Abstract. University Examination timetabling problem (UETP) is a hard combinatorial scheduling and it is fully automated. Visualization and steering of critical processing phases like examination clashes identification are essential for effective data analysis by examination timetabling research community. The visual analytics combined steering framework with interactive visualization techniques needs to take into account for both the inputs of the researcher (human timetabler) and the critical needs of the application including simulations and continuous visualization of significant events. In this work, we have developed an integrated problem solving environment (PSE) a computational steering mechanism for user-driven and automated steering interactions. The well sophisticated *ExamViz* user interface is for simulations, and analysis for critical conflict and reconciliation with visual cues. *ExamViz* provides the user steering control visualization over various parameters of the inputs, including the initial raw data solutions. The user are able to interact with conflicting data while the decision making algorithm in execution with visual cues to reconciliate the clashes and guide the user throughout the improvement of the solution. We have evaluated our approach for 13-real world examination timetabling problem (Carter et., al 1996) and the result are reported.

Keywords: Computational steering, Examination timetabling, Clash reconciliation.

1 Introduction

Automated timetabling schedulers simulation still requires hours or days of researcher's time to reach the optimality of results. Time and effort required for data input, output and further complicates the process. Computational steering is an investigative paradigm whereby the parameters of a running program can be altered according to what is seen in the currently visualized results of the simulation. Although these thoughts were expressed 25 years ago, they express a very simple idea that scientists want more interaction than is currently present in most simulation codes.

T. Huang et al. (Eds.): ICONIP 2012, Part IV, LNCS 7666, pp. 297–304, 2012.
© Springer-Verlag Berlin Heidelberg 2012

Computational steering has been defined as "the capacity to control the execution of long-running, resource-intensive programs" [1].

The modified definition suits to interactive visualization programs. As such, steering provides techniques to iteratively enhance or evaluate parameter changes in a complex computational environment. Hence, the paper focus on NP-hard optimization problems like resource-intensive programs with computationally complex constraint oriented examination timetabling problems.

The processing phases are (a) Pre processing phase (*Pre-P*) in this phase, an initial random solution is generated from the raw data in which we assume it as a set of unscheduled examination and the potency of future solutions are compared based on this initial solution. Naturally, as each of the examinations are assigned to a timeslot at random, it tends to generate the existence of possible conflicts where some examinations being assigned to a same timeslot, or multiple consecutive examinations. Therefore, with the adaptation of visually interactive elements (where data objects and relationships are represented in a simple, interactive and yet effective manner), users are allowed to directly interact with the data (schedule of events) graphically to modify as they like or perform reconciliation of conflicts. (b) During the processing phase (Due-P) has equipped with evolutionary algorithm (EA) and tuned dataset from Pre-P as input. The human timetablers are able to interact with the conflicting data which was represented in parallel coordinates style of visualization workspace while the algorithm is in execution. This sophisticated process of navigation allows the researcher to virtually "steer" through the computations and additionally, the computational steering mechanism is enhanced with an advanced analysis feature which displays the immediate results of every possible choices to further ease the user on clash reconciliations until the stopping criteria is invoked.

2   Related Work

Computational steering has been extensively studied over the past several years [4, 6]. A variety of steering systems have emerged like SCIRun [4], CUMULVS[1] etc. A taxonomy of different kinds of steering and steering systems and tools can be found in [4]. Performance steering allows scientists to change application parameters to improve application performance. Algorithmic steering uses an algorithm to decide application parameters to improve system and application performance [1]. Computational steering has been applied to different kinds of applications like molecular dynamics simulation, biological applications, astrophysics, computational fluid dynamics, atmospheric simulations [3, 6]. These frameworks were mainly developed for exploratory steering in order to change simulation parameters interactively and thereafter, visualizing the simulation output with the new parameters. CUMULVS [1] provides the user with a viewer and steering interface to modify the application's computational parameters and improving application performance. In our previous work [2], we have shown the proposed application and system parameters based on the uncapaciated nature of examination timetabling problem. To the best of our knowledge, ours is the first work that considers computational steering of examination timetabling applications with performance requirements.

3   Visual Analytics Steering Problem Solving Framework

The integrated problem solving environment *(PSE)* supported with visual analytics model that determines the application configuration for examination timetabling generations based on constraints and user input, a simulation process that performs examination scheduling with different configurations on each of the processing phases. The following subsections describe the primary components in detail:

3.1   User Interface: Workspace

The user gives input through the user interface. In particular, the user can specify the problem set from the 13 real-world dataset, the selected input values are sent to the Visualization workspace, and mapped with student, course coordinate axis collected from the raw data and further allocate the values to rooms and timeslots by drawing lines to the multiple axis. The sophisticated GUI has equipped with the user-driven steering interaction capabilities with the preprocessing phase *(Pre-P)*.

3.2   Interaction Manager

The interaction manager is the primary component of ExamViz and acts as the bridge between algorithmic steering and user-driven steering. The preprocessing *(Pre-P)* periodically update the clash reconciliation values (data) as new tuned input for the evolutionary decision algorithm for during the processing *(Due-P)*. Here the process checks the feasibility of running the simulation with the user inputs, advises the users of alternate options if not feasible and let the user invokes the decision algorithm with the user input parameters if feasible.   More details on the reconciliation of the algorithmic and user-driven steering are given in the section 4.2.

3.3   Simulation Process

The simulation processes in the examination timetabling that simulate the scheduling events. It starts with the preprocessing phase whereby the user-driven steering occurred by drag-n-drop the lines. Then during the processing phase is invoked once the parameters are configured. While the algorithm executes and assigning the scheduling of events from course to room and timeslots, the user able to interact with the clashed lines which is highlighted in red color from the course axis to the timeslot axis.

4   ExamViz

ExamViz is an integrated problem solving environment which allows human timetabler and researchers to perform a variety of tasks related to scientific computation. There are strategies in preprocessing phase such as exploration of the conflicting data in the search space, with visual cues and interact then reconciliate the clashes with drag and drop strategy. In during the processing phase the user able to communicate with the conflicting data while the algorithm under execution and facilitate heuristic strategies on assignment of events. This process is supported by the evolutionary algorithm to generate solutions with user interaction until the convergence and improvement of solution.

4.1   User-Driven Steering: Scientific Discovery in Preprocessing Phase (*Pre-P*)

The preprocessing phase consists of raw data randomization, initial clash identification and minimization, and constraint violation of patterns. An initial population is created and iteratively improved via the processes of evaluation. In addition to that, the system also provides a certain level of interactivity besides just graphic elements.

The rectangle box has colored concentric circles represents the population of chromosomes. For example, hovering the cursor over a chromosome (colored circles) will actually allow the preview of the chromosome contents as well as visualizing the content order on the axial canvas. Green circle appears the best fitness values (Fitness 1.0) has considered with the number of clashes. We have tested with the entire 13-real world benchmarking problem set provided by [7].

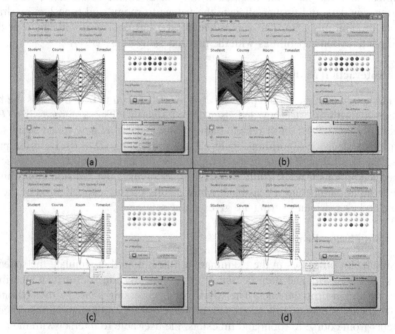

Fig. 1. User-Driven Steering on the preprocessing of examination timetabling

In Figure 1 (a) has shown the initial solutions with clashes and without clashes assignment of events. "Red" lines indicate clashes. "Grey" lines indicate the assignment without clashes. In Figure 1 (b) illustrates the detection of clash and resolve the clashes. The Room and Timeslot axial lines has enabled with "Blue" spots. The user click on the "Blue Spot" in the Room axis the corresponding Timeslot has steer the user by shown the available and not available timeslots symbolic with OPEN in "Green" and N/A "RED" symbols. This strategy is used as the primary method to reduce the clashes between Course to Rooms and Timeslot. In Figure 1 (c) the user interacts with the 10th Room and reduces the clashes from the 20th Timeslot. Currently the timeslot is shown as "not available (N/A)". Once the "RED" line moves to "available (OPEN)" timeslot the clashes are reduced. Figure 1 (d) shows the crossed lines are moved from 20th *Timeslot* N/A to 18th OPEN location thus minimizing the clashes from 783 to 778. This knowledge driven process are supported by a "tool-tip" information when the user moves the mouse cursor to a particular timeslot, a message appears and displays on how many number of students affected, what is the course ID, Room ID and slot ID that are involved. Moreover, the "Green" line visually draws the attention of the user on the affected student(s), course(s) to room and timeslot.

4.2 Algorithmic Steering and User-Driven Steering: Scientific Discovery in during the Processing Phase (*Due-P*)

The evolutionary algorithm makes a decision by determining the number of processors and the generations of schedules of examination timetable data based on the given problem set for simulation. The algorithm also takes as input (data) the execution times for different number of simulation. The execution considers the lower bound for students or upper bound for clashes between *rooms, courses* conflicts with which the human timetabler wants to visualize the allocation of events and conflicts. The objective of the evolutionary decision algorithm is to maximize the rate of simulations and to enable continuous visualization with maximum conflicts raised due to the violation of hard constraints. The traditional scheduling algorithms often minimizes the execution times while our algorithm may sometimes "slow- down" the simulations lead to faster consumption of visual elements satisfying the hard constraints over the problem set used for the particular simulation. Thus our problem is as an optimization problem that attempts to maximize simulation. In addition, as we have considered on the practicality of the application, we have added important hard constraints in order to ensure the quality of the visualization. Examinations are given priority, i.e. the number of free slots is available for the exam (saturation degree). The process begins with the creation of crossed line for a single value and the overall problem (hec-S-92) are represented as depicted in Figure 2. Now the users are able to click on the room(s) from the axis and identify which course or student containing the clash to *room* and to *timeslot*. If the maximum of allowed slots has already been reached the slot cannot occur and it randomly allocated to an existing *timeslot*.

4.3 Clash Reconciliation Algorithmic and User-Driven Steering Strategies

Initially, the interaction manager specifies default values for simulation on setting the algorithmic filters or configuration elements through the interface. The interaction manager then determines the events of output using the evolutionary decision making algorithm based on the constraints for the given problem set simulation. The simulation is started with these values. The user at the visualization "Workspace" can change the simulation parameters during execution through the user interface and this can be seen on the bottom right hand side with tab strips controls. The visualization workspace allows the user to specifically interact the declared rooms, timeslots which is steered by the "tool-tip" information to drag-n-drop the conflicted lines (RED) to the available (OPEN) timeslots.

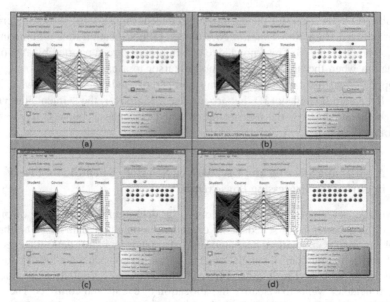

Fig. 2. User-driven and Algorithmic steering in during the processing phase

When a user specify/touch/interact a room (BLUE circle) the application proactively checks the feasibility with the conflicted constraints highlighted in "RED" lines between rooms and timeslots. However the visualization has shown various related information to course to rooms, timeslots and number of students affected. In Figure 2(d) explains second strategy of another computational steering mechanism with interval values or "Marvel Values" which gives visual cues to the user together with the OPEN and N/A cues. This Marvel Values is of integer values, whereby the user moves the conflicted line to the highest -ve (negative) values that weighted values of constraints are reduced and if they move conflicted line to the lowest +ve(positive) the conflicts has increased. In other words, as to why highest -ve (negative) values are presumed as the most favorable is simply because the result of the interaction will reduce the overall clashes significantly! These steering mechanisms are controlled by the framework and it assists the user towards the improvement of the solutions.

5   Evaluation

The performance of adapting visual analytics with steering visual representations is experimentally studied using benchmarking dataset [7]. The Carter dataset used in this study consists of 13 datasets, which reflect the real-world examination timetabling problems. For the purpose of our study, 12 datasets circulated in the literature were used. The characteristics of Carter datasets, varying in size and complexity, are shown in Table 1. The conflict matrix in the last column illustrates density, which is the ratio between the numbers of students share the exams and the total number of elements in the conflict matrix [5].

Table 1. Characteristics of the uncapacited examination timetabling dataset

Dataset	Time-periods	Exams	Students	Density
CAR-S-91-I	35	682	16925	0.13
CAR-F-92-I	32	543	18419	0.14
EAR-F-83-I	24	190	1125	0.27
HEC-S-92-I	18	81	2823	0.42
KFU-S-93	20	461	5349	0.06
LSE-F-91	18	381	2726	0.06
RYE-S-93	23	481	11483	0.07
STA-F-83-I	13	139	611	0.14
TRE-S-92	23	261	4360	0.18
UTA-S-92-I	35	622	21266	0.13
UTE-S-92	10	184	2750	0.08
YOR-F-83-I	21	181	941	0.29

Table 2. Effectiveness of the computational steering process in exam timetabling data

			Algorithmic Steering and User-driven steering experiments							
			Configuration in ExamViz					Computational Steering		
Data set	Size	cases	No. of Rooms	No. of Timeslots	Initial Clashes	Density	overflow	Power of (2)	User Steering reduces	Algorithmic Steering reduces
Car-S-91		1	20	50	544	0.54	-	5	455	351
		2	10	50	364	1.09	43	5	300	255
		3	30	40	685	0.45	-	5	600	555
	Large	4	50	50	472	0.22	-	5	400	400
		5	50	40	702	0.27	-	5	677	557
		6	40	30	941	0.45	-	5	850	804
		7	50	20	1005	0.36	-	5	823	759
ear-f-83		1	20	50	236	0.19	-	3	109	93
		2	10	50	216	0.38	-	3	199	95
		3	30	40	300	0.16	-	3	255	192
	Medium	4	50	50	220	0.08	-	3	185	175
		5	50	40	240	0.1	-	3	100	93
		6	40	30	429	0.16	-	3	359	299
		7	50	20	625	0.19	-	3	525	422
		6	40	30	162	0.32	-	3	142	123
		7	50	20	327	0.38	-	3	267	253
yor-s-92		1	20	50	173	0.18	-	2	156	134
		2	10	50	163	0.36	-	2	157	135
		3	30	40	232	0.15	-	2	200	146
	Small	4	50	50	169	0.07	-	2	145	122
		5	50	40	203	0.09	-	2	187	136
		6	40	30	300	0.15	-	2	256	244
		7	50	20	444	0.18	-	2	400	348

Table 2 investigates the influence of visual interactions with the steering significant to the user to quickly construct the initial solutions. Figure 3 has shown the performance of the user-driven steering versus the algorithmic steering on the reconciliation of constraints which was simulated, interacted in each of the 13 real world dataset.

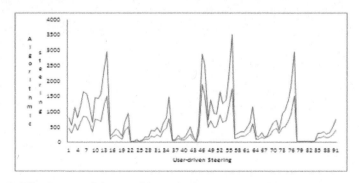

Fig. 3. Effectiveness of the user-driven steering versus algorithmic steering analysis

6 Conclusion

In this paper, we have described our integrated visual analytics steering framework, *ExamViz* that combines user-driven steering with automatic tuning of application parameters based on constraints and the needs of the examination timetabling application to determine the clash reconciliation. *ExamViz* proactively analyzes the impact of user inputs, advises the user on violations, guides with alternate options, and arrives at the initial solution with multiple steering strategies in preprocessing phase and during the processing phase of the examination timetabling for the improvement of best solutions. We have followed the similar heuristic ordering strategies same as the timetabling community.

References

1. Gu, W., Vetter, J., Schwan, K.: An Annotated Bibliography of Interactive Program Steering. ACM SIGPLAN Notices (1994)
2. Thomas, J.J., Khader, A.T., Belaton, B.: A Parallel Coordinates Visualization for the Uncapaciated Examination Timetabling Problem. In: Badioze Zaman, H., Robinson, P., Petrou, M., Olivier, P., Shih, T.K., Velastin, S., Nyström, I. (eds.) IVIC 2011, Part I. LNCS, vol. 7066, pp. 87–98. Springer, Heidelberg (2011)
3. Kohl, J.A., Wilde, T., Bernholdt, D.E.: Cumulvs: Interacting with High-Performance Scientific Simulations, for Visualization, Steering and Fault Tolerance. International Journal of High Performance Computing Applications (2006)
4. Parker, S., Miller, M., Hansen, C., Johnson, C.: An Integrated Problem Solving Environment: the SCIRun Computational Steering System. In: Proceedings of the Thirty-First Hawaii International Conference on System Sciences (1998)
5. Qu, R., Burke, E., McCollum, B., Merlot, L., Lee, S.: A survey of search methodologies and automated system development for examination timetabling. Journal of Scheduling 1, 12(1), 55 (2009)
6. Wright, H.R., Crompton, H., Kharche, S., Wenisch, P.: Steering and Visualization: Enabling Technologies for Computational Science. Future Generation Computer Systems 26, 506–513 (2010)
7. Carter, M., Laporte, G.: Recent Developments in Practical Examination Timetabling. Practice and Theory of Automated Timetabling 121 (1996)

A Data Gathering Scheme in Wireless Sensor Networks Using a Spiking Neural Network with Simple Local Information

Ikki Fujita, Hidehiro Nakano, and Arata Miyauchi

Tokyo City University
1-28-1,Tamazutsumi,Setagaya-ku,Tokyo,158-8557 Japan
{fujita,nakano,miyauchi}@ic.cs.tcu.ac.jp

Abstract. In wireless sensor networks (WSNs), compact wireless sensor nodes are deployed in observation area, form an impromptu network and gather sensing data periodically. Thereby, the environmental observation of large-scale area can be realized remotely. In the synchronization-based data gathering scheme, transmission timings for sensing data are synchronized by using spiking neural oscillators. Using this scheme, the number of the transmissions and receptions, and node power consumption can be reduced. However, in the conventional scheme, the duplicate transmissions of the same sensing data by plural wireless sensor nodes should be improved.

In this paper, a new data gathering scheme reducing duplicate transmissions of sensing data is proposed. In the proposed scheme, simple local information is used for transmissions and receptions of sensing data. Then, the traffic of the whole network can be significantly reduced. In the simulation experiments, the effectiveness of the proposed scheme is verified.

1 Introduction

In WSNs, compact wireless sensor nodes are deployed in an observation area, form an impromptu network and gather sensing data periodically[1][2]. Thereby, the environmental observation of large-scale area can be realized remotely. Each node consists of a sensor which measures a state (temperature, humidity, movement) for the observation, a limited processing function and a simple wireless transceiver. In addition, it usually operates by a battery. Therefore, efficient data gathering schemes saving node power consumption are required if data gathering is assumed to be performed by a lot of sensor nodes. Various data gathering schemes have been proposed in the conventional researches. In the synchronization-based data gathering scheme, transmission timings for sensing data are synchronized by using spiking neural oscillators[3]-[6]. If synchronization is achieved, each wireless sensor node can switch off the power supply of the wireless transceiver when it does not transmit sensing data. Then, the number of the transmissions and receptions, and node power consumption can be reduced.

T. Huang et al. (Eds.): ICONIP 2012, Part IV, LNCS 7666, pp. 305–312, 2012.
© Springer-Verlag Berlin Heidelberg 2012

However, in the conventional scheme, the duplicate transmissions of the same sensing data by plural wireless sensor nodes should be improved.

In this paper, a new data gathering scheme reducing duplicate transmissions of sensing data is proposed. In the proposed scheme, three values in each node are used for transmission and reception of sensing data. They are hop count to a sink node, sender ID and receiver ID. Sensing data is transmitted and received only if these IDs match to each other. By using the proposed scheme, the duplicate transmissions can be reduced. Then, the traffic of the whole network can be reduced. In the simulation experiments, the effectiveness of the proposed scheme is verified.

2   Synchronization-Based Data Gathering Scheme

First, a synchronization-based data gathering scheme presented in Ref.[5] is explained. A wireless sensor network consisting of N wireless sensor nodes is considered. Each wireless sensor node S_i $(i = 1, \cdots, N)$ has a timer which controls timing to transmit and receive sensing data. The timer in S_i is characterized by a phase $\phi_i \in [0,1]$, an internal state $x_i \in [0,1]$, a continuous and monotone function f_i, a nonnegative integer distance level $l_i > 0$, and an offset time δ_i. When there are not the couplings between wireless sensor nodes, the phase ϕ_i of a sensor node S_i changes as the following equation.

$$\frac{d\phi_i(t)}{dt} = \frac{1}{T_i}, \quad \text{for } \phi_i(t) < 1 \tag{1}$$

$$\phi_i(t^+) = 0, \quad \text{if } \phi_i(t) = 1 \tag{2}$$

where T_i denotes a period of the timer in S_i. That is, if the phase ϕ_i reaches the threshold 1, S_i is said to fire, and the phase ϕ_i is reset to 0 based on Equation(2), instantaneously. The internal state x_i is determined by the continuous and monotone function $f_i(\phi_i)$ where $f_i(0) = 0$ and $f_i(1) = 1$. Increase of the phase ϕ_i causes increase of the internal state x_i. If x_i reaches the threshold 1, x_i is reset to the base state 0, instantaneously.

The couplings between each wireless sensor node are realized by the following manner. Let S_j be one of the neighbor wireless sensor nodes located in the radio range of a wireless sensor node S_i. The wireless sensor node S_i has a nonnegative integer distance level l_i characterized by the number of hop counts from a sink node. The wireless sensor node S_i transmits the own distance level l_i and the own sensing data as a stimulus spike signal. If S_j receives the signal from S_i, S_j compares the received distance level l_i with the own distance level l_j. If $l_i < l_j$ is satisfied, S_j is said to be stimulated by S_i, and the phase and internal state of S_j change as follows:

$$x_j(t^+) = B(x_j(t) + \epsilon_j) \tag{3}$$

$$B(x) = \begin{cases} x, & \text{if } 0 \leq x \leq 1 \\ 0, & \text{if } x < 0 \\ 1, & \text{if } x > 1 \end{cases} \tag{4}$$

$$\phi_j(t^+) = f_j^{-1}(x_j(t^+)) \tag{5}$$

where ϵ_j denotes a strength of the stimulus. After S_j is stimulated, S_j does not respond to all stimulus signals from the neighbor wireless sensor nodes during an offset time δ_j. That is, each wireless sensor node has a refractory period corresponding to the offset time.

The stimulus signals are transmitted by the following manner. A wireless sensor node S_i broadcasts stimulus signals offset time δ_i earlier than the own firing time. That is, S_i broadcasts the stimulus signals if the following virtual internal state x_i' considered the offset time δ_i reaches the threshold 1.

$$\phi_i' = \phi_i + \delta_i \pmod{1} \tag{6}$$

$$x_i' = f_i(\phi_i') \tag{7}$$

Distance levels of each wireless sensor node are adjusted by the following manner. Initially, distance levels of each wireless sensor node are set to sufficiently large values, and that of the sink node is set to 0. A sink node broadcasts "level 0" as a beacon signal. If wireless sensor node S_j received a stimulus signal from other sensor node S_i, S_j compares the own distance level l_j with the received distance level l_i. If $l_i < l_j$, S_j receives a stimulus, and x_j is changed and updates l_j to $l_i + 1$. In other cases $l_i = l_j$ or $l_i > l_j + 1$, all the received stimulus signals are disregarded. As a result, each wireless sensor node has a distance level as corresponding to hop counts to its nearest sink node.

Sensing data is transmitted and received as the following. S_j receives sensing data from its neighbor wireless sensor node S_i if $l_i = l_j + 1$ is satisfied. Then, S_j aggregate the received sensing data and own sensing data. After that, S_j transmits the aggregated sensing data. Sensing data are transmitted and received in each firing period.

If a synchronization is achieved by the above explained manner considered offset time, sensor nodes with larger levels can transmit sensing data earlier than sensor nodes with smaller levels. By setting the offset time to sufficiently large value considered a collision of the MAC layer, transmissions begin from wireless sensor nodes far from a sink node sequentially.

3   Proposed Scheme

In this section, a new data gathering scheme reducing duplicate transmissions of sensing data is explained. As the same as the conventional scheme, a wireless sensor network consisting of N wireless sensor nodes are considered. Each wireless sensor node S_i $(i = 1, \cdots, N)$ has a timer. When there are not the couplings between wireless sensor nodes, the phase ϕ_i and the internal state x_i of a sensor node S_i change using Equations (1) and (2).

The couplings between each wireless sensor node are realized by the following manner. Let S_j be one of the neighbor wireless sensor nodes located in the radio range of a wireless sensor node S_i. As compared with the conventional scheme, the wireless sensor node S_i has not only a nonnegative integer distance level l_i

but also a sender ID ns_i and a receiver ID nr_i representing the relations between neighbor wireless sensor nodes (see Fig.1). The wireless sensor node S_i transmits the own values of l_i, ns_i and nr_i, as a stimulus signal. If S_j receives the signal from S_i, S_j compares the received l_i, ns_i and nr_i with the own l_j, ns_j and nr_j. If $l_i < l_j - 1$ is satisfied or if $l_i = l_j - 1$ and $nr_i = ns_j$ are satisfied, S_j is stimulated by S_i, and the phase and internal state of S_j change as Equations (3)-(5). In other cases, all the received stimulus signals are disregarded. After S_j is stimulated, S_j does not respond to all stimulus signals from the neighbor wireless sensor nodes during an offset time δ_j. The stimulus signals are transmitted based on Equations (6) and (7).

Distance levels, sender IDs, and receiver IDs of each wireless sensor node are adjusted by the following manner. Initially, distance levels, sender IDs and receiver IDs of each wireless sensor node are set to sufficiently large values. A sink node broadcasts "level 0" and "receiver ID 0" as a beacon signal.

If wireless sensor node S_j received a stimulus signal from other sensor node S_i, S_j compares the received l_i, ns_i and nr_i with the own l_j, ns_j and nr_j. If $l_i < l_j - 1$ is satisfied, S_j updates l_j to $l_i + 1$. Furthermore, ns_j and nr_j are updated to the value of nr_i as shown in Fig.1(a). If $l_i = l_j$ and $nr_i = nr_j$ are satisfied, nr_j is updated to $nr_j + 1$ as shown in Fig.1(b).

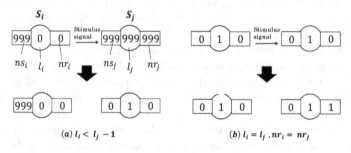

Fig. 1. Update of the distance levels, sender IDs and receiver IDs

As a result, each wireless sensor node has a distance level as corresponding to hop counts to its nearest sink node, and sender ID and receiver ID as corresponding to relations between neighbor wireless sensor nodes.

Sensing data is transmitted and received as the following. S_j receives sensing data from its neighbor wireless sensor node S_i if $l_i = l_j + 1$ and $ns_i = nr_j$ are satisfied. Then, S_j aggregate the received sensing data and own sensing data. After that, S_j transmits the aggregated sensing data. Sensing data transmitted and received in each firing period.

If data reception is performed on the above conditions, the communication paths of sensing data can be constructed as shown by the solid lines in Fig.2. On the other hand, in the conventional scheme, redundant paths can not be reduced as shown by not only the solid lines but also the dashed lines in Fig.2. Then, the duplicate transmissions of the same sensing data may occur and the network lifetime may be reduced. Note that the proposed scheme does not use complex routing functions but introduces the simple local information, sender and receiver IDs. Using the proposed scheme, the duplicate transmissions can be reduced.

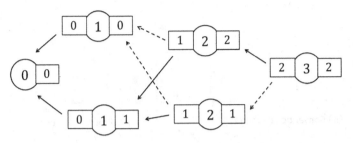

Fig. 2. Example of communication paths of sensing data

4   Experiments

In order to confirm the effectivity of the proposed scheme, numerical simulations are performed. First, a simulation was carried out in the wireless sensor network model as shown in Fig.3. In the figure, 6 wireless sensor nodes are deployed at (8,13),(8,8),(13,13),(13,8),(18,13),(18,8), and sink node is placed on (3,10). The radio ranges of each wireless sensor node and sink node are set to 8. Initial values of internal states in each wireless sensor node are set to random values. The parameters are fixed as follows.

$$\epsilon = 0.4, \ \delta = 0.4.$$

Fig.4 shows the finally decided communication paths of sensing data for the conventional and proposed schemes.

Next, a simulation was carried out in the wireless sensor network model as shown in Fig.5. In the figure, 100 wireless sensor nodes are deployed at random locations on 4 concentric circles whose centers are (0,0), and sink node are

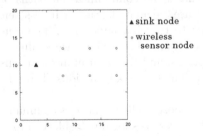

Fig. 3. A model of a wireless sensor network 1

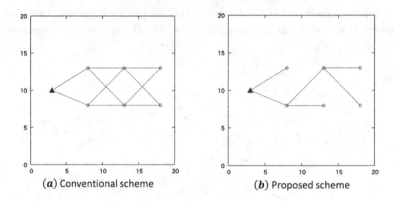

(a) Conventional scheme   (b) Proposed scheme

Fig. 4. Communication paths for the network model 1

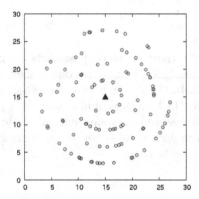

Fig. 5. A model of a wireless sensor network 2

located on the center. The radio ranges of each wireless sensor node and sink node are set to 5. The radii of each concentric circle are set to 3, 6, 9 and 12, respectively. $10n$ wireless sensor nodes are located on the n-th concentric circle from the center. Fig.4 shows the communication paths of sensing data. Next, the following quantities are calculated. Each wireless sensor node is assumed to transmit sensing data to the sink node once. The sensing data are relayed by multi-hop wireless communication. Then, we evaluate the total number of transmissions and receptions, and the total number of sensing data gathered at the sink node considered duplicate transmissions. Table 1 shows the simulation results.

Fig.4 shows that the proposed scheme has less communication paths than the conventional scheme. Moreover, as shown in Table 1, duplicate transmissions for the same sensing data can be significantly reduced.

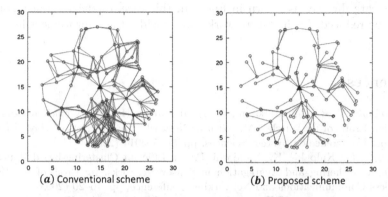

(*a*) Conventional scheme   (*b*) Proposed scheme

Fig. 6. Communication paths for the network model 2

Table 1. Results for the duplicate transmissions

	conventional scheme	proposed scheme
# of transmissions and receptions	10694	3794
# of the sensing data gathered by the sink node	3347	104

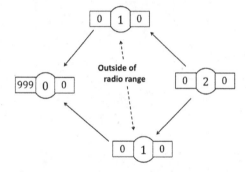

Fig. 7. Duplicate transmissions for the same sensing data in the proposed scheme

5   Conclusion

In this paper, a new data gathering scheme reducing duplicate transmissions of sensing data has been proposed. Through numerical simulations, it has been shown that the proposed scheme can reduce the redundant communication paths, and duplicate transmissions for the same sensing data.

In the proposed scheme, sensor nodes with the same distance level have different receiver IDs if they can communicate directly. Then, duplicate transmissions can be reduced. However, they can have the same receiver IDs if they can not communicate directly as shown in Fig.7. In this case, duplicate transmissions can not be reduced. In the future works, we consider the improvement for such a case.

References

1. Heinzelman, W.R., Chandrakasan, A., Balakrishnan, H.: Energy-efficient Communication Protocol for Wireless Microsensor Networks. In: Proc. of the Hawaii International Conference on System Sciences, pp. 3005–3014 (2000)
2. Dasgupta, K., Kalpakis, K., Namjoshi, P.: An Efficient Clustering-based Heuristic for Data Gathering and Aggregation in Sensor Networks. In: Proc. of the IEEE Wireless Communications and Networking Conference, pp. 16–20 (2003)
3. Mirollo, R.E., Strogatz, S.H.: Synchronization of Pulse-coupled Biological Oscillators. SIAM J. Appl. Math. 50, 1645–1662 (1990)
4. Catsigeras, E., Budelli, R.: Limit Cycles of a Bineuronal Network Model. Physica D 56, 235–252 (1992)
5. Wakamiya, N., Murata, M.: Synchronization-based Data Gathering Scheme for Sensor Networks. IEICE Trans. Communications E88-B(3), 873–881 (2005)
6. Nakano, H., Utani, A., Miyauchi, A., Yamamoto, H.: Data Gathering Scheme Using Chaotic Pulse-Coupled Neural Networks for Wireless Sensor Networks. IEICE Trans. Fundamentals E92-A(2), 459–466 (2009)

Utilizing Symbolic Representation in Synergistic Neural Networks Classifier of Control Chart Patterns

Kittichai Lavangnananda and Pantharee Sawasdimongkol

Data and Knowledge Engineering Lab., School of Information Technology (SIT)
King Mongkut's University of Technology Thonburi (KMUTT)
Bangkok, Thailand
{Kitt,Pantharee.boom}@sit.kmutt.ac.th

Abstract. Control Chart Patterns (CCPs) can be considered as time series. Industry widely used them in their process control. Therefore, accurate classification of these CCPs is vital as abnormalities can then be detected at the earliest stage. This work proposes a framework for neural networks based classifier of CCPs. It adopts a symbolic representation technique known as Symbolic Aggregate ApproXimation (SAX) in preprocessing. It was discovered that difficulty in classifying CCPs with high signal to noise ratio lies in differentiating among three very similar categories within their six categories. Synergism of neural networks is used as the classifier. Classification comprises two levels, the super class and individual category levels. The recurrent neural network known as Time-lag network is selected as classifiers. The proposed method yields superior performance than any previous neural network based classifiers which used the Generalized Autoregressive Conditional Heteroskedasticity (GARH) Model to generate CCPs.

Keywords: Classification, Control Chart Patterns (CCPs), Neural Networks, Process control, Symbolic Aggregate Approximation (SAX), Symbolic Representation, Time Series.

1 Introduction

A time series is a sequence of data points, measured typically at successive time instants spaced at uniform time intervals. Many fields of science ranging from bioinformatics to manufacturing to telecommunication are involved with time series information. Therefore, ability to detect, monitor and predict their time series information is beneficial. This is especially critical in manufacturing process where detecting abnormalities at early stage is crucial. This paper is concerned with classification of time series in manufacturing knows as Control Chart Patterns (CCPs). Section 2 briefly introduces this particular time series. Various methods for classification CCPs and related previous work are presented in Section 3.This work utilized a symbolic representation technique in preprocessing of CCPs prior to their classification by Synergistic neural networks classifier. The proposed method yielded the better accuracy than previous works. The details of the proposed method and results are explained in

T. Huang et al. (Eds.): ICONIP 2012, Part IV, LNCS 7666, pp. 313–321, 2012.
© Springer-Verlag Berlin Heidelberg 2012

Sections 4-9. Section 10 discusses the merits of the work while Section 11 concludes the work and suggests directions for future investigation.

2   Control Chart Pattern (CCPs)

CCPs are time series that show the level of a machine parameter plotted against time. In reality, CCPs can be different and time consuming to obtain, hence most previous researches relied on mathematical model. The most popular one is GARH (Generalized Autoregressive Conditional Heteroskedasticity Model)[1] as they represent most commonly occur pattern in manufacturing process. These patterns are normal, cyclic, increasing trend, decreasing trend, upward shift and downward shift. Figure 1 illustrates these patterns.

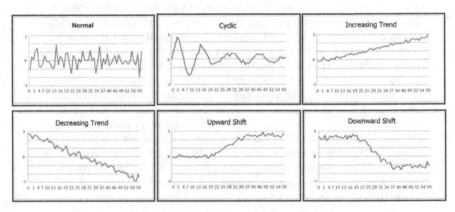

Fig. 1. Example of six Control Chart Patterns

The GARH (Generalized Autoregressive Conditional Heteroskedasticity Model) can be described as given below.

1. Normal: $y(t) = \mu + r(t)\sigma$ (1)
2. Cyclic: $y(t) = \mu + r(t)\sigma + a\sin(2\pi t/T)$ (2)
3. Increasing trend: $y(t) = \mu + r(t)\sigma + gt$ (3)
4. Decreasing trend: $y(t) = \mu + r(t)\sigma - gt$ (4)
5. Upward shift: $y(t) = \mu + r(t)\sigma + ks$ (5)
6. Downward shift: $y(t) = \mu + r(t)\sigma - ks$ (6)

where

$y(t)$	=	Time Series value	a =	Amplitude of cyclic variations
μ	=	Mean value	g =	Magnitude of gradient trend
$r(.)$	=	Normally distributed	k =	Determines shift position
t	=	Time	s =	Shift magnitude
σ	=	Standard deviation	T =	Period of cycle

From equations (1) – (6), σ is the most critical parameter as it dictates the level of noise which contain on them (i.e. higher value of σ generates patterns which higher degrees of signal to noise ratio). In this work, a pattern comprises of 60 interval values (i.e. t_1 to t_{60}) as use in most previous works which adopted this model.

3   Related Work

There exist several statistical methods in classification of CCPs. The very recent statistical approach[2] classified CCPs based on Bayesian inference and Maximum Likelihood Estimation by assuming its existence, the Maximum Likelihood Estimator of pattern parameters were obtained and then a measure called Belief was then determined. These works could classify CCPs in satisfactory results.

Earliest work which attempted to study the classification performance of neural networks on GARH model and was proven superior to the use of expert systems and statistical methods is in [3]. A systematic study of neural network capability to classify these CCPs was carried out in [4].

To improve classification, preprocessing techniques were introduced. One of the more promising approaches is to extract useful features from the signals. This advantage was affirmed in [5]. The first attempt was carried out by [6] where four features (mean, standard deviation, skewness and kurtosis) used in image processing were extracted from the original pattern. This work was extended later by introducing synergy of neural networks together with two more features (slope and Pearson correlation coefficient) and additional transformations of original patterns [7]. To date, it achieved the highest accuracy in classification of CCPs by means of statistical features extraction, together with additional transformation and synergism of neural networks.

The difficulty in learning or mining information from time series is due to its nature as data in time series are real value and homogeneous, especially if they are used as input to learning algorithms.

Therefore, many works which involves time series had attempted to transforms them into several suitable forms, these include Discrete Fourier Transform, Piecewise Linear Approximation, Max. Wavelet Transform and Piecewise Constant Approximation [8] depending on the application intended. This approach can be applied to preprocessing of CCPs too. Recent approach to the preprocessing is to transform the CCPs into a symbolic representation. Among a few symbolic representations exist, a recent and promising symbolic representation known as Symbolic Aggregate Approximation (SAX) [9] had been used in preprocessing for an evolutionary classifier known as Self-adjusting Association Rules Generator (SARG)[10] with some satisfactory results. The work also identified suitable parameters of SAX (this aspect is described in Section 4) for various σ values of the GARH model. The major drawback to the work is that application of SARG is time consuming, especially when compared with neural network classifiers.

The work in [10] also suggests that using SAX in preprocessing is likely to transform the CCPs into a more suitable format for a classifier than features extraction approach in previous attempts. As training neural networks is considerably less time consuming and demand less resources (i.e. memory) than SARG. A neural network based classifier for CCPs where SAX was utilized in preprocessing was implemented [11]. The work yielded the best performance for neural network based classifiers of CCPs using to date.

4   Symbolic Aggregate Approximation   (SAX)

Symbolic Aggregate ApproXimation (SAX) was first introduced in [12]. SAX has the advantage of dimensionality and noise reductions, and it also allows distance measures to be defined by symbolics that lower bound corresponding distance measures defined on the original series. SAX allows a time series of arbitrary length n to be reduced   to a string of arbitrary length l, ($l<n$, typically $l<<n$).

In SAX, the length of X-axis is represented as Piecewise Aggregate Approximation segments (l) is defined to reduce dimensions of the time series.The length of Y-axis represented as Alphabet Size (s), the value can be any arbitrary integer of at least >2. Hence, a time series data T of length n can now be represented in a l-dimensional space by a vector $\bar{T} = \bar{T}_1, ..., \bar{T}_l$. The ith element of $\bar{T}$ is calculated by following equation :

$$\bar{T}_i = \sum_{j=\frac{n}{l}(i-1)+1}^{\frac{n}{l}i} T_j \tag{7}$$

Table 1. Summarizes the Notations used in SAX

T	A time series $T = T_1, ..., T_2$
$\bar{T}$	A PAA of time series $\bar{T} = \bar{T}_1, ... \bar{T}_l$
$\hat{T}$	A symbol representation of time series $\hat{T} = \hat{T}_1,..\hat{T}_l$
l	The number of PAA segments representing time series T
s	Alphabet size (e.g., for the alphabet = {a,b,c} , $s = 3$

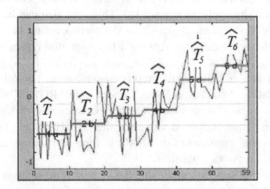

Fig. 2. An Example of SAX representation

In summary, to reduce the time series from n to l dimensions, the time series is divided into l equal sized 'frames'. The representation can be visualized as an attempt to approximate the original time series with a linear combination of box basis functions. For example, Figure 2 illustrates SAX representation of an upward shift control chart for $n = 60$, $l = 6$ and $s = 4$.

The result of the SAX representation for the upward shift above is '1a2b3b4b5d6d'. Empirical study in[10]had confirmed that good combination for PAA(l) and Alphabet Size(s) values for SAX is vital for its successful application. The work also suggested suitable combinations of PAA(l) and Alphabet Size(s) for σ = 5 to 10.

5   Major Difficulty in Control Chart Patterns Classification

Level of noises (or signal to noise ratio) in CCPs has a direct influence to degrees of accuracy in a classification of CCPs regardless of classifier used. Nevertheless, careful analysis and empirical study of these six patterns reveals a useful fact which can further improve the classification. Figure 3(a) depicts an ideal situation where all six patterns contain no noise in them.

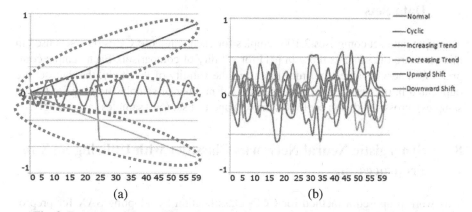

Fig. 3. Examples of Control Chart Patterns of noise free (a) and high level of noise (b)

From Figure 3(a), it can be seen that all six patterns can be superficially grouped into 3 classes as shown in their respective contours. Nothing that the mean and standard deviation values of each group are likely to be very close. These three super classes can be summarized as follows:

Super class1: These are patterns which belong to normal and cyclic categories.
Super class2: These are patterns which belong to increasing trend and upward shift categories.

Super class3: These are patterns which belong to decreasing trend and downward shift categories.

When CCPs are corrupted with undesirable signals (i.e. noises), then separating the-minto3super class are no longer obvious. Figure 3(b) depicts all six patterns in CCPs which contain high level of noise.

6   Neural Networks under Consideration

Previous literatures had reported that Recurrent Neural Network (RNN) is more suitable for prediction and classification of time series data [13]. This is also affirmed in [11]. Synergy of neural networks has been employed in various applications, ranging from design to diagnosis but its application in classification is prolific. The first systematic investigation attempt to study synergism in neural networks was carried out in [14]. As mention in Section 3, the best neural network based classifier [11] adopts the RNN known as Time-lag network. Therefore, this work also adopts the same network as used in [11] for validity in comparison of the results.

7   Data Sets

The input data set comprises 2,100 samples for each value ofσ.CCPs data sets used in this work are the same as those in [11] for validity of comparison. Each sample comprises 60 interval values. In order to ensure the validity of the study, the data set are duplicated into 5 sets. Each contains equal for number of samples training set (1,200 samples), cross-validation set (300 samples) and test set (600 samples) accordingly.

8   Synergistic Neural Networks Classifier with Utilizing SAX in Preprocessing

This work proposed a method for CCPs classification by adopting SAX for preprocessing and synergism of neural network classifiers. CCPs data sets are generated by using GARH model as described in Section 2. In preprocessing, each pattern is transformed into symbolic representation using SAX. The PAA (l) and Alphabet size (s) for SAX used in this work (i.e. for $\sigma = 5$ to 10) are the same as suggested in [10]. These symbolic become input vectors to Time-lag neural network classifiers. Classifications are performed at two level, the super class and individual category level. In super class level, CCPs (in SAX format) are classified into 3 super classes (as described in Section 5). Results from super class level become input vectors to next layer of classifiers to determine individual category. Figure 4 depicts the framework for synergistic neural networks classifier which utilizes Sax in preprocessing in this work.

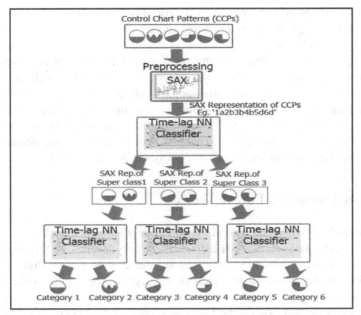

Fig. 4. Synergistic Neural Networks Classifier with SAX in Preprocessing

9   Results

In experimentation, each trail was trained and tested by 5 sample data sets, the result is taken to be the average of five trials. In preprocessing, the value for PAA (l) and Alphabet Size (s) for SAX were the same as suggested in [10].Table 2 reveals the comparisons and differences (in percentile) of the results in this work and the best neural network based classifier to date [11].

Table 2. Comparison Among Previous work [11] and Proposed Framework

	Accuracy (%)					
σ	5	6	7	8	9	10
Previous Work [11]	98.50	91.20	72.50	65.80	64.30	63.20
Proposed Framework	99.30	96.50	91.50	87.30	85.60	85.50
Improvement	1.00	5.30	19.00	21.50	21.30	22.30

10   Discussion

The results in Table 2 revealed that the proposed framework in this study yields superior performance than the best previous work [11] that used single neural network. When single neural network classifier could yield considerable high accuracy (e.g. when σ are 5 or 6), improvement by adopting synergism of neural networks is marginal. However, in situations where single neural network classifier could not yield high accuracy and more improvement is beneficial (e.g. when σ are 7 to 10), the synergistic framework proposed in this work manages to improve the accuracy in the classification considerably.

Note that selecting suitable combinations for PAA (*l*) and Alphabet Size (*s*) are crucial, in classification of CCPs, it was proven that there is no single combination suitable for all values of σ. The central idea for selecting these values is to retain or extract original characteristic. A guideline for selecting SAX parameters had been reported in [15]. Transforming CCPs into symbolic representation is superior than features extraction in preprocessing. This had been proven in single neural network classifier and this works affirms this further in synergistic framework.

11   Conclusion and Future Work

The framework of synergistic neural networks adopt in work comprises 2 levels of classification, the super classes and individual category levels. SAX representations of CCPs are classified into three categories at super classes. Output from each super class is then further classified into individual category. The results of the proposed framework are superior than the best previous work which was single unit neural network based classifier. This is the first neural network based classifier which could classify the GARH model of CCPs with σ value of 7 with greater than 90 % accuracy. The proposed framework ought to be applicable in other similar applications with multi-category classification.

Future work can be carried out in several aspects. In preprocessing, other symbolic representations may be investigated, a recent MDL-based Histogram Density Estimation [16] merits an experimentation. Experiments may be carried out on other types of noisy time series. This may reveal may be some useful facts for further study.

Acknowledgements. The authors gratefully acknowledge SIT for partial scholarship for Ms. Pantharee Sawasdimongkol and KMUTT for the support of this work.

References

1. Bollerslev, T.: Generalized Autoregressive Conditional Heteroskedasticity. Journal of Econometrics 31, 307–327 (1986)
2. Naeini, M.K.: A New Statistical Method for Recognition of Control Chart Patterns. In: Proc. of the 2011 IEEE Int. Conf. on Quality and Reliability (ICQR 2011), Bangkok, Thailand, pp. 609–612 (2011)
3. Pham, D.T., Oztemel, E.: Control Chart Pattern Recognition Using Neural Networks. Journal of Systems Engineering 2(4), 256–262 (1992)
4. Lavangnananda, K., Kasikitsakulphol, S.: Systematic Study to Assess Neural Networks' Capability in Classifying Control Chart Patterns. In: Proc. of the 2010 IEEE Int. Conf. on Intelligent Computing and Intelligent Systems (ICIS 2010), Xiamen, China, vol. 2 (2010)
5. Gauri, S.K., Chakraborty, S.: Recognition of Control Chart Patterns Using Improved Selection of Features. Computers and Industrial Engineering 56(4), 1577–1588 (2004)

6. Alcock, R.J., Manolopoulos, Y.: Time-Series Similarity Queries Employing a Feature-Based Approach. In: Proc. of 7th Hellenic Conference on Informatics, Ioannina, Greece, pp. 3.1–3.9 (1999)
7. Lavangnananda, K., Piyatumrong, A.: Image Processing Approach to Features Extraction in Classification of Control Chart Patterns. In: Proc. of 2005 IEEE Mid-Summer Workshop on Soft Computing in Industrial Applications (SMCia 2005), Espoo, Finland, pp. 85–90 (2005)
8. Mark, L., Abraham, K., Horst, B.: Data Mining in Time Series Databases, vol. 57, pp. 7–8. World Scientific Printers(S) Pte., Ltd. (2004)
9. Lin, J., Keogh, E., Lonardi, S., Chiu, B.: A Symbolic Representation of Time Series, with Implications for Streaming Algorithms. In: Proc. of 8th ACM SIGMOD Workshop on Research Issues in Data Mining and Knowledge Discovery, San Diego, CA, USA (2003)
10. Lavangnananda, K., Sawasdimongkol, P.: Capability of Classification of Control Chart Patterns Classifiers Using Symbolic Representation Preprocessing and Evolutionary Computation. In: Proc. of the 23rd IEEE Int. Con. on Tools with Artificial Intelligence (ICTAI 2011), Florida, USA, pp. 1047–1052 (2011)
11. Lavangnananda, K., Sawasdimongkol, P.: Neural Network Classifier of Time Series: A Case Study of Symbolic Representation Preprocessing for Control Chart Patterns. In: Proc. of the 2012 8th Int. Conf. on Natural Computation (ICNC 2012), Chongqing, China, pp. 351–356 (2012)
12. Keogh, E., Chakrabarti, K., Pazzani, M., Mehrotra, S.: Locally Adaptive Dimensionality Reduction for Indexing Large Time Series Databases. In: Proc. of ACM SIGMOD Conf. on Management of Data, Santa Barbara, CA., USA, pp. 151–162 (2001)
13. Pham, D.T., Liu, X.: Neural Networks for Identification, Prediction and Control, 238 p. Springer, London (1995)
14. Lavangnananda, K., Tengsriprasert, O.: Classification of TimeSeries Data: A Synergistic Neural Networks Approach. In: Proc. of the 9th Int. Conf. on Neural Information Processing (ICONIP 2002), Singapore, vol. 1, pp. 179–183 (2002)
15. Bing, H., Thanawin, R., Yuan, H., Scott, E., Stefano, L., Eamonn, K.: Discovering the Intrinsic Cardinality and Dimensionality of Time Series using MDL. In: Proc. of 2011 11th IEEE Int. Con. on Data Mining (ICDM 2011), Vancouver, Canada, pp. 1086–1091 (2011)
16. Kameya, Y.: Time Series Discretization via MDL-based Histogram Density Estimation. In: Proc. of the 23rd IEEE Int. Conf. on Tools with Artificial Intelligence (ICTAI 2011), Florida, USA, pp. 732–739 (2011)

Semantic Analysis of FBI News Reports

Sarwat Nizamani[1,2] and Nasrullah Memon[1]

[1] Mærsk McKinney Møller Institute, University of Southern, Denmark
[2] University of Sindh, Pakistan
{saniz,memon}@mmmi.sdu.dk

Abstract. In this paper we present our work on semantic analysis of FBI News reports. In the paper we have considered the News which are of the immense significance for the analyst who want to analyze the News of specific area. With this definite analysis we are able to extract critical events or concepts described in News along with entities involved in the event. These entities include important actors of the event or concept, with location and temporal information. This information will help News analyzers to retrieve the information of interest efficiently.

Keywords: Critical events, Concepts, FBI News reports, Information retrieval, Semantic analysis.

1 Introduction

In this paper we present the study of semantic analysis of FBI News reports. Through this analysis one can extract critical events present in the News with the key actors, temporal and location information efficiently. A News archive may contain a huge amount of News reports. Extracting information of interest from such a vast archive may be an intricate task. The simplest form of information retrieval would be a keyword based information retrieval. However, keyword based approach may ignore context information. In keyword based approach each word is treated equally, with no distinction of its context in the sentence. If a News is about killing incident that, 'person A killed person B', the keyword approach will just consider them as person A and person B without distinction of suspect and victim. This may not contain the information that who caused certain event and who was affected, where and when event occurred. This analysis has been performed by adopting predicate argument structure [3]. We perform this analysis by first identifying action words which is predicate in the News sentence, then based on that action word information, actors (entities) that are arguments of that action, are identified, with location and time information if any is present in the News. Once, the entities are identified, then based on context information of the entities, such as position of the entities in relation to action word, type of action word and sense of the action word, specific role (either entity caused the action or affected by the action) is assigned to that entity. The predicate mostly is the main verb of sentence and in English a word/ verb may have more than one senses/ meaning. The number and type of entities attached to the predicate vary

T. Huang et al. (Eds.): ICONIP 2012, Part IV, LNCS 7666, pp. 322–329, 2012.

from sense to sense. Therefore, in the analysis we also attached the sense information as number. Once the analysis has been carried out the result can be viewed by highlighting each critical part of the News. Therefore, contribution of the paper can be demonstrated as

- Efficient News analysis method
- Highlights prominent concepts/ events
- Highlights victims or suspects of events
- Highlights temporal and location information

We continue our discussion by first describing the dataset in the subsequent subsection, then discussing problem statement and outlining the paper structure. Following the paper outline, we continue our discussion to other sections.

1.1 FBI Dataset

The dataset comprises of the part of FBI News reports[10]. These News reports are from period 2001 to 2011. The full dataset comprises of 4397 files belonging to various divisions. We have taken a sample of 1056 files, which are national News in the dataset, each reporting the News of particular day. There is not a News report for each day but only the News if something special related to FBI happened is reported. The original dataset comprises of XML files and each XML file contains three elements, namely; title, date of publication and text. Initial preprocessing was required to extract text, and split text into sentences. This was implemented by using simple XML parser. Once the text is available as sentence by sentence, then News are ready for semantic analysis.

1.2 Problem Statement

We have an archive of News reports
Archive (News)= n_1, n_2,...n_m
We want to analyze News (n) such that, each News report is split as

- event/action/concept
- entity caused event /action/concept
- entity affected by event/ action/concept
- location event/action took place
- time event/action took place

Rest of the paper is organized as follows: Section 2 presents related work whereas Section 3 describes semantic analysis technique. Methodology is discussed in Section 4 and Concept extraction is demonstrated in Section 5. The results are illustrated in Section 6 whereas conclusion with future directions is given in Section 7.

2   Related Work

We present related work in this section by elaborating the use of the semantic analysis technique for various tasks. We do not include any comparative work to our article, because to best of our knowledge, we did not find any study on the dataset we used.

Shen and Lapata[1] have investigated the use of semantic role labeling for question and answering problem. The authors experimentally show that the use of SRL improves the performance of question and answering task over the state of art. The authors [2] applied SRL for machine translation from morphologically poor to morphologically rich languages, such as from English to Greek and Czech. Authors claim that they achieved error reduction in verb conjugation and noun case agreement. The article [3] elaborates the use of predicate argument structure which is the basis of SRL for information extraction. Authors mention that information extraction task can remarkably be improved using the approach adopted. Kulick et al. [4] have applied an approach to integrate biomedical text annotation using predicate argument structure of Propbank[1] and syntactic structure of Treebank[2]. The importance of Semantic Role Labeling can be realized from the fact that two consecutive CoNLL shared tasks [5] were dedicated to SRL in 2004 and 2005. The SRL task since then caught attention of NLP researcher and its importance was realized for the languages other than English. In this regard CoNLL shared task [9] in 2009 was dedicated to multilingual SRL.

Semantic roles have a great importance in language understanding [6]. SRL has a number of application including dialog understanding, question answering, machine translation, information extraction, dialogue understanding, word sense disambiguation and many more [6].

The authors [8] have emphasized on the use of SRL for Biomedical information extraction. The relation extraction is an important task for Biomedical text extraction. SRL can efficiently extract these relations which are ignored by other statistical NLP tasks. The SRL considers all the relations for an event such as what, when, how, extent and so on [8].

As it can be realized from above discussion, SRL can be a very handy technique for analysis of the text. In this regard we conducted the study of News analysis using SRL. We discuss the way we have adopted SRL technique in the study of News analysis in the following section.

3   Semantic Analysis Technique

For semantic analysis of the dataset we adopted semantic role labeling (SRL) technique. SRL is aimed to analyze the text at sentence level. Each sentence is then investigated against predicate argument structure. SRL answers a number of questions regarding a sentence. It analyzes the sentence by extracting pieces of information such that who did what to whom, why, how, when and where. For any News report, one

---

[1] Propbank is corpus containing annotations for predicate argument structure.

[2] Treebank is a parsed corpus containing syntactic parsing in tree structure.

expects all of these answers. For example if someone is interested to extract News regarding kidnap, SRL will return the query by providing information such as, who was kidnapped, who was the kidnapper, when and where kidnapping happened. This means that we can extract suspects, victims, instrument, location and temporal information about an event. SRL is based on predicate argument structure in which predicate mostly is a main verb of sentence, that describes the major concept of the sentence. Each predicate is attached to certain roles which provide detailed information on that action. We carried out the task of SRL on News reports using open source SRL tool[3]. The article [7] discusses the complete SRL model used in the tool.

Below we illustrate an example sentence from the dataset which is analyzed using proposed approach.

```
On Wednesday, November 18, 2009, at approximately 12:40
p.m. the Chase Bank, located at 16861 Bernardo Center
Drive, San Diego, California was robbed by an unknown
male.
```

The above sentence describes about robbery event, which took place in Chase Bank, by unknown male at location 1686 Bernardo Center Drive, San Diego, on time Wednesday, November 18, at approximately 12:40 p.m. As it can be seen that the event of robbery took place, the victim of event was Chase Bank, suspect is unknown male, location is 1686 Bernardo Center Drive, San Diego and time is Wednesday, November 18, at approximately 12:40 p.m. By the proposed analysis following pieces of information will be returned

- *Who?*   *unknown male*
- *What ?*   *robbery*
- *Whom?*   *Chase Bank*
- *Where?*   *1686 Bernardo Center Drive, San Diego*
- *When?*   *Wednesday, November 18, at approximately 12:40 p.m.*

The methodology of the research is discussed in the following section.

4   Methodology

The research presented in the paper is based on analysis of text in order to extract critical information from News reports in comprehensible pieces. After the analysis has been performed on News text sentence, one can precisely conclude the report. The analysis can also help querying the reports against certain events/ actions, suspect, locations etc.

The process of the News analysis is depicted in Figure 1. As it can be seen in the figure, initially we extracted the News in text, which were provided in XML format. News are analyzed sentence by sentence. From each sentence essential concepts are

---

[3] http://code.google.com/p/mate-tools/

extracted with all the arguments. Initially a predicate is extracted, which is the core concept of the News. Then sense of the concept is disambiguated, afterwards entities related to that concept are identified and then theses entities are assigned argument types such as A0, A1 and so on. The task of assigning roles to arguments of predicate makes the News analyses task robust as compared to keyword retrieval. Although the same can also be retrieved using keyword based approach but, the interesting idea of analyzing the News using proposed approach is to distinguish among the key event, the causer of the event   and entity who affected by the event.

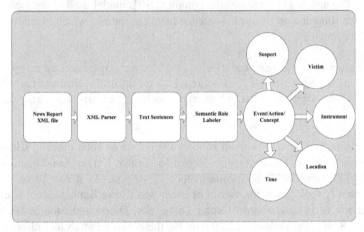

Fig. 1. Semantic analysis process of FBI News reports

5   Concept Extraction from News

As we previously mentioned that the predicates are considered as the events described in News report. The events of interest which are extracted by semantic analysis of the part of the dataset are given below in the table with their senses. We extracted those predicates which may be of great interest of security informatics personals who want to analyze the law and order situation of specific area. We have generalized the arguments of the predicate, which are the related entities of a concept/ event.

In the table 1, we have only included three arguments specific to the predicate , some predicates have two, others have three specific arguments and some may have up to five, but the predicates we used in our study from the dataset have maximum three arguments. As it can be observed in the table below that the predicates extracted from the dataset show the actions which tell about the theme of the News. Each action is related to certain entities, which are the causer or affected of that particular action. For example when concept of *murdering* is found in the News, then there is an entity *murderer*, who caused the *murder* which is also the *suspect of murder* and the entity *murdered*, who is affected by murdering action which is also the *victim of murder*, some other information can also be highlighted in the News such as *instrument* used for murdering, *location* information where murder took place and *temporal* information of the murder. It can be automatically answered from the proposed analysis method, that 'who murdered, whom, with what, when, where and why? If all such pieces of information are found in the News report. In the table below A0 represents the agent argument or causer of an action, argument A1 is the patient or the entity affected by the action. In some predicates the affected of the action is represented as A2,

Table 1. Interesting predicates found in the dataset

Predicate	Causer	Affected	Other Argument
assassinate.01	assassin agent (A0)	person assassi-nated(A1)	- - -
attack.01	Attacker (A0)	entity attacked (A1)	Attribute (A2)
bomb.01	bomb attacker (A0)	Entity attacked by bomb(A1)	- - -
commit.02	Criminal	Entity affected (A2)	Crime(A1)
kidnap.01	Kidnapper(A0)	Person (s) kidnapped (A1)	- - -
kill.01	Killer(A0)	Person killed (A1)	Instrument
testify.01	Witness(A0)	Witness against (A2)	Evidence(A1)
surrender.01	Person surrendering (A0)	Surrendered to (A2)	Surrendered for (A1)
threaten.01	Agent making threat (A1)	Threat given to (A2)	Threat (A1)
mislead.01	Liar(A1)	Lied to(A2)	False statement (A1)
damage.01	Agent damager (A0)	Entity damaged (A1)	Instrument (A2)
murdering.01	Murderer (A0)	Person(s) murdered (A1)	Instrument (A2)
victimize.01	Victimizer (A0)	Victim(A1)	Grounds for victimization (A2)
violate.01	Violator (A0)	- - -	Rule violated(A1)
destroy.01	Destroyer (A0)	Entity destroyed (A1)	Instrument for destruction (A2)
detonate.01	Exploder (A0)	Exploded entity (A1)	- - -
offence.01	Offender (A0)	Offended (A1)	- - -
endanger.01	Exposer to dand-er(A0)	Entity in danger (A1)	- - -
evasion.01	Avoider(A0)	- - -	Thing avoider(A1)
warn.01	Entity giving warn-ing (A0)	Warning giving to (A2)	Warning (A1)
prosecution.01	Prosecutor (A0)	Defendant (A1)	Law justifying case(A2)
mutilate.01	Mutilating agent (A0)	Entity mutilated de-stroyed (A1)	- - -
sabotage.01	Saboteur/ destroyer (A0)	Entity wrecked (A1)	Instrument (A2)
steal.01	Thief (A0)	Thing stolen from (A2)	Thing stolen (A1)
rob.01	Robber (A0)	Entity robbed (A1)	- - -
fake.01	Faker (A0)	Entity faked (A1)	- - -
hurt.01	Entity causing dam-age (A0)	Entity damaged (A1)	Instrument (A2)

because for these predicates A1 represent the certain action. For example in predicate *threat,* the entity who threatened is A0 or *agent* of threat, while threat itself is the argument A1 and the entity *threat given to* is A2. The other predicate such as *attack,* in which the *attacker* is the causer of attack which is the argument A0, the *entity attacked* is the A1 and the *instrument* used for attack is A2. The other arguments such as location and temporal are common to all predicates.

6   Results

We have statistically analyzed the dataset by providing the frequency of occurrence of each predicate in the dataset showing the event of the News. We also extracted the frequencies of each predicate as keyword in the dataset. The results of comparison of the predicates to the keyword retrieval is presented in the Figure 2. As it can be observed in the figure that the frequency of each keyword is much higher than the SRL predicates, because SRL extracts only if certain word is used as event in the text, while keyword extracts the every occurrence of the word without its context. As in English verbs have different forms, not each form represent the action.

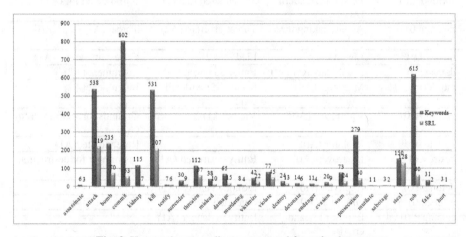

Fig. 2. Keywords vs. predicates extracted from the dataset

7   Conclusion and Future Directions

In the paper we presented a novel method of News reports analysis. The proposed method of News analysis can efficiently highlight the attention-grabbing aspects of the News report. Instead of going through the full News story, by the proposed method the core concept of the News can be scrutinized promptly and effectively. The proposed method of News analysis can help News analysts, security informatics personal and other concerned to glance the News promptly and justifiably. The News can be presented in core concept, entities causing the concept (event) or affected by the concept(event) as well as location, temporal and instrument information if any. We demonstrated the major concepts/ events found in FBI News with the type of entities related to those concepts. We statistically analyzed the occurrences of the predicates in the dataset as well as occurrences of these predicates as keywords. Currently we did not manually analyze the context of the keywords in order to further compare them with the SRL extracted predicates. This problem is left as future direction and we aim to manually analyze keywords to further refine the results. In current study we did not consider live News analysis so we consider it as another point for future direction. Analysis of online News / headlines can be handy for an ordinary News reader who wants to take a quick look on the News.

Acknowledgements. Authors would like to thank Anita Miller and James R. (Bob) Johnson ADB Consulting, LLC. University of Texas at Dallas, for providing FBI News dataset. Authors would also like to thank the Masters students Morten Gill Wollsen, Emil Nissen Gaarsmand and Rasmus Frostholm Petersen for parsing FBI dataset using SRL tool.

References

1. Shen, D., Lapata, M.: Using Semantic Roles to Improve Question Answering. In: Proceedings of Joint Conference on Empirical Methods in Natural Language Processing and Computational Natural Language Learning, pp. 12–21. Association for Computational Linguistics (2007)
2. Avramidis, E., Koehn, P.: Enriching Morphologically Poor Languages for Statistical Machine Translation. In: Proceedings of ACL 2008, pp. 763–770. Association of Computational Linguistics (2008)
3. Surdeanu, M., Harabagiu, S., Williams, J., Aarseth, P.: Using Predicate-Argument Structures for Information Extraction. In: Proceedings of 41st Meeting on Association of Computational Linguistics, pp. 8–15 (2008)
4. Kulick, S., Bies, A., Liberman, M., Mandeland, M., McDonald, R., Palmer, M., Schein, A., Ungar, L.: Integrated Annotation for Biomedical Information Extraction. In: HLT-NAACL 2004 Workshop: Biolink 2004, pp. 61–68 (2004)
5. Carreras, X., Márquez, L.: Introduction to the CoNLL-2005 Shared Task: Semantic Role Labeling. In: Proceedings Conference on Computational Natural Language Learning (CoNLL 2005), pp. 152–164. Association for Computational Linguistics (2005)
6. Gildea, D., Jurafsky, D.: Automatic Labeling of Semantic Roles. Computational Linguistics 28(3), 245–288 (2008)
7. Björkelund, A., Bohnet, B., Hafdell, L., Nugues, P.: A High-performance Syntactic and Semantic Dependency Parser. In: Coling 2010: Demonstration Volume, pp. 33–36 (2010)
8. Tsai, R.T.H., Chou, Y.C., Su, Y.S., Lin, Y.C., Sung, C.L., Dai, H.J., Yeh, I.T.H., Ku, W.: S.T.Y., Hsu, W.L.: BIOSMILE: A Semantic Role Labeling System for Biomedical Verbs Using a Maximum-entropy Model with Automatically Generated Template Features. BMC 8(1) (2007)
9. Haji, J., Ciaramita, M., Johansson, R., et al.: The CoNLL-2009 Shared Task: Syntactic and Semantic Dependencies in Multiple Languages (2009)
10. http://www.fbi.gov/news/stories

Regularized Signal Deconvolution Based on Hybrid Swarm Intelligence: Application to Neutron Imaging

Slami Saadi[1], Maamar Bettayeb[2], Abderrezak Guessoum[1], and M.K. Abdelhafidi[3]

[1] Department of Electronics, LATSI Laboratory, University Saad Dahlab of Blida, Algeria
saadislami@yahoo.fr, abderguessoum@yahoo.com
[2] Department of Electrical and Computer Engineering, University of Sharjah, UAE
maamar@sharjah.ac.ae
[3] LMetallic and Semiconducting Materials, University of Biskra, Algeria
abkamel18@yahoo.fr

Abstract. In this work, we introduce a new approach for the signal deconvolution problem, which is useful for the enhancement of neutron radiography projections. We attempt to restore original signals and get rid of noise present during acquisition or processing, due to gamma radiations or randomly distributed neutron flux. Signal deconvolution is an ill-posed inverse problem, so regularization techniques are used to smooth solutions by imposing constraints in the objective function. This paper proposes a new approach based on a synergy of two swarm intelligence algorithms: particle swarm optimization (PSO) and bacterial foraging optimization (BFO) applied for total variation (TV) minimization, instead of the standard Tikhonov regularization method. We attempt to reconstruct or recover signals using some a priori knowledge of the degradation phenomenon. The truncated singular value decomposition and the wavelet filtering methods are also considered in this paper. A comparison between several powerful techniques is conducted.

Keywords: Deconvolution, TV, Regularization, Swarm intelligence, Hybrid.

1 Introduction

By signal deconvolution, we seek to recover the original smooth signal using a mathematical model of the degradation process. Due to various unavoidable errors in the recorded signal, we cannot recover the original signal exactly. We can broadly classify deconvolution techniques into two classes: the filtering reconstruction techniques and the algebraic techniques. The first ones are rather classical and they make use of the fact that noise signals usually have higher frequencies than information signals. By selecting the proper filter, one can get a good estimate of the original signal. Example of such filters is the general linear model filter, in which the transfer function of the degraded system is inverted to produce a restored signal. It has been demonstrated that wavelets produce excellent results in signal de-noising. Shrinkage methods for noise removal, first introduced by Donoho in 1993 [1], have led to a variety of approaches combining wavelets with probabilistic concepts leading to new

T. Huang et al. (Eds.): ICONIP 2012, Part IV, LNCS 7666, pp. 330–338, 2012.

efficient de-noising procedures. The ill-posedness of this problem arises from the fact that the kernel of the blurring function is badly conditioned and the degraded signal contains noise. As a result, small perturbations in additive noise may lead to significant oscillations in the inversion result when using matrix inversion solution. Therefore, to correctly recover the unknown original signal, regularization is necessary. The basic regularization theory was first proposed by Tikhonov and Arsenin in (1977). The solution methods for this optimization problem include singular value decomposition (SVD) based direct method [2] and Newton and quasi-Newton methods [3]. Total variation is a regularization approach that performs edge preserving image restoration, but at a high computational cost. Iterative techniques have a common problem: the error starts increasing after it reaches a minimum. Most of the optimal techniques that have been proposed in literature over the past few decades to solve such problem by iterative optimization procedures are computationally demanding and time consuming. The novel approach introduced in this paper is to take advantage of swarm intelligence by synergy of two efficiently algorithms, PSO and BFO. For our test signals, we used the famous "blocks" test data credited to Donoho and Johnstone; and we consider the physical meaning of the widely used 8-bit images with pixel values varying from 0 to 255, we take the checkerboard image. We then reformulate the optimization problem by imposing nonnegative constraints.

2   Brief Review of Signal Deconvolution Techniques

2.1   Deconvolution Using a General Linear Model

We model signal degradation as a linear process characterized by a degradation matrix H of dimensions NxN, with $N=mxn$ and an observed g which, in vector form, are related by the following equation, f is the original signal:

$$g = Hf \tag{1}$$

Obtaining f from Equation (1) is not a straight forward task since, in most cases of interest, the matrix H is ill-posed [2].

2.2   Nonlinear Image Filtering

Nonlinear filters are often designed to remedy deficiencies of linear filtering approaches. They are usually defined by local operations on segments or windows of data. The size of the segments determines the scale of the filtering operation [4]. Larger segments produce more coarse scale representations, eliminating fine details.

2.3   Singular Value Decomposition (SVD)

Can be used to understand the ill-posed inverse problem and for describing the effect of the regularization method. Among all unitary transformations, SVD is optimal for a given signal in the sense that the energy packed in a given number of transformation coefficients is maximized. However, SVD requires a large number of computations for calculating singular values and singular vectors of large datasets [2].

2.4   Denoising with Wavelets

Wavelet transform is defined as a mathematical technique in which a particular signal is analyzed (or synthesized) in the time domain by using different versions of a dilated and shifted basis function called the wavelet prototype or the mother wavelet [1]. The original and simpler way to remove noise from a contaminated signal consists in modifying the wavelets coefficients in a way such that the "small" coefficients associated to the noise are basically neglected. The updated coefficients can thus be used to reconstruct the original function free from noise.

2.5   Total Variation Regularization (TV)

It is often used for signal filtering and restoration. TV was introduced by Rudin, Osher, and Fatemi [5]. It is an effective filtering method for recovering piecewise-constant signals. The famous implementations of TV approach are using Chambolle algorithm [6] and split Bregman methods [7]. The derivation in Chambolle algorithm is based on the min-max property and the majorization-minimization procedure.

3   Swarm Intelligence Approaches

3.1   Particle Swarm Optimization

The particle swarm optimization (PSO) algorithm was first described in 1995 by James Kennedy and Russell C. Eberhart [9]. In PSO, a swarm of individuals (particles) fly through the search space. Each particle represents a candidate solution to the optimization problem. The performance or quality of each particle (i.e., how close the particle is from the global optimum) is measured using a fitness function.

3.2   Bacterial Foraging Optimization Algorithm

Foraging means finding, handling, and ingesting food. Animals that have successful foraging strategies are privileged since they obtain enough food to enable them to reproduce. This has led scientists to model the activity of foraging as an optimization process. In [10], the author explains the biology and physics underlying the chemotactic (foraging) behavior of E.coli bacteria and gives a computer program that emulates the distributed optimization process represented by the activity of social bacterial foraging and applies that in adaptive controllers.

3.3   Hybrid Implementation: BFO-PSO

In the proposed hybrid approach, after undergoing a chemo-tactic step to perform a local search, each bacterium gets mutated by a PSO operator to accomplish a global search over the entire space [11]. BFO is changed by directing positions of bacteria and updating their velocities from the first chemotactic step using the power of PSO reaching the global solution. This hybridization improved the convergence speed and accuracy of solutions obtained by the classical BFO, however, what is requested is to attain a best approach to the original by finding the best solution, which is accomplished by this hybridization.

4   Simulation Results

In our application we consider a common additive noise model that comes essentially from the following sources:

- Photoelectric noise of background photons, from gamma sources. Noise from electronics used in processing, modelled usually by a white Gaussian noise.
- Film grain noise, from the randomness of silver halid grains in the film used for recording.
- Quantization noise which occurs during image digitization.

The simplest approach is to solve the least squares problem:

$$\min(\|h \otimes f - g\|_1) \qquad (2) \qquad \text{Where } \otimes \text{ is the convolution operator}$$

In practice the results obtained with this simple approach tend to be noisy, because this term expresses only the fidelity to the available data g. To compensate for this, the following regularization term is added to improve smoothness of the estimate:

$$0.004 \bullet \|Lf\|_1 \qquad (3)$$

L is the discrete *Laplacian*, which relates each data element to its neighbors. L is a 2D discrete approximation of:

$$\frac{\nabla^2 f}{2N} = \frac{1}{2N}\left(\frac{d^2 f}{dx^2} + \frac{d^2 f}{dy^2}\right) \qquad (4)$$

Where f is the estimated matrix. The matrix L has the same size as f with each element equal to the difference between an element of f and the average of its neighbors. For gray images, we also impose the constraint that the elements of f must fall between 0 and 255. To obtain the restored signal, we want to solve for X:

$$\min(\|h \otimes f - g\|_1 + 0.004 \bullet \|Lf\|_1 \qquad (5)$$

We carried out computer simulations to validate the applicability of this algorithm. In Fig.1, we used a sine test signal corrupted with additive noise and try to restore the original signal by applying soft heuristic SURE (Stein's Unbiased Risk Estimate) thresholding on detail coefficients obtained from the decomposition of the signal, at level 3 by sym8 wavelet then using our hybrid swarm intelligence implementation in which we got better results, Fig.1 (d) and (e). In Fig.2, we used the irregular blocks test signal corrupted with a multivariate Gaussian white noise exhibiting strong spatial correlation. We used four restoration algorithms for comparison. We see the improvement in signal enhancement in Fig.2 (g).

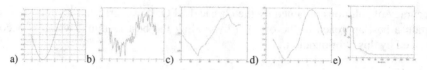

Fig. 1. Signal enhancement, a) original, b) degraded, c) soft heuristic thresholding in wavelet decomposition, d) hybrid PSO-BFO, e) objective function evolution

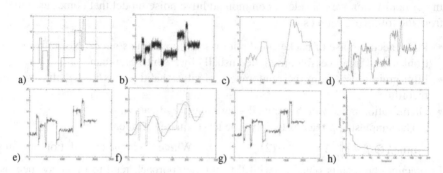

Fig. 2. a)Original, b)degraded, c)using Tikhonov regularization, d)using TV, e)using soft heuristic thresholding in wavelet decomposition, f)using TSVD, g)using hybrid BFO-PSO, h)objective function evolution

In Fig.3, restoration of a checkerboard test image (a) that has been blurred with motion blur function and noise is added (b), together with three methods: linear inversion filter (c), FFT with zero padding in frequency domain (d) and TSVD (e). In the TSVD, the condition number cond(A) = $\sigma 1/\sigma N$ was found to be 7.337638×10^4. Fig.4 below shows restorations of the previous blurred/noisy image (Fig.3a) using Tikhonov (a), R.O.F (b), Chambole Algorithm (c) and split Bregman method (d).

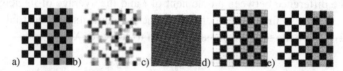

Fig. 3. a)Original Image, b)Blurred and Noisy Image, c)Restored with Inverse Filter, d)Restored with FFT and zero padding, e)Restored with TSVD

4.1   Using Chambole Algorithm

Chambolle describes in [6] an iterative algorithm for the resolution of TV restoration problem. This algorithm exploits a dual formulation of the minimization problem, and uses a fixed point iteration. Chambolle proves that these iterations are contracting, and thus converge to a solution with linear speed.

4.2   Using Split Bregman Method

The split Bregman iterative method is a concept that is used to optimize convex functions [7]. It was first used in image processing to the R.O.F model for TV denoising
[8]. It makes extensive use of Gauss-Seidel and Fourier transform methods.

Fig. 4. a)Original Image, b)Blurred and noisy Image, c)Tikhonov Regularization, b)R.O.F
method, c) Chambole Algorithm, d) split Bregman deconvolution

4.3   TV Minimization Using Hybrid PSO-BFO

It is very important that a reasonably effective set of algorithm parameters is chosen,
so that the resultant signal is accepted enough. For example, if we choose the bacteria
size S=30, we get better cost function and minimum MSE, but this had to be achieved
at the expense of more computational complexity. After fixing S=20, we change Nc to
reach the minimum cost found before, i.e., with the same computational resources.
After fixing S=20 and Nc=50, we increase Ns until stable values of MSE, PSNR and
cost function minimum; Ns is found to be 50. The rest parameters are also selected
following similar logic.

PSO parameters: C1=1.5; C2=4-C1; ω1=0.3; ω2=0.95; Swarm Size=120; Maximum Iterations=200
BFO parameters: s=20; Nc=50; Ns=50; Nre=80; Ned=1,2; Sr=s/2; Ped=0.25; c(i)=0.05

The hybridization of BFO with PSO improved the convergence speed and accuracy of
solutions. In Table 1, we compare the three algorithms, PSO, BFO and Hybrid BFO-
PSO. In Fig.5, we used four restoration algorithms for comparison and we see the
improvement in image enhancement in Fig.5(e).

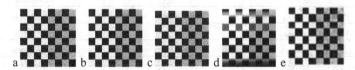

Fig. 5. Quality comparison between swarm approach and four restoration methods: a) Tikhonov, b) TSVD, c)Tikhonov (Sobolev), d)Iterative Least Square TV regularization and e) Hybrid
BFO-PSO

In Fig.6, we model gamma radiations by additive impulse noise and attempt to get
rid of such noise using the best adapted median filter.

Fig. 6. a)Original, b)impulse noise, c)median Filter, d)wavelet decomposition, e)split Bregman
denoise, f)hybrid BFO-PSO, g)regularized hybrid BFO-PSO

The PSNR throughout chemotactic steps increase, reveals the good choice of such scheme, as shown in Fig.7 and Table 1:

$$MSE = \frac{1}{mxn}\sum_{i=1}^{i=m}\sum_{j=1}^{j=n}[original\ (i,j) - restored\ (i,j)]^2 \qquad RMSE = \sqrt{MSE}$$

$$PSNR = 20.\log_{10}\left(\frac{255}{RMSE}\right)$$

In Table 2, we applied five methods in addition to the swarm intelligence approach to reconstruct the test image; numerical results show that these methods are promising for large-scale image deconvolution problems.

Table 1. RMSE and PSNR values of the three Algorithms reached with the identical computation time and population size

	Degraded	Restored with PSO	Restored with BFO	Hybrid BFO-PSO
RMSE	0.57	0.05	0.09	0.04
PSNR	52.99dB	75.01dB	68.68dB.	75.50dB

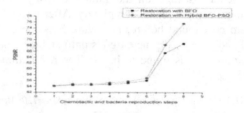

Fig. 7. PSNR with increasing number of bacteria, chemotactic and reproduction steps

Table 2. Quality comparison based between known methods and the proposed Implementation

	RMSE	PSNR
Blurred	0.57	52.99dB
TSVD Restoration	0.08	69.86 dB
Tikhonov Restoration	0.06	71.92 dB
Tikhonov (sobolev)	0.12	66.83 dB
TV LS Regularization	0.24	60.41 dB
Soft thresholding in wavelet	0.07	71.03 dB
PSO Restoration	0.05	74.01dB
BFO Restoration	0.09	68.68dB
Hybrid BFO-PSO	**0.04**	**75.50dB**

4.4   Application to Neutron Radiography Images Restoration

Neutron radiography projections are images acquired by a neutron imaging installation around a nuclear research reactor. The gray level value represents the relative linear attenuation coefficient of the object. Each of these gray images has 8-bit representations of their intensity levels. Hence, there are 256 gray levels. Fig.8-10.

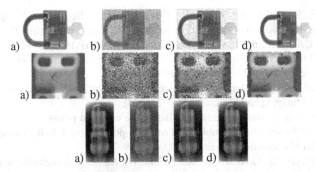

Fig. 8. a) Original image, b) with added noise, c) using median filter, d) using hybrid BFO-PSO

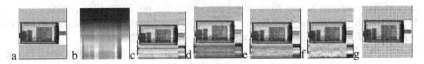

Fig. 9. a)Original image of an electrical relay, b) blurred image, c) TSVD, d) Tikhonov regularization, e) Tikhonov (sobolev), f) TV solved by split Bregman method,   g) Hybrid BFO-PSO

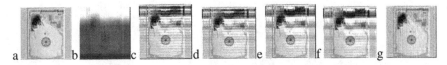

Fig. 10. a) Original image of computer hard disk, b)blurred image, c) TSVD, d) Tikhonov regularization, e) Tikhonov (sobolev), f) TV solved by split Bregman method, g) Hybrid BFO-PSO

5   Conclusion

In this work, we investigated a signal deconvolution approach in the aim to implement an enhancement method for neutron radiography projections. This approach is the synergy of two swarm algorithms to solve the ill-posed inverse problem. Computer simulations and visual inspection of the produced images illustrate that it yields optimistic results and good efficiency compared to other classical techniques.

References

1. Mallat, S.: A theory for multiresolution signal decomposition: the wavelet representation. IEEE Transactions on Pattern Recognition and Machine Intelligence 11(7), 674–693 (1989)
2. Hansen, P.C., James, G.: Deblurring Images: Matrices Spectra and Filtering. SIAM, Society for Industrial and Applied Mathematics, Philadelphia (2006)
3. Vogel, C.R.: Computational Methods for Inverse Problems. SIAM, Philadelphia (2002)

4. Undrill, P.E., Delibassis, K.: Stack Filter Design for Image Restoration Using Genetic Algorithms. In: Proceedings of the IEEE International Conference on Image Processing, Santa Barbara, CA, vol. 2, pp. 486–489 (1997), doi:10.1109/ICIP.1997.638814
5. Rudin, L., Osher, S., Fatemi, E.: Nonlinear Total Variation based noise removal algorithms. Physica D. Journal 60, 259–268 (1992)
6. Chambolle, A.: An algorithm for Total Variation minimization and applications. Journal of Mathematical Imaging and Vision 20, 89–97 (2004)
7. Bregman, L.: The relaxation method of finding the common points of convex sets and its application to the solution of problems in convex optimization. USSR Computational Mathematics and Mathematical Physics 7, 200–217 (1967)
8. Osher, S., Burger, M., Goldfarb, D., Xu, J., Yin, W.: An iterative regularization method for total variation-based image restoration. MMS 4, 460–489 (2005)
9. Kennedy, J., Eberhart, R.C.: Particle swarm optimization. In: Proceeding of IEEE International Conference on Neural Networks, Perth, Australia, pp. 1942–1948 (1995)
10. Passino, M.K.: Biomimicry of Bacterial Foraging, for Distributed Optimization and Control. 0272-1708/02/. IEEE Control Systems Magazine (June 2002)
11. Biswas, A., Dasgupta, S., Das, S., Abraham, A.: Synergy of PSO and Bacterial Foraging Optimization: A Comparative Study on Numerical Benchmarks. In: Innovations in Hybrid Intelligent Systems. ASC, vol. 44, pp. 255–263. Springer, Heidelberg (2007)

Artificial Bees Colony Optimized Neural Network Model for ECG Signals Classification

Slami Saadi[1], Maamar Bettayeb[2], Abderrezak Guessoum[1], and M.K. Abdelhafidi[3]

[1] Department of Electronics, LATSI Laboratory, University Saad Dahlab of Blida, Algeria
saadislami@yahoo.fr, abderguessoum@yahoo.com
[2] Department of Electrical and Computer Engineering, University of Sharjah, UAE
maamar@sharjah.ac.ae
[3] LMetallic and Semiconducting Materials, University of Biskra, Algeria
abkamel18@yahoo.fr

Abstract. The ECG signal is a representation of bioelectrical activity of the heart's pumping action. The doctor regularly uses a temporal recording of ECG and waveforms characteristics to study and diagnose the overall heart functioning. In some heart diseases, the correct diagnosis in an early time is essential for the patient survival. This need leads to the necessity to automate normal beat signals discrimination from abnormal beat signals. In our study, we have chosen the Multilayer Perceptron (MLP) as a classifier for this type of signals into two categories: normal (N) and pathological (V). To train this network, we used the database "MIT BIH arrhythmia database." This training is improved using a novel swarm optimization algorithm called Artificial Bees Colony (ABC) inspired from the foraging intelligence of honey bees. The (ABC) has the advantage of using fewer control parameters compared to other swarm optimization Algorithms. We propose several algorithms to filter, detect R peaks and extract the features of cardiac cycles to get it ready to be classified.

1   Introduction

The propagation of cardiac electrical activity causes the appearance of potential differences on the body surface, which can be stored in reference points. This is called the electrocardiogram (ECG). From a signal analysis point of view, the ECG is a non-stationary random signal, structured by the succession of waveforms and intervals (P, Q, R, S, T) as shown in Fig.1. Any temporal or morphological change on these events is a cardiac pathology.

Detection of cardiac arrhythmia is a race against time, there are varying ways to address them. The correct diagnosis in an early time is essential for patient survival. A number of research works have been successfully performed to demonstrate arrhythmias. For example, PVC (premature ventricular contractions) arrhythmias detection is used with different methods and different parameters [1, 2]. In [3] only the RR interval is used to find several arrhythmias. Several classification methods have been proposed in the literature to classify (N) or (V) beats for some patients [1, 4] and according to different rhythms (TV, BV...etc.) for other patients [5]. Actually, the

T. Huang et al. (Eds.): ICONIP 2012, Part IV, LNCS 7666, pp. 339–346, 2012.
© Springer-Verlag Berlin Heidelberg 2012

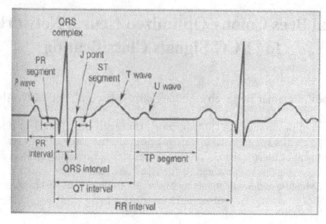

Fig. 1. A cycle of an ECG signal

computer-aided detection and classification of ECG signals is not new, including various techniques such as statistical methods, decision trees, fuzzy logic, expert systems and neural networks [5, 6, 7, 8]. In our study we adopted the (MLP) approach, which has a high accuracy testified proved by the large number of published works, for classifying heartbeats from ECG signal into normal (N) and pathological (V) cases. In [4], they combine the morphological and temporal information to classify the PVC with a robust neural network with performances estimated excellent between 95.16 and 96.82%. The separation between heart rhythms requiring shock or not, is studied in [8] using morphological and frequency criteria, the average score ranking for several arrhythmias exceeds 93%. The implementation of an intelligent pacemaker is our real objective. This is why we propose here the automatic detection using artificial neural network (ANN). Authors in [5] show the superiority of ANNs in detecting heart disease compared to traditional techniques. The (ABC) algorithm is a swarm based meta-heuristic algorithm that was introduced by Karaboga in 2005 [9] for optimizing numerical problems. The (ABC) algorithm was first applied to numerical optimization [10]. The (ABC) was applied to train neural networks [10-11]. In our application, we will enhance the model by applying this algorithm in the training step in order to get an optimized mode.

2   The Approach

Our procedure for automatically detecting cardiac arrhythmias includes mainly the selection of appropriate ECG signals and filtering for accurate extraction of pathologies indicators. There are four temporal and five morphological parameters. These parameters form the classifier input vectors (9 components).

2.1   Data Base Description

The signals used for ANN training and testing are extracted from the database: "MIT BIH arrhythmia database" available in the PhysioNet website [12]. It contains 48 recordings that are normal and pathological cases of about 30 minutes duration and sampling frequency of 360 Hz. Each cardiac cycle is characterized by peak temporal position R and labeled by a letter (V, S, etc...) called pathological symbol. Table.1 displays these records.

Table 1. ECG Records

ECG	N	V	ECG	N	V	ECG	N	V	ECG	N	V	ECG	N	V
100	2271	1	116	2302	109	200	1772	826	209	3003	1	223	2117	473
106	1506	520	118	2261	16	201	1752	208	210	2444	195	228	1690	362
108	1742	18	119	1542	444	202	2115	19	213	2668	220	230	2254	1
109	2491	38	121	1861	1	203	2530	444	214	2002	256	231	1568	2
111	2122	1	123	1514	3	207	1649	210	215	3197	164	233	2236	831
114	1831	43	124	1561	52	208	1587	992	219	2088	64	234	2749	3

	N	V
Total	62425	6517

2.2   Preprocessing of ECG Signal

Classification performances of an ANN are strongly related to pathologies extraction quality indicators. Therefore, pretreatment of the ECG signal is necessary before extracting these parameters accurately to remove, by filtering, frequencies (noises) that distort the detection, Fig. 2.

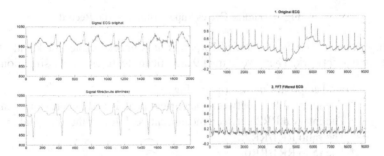

Fig. 2. Savitzky-Golay and FFT Signal Filtering

3   Parameter Extraction

3.1   Detection of R Peaks

After noise elimination and baseline correction, it is easier to detect R peaks. The best known algorithm to perform this task is that of Pan and Tompkins [13]. Therefore, we adopted Keselbrener method [14]. Each point of the signal is scanned and checked if it is above the threshold or not, Fig. 3.

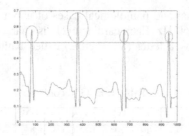

Fig. 3. Peaks Detection Illustration

3.2   Feature Extraction

After detection of R peaks, we take 75 points for each cardiac cycle in which R is at 25 points on the left side and 50 points on the right side. Fig.4 represents the superposition of these cycles record 100. Then, cycles are normalized between 0 and 1 by:

$$Y(n) = (X(n) - \min(X))/(\max(X) - \min(X))$$

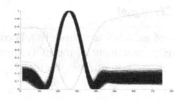

Fig. 4. Cardiac Cycles Superposition of 100 Records

Although each cycle is a vector of 75 points, it is unacceptable to take all of them as input to our network. Then, we extract 9 significant features for classification [15]:

1. Four Temporal Characteristics, Fig.5:
    * The interval RR1 between the current peak R and the previous peak.
    * The interval RR2 between the current peak R and the next peak.
    * The ratio RR1/RR2.
    * The ratio RR1/RR0.

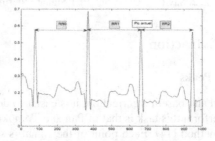

Fig. 5. Temporal Intervals Between Peaks

2. Five Morphological Features:
    * The cross-correlation between the current cycle and previous cycle.
    * The cross-correlation between the current cycle and the next cycle.
    * The percentage of points above 0.2.
    * The percentage of points above 0.5.
    * The percentage of points above 0.8.

The cross-correlation is given by the formula:

$$\rho = \frac{\langle [x_n - \mu_x][y_n - \mu_y] \rangle}{\sigma_x \sigma_y}$$

Where x, y are vectors of two successive cycles. μ_x, μ_y are their means respectively and ρ_x, ρ_y are their standard deviations respectively. We get a matrix P with:

- 9 columns representing the 9 input features.
- 68942 lines representing the number of cardiac cycles for all data, Table.1
- We add an output vector T of 68942 lines which takes 0 for (N) and 1 for (V).

4   Classification

4.1   Network Architecture

The MLP belongs to the family "feed-forward". It is widely used in approximation and classification problems. Our MLP consists of two fully connected layers of neurons: one hidden layer and one output layer, Fig. 6.

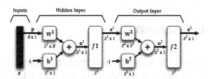

Fig. 6. The Multi Layer Perceptron

4.2   Neural Network Design Steps

a. Data Preparation: the input matrix P and the output vector T.

b. Development of the MLP Structure: the number of inputs is R=9 and the number of neurons in the output layer is S2 = 1. The number of neurons in the hidden layer will be determined during training.

c. Training: we enhance training by introducing a powerful swarm intelligence optimization algorithm: Artificial Bees Colony (ABC). The fitness value of each solution is the value of the error function evaluated at this position. To achieve the same error goal both with Levenberg-Marquaret and ABC swarm algorithm, we find that the ABC requires less number of computations; which is an observed performance. The

MLP structure trained using ABC algorithm by minimizing the mean squares of errors function [16]. The general scheme of the training by ABC algorithm is:

```
1. Initialize all the weights (w) and bias (b) of the
network to small random values.
2. For each association (P, T) in the training set:
Initialization Phase
      REPEAT
               Employed Bees Phase
               Onlooker Bees Phase
               Scout Bees Phase
               Memorize the best solution achieved so far
      UNTIL (Cycle=Maximum Cycle Number or a Maximum CPU
time)
3. Update the weights and biases.
4. If the stopping criterion is reached, then stop.
5. Otherwise, swap the order of the associations of the
training set.
6. Repeat step 2.
```

We divide data randomly into three distinct subsets: training data (60%), validation (20%) and test data (20%).

d. The transfer function: The transfer function of our neural network is the hyperbolic tangent (tansig) described by the following function.

$$a = \frac{e^n - e^{-n}}{e^n + e^{-n}}$$

5   Results and Discussion

We must test the classifier after training and measure its performance. We apply three statistical laws: Sensitivity (SE), Specificity (SP) and Classification rate (TC). It remains to specify the number of neurons in the hidden layer so that the MLP is complete. Only the experimental method can determine their number. Table.2 justifies the choice of five neurons.

Table 2. Optimizing the Neural Network Model

Number of Neurons in the hidden layer	Number of iterations	SE (%)	SP (%)	TC (%)
6	122	94.34	98.33	97.37
7	22	91.17	99.07	98.11
8	78	69.85	92.92	91.69

There are variants of the back-propagation algorithm [5,7] which accelerate the learning. A comparative study between some of them is summarized in Table.3 which shows that the ABC swarm algorithm is most suitable with a TC = 98.11 after only 22 iterations.

Table 3. Comparison between different training algorithms

Training Algorithm	Number of iterations	SE(%)	SP(%)	TC(%)
trainglm	265	86.32	96.04	96.32
trainbfg	185	87.01	98.71	98.09
traincgb	101	91.88	98.93	98.03
trainscg	279	91.75	98.57	97.87
traincgf	57	91.73	98.04	96.12
trainoss	186	90.12	98.93	96.14
traincgp	154	91.54	99.52	98.49
traingda	202	90.06	96.57	96.17
ABC	22	93.17	99.07	98.11

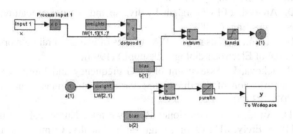

Fig. 7. ABC Training performance evolution & comparison with LM Algorithm

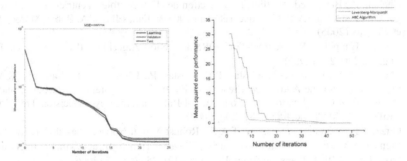

Fig. 8. The resultant Simulated Neural Network

Fig.7 shows that after only 22 training iterations, the error tends to 0.0072154. We can see that the results are reliable and represent an improvement to those published in references. The MLP classifier described in Fig.9 and Table.4 is very powerful.

Table 4. ECG signals classifier

Type of classifier	MLP
Transfer function	tansig
Number of inputs	9
hidden layer neurons	7
Output layer neurons	1
Weights number	70
Biais number	8

6   Conclusion

Our goal is to automatically classify cardiac arrhythmias using an optimized MLP. The performances obtained are very satisfactory when classifying patients with or without abnormalities. The ABC optimized MLP with one hidden layer of five neurons is the most appropriate classifier if we want a reduced learning time (22 iterations). It is still possible to gain time by implementing the network in hardware.

References

1. Chalabi, Z., Berrached, N., Ilies, L.: Détection des Extrasystoles Ventriculaires par les Algorithmes SOM & LVQ. Colloque Télécom 2005 & 4èmes JFMMA, Rabat, Maroc, 23, 24 et 25 Mars (2005)
2. Pan, J., Tompkins, W.: A Real-Time QRS Detection Algorithm. IEEE Trans. on Biom. Eng. 32, 230–236 (1985)
3. Themis, P.E., Markos, G.T., Costas, P.E., Costas, P., Dimitrios, I.F., Lampros, K.M.: A methodology for the Automated Creation of Fuzzy Expert Systems for Ischemic and Arrhythmic Beat Classification based on a Set of Rules obtained by a Decision Tree. Journal Artif. Intelli. Medicine 40(3), 187–200 (2007)
4. Omer, T.I., Laurent, G., Gregory, T.A.K.: Robust Neural Network based Classification of Premature Ventricular Contractions using Wavelet Transform and Timing Interval Features. Journal IEEE Trans. on Biom. Engin. 53(12), 2507–2515 (2006)
5. Silipo, R., Marchesi, C.: Artifical Neural Networks for Automatic ECG Analysis. IEEE Trans. on Signal Processing 46(5) (May 1998)
6. Übeyli, E.D.: Analysis of ECG signals by diverse and composite features. Journal of Electrical & Electronics Eng. (2007)
7. Giovanni, B., Brohet, C., Fusaro, S.: Possibilities of using neural networks for ECG classification. Journal of Electrocardiology 29(suppl.) (1996)
8. Krasteva, V., Jekova, I.: Assessment of ECG Frequency and Morphology parameters for automatic classification of life-threatening cardiac arrhythmias. Journal of Physio. Measur. 26(5), 707–723 (2005)
9. Karaboga, D.: An Idea Based on Honey Bee Swarm for Numerical Optimization. Technical Report-TR06, Erciyes University, Engineering Faculty, Computer Engineering Department (2005)
10. Karaboga, D., Akay, B.: An artificial bee colony (abc) algorithm on training artificial neural networks. In: 15th IEEE Signal Processing and Communications Applications, SIU 2007, Eskisehir, Turkiye, pp. 1–4 (June 2007)
11. Karaboga, D., Akay, B., Ozturk, C.: Artificial Bee Colony (ABC) Optimization Algorithm for Training Feed-Forward Neural Networks. In: Torra, V., Narukawa, Y., Yoshida, Y. (eds.) MDAI 2007. LNCS (LNAI), vol. 4617, pp. 318–329. Springer, Heidelberg (2007)
12. PhysioNet: Research Resource for Complex Physiolo- Signals, http://www.physionet.org
13. Tompkins, W.J.: Biomedical digital signal processing. University of Wisconsin-Madison (2000)
14. Keselbrener, L., Keselbrener, M., Akselrod, S.: Non-linear high pass for R-wave detection in ECG signal. Med. Eng. Phys. 19(5), 481–484 (1997)
15. Clifford, G.D., Azuaje, F., Patrick, E.: Advanced methods and tools for ECG Data Analysis. Artech house (2006)
16. Akay, B., Karaboga, D.: A modified Artificial Bee Colony algorithm for real-parameter optimization. Inform. Sci. (2010), doi:10.1016/j.ins.2010.07.015

Single-Trial Multi-channel N170 Estimation Using Linear Discriminant Analysis (LDA)

Wee Lih Lee[1], Tele Tan[2], Torbjörn Falkmer[3,4,5,6], and Yee Hong Leung[1]

[1] Department of Electrical and Computer Engineering
{Weelih.lee,Y.Leung}@curtin.edu.au
[2] Department of Computing
T.Tan@curtin.edu.au
[3] School of Occupational Therapy and Social Work, CHIRI
Curtin University, Perth, Australia
T.Falkmer@curtin.edu.au
[4] Rehabilitation Medicine, Department of Medicine and Health Sciences (IMH),
Linköping University & Pain and Rehabilitation Centre, Linköping, Sweden
[5] Department of Rehabilitation, Jönköping University, Jönköping, Sweden
[6] School of Occupational Therapy, La Trobe University, Melbourne, Australia

Abstract. N170 is one of the event-related potentials (ERPs) that have been extensively used to study the neurological response of a subject when presented with a visual face stimulus. Although many N170 experiments have been performed with multi-channel EEG recordings, most of the N170 analyses are still based on trial-averaged signals from one or a few selected electrodes. Not only does this method under utilise the information from all available electrodes, it inhibits a trial-to-trial analysis of N170. We address this issue by proposing a single-trial multi-channel N170 estimation method that estimates the latency, amplitude and scalp projection of N170 using a linear classifier-based approach.

Keywords: Event-Related Potentials, Linear Discriminant Analysis, Multi-Channel, N170, Single-trial.

1 Introduction

Event Related Potentials (ERPs) are brain activities generated in response to external stimuli. In electroencephalography (EEG), ERPs can be observed as a series of positive and negative peaks, where each peak is characterised by a specific amplitude, latency and scalp distribution pattern [1].

ERP plays an important role in the understanding of neurocognitive processes and also serves as a diagnostic tool for assessing neurological disorders. One of the important ERPs, which has been extensively studied, is N170 [2]. N170 is a negative signal which occurs around 170 ms after stimulus onset and is strongest in the occipital-temporal region. Previous studies have shown that N170 is often associated with the neural processing of face and object images, whereby the N170 amplitude is larger when faces are presented as stimuli [2][3]. Accordingly, N170 has been extensively used in the study of face processing deficits that are

T. Huang et al. (Eds.): ICONIP 2012, Part IV, LNCS 7666, pp. 347–355, 2012.

characteristic of Autism Spectrum Disorder (ASD) [4]. For instance, differences in the N170 latency are often observed between ASD and a control group [1][4]. Accurate amplitude and latency estimation are thus crucial. However, due to the stochastic nature of EEG and low signal-to-noise ratio (SNR) of ERP signals, the analysis of N170 and other ERPs remains a challenge.

Traditional methods for studying ERP involve electrode selection, trial-averaging and manually picking a peak within a time region where the ERP is most likely to occur. These methods have several drawbacks. Firstly, most of the EEG equipment is capable of recording signals simultaneously from multiple electrodes where up to 32 electrodes is common. Thus, studying time varying ERP waveforms solely from single channel is no longer effective, since this method does not utilise valuable information from all available electrodes. Secondly, trial-averaging, which is a classical procedure that has been adopted to improve the SNR, assumes that the ERPs are time-locked from trial-to-trial. In real life, the amplitude and latency of ERPs change according to a person's physiological conditions, such as mental fatigue, habituation and attentiveness. Lastly, trial-averaging techniques inhibit trial-to-trial analysis where important information, which may reflect the changing pattern of ERP amplitude and latency across trials, is lost during the averaging process.

To overcome these weaknesses, many solutions, notably single-trial multichannel ERP classification approaches, have been proposed. For example, some techniques extract ERP sources through Principal Component Analysis (PCA) and Independent Component Analysis (ICA). However, their main problem is that they do not extract the desired ERP in a straightforward manner. Often, many irrelevant sources are extracted at the same time and a related source has to be chosen through further efforts [5]. Another approach in ERP estimation is to use a predefined template [6][7]. Given that an ERP response looks similar to a certain waveform, a template is shifted across specific time range to find the best matching ERP. The drawback of this method is that in a given trial, the ERP signal may deviate significantly from the template. Moreover, often different ERPs share a similar waveform.

Single-trial N170 estimation can be treated as a two-stage task. This often involves the time localisation of N170 activities and the estimation of N170 parameters. To achieve good estimation, accurate time localisation of the N170 ERP is important. In this paper, we propose a single-trial N170 estimation method based on a linear classifier, whereby the classifier is trained to discriminate N170 and non-N170 time regions based on scalp patterns. The two main advantages of this approach are: (1) Classification is made based on the statistical properties of N170 that were learned from previous EEG samples; (2) Instead of locating N170 using a template waveform pattern, time localisation is achieved by finding the relevant scalp pattern as suggested by [1].

The rest of this paper is organised as follows. Section 2 describes how N170 parameters can be estimated through application of the proposed linear classifier. In Section 3, the performance of the proposed method was evaluated on both simulated and real data. The conclusion are drawn in Section 4.

2   Methodology

As shown in Fig. 1, the proposed method for N170 analysis involves three steps: training, detection and estimation. During training, from each trial, the time segment that contains N170 is assigned to a positive class, while the remaining time segment is assigned to a negative class. Based on these training classes, a binary linear classifier is then built and used to discriminate N170 and non-N170 time regions in an EEG trial. Once a N170 time region has been determined, estimation of N170 amplitude, latency and scalp distribution then follows.

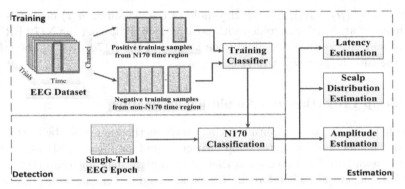

Fig. 1. Proposed single-trial multi-channel N170 analysis method

2.1   Linear Discriminant Analysis (LDA) for Locating N170

Many linear classifiers are available in the literature. In this study, LDA is chosen because of its simplicity. The procedure to build an LDA classifier is summarised as follows. Suppose the EEG equipment has D measurement channels, they are sampled synchronously at a sampling frequency of f_s Hz, and the measurements captured at time index t during the k-th trial are stored in the $D \times 1$ vector $\mathbf{x}_k(t)$. Let T be the number of snapshots taken in each trial, K be the number of trials, and the sets $\mathcal{T}_1$ and $\mathcal{T}_2$ contain, respectively, the time indices of the positive class and those of the negative class. (e.g. $\mathcal{T}_1 = \{t : 140ms < t/f_s < 190ms\}$). Let T_1 and T_2 be the number of elements in $\mathcal{T}_1$ and $\mathcal{T}_2$, respectively. To enable the classifier to learn the N170 scalp distribution pattern, we first compute the following two $D \times 1$ vectors which represent the mean scalp pattern of the positive and negative classes

$$\boldsymbol{\mu}_j = \frac{1}{T_j K} \sum_{t \in \mathcal{T}_j} \sum_{k=1}^{K} \mathbf{x}_k(t), \; j{=}1{,}2 \tag{1}$$

and the $D \times D$ average covariance matrix of both classes.

$$\mathbf{S} = \frac{1}{T-2} \sum_{j=1}^{2} \sum_{t \in \mathcal{T}_j} \sum_{k=1}^{K} (\mathbf{x}_k(t) - \boldsymbol{\mu}_j)(\mathbf{x}_k(t) - \boldsymbol{\mu}_j)^T \tag{2}$$

The LDA is given by the discriminative vector $\mathbf{w}$ which maximises the separability of the positive and negative classes. If the training samples from both classes can be assumed to be Gaussian distributed with equal covariance, then $\mathbf{w}$ is given by

$$\mathbf{w} = \mathbf{S}^{-1}(\boldsymbol{\mu}_1 - \boldsymbol{\mu}_2) \tag{3}$$

To determine whether a time region is correlated to N170 or not, the discriminative vector $\mathbf{w}$ is applied to an EEG trial and the classifier output $y_k(t)$ is calculated as follow:

$$y_k(t) = \mathbf{w}^T \mathbf{x}_k(t) - b \tag{4}$$

where $b = \mathbf{w}^T(\boldsymbol{\mu}_1 + \boldsymbol{\mu}_2)/2$ is the threshold of the classifier. $y_k(t)$ describes the possibility of an N170 occurrence with positive sign indicating the N170 ERP is present and vice versa. Accordingly, the latency of the N170 ERP can be taken to be the time instant when $y_k(t)$ is largest.

2.2   Scalp Projection and Amplitude Estimation

For scalp projection and amplitude estimation, a method described by Li *et al.* [6] is used. This method estimates scalp projection and amplitude with a Gaussian function template. Its drawback is that the template may not characterise the N170 ERP accurately. Therefore, in this work, a modified classifier output is used instead to guide the selection. The modified output $s_k(t)$ is defined below where $\mathbf{y}_k = [y_k(1), y_k(2), ..., y_k(T)]^T$. As can be seen, it makes no prior assumptions about the N170 waveform and is motivated by the observation that the activation of N170 is captured when $y_k(t)$ is positive.

$$s_k(t) = \begin{cases} \frac{y_k(t)}{max(\mathbf{y}_k)} & \text{if } y_k(t) > 0 \\ 0 & \text{if } y_k(t) \leq 0 \end{cases} \tag{5}$$

Scalp projection estimation consists of two steps. Firstly, assuming the N170 ERP is uncorrelated with other ERPs and background noise, the scalp projection $\mathbf{a}_k$ of an N170 source during the k-th trial is estimated with

$$\mathbf{a}_k = (\mathbf{s}_k^T \mathbf{s}_k)^{-1} \mathbf{X}_k \mathbf{s}_k \tag{6}$$

where $\mathbf{s}_k = [s_k(1), s_k(2), ..., s_k(T)]^T$ is a $T \times 1$ vector representing the time course of the N170 ERP, $\mathbf{X}_k = [\mathbf{x}_k(1), \mathbf{x}_k(2), ..., \mathbf{x}_k(T)]$ is a $D \times T$ matrix representing the scalp pattern across time, and $\mathbf{a}_k$ is a $D \times 1$ vector representing the magnitude of the source in different channels. Secondly, although the estimated scalp projection varies across trials, there exists a scalp projection pattern that is common to all trials. This common scalp projection, $\mathbf{a}_0$, is useful for understanding the source origin of the N170 ERP and is given by

$$\mathbf{a}_0 = \frac{1}{K} \sum_{k=1}^{K} \frac{\mathbf{a}_k}{\| \mathbf{a}_k \|} \tag{7}$$

For amplitude estimation, it was observed that the EEG measurements are corrupted by noise and there are trial-to-trial variability in the latency and amplitude of the N170 ERP. Therefore, to improve the statistical reliability of the estimated amplitude, a least-squares fit of the estimated scalp projection $\mathbf{a}_k$ to the common scalp projection $\mathbf{a}_0$ is performed. That is, we find the scaling factor σ_k that minimises the objective function $\parallel \mathbf{a}_k - \sigma_k \mathbf{a}_0 \parallel_2^2$, which has solution

$$\sigma_k = \mathbf{a}_0^T \mathbf{a}_k \qquad (8)$$

3 Experimental Results and Discussion

3.1 Test with Simulated Data

In this section, preliminary assessment of the proposed method was performed on simulated data that resemble the P1-N170-P2 complex often encountered in N170 datasets [9]. To evaluate the method under a possible worst case scenario, the data were generated to include dynamic latency variations in each ERP and the possibility of overlaps between the ERPs. For the test, 500 simulated trials were created using the linear generative EEG model introduced in [6] and [8], which is constructed as $\mathbf{X} = \sum_{i=1}^{3} \mathbf{a}_i \mathbf{s}_i^T$ where $s_i(t) = exp(-(t - \tau_i)^2/\delta_i^2)$ is a Gaussian function that resembles each ERP source with peak at latency τ_i and width of δ_i while $\mathbf{a}_i$ represents the scalp projection of each ERP source. On each simulated trial, P1, N170 and P2 latencies were chosen to be Gaussian distributed with the following values: 100 ± 25 ms, 170 ± 25 ms and 200 ± 25 ms while δ_i was fixed at 0.025. The scalp distribution of each ERP was also fixed and selected to mimic the findings of Kuefner et al. [3] and Boutsen et al. [9].

As latency estimation is crucial to the accurate estimation of amplitude and scalp projection, the goal of this study was to examine the accuracy of latency as estimated by our proposed method. Our results were also compared to the existing template-matching technique by Li et al. [6] for single-trial ERP estimation. The parameters used in our proposed method, and the template-matching technique, are described as follows. For our method, the data were first divided into training data and test data using the leave-one-out method. In this method, all except one trial was taken for classifier training, while the remaining trial was taken for testing. The time region for positive training samples were set from 150 ms to 190 ms. During the estimation phase, the process described in Section 2.1 was performed. These steps were repeated until every single trial was estimated. For the template-matching technique, an exact Gaussian template which matches the N170 waveform was used. To determine the latency, the predefined Gaussian template was scanned across time. The time where the shifted template best matches the signal using the Least-Square criterion was selected as the N170 latency. In addition, a latency correction method [6] was applied where the N170 latency was taken as the first local minima that occurred after 120 ms.

Fig. 2 compares the latency estimated from the our proposed method and the template-matching technique. As can be seen, our proposed method outperforms the existing template-matching technique where the latency estimated by our proposed method follow closely the actual N170 latency. It was observed the template matching technique failed when the P1 peak occurred after 120 ms or when overlapping occurred between the P1 and N170 ERPs. However, the proposed method managed to detect the N170 ERP in most cases except in situations where all the ERPs overlapped.

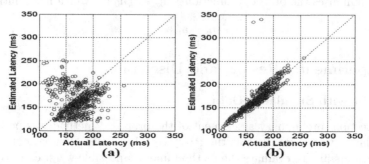

Fig. 2. Scatterplots of estimated latency versus actual latency of N170 on simulated data using (a) template-matching method [6] and (b) proposed method

3.2 Test with Real Data

We next evaluated the proposed method using part of the multi-modal face dataset contributed by Henson *et al.* [10]. The original dataset contains EEG, MEG and fMRI recordings of a subject for the purpose of measuring and studying the neurocognitive responses when performing face perception tasks. Two sessions were conducted on the same subject using the same stimuli presentation paradigm, based on Phase 1 of a previous study by Henson *et al.* [11]. In each session, a total of 86 faces and 86 scrambled faces were randomly presented [10]. The measurements were made by a 128-channel ActiveTwo EEG system with a sampling frequency of 2048 Hz. All electrodes were later re-referenced to the common-average reference.

For our study, we combined these two sessions. For each trial, a 400 ms segment of EEG signals was extracted starting from stimulus onset. Baseline correction was performed using a 200 ms segment of the EEG signal prior to stimulus onset. Any trial which had voltage exceeding 200 μV was rejected prior to analysis. A total of 166 trials were left in both the faces and scrambled faces dataset after this process. The trial-averaged waveform for faces and scrambled faces are shown in Fig. 3.

Part of the aim of this study was to estimate the N170 latency using the LDA classifier mentioned in Section 2.1 on the EEG recording for both the faces and scrambled faces stimuli presentation conditions. To demonstrate that the classifier can detect and estimate the N170 latency under both conditions, the classifier was trained using trial recordings only from the face stimuli dataset. The segments between 150 ms and 190 ms were assigned to the positive class. To avoid bias, the N170 estimation was performed using the leave-one-out method with the procedure described in Section 2. The trained classifier was then used to classify the N170 time region in the leftover trial and also one trial from the scrambled faces group. During latency estimation, the peak of the classifier output was selected from between 100 ms and 250 ms. This procedure was repeated until all trials were estimated.

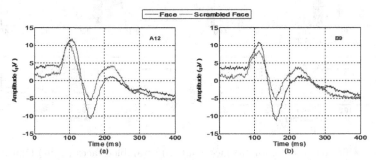

Fig. 3. Trial-averaged faces (blue) and scrambled faces (red) waveforms from two different electrodes which are located (a) at site A12 from the left occipital-temporal region and (b) at site B9 from the right occipital-temporal region

Fig. 4 shows a visual plot of the classifier output for all classified faces and scrambled faces trials. The circles indicate the position with the maximum classifier response between 100 ms to 250 ms for each trial. This is used to represent the N170 latency from the stimulus onset. There were 9 trials from scrambled faces condition that do not have positive $y_k(t)$ within the defined time region and were removed from the analysis. Although the classifier was trained using only samples from the faces stimuli, yet the classifier response for both stimuli conditions were similar. This indicated there was no overfitting in the LDA classifier and that the latency estimation was effective regardless of the stimuli condition. Besides that, Fig. 4 suggests that N170 is readily observed in single trial and the N170 latency is stable around 166 ms in both conditions. By applying the independent t-test to the estimated latency from both conditions, there was no significant difference in the latency for face (M=166.4, SD=7.6) and scrambled face (M=166.5, SD=10.1) conditions at single trial level ($t(321)$=-0.053, p =0.957). This can also be observed in the histogram of Fig. 5(a).

Fig. 6 shows the common scalp projection estimated from both faces and scrambled faces trials. It reveals scalp projections from both conditions have occipital-temporal distributions, which matches the literature findings [3]. Moreover, the results showed both conditions are similar to each other with correlation coefficient of 0.9729. This observation is, however, in contrast to the findings in Henson's previous study, which found that across subjects, N170 from face stimuli is usually more profound in the right hemisphere than in the left hemisphere [11]. Maybe this is due to the fact that the given subject have bi-laterised N170 in both conditions. Since scalp projection from both conditions for the given dataset are similar, their average was taken as common scalp projection for amplitude estimation.

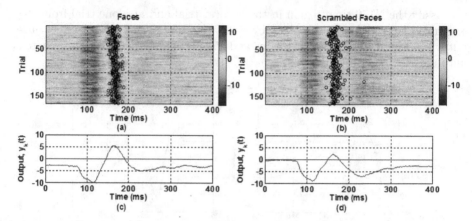

Fig. 4. Classifier output, $\mathbf{y}_k$ for (a) faces and (b) scrambled faces (right) trials. Their average were shown in (c) and (d). The circles represent the estimated N170 latency which was selected by taking the largest $y_k(t)$ between 100ms to 250ms in each trial.

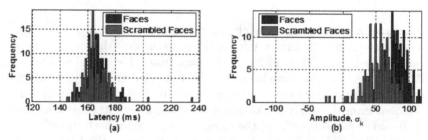

Fig. 5. Histogram of (a) estimated latency and (b) amplitude

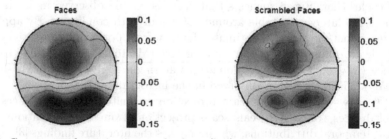

Fig. 6. Common scalp projection, $\mathbf{a}_0$ for faces (left) and scrambled faces (right) stimuli

Fig. 5(b) shows the histogram of estimated amplitude for both face and scrambled face trials. By applying the independent sample t-test on these estimated amplitudes, there was a significant difference between the amplitude for face (M=77.75, SD=17.36) and scrambled face (M=53.72, SD=26.52) conditions at single trial level ($t(266.694)=9.579$, $p <0.001$). These results support the established findings that N170 amplitude is larger in face stimulus as compared with non-face stimulus [1][3][11].

4   Conclusion

An LDA-based amplitude, latency and scalp distribution pattern estimatior targeting the N170 ERP was proposed and successfully tested using both synthetic and real EEG data. Our results show that the proposed method can be an effective technique to classify and estimate the N170 ERP from a single-trial multi-channel EEG recordings. The preliminary observations from this work will lead to better understanding and techniques for the analysis of N170 ERPs to support further work in understanding the facial processing characteristics of people afflicted with ASD.

References

1. Nelson, C.A., McCleery, J.P.: Use of Event-Related Potentials in the Study of Typical and Atypical Development. Journal of the American Academy of Child and Adolescent Psychiatry 47, 1252–1261 (2008)
2. Eimer, M.: The Face-Sensitive N170 Component of the Event Related Brain Potential. In: Calder, A.J., Rhodes, G., Johnson, M., Haxby, J. (eds.) The Oxford Handbook of Face Perception, pp. 329–344. Oxford University Press, Oxford (2011)
3. Kuefner, D., Heering, A.D., Jacques, C., Palmero-Soler, E., Rossion, B.: Early Visually Evoked Electrophysiological Responses over The Human Brain (P1, N170) Show Stable Patterns of Face-Sensitivity from 4 years to Adulthood. Frontiers in Human Neuroscience 67(3), 1–22 (2010)
4. Dawson, G., Webb, S.J., McPartland, J.: Understanding the Nature of Face Processing Impairment in Autism: Insights From Behavioral and Electrophysiological Studies. Developmental Neuropsychology 27(3), 403–424 (2005)
5. Onton, J., Westerfield, M., Townsend, J., Makeig, S.: Imaging Human EEG Dynamics using Independent Component Analysis. Neuroscience and Biobehavioral Reviews 30, 808–822 (2006)
6. Li, R.J., Principe, J.C., Bradley, M., Ferrari, V.: Single-trial ERP Estimation Based on Spatio-Temporal Filtering. In: IEEE EMBS Conference on Neural Engineering, Hawaii, pp. 538–541 (2007)
7. Spyrou, L., Sanei, S., Took, C.C.: Estimation and Location Tracking of The P300 Subcomponents From Single-Trial EEG. In: IEEE Conference on Acoustics, Speech and Signal Processing, Hawaii, pp. 1149–1152 (2007)
8. Parra, L.C., Spence, C.D., Gerson, A.D., Sajda, P.: Recipes for the Linear Analysis of EEG. NeuroImage 28, 326–341 (2005)
9. Boutsen, L., Humphreys, G.W., Praamstra, P., Warbrick, T.: Comparing Neural Correlates of Configural Processing in Faces and Objects: An ERP study of the Thatcher illusion. NeuroImage 32, 352–367 (2006)
10. N170 EEG Dataset Website, http://www.fil.ion.ucl.ac.uk/spm/data/mmfaces/
11. Henson, R.N., Goshen-Gottstein, Y., Ganel, T., Otten, L.J., Quayle, A., Rugg, M.D.: Electrophysiological and Haemodynamic Correlates of Face Perception, Recognition and Priming. Cerebral Cortex 13, 793–805 (2003)

EEG Based Foot Movement Onset Detection
with the Probabilistic Classification Vector Machine

Raheleh Mohammadi[1], Ali Mahloojifar[1], Huanhuan Chen[2], and Damien Coyle[3]

[1] Department of Biomedical Engineering, Tarbiat Modares University, Tehran, Iran
{raheleh.mohammadi,mahlooji}@modares.ac.ir
[2] School of Computer Science, University of Birmingham, Birmingham B15 2TT, UK
H.Chen@cs.bham.ac.uk
[3] Intelligent Systems Research Center, University of Ulster, Derry, UK
dh.coyle@ulster.ac.uk

Abstract. A critical issue in designing a self-paced brain computer interface (BCI) system is onset detection of the mental task from the continuous electroencephalogram (EEG) signal to produce a brain switch. This work shows significant improvement in a movement based self-paced BCI by applying a new sparse learning classification algorithm, probabilistic classification vector machines (PCVMs) to classify EEG signal. Constant-Q filters instead of constant bandwidth filters for frequency decomposition are also shown to enhance the discrimination of movement related patterns from EEG patterns associated with idle state. Analysis of the data recorded from seven subjects executing foot movement using the constant-Q filters and PCVMs shows a statistically significant 17% (p<0.03) average improvement in true positive rate (TPR) and a 2% (p<0.03) reduction in false positive rate (FPR) compared with applying constant bandwidth filters and SVM classifier.

Keywords: Constant-Q filter, Probabilistic classification vector machines, Self-paced BCI.

1 Introduction

In recent years, self-paced BCI systems have received increasing attention in BCI research [1-8]. BCI systems can be divided into two main classes based on their operation mode: synchronous and asynchronous (self-paced). In a synchronous BCI the onset of the mental activity is known in advance and the system analyzes and classifies the mental tasks in predetermined time intervals. In contrast self-paced BCIs allow the user to control the system when desired i.e., the communication period is not time-locked to a specific time window. Continuous classification of the EEG signal for detecting specific EEG oscillations or dynamics associated with a specific mental task makes it possible to design a self-paced BCI system. One of the challenges of these systems is mental task onset detection, i.e. detecting when the user shifts from the non-control state (when he/she is not executing any of the predefined mental tasks) to executing a mental task. Due to the movement or imagination of the

T. Huang et al. (Eds.): ICONIP 2012, Part IV, LNCS 7666, pp. 356–363, 2012.

movement, the EEG signal energy in specific frequency bands and also in specific regions of brain can be modulated producing an event related desynchronization (ERD) before and during movement and event related synchronization (ERS) in the beta frequency band after termination of the movement [9].

BCIs capable of detecting only one brain pattern from the continuous EEG signal is referred to as a brain switch and is suitable for controlling different applications [7]. Numerous signal processing and pattern recognition techniques have been developed for designing a movement based brain switch. For example in [1], a local neural classifier was used to reject non motor imagery signals. The low-frequency asynchronous switch design (LF-ASD) [3] is one of the first self-paced BCI systems. In the last design of the LF-ASD [4] features extracted from three neurological phenomena were used to train a set of SVM classifiers to detect right index finger flexion movement from idle states. In [5] the changes in average power spectral density (PSD) features are used to enhance classification of continuous EEG for onset detection. An unsupervised classification method based on Gaussian Mixture Model (GMM) was applied in [6] to detect the onset of the right hand movement. Another brain switch designed in [7, 8] proved the suitability of one single Laplacian derivation for detecting foot movement from ongoing EEG. Two distinct SVM classifiers were used to detect ERD and ERS patterns separately.

Kernel based methods such as SVM have been considered the state–of–the-art and successful methods for self-paced BCI. SVMs generalize well; however suffer from several disadvantages. In order to address the problems of SVMs probabilistic classification vector machines (PCVM) have been proposed [10]. The focus of this work is to apply PCVM for detection of ERS from EEG signal as a neurological phenomena representing foot movement. We also use constant-Q filters for frequency decomposition of the signal in the frequency range from 6 to 36 Hz which covers mu, beta and lower gamma frequency bands. We show that this feature of constant-Q filters contributes to defining more precisely the correlates of movement onset within the EEG signal and significantly improves the performance of the brain switch. The rest of the paper is organized as follows. Section 2 outlines data acquisition and the methodology of feature extraction, classification and performance evaluation. Results and conclusion are presented in section 3 and 5 respectively.

2   Method

A block diagram of the system built for this study is shown in Fig. 1.

2.1   Data Description

Our analysis is performed on the data provided by the laboratory of Brain Computer Interfaces (BCI-Lab), Graz University of Technology [7]. Seven healthy subjects participated in this study. Each subject performed 3 runs (each run comprised 30 trials). The subjects performed a brisk movement of both feet after the presentation of the cue. At the beginning of the trial (t=0) a "+" was presented; then at t=2 the presentation of an arrow pointing downwards cued the subject to perform a brisk foot movement of both feet for about 1 second duration. The cross and cue disappear at t=3.25s and at t=6s, respectively. At t=7.5 the trial ends. The sampling frequency was 250 Hz. Our analysis is performed on a single small Laplacian derivation over the Cz electrode.

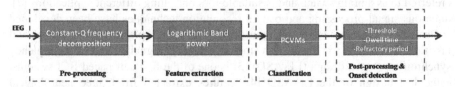

Fig. 1. Block diagram of the built system for this study

2.2   Feature Extraction

In this work we consider ERS features as a stable and detectable movement related pattern from spontaneous EEG signals. Due to the poor and low signal to noise ratio of ERD/S, its feature extraction is a challenging task. In this paper, band power which reveals the energy or power fluctuations of the signal in specific frequency bands is employed for feature extraction. However, the precise frequency band which is responsive to movement execution (ERS) can vary among subjects. From the viewpoint of learning, multiple possible frequency bands have to be examined or the optimum band has to be selected in the training phase for each subject. We process the EEG signal using an array of filter banks, each of them is a particular band-pass filter and they all together cover a frequency range of mu and beta rhythms. Twenty eight fifth order Butterworth band pass constant-Q filters with center frequencies at 6, 6.9, 7.8, 9, 10.2, 11.7, 13.4, 15.3, 17.5, 20.0, 22.8, 26.1, 29.8 and 33.5 Hz for two different Q ratios (Q=2 & Q=3) as suggested in [11,12] . In constant-Q filters, the ratio of center frequency to bandwidth for all the filters is the same and equals to Q. In other words, for low frequencies the frequency resolution is better while for high frequencies the time resolution is better.

Twenty eight logarithmic band power features are extracted from time segments of 1s length as follows: (i) band-pass filtering using 28 filters described above, (ii) squaring the value of each sample, (iii) averaging all samples within the time segment and (iv) applying the logarithm function. Each segment has 250 samples with an overlap of 125 samples between adjacent segments.

Feature vectors extracted from t=4-5s in each trial is labeled class 1 and is related to the ERS patterns after movement while the other feature vectors are labeled as class 0 or non-control states.

2.3   Classification (Probabilistic Classification Vector Machines)

The SVM classifier is known to be quite successful in various applications. However, there are a number of significant disadvantages that exist with the use of this method: 1) The number of support vectors grows linearly with the size of the training set, 2)

Predictions are not probabilistic and 3) SVM requires a cross-validation procedure to estimate the kernel parameter which is a waste of data and computation time. All of these problems can be rectified by using relevance vector machines (RVMs) developed by Tipping [13]. The RVMs approach is a Bayesian sparse kernel technique for regression and classification that shares many of the characteristics of the SVM whilst avoiding its principal limitations. Additionally, it typically leads to much sparser models resulting in correspondingly faster performance on test data whilst maintaining comparable generalization error. Requiring fewer basis functions could lead to significant reduction in the computational complexity of the decision function, therefore, making RVM more suitable for real-time applications. In RVMs [13] the zero-mean Gaussian prior are adopted over weights for both positive and negative classes, therefore   some of the training points belonging to positive class may have negative weights and vice versa. This formulation may lead to the situation that the decision of RVMs is based on some unreliable vectors, and thus is sensitive to the kernel parameter. In order to address this problem within RVMs and propose an appropriate probabilistic model for classification problems, probabilistic classification vector machines (PCVMs) was proposed in [10], which introduce different priors over weights for training points belonging to different classes, i.e., the nonnegative, left-truncated Gaussian for the positive class and the nonpositive, right-truncated Gaussian for the negative class . PCVMs also implement a parameter optimization procedure for kernel parameters in the training algorithm, which is proven to be effective in practice.

PCVM Algorithm. In supervised learning, we are given a set of input-target training pairs $\{X_i, y_i\}_{i=1}^{N}$ where $y_i = \{-1, +1\}$. Based on the training set, we wish to choose a learning model $f(X;W)$ which is controlled by some unknown parameters W. A flexible and popular set of candidates for $f(X;W)$ is that of the for

$$f(X;W) = \sum_{i=1}^{N} w_i \phi_{i,\theta}(X) + b = \Phi_\theta(X)W + b \tag{1}$$

where the prediction $f(X;W)$ is a linear combination of N basis functions $\{\phi_{1,\theta}(X), ..., \phi_{N,\theta}(X)\}$, (wherein θ is the parameter vector of the basis function), $W = (w_1, ..., w_N)^T$ is a parameter of the model and b is the bias. In the current research, the radial basis function (RBF) is chosen as the basis function. In order to map the linear outputs of $f(X;W)$ to the binary outputs, the probit link function $\Psi(x) = \int_{-\infty}^{x} N(t|0,1)dt$ is used for PCVMs and the model becomes:

$$l(X;W,b) = \Psi\left(\sum_{i=1}^{N} w_i \phi_{i,\theta}(X) + b\right) = \Psi(\Phi_\theta(X)W + b) \tag{2}$$

A prior distribution over each weight w_i is a truncated Gaussian and over the bias b is a zero-mean Gaussian as follows:

$$p(W|\alpha) = \prod_{i=1}^{N} p(w_i|\alpha_i) = \prod_{i=1}^{N} N_t(w_i|0,\alpha_i^{-1}) \tag{3}$$

$$p(b \mid \beta) = N(b \mid 0, \beta^{-1}) \tag{4}$$

where $N_t(w_i \mid 0, \alpha_i^{-1})$ is a truncated Gaussian function, α_i represents the precision of the corresponding parameter w_i, and β is the precision of the normal distribution of b. When $y_i = +1$, the truncated prior is a nonnegative, left-truncated Gaussian, and when $y_i = -1$, the prior is a nonpositive, right-truncated Gaussian. This can be formalized in (5):

$$p(w_i \mid \alpha_i) = \begin{cases} 2N(w_i \mid 0, \alpha_i^{-1}), & if \;\; y_i w_i \geq 0 \\ 0, & if \;\; y_i w_i < 0 \end{cases} \tag{5}$$

The gamma distribution is also used as the hyperprior of α and β. The expectation maximization (EM) algorithm is used to specify the model parameters such as w, bias b, and kernel parameters θ. The EM algorithm is a general algorithm for maximum a posteriori (MAP) estimation where the data are incomplete or the likelihood or prior functions involve latent variables. For more details of the EM algorithm in defining the PCVM parameter the reader is referred to [10].

Initial Parameter Selection and Training. The PCVM has only one parameter θ, which can be automatically optimized in the training process. However, the EM algorithm is sensitive to the initialization point and might get trapped in local maxima. The usual approach to avoid the local maxima is to choose the best initialization point. For each subject there are 3 runs each consisting of 30 trials. The patterns from 2 runs out of 3 runs are used as the training data. Therefore there are 60 training trials of two runs available for selecting the initialization points of PCVMs parameters. We train a PCVM model with different initializations (nine initializations 2 to 6 with step of 0.5) over the first five training folds of the data set. Hence, we obtain an array of parameters of dimensions 9×5 where the rows are the initializations and the columns are the folds. For each column, we select the results that maximize $TF = TPR - FPR$ (the difference between sample by sample true positive rates and sample by sample false positive rates), so that the array reduces from 40 to only five elements. Then, we select the median over those parameters.

Testing. The remaining run is used for testing the trained PCVMs. In order to simulate an online asynchronous system, we continuously compute logarithmic band power features applying a 1 second moving window at the rate of the sampling interval.

2.4   Performance Evaluation

Performance measurement of the online self-paced paradigm is carried out in an event by event manner. The event class posterior probability of PCVMs is post processed using threshold, dwell time and refractory period [14]. ROC analysis over the threshold value was conducted and the values for the dwell time (100 samples) and for the refractory period (750 samples) were picked by hand. For evaluation, the time interval from t=3 to 5.5 seconds of each is considered as the intentional control period. This interval is the same for all the subjects. All results reported in this paper were obtained from the ROC curves as the maximum TPR associated with a FPR≤0.1. The event by event TPR and FPR are calculated the same as [7].

3   Results

The results of foot movement onset detection using two different kinds of filter (constant bandwidth and constant-Q) for preprocessing and two classifiers (PCVMs and SVMs) are summarized in Table 1 and Table 2. In these tables for each subject TPR and FPR values are reported in the form of mean ± standard deviation. For each subject, 3 different combinations of train/test runs are possible. The average of TPR and FPR values of all 3 combinations for each subject are presented in the tables. In order to contrast the performance improvement of the self-paced system using constant-Q filters the results of applying constant bandwidth filters (a set of fifth order Butterworth filters with 2 Hz bandwidth and 1 Hz overlap between 6 to 36 Hz) are pr esented in table 1 and the results of applying constant-Q filters (Q=2 & Q=3 and 14 center frequencies from 6 to 36Hz) are presented in table 2. In addition the results of applying two different classifiers SVMs and PCVMs are illustrated in the left and right column of each table, respectively.

For selecting the parameters of SVMs, in each combination of train/test runs for each subject, we use a "grid-search" on C and σ with a 10-fold cross-validation. C and σ are varied from 2^{-8} to 2^{1} while for each step the values of parameters are doubled. Various pairs of (C, σ) are tried and the pair which maximize $TF = TPR - FPR$ are selected to train a final SVM [7].

According to the results of table 1 the average TPR achieved by applying PCVMs is significantly better than SVMs when the constant bandwidth filters are used for frequency decomposition of the EEG signal. According to this table the performance improvement is especially obvious for subjects s2, s6 and s7. PCVMs not only increase the average TPR of the brain switch (around 10% improvement) but also decrease the average FPR (around 1% improvement).

Table 2 shows the results of the brain switch when the constant-Q filter is applied. Comparing the results of SVMs in Table 1 and Table 2 shows the significant improvements achieved by constant-Q filters in detecting the foot movement from continuous EEG. The improvement originates from the characteristic feature of the constant-Q filter, its variable frequency resolution and time resolution according to the center frequency of the filter. The results of applying PCVMs and SVMs with constant-Q.

Table 1. Comparison of SVM and PCVM when constant bandwidth filters are applied for frequency decomposition

Subject ID	SVM		PCVM	
	TPR(Mean±SD)	FPR(Mean±SD)	TPR(Mean±SD)	FPR(Mean±SD)
S1	**94±5**	2±2	91±12	3±2
S2	54±20	10±2	**79±25**	**8±1**
S3	**97±4**	3±3	96±2	3±2
S4	80±7	6±4	**80±23**	**5±3**
S5	56±23	10±3	**60±11**	**7±1**
S6	65±7	7±4	**85±12**	7±3
S7	61±2	7±2	**91±2**	**5±3**
Average	73±13	6±3	**83±15**	5±2

Table 2. Comparison of SVM and PCVM when constant-Q filters are applied for frequency decomposition

Subject ID	SVM		PCVM	
	TPR(Mean±SD)	FPR(Mean±SD)	TPR(Mean±SD)	FPR(Mean±SD)
S1	98±4	2±2	98 ±4	**1±2**
S2	62±12	8±1	**87±9**	9±1
S3	98±2	2±1	98±2	2±2
S4	93±4	6±3	93±0	**3±1**
S5	80±3	8±1	80±3	8±2
S6	90±12	7±3	91±2	**4±3**
S7	**86±10**	5±2	80±6	**4±3**
Average	87±8	5±2	**90±4**	**4±2**

filters shows that TPR of the brain switch designed for most of the subjects are the same but the FPR is significantly lower in the case of using PCVMs. For subject s2 the PCVMs show the higher TPR compared with SVM. A two-sided Wilcoxon signed rank between the accuracies obtained by our proposed method (constant-Q filters +PCVMs) and the method proposed in [7] (constant bandwidth filters +SVMs) shows a significant improvements in the TPR (p =0.015), and FPR (p=0.015).

4   Conclusion

In this paper a novel approach for designing a foot movement-based brain switch was introduced. Constant-Q filters for frequency decomposition of the EEG signal in a broad frequency range (6 to 36 Hz) were used to extract the band power of the continuous EEG signal in 28 different frequency bands. A probabilistic algorithm, probabilistic classification vector machines (PCVMs) was applied for the first time in the context of BCI for classification of the movement related patterns (ERS) from noisy background EEG signals. Our results provide confirmation that using constant-Q filter along with PCVM improves the performance of the foot movement based brain switch significantly. Future work will include optimization of dwell time and threshold in the training session to make a system more suitable for online applications.

Acknowledgements. The authors are grateful to Prof. G. Pfurtscheller and Mr. T. Solis-Escalante of the laboratory of Brain Computer Interface (BCI-Lab), Graz University of technology for making their data available.

References

1. Millan, J.R., Mourino, J.: Asynchronous bci and local neural classifiers: An overview of the adaptive brain interface project. IEEE Trans. Neural Syst. Rehabil. Eng. 11, 159–161 (2003)
2. Scherer, R., Schloegl, A., Lee, F., Bischof, H., Jansa, J., Pfurtscheller, G.: The self-paced Graz brain–computer interface: methods and applications. Comput. Intell. Neurosci., 79826 (2007)
3. Mason, S.G., Birch, G.E.: A brain-controlled switch for asynchronous control applications. IEEE Trans. Biomed. Eng. 47, 1297–1307 (2000)
4. Fatourechi, M., Ward, R.K., Birch, G.E.: A self-paced brain–computer interface system with a low false positive rate. J. Neural Eng. 5, 9–23 (2008)
5. Galán, F., Oliva, F., Guardia, J.: Using mental tasks transitions detection to improve spontaneous mental activity classification. J. Med. Biol. Eng Comput. 45(6), 603–612 (2007)
6. Hasan, B.A.S., Gan, J.Q.: Unsupervised movement onset detection from EEG recorded during self-paced real hand movement Med. Biol. Eng. Comput. 48, 245–253 (2010)
7. Solis-Escalante, T., Muller-Putz, G.R., Pfurtscheller, G.: Overt foot movement detection in one single Laplacian EEG derivation. J. Neurosci. Methods 175, 148–153 (2008)
8. Pfurtscheller, G., Solis-Escalante, T.: Could the beta rebound in the EEG be suitable to realize a "brain switch"? Clin. Neurophysiol. 120, 24–29 (2009)
9. Pfurtscheller, G., Lopes da Silva, F.H.: Event-related EEG/MEG synchronization and desynchronization: basic principles. Clin. Neurophysiol. 110, 1842–1857 (1999)
10. Chen, H., Tiňo, P., Yao, X.: Probabilistic Classification Vector Machines. IEEE Trans. Neural Net. 20, 901–914 (2009)
11. Wang, T., Deng, J., He, B.: Classifying EEG-based motor imagery tasks by means of time–frequency synthesized spatial patterns. Clin. Neurophysiol. 115, 2744–2753 (2004)
12. Yamawaki, N., Wilke, C., Liu, Z., He, B.: An enhanced time-frequency-spatial approach for motor imagery classification. IEEE Trans. Neural Syst. Rehabil. Eng. 14, 250–254 (2006)
13. Tipping, M.E.: Sparse Bayesian learning and the relevance vector machine. J. Mach. Learn. Res. 1, 211–244 (2001)
14. Townsend, G., Graimann, B., Pfurtscheller, G.: Continuous EEG classification during motor imagery-simulation of an asynchronous BCI. IEEE Trans. Neural Syst. Rehabil. Eng. 12, 258–265 (2004)

Neural Networks Based System
for the Supervision of Therapeutic Exercises

Sven Nõmm[1,*], Alar Kuusik[3,**], Sergei Ovsjanski[2],
Ines Malmberg[4], Marko Parve[4], and L. Orunurm[4]

[1] Institute of Cybernetics at Tallinn University of Technology,
Akadeemia tee 21, 12618 Tallinn, Estonia
sven@cc.ioc.ee
[2] T.J. Seebeck Institute of Electronics, Tallinn University of Technology,
Ehitajate tee 5, 19086 Tallinn, Estonia
sergiorus@gmail.com
[3] Competence Center ELIKO, Teaduspargi 6/2, Tallinn 12618, Estonia
alar.kuusik@eliko.ee
[4] East-Tallinn Central Hospital, Ravi 18, 10138 Tallinn, Estonia
{Ines.Malmberg,Marko.Parve,Lii.Orunurm}@itk.ee

Abstract. Present contribution describes application of the neural networks based models to detect incorrectly performed therapeutic exercises within the frameworks of wearable supervision system. Electronic accelerometers and gyroscopes attached to the human upper and lower limbs gather information about performed exercise in real time. Trained, on the data describing correctly done exercises, neural network based dynamic model of the limb is used to find the difference between the actual and "ideal" performances and judge if exercises are performed in a correct way or not.

Keywords: Neural networks, dynamic model, NN-ANARX model, medical system, rehabilitation.

1 Introduction

Correctness in doing therapeutic exercises may play a key role for achieving complete rehabilitation of the patient's motor functions. Most practitioners contend that for positive results, exercises must be specific, of the proper intensity, and performed consistently with the correct technique. Techniques are often taught by a physical therapist or by use of various media, such as brochures, audiotapes, and videotapes. The effectiveness of exercise is most likely related to the quality of exercise performance [1]. Usually such exercises are performed either under the supervision of physiotherapists or on robotic simulators. Both cases lead rises

---

* This research was supported by the Estonian Science Foundation ETF through the state funding project SF0140018s08 and research grand ETF 8365.
** This research has been partially supported by European Regional Development Fund through the Competence Centre program of Enterprise of Estonia.

T. Huang et al. (Eds.): ICONIP 2012, Part IV, LNCS 7666, pp. 364–371, 2012.

in rehabilitation costs. While in the cases of more complicated traumas or on early stages of rehabilitation such costs are justified, in less complicated cases or on later stages of the rehabilitation those costs may be significantly reduced by using inexpensive wearable motion capture systems. Usual session lasts for approximately one hour, therefore employment of automated supervision system may lead to a significant time savings of medical stuff and relive personal from the routine operations.

Present paper concentrates its attention on the application of Neural Networks based Additive Nonlinear Auto Regressive eXogenous (NN-ANARX)[2] models as computational tool of the supervision system for therapeutic exercises. Different motion capture systems have been extensively used in different medical areas. For example infrared camera based systems are more popular in sportive medicine and in prothesis development areas. Also recent developments shows increasing attentions to such system from the area of medical robotics [3]. Field based systems are less popular due to the possible problems magnetic field may cause to other equipment. Systems based on wearable electronic sensors become more and more popular in different areas because of their portability and relatively low cost [4],[5]. Wearable motion capture system proposed in this paper consists of three sensors providing information about their orientation with respect to the horizon and linear accelerations along each of the axis in three-dimensional space.

Section 2 describes requirements and desired functionality expected from the supervision system. Both the mathematical tools and hardware elements of proposed system are described in Section 3. Application of the system to supervise therapeutic exercises described in Section 4. Conclusions are drawn in the last section.

2   Problem Statement

To develop light motion capture system and software for supervision of therapeutic exercises. One may separate requirements following from this target into three categories.

- **Ergonomics:** Number and weight of the sensors should be kept low, the sensors are expected to be easy to attach and detach and entire system should be easy to operate, preferably wireless.
- **Hardware:** Precision and sampling of each sensor should allow to detect "wrong" movements of the limbs but at the same time provide certain robustness with respect to the noise.
- **Software:** Allows fast identification of the patient specific parameters. Provides real time supervision of the exercises and informs patient if something is done in a wrong way. Ideally should be able to determine what was done wrong.

While therapeutic exercises contain both dynamic (including strengthening and ROM (range of motion) exercises) and static exercises within the frameworks of present contribution only ROM exercises (precisely AROM (active range of motion) are considered. AROM is movement of a segment within the unrestricted ROM that is produced by active contraction of the muscles crossing that joint [6]. Those exercises require patient to move their body part, following certain trajectories with a certain restrictions. Usually exercises performed in repeating cycles. For example one may require the hand to hold right angle in elbow, while turning it around humerus whereas such cycle should be repeated number of times.

3   Proposed System

Hardware part of the proposed system consist of three sensors, each equipped with electronic accelerometer and gyroscope, and wired or wireless controller allowing communication with a PC. The system architecture allows to add sensors, if needed for more complicated exercises. Software consists of two applications, the first one collects the data, performs off-line training of the neural network based models and allows to choose liminal values or thresholds to define required precision. The second one functions in real time, it collects the data performs analysis and makes decision about correctness of the performed exercise. Schematic diagram of marker placement and hand motions during on of the exercises is depicted in Figure 1. Main idea of the software may be described in the following way. First model of correct motion is produced off line on the basis of the data set obtained during patient's training on the robotic simulator or under supervision of the therapist. During the exercise system compares patient's actual motions with the model of correct motions and if the error exceeds liminal value, defined during off-line training, error message appears.

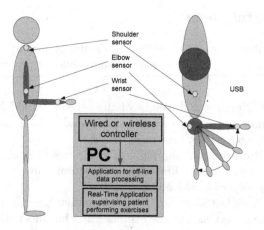

Fig. 1. General structure of the system

3.1   Mathematical Tools

Requirements imposed on the software in Section 2 lead following prerequisites to the models employed in the system. Obviously models required to be sensitive enough to recognize motions performed too fast or incorrect positions of upper and lower limbs. NN-based models represent one of possible choices to model motions of human hands and legs. One may suggest to apply some known form the literature model of human upper and lower limbs. Such approach is an interesting alternative but its main drawback that in this case one will have to use at least two structures for the upper limbs and two for the lower limbs in comparison with just one structure required in the case of NN-based models. Each of three sensors provide six data channels (three from acceleremeter and three from gyroscope) which leads totally eighteen data channels. Limbs required to move in one exercise may be required to be nearly motionless in the others, therefore informativeness of each sensor depends on the exercise. Latest leads the idea to associate with each exercise set of the most informative data channels, reducing number of NN inputs and eventually ease computational power required.

In order to model motions (gestures) of human upper and lower limbs NN-based Additive Nonlinear Auto Regressive eXogenous (NN-ANARX) models were employed [2]. In order to make this paper self-sufficient brief description of the NN-ANARX structure is provided. NN-ANARX is a subclass of a large NN-NARX models, which differs from its parent by separation of different time instances, leading restricted connectivity neural networks depicted in Figure 2. NN-ANARX models are usually described by high order input-output difference equation of the following form

$$y(t) = \sum_{i=1}^{n} \sum_{k=1}^{m} c_{i,k} \phi_i \left(w_{i,k}^1 y(t - n + i - 1) + w_{i,k}^2 u(t - n + i - 1) \right). \tag{1}$$

where $y(t)$ denotes system output, $u(t)$ - system input, c_k and $w_{i,k}^j$ are the synoptic weights, m is the number of neurons on hidden layer, n is the model order and ϕ_i are the saturation type smooth nonlinear functions.

One may easily see restrictions imposed on the network connectivity. While in theory restricted connective NN-model may be less accurate compared to the fully connected NN in practice the difference between fully connected NN and NN-ANARX were negligible which were demonstrated on numerous examples [7]. Also NN-ANARX models have less weights and therefore less computational power needed for training and simulation. The structure of the NN-ANARX models allows the order of the model to be identified online [7]. Unlike classical NN-ANARX in the case of modeling human hand or leg motions, NN-inputs are the data returned by accelerometers and gyroscopes of three sensors. One have relative freedom choosing the output. Namely any of the sensors outputs may play the role of the NN-output. Within the frameworks of present contribution mostly outputs of the gyroscopes are modeled because they are less affected by the noise level. Obviously in those cases when greater precision is required one may model number of outputs and then employ some multi-criteria and/or intelligent technique to combine the results.

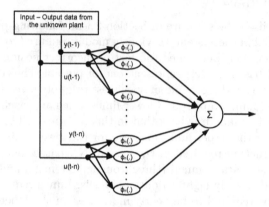

Fig. 2. Neural network corresponding to the ANARX structure

3.2   Hardware Description

For the testing, monitoring and gathering statistical data two prototypes wired and wireless, of wearable system were developed. Both prototypes consist of 3 micromechanical movement sensor modules and an USB host bridge device. Each sensor module contain a 3D accelerometer (Analog Devices ADXL345) and 3D gyroscope (Invensense ITG3200) connected to Atmel microcontroller. Relatively low sensor sampling rate of 20Hz were used, which is a tradeoff between signal noise suppression and required data throughput. Sensor boards have dimensions of 2.5cm by 4cm and can be tightly fixed to limbs with orthopedic stripes. Wireless sensor system is using Zigbee radio communication.

4   Application

Actual application of the system consist of two stages. On the first stage patient under the therapist supervision performs required exercises. As a result therapist gets data for each exercises saved in PC. Then it is required to identify the model. Since during NN training initial weights are selected randomly it is wise to identify a couple of models and then choose one on the basis of its performance.

4.1   Model Identification

Following the idea of reducing number of data channels genetic algorithm based method similar to [8] was applied to find combination of data channels leading accurate model for each given type of the exercise. As a result a table associating data channels to the therapeutic exercises were constructed. The model structure for all the exercises remains the same, but the meaning of the NN-inputs would vary. Since sampling time was about 0.2 sec and in average one cycle of the exercise took between two and four seconds to repeat. The latter means that

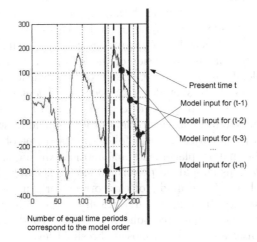

Fig. 3. Data acquisition technique

one cycle of the exercise is described by at least 40 time-instances. Which leads that to model just single output of the sensor on the basis of previous circle or even half circle one needs model of a very high order which is not realistic for proposed system. In order to overcome this problem and keep model order within reasonable limits, model (neural network) inputs are taken not from consequent past samples but with certain equal time intervals As shown in Figure 3. Fore example if one will choose 14 time instances to be the length of the interval between consequent samples then model of order 5 will make a prediction on the basis of previous 70 time instances which is equivalent to time period of 3.5 seconds. While the length of the interval between two instances requires to be defined it also provides one more possibility to tune the model. Of course one may suggest to reduce sampling rate but in a such way some information will be simple lost. Note that to model the value of time instance $t + 1$ one have to take time instances next to those taken to model value at time instance t, therefore no data will be lost. In the framework of present contribution 7 was found to be an optimal model order and length of the interval of 8 time instances. Neural networks with 5 neurons on hidden layer and *logsig* activation function were chosen. Training was performed with Levenberg-Marquardt algorithm. In average training took 1000 epochs to converge.

4.2   Supervision of Therapeuic Exercises

Once the model is identified one last step required on behalf of the therapist. Namely individual threshold levels should be determined for each exercise. Observing the curve representing error between NN-based prediction and actual output therapist may choose the value or it may be determined automatically. For each session patient should just attach the sensors and chose the name of the exercise. Numerous experiments has demonstrated that positions of the sensors have crucial importance for acquiring data suitable for NN-training. For example there is a big difference in attaching the wrist sensor. Attached to the place where usually watch is attached sensors gives more noisy compared to the case when sensor is attached on the write side of the wrist. Let us demonstrate entire process on example of simple exercise. One of the terapeutic exercises for increasing external rotation of the shoulder joint requires patient to hold right angle in elbow such that arm is kept pressed to the body and forearm put forward. Then patient requires to move forearm (right or left depending which hand is exercised) such that wrist's trajectory will follow 90° segment of the circle. Schematic diagram of the marker placements and motions during the exercise depicted in Figure 1. Motions should be performed slowly keeping right angle and arm pressed to the body. Once model of the correct performance is ready patient may start the exercise. In Figure 4 first graph represents actual output (blue dotted line) and model output (red solid line). The second graph is the normalized error, horizontal line shows proposed threshold level to distinguish between the correctly and incorrectly done cycles. In the third graph dashed blue line shows which circles in the exercise were performed correctly have value 1 and where incorrectly value 0. Solid red line represents detection made by the supervision system. Except initialization period of first hundred time instances, supervision system was able to detect incorrectly done exercise circles with a delay corresponding to the model order multiplied by the length of interval between consequent samples. Once therapist confirms the threshold patient may commence practicing exercises independently.

Initial testing has indicated that proposed system demonstrate better results to detect motions performed "too fast", namely about 85% of such errors were detected. At the same time only 75% of errors corresponding to the wrong angles were detected which leads necessity to improve detection of such errors. Due to limited space the table associating required data channels with exercises and testing results are omitted, but available from the authors upon request.

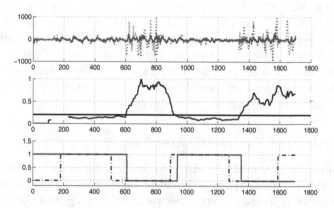

Fig. 4. Detection results

5 Conclusions

NN-based wearable system for the supervision of therapeutic exercises was proposed in this paper. NN-based modeling of the human upper and lower limbs motions and gestures was described in detail. Further research will be constitute by development of more precise models and mapping of therapeutical exercises.

References

1. Friedrich, M., Cermak, T., Maderbacher, P.: The effect of brochure use versus therapist teaching on patients performing therapeutic exercise and on changes in impairment status. Physical Therapy (Journal of the American Physical Therapy Association) 76, 1082–1088 (1996)
2. Kotta, Ü., Chowdhury, F., Nõmm, S.: On realizability of neural networks-based input-output models in the classical state space form. Automatica 42(6), 1211–1216 (2006)
3. Nõmm, S., Petlenkov, E., Vain, J., Yoshimitsu, K., Ohnuma, K., Sadahiro, T., Miyawaki, F.: Nn-based anarx model of the surgeon's hand for the motion recognition. In: Proceedings of the 4th COE Workshop on Human Adaptive Mechatronics (HAM), pp. 19–24. Tokyo Denki University, Tokyo (2007)
4. Zhou, H., Hu, H.: A Survey - Human Movement Tracking and Stroke Rehabilitation. Technical report, CSM-420, University of Essex (2004) ISSN 1744 - 8050
5. Lorincz, K., Chen, B., Welsh, M.: Mercury: a wearable sensor network platform for high-fidelity motion analysis. In: Proc of the 7th ACM Conference on Embedded Networked Sensor Systems, SenSys 2009, Bercley, pp. 183–196 (2009)
6. Kisner, C., Colby, L.: Therapeutic Exercise, Foundations and Techniques, 5th edn. F.A.Davis Company, Philadelfia (2007)
7. Nomm, S., Belikov, J.: Online identification of unknown plant by nn-based anarx model. In: 2011 International Conference on Adaptive and Intelligent Systems, Klagenfurt, Austria, pp. 5–15 (2011)
8. Nomm, S., Vassilejva, K., Beliokv, J., Petlenkov, E.: Structure identification of nn-anarx model by genetic algorithm with combined cross-correlation-test based evaluation function. In: 2011 9th IEEE International Conference on Control and Automation (ICCA), Santiago,Chile, pp. 65–70 (2011)

Fuzzy Model for Detection and Estimation of the Degree of Autism Spectrum Disorder

Wafaa Khazaal Shams[1], Abdul Wahab[1], and Uvais A. Qidwai[2]

[1] Computer Science Department, International Islamic University, Malaysia
wafaa_dth@yahoo.com, abdulwahab@iium.edu.my
[2] Computer Science and Engineering Department, Qatar University, Qatar
uqidwai@qu.edu.qa

Abstract. Early detection of autism spectrum disorder (ASD) is of great significance for early intervention. Besides, knowing the degree of severity in ASD and how it changes with the intervention is imperative for the treatment process. This study proposes Takagi- Sugeno-Kang (TSK) fuzzy modeling approach that is based on subtractive clustering to classify autism spectrum disorder and to estimate the degree of prognosis. The study has been carried out using Electroencephalography (EEG) signal on two groups of control and ASD children age-matched between seven to nine years old. EEG signals are quantized to temporal-time domain using Short Time Frequency Transformation (STFT). Spectrum energy is extracted as features for alpha band. The proposed system is modeled to estimate the degree in which subject is autistic, normal or uncertain. The results show accuracy in range (70-97) % when using fuzzy model .Also this system is modeled to generate crisp decision; the results show accuracy in the range (80-100) %. The proposed model can be adapted to help psychiatrist for diagnosis and intervention process.

Keywords: Autism Spectrum Disorder, Classification, EEG signals, Takagi-Sugeno-Kang fuzzy approach.

1 Introduction

Electroencephalography signals have been used widely to diagnose autism spectrum disorder as well as to study and understand its symptoms [1-6]. Autism spectrum disorder is a neurodevelopment disorder characterized by impairment in social interaction, deficits in communication, restricted and repetitive actions and other co-occurrence deficits such as motor and imitation. Those characteristics revealed by Diagnostic and Statistical Manual of Mental Disorder 4th.edition (DSM-IV) [7]. Autism is known to be diagnosed using psychology tests based on the social symptoms that appear in age three. The severity of autism varies from high to low symptoms that reflect the nature of the intervention. Thus, it is necessary to determine the degree of disorder and to see how it responds to the process of treatment. The remarkable affect of quantified EEG signals appears and can be appreciated when its extracted information helps in early detection of ASD. Therefore EEG signals are employed in this study to detect and estimate the degree of autism.

T. Huang et al. (Eds.): ICONIP 2012, Part IV, LNCS 7666, pp. 372–379, 2012.

Detection of autism signal is one of pattern recognition problem that is influenced by two factors: extracted features and classifier; where some of those extracted features facilitate discrimination process of classes which in turns normal classifier like different in mean distance is sufficient. In other cases there is a need for compatibility of features and classifier to get good results. However; pattern recognition problem is sort of decision making process. Most of the studies have been done concerning on the crisp decision making that tells whether the subject is autistic or normal. However, it is imperative to determine level of autism and how close autism subject is from normal one. This can be called the response of treatment.

Fuzzy system [8] is used to detect and estimate the degree of disorder. TSK fuzzy modeling approach based on subtractive clustering is used in this work due to its generalization and simplicity to model the complex systems [9],[10]. EEG signals are quantized to temporal-frequency domain using STFT to capture the dynamic change of signal with time. This model has been designed to give both crisp and fuzzy decision. The limitation of this work is the lack of psychology part which can be added and modified in future to be suitable for clinical purpose.

2   Method and Materials

2.1   EEG Data Descriptions

The data was collected from six typical children with age ranging from 7 to 9 from preschool (Malaysia International Islamic preschool) and six autistic children, ages (7 to 9) from the National Autistic Society of Malaysia (NASOM). Autism subjects were diagnosed based on DSM-IV [7]. Each subject is asked to sit in rest condition with open eyes looking at the screen at 75 cm distance .The EEG signals were collected using BIMEC machine with a sampling frequency of 250 hertz. These signals were recorded using eight channels (F3, F4, C3, C4, P3, P4, T3, and T4), placed according to the international 10-20 system with Cz as reference.

2.2   Data Process

EEG signals are affected by noise from child movement, environment and electrode connection. Therefore, it is necessary to pre-process EEG data to reduce the artefact. The mean value was subtracted from each channel then EEG data were filtered using Infinite Impulse Response(IIR) pass band filter (0-60) Hz. Another filter was performed to get alpha band (8-13) Hz, using pass band filter, Chebyshev (II). Meanwhile low quality EEG segments were manually excluded from the analysis. All EEG data were normalized between (0 to 1) after filter process.

2.3   Features Extracted

Temporal –frequency presentation for EEG signals provide important information for detection process. Short Time-Frequency Domain is one of the techniques that are widely used in bio-signal processing [11],[12].It is based on the assumption that within short time window EEG signal can be considered as a stationary and then Fourier transformation can be applied with it . EEG signals then can be defined as [12] .

$$STFT(f,t) = \int_{-\infty}^{\infty} X(t).W(t-\tau) \, e^{-j2\pi ft} \tag{1}$$

Where, X(t) is the EEG signal , t is the time frame , f is the frequency bin and W is the hamming time window. In this work STFT was applied to alpha band using 1 sec windows with 50% overlap. Each channel is presented by t time frame and f frequency bin which is in this work (117,512). The energy E of the frequency in each frame is computed as:

$$E(t) = \sum_{f=0}^{n} F(f)^2 \tag{2}$$

F is the spectrogram at frequency f, n is the bin number of frequency.

2.4   Fuzzy Logic Model

Fuzzy models are effective techniques to model complex, nonlinear and uncertain system that classical methods encounter difficulties to apply because lack of knowledge. Fuzzy approaches based on using the input-and output data to generate rules determined the behavior of the system. These rules extracted from the data structure that are identified using different cluster methods as well as those rules can be modified according to the system. The input data associated with the clusters based on membership function.

Takagi, Sugeno and Kang [8],[9] , introduced fuzzy model which is known (TSK) fuzzy system. This system is widely used in many theoretical and control systems. TSK fuzzy model based on using simple rules generates from the input –output data. These rules consequences with a simple linear regression model to predicate the output. In TSK approach, subtractive cluster methods [10] are used to cluster the data input by finding the center of each cluster which represents the point with highest number of neighborhood, consequently, the second cluster will be the second point of highest neighborhood. After using the subtractive cluster to identify number of cluster and its location, the rules for TSK fuzzy are extracted from training data. For example, the rules of j cluster can be expressed as:

IF x1 is in Aj1 and x2 is in Aj2 and x3 is in Aj3 ...... xn   is in Aj n

Then: $y^j = p_0 + p_1^j x_1 + p_2^j x_2 + p_3^j x_3 + \cdots + p_n^j.$ \hfill (3)

Where x represents the input variables from 1 to n, y is the output variable, while Ajn is the membership function for the cluster j and pjn is the   regression parameters for jth rules.

For the sake of this study, the input variables are the extracted features that stated previously, output are labeled by two values to indicate two groups (0=autism group, 1 = typical group). Then TSK fuzzy-subtractive approach was applied to the input-output variables to cluster the data and model the memberships that are associated with each variable and cluster.

2.5   Estimate the Degree of Autistic Child and Classification Process

The procedures can be explained as:

1. Features are extracted for each groups as explained in section 2.3
2. Using TSK fuzzy system with subtractive cluster to cluster the data and get the fuzzy inference system FIS .The input data was the extracted features for both

groups and the output data was labelled (0)'for autism group and (1) for normal groups.

3. Test the same data to evaluate the model. Estimate 0.5 as uncertain region between autism and normal groups. Thus, three classes are proposed autism, normal and uncertain region.

4. Test each subject separately and compute the distance between each output and autism, normal and the uncertain region. This can be done by subtracted the mean values of the output for each subject from the label of each groups. The results are shown in the next section.

5. In this system, crisp classification can be obtained by put 0.5 as the threshold for autism and normal children, instead of considering the value within 0.5 as uncertain region. The predicate value (output value) > 0.5   then features belong to normal group and if the predicate values < 0.5 the values belong to autism group.

3   Results and Discussion

For each group subjects, features are computed from STFT subspace for each channels and combined together. Fig. 1 shows the scattered distribution of these features of ASD and TP groups for channel C3. It can be seen that the data distribution has considerable overlap.

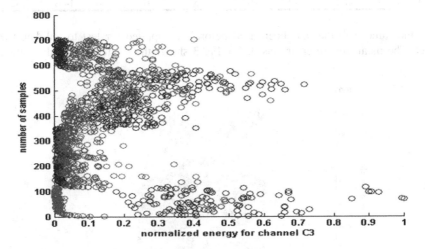

Fig. 1. Scatter plot of features for both autism subjects (red points) and typical subjects (blue points) for channels C3

It can be seen that the behavior of other channels is similar to Fig. 1, where there is a significant overlap between the two groups. TSK fuzzy system was used to predict the output values of two groups of data i.e. autism and typical groups. The input data, matrix of two groups with 8 column and 1404 rows, is clustered using subtractive clustering method as explained above .Two classes are identified for each variable (EEG-channel). Then these clusters were translated to TSK rules. Fig .2 shows the membership functions for variable 1.

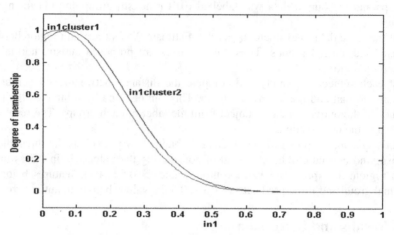

Fig. 2. Plot of the memberships for three classes for first input (in1), represent the C3 channel

Table 1. A rule base of TSK model

	P_0	P_1	P_2	P_3	P_4	P_5	P_6	P_7	P_8
R1	-3.02	-0.667	0.835	-4.55	-0.45	0.035	-1.89	0.234	7.52
R2	1.337	-1.1	-34.5	-6.72	-4.37	-3.91	-5.60	14.8	-4.82

Next, the same data that has been used before was applied for test to validate the model. The mean square error was 0.226. Fig 3 shows the scatter plot of the output values.

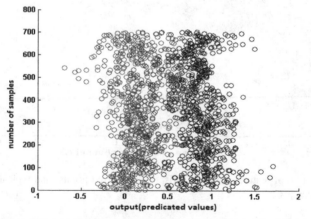

Fig. 3. Scatter plot of the predicate values for both autism subjects (red points) and typical subjects (blue points)

Clearly, the two groups are separated considerably, with clear overlap in some features at range 0.5. This model can be used to diagnose autistic and normal subjects as well as to determine the response of treatment for autism subject.

From TSK model, autism data presented by Gaussian distribution with mean equal (0) as well as normal groups have Gaussian distribution with mean values equal (1). Therefore, subject has distribution with output near to (0.5) will be in uncertainty. The mean value of output for each subject was computed .Then the distance between these values and the three classes were computed as explained in above. The results are shown in Table 2, presenting the closeness degree of each subject to each group. Besides, it gives us prediction of the degree of severity for each subject. Subject 1 in autism group is the highest, while subject 4 shows significant closeness to the uncertain region. This may be interpreted as the degree of response to the treatment .The results should be compared and evaluated with psychology profile of the subjects.

Table 2. The mean of output fuzzy model for each subjects and the distance from each class in percentage

	Mean values	Near from autism %	Near from normal %	Near from the uncertain region %
Autism(1)	0.072	92.723	7.276	57.27
Autism(2)	0.243	75.614	24.385	74.385
Autism(3)	0.231	76.818	23.181	73.181
Autism(4)	0.300	69.956	30.043	80.043
Autism(5)	0.123	87.632	12.367	62.367
Autism(6)	0.128	87.164	12.835	62.835
Typical(1)	0.970	2.926	97.073	52.926
Typical(2)	0.822	17.782	82.213	67.786
Typical (3)	0.822	17.786	82.213	67.786
Typical (4)	0.700	29.975	70.024	79.975
Typical (5)	0.779	22.035	77.964	72.035
Typical (6)	0.804	19.581	80.4182	69.581

Table 3. The accuracy of detect typical and autism subject

	Autism %	Typical %
Autism(1)	100	0
Autism(2)	94.87	5.13
Autism(3)	94.02	5.98
Autism(4)	83.76	16.42
Autism(5)	89.	10.62
Autism(6)	88.89	11.11
Typical(1)	97.44	2.5
Typical(2)	94.78	5.13
Typical (3)	94.78	5.13
Typical (4)	77.78	22.22
Typical (5)	89.74	10.26
Typical (6)	94.87	5.1

In typical group, most of the subjects are in the acceptable range of normal except subject 4 is diagnosed in uncertain region and subject 5 is strongly close to uncertain region. Thus, these subjects need further investigation and further test. Table 3 shows the accuracy of classification for autism and normal groups by using the same Fuzzy TSK model with 0.5 as threshold as previously explained.

By comparing this result with the Table 1, it can be noticed that the previous model of fuzzy output is more practical for analysis and to predicate the degree of disorder.

4   Conclusion and Future Work

In this work, fuzzy inference system is proposed to detect autism spectrum disorder and to estimate how close a child is to being normal or autistic using EEG signals in open eyes-condition. The system was built using TSK fuzzy based on subtractive clustering. First, the system was used to show the probability of a child being autistic, normal or in uncertain region and compute how a subject is close to or far from those classes. Next, the system was modified to have crisp decision for a subject being autistic or normal. The system shows correct decision for autism around 91% and around 96% for control subject except two control subjects were close to the uncertain region. Besides, the results show, one autistic subject is near to the uncertain region that may be identified as the response of autism to the treatment. The future work, hopes to design control system to diagnose autism, degree of autism and determine autism response to learning process as hop case or hopeless case assessment with their neuropsychological and psychiatric reports.

References

1. Sheikhani, A., Behnam, H., Mohammadi, M., Noroozian, M., Golabi, P.: Analysis of Quantitative Electroencephalogram Background Activity in Autism Disease Patients with Lempel-Ziv Complexity and Short time Fourier Transform Measure. In: 4th IEEE-EMBS International Summer School and Symposium on Medical Devices and Biosensors, pp. 111–114. IEEE Press, Cambridge (2007)
2. Sheikhani, A., Behnam, H., Mohammadi, M.R., Noroozian, M., Mohammadi, M.: Detection of Abnormalities for Diagnosing of Children with Autism Disorders using of Quantitative Electroencephalography Analysis. J. Med. Syst. 36, 957–963 (2012)
3. Ahmadlou, M., Adeli, H., Adeli, A.: Fractality and Awavelet-Chaos-Neural Network Methodology for EEG-based Diagnosis of Autistic Spectrum Disorder. Journal of Clinical Neurophysiology 27, 328–333 (2010)
4. Shams, W.K., Wahab, A.: Characterizing Autistic Disorder based on Principle Component Analysis. Australian Journal of Basic and Applied Sciences 6, 149–155 (2012)
5. Oberman, L.M., Hubbard, M.E., McCleery, J.P., Altschuler, L.P., Ramachandran, V.S., Pineda, J.A.: EEG Evidence for Mirror Neuron Dysfunction in Autism Spectrum Disorders. Cognitive 24, 190–198 (2005)
6. Bosl, W., Tierney, A., Tager-Flusberg, H., Nelson, C.: EEG Complexity as a Biomarker for Autism Spectrum Disorder Risk. BMC Medicine, 9–18 (2011)

7. American Psychiatric Association: Diagnose and Statistical Manual of Mental Disorder DSM-IV-IR. American Psychiatric Publishing, Inc., Washington DC (2000)
8. Zadeh, L.A.: Outline of a New Approach to the Analysis of Complex Systems and Decision Processes. IEEE Transaction on Systems, Man, and Cybernetics 3, 28–44 (1973)
9. Tagaki, T., Sugeno, M.: Fuzzy Identification of Systems and Its Application to Modelling and Control. IEEE Transactions on Systems, Men and Cybernetics 15, 116–132 (1985)
10. Chiu, S.L.: Fuzzy Model Identification based on Cluster Estimation. Journal of Intelligent Fuzzy Systems 2, 267–278 (1994)
11. Sanei, S.: Chambers: EEG Signal Processing. John Wiley & Sons Ltd, England (2007)
12. Subha, D.P., Joseph, P.K., Acharya, U.R., Lim, C.M.: EEG Signal Analysis: A survey. J. Med. Syst. 34, 195–212 (2010)

Detecting Different Tasks
Using EEG-Source-Temporal Features

Wafaa Khazaal Shams[1], Abdul Wahab[1], and Uvais A. Qidwai[2]

[1] Computer Science Department, International Islamic University, Malysia
wafaa_dth@yahoo.com, abdulwahab@iium.edu.my
[2] Computer Science and Engineering Department, Qatar University, Qatar
uqidwai@qu.edu.qa

Abstract. This study proposes a new type of features extracted from Electroencephalography (EEG) signals to distinguish between different tasks. EEG signals are collected from six children aged between two to six years old during opened and closed eyes tasks. For each time-sample, Time Difference of Arrival (TDOA) is applied to EEG time series to compute the source-temporal-features that are assigned to x, y and z coordinates. The features are classified using neural network. The results show an accuracy of around 100% for eyes open task and around (83%-95%) for eyes closed tasks for the same subject. This study highlights the use of new types of features (source-temporal features), to characterize the brain functional behavior.

Keywords: EEG signals, Classification, TDOA approach, Source-Temporal features.

1   Introduction

One of the important issues is how to present the EEG signals by features that characterize the brain behaviour. For more than two decades, power spectrum of EEG signals are commonly used to identify EEG wave bands: delta, theta, alpha, beta and gamma [1-3]. Even though Fourier transform are sensitive to noise, it is widely used to characterize EEG brain activity. Also time domain representation of EEG signal is found to be efficient and used widely for classification problem [1-3]. Most of the EEG features that are extracted using time-domain or frequency domain are a form of energy of the signal. In this study, source-temporal features that present x, y and z coordinates are proposed. Time Difference of Arrival approach is utilized for features extraction from EEG time series. Up to our knowledge, this is the first time TDOA method is applied on EEG signals. It is important to mention that this model deals with the dynamic behaviour of the signals to enable features extraction for each time-sample. Apart, EEG records the activities of brain; however there is a challenge of sensitivity of brain's activities to the task conditions, the complexity of the task and the environment. Therefore, in this study activities such as eyes open and eyes closed are chosen to test the model. As a result those tasks do not require much effort from subjects as well as the EEG waves is relativity uniform source.

T. Huang et al. (Eds.): ICONIP 2012, Part IV, LNCS 7666, pp. 380–387, 2012.

2   Time Difference of Arrival Approach

A Time difference of arrival (TDOA) is one of the known methods that it is widely used in wireless location system to find the source of the emitted signals as well as the location of the receiver [4], [5]. TDOA is based on estimating the delay in the arrival times between two synchronized receiver nodes where the signals transfer at the speed of light. By estimating the difference in distance between the source and the two node of receiver, the location of source can be determined by this difference and can be represented as hyperboloids .The intersection of these hyperboloids is the estimated location. To locate the source within three-dimensional (3D) space, three receivers must be used at least. In this work, Chan's method applied to find the source location for the active region in brain without using reference point but using four nodes [5]. Let (x, y, z) represents the location of source, and (x_i, y_i, z_i) represents the location of electrodes where $i=1…n$; n is the number of the receivers (electrodes).The difference in distance D_i can be estimated as :

$$D_i = \sqrt{(x_i - x)^2 + (y_i - y)^2 + (z_i - z)^2} \tag{1}$$

The distance between the source and the channel j

$$D_j = \sqrt{(x_j - x)^2 + (y_j - y)^2 + (z_j - z)^2} \tag{2}$$

The different in distance can be calculated as:

$$D_1 - D_2 = v * t_{12} \tag{3}$$

Where, t_{12} is the time delay between channels at locations 1 and 2. By finding (D_1-D_2), (D_1-D_3) and solving equation for x, y, z variables, the resultant equations is as follow:

$$R = Ax + By + Cz \tag{4}$$

Where

$$R = v * t_{12} - v * t_{11} + \left(\frac{(x_1^2 + y_1^2 + z_1^2)}{v * t_{12}} - \frac{(x_2^2 + y_2^2 + z_2^2)}{v * t_{12}} \right) \tag{5}$$

$$- \left(\frac{(x_1^2 + y_1^2 + z_1^2)}{v * t_{12}} - \frac{(x_3^2 + y_3^2 + z_3^2)}{v * t_{12}} \right)$$

$$A = 2[\frac{(x_2 - x_1)}{v * t_{12}} - \frac{(x_3 - x_1)}{v * t_{13}}] \tag{6}$$

$$B = 2[\frac{(y_2 - y_1)}{v * t_{12}} - \frac{(y_3 - y_1)}{v * t_{13}}] \tag{7}$$

$$C = 2[\frac{(z_2 - z_1)}{v * t_{12}} - \frac{(z_3 - z_1)}{v * t_{13}}] \tag{8}$$

x_1, y_1, z_1 are the coordinates of the electrode in locations 1. x_2, y_2, z_2 and x_3, y_3, z_3 are the coordinates of the electrodes in   locations 2 and 3 respectively,t_{13} is the time delay between EEG signals at locations 1 and 3.

By computing the differences of distance i.e. D_1-D_2, D_3-D_2, D_1-D_3, D_1-D_4, three linear equations are obtained. This can be solved by using Gaussian elimination or iteration methods [6].

3   Method and Material

3.1   EEG Data Descriptions

Six subjects participated in this study; aged (4-6) years .The EEG signal was recorded from eight channels based on the EEG International (10-20) Standard System by using BIMEC EEG machine with sampling frequency of 250 hertz. The eight channels represented as C3, C4, F3, F4, T3, T4, P3, P4 with Cz as reference. These channels collect data from cortex, frontal and parietal regions of brain. The subject was asked to sit in reset condition open eyes for one min then asked to close eyes for another one min.

3.2   Data Pre-process

EEG data were pre-processed to reduce the noise embedded as a result of body movement and from electrode connection. Thus the mean value was subtracted from each channel .Then EEG data was filtered by using Infinite Impulse Response IIR , Chebyshev (II) , to get the required band (8-13) Hz that represents the potential of alpha band. Low quality EEG segments (artifacts) were manually excluded from the analysis process.

3.3   Estimate Alpha Wave Speed

The speed of electromagnetic waves within any medium depends on materials properties represented by permeability and the permittivity and can be computed by this equation [7] :

$$v = 1/(\mu\varepsilon)^{1/2}$$ (9)

Where μ is the permeability and ε is the permittivity and given by:

$$\mu = \mu_0\mu_r$$ (10)

$$\varepsilon = \varepsilon_0\varepsilon_r$$ (11)

μ_0 and ε_0 are the permeability and permittivity in free space and μ_r , ε_r are the relative permeability and permittivity respectively.

Many researches have been conducted to estimate the permittivity and permeability of tissues in the brain [8-10]. From literature, the relative permittivity of tissues in the brain is 10e8 as estimated from Gabriel curves [8]. The values for skull, scalp and cerebrospinal fluid (CSF) are respectively, 5.2e5, 4e3 and 11e3 [11]. These values are employed in this study.The permeability of tissue is assumed to be equal to that of air (μ=1) [12] .Then by using the permittivity in free space which is (ε_0 :8.852 e-12 F/m)[7] and applying equation (9), the estimated speed of alpha wave inside the brain will be: 10.8 m/sec.

3.4   Features Extraction

Spatial features are extracted by applying the TDOA technique as explained above. The coordinates of each channel are taken from head spherical model [13].The time delay between each pair of channels was computed based on Generalized Cross-Correlation with Phase Transform method (GCC-PHAT) [14]. Then by applying eq. (3),(5),(6),(7),(8) and solving them using Gaussian elimination method, the x, y, z coordinates features for sources within the active regions can be computed. The choice of the geometry of the sensors (electrodes) was done based on 4 points to have real time difference among the sites. However this work was carried out without any specific geometrical path; it was selected arbitrarily. In this study, 20 iterations were applied for each time-sample, and for each iteration different 4 sites of electrodes were used. Fig.1shows the diagram of the proposed model.

3.5   Classification Process

Multiplayer neural network are widely used for bio-signal classification. In this study we use it to discriminate between two classes   ;the supervised classification was done using label (1) for eyes-open task and label (0) for eyes-close task. The network consist of   three layers- input layer , hidden layers   and the output layer which represents one node [15].According to pervious experimental results , it was found that 10-nodes in hidden layers give good results.

4   Results and Discussion

In this study, five-second segment was chosen from each channel to reduce the computation. The input EEG data for each subject is matrix with 8 channels and 1250 time samples. The size of time window used for computing time difference among the channels was 1 sec (having 250 samples) with a shift of one time sample. By applying the TDOA localization equations with the constant speed as assumed and explained in section 3.4, the output was matrix with 1000 time samples and 20 features for each dimensions x, y, z and for each subject within the task. This process has been done for eyes open and eyes closed tasks. Fig.2 shows the features extracted for both tasks open and closed eyes in x , y and z dimension and are demonstrated in Fig.3 and Fig.4 respectively.

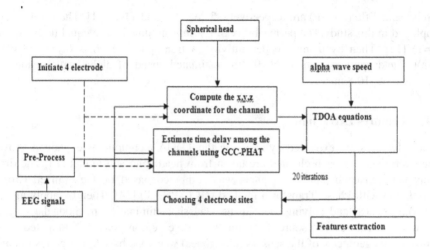

Fig. 1. Feature extracted model for one time-sample

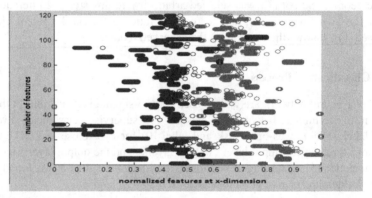

Fig. 2. Scatter plot of the features at x-dimension for open eyes task (blue points) and close eyes task (red points) for six subjects

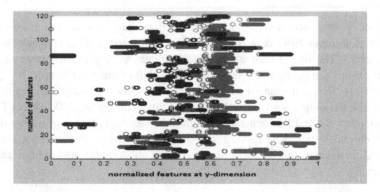

Fig. 3. Scatter plot of the features at y-dimension for open eyes task (blue points) and close eyes task (red points) for six subjects

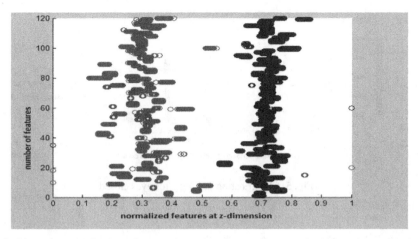

Fig. 4. Scatter plot of the features at z-dimension for open eyes task (blue points) and close eyes task (red points) for six subjects

It is clear that z-dimension shows significant features separation between the two tasks. In classification process, the features in the three dimensions x, y and z are combined and 8000 samples are chosen randomly from both classes and passed to the training process. Then the efficiency of the network has been tested for all subjects for both classes individually. The accuracy is shown in Table 1, and further illustrated in Fig.5 and Fig.6. The accuracy was around 100% for eyes open task for all subjects while it was around 80% in eyes closed tasks.

Table 1. The accuracy of detection eyes open and eyes closed in percentage

Eyes open task	Eyes open (%)	Eyes closed(%)
Subject 1	100	0
Subject 2	100	0
Subject 3	100	0
Subject 4	100	0
Subject5	100	0
Subject 6	100	0
Eyes closed task		
Subject 1	88.4	11.6
Subject 2	95.9	4.1
Subject 3	86.7	13.3
Subject 4	83.6	16.4
Subject5	88	12
Subject 6	83.1	16.1

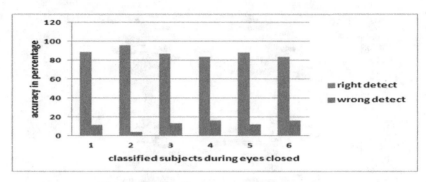

Fig. 5. Classification accuracy for eyes closed

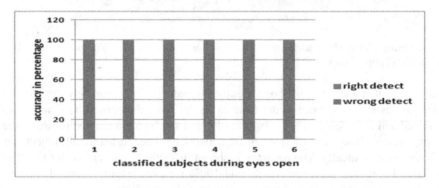

Fig. 6. Classification accuracy for eyes open

5   Conclusion and Future Work

In this study new kind of features are used to characterize the unique information from the EEG signals. These features present the x, y, z locations for the estimated active regions inside the brain. The results show high discrimination between both of tasks: eyes open and eyes closed. Using TDOA technique for feature extraction shows significant results. Particularly, this technique has computation complexity. Thus, it is used for short time periods. Future work can be done using different tasks and with different range of bands.

References

1. Sanei, S.: Chambers: EEG signal Processing. John Wiley & Sons Ltd., England (2007)
2. Subha, D.P., Joseph, P.K., Acharya, U.R., Lim, C.M.: EEG signal analysis: A survey. J. Med. Syst. 34, 195–212 (2010)
3. Thomas, H., Budzynski, H.K., James, R.E.: Introduction to Quantitative EEG and Neurofeedback, 2nd edn. Advance Theory and Application. Academic Press, Elsevier (2008)
4. Schmidt, R.O.: A New Approach to Geometry of Range Difference Location. IEEE Trans.on Aerospace and Electronics System 8, 821–835 (1972)

5. Chan, Y.T., Ho, C.K.: A Simple and Efficient Estimator for Hyperbolic Location. IEEE Transaction on Signal Processing 42, 1905–1915 (2004)
6. Atkison, K.A.: An Introduction to Numerical Analysis, 2nd edn. John Wiley & Sons, New York (1989)
7. Baden Fuller, A.J.: Microwaves, Oxford. Pergamon Press, New York (1979)
8. Gabriel, S., Lau, R.W., Gabriel, C.: The dielectric properties of biological tissues:III. Parametric Models for the Dielectric Spectrum of Tissues. Phys. Med. Biol. 41, 2271–2293 (1996)
9. Geddes, L.A., Baker, L.E.: The specific resistance of biological material—A compendium of data for the biomedical engineer and physiologist. Medical and Biological Engineering and Computing 5, 271–293 (1976)
10. Grave de Peralta Menendez, R., Gonzalez Andino, S.L., Morand, S., Michel, C.M., Landis, T.: Imaging the electrical activity of the brain: ELECTRA. Hum. Brain Mapping 9, 1–12 (2000)
11. Rafiroiu, D., Vlad, S., Cret, L., Ciupa, R.V.: 3D Modeling of the Induced Electric Field of Transcranial Magnetic Stimulation. In: Vlad, S., Ciupa, R., Nicu, A.I. (eds.) International Conference of Advancements of Medicine and Health Care through Technology. IFMBE Proceedings, vol. 26, pp. 333–338. Springer, Heidelberg (2009)
12. Sumi, C., Hayakawa, K.: Mathematical expressions of Reconstructions of Conductivity and Permittivity from Current Density. International Journal of Bioelectromagnetism 9, 103–104 (2007)
13. Kayser, J., Crig, E., Tenke: Principle Components Analysis of Laplacian Waveforms as Generic Method for Identifiying ERP Generator Patterns: Evaluation with Auditory Oddball Tasks. Clinical Neurophysiology 117, 348–368 (2006)
14. Knapp, C., Carter, G.: The Generalized Correlation Method for Estimation of Time Delay. IEEE Transactions on Acoustics, Speech, and Signal Processing 4, 320–327 (1976)
15. Lerner, B., Guterman, H., Dinstein, I., Romen, Y.: Learning Curves and Optimization of a Multilayer Perceptron Neural Network for Chromosome Classification. World Congress on Neural Network 3, 248–253 (1994)

Artificial Neural Network Classification Models for Stress in Reading

Nandita Sharma and Tom Gedeon

Research School of Computer Science
Australian National University, Canberra, Australia
{Nandita.Sharma,Tom.Gedeon}@anu.edu.au

Abstract. Stress is a major problem facing our world today and it is important to develop an understanding of how an average person responds to stress in a typical activity like reading. The aim for this paper is to determine whether an artificial neural network (ANN) using measures from stress response signals can be developed to recognize stress in reading text with stressful content. This paper proposes and tests a variety of ANNs that can be used to classify stress in reading using a novel set of stress response signals. It also proposes methods for ANNs to deal with hundreds of features derived from the response signals using a genetic algorithm (GA) based approach. Results show that ANNs using features optimized by GAs helped to select features for stress classification, dealt with corrupted signals and provided better classifications. ANNs using GAs were generated to exploit the time-varying nature of the signals and it was found to be the best method to classify stress compared to all the other ANNs.

Keywords: stress classification, artificial neural networks, genetic algorithms, physiological signals, physical signals, reading.

1 Introduction

Stress was coined by Hans Selye and he defined it as "the non-specific response of the body to any demand for change" [1]. It is the body's reaction or response to the imbalance caused between demands and resources available to a person. Stress is seen as a natural alarm, resistance and exhaustion [2] system for the body to prepare for a fight or flight response to protect the body from threats and changes. When experienced for longer periods of time and not controlled, stress has been widely recognized as a major growing concern in our age adversely impacting society due to its potential to cause chronic illnesses (e.g. cardiovascular diseases, diabetes and some forms of cancer) and high economic costs in societies (especially in developed countries [3, 4]). Benefits of stress research range from improving personal operations, through increasing work productivity to benefitting society - motivating interest, making it a socially beneficial area of research and posing technical challenges in Computer Science. Numerous computational methods have been used to objectively define and classify stress to differentiate conditions causing stress from other conditions. The methods used simplistic models formed from techniques like

T. Huang et al. (Eds.): ICONIP 2012, Part IV, LNCS 7666, pp. 388–395, 2012.

Bayesian networks [5], decision trees [6] and support vector machines [7]. These models have been built from a relatively smaller set of stress features than the sets used in the models in this paper.

The human body's response signals obtained from non-invasive methods that reflect reactions of individuals and their bodies to stressful situations have been used to interpret stress levels. These measures have provided a basis for defining stress objectively. Stress response signals used in this paper fall into two categories - physiological and physical signals. The physiological signals are the galvanic skin response (GSR), electrocardiogram (ECG) and blood pressure (BP). Unlike physiological signals, we define physical signals as a time-varying characteristic where changes can be seen by humans without the need for equipment and tools that need to be attached to individuals to detect general fluctuations. However, sophisticated equipment and sensors using vision technologies are still needed to obtain physical signals at sampling rates sufficient for data analysis and modeling. The physical signals used in this paper are eye gaze positions and pupil dilation. GSR, ECG, BP, eye gaze tracking and pupil dilation have been used to detect stress in literature [5, 8, 9] but this combination is novel to stress research. In this paper, we refer to the physiological and physical signals as primary signals for stress.

Artificial neural networks (ANNs) have been successfully used for emotion-based stress classification for playing video games [10] and preliminary work in stress classification for reading [11]. This paper uses ANN models to recognise *stressed* and *non-stressed* reading based on features derived from primary measures for stress.

Hundreds of stress features can be derived from primary signals for stress to classify stress classes for the different types of reading. If an ANN used all these features as inputs then it would result in quite a large network. This can lead to the issue where the ratio for number of samples to the number of connections and weights in an ANN could be relatively small, which could affect classifications. Furthermore, in order to achieve a good classification model, it should be robust to input tuples or features that suffer from corruption. A genetic algorithm (GA) could help solve these problems by selecting subsets of features for optimizing ANN stress classifications. A GA is based on the concept of natural evolution and has been commonly used to solve optimization problems [12]. It evolves a population of candidate solutions using crossover, mutation and selection methods in search for a population of a better quality. The quality for each individual or chromosome in the population is defined by some fitness function. GAs have been successfully used to select features derived from physiological signals to classify emotions [13, 14].

ANNs have been used in microarray studies for gene expression levels for classification. Like the stress feature space used in this paper, they also faced the issue of dealing with a relatively large feature space [15], tens of magnitude larger than our feature space and with fewer samples. A GA was used to optimize the inputs for an ANN by maintaining features that produced better classifications [15]. The size of a chromosome in the GA, which mirrored the number of features and the number of inputs to the ANN, was set based on what was previously proposed in the literature regarding a similar type of problem [16]. The chromosome size was fixed. For our stress features, we need to discover the number of features that is sufficient for stress classification. This paper investigates an ANN with all features as inputs (which is referred to as ANN-AllInputs), another ANN with inputs selected by a GA to use features that produced better classifications (ANN-GAInputs) and an ANN with relatively fewer feature inputs and inputs defined to exploit the time-varying characteristic for features (ANN-3Seg-GAInputs). The performance of the ANNs will not only provide how well stress can be classified from the features derived from the primary signals for stress but also provide an insight to irrelevant features that exist in the feature space.

This paper presents the method for data collection from the reading experiment. Various primary signals for stress were recorded during the experiment, which were GSR, ECG, BP, eye gaze tracking and pupil dilation signals. The paper then describes the features obtained from the signals that were used by the different classification models to learn patterns to classify the different types of reading. As mentioned above, the three different types of classification models were ANN-AllInputs, ANN-GAInputs and ANN-3Seg-GAInputs. Each model made use of different facets of stress feature signals from determining features that produced better stress classifications to exploiting time-varying characteristics for signals. Further, the paper provides results, analyses the results and presents a discussion for the results. It concludes by summarizing the work and proposes work that can be done in future to extend the research.

2   Data Collection from Reading Experiment

Thirty-five Undergraduate Computer Science students, compromising 25 males and 10 females, over the age of 18 years old were recruited as experiment participants (after obtaining Ethics Approval from the Australian National University Ethics Committee). Each participant had to understand the requirements of the experiment from written experiment instructions with the guidance of the experiment instructor before they filled in the experiment consent form. Afterwards, physiological stress sensors were attached to the participant and physical stress sensors were calibrated. The instructor notified the participant to start reading, which triggered a sequence of text paragraphs. After finishing the reading, participants had to do an assessment. The experiment process was an extension to the experiment process done in [11] in that the experiment was extended to include a wide range of stress sensors whereas the earlier experiment only had the GSR sensor. An outline of the process of the experiment for an experiment participant is shown in Fig. 1.

Fig. 1. Process followed by participants during the reading experiment

Each participant had physiological and physical measurements taken over the 12 minutes reading time period. During the reading period, a participant read *stressed* and *non-stressed* types of text. Stressed text had stressful content in the direction towards distress, whereas the non-stressed text had content that created an illusion of meditation or soothing environments. Each type of text had the same number of paragraphs and each paragraph was displayed on a computer monitor for participants to read. For consistency, each paragraph was displayed on a 1050 x 1680 pixel Dell monitor, displayed for 60 seconds and positioned at the same location of the computer screen for each participant. Each line of the paragraph had 70 characters including spaces.

Feature values were derived from physiological and physical signals. Biopac ECG100C, Biopac GSR100C and Finapres Finger Cuff systems were used to take ECG, GSR and blood pressure recordings at a sampling rate of 1 kHz. Eye gaze and pupil dilation signals were obtained using Seeing Machines FaceLAB system with a pair of infrared cameras at 60 Hz. Other signals were derived from primary signals such as, heart rate variability, which was calculated from consecutive ECG peaks and another popular signal used for stress detection [17, 18]. Statistics (e.g. mean and standard deviation) were calculated for the signal measurements for each 5 second interval during the stressed and non-stressed reading. Measures such as the number of peaks for periodic signals, the distance an eye covered, the number of forward and backward tracking fixations, and the proportion of the time the eye fixated on different regions of the computer screen over 5 second intervals were also obtained. The regions of the computer screen are shown in Fig. 2. The statistic and measure values formed the stress feature set. In total, there were 215 features.

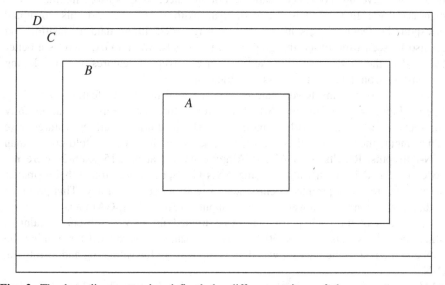

Fig. 2. The bounding rectangles defined the different regions of the computer screen to determine a subset of eye gaze features. The bounding rectangles show regions A, B, C and D. Region A contained the text area where text was displayed to a participant. Region B is the region without A and regions C and D are defined similarly. Region D had the application menu-bar and the tool-bar. Note that the diagram is not drawn to scale.

3   Artificial Neural Network Stress Classifiers

ANNs, inspired by biological neural networks, have characteristics for learning patterns to classify input tuples into classes. It is made up of interconnected processors, known as *artificial neurons*, which are connected by weighted links that pass signals between neurons. In this paper, feed-forward ANNs trained using backpropagation were used. Three topologies were used to classify stress in reading. For two ANNs, a GA was used to select stress features, which formed inputs to the ANNs. As a consequence, the three ANNs differed only on the number of inputs used. The ANNs were:

ANN-AllInputs: A feed-forward ANN was generated for each participant. The ANN had 215 inputs, which mirrored the number of features derived from the primary signals for stress.

ANN-GAInputs: This ANN was similar to ANN-AllInputs but the number of inputs was based on the features selected by a GA. The GA selected features to improve the classification rate for stressed and non-stressed states.

ANN-3Seg-GAInputs: An ANN was generated for each participant with at most 10 features with 3 sequential time segments for each feature as inputs to the ANN. Like ANN-GAInputs, the GA selected features to improve the classification rate for stressed and non-stressed states.

GAs were not only used to improve the classification by selecting relevant features but also to leave out corrupted features. For instance, ECG sensors malfunctioned when acquiring ECG signals for participant with index 19, and this corrupted corresponding features. This information was recorded during data acquisition and can also be seen from observing graphs of raw signals. With a GA, there is a better chance that the ANN classifiers will not use corrupted features for developing relationships from features for stress classification.

A reason for selecting fewer features than the total number of features for ANN-3Seg-GAInputs was to determine whether fewer features can be used to successfully represent the feature space with a smaller network. In addition, if all the features were used as inputs then, in total, the number of inputs would have been 3-fold greater than ANN-AllInputs. Results from ANN-GAInputs show that all 215 features were not needed for an ANN classification. Using ANN-GAInputs, it was found that a smaller subset of features can produce a classification with a better accuracy. This provided another motivation to use fewer features as inputs in ANN-3Seg-GAInputs.

The data sets for each participant was divided up into 3 subsets – training, validation and test sets – where 50% of the data samples were used for training the ANN and the rest of the data set was divided up equally for validating and testing the ANN.

MATLAB was used to implement and test the ANNs. The MATLAB adapt function was used for training the ANN on an incremental basis. Each network was trained for 1000 epochs using the Levenberg-Marquardt algorithm. The network had 7 hidden neurons and one neuron in the output layer. Future work could investigate optimizing the topology of the ANN for stress classification on the reading data set. The accuracy, sensitivity, specificity and the F-Score were calculated to determine the quality of the classification.

GAs were used to select features in ANN-GAInputs and ANN-3Seg-GAInputs based on quality measures for stress classification from ANNs. A GA is a global search technique and has been shown to be useful for optimization problems. Given a population of subsets of features, the GAs evolved the feature sets by applying crossover, mutation and selecting feature sets during each iteration of the search to determine sets of features that produced better quality ANN classifications. The initial population for the GAs was set up to have all the features. The number of features in the chromosomes varied but the chromosome length was fixed. The length of a chromosome was equal to the number of features in the feature space. A chromosome was a binary string where the index for a bit represented a feature and the bit value indicated whether the feature was used in the ANN. For ANN-GAInputs, the initial population had features varying in numbers equally distributed from 1 to 215 features. On the other hand for ANN-3Seg-GAInputs, each chromosome in the initial population had 5 features with feature values for 3 consecutive time segments. The other parameters for the GA were set as provided in Table 1.

Table 1. Settings for GA parameters for ANN-GAInputs and ANN-3Seg-GAInputs

GA Parameter	Value/Setting
population size	100
number of generations	2000
crossover rate	0.8
mutation rate	0.01
crossover type	MATLAB's Scattered Crossover
mutation type	MATLAB's Uniform Mutation
selection type	MATLAB's Stochastic Uniform Selection

4   Results and Discussion

Stressed and non-stressed states for participants were classified using ANN-AllInputs, ANN-GAInputs and ANN-3Seg-GAInputs. The accuracy values for the classification are provided in Fig. 3 for each experiment participant. The best accuracy values were obtained from ANN-3Seg-GAInputs. The accuracy values were significantly different for the three ANNs based on the Student's T-test ($p < 0.01$).

ANN-AllInputs performed the worst compared to the other two ANNs. This shows that the ANN is susceptible to features that were irrelevant and redundant for stress classification. The performance for ANN-GAInputs, which was better than ANN-AllInputs, confirmed this. It used features that were optimized for stress classification. The ANN that exploited the time-varying nature associated with stress features gave the best stress classification. Results were significantly better for ANN-3Seg-GAInputs. This showed that feature values for stress has information in the patterns over time – a significant observation derived directly from this ANN result.

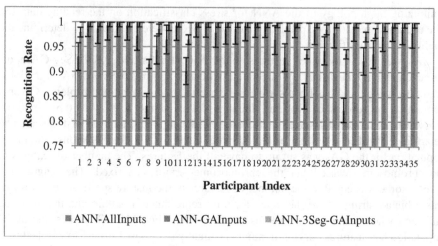

Fig. 3. Recognition rates obtained from cross-validation on the different ANNs

The classification results from the ANNs mirrored the participant reported responses for stress for the different types of text. Participants classified stressed and non-stressed text where classifications were significantly different based on the Chi-test ($p < 0.01$).

5   Conclusion and Future Work

Stress in reading has been successfully classified using an artificial neural network that was built on features derived from an individual's stress response signals. Features can become voluminous and this may increase chances for an ANN to use irrelevant and redundant input features. As a consequence, a genetic algorithm was developed for selecting relevant features for the ANN classification. The GA and ANN hybrid provided better quality classifications. Moreover, by exploiting the time-varying nature of features and incorporating it in the hybrid, the quality of the stress classifications improved and formed a more robust stress classification model. Investigations in the future could examine how different ANN models and topologies may influence stress classification, and to investigate alternative optimization techniques.

References

1. Selye, H.: The stress syndrome. The American Journal of Nursing 65, 97–99 (1965)
2. Hoffman-Goetz, L., Pedersen, B.K.: Exercise and the immune system: a model of the stress response. Immunology Today 15, 382–387 (1994)
3. The-American-Institute-of-Stress. America's No. 1 Health Problem - Why is there more stress today? (August 05, 2010), http://www.stress.org/americas.htm

4. Lifeline-Australia (Stress Costs Taxpayer $300K Every Day (2009),
   http://www.lifeline.org.au
5. Liao, W., et al.: A real-time human stress monitoring system using dynamic bayesian network. In: Computer Vision and Pattern Recognition - Workshops. CVPR Workshops (2005)
6. Zhai, J., Barreto, A.: Stress recognition using non-invasive technology. In: Proceedings of the 19th International Florida Artificial Intelligence Research Society Conference FLAIRS, pp. 395–400 (2006)
7. Dou, Q.: An SVM ranking approach to stress assignment. University of Alberta (2009)
8. Labbé, E., et al.: Coping with stress: the effectiveness of different types of music. Applied Psychophysiology and Biofeedback 32, 163–168 (2007)
9. Healey, J.A., Picard, R.W.: Detecting stress during real-world driving tasks using physiological sensors. IEEE Transactions on Intelligent Transportation Systems 6, 156–166 (2005)
10. Scherer, S., et al.: Emotion recognition from speech: Stress experiment. In: Proceedings of the 6th International Language Resources and Evaluation (LREC 2008), Marrakech, Morocco (2008)
11. Sharma, N., Gedeon, T.: Stress Classification for Gender Bias in Reading. In: Lu, B.-L., Zhang, L., Kwok, J. (eds.) ICONIP 2011, Part III. LNCS, vol. 7064, pp. 348–355. Springer, Heidelberg (2011)
12. Goldberg, D.E.: Genetic algorithms in search, optimization, and machine learning. Addison-Wesley (1989)
13. Park, B.J., et al.: Feature selection on multi-physiological signals for emotion recognition. In: 2011 International Conference on Engineering and Industries (ICEI), Korea, pp. 1–6 (2011)
14. Niu, X., et al.: Research on genetic algorithm based on emotion recognition using physiological signals. In: 2011 International Conference on Computational Problem-Solving (ICCP), pp. 614–618 (2011)
15. Yacci, P.: Feature selection of microarray data using genetic algorithms and artificial neural networks. Rochester Institute of Technology (2009)
16. Golub, T.R., et al.: Molecular classification of cancer: class discovery and class prediction by gene expression monitoring. Science 286, 531–537 (1999)
17. Ferreira, P., et al.: License to chill!: how to empower users to cope with stress. In: Proceedings of the 5th Nordic Conference on Human-Computer Interaction: Building Bridges, pp. 123–132 (2008)
18. Dishman, R.K., et al.: Heart rate variability, trait anxiety, and perceived stress among physically fit men and women. International Journal of Psychophysiology 37, 121–133 (2000)

Performance Evaluation of TCP and UDP Traffic in IEEE 1451 Compliant Healthcare Infrastructure

Junaid Ahsenali Chaudhry[1], Uvais A. Qidwai[1], and Mudassar Ahmad[2]

[1] Department of Computer Science and Engineering, Qatar Univesrity, Qatar
[2] Department of Information and Communication, Universiti Teknologi Malaysia
{junaid,uqidwai}@qu.edu.qa, amudassar2@utm.my
http://www.qu.edu.qa, http://www.utm.my

Abstract. Wireless ad-hoc and intrastructure networks have become very popular during the last ten years. Due to this reason, wireless security has become a very serious issue, specially in healthcare intrastructure networks. The aim of this research is to quantify the impact of security on the performance of healthcare networks. Many experiments are conducted by using TCP and UDP traffice to analyse the security behavior on network performance. Results show that security implementation affact the performance of wireless healthcare centers. At the end we will show that how our proposed approach will achive the maximum performance.

Keywords: Healthcare, TCP, UDP, Security, Healthcare Monitoring.

1 Introduction

The market for wireless healthcare infrastructure has made incredible growth during the past few years. Medical and the computer industry have an important place in Wireless technologies [1]. Network flexibility and mobility are the major benefits of such wireless infrastructures. In mobile network data can be easily accessed from anywhere because there is no headache of wires, as in case of Wired Network [2]. The medium of transmission between sender and receiver in wireless communication is air, hence makes the data insecure. Hence, security is a very important issue in this sense because insecure data cause a great loss for organizations. Several security algorithms have been discovered to solve the IEEE 802.11 security issues .Performance reduction is the drawback of these security algorithms. Effect of these algorithms is being studying [3]. This research unveils and compares the effects of WEP, 802.11x, and WPA on 802.11g wireless network performance. In this search we use traffic type, transmission power and security mechanisms as metrics to analyze the performance of wireless network. The experiments were conducted using a single celled network in a clean environment proving that 802.11g wireless network. The rest of paper is organized as follows. In section 2, we discuss the related work. Section 3 describes the experimental

T. Huang et al. (Eds.): ICONIP 2012, Part IV, LNCS 7666, pp. 396–403, 2012.

setup. Performance evaluation methodology is explained in section 4. The results obtained are described in section 5 we present our conclusion and discuss future work in section 6.

2   Experimental Setup

The test bed configuration was based on the traditional client/server architecture but using wireless connections. The only difference was that no real-time data traffic was used in the experiments. Data traffic was generated by a traffic generating tool iperf [4]. The laboratory was designed as a clean environment ; that is, no background noise or other interferences was present. It was a single cell environment that included one server, one client and one 802.11g access point. The server P-3 Toshiba Satellite 800 Laptop was connected to access point with 100 Mbps wired connection. The client P-3 IBM ThinkPad Laptop T22 was placed at a fixed distance of 7 meters from access point with 54 Mbps wireless connection. The network traffic was generated by the Iperf installed on both the server and the client laptops. The bandwidth of the Ethernet connection between the Access Point and the server was equal to 100 Mbps. Between the access point and the client, the bandwidth was 54 Mbps. In the experiment, Windows-based operating systems were used because Windows XP and Windows 2003 Server have a built in implementation of the IEEE 802.11 security mechanisms and 802.1x authentication protocol such as: PEAP. Fig 1 shows a graphical setup representation of the single cell network. Experiments were conducted in a single-celled environment. Iperf was installed on both client and server laptops, to generate TCP and UDP traffic of different packet size. Total amount of data sent per session was kept constant i.e. 15 MB. Performance evaluation was characterized on the basis of Response time and Throughput by varying Security, Transmit Power and TCP and UDP window size.

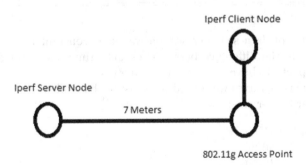

Fig. 1. Experimental Setup, TestBed

3   Performance Evaluation Methodology

Performance was evaluated by applying 15 different security mechanisms on UDP & TCP traffic types and changing the transmit power of AP. Traffic was generated using Iperf. The results were analyzed using statistical analysis tool SPSS. The experiments were conducted for a single client, changing transmit power of AP and were repeated for different packet sizes in case of UDP and different window sizes in case of TCP. Two performance metrics were used i.e. Throughput in Mbits/sec and Response time in seconds. The Iperf being used allowed us to set up the traffic model types. UDP and TCP are the two traffic models that were used in our experiments. Both of the traffic types affect the performance of the network in different ways. I used 15 different security mechanisms to check the impact of security protocols on the performance of different traffic types. Total data sent per interval was kept constant i.e. 15 M, in case of TCP window size was changed 1K, 500K and 1000K and in case of UDP window size was kept constant i.e. 500K while packet size was changed between 300 bytes, 600 bytes and 900 bytes. The results obtained for the mean throughput and response time regarding security protocols are analyzed. Response time and throughput of two traffic types UDP & TCP were measured using Iperf. , against the following parameters show in table 1.

Table 1. Enumirology of Security Mechanisms

Security Mechanisms
No security with SSID
MAC address authentication
Open System Authentication with 64-bit WEP Encryption
Open System Authentication with 128-bit WEP Encryption
Open System Authentication with 152-bit WEP Encryption
Shared Key Authentication with 64-bit WEP Encryption
Shared Key Authentication with 128-bit WEP Encryption
Shared Key Authentication with 152-bit WEP Encryption
Open System / Shared Key Authentication with 64-bit WEP Encryption
Open System/ Shared Key Authentication with 128-bit WEP Encryption
Open System / Shared Key Authentication with 152-bit WEP Encryption
WPA-PSK Authentication with AES Encryption
WPA-PSK Authentication with TKIP Encryption
WPA-EAP-PEAP Authentication with AES Encryption
WPA-EAP-PEAP Authentication with TKIP Encryption

Total amount of data sent per session was kept constant i.e. 15 Mbytes, in order to measure the difference between response time and throughput values due to the impact of above mentioned parameters.

Following iperf syntaxes were used to measure the traffic.

Iperf Syntax at Server side:

```
iperf s D n 15M w 1K
iperf s D n 15M w 500K
iperf s D n 15M w 1000K
iperf s D n 15M w 500K u l 300
iperf s D n 15M w 500K u l 600
iperf s D n 15M w 500K u l 900
```

Iperf Syntax at Client side:
iperf c ms-client1 n 15M w 1K
iperf c ms-client1 n 15M w 500K
iperf c ms-client1 n 15M w 1000K
iperf c ms-client1 n 15M w 500K u b 54M l 300
iperf c ms-client1 n 15M w 500K u b 54M l 600
iperf c ms-client1 n 15M w 500K u b 54M l 900

Fig 2 shows that the result of one-way ANOVA at 95% confidence interval is
highly significant, thus it is proved that the performance of both TCP and UDP
traffic degrades with enhancement in security. Our results showed that for all the
security mechanisms TCP performance is almost half of the UDP performance.
Therefore it is confirmed that UDP utilizes maximum possible bandwidth as
presented by [3]. The results obtained for response time are opposite to those
of throughput here TCP values were higher than UDP which is almost half of
TCP. The Response time of UDP traffic remains same but that of TCP traffic
is highest at minimum transmit power. The transmit power has no effect on the
performance of UDP throughput while in case of TCP performance degrades
when the transmit power is minimum.

ANOVA

		Sum of Squares	df	Mean Square	F	Sig.
TCP Throughput in Mbits/sec	Between Groups	58.939	14	4.210	7.900	.000
	Within Groups	71.939	135	.533		
	Total	130.878	149			
UDP Throughput in Mbits/sec	Between Groups	3294.100	14	235.293	132.154	.000
	Within Groups	240.360	135	1.780		
	Total	3534.460	149			
TCP response time in seconds	Between Groups	4.809	14	.344	6.816	.000
	Within Groups	6.804	135	.050		
	Total	11.613	149			
UDP response time in seconds	Between Groups	17.187	14	1.228	105.833	.000
	Within Groups	1.566	135	.012		
	Total	18.753	149			

Fig. 2. Analysis of Variance TCP and UDP Throughput and RTT

Different types of traffic were generated using Iperf. TCP traffic was generated
using three different window sizes in order to understand the impact of window
size on network performance.

The fig 3 shows that larger window size gives better TCP throughput and
lower response time while smaller window size gives lower throughput and high
response time value. Proving as the window size is increased performance is
increased. The fig 4 shows that UDP throughput increases with the increase in
packet size while its response time decreases as it takes more time to deliver a
larger size of packet.

The experiments were conducted by changing the transmit power of AP from
full to half and from half to minimum. All these transmit powers effect the
network performance in different ways. The results obtained shows that transmit

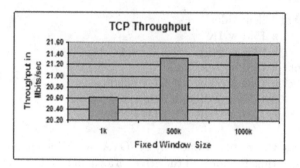

Fig. 3. TCP Throughput Chart

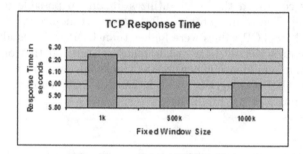

Fig. 4. TCP Response time Chart

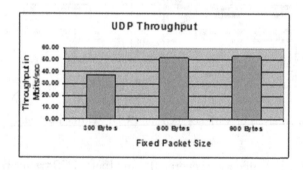

Fig. 5. UDP Throughput Chart

power has no effect on the performance of UDP throughput while in case of TCP performance degrades when the transmit power is minimum. The Response time of UDP traffic remains same but that of TCP traffic is highest at minimum transmit power.

Results show that in case of 1K window size low throughput and high response time was noticed from security mechanism 10 to security mechanism 15. High throughput and low response time was achieved from security mechanism 6 to security mechanism 9. Lowest throughput and highest response time was achieved

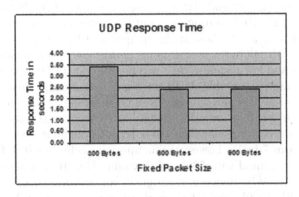

Fig. 6. UDP Response time Chart

at 14. With 500K overall good throughput was noticed from security mechanism 3 to 13. Here lowest throughput and highest response time was achieved at mechanism 2. Hence it gives good performance as compared to 1K window size. Overall high performance was measured with 1000K TCP window size. This is because with larger window size, security mechanisms work efficiently, and transfer more data in less time efficiently. This means that when we want security in our WLANs, then we must use such mechanism where TCP will send data with larger windows size to have larger throughput and low response time. Exceptions are WPA-EAP-AES-PEAP and WPA-EAP-TKIP-PEAP which gives better performance at 500K. Results show that in case of 300byte datagram size low throughput and high response time were noticed from security mechanism 3 to security mechanism 15. High throughput and low response time was achieved from security mechanism 1 and 2. With 600byte overall good throughput was noticed in all security mechanisms. Overall high performance was measured with 900bytes datagram size. This is because with larger packet size, security mechanisms work efficiently, and transfer more data in less time efficiently. This means that when we want security in our WLANs, then we must use such mechanism where UDP will send data with larger datagram size to have larger throughput and lowest response time.

Results shows that in case of 1K window size low throughput and high response time was noticed at security mechanism 3, 8, 10, 14 and 15. High throughput and low response time was achieved at security mechanism 1, 2, 6 and 7. Lowest throughput and highest response time was achieved at 14. Overall high performance was measured with 500K TCP window size. Here lowest throughput and highest response time is achieved at 5 (Shared key 152-bit). Hence it gives good performance as compared to 1K window size. With 1000K overall good throughput was achieved, low throughput was achieved at security 4, 14 and 15. All other mechanisms gave good performance value but overall effect of 500K gave better performance than 1000K in case of Half transmit power. Giving the result that with half transmit power larger throughput and lowest response time is achieved at 500K window size. Results shows that in case of 300byte packet size low throughput and high response time was noticed between security mechanism 3 and 15. High throughput and low response time was achieved at security mechanism 1 and 2. Lowest throughput and highest response time was achieved at 4. With 600bytes overall good performance was achieved. Overall high throughput and low response time was measured with 900byte UDP packet size, giving the impression that UDP does not have much impact of security with large packet size. Its performance does not degrade as that of TCP.

In this case our results shows that in case of 1K window size low throughput was noticed from security mechanism 1to 6. High throughput was achieved from security mechanism 7 to 15. Lowest throughput was achieved at 4. Overall high throughput was measured with 500K TCP window size. Here lowest throughput is achieved at 2 (MAC). Hence it gives good performance as compared to 1K window size. With 1000K highest throughput was achieved at 1, 5 ,9 ,12 and 13 security mechanisms, low throughput was achieved at security 2, 4, 6, 8, 14 and 15. At security 3, 10 and 11 1000K and 500K gave same good throughput values. Experimental results shows that in case of 1K window size high response time was noticed at security mechanism 1, 6 and 15. Low response time was achieved from security mechanism 5 to 14. Highest response time was achieved at 4. Overall low response time was measured with 500K TCP window size. Here highest response time is achieved at 1. Hence it gives good performance as compared to 1K window size with enhanced security. With 1000K lowest response time was achieved at 7 , 8, 11 and 12 security mechanisms, high response time was achieved at security 1, 4, 6, and 15. At security 3, 10 and 11 1000K and 500K gave same response time values. Results shows that in case of 300byte packet size low throughput and high response time was noticed from security mechanism 3 to 15. High throughput and low response time was achieved at security mechanism 1 and 2. Lowest throughput and highest response time was achieved at 12. With 600bytes overall good performance was achieved. Overall high throughput and low response time was measured with 900byte UDP packet size, giving the impression that UDP does not have much impact of security with large packet size. Its performance does not degrade as that of TCP.

4   Conclusion

Overall research concluded that security mechanisms effect differently if other factors like traffic types, packet size, window size, and transmit power are taken into consideration. It is proved that UDP utilizes more bandwidth as compared to TCP and is less affected by other factors. Transmit power has more clear effect on TCP as with minimum 12.5% transmit power security effect is reversed in case of TCP. To get maximum performance for TCP traffic use shared key authentication with 128 bit WEP encryption, half transmit power and medium packet size. This could be left another area of research. All above results prove that good network performance of TCP traffic with security is achieved with when transmit power of AP is Full and TCP window size is large. UDP throughput with security is not much affected by transmit power, it gives overall good performance than TCP traffic. WPA-EAP-PEAP Authentication with AES Encryption can be used for UDP traffic for all transmit powers and large packet size. For future work other standards 802.11n can be compared with 802.11g. This research was conducted with no interference another research could take this factor consideration.

Acknowledgment. This publication was made possible by a grant from Qatar National Research Fund under its National Priority Research Program, for projects NPRP 09-292-2-113. Its contents are solely the responsibility of the authors and do not necessarily represent the official views of Qatar National Research Fund.

References

1. Stallings, W.: IEEE 802.11: Moving Closer to Practical Wireless LANs. in IT Professional IEEE 3(3), 17–23 (2001)
2. Hunt., B.N.a.R.: An Experimental Study of Cross-Layer Security Protocols. In: IEEE Public Access Wireless Networks, Globecom 2005, St. Louis, USA (2004)
3. Senat, N.J.: Performance Study on IEEE 802.11 Wireless Local Area Network Security (2006)
4. Barka, E., Boulmalf, M.: On The Impact of Security on the Performance of WLANs. Journal of Communications Proceedings of the IEEE 2(4), 10–17 (2007)
5. Narayan, S., Feng, T., Xu, X., Ardham, S.: Network Performance Evaluation of Wireless IEEE802.11n Encryption Methods on Windows Vista and Windows Server 2008 Operating Systems. IEEE Performance Evaluation, 1–5 (2009)
6. Narayan, S., Feng, T., Xu, X., Ardham, S.: Impact of Wireless IEEE802.11n Encryption Methods on Network Performance of Operating Systems. In: Second International Conference on Emerging Trends in Engineering and Technology, vol. 12, pp. 1178–1183 (2009)
7. Kolahi, S.S., Singla, H., Ehsan, M.N., Dong, C.: Performance of IPv4 and IPv6 Using 802.11n WLAN in Windows 7- Windows 2008 environment. In: Baltic Congress on Future Internet and Communication, pp. 50–53 (2011)
8. Li., K.S.S.A.P.: Evaluating IPv6 in Peer-to-Peer 802.11n Wireless LANs, pp. 70–74. IEEE Computer Society (2011)
9. Murty, M.S., Veeraiah, D., Rao, A.S.: Performance Evaluation of Wi-Fi comparison with WiMAX Networks. International Journal of Distributed and Parallel Systems (IJDPS) 3(1), 321–329 (2012)
10. Likhar, P., Yadav, R.S., Keshava Rao, M.: Securing IEEE 802.11G Wlan Using Open Vpn and ITS Impact Analysis. International Journal of Network Security and its Applications (IJNSA) 3(6), 97–123 (2011)
11. iperf, Network Bandwidth Measuring Tool

Power and Task Management in Wireless Body Area Network Based Medical Monitoring Systems

Robert G. Rittenhouse[1], Malrey Lee[2], Junaid Ahsenali Chaudhry[3],
and Uvais A. Qidwai[3]

[1] Keimyung Adams College, Keimyung University
Daegu, 704-701, South Korea
rrittenhouse@acm.org
[2] The Research Center for Advanced Image and Information Technology,
School of Electronics & Information Engineering, ChonBuk National University
JeonJu, ChonBuk, 561-756, South Korea
mrlee@chonbuk.ac.kr
[3] Department of Computer Science and Engineering,
Qatar University, Doha, Qatar
{junaid,uqidwai}@qu.edu.qa

Abstract. Intelligent healthcare systems incorporating wireless sensors in a ubiquitous computing environment have the potential to revolutionize outpatient care. Such systems must balance effective and timely reporting of results with power requirements and available communication methods. We intend to design a system based on a wireless body area network (WBAN) which will effectively detect medical problems, reduce the time lag between detection of a medical problem as well as manage power and communications. We plan to develop an algorithm that can find an optimal solution, within an acceptable time, and be faster than current algorithms in assigning tasks and processing and transmitting sensor data such that the system end-to-end delay is minimized while guaranteeing required system battery lifetime and availability. This system will also be able to analyze a patient's health data and report the results to the user while simultaneously converting the sensor data to the standard HL7 (Health Level Seven) format and transmitting it to a healthcare server.

Keywords: Healthcare, Power Management, Wireless Networks, Body Area Networks.

1   Justification for the Research

Many countries will be confronted with challenges supporting quality health care. The need for home health care is being driven by several factors including demographic trends, particularly the aging population; the desire of patients to remain in their homes and, the need to dramatically reduce the costs of health care delivery while still providing quality care [1].

T. Huang et al. (Eds.): ICONIP 2012, Part IV, LNCS 7666, pp. 404–414, 2012.
© Springer-Verlag Berlin Heidelberg 2012

The traditional methods of taking care of outpatients alone at home include periodic visits by the patient to medical facilities providing visiting nursing services in the home or a live in caretaker. There are drawbacks to all these approaches. Patients with limited mobility may find it difficult to get to patient care facilities and providing visiting nursing care is expensive. Neither approach will serve patients whose condition requires constant monitoring. Live in caretakers are prohibitively expensive and patients may resent the loss of privacy. One way to address this difficulty is to enable people to become outpatients while still having their condition monitored.

In modern healthcare environments patients wear sensors to monitor their health. These devices measure the patient's physiological data such as skin temperature, respiration rate, heart rate, blood pressure, etc. The raw data gathered by the sensors must be transferred to a computer for further analysis. This healthcare server holds all the pertinent information about the patient. Much existing research has focused on the exploitation of ubiquitous systems for increased monitoring in the outpatient environment [1–3].

These systems must balance power requirements, monitoring requirements based on the patient's condition and communications bandwidth. We intend to design a system based on sensors combined with a Wireless Body Area Network (WBAN) and communication technologies, to reduce the time lag between the detection of a problem and the time that care is administered and manage in cases where the patient is away from the hospital premises.   As part of the system we are developing an algorithm to find an optimal solution that will be faster than current algorithms in assigning tasks, processing sensor data and transmitting it to a central location. The algorithm will minimize system end-to-end delay while ensuring required system battery lifetime and availability. This system will also analyze a patient's health data and report the results to the user. At the same time the system will convert raw patient data from the sensors to the standard HL7 (Health Level Seven) format for further processing and record keeping.

For example, if the signal processing algorithm on the mobile device detects an imminent health problem the patient can be warned via the mobile device. At the same time an alarm and the patient's GPS position and diagnostic data can be sent to the healthcare center. Thus an attempt is made to supervise the dynamic situation by using agent based ubiquitous computing and to find the appropriate solution for emergency circumstances, by providing correct diagnosis and appropriate treatment in time.

The results of this research should be helpful to high risk patients that may need emergency medical treatment as well as monitoring and advising services for patients with chronic conditions. Examples include people suffering from COPD, heart disease, diabetes and similar conditions. For these patients speedy emergency medical treatment can prevent sudden death. Systems built using this research may also help physicians obtain online access to a patient's data including information on the events immediately prior to the alert, medication and service priority (e.g. emergency case).

2 · Research Objectives and Content

As computing power, network infrastructure and device technology have developed, research has been directed at providing ubiquitous technology. The goal of ubiquitous computing is to provide the 5-Any services "Anytime, Anywhere, Any Network, Any Device, Any Service" [4]. Combining ubiquitous computing with healthcare yields ubiquitous healthcare (U-healthcare). U-healthcare enables patients to utilize healthcare services such as diagnostic services, emergency management services and monitoring services anytime, anywhere. [5, 6]

To provide U-healthcare service for a patient a ubiquitous healthcare infrastructure needs to be constructed. The U-healthcare infrastructure consists of sensing devices, network, the patient's mobile system and a healthcare server (See Figure 1 for more information).

In *Level 1* of a ubiquitous healthcare infrastructure the patient will wear one or more wireless medical sensors. Each wireless sensor can sense, sample, and process one or more physiological signals. For example, an electrocardiogram sensor (ECG) can monitor heart activity, a blood pressure sensor can monitor blood pressure, an electroencephalogram sensor (EEG) can monitor brain electrical activity, a breathing sensor can monitor respiration and motion sensors can help determine the patient's status and estimate her or his level of activity[3].

Level 2 infrastructure includes a Personal Server (PS) application running on a smart phone or similar device. The PS is responsible for a number of tasks such as providing a transparent interface to the wireless medical sensors, an interface to the user and an interface to the healthcare server [3]. The interface to the Wireless Body Area Network (WBAN) includes network configuration and management. Network configuration includes sensor node registration (type and number of sensors), initialization (e.g., specifying sampling frequency and mode of operation), customization (e.g., running user specific calibration or signal processing procedure upload) and setup of a secure communication channel (key exchange). During operation the PS must monitor system power and communication availability and reconfigure the WBAN appropriately. The combination of Level 1 and Level 2 makes up the Outpatient Monitoring System (OMS).

After the WBAN network is configured, the PS application performs network management including channel sharing, time synchronization, data retrieval and processing and aggregating data. Based on the analysis of information from multiple medical sensors the PS application determines the patient's state and health status and provides feedback through a user-friendly and intuitive graphical or audio user interface. Finally, if a communication channel to the medical server is available, the PS establishes a secure link to the medical server and sends reports in HL7 format that can be integrated into the patient's medical record. If a link between the PS and the medical server is not available the PS will store data locally and initiate data uploads whenever a link becomes available. The PMS should be capable of some degree of autonomy when communication with the healthcare server is unavailable.

Level 3 includes a medical server or servers accessed via the Internet or cell phone network. In addition to the medical server the last level may encompass other servers and actors, such as informal caregivers, commercial health care providers, and even emergency services. The medical server typically runs a service that sets up a communication channel to the patient's PS, collects reports and integrates the data into the patient's medical record. The service can issue recommendations, and even issue alerts, if reports seem to indicate an abnormal condition. More details about this architecture and services can be found in [1].

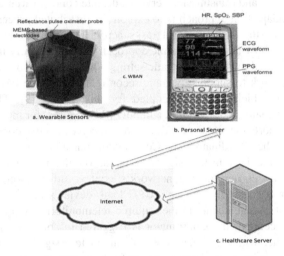

Fig. 1. Ubiquitous Healthcare Infrastructure

2.1   Research Strategies and Methods

Our primary research strategy is to implement an API for smart phones or similar devices that can monitor patients' health. Our research also includes the conversion of the data from sensors into a standard HL7 format and its subsequent transfer to a healthcare server [2]. For the implementation of the system, there are four tasks that need to be implemented.

In our discussion below *Part i* deals with the Task Assignment Algorithm used to assign a task for a particular sensor from the PS. *Part ii* deals with the PS collecting data from each sensor node in the WBAN. *Part iii* discusses processing and analysis of the raw data. *Part iv* deals with the conversion of the sensor data to HL7 format and its subsequent transfer to a remote healthcare server from the PS.

i. Task Assignment. For implementation of the system, we need to transfer the wireless sensor data from a sensor located on a patient's body to the PS. We refer to the underlying computation and communication resources of a PMS as an m-health platform. On top of this platform, various telemonitoring applications can operate continuously (24/7). Examples of telemonitoring applications include safety-critical applications such as detection of falls, heart attacks, fetal distress and premature labor [7].

As with other applications operating in a mobile environment, a PMS can be greatly affected by context changes and scarcity of resources, e.g. network bandwidth, battery power and computational power of handhelds. For example, a drop in network bandwidth due to a patient's mobility can result in transmitted bio-signal loss or excessive delay. When the mismatch between application demand and resource supply exceeds a tolerable level, the entire PMS may fail to respond accurately and in a timely manner to an emergency situation [8]. Thus, the success of a PMS relies heavily on whether the system can adapt itself and provide adequate and continuous bio-signal processing and transmission services, despite context variations.

One possible adaptation approach is to exploit the distributed processing paradigm of PMS and adjust the assignment of tasks across available devices at run-time. Simulation results show that this dynamic approach can significantly improve system performance compared to the current static setting. Other related work proposes a task distribution framework to support dynamic reconfiguration of the PMS by means of task redistribution which we have adopted for this project [9]. This framework consists of a Coordinator and a set of Facilitators (See Figure 2 for more information).

A Facilitator reports the status of a device and receives control commands on task management from the Coordinator. The Coordinator runs a task assignment algorithm that can identify the best task assignment based on the required telemonitoring application and the current device network's context information, e.g. available devices and their connectivity, device's CPU load, device's remaining battery, etc. Once a significant change occurs in the required telemonitoring application or in the device network's context, the Coordinator is triggered to compute the optimal task assignment, together with a few near-optimal candidate assignments, under the new situation. These candidate assignments are further ranked subject to both their performance enhancements and their reconfiguration cost. If the identified best assignment is different from the current one deployed in the system, the Coordinator constructs a reconfiguration plan and controls the Facilitators to deploy the new task assignment, by means of task redistribution [10].

Task assignment algorithms targeting some special topologies of PMS have been previously studied [11, 12]. Task assignment (also referred to as allocation, mapping or partitioning) is a well-known NP-hard problem in its general form. A good survey describing the basic concepts and models of task assignment problem can be found in [13]. This problem has been studied extensively within various application areas. For example, it is an essential problem in the field of wireless sensor networks ([14, 15]) grid computing ([16]), and distributed databases ([17, 18]). Except for the earlier work previously cited we did not locate other research which addresses the task assignment problem in PMS. Here, we compare our work with other research in somewhat related fields.

Genetic algorithms are based on simplified evolutionary processes, using directed selection to achieve optimal results. The selection algorithms evaluate components of random sets of solutions to a problem. The solutions that come out on top are then recombined and mutated and run through the process again. This process is repeated until an acceptable solution is discovered. Genetic Algorithms are described more thoroughly in ([19, 20]).

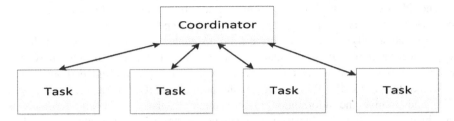

Fig. 2. Task distribution framework

In this research, we intend to implement Task assignment using the genetic algorithm approach. The problem we face is determining an optimal solution to a problem with multiple sometimes conflicting objectives which vary dynamically. Genetic algorithms are well suited for such tasks [20].

The heterogeneous m-health platform in a PMS exhibits three special properties that require special attention. *First*, it is modeled as an arbitrary network and has a random topology. *Second*, the network connections are asymmetric, thus channel directions have to be modeled. *Third*, devices can serve as relays for data streams. These aspects have been addressed in the past individually but not in combination.

The task assignment problem in arbitrary processor networks was studied in ([15]). A heuristic algorithm was proposed that can assign tasks to the most suitable processor by taking into account the network topology. However, our experiments show this kind of greedy algorithm suffers from the risk of finding no suitable task assignment in the PMS.

Hu and Marculescu and Alsalih, Akl and Hassanein modeled explicitly channel directions in a processor network [21, 22]. The objectives of their studied task assignment algorithms were to minimize total system energy consumption. Hu and Marculescu proposed a Branch-and-Bound algorithm [21] and Alsalsalih, Akl and Hassanein suggested a greedy approach [22].

Lee and Shin considered the possibility of relaying data streams by devices [23]. However, their performance estimation of a task assignment did not include the extra resource consumption at a relaying device, caused by a relayed data stream [10].

Thus, another unique aspect of the task assignment problem in PMS is that all these special features must be taken into account. In this research, we use a new genetic algorithm model, propose specific assignment constraints and use expansion rules to tackle these challenges.

ii. Collecting Data from the WBAN. Once the task assignment algorithm assigns the task to a particular sensor the Personal Server (PS) is solely responsible for collecting data and events from the WBAN. The PS provides the user interface, controls the WBAN, combines data and events, and creates unique session archive files. The software we are developing will be implemented in Java for the Android Mobile Operating System. It should run on either an Android smart phone or similar devices. Based on our communication protocol and the size of the super frame, the system should be able to support as many as nineteen sensor nodes, although in practice typical WBAN research involves fewer.

The PS begins a health monitoring session by wirelessly configuring sensor parameters such as sampling rate, selecting the type of physiological signal of interest, and specifying events of interest. For example, motion sensors are capable of step detection if placed on the ankles, upper body tilt if placed on the chest, and energy estimation depending on location. Sensors, in turn, transmit pertinent event messages to the PS. The PS must aggregate the multiple data streams, create session files and archive the information in the patient database. Real-time feedback is provided through the user interface. The patient can monitor his /her vital signs and be notified of any detected warnings or alerts [24].

Each sensor node in the network is sampling, collecting and performing basic processing of data. Depending on the type of sensor and the degree of processing specified at configuration, a variety of events will be reported to the PS. An event log is created by aggregating event messages from all the sensors in the WBAN; the log is then stored in a session archive file. The PS must recognize events as they are received and make decisions based on the severity of the event. For example normally heartbeat events do not create alerts and are only logged in the event log. However, the PS must recognize when the heart rate exceeds threshold values and issue alerts.

Although all sensors in the system perform on-sensor processing and event detection, there are events where processed and summary events are not sufficient and real-time raw signal capture is necessary. In a deployed system, where intelligent sensors analyze raw data, process and transmit application event messages, there may be cases where it is necessary to transmit raw data samples. A case in point is a deployed ECG monitor. When embedded signal processing routines detect an arrhythmic event, the node should send an event message to the PS, which will then be relayed to the appropriate medical server. The medical server, in turn, will provide an alert to the patient's physician. However, a missed heart beat can also be caused by electrode movement. Therefore, it would be useful to augment this event with actual recording of the fragment of unprocessed ECG sensor data. The recording can be used by a physician to evaluate the type and exact nature of the event or to dismiss it as a recording artifact. In this case, the embedded sensor would stream real-time data to the personal server, during a predefined time period   [24].

iii. Processing and Analyzing the Raw Data. After the raw data has been transferred to the PS the data has to be processed and analyzed. The raw data is directed to a streaming database with processing nodes. These computation nodes will transform, in real–time, the raw data streams into more usable information such as heart rate (HR) [25]. ECG algorithms must meet certain requirements in order to be useful for this system. As patients walk around or shift artifacts from sensor displacement, muscle noise and baseline wander appear in the signal. The scalable premise implies that the processing must be fast, in order to accommodate many patients and caregivers, and allocate time to other tasks such as location tracking, alarm handling, storage and retrieval of historical data, and resource management. Also, system dependability is tightly tied to an accurate screening of alarms. Response times can improve only if false alarms are kept to a minimum. The selected processing algorithm must be fast, noise resistant and accurate to conform to all these requirements.

Other sensor data can be processed and analyzed in a similar manner. As the research goes forward, we will try to process and analysis other sensor data using efficient algorithms.

iv. HL7 Conversion of Sensor Data. Along with the processing and analysis of the raw data, the PMS needs to convert the raw data from the sensors in to a standard HL7 format in order to transmit it to a remote healthcare server through the mobile device's WI-FI or GPRS internet connection. This way the care giver can have access to the user data and have the information required to better administer care to the patient when required. In this section we discuss our proposed middleware which can connect to commercial monitoring devices in order to transmit vital signs, together with patient demographics. Later, data packets from monitoring devices are parsed to extract HL7 compatible data. We have developed middleware to support data transmission from medical devices to arbitrary information systems [2].

3   The Use of Performance Measures and Expected Results

3.1   Task Assignment

As with other applications operating in a mobile environment a PMS could be greatly affected by context changes and scarcity of m-health platform resources, e.g. network bandwidth, battery power and computational power of handhelds. Dynamic context-aware adaptation mechanisms are required in order to meet the stringent requirements of such mission critical applications. The core of the adaptation mechanism is a decision-making component that can calculate/select the optimal task assignment to be enforced by taking into account the reconfiguration costs.

This research first studied the performance requirements of. We identified three key performance measures that are critical to the success of system: end-to-end delay, system battery lifetime and availability level. Secondly, using an intelligent system model, we proposed a genetic algorithm based task assignment algorithm that minimizes the system end-to-end delay while guaranteeing required performance. We intend to evaluate the algorithm performance using experiments and provide recommendations for further improvements. Our estimated results show that this dynamic approach can significantly improve system performance compared to the current static setting.

3.2   Collecting Data from the WBAN

The proliferation of wireless devices and recent advances in miniature sensors supports the technical feasibility of a ubiquitous heath monitoring system. However, WBAN designers face a number of challenges in an effort to improve user's compliance that depends on the system's ease of use, size, reliability, and security. In order to address some of these challenging tasks, we have designed a WBAN prototype that includes an ECG sensor, a blood pressure sensor and a smart phone or similar mobile device based personal server. In this research we describe both the hardware and software architecture of our prototype. Our hardware architecture leverages off-the-shelf commodity sensor platforms. Similarly, our software architecture builds upon the Android OS, a widely used open-source operating system for embedded sensor networks [24].

3.3   HL7 Conversion of Sensor Data

We have previously developed and implemented HL7 compliant middleware [2]. The preliminary results are satisfactory and encouraging. Given a set of IEEE 1451 data formats, the middleware will convert the data into HL7 format, which can be transmitted to any information systems in general or healthcare server. We will continue to experiment in order to accomplish a general-purpose interfacing facilitator, to deal with many different kinds of patient care devices being developed. We hope that the middleware library package will evolve to become embedded software that can be used in ubiquitous healthcare devices.

Acknowledgements. Dr. Rittenhouse is supported by a Bisa Research Grant of Keimyung University in 2011. Professor Lee was financially supported by the Ministry of Education, Science Technology (MEST) and National Research Foundation of Korea (NRF) through the Human Resource Training Project for Regional Innovation.

References

1. Gatton, T.M., Lee, M.: Fuzzy Logic Decision Making for an Intelligent Home Healthcare System. In: 2010 5th International Conference on Future Information Technology, pp. 1–5. IEEE, Busan (2010)
2. Lee, M., Gatton, T.M.: Wireless Health Data Exchange for Home Healthcare Monitoring Systems. Sensors 10, 3243–3260 (2010)
3. Milenkovic, A., Otto, C., Jovanov, E.: Wireless sensor networks for personal health monitoring: Issues and an implementation. Computer Communications 29, 2521–2533 (2006)
4. Lee, M., Gatton, T.M., Lee, K.-K.: A Monitoring and Advisory System for Diabetes Patient Management Using a Rule-Based Method and KNN. Sensors (2010)
5. Kirn, S.: Ubiquitous healthcare: The onkonet mobile agents architecture. Objects, Components, Architectures, Services, and Applications for a Networked World, 265–277 (2009)
6. Ko, E.J., Eee, H.J., Eee, J.W.: Ontology-Based Context-Aware Service Engine for U-HealthCare. In: 2006 8th International Conference Advanced Communication Technology, pp. 632–637 (2006)
7. Jones, V., Halteren, A., van, W.I., Dokovsky, N., Koprinkov, G., Bulta, R., Konstantas, D., Herzog, R.: MobiHealth: Mobile Health Services based on Body Area Networks. In: Istepanian, R.H., Laxminarayan, S., Pattichis, C.S. (eds.) M-Health: Emerging Mobile Health Systems, pp. 219–236 (2006)
8. Jones, V., Incardona, F., Tristram, C., Virtuoso, S., Lymberis, A.: Future challenges and recommendations. M-Health, 267–270 (2006)

9. Mei, H., van Beijnum, B.-J., Pawar, P., Widya, I., Hermens, H.: Context-Aware Dynamic Reconfiguration of Mobile Patient Monitoring Systems. In: 2009 4th International Symposium on Wireless Pervasive Computing, pp. 1–5 (2009)
10. Mei, H., Beijnum, B.-J., van, P.P., Widya, I., Hermens, H.: A*-Based Task Assignment Algorithm for Context-Aware Mobile Patient Monitoring Systems. IEEE (2009)
11. Pawar, P., Mei, H., Widya, I., van Beijnum, B.-J., van Halteren, A.: Context-aware task assignment in ubiquitous computing environment - A genetic algorithm based approach. In: 2007 IEEE Congress on Evolutionary Computation, pp. 2695–2702 (2007)
12. Mei, H., Pawar, P., Widya, I.: Optimal Assignment of a Tree-Structured Context Reasoning Procedure onto a Host-Satellites System. IEEE (2007)
13. Norman, M.G., Thanisch, P.: Models of machines and computation for mapping in multicomputers. ACM Computing Surveys 25, 263–302 (1993)
14. Gu, Y., Tian, Y., Eylem, E.: Real-time multimedia processing in video sensor networks. Signal Processing: Image Communication 22, 237–251 (2007)
15. Zhao, B., Wang, M., Shao, Z., Cao, J., Chan, K.C.C., Su, J.: Topology Aware Task Allocation and Scheduling for Real-Time Data Fusion Applications in Networked Embedded Sensor Systems. IEEE (2008)
16. Cooper, K., Dasgupta, A., Kennedy, K., Koelbel, C., Mandal, A., Marin, G., Mazina, M., Mellor-Crummey, J., Berman, F., Casanova, H., Chien, A., Dail, H., Liu, X., Olugbile, A., Sievert, O., Xia, H., Johnsson, L., Liu, B., Patel, M., Reed, D., Deng, W., Mendes, C., Shi, Z., YarKhan, A., Dongarra, J.: New grid scheduling and rescheduling methods in the GrADS project. IEEE (2004)
17. Paré, G., Jaana, M., Sicotte, C.: Systematic review of home telemonitoring for chronic diseases: the evidence base. Journal of the American Medical Informatics Association: JAMIA 14, 269–277 (2007)
18. Pietzuch, P., Ledlie, J., Shneidman, J., Roussopoulos, M., Welsh, M., Seltzer, M.: Network-Aware Operator Placement for Stream-Processing Systems. In: 22nd International Conference on Data Engineering (ICDE 2006), pp. 49–49. IEEE (2006)
19. Lee, M.: Evolution of behaviors in autonomous robot using artificial neural network and genetic algorithm. Information Sciences 155, 43–60 (2003)
20. Konak, A., Coit, D.W., Smith, A.E.: Multi-objective optimization using genetic algorithms: A tutorial. Reliability Engineering & System Safety 91, 992–1007 (2006)
21. Hu, J., Marculescu, R.: Energy- and performance-aware mapping for regular NoC architectures. IEEE Transactions on Computer-Aided Design of Integrated Circuits and Systems 24, 551–562 (2005)
22. Alsalih, W., Akl, S., Hassanein, H.: Energy-Aware Task Scheduling: Towards Enabling Mobile Computing over MANETs. In: 19th IEEE International Parallel and Distributed Processing Symposium, p. 242a. IEEE (2005)
23. Lee, C.H., Shin, K.G.: Optimal task assignment in homogeneous networks. IEEE Transactions on Parallel and Distributed Systems 8, 119–129 (1997)
24. Otto, C., Milenkovic, A., Sanders, C., Jovanov, E.: System architecture of a wireless body area sensor network for ubiquitous health monitoring. Journal of Mobile Multimedia 1, 307–326 (2006)
25. Drew, B.J., Califf, R.M., Funk, M., Kaufman, E.S., Krucoff, M.W., Laks, M.M., Macfarlane, P.W., Sommargren, C., Swiryn, S., Van Hare, G.F.: Practice standards for electrocardiographic monitoring in hospital settings: an American Heart Association scientific statement from the Councils on Cardiovascular Nursing, Clinical Cardiology, and Cardiovascular Disease in the Young: endorsed by the Inte. Circulation 110, 2721–2746 (2004)

26. IEEE Standards Organization: IEEE Standard for a Smart Transducer Interface for Sensors and Actuators - Common Functions, Communication Protocols, and Transducer Electronic Data Sheet (TEDS) Formats (2007)
27. IEEE Standards Organization: IEEE Standard for a Smart Transducer Interface for Sensors and Actuators Wireless Communication Protocols and Transducer Electronic Data Sheet (TEDS) Formats (2007)
28. IEEE Standards Organization: An Overview of IEEE 1451. 4 Transducer Electronic Data Sheets (TEDS)
29. IEEE Standard Organization: IEEE Standard for a Smart Transducer Interface for Sensors and Actuators - Mixed-Mode Communication Protocols and Transducer Electronic Data Sheet (TEDS) Formats (2007)

Feature Salience for Neural Networks: Comparing Algorithms

Theodor Heinze, Martin von Löwis, and Andreas Polze

Hasso-Plattner-Institute for Software Systems Engineering
{Theodor.Heinze,Martin.vonLoewis,Andreas.Polze}
@hpi.uni-potsdam.de

Abstract. One of the key problems in the field of telemedicine is the prediction of the patient's health state change based on incoming non-invasively measured vital data. Artificial Neural Networks (ANN) are a powerful statistical modeling tool suitable for this problem. Feature salience algorithms for ANN provide information about feature importance and help selecting relevant input variables. Looking for a reliable salience analysis algorithm, we found a relatively wide range of possible approaches. However, we have also found numerous methodological weaknesses in corresponding evaluations. Perturb [11][7] and Connection Weight (CW) [1] are two of the most promising algorithms. In this paper, we propose an improvement for Connection Weight and evaluate it along with Perturb and the original CW. We use three independent datasets with already known feature salience rankings as well as varying topologies and random feature ranking results to estimate the usability of the tested approaches for feature salience assessment in complex multi-layer perceptrons.

Keywords: Feature Salience, Sensitivity Analysis, Neural Networks, Machine Learning, Telemedicine.

1 Introduction

Working in the field of telemedicine, one of the key problems is the estimation of patient's health based on the history of incoming non-invasively measured vital data. Artificial Neural Networks (ANN) are a powerful statistical model suitable for modeling this type of non-linear problems. One drawback is the "black box" outward appearance of ANN since contributions of each input variable in the prediction process are not trivially determinable. To address this issue, a wide range of so called feature salience/sensitivity analysis algorithms has been introduced already. A recent study [5], comparable to our current work, has employed many of them to estimate the importance of input features in the analysis of vital measurement data. Also, [5] hints that current feature salience algorithms are not reliable enough to be used stand-alone and considers ensemble results.

Since the 90ies, sensitivity analysis and its variants have been the dominating paradigm [6][4][10][12]. Garson's algorithm [8], a widely used approach, was found to be quite weak by several comparisons [1][11][5]. It has been the field of ecology in

T. Huang et al. (Eds.): ICONIP 2012, Part IV, LNCS 7666, pp. 415–422, 2012.

particular, in which many different methods have been developed and tried [3][9][18][1]. Unfortunately, some of ecology-based evaluations employ datasets with effectively unknown true feature salience. Expert opinion is used to rank the input variables and then proposed algorithms are tested for compliance [11]. Also, in many cases only one topology of ANN is used, usually a very small one [1][13]. Another frequent evaluation weakness is the employment of only one dataset.

The main contribution of this paper is the introduction of a ranking-based testing methodology for feature salience algorithms and applying it to the two most promising techniques: Perturb [11][7] and Connection Weight [18][1]. Also, we introduce an improvement for the latter. Three independent datasets with already known feature salience ranking are used. Also, we try several topologies and varying training quality to analyze algorithm behavior under different conditions and compare its quality with random ranking. We also show how cumulative analysis of a set can distinctively improve ranking performance.

2   Feature Salience Algorithms

Feature salience algorithms are used to analyze a trained Neural Network and determine the importance of each input feature. Usually, *real* $[+\infty...-\infty]$ values are assigned to all input features so that a ranking can be drawn. If several trained neural networks are available, it is possible to sum the assigned values corresponding to the input features over several networks and compute a cumulative ranking then. Assigned values have to be normalized each step, so that the contribution of the analysis of one single network does not dominate the whole set with its high values.

2.1   Perturb

Perturb seems to be the method of choice for many researchers. In fact, its concept is very logical and we have not seen one single review where its results have been questioned. Perturb measures the change in the root mean squared error (RMSE) of the network, while noise is added to the input variables successively. The input features are then ranked by the RMSE change they induce. A recent analysis [7] shows that the optimum input perturbation ratio range is around [-20%, 20%].

```
given: a dataset and a trained neural network
initialize Array RMSEchangeSum[input features]
FOR all samples in dataset
  Measure output RMSE error
  FOR all input features
    perturb feature and measure change in RMSE
    add RMSE change to RMSEchangeSum[this_input_feature]
  NEXT input feature
NEXT sample
Rank input feature salience by values in RMSEchangeSum
[Pseudo-code for the Perturb Algorithm]
```

2.2   Connection Weight

Connection Weight (CW) is another powerful and also logical approach for feature salience. It "calculates the product of the raw input-hidden and hidden-output connection weights between each input neuron and sums the products across all hidden neurons" [1]. A comprehensive introduction and review is offered by [1]. This method is similar to the formerly very popular Garson's algorithm [8], though is reported to perform much better [11][5].

2.3   Connection Impact

Our algorithm exploits ideas used in already known methods, mainly CW. Though, we do not use the raw connection weights, rather for every data sample and every weight, the absolute "impact" is computed, which describes the percentage of contribution of the connection to the target neuron. The impact is computed by multiplying the raw connection weight with the activation of the source neuron and then normalizing. The impact values are used to backtrack the square error from network output to the input neurons, where it is then summed over the data samples.

```
given: a dataset and a trained neural network
initialize Array Salience[input neurons]
FOR all samples in dataset
   process sample with trained network, compute output
   map input range to -1 ... 1  (= activation input layer)
   FOR every weight:
       Impact[weight] = weight * activation(source neuron)
   NEXT weight
   Normalize Impacts, so that for every target neuron,
       incoming impacts (absolute values) sum up to 1
   FOR all neuron layers starting with output layer
      FOR all neurons
        IF output layer THEN
           neuron_value = (correctOutput - computedOutput)²
           CONTINUE with next neuron
        ELSE neuron_value = ∑(outgoingImp *
                                   neuron_value(target))
        IF input layer THEN
                Salience[neuron] += absolute(neuron_value)
      NEXT neuron
   NEXT layer
NEXT data sample
Rank input feature salience by values in Salience
```

[Pseudo-code for the Connection Impact algorithm]

3   Testing Methodology

For every dataset we have trained a set of neural networks, which are then analyzed by the assessed algorithms. In single network analysis, an algorithm computes a ranking from one dataset/network combination, while in cumulative analysis a dataset and a (sub)set of its trained networks are employed to calculate one cumulative ranking. Since all tested algorithms work with accumulations of values corresponding to the impact of input features, we will test if accumulating over several trained networks affects ranking accuracy. The ranking error of all input features is computed by calculating the Euclidean distance between the true and the estimated ranking vector.

If e.g. out of three input features the 2nd feature is the most important and the 1st feature is the least important, then the true ranking is {3, 1, 2}. If an algorithm estimates a wrong ranking {2, 1, 3}, then the ranking error is (1).

$$\sqrt{(3-2)^2 + 0 + (2-3)^2} = \sqrt{2} = 1,414 \qquad (1)$$

The maximum ranking error for a set of n features is (2).

$$maxErr = \sqrt{\frac{n^3-n}{3}} \qquad (2)$$

For better comparability between problem sizes, we normalize the actual ranking error with this maximum.

3.1   Artificial Neural Networks and Testing Procedure

In our experiments we have used standard feed-forward multi-layer neural network design with sigmoid activation function and squared error function. Network topology has been varied over the datasets to account for different complexity levels of the data. Training was performed with the standard-backpropagation approach [16] along with RPROP enhancement [15] for weight adjustment. We have used a training and a validation set for training and a test set to determine the true performance of the trained network. Input/output normalizing, early stopping techniques and error measures from [14] have been used to standardize our results.

3.2   Datasets

We have used three different datasets to review the considered algorithms. Our main concern was to have more than one dataset and also try different topologies to test resulting performance changes. Also, it is crucial to know the true feature salience in advance.

The first dataset was *Demosaicing* with 36 input and 12 output variables. Demosaicing is an image processing technique which is used to reconstruct three channel (red, green and blue) 2x2 pixel matrix information from digital sensor data which only captures information of one channel (red, green or blue) for each pixel location. Thus, the goal is the estimation of the missing color values.

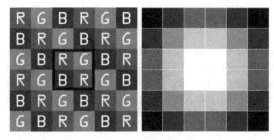

Fig. 1. Demosaicing problem and feature salience distribution (brighter = better)

The input vector corresponds to a 6x6 crop from the digital sensor around the 2x2 matrix (output vector). We have used an imaging model described in [2] to artificially simulate a digital sensor and thus generate 10.000 data samples from 400 digital images.

We had used a very similar approach before [17] to train a neural network which was then able to properly render real-world images from data generated by a real digital imaging sensor. Obviously, due to the distribution of light, input features in the center of the sensor crop are more relevant for the restoration of the true color values than the pixels on border and edges of the crop. We used the Euclidean distance function in true feature ranking to account for that (Figure 1).

Our second dataset, Tic-Tac-Toe (TTT), is taken from the UCI Machine Learning Repository [19]. TTT contains a of 958 endgame positions from the homonymous game (sometimes also called *Noughts and Crosses*) classified as *won* or *not-won* for the first player. Here, the 9 input features correspond to the occupation of the 9 fields: *cross*, *nought* or *empty*.

Fig. 2. TTT: typical endgame position (left) and feature salience distribution (brighter = better)

Since from the center field there are four possibilities to build a row and only three and two from edges and border accordingly, the true ranking of feature salience is pretty clear (Figure 2).

The third dataset, simulated data, is based on a artificially generated function with 20 input variables which correlate with the output variable in a range between 0.1 and 0.9, similarly to [1]. This dataset contains 1000 samples.

4   Results

For all datasets, we have trained a set of 100 networks. Network topology was set to one hidden layer (10 neurons) for TTT and the artificial dataset, while for Demosaicing it was 2 hidden layers (25 + 16). RMSE [14] was fairly low, as shows Figure 3. For Tic-Tac-Toe, we picked 100 ANN samples with perfect training (≈zero error).

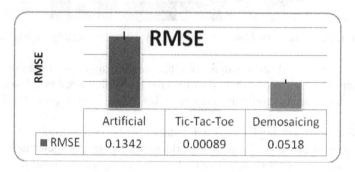

Fig. 3. RMSE and standard deviation, 100 samples

4.1   Single Network Analysis

Figure 4 shows the mean ranking errors (% of max. error) for every algorithm/set combination. We have added random ranking error values to give the impression how useful the results are. All algorithms performed reasonably well analyzing networks trained with the artificial dataset. Unlike in [1], Perturb performed better than CW. We suppose the reason is the increased number of input features (20 vs. 10) and the more complex network topology. Perturb also performed clearly best on the rest of the tested sets and was in some cases even able to determine the perfect ranking for the TTT dataset, while CW showed rankings not so much better than random.

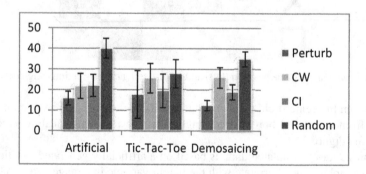

Fig. 4. Single network analysis: Mean errors and standard deviation, 100 samples

4.2   Cumulative Analysis

In many cases there are several trained networks available for a dataset, so why not use that extra information for feature salience analysis? To test algorithm performance in this case, we partitioned the available trained networks into 10 subsets with 10 networks each. Then we calculated the cumulative ranking for every subset/algorithm combination (Figure 5). Error rates improved in most cases. Considering accuracy and precision, Perturb is still the clear winner. Especially remarkable is its performance on the TTT set, where it was able to perfectly rank the features in 7/10 cases. In the last experiment, the whole set of 100 networks is employed to compute the cumulative ranking (Figure 6). Perturb and CI were able to perfectly rank the features of the TTT dataset. CI slightly outperformed Perturb on the artificial dataset.

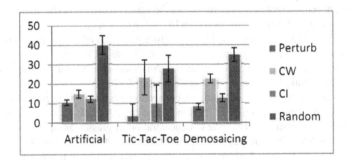

Fig. 5. Cumulative analysis: Mean errors and standard deviation, 10 samples

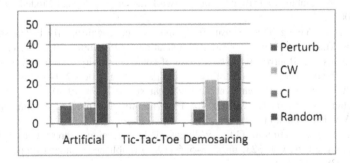

Fig. 6. Cumulative analysis: Ranking error

5   Conclusion and Outlook

The original Connection Weight performed worst, e.g. calculating a ranking only slightly better than Random for Demosaicing. CI did better overall, but was also not able to compete against Perturb which consistently produced best results and generated usable (clearly better than Random) rankings. With cumulative analysis, situation improved even further. The choice of a feature salience algorithm for our future work is now clear.

References

1. Olden, J.D., Joy, M.K., Death, R.G.: An accurate comparison of methods for quantifying variable importance in artificial neural networks using simulated data. Ecological Modelling 178, 389–397 (2004)
2. Farsiu, S., Elad, M., Milanfar, P.: Multi-frame demosaicing and super-resolution of color images. IEEE Trans. on Image Processing 15, 141–159 (2006)
3. Lek, S., Delacoste, M., Baran, P., Dimopoulos, I., Lauga, J., Aulagnier, S.: Application of neural networks to modelling nonlinear relationships in ecology. Ecological Modelling 90, 39–52 (1996)
4. Wang, W., Jones, P., Partridge, D.: Assessing the Impact of Input Features in a Feedforward Neural Network. Neural Computing & Applications 9, 101–112 (2000)
5. Cui, X.R., Abbod, M.F., Liu, Q., Shieh, J.S., Chao, T.Y., Hsieh, C.Y., Yang, Y.C.: Ensembled artificial neural networks to predict the fitness score for body composition analysis. The Journal of Nutrition Health Aging 15, 341–348 (2011)
6. Tchaban, T., Taylor, M.J., Griffin, J.P.: Establishing impacts of the inputs in a feedforward neural network. Neural Computing & Applications 7, 309–317 (1998)
7. Bai, R., Jia, H., Cao, P.: Factor Sensitivity Analysis with Neural Network Simulation based on Perturbation System. Journal of Computers 6 (2011)
8. Garson, G.D.: Interpreting neural-network connection weights. AI Expert 6, 46–51 (1991)
9. Dimopoulos, I., Chronopoulos, J., Chronopoulou-Sereli, A., Lek, S.: Neural network models to study relationships between lead concentration in grasses and permanent urban descriptors in Athens city (Greece). Ecological Modelling 120, 157–165 (1999)
10. Montaño, J.J., Palmer, A.: Numeric sensitivity analysis applied to feedforward neural networks. Neural Computing & Applications 12, 119–125 (2003)
11. Gevrey, M., Dimopoulos, I., Lek, S.: Review and comparison of methods to study the contribution of variables in artificial neural network models. Ecological Modelling 160, 249–264 (2003)
12. Cheng, A.Y., Yeung, D.S.: Sensitivity analysis of neocognitron. IEEE Transactions on Systems, Man and Cybernetics, Part C (Applications and Reviews) 29, 238–249 (1999)
13. Dimopoulos, Y., Bourret, P., Lek, S.: Use of some sensitivity criteria for choosing networks with good generalization ability. Neural Processing Letters 2, 1–4 (1995)
14. Prechelt, L.: Proben 1. Presented at the Technical Report 21/94 (1994)
15. Riedmiller, M., Braun, H.: RPROP - A Fast Adaptive Learning Algorithm. In: Proc. of ISCIS VII, Universitat (1992)
16. Rumelhart, D.E., Hinton, G.E., Williams, R.J.: Learning representations by back-propagating errors, vol. 323, pp. 533–536 (1986) (published online: October 09, 1986), doi:10.1038/323533a0
17. Heinze, T., von Lowis, M., Polze, A.: Joint multi-frame demosaicing and super-resolution with artificial neural networks. In: 2012 19th International Conference on Systems, Signals and Image Processing (IWSSIP), pp. 540–543 (2012)
18. Olden, J.D., Jackson, D.A.: Illuminating the "black box": a randomization approach for understanding variable contributions in artificial neural networks. Ecological Modelling 154, 135–150 (2002)
19. UCI Machine Learning Repository,
   http://archive.ics.uci.edu/ml/datasets.html

Adaptive Modeling of HRTFs
Based on Reinforcement Learning

Shuhei Morioka[1,*], Isao Nambu[1], Shohei Yano[2], Haruhide Hokari[1],
and Yasuhiro Wada[1]

[1] Nagaoka University of Technology, 1603-1 Kamitomioka Nagaoka Niigata, Japan
[2] Nagaoka National College of Technology, 888 Nishikatagai Nagaoka Niigata, Japan
`morioka@stn.nagaokaut.ac.jp`

Abstract. Although recent studies on out-of-head sound localization technology have been aimed at applications in entertainment, this technology can also be used to provide an interface to connect a computer to the human brain. An effective out-of-head system requires an accurate head-related transfer function (HRTF). However, it is difficult to measure HRTF accurately. We propose a new method based on reinforcement learning to estimate HRTF accurately from measurement data and validate it through simulations. We used the actor-critic paradigm to learn the HRTF parameters and the autoregressive moving average (ARMA) model to reduce the number of such parameters. Our simulations suggest that an accurate HRTF can be estimated with this method. The proposed method is expected to be useful for not only entertainment applications but also brain-machine-interface (BMI) based on out-of-head sound localization technology.

Keywords: HRTF, Actor-critic, Reinforcement learning, ARMA.

1 Introduction

A brain-machine-interface (BMI) is a system that exchanges information between the human brain and computers. Most BMI research is aimed at developing communication tools for motion-impaired persons such as amyotrophic lateral sclerosis (ALS) patients. A representative non-invasive BMI is the P300 speller which uses the event-related potential that is evoked by visual stimuli. However, BMIs that use visual stimuli cannot be used by visually impaired users.as

Schreuder et al.[1] proposed an auditory BMI using speakers to estimate the user's intended direction. Much information can be sent to the computer by using this auditory BMI. However, this system requires a large space in order to locate multiple speakers. In contrast, Ebisawa et al. [2] has proposed an auditory BMI using out-of-head sound localization technology in which 3D sounds are reproduced from headphones. The BMI in this system does not require a large space or a lot of devices to present stimuli to the user because the system can only give the location information of the sound through the headphones.

---

* Corresponding author.

T. Huang et al. (Eds.): ICONIP 2012, Part IV, LNCS 7666, pp. 423–430, 2012.

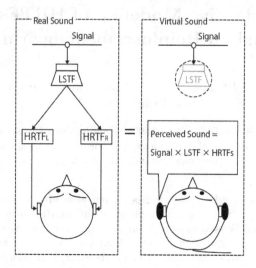

Fig. 1. Principle of out-of-head localization

The principle of out-of-head sound localization is depicted in Fig. 1. Out-of-head sound localization is achieved by equating the auditory stimulus produced by the speakers and received on the ear drums to the one produced by the headphones. Specifically, if the measured impulse responses of the loudspeaker transfer function (LSTF) and the head-related transfer function (HRTF) are convoluted with the sound signal and reproduced on the headphones, the user can perceive the sound as if it is coming from the speakers.

In order to achieve an effective out-of-head sound localization system, an accurate measurement of the head-related transfer function must be obtained. However, this can prove to be difficult since it requires a large scale measurement environment such as an anechoic chamber [3]. Thus far, techniques that correct the ear canal transfer function and change the direction of the convolution HRTF according to the movement of the user's head have been proposed. However, a method to improve the HRTF itself has not been studied sufficiently. We propose a method to increase the accuracy of out-of-head sound localization by obtaining an accurate HRTF using a measured HRTF.

This method incorporates reinforcement learning. An agent learns an HRTF adaptively based on an evaluation of accuracy of sound localization decided by the user. When the learning starts, an accurate HRTF is unknown. So, reinforcement learning that can obtain a correct control law without a supervised signal is employed. We adopted the actor-critic method, which is a type of temporal difference learning [4]. The number of parameters that are dealt with in reinforcement learning is enormous, if an agent learns using all the sample points of the HRTF. Therefore, in our research, the head-related impulse response (HRIR), an impulse response of the HRTF, was modeled by means of the autoregressive moving average (ARMA) model [6]. This makes it possible to reduce the number of parameters to be learned.

Section 2 describes the basic reinforcement learning method and the ARMA model, and gives the sequence of the proposed method. Section 3 explains how the validity of the proposed method was verified through simulation. Finally, in sections 4 and 5, we discuss the simulation results and the conclusions of our work respectively.

2   Method

2.1   Autoregression Moving Average Model

The ARMA model that models the HRIR is shown in equation (1).

$$\hat{h}(k) = \sum_{k=1}^{P} a(k)h(n-k) + \sum_{k=0}^{Q} b(k)\delta(n-k). \tag{1}$$

In this equation, $\hat{h}$ is the impulse response modeled by the ARMA model, δ is the Dirac delta function, and h is the 128-point measured impulse response, a_k is coefficient of AR parameter, and b_k is coefficient of MA parameter. The order of the ARMA parameters was determined using a criterion similar to that of Haneda et al.[6]. The 128 samples of the impulse response can be represented by about 50 ARMA parameters. Figure. 2 plots the HRIR of the right ear at 30 degrees forward on the right side of the head. This HRTF was measured in similar condition to that of Ebisawa et al. [2]

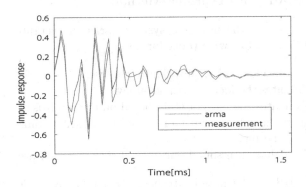

Fig. 2. HRIR at 30 of right ear. This model consists of 5 AR parameters and 47 MA parameters.

2.2   Actor Critic

The framework of the actor-critic method employed in this paper is shown in Fig. 3. In this method, the critic learns to predict the correct evaluation value

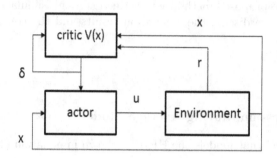

Fig. 3. Actor-critic scheme

$V(\mathbf{x}(t))$ for the current state. The evaluation value $V(\mathbf{x}(t))$ is the predicted value of the sum of future rewards r, and it is the job of the actor to learn the policy that maximized this value. The $\mathbf{x}$ is the vector of current ARMA parameters, and $\mathbf{u}$ is the vector of ARMA parameters that were updated by the actor. The critic calculates temporal difference (TD) error between the current evaluation value and past values. Both the actor and the critic learn based on TD error. In this study, continuous state-space was achieved by using both the actor-critic model and INGnet[5].

2.3   Proposed Adaptive Learning Method

Figure 4 illustrates how the proposed system learns an accurate HRTF. This system learns using the following procedure.

1. Present out-of head sound localization to the user.
2. The user evaluates the localization accuracy.
3. The critic evaluates the current state based on the rewards.
4. The actor learns how to output more accurate ARMA parameters based on the critic's evaluation.
5. Repeat 1 through 4 until sound image localization is accurate.

In our method, HRIR, the impulse response of HRTF, is modeled using the ARMA model, and ARMA parameters are learned. Thereby, this system obtains more accurate HRTFs than measured HRTFs. This paper also discusses the validity of the proposed algorithm by using a simulation without an evaluation by the user. The evaluation of sound image localization accuracy is described later.

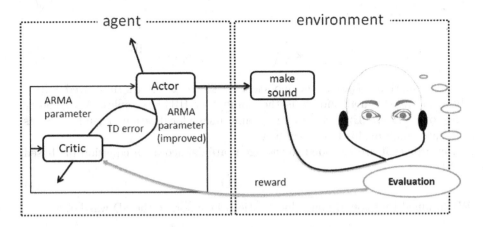

Fig. 4. Adaptive HRTF modeling system using reinforcement learning

3   Simulation

A simulation was done to validate the proposed learning method. When this system learns parameters in a real application, both of the AR and MA parameters should be learned. However, only 20 points of MA coefficients which are great changing depending on each direction were multiplied coefficient which is a Gaussian noise in order to simplify the problem. Then, MA parameters were checked that it can obtain accurate parameters. The initial values in the simulation, the method used to evaluate the sound image localization, and the learning conditions are described in this section.

3.1   Setting the Initial Value

We assumed that the HRTF could not be measured accurately, in other words, the measured HRTF includes noise. The initial value of the MA parameter $MA_{initial}$ was set as follows in the simulation.

$$MA_{initial} = MA_{measured} \times coefficient \qquad (2)$$

In this equation, $MA_{measured}$ is an MA parameter that models the measured HRIR. Coefficient is a Gaussian noise with a variance of 0.2 and a mean of 1.0.

3.2   Evaluation

In this simulation, we adopted spectrum distortion (SD) as a reward. SD is an indicator that expresses the modeling accuracy of HRTF, and is defined as in equation (3).

$$SD = \sqrt{\frac{1}{N} \sum_{k=1}^{N} \left(20 \cdot \log_{10} \frac{|H(\omega_k)|}{|\hat{H}(\omega_k)|}\right)^2} \text{ [dB]} \qquad (3)$$

In this equation, $H(\omega)$ is the modeled HRTF, $\hat{H}(\omega)$ is the improved HRTF obtained by learning, and N is the number of HRTF samples. A frequency range of 500Hz-15kHz is required for sound image localization. Therefore, SD is calculated only in this frequency band.

Rewards that are provided to the agent are defined as in equation (4) based on SD.

$$r = -\mathbf{SD} \qquad (4)$$

We assumed that sound image localization was poor, so the SD was large.

3.3   Conditions

In this simulation, learning was performed using a method similar to one previously reported [5]. The learning coefficients of the actor and critic were 0.15 and 0.12, and the variance of radial basis function (RBF) that was added to the hidden layer was 0.4. The other parameters were the same as in [5].

Each MA parameter is normalized to configure simply the parameters of reinforcement learning. The amplitude of the HRIR differs by about 10 times in the sound source and its opposite side. Thus, normalization is divided into some courses. Specifically, HRTFs that were measured for 15 degrees in 24 directions were divided into 6 intervals (from the front in the clockwise direction 0-60, 60-120, 120-180, 180-240, 240-300, and 300-360), and MA parameters were normalized so as to become 1.0 to maximum the absolute value of MA parameters for each direction.

4   Result

We confirmed through this simulation that the proposed method was effective for learning accurate HRTFs in many cases. These results are shown in Fig. 5. Fig. 5(a) shows that the learned HRTF is in accord with the measured HRTF. Early in learning, SD is 1.03 [dB], and after learning, SD is 0.07 [dB]. The SD decreased to less than 0.2 [dB] after approximately 50,000 trials from the start of learning, as shown in Fig. 5(b). The pre- and post-learning SDs across all directions are shown in Fig. 5(c). In this figure, starting from the front (user's face), positive values increase from 0 to 165 degrees in a clockwise direction and represent the right side. Negative values decrease from 0 to -180 degrees in a counterclockwise direction and denote the left side. Differences between each direction were observed, but SD were less than 0.2 [dB] in many directions. However,at 75, -120, and -180 degrees, SD were greater than 0.4 [dB] , even when the initial value and exploratory noise were changed.

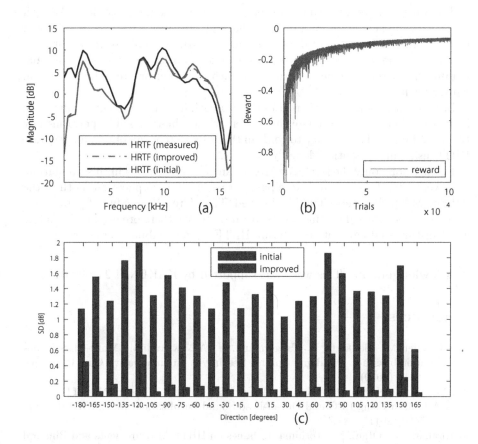

Fig. 5. Simulation results (a) HRTFs at 30 degrees of right side. Blue line is measured HRTF. Broken red line is learned HRTF. Black line is the HRTF before learning. (b) Reward in learning at 30 degree of right side. (c) The pre- and post-learning SD across all directions.

5   Discussion

As can been seen in Figure 5(c), these results indicate that the spectral distortion could be improved across all directions with the proposed method and suggest that an effective sound localization can be achieved through reinforcement learning.

However, at 75, -120, and -180 degrees, SD were greater than 0.4 [dB]. The reasons are though to be the initial value, exploratory noise, number of ARMA parameters , and parameters of reinforcement learning such as learning coefficient or variance of RBF function. Each HRIR was modeled with the optimal order of parameters, and the orders were different in each direction. At the degrees where a decrease in SD could not be confirmed and the degrees that where accurate HRTFs were obtained, the orders were not different. We varied each condition and conducted more simulation. No decrease in SD could not be confirmed even if initial value and exploratory noise were changed. However, a decrease in SD could be confirmed if parameters of reinforcement learning were optimized at each direction. Thus it is necessary to propose a method that searches the optimal parameters of reinforcement learning at each direction or are able to learn in parameter free.

In this simulation, SD was used in the evaluation of sound image localization. However, in a real application of this system, the evaluation will be performed by the user. Thus, it is necessary to confirm that our system can obtain an accurate HRTF using an evaluation done by humans.

Furthermore, only the right side MA parameters are learned in this simulation. However, when this system learns parameters in a real application, both of the AR and MA parameters should be learned. Therefore, the number of parameters that should be learned will increase substantially. We therefore need to formulate an algorithm that can obtain accurate HRTFs in a small number of trials.

Acknowledgement. This work was supported by KAKENHI 24300051.

References

1. Schreuder, M., Blankertz, B., Tangermann, M.: A New Auditory Multi-Class Brain-Computer Interface Paradigm: Spatial Hearing as an Informative Cue. PLos One 5(4), e9815 (2010)
2. Ebisawa, M., Kogure, M., Yano, S., Matsuzaki, S., Wada, Y.: Estimation of Direction of Attention Using EEG and Out-of-head Sound Localization. In: EMBC 2011, pp. 7417–7420 (2011)
3. Hirahara, T., Otani, M., Toshima, I.: Issues on HRTF Measurements and Binaural Reproduction. Fundamentals Review 2(4), 68–85 (2009)
4. Barto, A.G., Sutton, R.S., Anderson, C.W.: Neuronlike Adaptive Elements That Can Solve Difficult Learning Control Problems. IEEE SMC 13(5), 834–846 (1983)
5. Morimoto, J., Doya, K.: Learning Dynamic Motor Sequence in High-Dimensional State Space by Reinforcement Learning - Learning to Stand Up -. IEICE J82-D-II(11), 2118–2131 (1999)
6. Haneda, Y., Makino, S., Kaneda, Y.: Common-Acoustical-Pole and Zero Modeling of Head-Related Transfer Functions. IEEE Trans. Speech Audio Processing 7(2), 181–196 (1999)
7. Morikawa, D., Shimakura, N., Hirahara, T.: Threshold of Hearing and Signal Bandwidth Necessary for Horizontal Sound Localization. IEICE Technical Report EA2009(70), 83–88 (2009)
8. Huang, Q., Liu, K.: A Reduced Order Model of Head-related Impulse Responses based on Independent Spatial Feature Extraction. In: IEEE International Conference on Acoustics Speech and Signal Processing, pp. 281–284 (2009)

Damage Pattern Recognition of Refractory Materials Based on BP Neural Network

Changming Liu[1], Zhigang Wang[1], Yourong Li[1], Xi Li[1], Gangbing Song[1,2], and Jianyi Kong[1]

[1] Key Laboratory of Metallurgical Equipment & Control Technology, Ministry of Education, Wuhan University of Science and Technology, Wuhan, China
{lliuchangming,wzgwy}@126.com,
{371373011,459320376}@qq.com,
xbs@wust.edu.cn
[2] Department of Mechanical Engineering, University of Houston, Houston, USA
gsong@uh.edu

Abstract. The determination of the damage mode and the quantitative description of the damage of the clustered acoustic emission (AE) signal of the refractory materials based on the BP (back propagation) Neural Network are the subjects of this paper. In this paper, a large number of AE signals in the process of a three-point bending test were studied and the pattern recognition system of refractory materials based on BP neural network was established with the AE characteristic parameters such as amplitude, counts, rise time, duration and centroid frequency etc. The results show that the total recognition rate of material damage types with this method is as high as 97.5%, and the prediction error of the extent of the damage is about 5%, which indicates that this method has the value of application and dissemination in the aspect of micro-damage pattern recognition and extent prediction of the damage.

Keywords: Acoustic emission, Refractory materials, BP neural network, Pattern recognition.

1 Introduction

Refractory materials, which are the key for the safe operation of the high temperature equipment, are widely used in the metallurgical industry. Meanwhile, they are also the weakest vulnerable link in the high temperature furnace lining structure, which might cause irreplaceable casualties and property losses. So far, the acoustic emission (AE) technique is the only method for the real-time tracking of the generation and development of defects [1]. The purpose of the processing AE data is to clearly recognize of the damage source. So far, the artificial neural network is one of the major methods in the pattern recognition field, which therefore plays a major role in the processing methods of AE signals [2].

The commonly used methods in damage pattern recognition include amplitude discrimination, frequency discrimination, statistical pattern recognition and artificial neural network, etc [3]. While the first three methods could be used in recognition in

T. Huang et al. (Eds.): ICONIP 2012, Part IV, LNCS 7666, pp. 431–440, 2012.

practical application, rich background knowledge, as well as long-term AE testing and data analysis experience is needed during actual operation [4]. Compared to other methods, the pattern recognition method based on the BP neural network has a strong adaptive learning ability, which could be used in complex classification of the feature space and suitable for a high speed parallel processing system. The BP neural network is one of the most commonly used methods in the pattern recognition area, has and is relatively mature in theory and application.

The pattern recognition of the refractory materials consists of the classification of the damage mode and the recognition of the extent of the damage. Clustering of the AE signals could be used to identify the micro-damage mechanisms of refractory materials and determine the AE characteristics corresponding to the different micro-damage mechanisms. The k-means algorithm was used to divide the AE signals collected during the three-point bending test into two types in previous studies [3-5]. Combined with the analysis of AE parameters, the micro-damage pattern of the refractory materials could be distinguished. In this paper, the k-means clustered AE signals of the refractory materials were first preprocessed with the principal component analysis and normalization processing and used as the sample data for the training, testing and validation of the BP neural network. Then a three-layer BP neural network was constructed to carry out the anti-normalization to predict the results of the classification of the damage mode and the recognition of the extent of the damage, which could not only be used to predict the micro-damage of the refractory materials, but also to establish the pattern recognition system of refractory materials based on the BP neural network.

2   BP Neural Network

The error back propagation neural network is called the BP neural network, which is composed of three neuron layers. Fig. 1 is the structure schematic diagram of the BP neural network. The left is the input layer, the middle is the hidden layer and the right is the output layer.

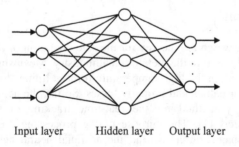

Input layer   Hidden layer   Output layer

Fig. 1. BP neural network

The learning process of the BP neural network is composed of the forward and backward propagation. The input information passes from the input layer and the hidden layer to the output layer during forward propagation. The condition of the next layer is only affected by the previous one. If the wanted output could not be achieved, the propagation will turn back along the original path and the error signals will thus be returned. The neural connection weights will be modified one after another in the returning process and this process is iterative until the error is within satisfactory levels [5].

3   Experiment

3.1   Presentation of the Refractory Materials

The refractory materials tested in the study are composed of magnesia aggregates and a binder containing graphite and/or additional metallic elements (e.g. aluminum, silicon). The components were pressed under high pressure (around 150 MPa) and insulated under 110°C for 24 hours. The dimensions of the materials were 125mm × 25mm × 25mm.

3.2   Equipment

Because of the low tensile strength of the refractory materials and the difficultly in the design of the clamp under tension, the three-point bending test was adopted for its good noise immunity and facility in the control of the damage. The three point bending tests were performed using a HMOR/STRAIN loading machine. At the same time, the DISP AE detection system from PAC company was used to collect the on-time AE signals, which consisted of rise time, peak value counts, counts, energy, duration, amplitude and average signal level values (ASL). The magnesia refractory materials were loaded until the fracture of the specimen. The crosshead speed of the machine was fixed linearly at 0.25N / (m2×s). The diagram of the experimental principle is shown in Fig. 2.

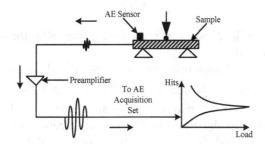

Fig. 2. Experimental setup used for AE measurements

4   Pattern Recognition with BP Neural Network

The parameters of the AE signals differ significantly between various stages of the damage. Moreover, the amount of energy carried by the AE signals corresponds to different extents of the damage. Therefore, the damage mode of the AE signals corresponding to the different damage stages could be predetermined and the pattern recognition of the AE signals could be processed by the BP neural network.

4.1   The Construction of the BP Neural Network

The BP network is composed of neurons such as the input layer, hidden layer and the output layer, as well as the activation function in each layer. Studies have shown that a three-layer BP network could approximate any continuous function with arbitrary accuracy [6]. Therefore, the BP network was constructed in a three-layer format, including an input layer, a hidden layer and an output layer .The number of the neurons of the input layer and the output layer was determined by the dimension of the input and output data respectively.

The Determination of the Nodes in Different Layers. In AE, the characteristics of composite materials, the ring count, amplitude, AE duration (under constant load), and the Felicity ratio are the major parameters in distinguishing the damage stages, type, and mechanical properties of the composite component [7]. Therefore, the amplitude, ring count, rise time, duration and centroid frequency were selected as the input layer neurons in the BP network.

The nodes of the hidden layer are usually determined based on experience and test data. Generally, the nodes of the hidden layer have a direct relationship with the requests and the number of the input and output units. In addition, it will lead to a long learning time if there are too many hidden nodes. At the same time, it will cause a bad fault tolerance and a low recognition capacity if there are too few nodes. Therefore, multi-disciplinary factors should be considered in the process of design.

Based on experience, the BP neural network for pattern recognition and classification could be designed referring to the following formula:

$$n = \sqrt{n_i + n_0} + a \tag{1}$$

Where n is the hidden layer nodes, n_i is the input nodes, n0 is the output nodes; a is the constant from 1 to 10.

According to the AE characteristic parameters introduced, the input nodes n_i=5, the output nodes n_0=2 and the hidden layer nodes $n = \sqrt{5+2} + a = 2.64575 + a$, which means the hidden layer may be chosen to contain   3 to 13 nodes. Figure 3 shows that compared to the Resilient Propagation method (RPROP) and Scaled Conjugate Gradient method (SCG) method, the Levenberg-Marquardt method (L-M) is good at reducing the mean square error, moves smoothly and is insensitive to the number of the hidden layer nodes. The mean square errors were all relatively small when the hidden layer nodes were set to 5 using the three methods. Therefore, the L-M training function was selected and the number of the hidden layer nodes was set to 5.

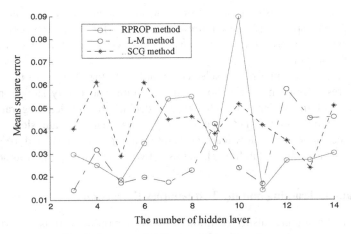

Fig. 3. Contrast of different training functions

The number of the nodes of the output layer depends on two aspects of the type and size of the data. The refractory pattern recognition includes the recognition of the damage and the identification of the extent of the damage. Previous studies have shown that the micro-damage type could be achieved by the K-means clustering method in the three point bending test and the extent of the damage could be expressed by $\frac{E_{current}}{E_{initial}} \times 100\%$, where the $E_{current}$ is the elastic modulus of the specimen in the current state of the damage and the $E_{initial}$ is initial elastic modulus of the specimen [8]. Therefore, the number of the output layer nodes of the BP neural network is selected as 2. The K-means clustering method results show that the damage types of the refractory materials were matrix damage and interface damage, corresponding to two types of damage. In data preprocessing, the first damage type (interface damage) was expressed by 0.2 and the second damage type (matrix damage) was expressed by 0.8.

Determination of the Initial Weight and Learning Rate. Because the system is nonlinear, the initial weights have a strong influence on whether the study reached the local minimum and when convergence can be achieved. An excessively large initial weight value will cause the weighted input to fall into the saturation region of the activation function, which will result in extremely small derivatives and break the adjustment. For the feedforward networks, there are two different initialization methods, which apply to the linear and nonlinear transforming system, respectively. The linear system initializes the weight matrix and bias according to the initialization parameters of each layer. However, in the nonlinear system the initial weights and bias values are usually brought forward by the Nguyen-Widrow method, which makes the activity area of each neuron roughly distributed in the input space. It has the following advantages compared to the random simple assignment to the weight and bias: (1) Reduce the waste of the neurons (for that, the activity area of all neurons is inside the input space). (2) Faster training velocity (for that, each region in input space is inside the range of active neurons). In this system, the former was selected to initialize the network.

Generally, the learning rate is selected according to the changes of the sum of the error squares after each training to avoid oscillation and overly slow convergence. The change of the weight in each circle is decided by the learning rate. A high learning rate destabilizes the system and the low learning rate slows down the convergence. In general, a low learning rate (always between 0.01 and .08) is always chosen to ensure system stability [5]. In this network the learning rate was set to 0.02.

4.2   Sample Training and the Test Samples Extraction

The principal and prerequisite condition of the BP neural network is to have adequate amount of high precision samples. Moreover, in order to avoid the over-fitting in the training process and estimate the performance and the generalization of the network model, the data collected should be randomly divided into the following three parts: training sample, checking sample (above 10%) and testing sample (above 10%). In addition, the balance between the damage modes of the samples in the data grouping process should be considered. Therefore, in order to enhance the generalization of the network, the k-means clustered AE signals of the refractory materials are first preprocessed with the normalization processing of the input and target vector. Then the predicted output vector will be anti normalized to achieve the final results.

Generally, a certain number of samples are needed in the training. Yet the collection of the sample is always limited by the objective conditions. Besides, when the sample reaches to a certain extent, the speed of the network will be affected. According to experiments, the number of the sample depends on the complexity of nonlinear mapping relationship between input and output. The more complex the mapping relationship is, the greater the size of the network and noise levels, and more samples are needed in order to guarantee the accuracy of the mapping. An empirical rule that the number of training samples is 5-10 times of the number of the connection weights in the network could be referenced in [9].

In order to enhance the training efficiency of the neural network, the sample data should be preprocessed properly. The sample data should first be normalized to make the input and target data obey a normal distribution. Then the normalized sample data is carried by the principal component analysis to eliminate the redundant components of the data and meet the dimension reduction goal. At the same time, in order to improve the generalization and identification ability, the early stop method was adopted in the training [10]. Therefore, the processed sample data were divided into the training sample set, the checking sample set and the testing sample set. There were 400 sample data points that were chosen in the process of the analysis. The checking sample and testing sample accounted for 20% of the total sample respectively and the others were selected as the training sample. Before training, the principal component analysis method was first used to classify the damage mode for the amplitude of the input layer, ringing counts, rising time, duration and centroid frequency. Then they were normalized so that the parameters remain unrelated.

5   Results and Discussion

The BP neural network constructed could be used in the classification of the damage mode and the recognition of the extent of the damage. Fig. 4 shows the error curve of network training.

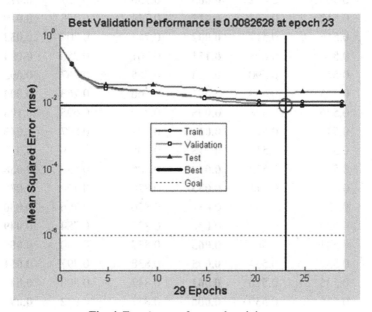

Fig. 4. Error curve of network training

Table 1. The classification of the damage mode (recognition rate)

Damage mode	sample	Correct number	Error type	Recognition rate (%)
Matrix damage	77	76	1 interface damage	98.7
Interface damage	3	2	1 matrix damage	66.7
Total recognition rate (%)	80	78	-	97.5

There were 400 sample data points and the data tested accounted for 20% of the overall sample. Seen from the type and the extent of the damage of the 80 tested data, the recognition rate of the BP neural network in the classification of the damage mode was as high as 97.5% and the relative error in the recognition of the extent of the damage was comparatively small, which indicated that the BP neural network could be used well in the classification of the damage mode and the recognition of the extent of the damage. Table 1 and Table 2 show the classification of the damage mode and prediction of the recognition of the extent of the damage respectively.

Table 2. The prediction of the recognition of the extent of the damage

Experiment	Prediction	Error	Experiment	Prediction	Error
0.500	0.443	**0.129**	0.735	0.718	**0.024**
0.503	0.472	**0.065**	0.736	0.727	**0.013**
0.512	0.361	**0.417**	0.744	0.758	**0.019**
0.516	0.517	**0.002**	0.744	0.708	**0.052**
0.516	0.608	**0.151**	0.761	0.758	**0.004**
0.516	0.590	**0.124**	0.765	0.767	**0.003**
0.525	0.559	**0.061**	0.766	0.763	**0.004**
0.526	0.528	**0.005**	0.770	0.695	**0.108**
0.526	0.515	**0.021**	0.771	0.807	**0.045**
0.527	0.519	**0.015**	0.805	0.779	**0.033**
0.527	0.532	**0.010**	0.805	0.837	**0.038**
0.527	0.511	**0.033**	0.820	0.754	**0.088**
0.528	0.502	**0.051**	0.820	0.796	**0.030**
0.529	0.549	**0.036**	0.822	0.784	**0.049**
0.531	0.567	**0.063**	0.837	0.790	**0.059**
0.532	0.534	**0.005**	0.838	0.797	**0.051**
0.535	0.570	**0.061**	0.847	0.808	**0.048**
0.535	0.539	**0.006**	0.847	0.932	**0.092**
0.602	0.634	**0.051**	0.853	0.798	**0.069**
0.609	0.927	**0.343**	0.864	0.823	**0.051**
0.647	0.675	**0.041**	0.865	0.805	**0.024**
0.656	0.653	**0.005**	0.866	0.799	**0.013**
0.656	0.676	**0.030**	0.870	0.863	**0.019**
0.661	0.650	**0.017**	0.870	0.776	**0.052**
0.664	0.693	**0.042**	0.871	0.824	**0.004**
0.664	0.724	**0.083**	0.878	0.837	**0.003**
0.666	0.681	**0.022**	0.879	0.838	**0.004**
0.667	0.684	**0.025**	0.879	0.836	**0.108**
0.667	0.708	**0.058**	0.882	0.834	**0.045**

Table 2. (*Continued*)

Experiment	Prediction	Error	Experiment	Prediction	Error
0.670	0.69	**0.029**	0.882	0.867	**0.033**
0.681	0.673	**0.012**	0.885	0.803	**0.038**
0.681	0.717	**0.050**	0.886	0.828	**0.088**
0.696	0.689	**0.010**	0.894	0.913	**0.030**
0.697	0.707	**0.015**	0.894	0.833	**0.049**
0.697	0.691	**0.009**	0.896	0.482	**0.059**
0.698	0.686	**0.017**	0.897	0.850	**0.051**
0.724	0.707	**0.024**	0.899	0.872	**0.048**
0.733	0.713	**0.028**	0.903	0.954	**0.092**
0.734	0.743	**0.012**	0.996	0.761	**0.069**
0.734	0.719	**0.020**	0.997	0.650	**0.051**

6   Conclusion

The AE technique could be used in the on-time tracking of the generation and development of the damage of the refractory materials. The pattern recognition system of refractory materials based on BP neural networks is established with the AE characteristic parameters such as amplitude, counts, rise time, duration and centroid frequency etc. The results show that the total recognition rate of material damage types with this method was as high as 97.5%, and the prediction error of the extent of the damage was about 5%, which indicates that this method could be used in the pattern recognition and damage extent prediction. Compared to other pattern recognition methods, the BP neural network is superior in the damage recognition rate and the prediction accuracy of the extent of the damage.

Acknowledgements. The authors would like to thank the National Natural Science Foundation of China (NSFC 51075310) for the financial support.

References

1. Huguet, S., Godina, N., Gaertnera, R., Salmonb, L., Villard, D.: Use of acoustic emission to identify damage modes in glass fibre reinforced polyester. Compos. Sci. Technol. 62, 1433–1444 (2002)
2. Marec, A., Thomas, J.H., Guerjouma, R.E.: Damage characterization of polymer-based composite materials: Multivariable analysis and wavelet transform for clustering acoustic emission data. Mech. Syst. Signal. Pr. 22, 1441–1464 (2008)

3. Godina, N., Huguet, S., Gaertner, R., Salmon, L.: Clustering of acoustic emission signals collected during tensile tests on unidirectional glass/polyester composite using supervised and unsupervised classifiers. NDT&E International 37, 253–264 (2004)

4. Kalogiannakis, G., Quintelier, J., De Baets, P., Degrieck, J., Van Hemelrijck, D.: Identification of wear mechanisms of glass/polyester composites by means of acoustic emission. Wear 264, 235–244 (2008)

5. Philippidis, T.P., Nikolaidis, V.N., Anastassopoulos, A.A.: Damage characterization of carbon/carbon laminates using neural network techniques on AE signals. NDT&E International 31, 329–340 (1998)

6. Rao, H.S., Mukherjee, A.: Artificial neural networks for predicting the macromechanical behaviour of ceramic-matrix composites. Comp. Mater. Sci. 5, 307–322 (1996)

7. Rajendraboopathy, S., Sasikumar, T., Usha, K.M., Vasudev, E.S.: Artificial neural network a tool for predicting failure strength of composite tensile coupons using acoustic emission technique. Int. J. Adv. Manuf. Technol. 44, 399–404 (2009)

8. Schmitt, N., Berthaud, Y., Poirier, J.: Tensile behaviour of magnesia carbon refractories. J. Eur. Ceram. 20, 2239–2248 (2000)

9. Bar, H.N., Bhat, M.R., Murthy, C.R.L.: Parametric Analysis of Acoustic Emission Signals for Evaluating Damage in Composites Using a PVDF Film Sensor. J. Nondestruct. Eval. 24, 121–134 (2005)

10. Soheil, A., Carey, J.S., Robert, W.B.: Application of neural networks to predict the transient performance of a Run-Around Membrane Energy Exchanger for yearly non-stop operation. Int. J. Heat Mass Tran. 20, 1–14 (2012)

Robust and Optimum Features for Persian Accent Classification Using Artificial Neural Network

Azam Rabiee[1] and Saeed Setayeshi[2]

[1] Department of Computer Science, Dolatabad Branch, Islamic Azad University, Isfahan, Iran
rabiee@iauda.ac.ir
[2] Department of Medical Radiation, Amirkabir University of Technology, Tehran, Iran
setayesh@aut.ac.ir

Abstract. This paper presents a classification model for regional accents of Persian. The model is based on a combination of the conventional speech coding and pattern recognition techniques. In this study, the well-known multilayer perceptron plays the role of the classifier. Moreover, a wide variety of speech coding techniques is utilized for feature extraction. Among them, we determine the robust and optimum features for this task by comparing the classification performance. The method is validated on a corpus containing recordings from ten speakers, five males and five females, for each accent. Results show that perceptual linear predictive (PLP), relative spectral transform PLP (Rasta PLP), and linear predictive coefficient (LPC) perform well under both clean and noisy conditions.

Keywords: accent classification, optimum feature, robust feature, Persian accent.

1 Introduction

Groups of people of a similar geographical, linguistic, social or cultural background can be considered to share various common patterns in their speech, resulting in an impression of a particular accent when they talk. In fact, the accent refers to the use of particular vowel and consonant sounds, and how these change when they are combined in words and groups of words [1]. In a subjective study, Ikeno and Hansen [2] have showed that how listener's accent background affects accent perception and comprehensibility. Obviously, in automatic speech recognition (ASR) systems, correct classification of speaker's accent can improve the performance. Thus, the accuracy of an ASR system is greatly reduced when the speaker's accent is different from that for which it is trained. Furthermore, the ability to estimate and characterize accents would provide valuable information in the development of more effective speech systems, such as speaker classification, channel monitoring, voice conversion, and audio stream tagging in spoken document retrieval.

Various studies for automatic accent classification have been proposed in the literature [1, 3–14]. Although, majority of the studies have examined English accents, limited studies have investigated the non-English ones [13–14]. In a previous study

T. Huang et al. (Eds.): ICONIP 2012, Part IV, LNCS 7666, pp. 441–449, 2012.

[14], we have presented a model for Persian accent identification, in which the well-known multilayer perceptron (MLP) classifies conventional features extracted from the accented speech signal. Our results in [14] emphasize that the MLP, as an adaptive classifier, outperforms the statistical support vector machine (SVM) and k-nearest neighbor (KNN) approaches in this task, especially when the number of accents increases.

In our previous study, we have extracted the second and third formants, Mel frequency cepstral coefficients (MFCCs), and energy as the relevant features for accent classification. Moreover, the experiments have been performed in clean condition. This paper develops the previous study for noisy condition. Furthermore, the robust and optimum features for this task are determined from a wide range of features contained within the speech signal. Section 2 presents our accent classification model in details. Section 3 explains the implementation and experimental results. Finally, conclusion is given in Section 4.

2   Model Overview

Our engineering model for Persian accent classification is based on a combination of the conventional speech coding and pattern recognition techniques. The block diagram of the model is depicted in Fig. 1. The model contains pre-processing, feature processing, and classification. In the pre-processing stage, a band-pass filter eliminates all irrelevant frequencies from the speech signal. A framing and windowing process converts the continuous speech signal into frames. Finally, the silence removal process eliminates all silence frames.

The second stage is feature processing comprising pre-emphasis, feature extraction, feature compression, and dimension reduction. The pre-emphasis makes the features distinctive. In the feature extraction stage, we extract the appropriate feature for the accent classification task. Eventually, feature compression and dimension reduction decrease the amount of data and subsequently, computational time for the next stage.

The last stage is the well-known MLP, as our artificial neural network classifier. Processed features prepare the input vector of the MLP. Moreover, each output neuron of the MLP belongs to one accent class. The details of the model are explained in the following subsections.

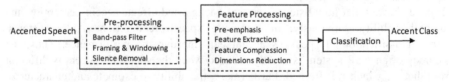

Fig. 1. The block diagram of the Persian accent classification model

2.1   Pre-processing

Band-Pass Filter. The input signal may include unknown signals and noise, which affect the classification performance. Hence, the frequency components that are out of the speech frequency range are eliminated using a band-pass filter. Instead of a band-pass filter, we utilize a high-pass and a low-pass filter. Similar to the MELP standard [15], our high-pass filter is a 4th order Chebychev type II, with a cutoff frequency of 60 Hz, and a stopband rejection of 30db. Furthermore, we use a 4.5 kHz 6th order Butterworth low-pass filter.

Framing and Windowing. The continuous speech signal is divided into overlapped frames. The length of the frames is 25 ms, and the steps are 10 ms; therefore, each frame has 15-ms overlap with the next and previous frames. Furthermore, before any further process, a Hamming window prevents the aliasing effect.

Silence Removal. The silence frames do not have appropriate features and should be removed. A silence frame is the frame whose energy is less than 0.15 of the average energy of the entire waveform.

2.2   Feature Processing

Pre-emphasis. The pre-emphasis stage contains the filter $1 - \alpha z^{-1}$, in which usually α is equal to 0.97. The filter is inspired from the lip model, and is equal to a high-pass filter that strengthens the high-frequency components. Utilizing the filter before the feature extraction helps the features to be extracted distinctly.

Feature Extraction. There is a wide range of features contained within the speech signal, which provide information concerning a particular speaker's characteristics, such as age, gender, emotion, stress, accent, dialect, and health [2]. Many studies used MFCC, energy, as well as second- and third- formant frequencies as the most relevant features of a speaker's accent [1, 4, 11, 12]. However, all these features may not be relevant for the accent classification task. In this subsection, we review a wide variety of features including both biologically inspired and engineering ones; but, for each experiment, only one feature set is employed in the feature extraction stage. Thus, we can find the robust and optimum feature set in this task by comparing the results.

Various types of features can be extracted from a speech signal including scalar and vector feature sets. The scalar features, such as energy, gain, jitter, pitch, mean, variance, and formants are lonely weak for accent classification; thus they are usually employed besides vector features. We explain the extraction processes of some vector features in the following.

Linear Predictive Coefficient (LPC). One of the most popular features is the LPC. The LPC extraction starts with the assumption that the nth sample of a speech signal $s(n)$ can be estimated using a linear combination of previous samples by

$$\hat{s}(n) = \sum_{k=1}^{p} a_k s(n - k) \tag{1}$$

where $\hat{s}(n)$ is the estimated sample and a_k are the linear predictive coefficients. These coefficients are obtained using the least mean square (LMS) algorithm to decrease the distance between $s(n)$ and $\hat{s}(n)$. It can be performed by autocorrelation or covariance method [16]. In Section 3, we refer to the LPC extracted by the autocorrelation method as LPC_Auto.

*Mel Frequency Cepstral Coefficient (MFCC).*The block diagram of the MFCC extraction is depicted in Fig. 2. The first step is a discrete Fourier transform (DFT). The result is fed to a Mel filter bank. Then, the logarithm of each Mel filter output is computed. Finally, the amplitude of the resulting discrete cosine transform (DCT) of the Mel log are the required features [17].

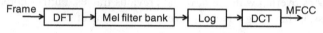

Fig. 2. The block diagram of the MFCC extraction

The log stage can be carried out on the amplitude or power of the frequency components, referring to them as AD_MFCC and PD_MFCC, respectively. As the other speech feature sets, we use the first and second derivatives of MFCC, D_xD_MFCC and DD_xD_MFCC, respectively. Moreover, a teager energy operator (TEO) [18] on the MFCCs features results a new feature set, named teager_MFCC.

Log Frequency Power Coefficient (LFPC). The LFPC is a nonlinear feature that is obtained by a logarithmic filter bank from 200 Hz to 4 kHz [19]. The logarithmic filter bank can be regarded as a model that follows the varying auditory resolving power of the human ear for various frequencies. Fig. 3 depicts the block diagram of the LFPC extraction.

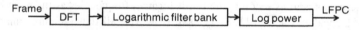

Fig. 3. The block diagram of the LFPC extraction

The feature can be extracted in both frequency and time domain, denoted by FD_LFPC and TD_LFPC, respectively. Their first and second derivatives, D_xD_LFPC and DD_xD_LFPC, respectively, are also regarded as speech features.

Wavelet Packet Cepstral Coefficient (WPCC). The block diagram of the WPCC extraction is shown in Fig. 4. The first step is decomposing the frame into a wavelet packet (WP) tree up to 6th level. An ordered set of 24 nodes of the WP tree is selected. Consequently, 24 normalized filter bank energies $S(k)$ are computed from the WP transform coefficients at each node or subband. Finally, from each frame, WPCC is computed by taking discrete wavelet transform (DWT) of $\log(S(k))$ [18]. In our model, we use the mother wavelets *db3*, *db6*, *db10*. In Section 3, the TEO of the WPCC in the frequency domain is denoted by tfWPCC.

Fig. 4. The block diagram of the WPCC extraction

Perceptual Linear Predictive (PLP). The PLP technique was designed to suppress speaker-dependent components in the features. Several basic properties of human hearing were integrated in PLP [20]. The block diagram of PLP extraction is shown in Fig. 5 with the solid arrows. The LPC extracted at the final stage of this process is called PLP_lpcas. Moreover, we employ the spectrum and cepstrum of these features, which are denoted by PLP_spectra and PLP_cepstra, respectively.

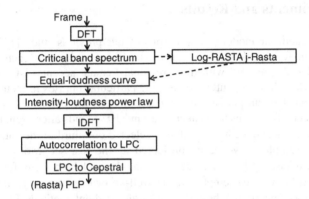

Fig. 5. The block diagram of the PLP and Rasta PLP extraction. Solid arrows show the pathway of the PLP extraction. Dashed arrows belong to the Ratsa PLP extraction.

RelAtive Spectral TrAnsform PLP (Rasta PLP). The Rasta PLP is based on certain temporal properties of human hearing. This technique uses the fact that linear distortions and additive noise in speech signal are shown as a bias in the short-term spectral parameters. Hence, the Rasta PLP is proposed based on filtering the temporal trajectories of the speech parameters [21]. The block diagram of the Rasta PLP extraction is depicted in Fig. 5. We also use the spectrum and cepstrum of this feature as the speech feature sets, named RPLP_spectra and RPLP_cepstra, respectively.

Feature Compression and Dimension Reduction. A feature compression is carried out based on averaging the features of every three neighbor frames. This is because every phoneme usually takes 3 to 10 frames. Later, polynomial approximation [22] is performed on the trajectories of various features in time domain. This process helps to decrease the unwanted noise in the feature extraction process due to the computational limitation. Moreover, principal component analysis (PCA) [23] helps us to remove irrelevant and redundant features, resulting in speeding up the learning process, and improving the model interpretability.

2.3   Classification

A two-layer feed-forward MLP is used to classify the features, same as our previous study [14]. The number of neurons in the input layer is equal to the feature vector length, and the number of neurons in the hidden layer is equal 5, determined by trial and error. Finally, the number of output neurons is equal to the number of accents. Each output neuron belongs to one accent. The desired value of each output neuron is 1, if the input pattern belonged to the same accent; otherwise, it is −1. We employ the *tangent sigmoid* function as the transfer function of the hidden neurons and the *linear* function for the outputs. The learning rate in our MLP is equal 0.05.

3   Experiments and Results

We have evaluated our model on the corpus of our previous study [14], containing recordings from ten speakers, five males and five females, for each accent. The speakers are 23-52 years old with three regional Persian accents: Isfahani, Tehrani, and Kermanshahi. They have uttered the same particular phrases in their own accent. We have selected Persian phrases that contain most of the phonemes. The length of the acoustic speech file for each speaker is around 10 seconds after silent removal.

In all the experiments, we have randomly selected one-third of the dataset for train. The remaining samples have been considered as the test set. Moreover, we have normalized the dataset to have a value between −1 and 1 before any further process. To train the MLP, we have employed the well-known *conjugate gradient* learning method, which is one of the best second-order gradient methods [24]. In all the experiments, we have employed a vector of 12 coefficients from only one type of the features explained in Section 2.2.

3.1   Clean Condition

The first experiment is performed in the clean condition. We have repeated the experiment for all the types of the feature sets. Fig. 6 depicts the classification

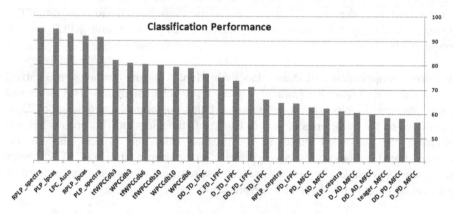

Fig. 6. Classification performance for each feature vector

performance for each feature set. The results are the average on the three classes of the Persian accents. As shown in the figure, the RPLP_spectra, PLP_lpcas, LPC, RPLP_lpcas and PLP_spectra show the best results, higher than 90%. The worst results have been observed when teager_MFCC, DD_PD_MFCC, and D_PD_MFCC are employed (50%-60%). it is worth mentioning that the minimum boundary for the classification performance belongs to the chance, as a classifier that shows 33% accuracy for the three classes.

3.2 Noisy Condition

Existence of various noises in the environment is unavoidable, and corrupted data reduces the performance of the system. Hence, the second experiment is to find the robust feature in the noisy condition, when the data is corrupted by the environmental noise.

We have employed two different environmental noises: (1) babble noise, which is the background sound from a party, and (2) siren or alarm. We have selected the babble and the siren because of their different characteristics. The babble noise contains almost all the frequency bins uniformly; while the harmonic siren includes some specific frequency bins.

We have repeated the accent classification experiment for all the types of the feature sets extracted from the corrupted signals. Fig. 7 and Fig. 8 show the average classification performance for babble and siren noises, respectively. In both figures, PLP_lpcas is the pioneer feature. The results of the both noises convince that the PLP_lpcas, LPC_auto, RPLP_spectra, and PLP_spectra are the most robust features for this task. Same as the clean condition, the Teager_MFCC feature shows the worst results in both noises.

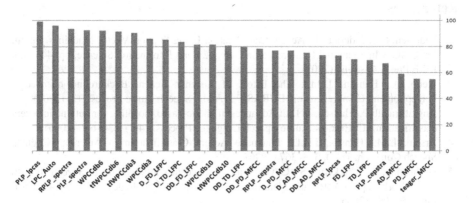

Fig. 7. Classification performance on corrupted data with babble (party) noise

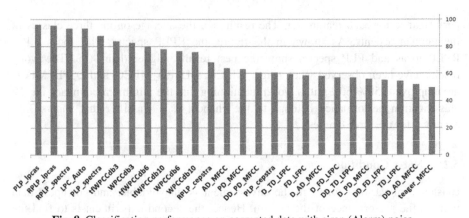

Fig. 8. Classification performance on corrupted data with siren (Alarm) noise

4   Conclusion

In this study, a Persian accent classification is carried out using a model including pre-processing, feature processing, and an adaptive classifier. Our model is a combination of speech coding and pattern recognition techniques. In this paper, we have reviewed a wide variety of biologically inspired and engineering features. Then we have found the robust and optimum feature sets for this task by comparing the results. Results have showed that the biologically inspired features PLP, Rasta PLP, and LPC perform well under both clean and noisy conditions. Although, the MFCC is the pioneer feature for ASR, the worst results of this study belong to the MFCC and its derivatives.

References

1. Pedersen, C., Diederich, J.: Accent Classification Using Support Vector Machines. In: 6th IEEE/ACIS International Conference on Computer and Information Science, pp. 444–449. IEEE Press, Australia (2007)
2. Ikeno, A., Hansen, J.H.L.: The Effect of Listener Accent Background on Accent Perception and Comprehension. EURASIP J. Audio Speech Music Process. (3) (2007)
3. Angkititrakul, P., Hansen, J.H.L.: Advances in phone-based modeling for automatic accent classification. IEEE Trans. Audio Speech Language Process. 14(2), 634–646 (2006)
4. Arslan, L.M., Hansen, J.H.L.: Language Accent Classification in American English. Speech Commun. 18(4), 353–367 (1996)
5. Deshpande, S., Chikkerur, S., Govindaraju, V.: Accent Classification in Speech. In: 4th IEEE Workshop on Automatic Identification Advanced Technologies, pp. 139–143. IEEE Press, Buffalo (2005)
6. Faria, A.: Accent Classification for Speech Recognition. In: Renals, S., Bengio, S. (eds.) MLMI 2005. LNCS, vol. 3869, pp. 285–293. Springer, Heidelberg (2006)
7. Fung, P., Kat, W.L.: Fast Accent Identification and Accented Speech Recognition. In: IEEE International Conference on Acoustics, Speech, and Signal Processing, pp. 221–224. IEEE Press, Arizona (1999)

8. Huang, R., Hansen, J.H.L., Angkititrakul, P.: Dialect/Accent classification using unrestricted audio. IEEE Trans. Audio Speech Language Process. 15(2), 453–464 (2007)
9. Kumpf, K., King, W.R.: Automatic Accent Classification of Foreign Accented Australian English Speech. In: 4th International Conference on Spoken Language Proceedings 3, pp. 1740–1743. IEEE Press, Pennsylvania (1996)
10. Tang, H., Ghorbani, A.A.: Accent Classification Using Support Vector Machine and Hidden Markov Model. In: Xiang, Y., Chaib-draa, B. (eds.) Canadian AI 2003. LNCS (LNAI), vol. 2671, pp. 629–631. Springer, Heidelberg (2003)
11. Ullah, S., Karray, F.: Speaker Accent Classification Using Distance Metric Learning Approach. In: 7th IEEE International Symposium on Signal Processing and Information Technology, pp. 900–905. IEEE Press, Egypt (2007)
12. Ullah, S., Karray, F.: Speaker Accent Classification System Using A Fuzzy Gaussian Classifier. In: International Conference on Information and Emerging Technologies, pp. 1–5. IEEE Press, Pakistan (2007)
13. Zheng, Y., Sproat, R., Guy, L., Shafranz, I., Zhouz, H., Suz, Y., Jurafsky, D., Starr, R., Yoon, S.: Accent Detection and Speech Recognition for Shanghai-Accented Mandarin. In: InterSpeech 2005, pp. 217–220. ISCA, Portugal (2005)
14. Rabiee, A., Setayeshi, S.: Persian Accents Identification Using an Adaptive Neural Network. In: 2nd International Workshop on Education Technology and Computer Science, pp. 7–10. IEEE Press, China (2010)
15. Supplee, L.M., Cohn, R.P., Collura, J.S., McCree, A.V.: MELP: the new Federal Standard at 2400 bps. In: International Conference on Acoustics, Speech, and Signal Processing, vol. 2, pp. 1591–1594. IEEE Press, Germany (1997)
16. Deller, J., Hansen, J.H.L., Proakis, J.: Discrete-Time Processing of Speech Signals. IEEE Press (2000)
17. Molau, S., Pitz, M., Schluter, R., Ney, H.: Computing Mel-frequency cepstral coefficients on the power spectrum. In: International Conference on Acoustics, Speech, and Signal Processing, pp. 73–76. IEEE Press, Utah (2001)
18. Kandali, A.B., Routray, A., Basu, T.K.: Vocal emotion recognition in five native languages of Assam using new wavelet features. Int. J. Speech Tech. 12, 1–13 (2009)
19. Nwe, T.L., Foo, S.W., De Silva, L.C.: Classification of stress in speech using linear and nonlinear features. In: International Conference on Acoustics, Speech, and Signal Processing, pp. II-9–II-12. IEEE Press, Hong Kong (2003)
20. Hermansky, H.: Perceptual linear predictive (PLP) analysis for speech. J. Acoust. Soc. Am. 87(4), 1738–1752 (1990)
21. Hermansky, H., Morgan, N., Bayya, A., Kohn, P.: RASTA-PLP speech analysis technique. In: International Conference on Acoustics, Speech, and Signal Processing, pp. 121–124. IEEE Press, California (1992)
22. Dusan, S., Flanagan, J.L., Karve, A., Balaraman, M.: Speech Compression by Polynomial Approximation. IEEE Trans. Audio Speech Language Process. 15(2), 387–395 (2007)
23. Jolliffe, T.: Principal Component Analysis, 2nd edn. Springer, New York (2002)
24. Gupta, M., Jin, L., Homma, N.: Static and Dynamic Neural Networks: From Fundamental to Advanced Theory. John Wiley & Sons Inc. (2003)

Identification of Factors Characterising Volatility and Firm-Specific Risk Using Ensemble Classifiers

Pascal Khoury[1,2] and Denise Gorse[1]

[1] Dept. of Computer Science, University College London, Gower Street,
London WC1E 6BT, UK
[2] Charlemagne Capital, 39 St. James's St., London SW1A 1JD, UK
{P.Khoury,D.Gorse}@cs.ucl.ac.uk

Abstract. Ensemble classifiers comprised of neural networks trained using particle swarm optimisation are used to identify characteristic factors of companies that have experienced high share price volatility either by an objective measure or relative to their peers, or whose calculated firm-specific equity risk exceeds a comparison value appropriate to their industry and region. Use is made of a novel training metric, the Matthews correlation coefficient, that is shown to better handle numerically unbalanced data sets. A comparison is made with results from input-output correlation analysis and it is noted that the factors derived from ensemble weightings appear to be more predictive.

Keywords: Ensemble classifiers, particle swarm optimisation, volatility.

1   Introduction

Volatility is the relative rate at which the price of a security moves up and down. For stocks it is usually calculated as the annualised standard deviation of daily returns (changes in closing price from one day to the next). Unexplained excess volatility is disliked by investors as it undermines the usefulness of the stock price as a reflection of a company's intrinsic value and hence it is desirable to discover the factors that might underlie such excess volatility.

While volatility can be in part explained by macroeconomic and geopolitical factors that could affect the stock market as a whole there may also be firm-specific factors that cause a particular stock price to behave differently to its industrial or geographical peers. The objective of the current work is to identify, via their use to classify companies into low/high risk groups, those financial factors most greatly associated with firm-specific stock price volatility.   Classification will be done using an ensemble of neural networks trained using particle swarm optimisation (PSO), making a novel use of the Matthews correlation coefficient as a fitness function.

The importance of individual factors will be assessed according to two different criteria and networks retrained on subsets of these factors, demonstrating both that a substantial number of candidate factors are redundant and that those key factors identified on the basis of neural network weight magnitudes are more predictive.

T. Huang et al. (Eds.): ICONIP 2012, Part IV, LNCS 7666, pp. 450–457, 2012.

2   Background

2.1   Particle Swarm Optimisation as a Training Method for Neural Networks

Particle swarm optimisation (PSO) [1] is a population-based search algorithm, based on observations of social behaviour in birds and other animals, that has been found effective in a wide variety of application areas, including finance [2,3].   While PSO can be used successfully as a training method for neural networks there are usually no strong grounds for preferring it to better known methods like error backpropagation other than the observation that PSO-based training can be faster and less prone to trapping in local minima [4]. However the real strength of the method is that it does not require a differentiable performance measure, a feature which will be important in the current work, while being at the same time simpler and less computationally demanding than the possible alternative use of a genetic algorithm [5].

PSO combines learning based on each particle's own past experience (*cognitive* contribution weighted by φ_1) and learning based on following the swarm's best performing member (*social* contribution weighted by φ_2). Every particle has a velocity $\underline{v}_i$ and position $\underline{x}_i$, where the latter will here correspond to the full list of weights possessed by the ith network. The equations used to update the velocity $\underline{v}_i$ and position are

$$\underline{v}_{i,t+1} = W\underline{v}_{i,t} + \varphi_1\beta_1(\underline{p}_{i,t} - \underline{x}_{i,t}) + \varphi_2\beta_2(\underline{g}_t - \underline{x}_{i,t}) \ , \tag{1a}$$

$$\underline{x}_{i,t+1} = \underline{x}_{i,t} + \underline{v}_{i,t+1} \ , \tag{1b}$$

where $\underline{p}_{i,t}$ ('personal best' or *pbest*) is the best position (weight set) found at time t by particle (net) i, $\underline{g}_t$ ('global best' or *gbest*) is the best   position found at this time by any particle, β_1, β_2 are random numbers chosen uniformly from the interval [0,1], and W is an iteration-decreasing *inertia weight* that balances the above forms of learning (exploitation of the search space) with random search (exploration).

2.2   Use of Matthews Correlation Coefficient (MCC) as a Fitness Metric

The *Matthews correlation coefficient* (MCC) [6] , sometimes also known as the *phi coefficient*, was introduced as a measure of classification success in bioinformatics, where there are many classification problems dominated by a majority type and where using a simple measure like root mean squared error (RMSE) a deceptively high score can be gained by assigning all examples to a single class. For a two-class problem the MCC is defined by

$$MCC = \frac{n_{11}n_{00} - n_{01}n_{10}}{\sqrt{(n_{11}+n_{01})(n_{11}+n_{10})(n_{00}+n_{01})(n_{00}+n_{10})}} \ , \tag{2}$$

where n_{00} (n_{10}) is the number of true (false) negative examples and n_{11} (n_{01}) is the number of true (false) positive examples. An MCC of 1 denotes perfect performance, a value of 0 either a random guess or the assignment of everything to one class.

The MCC contains within a single value similar information to that given by the joint measures *sensitivity* (the ability to correctly identify positive examples) and *specificity* (the corresponding ability to identify negative examples). It is more informative than the F1 score in that the latter does not take into account the true negative rate. These features suggest the use of the MCC not only as a measure of classification success but also as a training fitness function. While there are many instances of the use of the MCC to score training success, especially in biomedical application areas, to our knowledge it has not previously used as a training metric. A possible reason is that the MCC is a counting measure and as such is problematic for training methods such as error backpropagation requiring a differentiable performance metric; this is not however necessary for PSO, the form of learning used in the current work.

The results of Section 4.1 will demonstrate the superiority of the MCC over the more commonly used root mean squared error (RMSE) as a PSO training metric in the three volatility classifcation tasks considered here, especially for Task 3, the one for which the example classes are least numerically balanced.

3   Methods

3.1   Data Used

The data used are from 701 companies in five emerging markets regions (Asia, Latin America, Europe, Middle East and Africa, Global). There are in addition 49 industry groups, examples being 'Korean industrials' or 'Chinese healthcare companies'.

Table 1. Input factor summary for data used in the study

Index	Name	Comments
1	Book to price	Company book value per share divided by share price
2	Earnings yield	Total earnings per share divided by share price
3	Dividend yield	Total dividend per share divided by share price
4	CAGR	Compounded annual growth rate of 3-year earnings
5	Earnings growth	Earnings growth over last year
6	Size	Log of market capitalisation
7	Non-linear size	Cube of log of market capitalisation
8	Free cash flow yield	Indicates free cash flow on a per share basis
9	Net debt to assets	Net debts / total assets; an indicator of company leverage
10	Net debt to equity	Net debts / total equity; an indicator of company leverage
11	Liquidity	Average 3 months trading volume
12	RSI	3 months share price momentum
13	dRSI	Change in RSI as a proportion of last 3-month measurement

For each company thirteen factors (as defined in Table 1) considered potentially important in relation to firm-specific stock price volatility [7] are calculated on the basis of 2011-12 financial data; these will be the inputs to an ensemble classifier aiming to use them to separate companies into lower and higher risk groups.

For Tasks 1 and 2 (as defined below) a linear scaling was applied to the majority of the input factors to bring them roughly into $O(1)$, with the exception of factors 1, 5, 8, 11 to which a log transform was applied in order to minimise the effect of outliers.

For Task 3 (again defined below) the inputs were scaled in relation to others in the same region or industrial sector, or with respect to the full training list, as deemed most appropriate using domain knowledge from the investment industry.

3.2   Definition of the Three Classification Tasks

The broad objective in each case is to divide companies into lower and higher risk groups. However the tasks are different in their detailed definitions, given below:

Task 1. The objective here is to classify companies into a lower or higher risk group by a measure in which low volatility risk corresponds to an annualised volatility (defined here as the standard deviation of the stock price over the past year relative to its mean value) of less than 40%, a high volatility conversely.

Task 2. The objective here is to classify companies into high or low volatility relative to comparable peers. The lower risk examples in this case correspond to an annualised volatility (calculated as in Task 1) that is less than the group average, the higher risk cases to a volatility that is greater than the group average.

Task 3: For each company a ratio of 'firm-specific risk' to 'total risk' is calculated, with each component in this ratio being calculated using an equity risk model [8]. Companies with a value of less than 0.65 for this ratio are considered to have lower firm-specific risk, with values above this corresponding to higher risk.

Table 2. Class distributions of the data used in the three tasks of the study

	Task 1	Task 2	Task 3
Positive (higher risk)	218	299	187
Negative (lower risk)	483	402	514
Ratio majority/total	0.689	0.573	0.733

Table 2 shows how the data are distributed between the two classes for each of the tasks. Task 2 has a roughly even division into target types, as one might expect given the observation that the data are near-normally distributed so that roughly as many examples will be above the mean as below it. This is a favourable situation for a classifier. The situation for the other two tasks, especially the third, is less favourable as there is a clear majority class.   There is usually a tendency to assign all examples to a strong majority class; however in the current work this problem will be seen to be alleviated by the use of the Matthews correlation coefficient, described in section 2.2, as a training fitness measure.

3.3   Classifier Training Methodology and Assessment of Results

Each experiment consisted of ten runs, with each run consisting of the following:

- The entire data set is randomised and divided into five non-overlapping but roughly equally sized subsets which will be the five test sets for this run;

- For each such division the remaining four-fifths of the data is itself divided into five non-overlapping subsets and fivefold validation carried out, with the *gbest*s at each iteration also assessed on the validation data, the *gbest* weight set saved from each training/validation split being that which did best on the validation set;
- The five weight sets thus selected are used as a committee to classify the test set by taking an average of their outputs.

It was discovered that using multilayer nets as swarm/ensemble members did not improve classification performance on the validation sets. This is a not unexpected result given the amount of noise typically present in financial data, and was useful in that with single layer nets the magnitudes of input layer weights could then give an indication of the degree to which factors were involved in the classifier decision process (see Section 4.1), but it is an open question whether a larger amount of data would have allowed the use of more complex networks and it is planned to repeat these experiments with a larger number of companies and over a longer time period.

The PSO algorithm used 500 training iterations and a swarm size of 25 particles, with cognitive and social learning factors given by $\varphi_1 = \varphi_2 = 2$, with decreasing inertia weight W in the range [1, 0.2]. These settings would be considered good general-purpose choices in the PSO literature and were not optimised for this particular task.

Two fitness criteria were used for training and validation, the Matthews correlation coefficient (MCC) as described in Section 2.2 and in Section 4.1—where a comparison between training according to these two fitness criteria is carried out—also the root mean squared error (RMSE).

Test results (each of the 701 data points during a given run being used as a test point once and only once) were assessed according to two measures: the MCC and the *normalised percentage better than random* (NPBR), defined as

$$NPBR = \frac{(n_{11} + n_{00}) - R_{total}}{t - R_{total}} \times 100 , \tag{3a}$$

where $t = n_{00} + n_{01} + n_{10} + n_{11}$ (these confusion matrix entries as defined in Section 2.2) and R_{total} is the number of correct class assignments that could be expected by chance

$$R_{total} = \frac{(n_{11} + n_{01})(n_{11} + n_{10}) + (n_{00} + n_{01})(n_{00} + n_{10})}{t} \tag{3b}$$

For each experiment and assessment criterion (MCC, NPBR) considered, a mean value over the ten runs and standard deviation will be given.

4   Results

4.1   Predicting Volatility and Risk on the Basis of Company Performance

This section will consider whether, on the basis of the thirteen factors of Table 1, stock price behaviour during a given period could have been categorised into a more or less volatile class, and also whether this is better done using the MCC as the training fitness measure or the more usual RMSE. If a set of financial factors can be used in this way it is reasonable to regard those factors as an at least partial explanation for the excess volatility. We also assume a more compact explanation is preferred and will therefore in the following section consider two different ways to select those factors that appear most important and test this by retraining the system with progressively enlarged factor subsets (from highest to lowest ranked).

Table 3 summarises the test results obtained using both forms of training. As can be seen for Tasks 1 and 2 there is a 45-50% better than random assignment of test data to the two volatility classes. Task 3 is clearly more difficult; however the results for MCC-based training are considerably better than those from RMSE-based training. Comparing the NPBR values, p-values for Tasks 1 and 2 are < 0.01 and > 0.1, respectively, while for Task 3 $p < 0.001$. Thus for this last task individually the difference between MCC-based and RMSE-based training is statistically significant, and since combining the p-values for the three tasks gives $p < 0.0003$ it can also be said that overall MCC-based training is superior. However it should be noted that the bulk of the advantage derives from Task 3, the case in which positive and negative classes are most clearly unbalanced in size (Table 2), as one would expect given that the MCC measure specificially penalises excess assignment to a majority class.

Table 3. Performance on test data with ensemble classifiers trained either to maximise the Matthews correlation coefficient (MCC) or to minimise root mean squared error (RMSE). For each method both MCC and normalised percentage better than random (NPBR) are given.

	MCC-based training		RMSE-based training	
	MCC	NPBR	MCC	NPBR
Task 1	0.51±0.01	50.87±0.88	0.48±0.02	47.98±2.58
Task 2	0.45±0.02	45.26±1.96	0.45±0.02	45.17±1.52
Task 3	0.20±0.02	19.77±2.41	0.14±0.03	9.74±2.11

4.2   Ranking of Financial Factors Characterising Volatility and Risk

Section 4.1 has demonstrated the financial factors of Table 1 do provide information useful in characterising volatility; in this following section some of those factors most important for each of the tasks will be identified and discussed.

The first basis for factor selection, designated as 'WGT', will be the magnitude of the ensemble-averaged weights, where the ensembles are as described in Section 3.3; given that single layer nets are used it is not unreasonable to use such averaged weight magnitudes as an indicator of the degree to which factors are used. Selections made on this basis will be compared to those from a second method, 'XCR', in which factors are selected on the basis of single-variable input-output correlation.

Table 4 shows the most-to-least significant (left-to-right) ranking of input factors for the three tasks using both methods described above. Figs. 1 and 2 show how classification improves as factors are added, beginning with the most significant (highest magnitude averaged weighting or cross correlation value).

Table 4. Ranking of input factors for the three tasks according to the methods WGT and XCR. Leftmost factors are the most significant, with black-on-white indicating a positive weight or correlation, white-on-grey a negative value.

Task	Method												
Task 1	WGT	3	11	2	12	7	13	1	5	6	8	10	9	4
	XCR	1	7	6	3	2	12	9	11	13	8	10	4	5
Task 2	WGT	3	11	2	7	12	6	1	5	13	9	8	4	10
	XCR	7	1	6	2	12	3	9	13	5	8	4	11	10
Task 3	WGT	7	6	11	3	13	4	12	2	9	1	5	10	8
	XCR	7	6	11	12	13	1	10	5	2	3	9	4	8

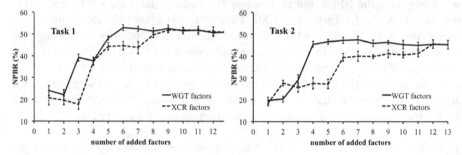

Fig. 1. Test data performance in terms of NPBR for Task 1 (left) and Task 2(right) as input factors are added in the orders given in the first four rows of Table 4

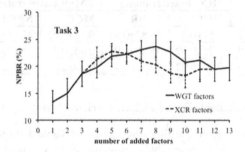

Fig. 2. Test data performance in terms of NPBR for Task 3 as input factors are added in the orders given in the last two rows of Table 4

From Figs.1 and 2, it appears that factors singled out by WGT are more useful in understanding the sources of excess volatility than those obtained by looking at individual input/output correlations (XCR), and that relatively few factors are needed in order to achieve the best obtainable classification performance (six factors for Task 1, four for Task 2, and eight for Task 3). A combination of Fig.1 and Table 4 yields a profile of a company with a more volatile one-year stock behaviour: among other characteristics these tend to be small (negative weight on factors 6 and 7 (6-,7-)), to not pay dividends (3-), but to generate high earnings (positive weight on factor 2 (2+)) and be consequently followed by the market (11+). A combination of Fig. 2 and Table 4 gives a comparable profile of a company with high firm-specific risk: small (6-,7-), not paying dividends (3-), but in this case not generating high earnings (4-) and so not well followed by the market (11-). The difference between these profiles is important as a volatile stock more clearly indicates an investment opportunity while a company with high firm-specific risk should be approached with caution.

5   Discussion

It has been demonstrated that for numerically unbalanced data sets there is an advantage in training an ensemble classifier with the Matthews correlation coefficient as a performance measure, something not feasible using a gradient descent method but achievable using instead particle swarm optimisation. Sufficiently good results were obtained to allow an investigation of the involvement of various financial factors in characterising high stock price volatility and firm-specific risk, and preliminary conclusions were drawn on the basis of associated weight magnitudes about the typical profile of companies exhibiting these characteristics. Future work will investigate the use of hybrid binary/continuous PSO to discover significant factors and the value of these characteristics as indicators of future stock returns.

References

1. Kennedy, J., Eberhart, R.: Particle Swarm Optimization. In: IEEE International Conference Symposium on Neural Networks, pp. 1942–1948. IEEE Press, New York (1995)
2. Poli, R.: An Analysis of Publications on Particle Swarm Optimisation Application. Technical report, Department of Computer Science, University of Essex (2007)
3. Banks, A., Vincent, J., Anyakoha, C.: An Review of Particle Swarm Optimization. Part II: Hybridisation, Combinatorial, Multicriteria and Constrained Optimization, and Indicative Applications. Natural Computing 7, 109–124 (2008)
4. Gudise, V.G., Venayagamoorthy, G.K.: Comparison of Particle Swarm Optimization and Backpropagation as Training Algorithms for Neural Networks. In: Swarm Intelligence Symposium, pp. 110–117. IEEE Press, New York (2003)
5. Lee, J.-s., Lee, S., Chang, S., Ahn, B.-H.: A Comparison of GA and PSO for Excess Return Evaluation in Stock Markets. In: Mira, J., Álvarez, J.R. (eds.) IWINAC 2005. LNCS, vol. 3562, pp. 221–230. Springer, Heidelberg (2005)
6. Matthews, B.W.: Comparison of the Predicted and Observed Secondary Structure of T4 Phage Lysozyme. Biochim. Biophys. Acta 405, 442–451 (1975)
7. Ang, A., Hodrick, R.J., Xing, Y., Zhang, X.: The Cross-Section of Volatility and Expected Returns. Journal of Finance 57, 259–299 (2006)
8. Fabozzi, F.J., Markowitz, H.M.: Multifactor Equity Risk Models. Wiley, NJ (2011)

Fuzzy Classification-Based Control of Wheelchair Using EEG Data to Assist People with Disabilities

Uvais A. Qidwai and Mohamed Shakir

Computer Science and Engineering Department, Qatar University, Qatar
{uqidwai,shakir}@qu.edu.qa

Abstract. Electroencephalography (EEG) will play an intelligent role our life: especially EEG based health diagnosis of brain disorder and Brain-Computer Interface (BCI) are growing areas of research. However, these approaches fall extremely short when attempting to design an automatic detection system and to use the same in BCI framework. The situation becomes even more difficult when the measurement system is being designed for a ubiquitous application for supporting people with disabilities, in which the patient is not confined to the hospital and the device is attached to him/her externally while the person is involved in daily chores. This paper presents a classification technique for one such system which is being built by the same team. Hence the presented work covers the initial findings related to some of the brain conditions in different scenarios that can be monitored in this setting and the detection system can produce control signals for the wheel chair movements that can be conveyed. Due to the compact nature of such systems, the detection and classification techniques have to be extremely simple in order to be stored in the small memory of the microcontroller of the ubiquitous system. The paper presents one such technique which is based on Fuzzy classifications of the EEG data using certain statistical features from the signal.

Keywords: Fuzzy Classifier, Wearable Devices, Electroencephalography (EEG), Ubiquitous computing, Motor controller, Brain Computer Interface, wheel chair Control.

1 Introduction

Detection of commands from EEG of people with disabilities is extremely important and useful since it can help them to achieve control over a number of aspects of things in their life such as mobility in a wheel chair. Many methods have been developed for the detection of EEG control signal in controlled environment. In this paper, different control "thought" EEG signals have been detected in various practical scenarios. These scenarios include different types of background activities such as talking over a phone, short thought vs. long thoughts, and background music, etc. and have been used with Fuzzy classifiers in order to obtain a better classification/detection of a specific type of thought-command. This will help in developing a more generalizable solution as a low cost wearable EEG monitoring and control system for the people with disabilities. Wheelchair control is a subset of such a system. A significant

T. Huang et al. (Eds.): ICONIP 2012, Part IV, LNCS 7666, pp. 458–467, 2012.

advantage of this system is that all the filtering and preprocessing is done by the main sensory unit, Emotiv headset and the SDK [1]. This helps in simplifying the process and reduces the cost in the initial stages of such study.

From initial demonstrations of EEG single-neuron-based device control, researchers have gone on to use electroencephalographic, intra-cortical, electrocorticographic, and other brain signals for increasingly complex control of cursors, robotic arms, prostheses, wheelchairs, and other devices [2]. A conventional approach for the movement of wheelchair for people with disabilities was to equip the wheel chair with sensors, joysticks, keyboard, switches to perform control, obstacle detection and localization [3]. When it comes to more advanced controllers, tools like, Image processing-Virtual computing, multi-channel BCI were used. As a part of BCI, statistical pattern recognition, Type-2 Fuzzy Logic, spatial analysis, artificial intelligence was implemented and tested [4-8]. When it comes to Type-2 fuzzy logic systems (T2 FLSs), it has been only used for modeling uncertain data, Interpretability and transparency of models are obtained with this approach [4]. In some methods, both frequency and amplitude information from an epoch of the EEG signal are extracted into a vector that is then compared to previously taught vectors representing the canonical features. When it comes to any fuzzy logic technique, the membership functions are calculated in each epoch; where the maximum degree of membership is scored and classified. This was implemented in software using the C programming language[9]. Some system utilizes signals from parietal region of the human brain to determine the direction of the wheelchair navigation [10]. Some devices follow the clinical protocol where the EEG signals are synchronized with external cues, using one of the common event-related potentials [11]. In some of the system, as the EEG signals are detected, spatial resolution is reduced while the noise level is increased. One of the main challenges of EEG-based BCI is insufficient decoding accuracy due to the low signal-to-noise ratio (SNR) of EEG signals, which are generated by synchronous activity of millions of cortical neurons[12]. In all the methodologies mentioned above require huge computational capabilities and would be very difficult to implement in near realtime ubiquitous system , moreover it will add cost especially if we intend to develop cost effective systems. Hence in this proposed system , we introduce low cost EEG controlled wheelchair by using Fuzzy logic interface to classify different directions on the wheelchair.

2   Overview of the System

A new system for this purpose is being built by the authors and their team to alleviate the existing system from the limitations related to the mode of on-board processing, range of access to the healthcare facilities, and correct control action. This paper presents one of the techniques in this context that has been implemented for detecting the control commands for the wheel chair locomotion using EEG signal in different environments and cases. Specifically, less noisy environment, short thought of command, long command thought, command thought while in another action and command action while talking. This gives a very good data EEG sets to evaluate the responsiveness of the system for Stop, Front, Back, Right and Left commands.

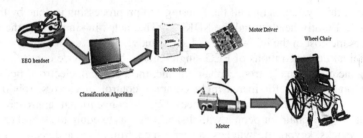

Fig. 1. Overview of the System

3   Experimental System

The intension of this experimental setup is to generate the waveform of EEG using an commercially available cheap EEG headset to product waveforms of different command signal from the subject related conditions and analyzing the waveform before trying to control the wheel chair . Although the research interface software from emotiv is capable of obtaining 14- lead data from this headset, but the actual wearable system will only have two channel of data and hence the acquired data from this experimental system was also restricted to the F8 and FC6 channel only.

Since the aim of the system is to control the wheel chair by EEG signal using fuzzy logic controller, in the first place we need to tap the EEG signal from the brain. For this purpose we are using Emotiv EPOC EEG research headset. This is used to identify a particular signature for a particular emotion, feelings or command in real-time to classify those brain patterns. The Research Edition SDK includes an Emotiv EEG Neuro headset: a 14 channel (plus CMS/DRL references, P3/P4 locations) high resolution, neuro-signal acquisition and processing wireless Neuro headset. Channel names based on the International 10-20 locations are: AF3, F7, F3, FC5, T7, P7, O1, O2, P8, T8, FC6, F4, F8, AF4. By exploiting this feature, we have used the research SDK for tapping the signal in different condition sets and export it to an excel file.

Once this is done, the file is accessed from MATLAB. We have found that from all the fourteen channels, the channel at the right fronto-temporal which is F8 and FC6 have shown good response. Hence we have taken data from this two channels and applied distribution techniques like Mean, Variance, Sum of elements, Standard deviation, Skewness, Kurtosis and Entropy. for all the data sets, we got a resultant matrix of [6] after applying all these filters.

We have found that Skewness, Kurtosis and Entropy showed better dynamics and variation with various command inputs from the EEG signal. This will help in determining the difference between the command signals. In this case, we have eliminated FC6 data and only considered F8 probe since it was also showing significant dynamics for all the cases listed in table 1.

Table 1. A rule base of TSK model

Control action command	State Function1 Command at less Noisy Environment	State Function2 Short command Thought	State Function 3 Long command Thought	State Function 4 Command thought while other action	State Function 5 Command thought while talking
STOP	✔	✔	✔	✔	✔
FRONT	✔	✔	✔	✔	✔
BACK	✔	✔	✔	✔	✔
RIGHT	✔	✔	✔	✔	✔
LEFT	✔	✔	✔	✔	✔

4   The EEG Signals

The patient was asked to wear the EEG probe headset for tapping the signal from the brain and sending the signal wirelessly to the EEG SDK in the computer. this file is then saved as an excel sheet. This this file is then accessed from the Matlab. The physical situations of the patient was in an sitting posture waveform, has been used as the reference state of the subject in the presented work.

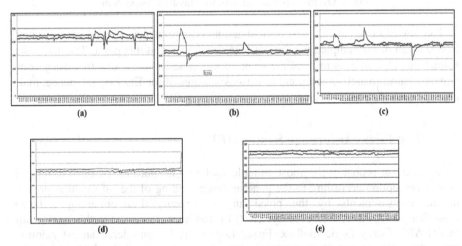

(a)   (b)   (c)

(d)   (e)

Fig. 2. EEG Signals of a patient: from probes F8 and FC6 (a) Back thought while talking, (b) Front thought while talking, (c) Left thought while talking, (d) Right thought while talking & (e) Stop thought while talking.

The high amount of noise indicates the electronic sampling and other such noise sources and the waveforms represent very raw form of the EEG dataset without any processing. The five control command cases selected for classification and identification.

5   The Algorithm

The Emotiv headset probe gives the raw data; in order to test our algorithm's capability to clearly distinguish the command signal even if the noise exist, will be one of the challenges for this paper. Figure 3 shows various components of the main algorithm presented in this paper for type classification for the EEG signals shown in Figure 2 (a)-(e).

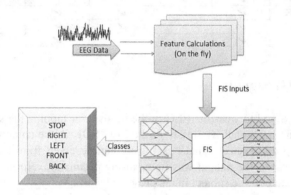

Fig. 3. Overall block diagram of the classification system

The raw EEG signals are first normalized and de-trended as a standard procedure. Hence mean of the sampled window of data is calculated at this point. The main idea behind this algorithm is to have the signal decomposed into various frequency components that ultimately correspond to various features of the EEG signal, hence, making it possible to classify them.

6   The Fuzzy Inference System (FIS)

Fuzzy logic is probably the most suitable tool for the purpose of quantifying the human perception according to the general understanding of the information present in the measurements by the physician or surgeon, thus obtaining a better classification. A Fuzzy Inference System (FIS) has been developed in this work using MATLAB's Fuzzy Logic toolbox. Fuzzy Logic may be considered an extension of binary logic theory that does not require crisp definitions and distinctions. Hence, the developed FIS is not only innovative in terms of its structure and functionality and its application in clinical practices, but also very powerful since it translates the heuristics from human experts into tangible quantitative data and consequently into useful estimates.

6.1   Feature Selection

Since the ultimate technique's usage is in a wheel chair system, hence, there is a need for measures or Features that can be calculated quite quickly and recursively. This led to the decision to use the statistical measures only instead of geometrical measures as used by most of the automated detection algorithms. As mentioned before, a set of 7 different features were tried out with various waveforms obtained from the filter banks. The features used are: 1. Mean, 2.Standard Deviation, 3.Variance, 4.Energy (sum squared values), 5. Skewness, 6. Kurtosis & 7.Entropy. Figure 4(a)-(c) shows the values of various waveforms from the filter bank. Based on the selectivity of various measures, it can be seen from Figure 4 (a)-(c) that certain measures are good in classifying the five control command signal from the EEG. After applying all the features to the EEG signals which defines all the control command on F8 and FC6 signal, we have found that features 5,6 and 7 showed better dynamic variation for all the 25 cases. This helped the system to further simplify the algorithm.

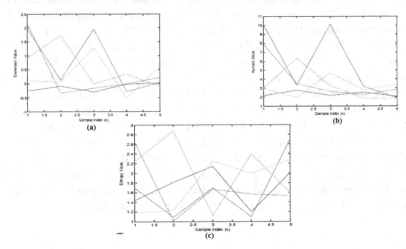

Fig. 4. Feature selection strategy: Plot of the feature values for filtered outputs from by using (a) Skewness feature (b) Kurtosis feature & (c) Entropy, for all the control command through cases

As can be seen that the features 5, 6, and 7 occur most frequently and hence were used for the feature space. Also, for the selected four classes of the outputs, it was found that only the EEG signal and the three feature selection technique output would be sufficient for the required classification.

The proposed FIS is shown in Figure 5, and is composed of the following:

- Three input membership descriptors (representing the feature space; SF0, SF1 and KF0). Here SF0 represent the Skewness , SF1 represents Kurtosis of the original signal, and KF0 represents Entropy.
- Five output membership descriptors for the control cases (STOP, FRONT, BACK, RIGHT and LEFT) and
- A set of 6 rules that represents the heuristical combination of the membership functions with historical understanding of the human user in the domain under study.

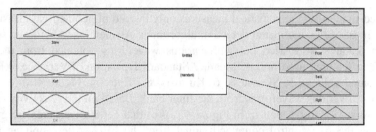

Fig. 5. The FIS structure for the ECG classifier

Figure 6(a)-(c) represents the input membership functions with individual grouping. First of all the data from the training signals was collected under each category of the five features. Then a Fuzzy c-mean clustering algorithm was applied to find feasible boundaries between these classes. These boundary values are then used in each class of input as Stop, Right, Left, Front, back and stop memberships so that forming rules can be accomplished easily. In each membership distribution, the boundary value represents the middle of the trapezoidal function with 10% gradient fall on either side. Hence, each of these degrees can now be represented by a mathematical function that will map the input value of the feature with its functional weights to produce the fuzzified version of the input data. The Output variable of the FIS (Figure 7) corresponds to the three degrees of membership representing five command control. Each one of these memberships is evenly distributed triangular distributions corresponding to "Un-likely", "Likely", and "Highly-likely" degrees. The x-axis in all of the curves shown in Figures 6 represents the input values for each input membership function. The y-axis is from 0 to 1 representing the overall probabilistic space.

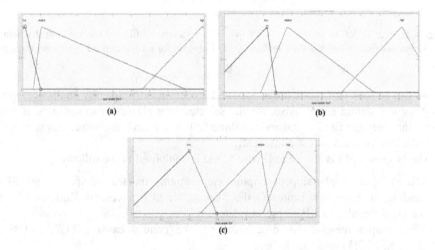

Fig. 6. Input Membership functions; (a) SF0, (b) SF1 and (c) KF0

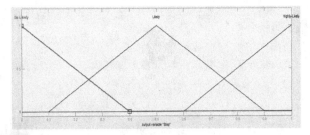

Fig. 7. Output Membership function for the class "Stop". Other three classes are also the same in structure.

6.2   Rule Base

A set of 6 rules was formed based on typical visual heuristics. This rule-base utilizes the membership degrees and their underlying values mapped from the membership functions to perform Boolean Logical inference for a particular set of inputs. For each rule, a decision bar is generated which, when combined with the other rules in a similar way, constitutes a decision surface. This set of rules is an intuitive collection of antecedents and their consequents which most physicians will agree with. Each rule represents a collection of possible heuristics combined together using AND operation. None of the rules actually utilize any form of statistical or algebraic decision boundary. These rules are shown below:

Table 2. Rule-base shown in the form of a Matrix

Skewness	Kurtosis	Entropy	Output
Low	Medium	None	Back
Medium	Medium	Medium	Left
Medium	Medium	High	Right
None	Medium	High	Stop
Low	Medium	Medium	Front
Low	High	Medium	Front

The actual rules can be read from the above matrix directly. For instance, the first rule can be stated in words as: "If (Skewness is low) and (Kurtosis is Medium) then ("BACK" is Highly Likely)". Obviously, there can be other possibilities or other combinations of these memberships for other output characterizations but were not exhaustively tested as part of the presented work. Once these rule-base implications are established, an overall decision surface is pre-calculated. For each set of input values, the centroid is calculated for the area of this decision surface that overlaps with the decision rules applicable to the input memberships. The centroid value is an indicator of the degree to which the inputs correspond to the rule-base and consequently provide a number that depicts the degree of output. The resulting decision surfaces are multidimensional and cannot be displayed as one hyper surface. However, some subset decision surfaces can be plotted and are shown in Figure 8.

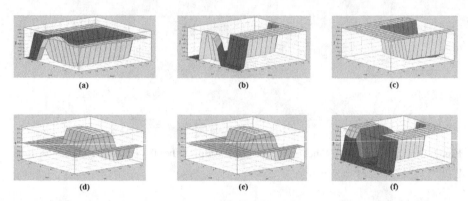

(a)   (b)   (c)

(d)   (e)   (f)

Fig. 8. Decision surface for (a) "BACK" between "Kurtosis" and "Skewnes", (b) "FRONT" between "Entropy" and "Skewness", (c) "FRONT" between "Kurtosis" and "Entropy", (d) "LEFT" between "Kurtosis" and "Entropy", (e) "RIGHT" between "Entropy" and "Skewness", and (f) "STOP" between "Entropy" and "Skewness"

The results have shown almost 100% correct detection for the five control command cases in various environment. Other cases can be similarly incorporated in the rule base and hence into the whole classification system.

7   Controlling the Wheel Chair

After the fuzzy classification is made, and when a clear differentiation is made, the data is passed to the LabVIEW function to call the variables to know the control command.

8   Conclusion

In this paper, an innovative strategy is presented to perform computationally low cost classification for the EEG signals. This technique can be used in real-time classification from the EEG data as it arrives. Hence the system can be used as classifier as well as a predictor for certain epileptic disorder conditions and can be enhanced to clinical applications in the future.

Actual hardware implementation of this system is underway for a small form factor which will make it feasible for wearable EEG monitoring and control system. While all the constituent modules are not really new, however, the overall strategy presented here is quite unique and has shown results which are very interesting and showing the usefulness of the technique. Detecting the brain conditions (for certain targeted commands) on the fly (i.e., as the data samples arrive) is an extremely embedded design. However, with the presented strategy, one can isolate the most useful part of the signal can be indirectly detected and hence it can be further exploited for predicting the successive conditions based on the current data.

Acknowledgment. This publication was made possible by a grant from Qatar National Research Fund under its National Priority Research Program, for projects NPRP 09-292-2-113. Its contents are solely the responsibility of the authors and do not necessarily represent the official views of Qatar National Research Fund.

References

1. http://www.emotiv.com/store/sdk/eeg-bci/research-edition-sdk/
2. Shih, J.J., Krusienski, D.J., Wolpaw, J.R.: Brain-Computer Interfaces in Medicine. Mayo Clinic Proceedings 87(3), 268–279 (2012)
3. Rebsamen, B., Burdet, E., Guan, C., Zhang, H., Teo, C.L., Zeng, Q., Laugier, C., Ang Jr., M.H.: Controlling a Wheelchair Indoors Using Thought. IEEE Intelligent Systems 22(2), 18–24 (2007)
4. Herman, P., Prasad, G., McGinnity, T.M.: Investigation of the Type-2 Fuzzy Logic Approach to Classification in an EEG-based Brain-Computer Interface. In: 27th Annual International Conference of the Engineering in Medicine and Biology Society, IEEE-EMBS 2005, January 17-18, pp. 5354–5357 (2006)
5. James, C.J., Jones, R.D., Bones, P.J., Carroll, G.J.: Spatial analysis of multi-channel EEG recordings through a fuzzy-rule based system in the detection of epileptiform events. In: Proceedings of the 20th Annual International Conference of the IEEE Engineering in Medicine and Biology Society, October 29- November 1, vol. 4, pp. 2175–2178 (1998)
6. Riddington, E.P., Wu, J., Ifeachor, E.C., Allen, E.M., Hudson, N.R.: Intelligent enhancement and interpretation of EEG signals. In: IEE Colloquium on Artificial Intelligence Methods for Biomedical Data Processing, April 26, pp. 11/1–11/7 (1996)
7. Herman, P., Prasad, G., McGinnity, T.M.: Design and on-line evaluation of type-2 fuzzy logic system-based framework for handling uncertainties in BCI classification. In: 30th Annual International Conference of the IEEE Engineering in Medicine and Biology Society, EMBS 2008, August 20-25, pp. 4242–4245 (2008)
8. Rao, R.P.N., Scherer, R.: Chapter 10 - Statistical Pattern Recognition and Machine Learning in Brain–Computer Interfaces. In: Oweiss, K.G. (ed.) Statistical Signal Processing for Neuroscience and Neurotechnology, pp. 335–367. Academic Press, Oxford (2010)
9. Hu, J., Knapp, B.: Electroencephalogram pattern recognition using fuzzy logic. In: 1991 Conference Record of the Twenty-Fifth Asilomar Conference on Signals, Systems and Computers, November 4-6, vol. 2, pp. 805–807 (1991)
10. Kaneswaran, K., Arshak, K., Burke, E., Condron, J.: Towards a brain controlled assistive technology for powered mobility. In: 2010 Annual International Conference of the IEEE Engineering in Medicine and Biology Society (EMBC), August 31-September 4, pp. 4176–4180 (2010)
11. Iturrate, I., Antelis, J., Minguez, J.: Synchronous EEG brain-actuated wheelchair with automated navigation. In: IEEE International Conference on Robotics and Automation, ICRA 2009, May 12-17, pp. 2318–2325 (2009)
12. Huang, D., Qian, K., Fei, D.-Y., Jia, W., Chen, X., Bai, O.: Electroencephalography (EEG)-Based Brain–Computer Interface (BCI): A 2-D Virtual Wheelchair Control Based on Event-Related Desynchronization/Synchronization and State Control. IEEE Transactions on Neural Systems and Rehabilitation Engineering 20(3), 379–388 (2012)
13. http://www.robokits.co.in/documentation/Dual_DC_Motor_Driver_20A.pdf

EEG-Based Emotion Recognition in Listening Music by Using Support Vector Machine and Linear Dynamic System

Ruo-Nan Duan[1], Xiao-Wei Wang[1], and Bao-Liang Lu[1,2,3,4,*]

[1] Center for Brain-Like Computing and Machine Intelligence
Department of Computer Science and Engineering
[2] MOE-Microsoft Key Laboratory for Intelligent Computing and Intelligent Systems
[3] Shanghai Key Laboratory of Scalable Computing and Systems
[4] MOE Key Laboratory of Systems Biomedicine
Shanghai Jiao Tong University, 800 Dongchuan Road, Shanghai 200240, China
bllu@sjtu.edu.cn

Abstract. This paper focuses on the variation of EEG at different emotional states. We use pure music segments as stimuli to evoke the exciting or relaxing emotions of subjects. EEG power spectrum is adopted to form features, power spectrum, differential asymmetry, and rational asymmetry. A linear dynamic system approach is applied to smooth the feature sequence. Minimal-redundancy-maximal-relevance algorithm and principal component analysis are used to reduce the dimension of features. We evaluate the performance of support vector machine, k-nearest neighbor classifiers and least-squares classifiers. The accuracy of our proposed method reaches 81.03% on average. And we show that the frequency bands, beta and theta, perform better than other frequency bands in the task of emotion recognition.

Keywords: emotion recognition, electroencephalogram, power spectrum.

1 Introduction

Emotional states significantly affect the cognition and behaviors of human. Scientific findings suggest an increasingly large number of important functions of emotion [1]. Emotion recognition based on information technology is an important research topic in the field of neural engineering.

Most previous studies on emotion recognition focused on extrinsic signals, such as facial expressions [2] and voice [3]. However, human may modify their appearance deliberately, and some disabled persons can not express their emotion in these extrinsic ways. Detecting emotional states of people through their physiological signals can solve this problem in some degree. Common approaches to recognize emotion based on physiological signals include detecting emotional states

---

* Corresponding author.

T. Huang et al. (Eds.): ICONIP 2012, Part IV, LNCS 7666, pp. 468–475, 2012.

through electroencephalography (EEG), electrocardiogram (ECG), electromyogram (EMG), skin resistance (SC), skin temperature, pulse and respiration signals. EEG signal is a typical central nervous system signal, which relates to emotion activities more closely than automatic nervous system signal, such as ECG and EMG signal.

Recently, many institutions have started to study the EEG-based emotion recognition. Aftanas *et al.* made use of Fourier transform to map the original EEG signal to theta, alpha and beta frequency bands, and calculated power spectral density of each electrodes as features to recognize emotional states [4]. Lin *et al.* classified emotional states into four categories using support vector machine, and found the 30 most relevant EEG features to the emotion recognition [5]. Nie *et al.* figured out the key brain regions and frequency bands of EEG signals according to the correlation coefficients of frequency-domain features in the emotion recognition task [6].

In this paper, we did experiments to collect EEG signals of subjects when they were listening to music with exciting or relaxing emotional elements. And we compared the performance of three kinds of frequency-domain features in five frequency bands in emotion classifying task. We adopted linear dynamic system (LDS) approach to remove the noise of the features, and employed support vector machine (SVM) as the classifier. At last, minimal-redundancy-maximal-relevance criterion (MRMR) and principal component analysis (PCA) were applied to reduce the dimension and speed up the classifying procedure.

2   Experiment

2.1   Stimuli

In order to evoke the emotion of the participants, we chose to use pure music segments in specific emotion style as stimuli, such as *Kiss The Rain* for relaxing emotion and *He's A Pirate* for exciting emotion. We used music with no lyrics for the reason that we had found that songs with lyrics in different languages impact the EEG of participants in different degrees [7]. We selected 16 pieces of pure music, each of them lasting 3 minutes.

2.2   Subjects

Five right-hand students were chosen as subjects who were all from Shanghai Jiao Tong University, aged between 21 to 24 years old, 3 males and 2 females. They had no history of brain damage and mental illness before. We informed them of the purpose and procedure of the experiment, and that the EEG recording device was harmless.

2.3   Procedure

Before the experiment, we recorded the information of the subjects and the environment. We played the stimuli using Stim 2 software, and ESI NeuroScan

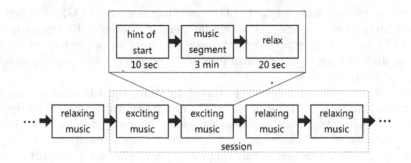

Fig. 1. Procedure of the stimuli playing

System with a 62-channel electrode cap was employed to record the EEG signals of subjects. The sampling rate was 500Hz. Those music segments were divided into 4 sessions, each session containing 2 pieces of exciting music and 2 pieces of relaxing music. We played those music segments in the order shown in Fig. 1.

3   Methods

3.1   Preprocessing

In order to speed up the computation, we down-sampled the EEG data with sample frequency 200Hz. Then we removed the electromyogram and other artifacts manually.

3.2   Feature Extraction

We used frequency-domain features and their combinations as features. At first, the 512-point short-time fourier transform (STFT) with a non-overlapped Hanning window of 1s was applied to the EEG data from each of the 62 channels to compute power spectrum. Then we averaged the spectral time series into five frequency bands (delta: 1-3Hz, theta: 4-7Hz, alpha: 8-13Hz, beta: 14-30Hz, gamma: 31-50Hz), and combined them to form 3 kinds of features named PS (power spectrum), DASM (differential asymmetry), RASM (rational asymmetry). We chose Fp1, F7, F3, FT7, FC3, T7, P7, C3, TP7, CP3, P3, O1, AF3, F5, F7, FC5, FC1, C5, C1, CP5, CP1, P5, P1, PO7, PO5, PO3, CB1 of the left hemisphere, and Fp2, F8, F4, FT8, FC4, T8, P8, C4, TP8, CP4, P4, O2, AF4, F6, F8, FC6, FC2, C6, C2, CP6, CP2, P6, P2, PO8, PO6, PO4, CB2 of the right hemisphere as the corresponding electrodes .

PS features were consisted of the average power spectrum of 62 scalp electrodes in the five frequency bands. RASM and DASM features were the average power spectrum ratios and differences of those 27 pairs of hemispheric asymmetry electrodes, respectively. Totally, we gain 580 features for one sample, and one sample per second. The number of features is shown in Table 1.

Table 1. Statistics of the feature number

Feature	Delta	Theta	Alpha	Beta	Gamma	Total
PS	62	62	62	62	62	310
RASM	27	27	27	27	27	135
DASM	27	27	27	27	27	135

3.3 Feature Smoothing

The features we extracted above contained many rapid fluctuations. As we all know, emotion usually varies smoothly, and the features relevant to emotional states should change relatively slowly. So the features that changed rapidly have nothing to do with the emotional states and we should smooth the features to remove those noise components.

We used a linear dynamic system (LDS) approach [8] to remove noise. We applied off-line LDS with the window of 20s to smooth the features and to remove noise in some degree.

3.4 Classification

Our classifying task is to use part of the EEG data of one experiment to train the model and the rest data recorded in the same experiment to test it. In practical applications, we can only use the data we have obtained in laboratory to classify the data we record later, so we used the data of the first three sessions to train the model, and the rest one session to evaluate it. Since each session contained 700 to 800 cases, we had about two thousands training samples and seven hundreds test samples.

We classified the EEG features using three kinds of algorithms, k-nearest neighbors (KNN) algorithm, least-squares (LS) classification and support vector machine (SVM). kNN model works in the way that classifying the test case to the most common category of the k nearest neighbors with known category. Least-squares classification tries to find a model that is a linear function with parameters which minimize the sum-of-squares error [10]. SVM projects the features onto another feature space using linear kernel function. Then SVM iteratively approaches the optimal hyperplane which has the maximal margins. We used LIBSVM [9] software to train the SVM classifier.

3.5 Dimension Reduction

To speed up the computing procedure and reduce the complexity of the model, we applied principal component analysis (PCA) and minimal-redundancy-maximal-relevance (MRMR) algorithm to reduce the dimension of the features. PCA projects data onto a new space with a lower dimension and maximizes the variance of the projected data. And MRMR algorithm uses a series of intuitive measures of redundancy and relevance to select optimal feature subsets [11].

4   Results

4.1   The Effect of the LDS Approach

We applied the LDS approach to smooth the features and remove the noise. A piece of feature sequence is cut to show the performance of LDS (Fig. 2).

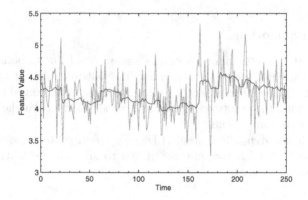

Fig. 2. Performance of LDS smoothing on feature sequence. The green curve represents the original feature values, and the red one stands for the feature values after smoothing. Feature sequence shown is part of the 24th PS feature in gamma frequency band of subject 4 in session 1.

In the experiment, we trained and tested SVM models using original features and features after LDS smoothing respectively. The features we adopted to evaluate the performance of LDS smoothing were PS in five frequency bands. The comparison of classification accuracies between original features and smoothed features is shown in Table 2. The accuracies indicate that LDS smoothing helps to increase the accuracy in some degree. And the effect of LDS smoothing becomes significant on unclean data.

Table 2. Classification accuracies of features with and without LDS smoothing

	Subject 1	Subject 2	Subject 3	Subject 4	Subject 5	Average
No smoothing	74.97	70.96	71.84	**98.97**	65.01	76.35
LDS smoothing	**76.48**	**71.95**	**85.86**	98.63	**67.47**	**80.08**

4.2   Performance of Features and Classification Methods

To compare the performance of different frequency bands and different features, we adopted linear-SVM to classify the features after smoothing them. The accuracies of classification is shown in Table 3. It is obvious that the frequency

Table 3. Classification accuracies using different kinds of features

Subject	Feature Name	Delta	Theta	Alpha	Beta	Gamma	Total
1	DASM	57.59	62.22	67.79	74.86	77.75	**84.13**
	RASM	57.82	64.54	56.32	76.38	80.30	**81.34**
	PS	70.10	63.38	72.54	77.98	**82.39**	76.48
2	DASM	59.07	78.72	72.46	77.85	**81.23**	78.60
	RASM	63.95	77.97	52.82	77.35	**80.48**	74.72
	PS	61.58	74.34	70.09	70.34	**74.72**	71.96
3	DASM	78.91	59.79	60.95	65.70	72.42	**84.82**
	RASM	80.53	56.32	57.94	67.09	75.55	**84.36**
	PS	78.68	63.27	59.10	70.10	84.01	**85.86**
4	DASM	60.96	83.45	76.03	98.06	**98.06**	97.95
	RASM	56.51	72.83	59.13	98.06	**98.17**	98.17
	PS	62.67	80.25	89.38	98.40	98.40	**98.63**
5	DASM	64.95	63.93	59.82	**66.78**	57.53	66.32
	RASM	61.99	63.13	64.50	59.36	57.19	**64.61**
	PS	**78.54**	61.19	65.98	65.53	65.30	67.47
Average	DASM	64.30	69.62	67.41	76.65	77.40	**82.36**
	RASM	64.16	67.18	58.14	75.65	78.34	**80.64**
	PS	70.31	68.49	71.42	76.47	**80.96**	80.08

bands beta and gamma perform better than frequency bands delta, theta and alpha. This result is consistent with previous results that EEG signals in high frequency bands are more relevant to the emotional states of human [12]. We find that features, RASM and DASM, gain relatively high classification accuracies with lower feature dimensions. This result reflects that the differences and ratios of EEG signals between left and right hemispheres are related to the human emotion in some degree.

To evaluate the performance of different classifiers, KNN algorithm, linear-squares classification and liner-SVM were used to classify LDS-smoothed EEG features. To compare the performance of the three classifiers, we show the accuracies of classifying PS features in all frequency bands in Table 4. According to the result, KNN classifier and SVM perform better than least-squares classification. The performance of KNN classifier and SVM are similar to each other. However, SVM works much faster than KNN classifier in this task because of the high dimension of the features.

Table 4. Classification accuracies of different classifiers

Classifier	Subject 1	Subject 2	Subject 3	Subject 4	Subject 5	Average
KNN	72.19	71.34	**87.72**	97.83	**69.41**	79.69
LS	**78.45**	62.08	79.49	88.47	59.02	73.50
SVM	76.48	**71.95**	85.86	**98.63**	67.47	**80.08**

4.3 Dimension Reduction

PCA and MRMR were used to reduce the dimension of the features. The original features were the combinations of PS, DASM and RASM in five frequency bands, so the original dimension of the features was 580. After the dimension was reduced to 510, 460, ..., and 10 with the intervals of 50, the performance of SVM classifier changes as shown in Fig. 3.

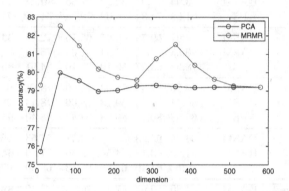

Fig. 3. Classification accuracies on features with different dimensions. The accuracies shown were the average accuracies of the five subjects by using SVM classifier and two reduction methods.

The figure indicated that the dimension reduction using PCA affected the accuracy in a small range when the dimension was larger than 100, whereas MRMR showed the possibility of increasing the accuracy. This increment was caused by the removing of emotion irrelevant features which annoyed the classifier. And this result implies that we can reduce the dimension of features to speed up the calculation.

5 Conclusion

Using the EEG data recorded from subjects when they were in some emotional state evoked by music, we obtained an efficient and stable emotion classifier. The accuracies of classification reached 81.03% on average. In our study, we showed that the frequency bands beta and gamma performed better than other frequency bands in the task of emotion recognition. In addition, PCA, MRMR and LDS can help to increase the speed and improve the stability of the classification procedure, respectively.

Acknowledgments. This research was partially supported by the National Basic Research Program of China (Grant No. 2009CB320901), and the European Union Seventh Framework Programme (Grant No. 247619).

References

1. Picard, R.W., Klein, J.: Toward Computers that Recognize and Respond to User Emotion: Theoretical and Practical Implications. Interacting with Computers 14(2), 141–169 (2002)
2. Anderson, K., McOwan, P.W.: A Real-Time Automated System for the Recognition of Human Facial Expressions. IEEE Transaction on System, Man, and Cybernetics, Part B: Cybernetics 36(1), 96–105 (2006)
3. Bruck, C., Kreifelts, B., Wildgruber, D.: Emotional Voices in Context: a Neurobiological Model of Multimodal Affective Information Processing. Physics of Life Reviews 8, 383–403 (2011)
4. Aftanas, L.I., Lotova, N.V., Koshkarov, V.I., et al.: Non-Linear Dynamic Complexity of the Human EEG During Evoked Emotions. International Journal of Psychophysiology 28, 63–76 (1998)
5. Lin, Y.P., Wang, C.H., Jung, T.P., et al.: EEG-Based Emotion Recognition in Music Listening. IEEE Transaction on Biomedical Engineering 57(7), 1798–1806 (2010)
6. Nie, D., Wang, X.W., Shi, L.C., et al.: EEG-Based Emotion Recognition During Watching Movies. In: Proceedings of IEEE EMBS Conference on Neural Engineering, pp. 667–670. IEEE Press (2011)
7. Shi, S.J., Lu, B.L.: EEG Signal Classification during Listening to Native and Foreign Languages Songs. In: Proceedings of 4th International IEEE EMBS Conference on Neural Engineering, pp. 440–443 (2009)
8. Shi, L.C., Lu, B.L.: Off-Line and On-Line Vigilance Estimation Based on Linear Dynamical System and Manifold Learning. In: Proceedings of 32nd International Conference of the IEEE Engineering in Medicine and Biology Society, pp. 6587–6590. IEEE Press (2010)
9. Chang, C.C., Lin, C.J.: LIBSVM: A library for support vector machines, http://www.csie.ntu.edu.tw/~cjlin/libsvm
10. Bishop, C.M.: Pattern Recognition and Machine Learning. Springer, New York (2006)
11. Peng, H., Long, F., Ding, C.: Feature Selection Based on Mutual Information: Criteria of Max-Dependency, Max-Relevance, and Min-Redundancy. IEEE Transactions on Pattern Analysis and Machine Intelligence 27(8), 1226–1238 (2005)
12. Muller, M.M., Keil, A., Gruber, T., Elbert, T.: Processing of Affective Pictures Modulates Right-Hemispheric Gamma Band EEG Activity. Clinical Neurophysiology 110(11), 1913–1920 (1999)

Analysis of Genetic Disease Hemophilia B by Using Support Vector Machine

Kenji Aoki[1], Kunihito Yamamori[2], Makoto Sakamoto[2], and Hiroshi Furutani[2]

[1] Information Technology Center, University of Miyazaki,
Miyazaki City, Miyazaki 889-2192, Japan
aoki@cc.miyazaki-u.ac.jp
[2] Faculty of Engineering, University of Miyazaki,
Miyazaki City, Miyazaki 889-2192 Japan
{yamamori,sakamoto,furutani}@cs.miyazaki-u.ac.jp
http://www.miyazaki-u.ac.jp/

Abstract. Hemophilia B is a genetic disease resulting from deficiency of factor IX. The database of mutations causing hemophilia B has been developed by the world wide collaboration. Most common mutations are amino acid changing substitutions, which we call missense mutations, and factor IX activity is closely related to the type and position of a substitution. In this study, we examined the relation between clotting level of factor IX and the type of a missense mutation by using Support Vector Machine (SVM). As parameters, we used four physical-chemical parameters of amino acids and a special flag variable representing amino acid cysteine. As a result, EGF(1st) and EGF(2nd) have the relationship in the prediction of serious or slight illness in hemophilia B. Cysteine substitution parameters influence the prediction in Activation region.

Keywords: Hemophilia B, Factor IX, Amino Acid, Support Vector Machine.

1 Introduction

In recent days, it becomes very important to study mutations in genes responsible for diseases. The Human Gene Mutation Database includes mutations causing or associated with human inherited disease (HGMD, http:www.hgmd.cf.ac.uk) [1]. This database was established for the study of mutational mechanisms in human genes. Now, over 250 journals are surveyed for articles describing mutations causing genetic diseases. However, it is a time-consuming, laborious and expensive task to distinguish a disease-causing mutation from neutral ones. To overcome such a problem, computational approaches have been developed by many researchers. We also applied a multiple regression model to predict the effect of a missense mutation in Factor IX gene of hemophilia B patient [2]. As an extension of this work, we carried out the same type of analysis using Support Vector Machine (SVM) [3,4]. We report results obtained in this analysis.

Genetic defects in blood coagulation proteins are associated with a bleeding disorder known as hemophilia. Hemophilia A accounts for about 85% of this

T. Huang et al. (Eds.): ICONIP 2012, Part IV, LNCS 7666, pp. 476–483, 2012.

disorder, while Hemophilia B for 10 − 12 % [5]. Hemophilia B is a hereditary, X-linked, recessive hemorrhagic disorder, caused by various types of mutations in factor IX gene [5,6]. Mutation in factor IX is made up of a majority of point mutation. Substitution of amino acid sequence is the most common form of point mutation. In general, substitution in important site and substitution to different character from original amino acid are supposed to drastic decrease in activity of factor IX. Cysteine, one of amino acids, has different properties from others. Cysteine contains sulfur, thus it makes S-S binding with another sulfur [6].

There have been reported a variety of defects in the factor IX gene from hemophilia B patients, and these are summarized in the hemophilia B database [7]. We analyzed amino acid changing mutations, or missense mutations in the database described with factor IX activity values. Among them, the cases of double and triple mutations and female patients were excluded, We adopted 1494 cases as total. We have introduced distances between 20 amino acids by using the following four physical-chemical properties: 1. Molecular volume, 2. Hydropathy, 3. Polar requirement, and 4. Isoelectric point.

We carried out an analysis of missense mutations in the database by using SVM. In addition, we studied the cases whose wild type amino acid is cysteine. This amino acid plays an important role in determining the structure of factor IX protein.

2   Hemophilia B Database

We adopted Hemophilia B Mutation Database [7] for the analysis. There are 2,891 entries in this database, and 34 double mutations and 1 triple mutation are included. Of the 2,891 patients listed, 962 show unique molecular events probably causing the disease. Most of them are point mutations, and the database contains 561 different amino acid substitutions.

The gene for factor IX contains eight exons and seven introns with an overall size of 33k base-pairs [6]. Factor IX is a glycoprotein of 461 amino acids essential for blood coagulation, and made up of seven regions: (1)Signal peptide, (2)Propeptide, (3)Gla, (4)EGF(1st), (5)EGF(2nd), (6)Activation, and (7)Catalytic. Signal peptide and propeptide are removed during biosynthesis, and remaining protein circulates in the blood as a mature factor IX.

Domain	Location	Number of mutants
Signal peptide	-46 to -18	18
Propeptide	-17 to -1	107
Gla	1 to 46	99
EGF(1st)	47 to 84	138
EGF(2nd)	85 to 127	95
Activation	128 to 195	241
Catalytic	196 to 415	795

Activity of factor IX in a patient's blood depends on a position of the substitution and combination of original and substituting amino acids. Classification of the disease is given by in vitro clotting activity, (1) Severe hemophilia B: <1 % factor IX, (2) Moderate hemophilia B: 1 % – 5 % factor IX, (3) Mild hemophilia B: > 5 % factor IX. We performed SVM discriminative analysis to classify severe hemophilia B from moderate and mild ones.

3   Methods

3.1   Support Vector Machine

Support vector machine (SVM)[3][4] can classify the samples x_i $(i = 1, \cdots, n)$ belonging to unknown class into two classes C_1 or C_2. The classification function $f(x)$ is defined as the Equation (1).

$$f(x) = sign\,(g(x)) = sign(w^t x + b),\qquad(1)$$

where w and b are parameters.

Let x_i belong to the class y_i $(= \{1, -1\})$, and if all the samples are correctly classified, Equation (2) will be satisfied.

$$\forall i,\ y_i \cdot \left(w^t x_i + b\right) - 1 \geq 0.\qquad(2)$$

When Equation (2) is satisfied, no samples exist between the $H_1 : (w^t x + b) = 1$ and the $H_2 : (w^t x + b) = -1$, and the distance between H_1 and H_2, called as *margin*, becomes $\frac{2}{||w||}$. To obtain the maximum margin, we minimize $\frac{1}{2}||w||^2$. In SVM, it is solved by a Lagrange-multiplier method. To maximize the margin, we rewrite the objective function as Equation (3) in subject to Equation (2),

$$L\,(w, b, \alpha) = \frac{1}{2}||w||^2 - \sum_{i=1}^{n} \alpha_i \left[y_i \left(w^t x + b\right) - 1\right].\qquad(3)$$

where $\alpha \geq 0$ denotes Lagrange-multiplier. Partial differentiations of Equation (3) by w and b are substituted for Equation (3), we obtain Equation (4).

$$L(\alpha) = \sum_{i=1}^{n} \alpha_i - \frac{1}{2} \sum_{i,j=1}^{n} \alpha_i \alpha_j y_i y_j x_i^t x_j,\qquad(4)$$

in subject to

$$\forall i,\ \alpha_j \geq 0,\ \sum_{i=1}^{n} \alpha_i y_i = 0.\qquad(5)$$

Here we denote α_i to maximize Equation (4) as α_i^*. The sample x_i with $\alpha_i^* > 0$ is called as Support Vector (SV), it exists on H_1 or H_2. The optimum of w denoted as w^* is obtained from the partial differentiations of Equation (3) and α_i^* by Equation (6).

$$w^* = \sum_{i=1}^{n} \alpha^* y_i x_i.\qquad(6)$$

The optimum of b denoted as b^* is obtained from the Equation (7) with any x_s $(s \in SV)$.

$$b^* = y_s - w^{*t}x_s. \tag{7}$$

Finally, we obtain the discriminant function of SVM for linearly separable problem as Equation (8).

$$f(x) = sign\left(\sum_{i \in SV} \alpha_i^* y_i x_i^t x + b^*\right). \tag{8}$$

We use kernel trick for non-linear problems in usual. The kernel function was radial basis function (RBF) in this study. We used the application program SVM^{light} to calculate the support vector machine [8]. The parameter C which is the trade-off between training error and margin is 1, and γ which is parameter in RBF kernel is 1.

3.2 Variables

We use a distance between amino acids for each four amino acid parameters (Molecular volume, Hydropathy, Polar requirement and Isoelectric poin). The k-th distance between amino acid A_i and A_j is defined as

$$D_{ij}^{(k)} = |f_k(A_i) - f_k(A_j)|, \quad (k = 1, 2, 3, 4). \tag{9}$$

We define that the value of cysteine parameter is 1 in the case of cystaine changed mutation, and 0 in the other case.

3.3 Dataset

We used data of Hemophilia B Mutation Database. The number of data in Activation, in EGF(1st) and in EGF(2nd) are 241, 138 and 95, respectively. These data was divided into training data and test data of SVM in each region. The ratio of training data and test data is 1 : 3. When we analyzed between regions, we used all data in each region. We considered serious illness with less than 1% of factor IX clotting activity, and slight illness with more than 1% of one. We predicted the serious or slight illness of Hemophilia B by SVM based on these data.

4 Results

4.1 Relationship between EGF(1st) and EGF(2nd)

In Fig. 1, Fig. 2, Fig. 3 and Fig. 4, the horizontal axis is the false positive rate, and the vertical axis is the true positive rate. False positive means that the predicted result is positive (serious), but observed result is negative (slight). True positive means that both the prediction result and observed result are positive.

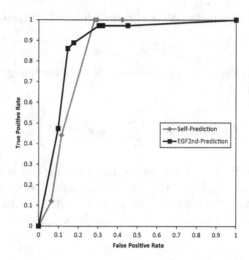

Fig. 1. The relationship between the false positive rate and the true positive rate in EGF(2nd) region. The diamond mark shows the result of prediction using EGF(2nd) region data only. The square mark shows the result of prediction using EGF(1st) region data as the training data and EGF(2nd) region data as the test data.

These figures are called ROC curve. ROC curve is used for a comparison of the inspection performance. The curve which stretched on the upper left are more superior performance.

In Fig. 1, diamond mark shows the result of prediction using EGF(1st) region data only. The square mark shows the result of prediction using EGF(2nd) region data for training and EGF(1st) region data for test. These two curves are similar.

In Fig. 2, the diamond mark shows the result of prediction using EGF(2nd) region data only. The square mark shows the result of prediction using EGF(1st) region data for training and EGF(2nd) region data for test. Although these tow curves are not completely same in a middle range, these are almost similar.

In Fig. 3, the diamond mark shows the result of prediction using Activation region data only. The square mark shows the result of prediction using EGF(2nd) region data for training and Activation region data for test. These two curves are not similar.

In FIg. 4, the diamond mark shows the result of prediction using Activation region data only. The square mark shows the result of prediction using EGF(1st) region data for training and Activation region data for test. This two curves are not also similar .

4.2   Influence of Cysteine

In Fig. 5 the horizontal axis is the false positive rate, and the vertical axis is the true positive rate. This figure shows result of prediction of serious or slight illness in Activation region. The diamond mark shows the result of prediction using 4

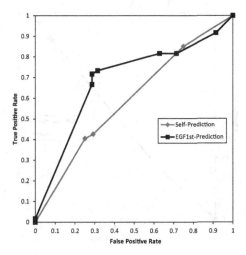

Fig. 2. The relationship between false the positive rate and the true positive rate in EGF(1st) region. The diamond mark shows the result of prediction using EGF(1st) region data only. The square mark shows the result of prediction using EGF(2nd) region data as the training data and EGF(1st) region data as the test data.

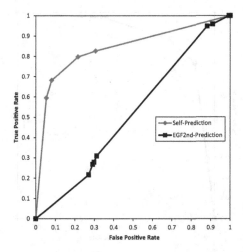

Fig. 3. The relationship between false the positive rate and the true positive rate in Activation region. The diamond mark shows the result of prediction using Activation region data only. The square mark shows the result of prediction using EGF(1st) region data as the training data and Activation region data as the test data.

parameters. The square mark shows the result of prediction using 5 parameters including cysteine parameter. The curve of result using 5 parameters is on more upper left. Therefore, the result using 5 parameters is more superior than the result using 4 parameters. In other region, the noticeable difference is not shown between the 5 parameters result and the 4 parameters result.

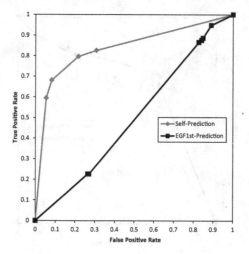

Fig. 4. The relationship between the false positive rate and the true positive rate in Activation region. The diamond mark shows the result of prediction using Activation region data only. The square mark shows the result of prediction using EGF(2nd) region data as the training data and Activation region data as the test data.

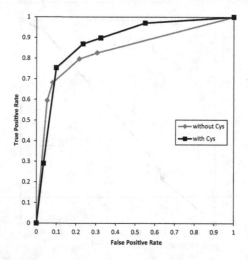

Fig. 5. The relationship between the false positive rate and the true positive rate in Activation region. The diamond mark shows the result of prediction using 4 parameters. The square mark shows the result of prediction using 5 parameters which included cysteine parameter.

5   Discussion

We can observe similar two curves in Fig. 2 and Fig. 1, respectively. In EGF region, we know that there is the highly homology of amino acid arrangement between EGF(1st) region and EGF(2nd) region. Therefore, we suppose that the model of SVM in EGF(1st) region can be applied to the prediction serious or slight illness in EGF(2nd) region, and the reverse is similar. These results supported our supposition. Furthermore, this result might apply to between genetic diseases with similar amino acid arrangements.

Fig. 5 shows that the prediction using the cystaine parameter is more superior in Activation region. Cysteine is the important amino acid to determine the structural arrangement of protein. We thought that cysteine substitution make the serious illness of disease. This result suggests that cysteine relate to the serious or slight illness in hemophilia B. However, we did not observe the similar relation in other region.

6 Conclusion

We analyzed the relatonship between EGF(1st) and EGF(2nd) by Sapport Vector Machine analysis. As a result, EGF(1st) and EGF(2nd) have the relationship in the prediction of serious or slight illness in hemophilia B. This result suggested that the prediction of serious illness is executed between proteins with the similar arrangement of amino acid. In addition, we examined the influence of the cystaine substitution to prediction of serious illness in hemophilia B. In the Activation region, we got the result that cystaine substitution parameters influence the prediction. This suggested that cysteine infrences serious illness of hemophilia B.

References

1. Sternson, P.D., Mort, M., Ball, E.V., Howells, K., Phillips, A.D., Thomas, N.S., Cooper, D.N.: The Human Gene Mutation Database: 2008 update. Genome Medicine 1, 13 (2009)
2. Utsunomiya, M., Sakamoto, M., Furutani, H.: Regression Analysis of Amino Acid Substitutions and Factor IX Activity in Hemophilia B. Artificial Life and Robotics 13, 531–534 (2008)
3. Schölkopf, B., Burges, C.J.C., Smola, A.J.: Advances in Kernel Methods. The MIT Press, London (1999)
4. Schölkopf, B., Tsuda, K., Vert, J.P.: Kernel Methods in Computational Biology. A Bradford Book, The MIT Press, London (2004)
5. Furie, B., Furie, B.C.: The Molecular Basis of Blood Coagulation. Cell 53, 505–518 (1988)
6. Yoshitake, S., Schach, B.G., Foster, D.C., Davie, E.W., Kurachi, K.: Nucleotide Sequence of the Gene for Human Factor IX. Biochemistry 24, 3736–3750 (1985)
7. Haemophilia B Mutation Database: A database of point mutations and short additions and deletions in the factor IX gene. version 13 (2004)
8. Jachims, T.: SVM^{light} Version 6.02 (2008), http://svmlight.joachims.org/

EEG-Based Fatigue Classification by Using Parallel Hidden Markov Model and Pattern Classifier Combination

Hui Sun[1] and Bao-Liang Lu[1,2,3,4,*]

[1] Center for Brain-Like Computing and Machine Intelligence
Department of Computer Science and Engineering
[2] MOE-Microsoft Key Lab. for Intelligent Computing and Intelligent Systems
[3] Shanghai Key Laboratory of Scalable Computing and Systems
[4] MOE Key Laboratory of Systems Biomedicine
Shanghai Jiao Tong University
800 Dongchuan Road, Shanghai 200240, China
bllu@sjtu.edu.cn

Abstract. Fatigue is the most important reason leading to traffic accidents. In order to ensure traffic safety, various methods based on electroencephalogram (EEG) are proposed. But most of them, either regression or classification, are focused on the relationship between feature space and observation values, so the changing patterns of features are ignored or discarded. In this paper, we propose a new fatigue classification method by using parallel hidden-Markov-model and pattern classifier combination techniques, where each model represents a particular fatigue-high-related feature. In the experiment, subjects are asked to accomplish some simple tasks, and both their fatigue states and their EEG signals are recorded simultaneously. Experimental results indicate that the mean error rate obtained by using our new method are 11.15% for classifying 3 states and 16.91% for classifying 4 states, respectively, while the existing approach can only reach 16.45% and 23.55% under the same condition.

Keywords: EEG, Fatigue Classification, PHMM, Classifier Combination, Fuzzy integral.

1 Introduction

In recent years, the traffic safety is worrying because of the frequently happened traffic accidents and the great damage to people and economy. Among the numerous factors, fatigue driving is the most significant one [6,12]. Therefore many researchers are trying to develop a fatigue on-line estimating system with high time resolution and high accuracy. They have tried various signals including pulse, facial movement or other physiological signals. Among these signals, electroencephalogram (EEG) has been proved to be high correlated, accurate and

---

* Corresponding author.

T. Huang et al. (Eds.): ICONIP 2012, Part IV, LNCS 7666, pp. 484–491, 2012.

reliable [7]. Meanwhile, many good characteristics of EEG have been pointed out, including the EEG spectrum whose principal component has the positive correlation with fatigue state. By using the EEG spectrum features, we proposed a method to solve these problems to some extent for fatigue classification.

The existing approaches suffer the following four main deficiencies: a) Most estimation models are statistical analysis models or off-line supervised learning models which can't be applied for practical applications. b) Most models just use single classifier, but the information about single classifier alone is always insufficient to get the best result. c) As most studies pointed out, the changing patterns are different in different fatigue states. But most of the existing methods haven't taken this knowledge into consideration. d) Fatigue transition is a dynamic process but we are still using the EEG temporal static features, the dynamic information is wasted.

To deal with the first problem, we introduce linear dynamic system (LDS) technique [11,9]. LDS combines both the spatial and temporal information of EEG features and keeps time series information when smoothing features in second-scale [10]. For the second and third problems, we adopt the parallel hidden Markov models (PHMM). HMMs have been applied in many fields including speech recognition, activity recognition and motor imagery, but there is few research using HMMs to classify fatigue [4]. As we know, HMMs are capable of differentiating different signals that their statistical properties changing over time, thus considering different fatigue state have different changing patterns, we construct corresponding Hidden Markov models to imitate the way of different fatigue states transition. In addition, we use fuzzy integral as a pattern classifier combination strategy in a more reasonable way [3]. For the fourth problem, the EEG temporal dynamic features are used to classify different EEG patterns.

This paper is organized as follows. In Sect 2, the proposed method and the main processes including data processing, dynamic feature extraction and classification scheme are described. In Sect 3 we introduce the experimental setup and the collected data forms. The experimental results and discussions are reported in Sect 4, followed by conclusions in Sect 5.

2 Methods

The overall flow chart of the proposed fatigue classification method is depicted in Fig 1. The original EEG signals and subject's performance are the inputs from the most left and the optimal solution from the right bottom is the output. From original EEG, we can extract the static and dynamic features. From subject's performance we can get the reference fatigue level. All the processes will be presented in this section.

2.1 Data Processing

We use the error rate of subject's reaction as the reference. This is based on the viewpoint that high fatigue level means making more mistakes. By averaging the error rate within a 2-minute window at 2-second step, the reference fatigue level is obtained.

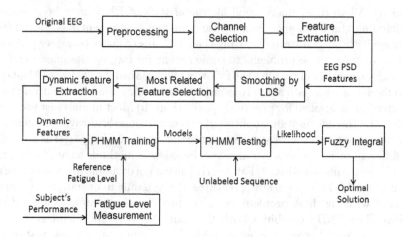

Fig. 1. Flow chart of the proposed EEG-based fatigue classification method

Since raw EEG signals contain much noise, we need to wipe off the EEG-unrelated signals and fatigue-unrelated signals, and remain the highly fatigue-related EEG signals.

In the preprocessing process, we discard the first few minutes EEG signals because subjects need to adapt the experiment in the very beginning. Then we remove the artifacts and use a band-pass filter (1Hz-16Hz) to remove the low-frequency and high-frequency noise. In the channel selection process, we select 6 EEG channels (P1, Pz, P2, Po3, Poz, Po4) from the posterior scalp which are proved to be the most fatigue-related region in our previous work [10]. In the feature extraction process, we use short-time Fourier transform (STFT) with a 2 seconds Hamming window and 0.5 seconds overlapping to calculate the log power spectral density of every single frequency in 1Hz-16Hz. Next we use a linear dynamic system (LDS) approach to smooth the features, because LDS-based filter has higher time resolution and ability to keep the spatial and temporal information in EEG signal smoothing. In the feature selection process, we use the correlation coefficients between features and reference fatigue level (the reference fatigue level is still continuous here) to measure the importance of features. We select the top Q high-correlation-coefficient features in training data, and select corresponding features in testing data.

2.2   Dynamic Feature Extraction

The dynamic features are calculated using the following first-order orthogonal polynomial formula [4,2]:

$$\Delta F(l) = \frac{\sum_{i=-K}^{K} i F(l-i)}{\sum_{i=-K}^{K} i^2}, K+1 \leq l \leq N-K \tag{1}$$

where $F(l)$ is the static feature, l is the sample index, N is the size of the sample set, K is the window length, which should be carefully chosen to balance the weights of the adjacent samples, and $\Delta F(l)$ is the dynamic feature of input $F(l)$. When $F(l)$ is static feature, $\Delta F(l)$ will be the first-order dynamic feature. While $F(l)$ is first-order dynamic feature, $\Delta F(l)$ will be the second-order dynamic feature.

Each static feature and its corresponding dynamic features are combined to form one feature vector. So the number of static features is equal to the number of feature vectors. All feature vectors will be processed independently later.

2.3   Classification Scheme

The flow chart of the classification scheme proposed in this paper is shown in Fig. 2, which is an example for 2 classes, Q is the number of feature vectors.

Before training PHMMs, the samples should be partitioned into fixed-length sequences. Then, the continuous reference fatigue level should turn into discrete value refer to different classes. Ranges of each class can be divided equally, or divided by artificial thresholds.

Suppose we have M classes and Q feature vectors. After training process, we get $M * Q$ HMMs. All sequences of each class and each feature vector will be used to train a HMM with mixture of Gaussians outputs. The fatigue is continuous varying, so the states of HMM can only transfer to their adjacent states and themselves. The state transition diagram with three states applied in our method is shown in Fig. 3.

The HMM parameters are initialized with k-means algorithm, and using the Baum-Welch algorithm in training [8,4]. When we have a new unlabeled sequence, we compute the likelihood of each HMMs based on the new sequence

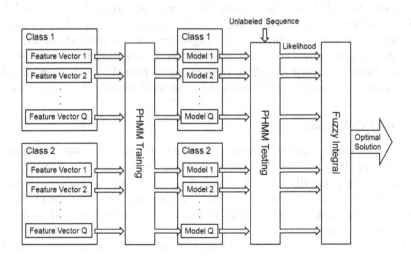

Fig. 2. Flow chart of the proposed classification scheme

Fig. 3. The state transition diagram of our HMM with three states

in the testing process. Then we use fuzzy integral algorithm to determine which class is the optimal solution.

We can also use some simple pattern classifier combination strategies like maximal, sum, voting and product rule. All the definitions can be found in [5]. Compared with fuzzy integral, these methods need no training process. However, they are more inaccuracy. We have implemented all these methods and the results will be shown in Sect 4.

3   Materials

There is no fatigue EEG database available online till now. So we use the database collected in our laboratory. The following sections present the experiment settings and the data forms.

3.1   Experiment

In our experiment, subjects need to complete a monotonous visual task in which subjects are asked to recognize the sign color and press the correct button on the response pad.

There are four colors in all, each has more than 40 different traffic signs. The NeuroScan Stim2 software presents 5~7 seconds black screen and 0.5 seconds random sign in each trial. The experiment room is soundproof and normal illuminated. Subjects sit in a comfortable chair, 2 feets away from LCD. In order to collect sufficient and useful data, the experiment starts from about half hour after lunch, lasting more than one hour, better if including the whole process from wakefulness to sleep. Totally 9 subjects aging from 19 to 28 years old have participated in our experiment, each subject at least twice.

3.2   Data Collection

There are two kinds of collected data. One is the EEG signal. A total of 62 channels EEG signals sampled with 500Hz are recorded by the NeuroScan system. The system will primarily filter the EEG signals between 0.1 and 100Hz to remove some high-frequency noise from surroundings and muscles. The electrodes are arranged based on extended 10/20 system. The reference electrode is located on the top of the scalp.

The other is the visual stimulus sequence and subjects' performance. According to these data, we can estimate subjects' reference fatigue level during experiment, which helps us training and evaluating our approach.

4   Results and Discussions

EEG signal is time sequence, so we can't randomly select training and testing data in one session. So we use two sessions of one subject to train and test separately. There are totally 18 pairs of training and testing sessions.

Table 1. The RMSE for different class numbers and different features

No. of class	static	1st order	2nd order
2	6.84 ± 0.03	6.54 ± 0.02	**6.06 ± 0.04**
3	11.31 ± 0.08	11.25 ± 0.02	**11.15 ± 0.10**
4	18.05 ± 0.14	17.84 ± 0.08	**16.91 ± 0.15**
5	23.64 ± 0.04	23.62 ± 0.04	**22.54 ± 0.07**

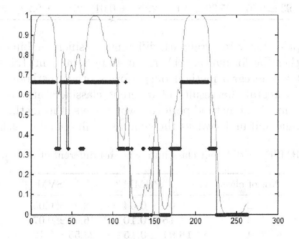

Fig. 4. Example of one testing result with 3 classes

In Table 1, we calculate the root mean squared error (RMSE) of all 18 testing error rates with different features and different class numbers. Fatigue is generally divided into 5 periods, so we choose 2 to 5 classes in our test. The second column to the fourth column refer to different features. The pattern classifier combination method is fuzzy integral and the classifier's parameters are tuned by trial and error. The best results are highlighted in bold.

We can see that, firstly, the RMSE increases quickly with the number of classes increasing. Because when the number of classes increases, the range of each class will be narrower, so the confused boundary will greatly affect the classification accuracy on adjacent classes. Secondly, the dynamic features perform better than static features, and 2nd order outperforms 1st order. This means the dynamic information contributes to classify fatigue levels.

Figure 4 shows a testing result with 3 classes, 5 Gausses, and 3 fatigue states. And the classifier combination method is fuzzy integral. The horizontal axis means sequence index. The vertical axis means the fatigue level value (lower value is lower fatigue). The green line is the continuous fatigue level value. The red line is the discrete fatigue level value, which is equally divided and valued as 0, 1/3 and 2/3. The blue points is the estimated discrete fatigue level value.

In Fig. 4, there are two main kinds of error points. One is at the junction of two adjacent classes with a bias towards the previous class. The other is at the fluctuate of continuous fatigue level with a bias towards the fluctuate direction.

Table 2. The RMSE for different class numbers and different classifier combination methods

No. of class	sum	voting	product	maximal	fuzzy integral
2	6.38 ± 0.06	6.17 ± 0.05	6.11 ± 0.07	7.44 ± 0.14	**6.06 ± 0.04**
3	11.80 ± 0.10	12.53 ± 0.04	11.53 ± 0.10	14.32 ± 0.15	**11.15 ± 0.10**
4	17.93 ± 0.13	18.37 ± 0.11	17.60 ± 0.03	19.31 ± 0.15	**16.91 ± 0.15**
5	23.93 ± 0.01	25.36 ± 0.15	22.82 ± 0.01	27.82 ± 0.22	**22.54 ± 0.07**

Next, we replace fuzzy integral with different classifier combination methods and compare their performance. The result is presented in Table 2. We can see that, the fuzzy integral method outperforms other methods. That means by using fuzzy integral the results of different classifiers are combined more reasonably. The maximal method performs worst, cause this method emphasizes the maximal value and undervalues the general results of all classifiers.

Table 3. The RMSE for different class numbers and different classification methods

No. of class	PHMM	SVM
2	**6.06 ± 0.04**	8.22 ± 0.05
3	**11.15 ± 0.10**	16.45 ± 0.07
4	**16.91 ± 0.15**	23.55 ± 0.12
5	**22.54 ± 0.07**	28.38 ± 0.09

Last, we compare PHMM with SVM. LibSVM is used to train and test models [1]. Here we replace the PHMM modules in Fig. 2 by the SVM methods, in which RBF kernel function is used and the optimal parameters are selected by Cross Validation. The classifier combination method is still fuzzy integral. The result is shown in Table 3. Obviously our method performs better than SVMs. Because PHMM can construct the internal transition regularity of different fatigue states. And SVMs only use the information of feature space.

5 Conclusions

This paper presents a novel EEG-based fatigue classification method by integrating PHMM and fuzzy integral. In this method, we introduce dynamic features to our feature extraction process. And the dynamic features are proved to be better than static features by our experimental results. In the classification scheme, we use PHMM to construct the changing patterns of different fatigue states and use fuzzy integral to combine the results of different classifiers. By comparison, PHMM outperforms SVMs, and the results show changing information is discriminative in EEG-based fatigue classification. We also evaluate the performance of different classifier combination methods (fuzzy integral, maximal, sum, product and voting). The results show that fuzzy integral outperforms others.

Acknowledgments. This work was partially supported by the National Basic Research Program of China (Grant No. 2009CB320901), and the European Union Seventh Framework Programme (Grant No. 247619).

References

1. Chang, C.C., Lin, C.J.: LIBSVM: A library for support vector machines. ACM Transactions on Intelligent Systems and Technology 2, 27:1–27:27 (2011), software, http://www.csie.ntu.edu.tw/~cjlin/libsvm
2. Deller, J., Proakis, J., Hansen, J.: Discrete-time processing of speech signals. Macmillan (2000)
3. Grabisch, M.: Fuzzy integral for classification and feature extraction. Fuzzy Measures and Integrals - Theory and Applications, 415–434 (2000)
4. Lederman, D., Tabrikian, J.: Classification of multichannel eeg patterns using parallel hidden markov models. Medical and Biological Engineering and Computing, 1–10 (2012)
5. Li, B., Lian, X., Lu, B.: Gender classification by combining clothing, hair and facial component classifiers. Neurocomputing, 18–27 (2011)
6. Molloy, R., Parasuraman, R.: Monitoring an automated system for a single failure: Vigilance and task complexity effects. Human Factors: The Journal of the Human Factors and Ergonomics Society 38(2), 311–322 (1996)
7. Oken, B., Salinsky, M., Elsas, S.: Vigilance, alertness, or sustained attention: physiological basis and measurement. Clinical Neurophysiology 117(9), 1885–1901 (2006)
8. Rabiner, L.: A tutorial on hidden markov models and selected applications in speech recognition. Proceedings of the IEEE 77(2), 257–286 (1989)
9. Shi, L., Lu, B.: Dynamic clustering for vigilance analysis based on eeg. In: Proc. of 32nd International Conference of the IEEE Engineering in Medicine and Biology Society, pp. 54–57 (2008)
10. Shi, L., Lu, B.: Off-line and on-line vigilance estimation based on linear dynamical system and manifold learning. In: Proc. of 32nd Annual International Conference of the IEEE Engineering in Medicine and Biology Society, pp. 6587–6590 (2010)
11. Shi, L., Yu, H., Lu, B.: Semi-supervised clustering for vigilance analysis based on eeg. In: Proc. of 20th IEEE International Joint Conference on Neural Networks, pp. 1518–1523 (2007)
12. Weinger, M.: Vigilance, boredom, and sleepiness. Journal of Clinical Monitoring and Computing 15(7), 549–552 (1999)

A Novel Approach to Protein Structure Prediction Using PCA or LDA Based Extreme Learning Machines

Lavneet Singh, Girija Chetty, and Dharmendra Sharma

Faculty of Information Sciences & Engineering
University of Canberra, Australia
{Lavneet.singh,giriza.chetty,dharmendra.sharma}@canberra.edu.au

Abstract. In the area of bio-informatics, large amount of data is harvested with functional and genetic features of proteins. The structure of protein plays an important role in its biological and genetic functions. In this study, we propose a protein structure prediction scheme based novel learning algorithms – the extreme learning machine and the Support Vector Machine using multiple kernel learning, The experimental validation of the proposed approach on a publicly available protein data set shows a significant improvement in performance of the proposed approach in terms of accuracy of classification of protein folds using multiple kernels where multiple heterogeneous feature space data are available. The proposed method provides the higher recognition ratio as compared to other methods reported in previous studies.

Keywords: Support Vector Machines (SVM), Extreme Learning Machines (ELM), Protein Folding, Principal Component analysis (PCA), Linear Discriminant Analysis (LDA).

1 Introduction

There is an urgent need of transforming the large harvested human genomic sequences data using effective and efficient computational algorithms to extract the biological knowledge. The structure of protein plays an important role in its biological and genetic functions [1]. Past studies tried to predict protein folding patterns based on an approach involving comparison of between the unknown protein sequences, and the known protein sequences by computing sequence similarities. In contrast, several machine learning methods have been introduced for feature selection and classification to predict protein structures in various folds. Ding and Dubchak [2] proposed the approach based on Support Vector Machines (SVM) and Neural Networks (NN) classifiers. Shen and Chou [3] proposed a model based on nearest neighbor algorithm and it's modified nearest neighbor algorithm called K-local hyperplane (H-KNN) and was implemented by Okun [4]. Nanni [5] also proposed model using Fishers linear classifiers and H-KNN classifiers. Eddy [6] used Hidden Markov Model for protein folding recognition. M Madera and Ch. Lampros [7] and [8] also used the Hidden Markov Models but on reduce state space with small architecture.

T. Huang et al. (Eds.): ICONIP 2012, Part IV, LNCS 7666, pp. 492–499, 2012.

The major challenge in above mentioned methods and approaches lies in the complexity of data, which involve a large number of folding classes with only small number of training samples and multiple heterogeneous feature groups, making it harder for pattern discovery in protein folding prediction to reach more than 60%, which is below than normal classification in pattern recognition problems.

To deal with this problem, many studies have been formulated using the fusion and ensemble classifiers for protein fold recognition problems [9][10][11][12][13][14][15] [21][22][23].

This ensemble fused classifiers are designed through combining a number of component classifiers on multiple set of features or attributes derived from different combinations of amino acids. For example, OET-KNN (Optimized Evidence-Theoretic K Nearest neighbors) classifier is modeled on each of the nine different feature groups in Shen and Chou [3]. Hence, the performance of ensemble classifiers can increase the classification accuracy of protein folding prediction.

In this paper, we propose a novel extreme learning machine algorithm using multiple kernel learning on preprocessed data using feature selection algorithms - the Principal Component Analysis (PCA) and the Linear Discriminant Analysis (LDA) for multi class protein fold recognition problem, where multiple heterogeneous feature space data are available. The results of our novel algorithm resulted in significant improvement in performance in terms of classification accuracy of protein folds.

2   Extreme Learning Machines

Unlike, traditional popular machine learning algorithms, new Extreme Learning Machine (ELM) [16] [17] [18] is based on single hidden layer feed forward neural networks (SLFNs) with additive neurons which randomly chooses the input weights and hidden neurons biases. Unlike traditional approaches such as Back Propagation (BP) algorithms, ELM doesn't need manual tuning parameters and local minima, providing fast and good classification accuracy rate. ELM doesn't have control parameters as learning epochs etc. and hidden neurons are chosen randomly based on Gaussian Probability Distribution and the output weights are calculated using simple generalized inverse method known as Moore-Penrose generalized pseudo inverse [19].

Given a series of training samples (x_i, y_i) $_{i=1, 2 ...N}$ and $\hat{N}$ the number of hidden neurons where $x_i = (x_{i1}, ... x_{in}) \in R^n$ and $y_i = (y_{i1}, ... y_{in}) \in R^m$, the actual outputs of the single-hidden-layer feed forward neural network (SLFN) with activation function $g(x)$ for these N training data is mathematically modeled as

$$\sum_{k=1}^{\hat{N}} \beta_k g\big((w_k, x_i) + b_k\big) = 0_i, \forall = i = 1, ..., N \qquad (1)$$

Where $w_k = (w_{k1}, ..., w_{kn})$ is a weight vector connecting the kth hidden neuron, $\beta_k = (\beta_{k1}, ... \beta_{km})$ is the output weight vector connecting the kth hidden node and output nodes. The weight vectors w_k are randomly chosen. The term (w_k, x_i) denotes the inner product of the vectors w_k and x_i and g is the activation function. The above N equations can be written as $H\beta = O$ and in practical applications $\hat{N}$ is usually much less than the number N of training samples and $H\beta \neq Y$, where

$$H = \begin{bmatrix} g((w_1, x_1) + b_1) & \cdots & g((w_{\hat{N}}, x_1) + b_{\hat{N}}) \\ \vdots & \ddots & \vdots \\ g((w_1, x_{1N}) + b_1) & \cdots & g((w_{\hat{N}}, x_N) + b_{\hat{N}}) \end{bmatrix}_{NX\hat{N}} \tag{2}$$

The matrix H is called the hidden layer output matrix. For fixed input weights $w_k = (w_{k1},.....,w_{kn})$ and hidden layer biases b_k, we get the least-squares solution $\hat{\beta}$ of the linear system of equation $H\beta = Y$ with minimum norm of output weights β, which gives a good generalization performance. The resulting $\hat{\beta}$ is given by $\hat{\beta} = H + Y$ where matrix H^+ is the Moore-Penrose generalized inverse of matrix H [14].

3   Support Vector Machines

The support vector machine (SVM) is a well-known large margin classifier proposed by Vapnik [13]. The basic concept of the SVM classifier is to find an optimal separating hyper- plane, which separates two classes. The decision function of the binary SVM is

$$f(x) = \sum_{i=1}^{N} \alpha_i y_i K(x_i, x) + b) \tag{3}$$

where b is a constant, $y_i \in \{-1,1\}$, $0 \leq \alpha_i \leq C$, $I = 1,2,...............N$ are non-negative Lagrange multipliers, C is a cost parameter, that controls the trade-off between allowing training errors and forcing rigid margins, x_i are the support vectors and $K(x_i, x)$ is the kernel function.

We use the software LIBSVM library for experiments. LIBSVM is a general library for support vector classification and regression, which is available at http://www.csie.ntu.edu.tw/~cjlin/libsvm. In this study, we also did comparisons with multiple kernels for better optimization.

3.1   The PCA or LDA Based Extreme Learning Machines and Support Vector Machines Algorithm

Step 1: Use PCA on training and testing dataset

$$\hat{S} = \frac{1}{N} \sum_{j=1}^{K} \sum_{i=1}^{N_j} (x_{ji} - \hat{\mu})(x_{ji} - \hat{\mu})^T$$

Step2: Use LDA on training and testing dataset

$$\hat{S}_W = \frac{1}{N} \sum_{j=1}^{K} \sum_{i=1}^{N_j} (x_{ji} - \hat{\mu})(x_{ji} - \hat{\mu})^T$$

$$\hat{S}_B = \sum_{j=1}^{K} N_j (\mu_j - \mu)(\mu_j - \mu)^T$$

$$J(T) = \frac{\left| \hat{S}_B \right|}{\left| \hat{S}_W \right|} \quad \textit{Where J(T)=Linear Discriminant Function}$$

Given a training set

$$\hat{S} = \{(x_i, y_i) \in \Re^{m+n}, y_i \in \Re^m\}_{i=1}^{N} \sum \quad , \text{ for activation function } g(x) \text{ and the}$$

number of hidden neurons $\hat{N}$;

Step2: For $k = 1,...,\hat{N}$ randomly assign the input weight vector $w_k \in \Re^n$ and bias $b_k \in \Re.$

Step3: Determine the hidden layer output matrix H .

Step4: Calculate H^+ .

Step5: Calculate the output weights matrix $\hat{\beta}$ by $\hat{\beta} = H^+ T$.

Step 6: Calculate the decision function of SVM is

$$f(x) = \sum_{i=1}^{N} \alpha_i y_i K(x_i, x) + b)$$

We have used RBF, Sigmoid (Sig) and Linear (Lin) as activation functions in ELM and SVM. In later section, a comparative study has been shown on multi kernels with ELM respect to their learning classification accuracy rate. The parameters C from Eq. (1) and g have certain parametric value. Both values has been experimentally chosen, which was done using a cross-validation procedure on the training dataset. The best recognition ratio was achieved using parameter values gamma = 0.5 and C = 300.

4   Experiments

4.1   Dataset

In our experiments, we have used the features described by Ding and Dubchak [2]. All feature vectors are standardize and normalized to the range of [-1; +1] before applying any classifiers. The proteins in both the training and test sets belong to 27 different protein folds corresponding to four major structural classes: α, β, α/β, α+β. In this study, we compare classification results of protein folds by overall accuracy Q, which is defined as the percentage of correctly recognized proteins to all proteins in the test dataset which can be expressed as Q=c/n where c is the number query proteins whose folds have been correctly recognized and n is the total number of proteins in the test dataset. Table 1 shows extracted parameters from protein sequence with 125 dimensional feature vector for each protein in this dataset.

Table 1. Six Extracted features with their dimensions from Protein Sequence

Protein Features	Dimension
Amino Acids Composition	20
Predicted Secondary Structure	21
Hydrophobicity	21
Van der Waals Volume	21
Polarity	21
Polarizability	21

All the simulations for PCA-ELM and LDA-ELM are carried out in Matlab 10.0 environment running in Pentium dual core, 201 GHZ CPU. Figure 1 demonstrates the principal components (eigen values) of feature vectors of protein sequences.

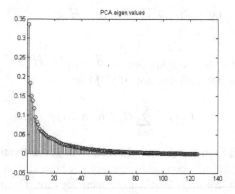

Fig. 1. PCA features of protein sequences

5   Results and Discussion

In this paper, we present a combined generative based classifier and use of feature selection algorithm for protein fold recognition. Table 2 explains the comparison between different methods used in terms of recognition ratio. In comparison with other methods, our proposed PCA and LDA based SVM classifier show higher accuracy rate up to 65.32% and 68.12%. Table 3 presents the recognition ratio of using multiple kernels with proposed PCA and LDA based SVM and PCA and LDA based ELM. It can be seen in Table 3 that in both classifiers, RBF kernels gives the higher accuracy rate comparative to other kernels. In addition, PCA and LDA based

ELM classifier from Table 3 shows more promising results in terms of recognition rate upto 82.45% after 50 trials using RBF kernels. As a result, it can be seen that from all classifiers including SVM and proposed SVM, our proposed Extreme Learning Machines shows promising results in terms of higher accuracy rate of protein folds. Table 4 depicts the ELM training time and testing time ranging from 31.2 ms to 67.5 ms as a function of number of hidden neurons (N). This indicates a fast training process unlike training of a gradient descent based BPN which usually gets trapped in multiple local minima's and thus waste time.

Table 2. Comparison among different methods

Method	Recognition Ratio (%)
SVM	63.75
H-KNN	57.4
Bayesian Naives	52.30
Random Forest	53.72
MLP	54.72
LDA-SVM	65.32
PCA-SVM	68.12
LDA-ELM	77.67
PCA-ELM	82.45

Table 3. Comparison among different methods using multiple kernels

Method	Sigmoid Kernels	RBF Kernel	Linear Kernel
LDA-SVM	57.23	65.32	52.11
PCA-SVM	65.78	68.12	51.89
LDA-ELM	72.34	77.67	68.90
PCA-ELM	78.56	82.45	71.35

Table 4. Training and Testing Time computed time spans for PCA/LDA based ELM

Number of Hidden Neurons (N)	ELM Training Time (Sec)	Testing Time (Sec)
20	0.312	0.456
40	0.312	0.545
60	0.675	0.521

6   Conclusions and Future Work

In this study, we compared various classifiers and proposed an improved and more accurate hybrid ensemble classifiers based on Support Vector Machines (SVM and Extreme Learning Machines for protein folding recognition. In contrast to protein folds prediction, it's very hard to classify its various folds with its different amino acids attributes due to the limited training data availability. Our proposed classifier involves dimensionality reduction using PCA and LDA prior to classification This results in significant improvement in recognition accuracy. The combined PCA-SVM and LDA-SVM classifiers results in an accuracy of (65.32%) and (68.12%), as compared to SVM (56.70%) and other above mentioned classifiers. The similar PCA/LDA feature selection for Extreme Learning Machines also results in further improvement in terms of recognition accuracy up to (77.67%) and (82.45%). We also compare our proposed classifiers using multiple kernels approach resulting in promising results. Table 3 reveals that RBF Kernel give the better classification accuracy rate for both ELM and SVM classifiers compared to other kernels. These results seem to be very promising for classification for protein sequences as compared to results reported in the previous work.

Additionally, all our experiments were done on the original protein features developed by Ding and Dubchak [2]. Further work can be done using different feature selection and classification algorithms on other high dimensional protein sequences to reduce the computational power, more accuracy and higher recognition rate. Although, the obtained results are very encouraging, further experiments using different approaches such as SVM with binary decision trees and other ensemble hybrid classifiers will be investigated.

References

1. Chan, H.S., Dill, K.: The protein folding problem. Physics Today February, 24–32 (1993)
2. Ding, C.H., Dubchak, I.: Multi-class protein folds recognition using support vector machines and neural networks. Bioinformatics 17, 349–358 (2001)
3. Shen, H.B., Chou, K.C.: Ensemble classifiers for protein fold pattern recognition. Bioinformatics 22, 1717–1722 (2006)
4. Okun, O.: Protein fold recognition with k-local hyperplane distance nearest neighbor algorithm. In: Proceedings of the Second European Workshop on Data Mining and Text Mining in Bioinformatics, Pisa, Italy, pp. 51–57 (2004)
5. Nanni, L.: A novel ensemble of classifiers for protein folds recognition. Neuro Computing 69, 2434–2437 (2006)
6. Eddy, S.R.: Hidden Markov models. Current Opinion in Structural Biology 6, 361–365 (1995)
7. Madera, M., Gough, J.: A comparison of profile hidden Markov model procedures for remote homology detection. Nucleic Acids Research 30(19), 4321–4328 (2002)
8. Lampros, C., Papaloukas, C., Exarchos, T.P., Golectsis, Y., IFotiadis, D.: Sequence-based protein structure prediction using a reduced state-space hidden Markov model. Computers in Biology and Medicine 37, 1211–1224 (2007)
9. Lampros, C., Papaloukas, C., Exarchos, K., IFotiadis, D.: Improving the protein fold recognition accuracy of a reduced state-space hidden Markov model. Computers in Biology and Medicine 39, 907–914 (2009)
10. Shen, H.B., Chou, K.C.: Hum-mPLoc.: An ensemble classifier for large-scale human protein subcellular location prediction by incorporating samples with multiple sites. Biochemical and Biophysical Research Communications 355, 1006–1011 (2007)

11. Ghanty, P., Pal, N.R.: Prediction of protein folds.: Extraction of new features, dimensionality reduction and fusion of heterogeneous classifiers. IEEE Transactions on Nanobioscience 8, 100–110 (2009)
12. Guo, X., Gao, X.: A novel hierarchical ensemble classifier for protein folds recognition. Protein Engineering, Design and Selection 21, 659–664 (2008)
13. Kechman, V., Yang, T.: Protein folds recognition with adaptive local hyperplane algorithm. In: Proceedings of IEEE Symposium on Computational Intelligence in Bioinformatics and Computational Biology, Nashville, TN, USA, pp. 75–78 (2009)
14. Wieslaw, C., Katarzyna, S.: A hybrid discriminative/ generative approach to protein fold recognition. Neurocomuting, 194–198 (2012)
15. Zhang, C.X., Zhang, J.S.: RotBoost: a technique for combining rotation forest and adaboost. Pattern Recognition Letters 29, 1524–1536 (2008)
16. Lin, M.B., Huang, G.B., Saratchandran, P., Sudararajan, N.: Fully complex extreme learning machine. Neurocomputing 68, 306–314 (2005)
17. Huang, G.B., Zhu, Q.Y., Siew, C.K.: Extreme Learning Machine: Theory and Applications. Neurocomputing 70, 489–501 (2006)
18. Huang, G.B., Zhu, Q.Y., Siew, C.K.: Real-Time Learning Capability of Neural Networks. IEEE Transactions on Neural Networks 17, 863–878 (2006)
19. Serre, D.: Matrices: Theory and Applications. Springer Verlag, New York Inc. (2002)
20. Vapnik, V.: The Nature of Statistical Learning Theory. Springer, New York (1995)
21. Anurag, M., Lavneet, S., Giriza, C.: A Novel Image Water Marking Scheme Using Extreme Learning Machine. In: Proceedings of IEEE World Congress on Computational Intelligence (WCCI 2012), Brisbane, Australia (2012)
22. Lavneet, S., Giriza, C.: Hybrid Approach in Protein Folding Recognition using Support Vector Machines. In: Proceedings of International Conference on Machine Learning and Data Mining (MLDM 2012), Berlin, Germany. LNCS. Springer (2012)
23. Lavneet, S., Giriza, C.: Review of Classification of Brain Abnormalities in Magnetic Resonance Images Using Pattern Recognition and Machine Learning. In: Proceedings of International Conference of Neuro Computing and Evolving Intelligence, NCEI 2012, Auckland, New-Zealand. LNCS(LNBI). Springer (2012)

Hierarchical Parallel PSO-SVM
Based Subject-Independent Sleep Apnea Classification

Yashar Maali and Adel Al-Jumaily

University of Technology, Sydney (UTS)
Faculty of Engineering and IT, Sydney, Australia
Yashar.Maali@student.uts.edu.au, Adel.Al-Jumaily@uts.edu.au

Abstract. This paper presents a method for subject independent classification of sleep apnea by a parallel PSO-SVM algorithm. In the proposed structure, swarms are separated into masters and slaves and accessing to the global information is restricted according to their types. Biosignal records that used as the input of the system are air flow, thoracic and abdominal respiratory movement signals. The classification method consists of the three main parts; feature generation, feature selection and data reduction based on parallel PSO-SVM, and the final classification. Statistical analyses on the achieved results show efficiency of the proposed system.

Keywords: Sleep apnea, particle swarm optimisation, parallel processing, support vector machines.

1   Introduction

Sleep apnea is one of the most common and important components of sleep disorders. Sleep apnea (SA) and sleep hypopnea are characterized by the repeated temporary cessation of breathing during sleep [1]. Clinically, apnea is defined as the total or near-total absence of airflow. This becomes significant once the reduction of the breathing signal amplitude is at least around 75% with respect to the normal respiration and occurs for a period of 10 seconds or longer. A hypopnea is an event of less intensity; it is defined as a reduction in baseline of the breathing signal amplitude around 30–50%, also lasting 10 seconds in adults [2]. Medical checking for the patients – who have symptoms of SA- need to be done through an overnight sleep study in a sleep centre to record related bio-signals and other associated information. The manual review and analyses by experts is highly cost and time consuming. Thus several efforts have been done to develop a systems that can automatically review and analyse the recorded data [3]. Many techniques are used in this area such as fuzzy rule based system [4] or genetic SVM [5] algorithms that are proposed in our previous studies.

In this paper we used a hierarchical multi swarm structure model [6] for classification of sleep disorder events to apnea or hypopnea. Previously we proposed this parallel structure for subject dependence sleep apneic detection [6], but in this study we investigate it for subject independent sleep apneic classification. The rest of this paper

T. Huang et al. (Eds.): ICONIP 2012, Part IV, LNCS 7666, pp. 500–507, 2012.
© Springer-Verlag Berlin Heidelberg 2012

is covers details of the parallel model in the second section. PSO-SVM classifier is introduced in the third section, followed by experimental results in the section four and the conclusion in section five.

2   Parallel PSO

Particle swarm optimisation (PSO), introduced by Kennedy and Eberhart in 1995 [7, 8] based on the movement of swarms and inspired by the social behaviour of birds or fishes. Similarly to the genetic algorithm, PSO is a population-based stochastic optimisation technique. In the PSO each member is named particle, and each particle is a potential solution to the problem. In comparison with the genetic algorithms, PSO updating the population of particles by considering their internal velocity and position, that are obtained by the experience of all the particles.

2.1   Hierarchical Parallel Structure

In this work by consideration of the big size of the search space (around 1300 data in training with 205 features and the SVM parameters that must be tunned), single PSO cannot have a good performance and leads to the local optimum with low accuracy. Therefore a new parallel structure is developed to perform better exploration and exploitation together in the search space.

In this study a new hierarchical multi swarm structure with a new cooperative strategy among swarms is used [6]. In the traditional multi PSO models, all of the swarms are in the same level and exchanging of the information is just based on the neighbourhood definition [9]. But, in this structure, swarms are classified into two different levels as masters and one (or more) slave(s). Master swarms have access to the best particle of others swarms while the slaves swarms have no access to others information; they actually just provide information for others. Sending the best local particle information among the masters and from the slave(s) to the masters can be performing in the each iteration or by a specified frequency.

In this hierarchical model, one of the master swarms is considered as the centre swarm. All of the swarms, masters or slaves, send the local best particle to the centre swarm. And the centre swarm computes the global best particle and sends it to the other master swarms. So all of the master swarms update their particles by the global best particle, but the slave swarms only use their own local best particles for updating process. Pseudo code of the proposed multi swarm PSO is as follows.

Begin

Select the number of master and slave swarms, number of the particles for each sub-warm and also the frequency for the sending of the information. Select one of the master swarms as the centre.

Initialize the position and velocity of each particle

Do in parallel until the maximum number of iterations has reached {

Evaluate the fitness value of each particle

Find out the local best particle in each sub-swarm

If meet sending condition

   Sending the local best particle (lp_{best}) from each swarm to the centre swarm.

Updating global best particle (Gp_{best}) in the centre swarm.

Sending the global best particle to the master swarms.

End If

Calculate the new velocity of each particle in each sub-swarm

Update the position of each particle in each swarm

   End Do}

Return the best solution (the global best particle) of the algorithm

End

3   Approach and Method

In this section the proposed algorithm for subject independent classification of the sleep apneic events to apnea or hypopnea is presented. This algorithm is based on using just three bio-signals, airflow, abdominal and thoracic movement, as the input. Detection of apneic events by the same set of input is considered by the authors previously [5, 6]. In this paper apneic events will be classified into the apnea or hypopnea. The proposed methodology is as follows:

- Feature generation: In the first stage 10 seconds after each apneic event is selected and then several statistical features are generated for these windows from the wavelet packet coefficients.
- Parallel PSO-SVM: in this stage, PSO algorithm is used for selecting a best training data and features subset interactively by the SVM. In this process SVM is used as the fitness evaluator. PSO also is used for tuning parameters of the SVM. This PSO-SVM is applied in parallel by the new architecture to achieve better performance and also to avoid of the local optimal solutions.
- Final classification: In this stage selected pattern is used for classification of the unseen validation data. The accuracy of this step is assumed as the final performance of the algorithm.

Details of these steps are as follows:

3.1   Features Generation

After each apneic event 10 seconds is considered as the time window, and then several statistical measures corresponding to the input signals are generated as features. To find these features, 3 levels "Haar" wavelet packet is applied on the input signals for each time window. Then several statistical measures are computed by attention to the wavelet coefficients related to each window. These features represent the inputs of the proposed PSO-SVM algorithm in the next step. Full list of the generated features are as Table 1. In these formulas x is representing coefficients of wavelet packet.

Table 1. List of statistical features, x is coefficients of wavelet packet

$\log(\text{mean}(x^2))$	$\text{kurtosis}(x^2)$	$\text{geomean}(|x|)$
$\text{std}(x^2)$	$\text{var}(x^2)$	$\text{mad}(x)$
$\text{skewness}(x^2)$	$\text{mean}(|x|)$	$\text{mean}(x^2)$
$\text{skewness}(x)$	$\text{kurtosis}(x)$	$\text{var}(x)$
$\text{geomean}(x^2)$	$\text{mad}(x^2)$	$\text{std}(x)$

3.2   Parallel PSO-SVM Algorithm

After generating features, events must be classified to the apnea or hypopnea. For this reason, the described hierarchical parallel PSO-SVM algorithm[6] is used for pattern selection and also tuning the parameters of the SVM.

In this study SVM is used to evaluate each particle to classify the validation set. Accuracy of this classification is considered as the fitness of that particle.

In the first step, total data from all of the samples are integrated as meta data then, the meta data is separated to the train and validation by the 0.9 and 0.1 ratios. Then two experiments are investigated.

Experiment 1: In the first experiment, the parallel PSO-SVM is used by 5*10 CV paradigm to select a best feature set from the training set. And then the whole training data by the selected features is used to classify the validation set.

Experiment 2: In the second experiment, training data is separated into the two sets as train and test. The parallel PSO- SVM is used to select a best feature set and also to select a best training data for classifying the test set from the train set. After finding the best features and training data, final output of the algorithm is as the result of the classification of the unseen validation set by the selected training data and features. Figure 1, represent these two experiments.

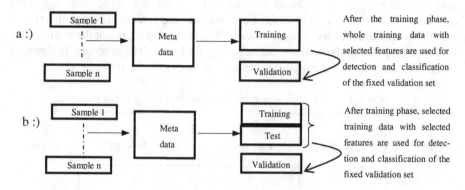

Fig. 1. a, is demonstration for the experiment 1 and b, is demonstration for the experiment 2

Briefly, in the experiment 1 just feature selection is performed, but in the experiment 2, feature selection and pattern selection, or training data selection is considered. So in the experiment 2, after the training phase by the PSO-SVM, we have a set of the best features and also selected training data from the training set.

PSO Structure
In this study constriction coefficient PSO is used [10]. In this approach the velocity update equation is as (1),

$$v_{ij}(t+1) = \chi \left[v_{ij}(t) + \phi_1 \left(y_{ij}(t) - x_{ij}(t) \right) + \phi_2 \left(\hat{y}_{ij}(t) - x_{ij}(t) \right) \right], \quad (1)$$

where y_{ij} is the particle best and $\hat{y}_{ij}$ is the global best particles. And also,

$$\chi = \frac{2k}{|2 - \phi - \sqrt{\phi(\phi - 4)}|}, \quad (2)$$

with,

$$\phi = \phi_1 + \phi_2, \quad \phi_i = c_i r_i \quad i = 1,2.$$

Equation (2) is used under the constraints that $\phi \geq 4$ and $k, r_i \in [0,1]$. The parameter k in the equation (2) controls the exploration and exploitation. For $k \sim 0$, fast convergence is expected and for $k \sim 1$ we can expect slow convergence with the high degree of exploration [10].

By attention to the proposed parallel structure, for the slave swarms, k considered as 0.8 and $c_1 = 2, c_2 = 4$, and for the master swarms k considered as 0.2 and $c_1 = 4, c_2 = 2$.

Particle Representation
In the first experiment, each particle consists of two arrays, the length of the first array is equal to the number of features and each cell can get a number between zero and one. If the value of a cell is higher than 0.5 then the corresponding feature is selected for classification. The second array is related to the gamma and cost as parameters of the SVM and has two cells and each of them can get a value between 2^{-5} to 2^5.

In the second experiment, each particle consists of three arrays, which two arrays are same as the first experiment. And the length of the third array is equal to the number of the train data. Each cell of the third array can get a number between zero and one. If the value of a cell is higher than 0.5 then the corresponding training data is selected for classification.

4   Results and Discussion

Experimental data consist of 20 samples which events of them are annotated by an expert were provided by the concord hospital in Sydney. RBF kernel is selected for the both of the master and slave swarms, and 4 slaves and two masters are selected and each swarm contain 20 particles. Frequency for changing information between swarms is set as 5 iterations.

In the first experiment 5 samples are considered as the validation and 15 samples make the training data. And in the second experiment, again the same 5 samples are considered as the validation, 5 samples make the test and 10 samples make the training data. It must be considered that to overcome the impact of validation set on the final result, 5 independent tests are runs (5*10 CV). Table 2 tabulates the number of sleep apnea or Hypopnea events in each of the validation set, train and test for these 5 runs.

Table 2. Diversity of classes in different runs

	Experiment 1				Experiment 2			
	Validation		Training		Train		Test	
	Apnea	Hypopnea	Apnea	Hy-popnea	Apnea	Hypopnea	Apnea	Hypopnea
#1	316	485	818	876	655	701	163	175
#2	310	532	824	829	655	685	169	144
#3	428	422	706	939	602	728	104	211
#4	289	548	845	813	668	665	177	148
#5	365	539	769	822	636	660	133	162

Accuracy of the proposed experiments are as Table 3, in this table also we mention to the obtained accuracy if we use all of the training data for classification of the validation set without any feature or data selection as appeared in full data column.

Table 3. Accuracies of the three different approaches

	Full data	Experiment 1	Experiment 2
#1	75.12	76.16	78.04
#2	76.41	79.79	83.13
#3	75.49	79.29	82.79
#4	76.45	76.89	83.53
#5	77.04	82.16	83.16

The average accuracies for these three methods are as 76.1, 78.86, 82.13, respectively. Also, for more reliable evaluation between results of these three methods, ANOVA test is performed to show if there is a significant difference in mean of these three groups. The p value of the ANOVA test is as 0.001 which shows there is a significant difference between these three methods; while to have another way to judge which method is better, pair t-test is used. The p value between full data and experiment 1 is as 0.03 and the p value between experiment 1 and experiment 2 is as 0.02. These statistical tests show that the results obtained by the experiment 2 are significantly better than the results of experiment 1, and also result of experiment 1 are better than the results that obtained when we used the whole training data.

Also, we consider the f-score as another performance measure to compare these methods. Table 4, tabulated the f-scores for these two experiments and also using the whole training data. ANOVA test and paired t-test show that experiment 2 is also better than other methods by considering the f-score measure.

Also it can be noticed that, in the experiment 2 generally 12% of training data are not selected. By attention to these facts that; by the implemented particle structure there is not any limit for the number of selected training data, or in another way it is possible for all of the training data to be selected. And by attention to the superiority of the experiment 2 to the experiment1 we can find that, in this study selecting the training data, together by the feature selection can improve the classification performance.

Table 4. F-scores of the three different approaches

	Full data	Experiment 1	Experiment 2
#1	0.76	0.78	0.79
#2	0.78	0.80	0.82
#3	0.74	0.80	0.81
#4	0.76	0.78	0.82
#5	0.78	0.83	0.84

But to implement the same paradigm for the general classification problems we must be careful about risk factors of this paradigm. For example, in this study we have plenty of data as training. So, deleting few of the training data works here. Also, if the length of the samples of classes are very far from each other and we have unbalance classes, then eliminating of some training data can be result in over classification towards the big class.

5   Conclusion

In this paper an algorithm proposed for subject independent classifying the apneic events. The proposed algorithm used a new parallel hierarchical PSO. This proposed structure is used for classifying apneic events to apnea or hypopnea. In this study two paradigms are experimented. In the first one just feature selection is performed but in the second one also elimination of some training data from the training set is considered. The second experiment shows it superiority in this study. But for using it in the general classification problems, that have small training data or unbalances classes, more thoughtfulness is needed.

References

1. Guilleminault, C., Hoed, J.V.D., Mitler, M.: Overview of the sleep apnea syndromes. In: Guilleminault, C., Dement, Wc. (eds.) Sleep Apnea Syndromes, pp. 1–12. Alan R Liss, New York (1978)
2. Flemons, W.: Sleep-related breathing disorders in adults: Recommendations for syndrome definition and measurement techniques in clinical research. Sleep 22(5), 667–689 (1999)
3. Chazal, D.P.: Automated processing of the single-lead electrocardiogram for the detection of obstructive sleep apnoea. IEEE Transactions on Biomedical Engineering 50(6), 686–696 (2003)
4. Maali, Y., Jumaily, A.A.: Genetic Fuzzy Approach for detecting Sleep Apnea/Hypopnea Syndrome. In: 2011 3rd International Conference on Machine Learning and Computing, ICMLC 2011 (2011)
5. Maali, Y., Jumaily, A.A.: Automated detecting sleep apnea syndrome: A novel system based on genetic SVM. In: 2011 11th International Conference on Hybrid Intelligent Systems, HIS (2011)
6. Maali, Y., Jumaily, A.A.: A novel partially connected cooperative parallel PSO-SVM algorithm: Study based on sleep apnea detection. In: 2012 IEEE Congress on Evolutionary Computation, CEC (2012)
7. Eberhart, R., Kennedy, J.: A new optimizer using particle swarm theory. In: Proceedings of the Sixth International Symposium on Micro Machine and Human Science, MHS 1995 (1995)
8. Kennedy, J., Eberhart, R.: IEEE, Particle swarm optimization. In: 1995 IEEE International Conference on Neural Networks Proceedings, vol. 1-6, pp. 1942–1948 (1995)
9. Fan, S.K.S., Changand, J.M.: Dynamic multi-swarm particle swarm optimizer using parallel PC cluster systems for global optimization of large-scale multimodal functions. Engineering Optimization 42(5), 431–451 (2010)
10. Clerc, M.: The swarm and the queen: towards a deterministic and adaptive particle swarm optimization. In: Proceedings of the 1999 Congress on Evolutionary Computation, CEC 1999 (1999)

Air Quality Monitoring and Prediction System Using Machine-to-Machine Platform

Abdullah Kadri[1], Khaled Bashir Shaban[2],
Elias Yaacoub[1], and Adnan Abu-Dayya[1]

[1] Qatar Mobility Innovations Center
Qatar Science and Technology Park, Doha, Qatar
{abdullahk,eliasy,adnan}@quwic.com
[2] Computer Science and Engineering Department
College of Engineering, Qatar University
khaled.shaban@qu.edu.qa

Abstract. This paper presents an ambient air quality monitoring and prediction system. The system consists of several distributed monitoring stations that communicate wirelessly to a backend server using machine-to-machine communication protocol. Each station is equipped with gaseous and meteorological sensors as well as data logging and wireless communication capabilities. The backend server collects real time data from the stations and converts it into information delivered to users through web portals and mobile applications. In addition to manipulating the real time information, the system is able to predict futuristic concentration values of gases by applying artificial neural networks trained by historical and collected data by the system. The system has been implemented and four solar-powered stations have been deployed over an area of 1 km^2. Data over four months has been collected and artificial neural networks have been trained to predict the average values of the next hour and the next eight hours. The results show very accurate prediction.

Keywords: Air quality monitoring and prediction, Artificial neural network, Machine-to-Machine communication.

1 Introduction

The effect of air pollution on human health is considered a major and serious problem globally, especially in countries where oil and gas industries are prevalent. Huge efforts are being done in order to improve air quality in both indoor and outdoor environments. According to the United States Environmental Protection Agency (US EPA) [1], the air quality is characterized by measuring specific gases that affect the health the most, out of which are: ground-level ozone (O_3), carbon monoxide (CO), and hydrogen sulfide (H_2S), [2]. Often, the temporal environmental data is reported within a time frame defined by the standard. For example, CO is reported either by 1-hour average or by 8-hour average; whereas O_3 and H_2S are reported by 8-hour average [1].

T. Huang et al. (Eds.): ICONIP 2012, Part IV, LNCS 7666, pp. 508–517, 2012.

Traditionally, bulky air quality monitoring stations are used for collecting various gases concentrations. These stations include many reference analyzers where each analyzer measures one gas. Although these analyzers produce measurements with high level of accuracy, such stations require frequent calibration and maintenance and they need access to power socket mainly for air conditioning. This inevitably limits their use on large scale. Nowadays, and because of the recent advances in micro-electro-mechanical (MEMS) systems, research and industrial bodies are focusing on developing new generation of sensing stations with low cost, smaller size, and more mobility features [3]. Variations of such stations are being used in different indoor and outdoor environments for both residential and industrial applications.

In [4], a smart sensor network for air quality monitoring applications has been discussed. The authors have shown that using multi-input, single-output artificial neural networks (ANNs) can solve inherent problems of the used sensors, namely the dependency on both ambient temperature and relative humidity. Models for air pollution concentrations as a function of the emission distribution have been investigated in [5]. In [6], an auto-calibration method for a dynamic gas sensor network for air pollution monitoring is proposed. The simulation results show that using this method a high accuracy can be achieved. More related work can be found in [8], [7], [9], [12], [10], and [11]. However, to the best of the authors' knowledge, none of the previous works has presented a comprehensive end-to-end system for: 1) real-time air quality monitoring using an M2M communication paradigm, and 2) predicting short-term concentration values of gases using ANNs.

In this paper, air quality monitoring and prediction (AQMP) system is presented. This system is based on utilizing multi-gas (MG) monitoring stations that communicate with a platform by the means of M2M communication. Each MG monitoring station includes gaseous sensing elements, data logging component, and wireless communication board and it is powered by the solar energy. The platform is located on a backend server where data cleaning and filtering operations are carried out. In addition, this platform converts the received data to useful information that are delivered to users through web portal and mobile applications. The platform also consists of a prediction component that is equipped with ANNs for estimation and forecasting purposes. This paper is organized as follows: in section 2, a description of the systems is given. The prediction module is explained in section 3. Section 4 discusses the results and the conclusions are drawn in section 5.

2 System Description

The system architecture of AQMP is shown in Fig. 1. The end-to-end system consists of two main subsystems: 1) the MG monitoring stations, and 2) the platform at the back-end server. The MG monitoring stations communicate with the backend server via M2M communication protocol. More detailed description of these subsystems is given in the following subsections.

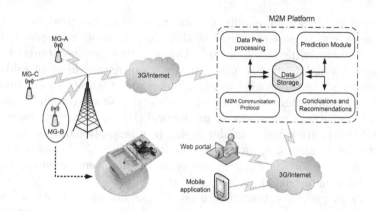

Fig. 1. Air Quality Monitoring and Prediction System Architecture

2.1   Multi-gas Monitoring Stations

An MG monitoring station consists of several gaseous and meteorological sensing elements, the data logging and wireless communication board, and the power supply system. the current system has three stations: MG-A, MG-B, and MG-C. The first two stations are equipped with O_3 and CO sensors, and the third station is only equipped with H_2S sensor. In future implementations, more stations equipped with more gaseous sensing elements for NO and $PM_{2.5}$ will be deployed. The first two stations are 750 m away.

Gaseous and Meteorological Sensing Elements. Normally, each MG monitoring station carries three gaseous sensing elements (for O_3, CO, and H_2S) and two sensing elements for the ambient temperature and relative humidity. The gaseous sensing elements are based on nanotechnology semiconductor concept where the gas concentration is evaluated by measuring the electrical conductivity of a thin metal-oxide layer. When a toxic gas touches this layer, it is absorbed and, consequently, its electrical conductivity changes. The gas concentration is a function of electrical conductivity variation. All these sensing elements are connected to the analog inputs of the data logging and communication board.

Data Logging and Wireless Communication Board. this board is based on Atmega 2560 microprocessor and houses an external MicroSD memory with 2 GB capacity. The main functions of this board are data acquiring, processing, logging, and transmitting. The board is equipped with a GPRS modem for wireless connectivity. The MG monitoring stations are configured to take a sample of all parameters every 1 min and then calculate and save the average of five readings. The station establishes an TCP/IP Internet connection through the GPRS modem with the M2M platform located at the backend server.

Solar Power System. the sensing elements and the data logging and wireless communication board are powered by a solar power system (a solar panel, battery, and charger). The solar system is designed in such a way to operate each station around the clock. The level of the internal battery is sent to the backend server along with measurement data.

2.2 M2M Platform

The M2M platform is operating on a backend server. The main modules of this platform are: an M2M communication module, data integrity module, data processing module, data storage, and prediction module. The M2M communication protocol operates over GPRS or 3G network and is responsible for connecting to all MG monitoring stations for data transfer. The stations are configured to initiate a TCP/IP connection with the platform. The communication stays for 1 min during which the station sends the data. Once received, the data is sent to the data integrity module which is responsible for handling missing, erroneous, and noisy data. The output of this module will be stored in the database. The data processing module applies basic statistical operations on the data before presenting to the user interface. A key module in this platform is the *Prediction* module which will be explained in the next section.

3 Prediction Module

This module is responsible for predicting futuristic measurements for all the gasses O_3, CO, and H_2S. ANNs are utilized in the prediction process and trained by historical and collected data over four months of the AQMP system operation. Average values of the next one-hour and eight-hours' time-windows are predicted. The following subsections detail the type and setups of the ANNs used in this module.

3.1 Artificial Neural Networks

Artificial neural networks are a family of techniques that have been inspired and developed to be analogous with the neural system in the humans. With simple processing units, called neurons, signals are collected from other neurons after being weighted in connection links [13]. Neural networks can differ based on the way their neurons are connected, the specific kinds of computations their neurons perform and the way they transmit patterns of activity throughout the network. ANNs have proven their efficiency in different applied pattern recognition problems [14].

A. Multi-layer Feed Forward Artificial Neural Networks. In this work, two, identical in structure, multi-layer feed forward artificial neural networks (MLFF-ANN) are used to learn the patterns in the time series readings of the different sensed gases. One MLFF-ANN is used for each gas to predict the average of the next hour and the next eight hours, i.e., six ANNs in total. Figure 2

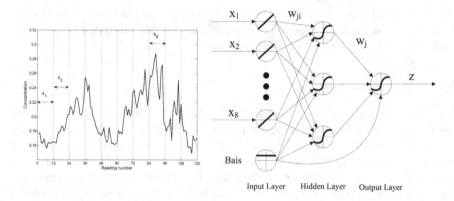

Fig. 2. The MLFF-ANN Structure used for Prediction

shows the MLFF-ANN topology used for prediction with three layers: the input layer, hidden layer and the output layer. Pre-processing of the time series signal is also depicted in the figure. The neural network used in this work has eight neurons in the input layer, one hidden layer with six neurons, and one neuron at the output layer. Each of the eight inputs, which are the mean values of the gas readings from three stations measured during a time window of one hour (i.e., the arithmetic mean of 12 five-minutes readings), are presented to one of the input neurons. The raw data is preprocessed to replace missing values and delete the repeated ones. It is found that about 2% of the readings were missing and therefore replaced with average values of the previous and next reading if the missing data does not exceed 12 readings (i.e. one hour). Otherwise, a copy of the same time segment is copied from a similar day from the previous week. The predicted gas concentration value, which is the average value of the next hour or the next 8 hours, appears at the output neuron. The input samples in the first layer are sent to the hidden layer through weighted connection links. The hidden layer nodes calculate their net activations as in the following equation:

$$net_j = \sum_{i=1}^{8} x_i w_{ji} + b \tag{1}$$

where 8 is the number of features $(x_1, x_2, ..., x_8)$, and w_{ji} represents the weights between the ith input neuron and the jth hidden neuron. b is a bias value for the network. The output of each node in the hidden layer, which is a nonlinear function of its net activation, is shown as the following:

$$y_j = f(net_j) \tag{2}$$

where y_j is the output of the hidden layer neuron j. The output layer single neuron calculates its net activation as in the following equation:

$$net = \sum_{j=1}^{6} y_j w_j + b \tag{3}$$

where 6 is the number of hidden neurons and w_j is the weights between the output neuron and the jth hidden neuron. The output layer yields an output as a nonlinear function of its net activation as shown in the following equation:

$$z = f(net) \tag{4}$$

where z is the output which is equal to one of the concerned measurements. Gradient descent learning is used here; therefore, the nonlinear sigmoid function used is continuous as defined in equation 5.

$$f(net_j) = \frac{1}{1 + e^{-net_j}} \tag{5}$$

The output of neural network can be expressed as a function of the inputs, the weights between input layer and the hidden layer and the weights between the hidden layer and the output layer as per the following equation:

$$g(x) = z = f\left(\sum_{j=1}^{6} w_j f\left(\sum_{i=1}^{d} x_i w_{ji} + b\right) + b\right) \tag{6}$$

B. Back-Propagation Training Method. The main objective behind using the back-propagation training method is to use training samples of inputs and outputs in the network to adjust the weights' values (w_{ji}, w_j) to minimize the difference between the predicted and the actual outputs. The optimum weights (w_{ji}, w_j) are learned by minimizing the training error given in the following equation:

$$J(w_{ji}, w_j) = \frac{1}{2}(t - z)^2 \tag{7}$$

where $J(w_{ji}, w_j)$ is the mean square error and t is the target output at the output z. Using gradient descent, the updated weights are calculated as the following:

$$w_{ji}^{t+1} = w_{ji}^t - \eta \frac{\partial J}{\partial w_{ji}} \tag{8}$$

$$w_j^{t+1} = w_j^t - \eta \frac{\partial J}{\partial w_j} \tag{9}$$

Using the chain rule, $\partial J/\partial w_{ji}$ and $\partial J/\partial w_j$ are calculated yielding the following expressions:

$$\frac{\partial J}{\partial w_j} = -(t - z)f'(net)y_j \tag{10}$$

$$\frac{\partial J}{\partial w_{ji}} = -f'(net_j)x_i(t - z)f'(net)w_j \tag{11}$$

4   Results for 1-Hour and 8-Hours' Time Window Prediction

In light of the limitation in the number of available samples and to increase the statistical significance of the results, round robin strategy is used in training and testing the neural network. On a 10–fold validation basis and cycling over all the samples, 18 samples of the measured gas concentrations data of two months are spared for testing while the rest of the samples are used for training the neural network. Figs. 3–6 show the measured values of the gases along with their

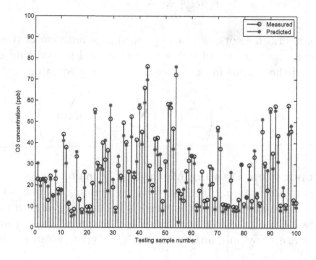

Fig. 3. Prediction of O_3 concentration for the next eight hours

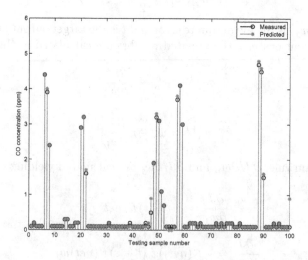

Fig. 4. Prediction of CO concentration for the next eight hours

predicted values for the next eight hours average values. To evaluate the performance of the prediction module, one performance parameter namely, the root mean squared error (RMSE), is used. For air quality monitoring applications, (RMSE) is good enough.

Fig. 3 shows the measured and predicted values for O_3. The RMSE of the O_3 prediction using the 10–fold strategy is calculated to be (0.848 ± 0.111) when predicting the next eight hours. The results of predicting the average of the next one hour and the next eight hours for CO are shown in Figs. 5 and 4. The RMSE value of the average for the next one hour is (0.346 ± 1.66) and for the average

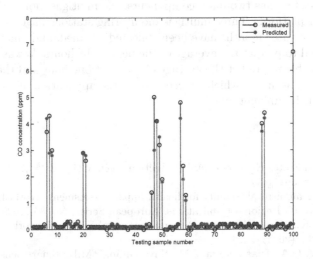

Fig. 5. Prediction of CO concentration for the next one hour

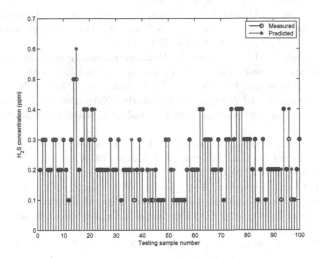

Fig. 6. Prediction of H_2S concentration for the next eight hours

of the next eight hours (0.043 ± 0.023) respectively. Fig. 6 shows the measured and predicted values for H_2S. The RMSE of the H_2S prediction using the same strategy is calculated to be (0.016 ± 0.012) when predicting the next eight hours.

The obtained results show that the maximum root mean squared error is (0.848 ± 0.111). For the application of predicting gaseous concentrations in ambient air, this is within the acceptable accuracy range.

5   Conclusions

In this paper, a system for ambient air quality monitoring and prediction is presented. The system has two main components, the multigas monitoring stations and the M2M platform. Three multigas monitoring stations have been deployed and the data of three months have been collected, cleaned, and analyzed. ANNs have been used to predict the average of the next eight-hours. It was shown that the system is able to predict the average of the next one hour and the next eight hours with high accuracy which is accepted for the application of predicting gas concentrations in ambient air.

References

1. The United States Environmental Protection Agency (US EPA),
   http://www.epa.gov/
2. Monitoring ambient air quality for health impact assessment. World Health Organization, WHO Regional Publications. European Series, No. 85 (1999)
3. Martinez, K., Hart, J.K., Ong, R.: Environmental Sensor Networks. IEEE Computer, 50–56 (2004)
4. Postolache, O.A., Dias Pereira, J.M., Silva Girão, P.M.B.: Smart Sensors Network for Air Quality Monitoring Applications. IEEE Transactions on Instrumentation and Measurements 58(9), 3253–3262 (2009)
5. Andò, B., Baglio, S., Graziani, S., Pitrone, N.: Models for Air Quality Management and Assessment. IEEE Transactions on Systems, Man, and Cybernetics–Part C: Applications and Reviews 30(3), 358–363 (2000)
6. Tsujita, W., Ishida, H., Moriizumi, T.: Dynamic gas sensor network for air pollution monitoring and its auto-calibration. Proceedings of IEEE Sensors 1, 56–59 (2004)
7. Andò, B., Cammarata, G., Fichera, A., Graziani, S., Pitrone, N.: A Procedure for the Optimization of Air Quality Monitoring Networks. IEEE Transactions on Systems, Man, and Cybernetics–Part C: Applications and Reviews 29(1), 157–163 (1999)
8. Khedo, K.K., Perseedoss, R., Mungur, A.: A Wireless Sensor Network Air Pollution Monitoring System. International Journal of Wireless & Mobile Networks (IJWMN) 2(2), 31–45 (2010)
9. Liu, J.-H., Chen, Y.-F., Lin, T.-S., Lai, D.-W., Wen, T.-H., Sun, C.-H., Juang, J.-Y., Jiang, J.-A.: Developed Urban Air Quality Monitoring System Based on Wireless Sensor Networks. In: 5th International Conference on Sensing Technology, pp. 549–554 (2011)
10. Lu, W., Wang, X., Wang, W., Leung, A.Y.T., Yuen, K.: A preliminary study of ozone trend and its impact on environment in Hong Kong. Environment International 28(6), 503–512 (2002)

11. Hou, A.S., Lin, C.E., Gou, Y.Z.: A Wireless Internet-Based Measurement Architecture for Air Quality Monitoring. In: Proceedings of the 21st IEEE Instrumentation and Measurement Technology Conference (IMTC 2004), vol. 3, pp. 1901–1906 (2004)

12. Tzeng, C.-B., Wey, T.-S.: Design and Implement a Cost Effective and Ubiquitous Air Quality Monitoring System Based on ZigBee Wireless Sensor Network. In: Second International Conference on Innovations in Bio-inspired Computing and Applications (IBICA), pp. 245–248 (2011)

13. Hayken, S.: Neural Networks a Comprehensive Foundation, 2nd edn. Prentice Hall, New Jersey (1999)

14. Fausett, L.: Fundamentals of Neural Networks: Architectures, Algorithms, and Applications. Prentice Hall, New Jersey (1993)

An Extension of the Consensus-Based Bundle Algorithm
for Group Dependant Tasks
with Equipment Dependencies

Simon Hunt, Qinggang Meng, and Chris J. Hinde

Loughborough University, Department of Computer Science, Loughborough, United Kingdom
{s.hunt,q.meng,c.j.hinde}@lboro.ac.uk

Abstract. This paper addresses the problem of multi-agent, multi-task assignment with multiple agent requirements on tasks for unmanned aerial vehicles by presenting the Consensus Based Grouping Algorithm. The algorithm is an extension of the Consensus Based Bundle Algorithm that converges to a conflict free, feasible solution of which previous algorithms are unable to account for. Furthermore the algorithm takes into account heterogeneous agents, deadlocking and a method to store assignments for a dynamical environment.

Keywords: CBAA, CBBA, UAV, Consensus, Task Allocation, Cooperation.

1 Introduction

With the rising use of Unmanned Aerial Vehicles (UAV) prevalent throughout the world, UAVs are finding valuable usage in performing military tasks that fall into the categories of the dull, dirty and dangerous [1]. As we progress the future of UAVs look increasingly towards civilian activities [2]. Common applications include surveillance of power lines or pipes [3], disaster monitoring [4] and search and rescue operations [5]. As the applications for UAVs increase so too does their need to cooperate to perform bigger and more complex tasks.

Creating a UAV to cover all situations and problems is difficult due to hardware and software limitations [11] thus it is easier to specialize UAVs to a precise problem. However doing so reduces the UAVs ability to solve a wide variety of tasks in a dynamic environment. With a diverse selection of UAVs that can form teams and work together to complete tasks we can solve the limitation of any one UAV. Using multiple UAVs will improve the efficiency at which a number of tasks can be performed by completing tasks in parallel.

Particularly within the area of UAV cooperation is the Task Assignment Problem (TAP) which assigns a finite number of agents to complete a finite number of tasks as efficiently as possible. This problem can be solved with a centralized or decentralized solution but current research looks at decentralized solutions, which provide a more feasible solution for real world adaption. Many researchers have solved the TAP using auction algorithms [6] [7] where agents make bids for tasks and

T. Huang et al. (Eds.): ICONIP 2012, Part IV, LNCS 7666, pp. 518–527, 2012.

receive assignments based on their bids by a single auctioneer. One such solution that makes use of auction algorithms is the Consensus Based Auction Algorithm (CBAA) [8]. The CBAA lets agents make bids towards each task where their bid is the reward the group would receive should that agent complete the task. Whilst task allocation for an individual agent is relatively simple, the difficulty comes with consensus between all agents when using a decentralized algorithm. The CBAA succeeds in giving us a conflict free solution that has superior convergence and performance than other auction algorithms.

The Consensus Based Bundle Algorithm (CBBA) was created to solve an extension of the TAP where agents are allowed to queue up tasks they'll complete. Individual agents take available tasks and compute every permutation given their current queue of tasks. The greatest increase in reward is used as the tasks bid. Agents continually add and remove tasks as other agents find higher valued sequences with that task. The CBBA gives a conflict free solution with a guaranteed 50% optimality to the multi-agent to multi-task assignment problem [8].

We can further extend the problem towards a realistic simulation by placing equipment requirements onto each task that would restrict which UAVs can complete specific tasks. After developing a structure for assigning multiple agents to a single task we can use cooperation to solve equipment limitations. Current algorithms are unable to account for the assignment and consensus when multiple agents are required for a single task. Using the framework set up by the CBBA we extend the algorithm to account for the new limitations on tasks. Tasks will require varying numbers of agents and equipment. This extension leads us to the Consensus Based Grouping Algorithm (CBGA).

2 Background

2.1 Task Assignment Problem

The task assignment problem is a combinatorial optimization problem that looks at finding the least-cost solution between two disjoint sets [9]. There is a set of agents $A = \{a_1,....,a_n\}$ and a set of tasks $T = \{t_1,....,t_m\}$.

An agent has a cost associated with it for completing each task. Let C_{ij} be the non-negative cost of assigning the i_{th} agent to the j_{th} task. The objective is to assign each task one agent in such a way as to minimize the overall cost of completing all the tasks. If we define a binary variable X_{ij} where $X_{ij} = 1$ to indicate agent a_i is assigned to task t_j. Otherwise $X_{ij} = 0$. Then the total cost of the assignment is equal to (1).

$$\sum X_{ij}*C_{ij} \text{ for } i = 1 \text{ to } n, j = 1 \text{ to } m. \tag{1}$$

With a valid assignment each agent $a \in A$ must be assigned to only one task and each task $t \in T$ must have exactly one agent assigned. For an assignment to be efficient we say the task allocation must be valid and the cost is minimized (2).

$$\text{Cost} = \min (\sum X_{ij}*C_{ij} \text{ for } i = 1 \text{ to } n, j = 1 \text{ to } m) \tag{2}$$

2.2   Restricted Task Assignment Problem

As we extend the TAP we are creating restrictions that limit which agents are valid depending on their equipment but we loosen the single assignment restriction.

Each agent a_i can be assigned to multiple tasks as part of the CBBA, conversely each task t_j can similarly have multiple agents assigned to it.

There is a set of sensor requirements $R = \{r_1,....,r_p\}$ that each task t_j has and correspondingly a sensor list that each agent a_i is equipped with. Where r_{pj} is the number of sensors p required for task j and r_{pi} is the existence of sensor p on agent i.

$$\sum r_{qj} * r_{qi} \text{ for } q = 1 \text{ to } p. \qquad (3)$$

Thus when (3) is non-negative agent i is capable of assisting on task j.

Although multiple agents can potentially be assigned to a single task, the cost function will stay the same, however, the algorithm wont limit $X_{ij} = 1$ to a single instance for each j, limiting it only to the maximum assignments to a specific task.

2.3   Consensus Based Bundle Algorithm

The primary contribution of this work is to extend the CBBA presented in [8]. The CBBA contains two distinct phases for controlling the allocation and consensus of tasks. During the first phase an agent internally builds a bundle b_i, with a path p_i. The path p_i, is an ordered sequence of tasks that agent i will perform. A new task is intersected into the current path at all possible locations to find the highest increase in reward. This increase is compared to the current winning bid list, y_i, and if a greater reward is provided agent i allocates itself the task with the corresponding path.

During the second phase, agents communicate their assignments and winning bid lists to reach consensus on the TAP. Conflicts are resolved using the winning bid list y_i, the winning agent list z_i, and a time stamp vector, s_i, which indicates the age of a teammate's information from each agent.

As conflicts are resolved and the information an agent has access to is updated an agents bundle and path will adjust itself to the other agents. The algorithm cycles between the two phases until eventually consensus is made and a solution is found where as many tasks have been assigned to agents as possible.

The CBBA provides a conflict free solution with a guaranteed 50% optimality, however once we expand on the simulation and increase the requirements of tasks the algorithm cannot complete these problems. The focus of this paper is to extend the CBBA to manage the increased complexity on requirements within the simulation.

3   Consensus Based Group Algorithm

3.1   Local Data

With the CBBA task and agent information are stored locally at the beginning of the simulation and are fixed for the duration of the simulation. Each agent stores two vectors a winning bids list y_i and the winning agent list x_i. With the CBGA we need to modify the storage of these values to allow multiple agent assignments. We can merge both vectors into an $m*n$ matrices where m is the number of tasks and n is the number of agents in the environment. Therefore X_{ij} is equal to the winning bid of agent i for task j or equal to 0 if no assignment has been made. The reward for task j can be calculated as $\sum X_{ij}$ for $i = 1$ to n, however the total instances of non-zero values in a row should never exceed the maximum number of agents assignable to that task.

In a dynamic system we cannot assume each agent will store data in the same order, therefore we cannot use the matrix index as an identifier for an agent or task. Agents therefore store an agent vector I that contains all agent IDs and a task vector J containing task IDs. These two vectors are used as lookups to the assignment matrix X. With this new matrix agents can store data dynamically and build up a list of agent to task assignments as they discover new agents or tasks in the system. When a new row (tasks) or column (agents) is created the ID is added to the appropriate vector and a new row or column is created. Agents can individually build up their assignment matrix in different orders but still store and exchange the data reliably. Update times from agents can continue to be stored in a vector s_i and are identified using I. Additionally each agent continues to store a bundle and path vector b_i and p_i respectively for calculating individual pathing and assignments.

3.2   Communication

The CBBA communicated three sets of data to nearby agents, the winning bids list y_i, the winning agent list x_i and the time stamp s_i. As each agent can order their assignment matrix differently the matrix cannot be directly communicated. A matrix $3*u$ where u is the number of assignments in X is sent to each agent. This matrix contains each assignment in the form $[j \ i \ x]$ which is used to reconstruct the senders assignment matrix X.

On initial communication with an agent their ID i and equipment list R is sent to enable agents to calculate when sensor requirements are met for a task. As a new communication is received the agent vector is updated and a new column is added to the assignment table and populated with 0's. As in the CBBA the time stamp vector s_i sent once per communication.

3.3   Bundle Construction

In phase 1 each agent constructs a bundle of tasks b_i and the ordered path for those tasks p_i. Bundle and path construction works as developed in the CBBA [8]. However task selection and costing works out differently with multi-agent tasks. The cost function for an agent k completing task j is worked in (4) where d_{kj} is the distance agent k is from task j and t_j is the time it takes to complete task j. The sum of these costs are taken away from the reward r_j of completing task j.

$$C_{kj} = r_j - (d_{kj} + t_j) \tag{4}$$

When the task is placed inside the path, d_{kj} takes the form d_{lj} where l is the previous tasks location. The value C_{kj} is used to work out whether a bid will be successful against another agent. However when working out the min cost in p_i we need to account for other agents involved in the task and their travel times, thus overall cost for task k is

$$C_{kj} = r_j - (\max(d_{mj} (\forall m \in M \; X_{kjm} > 0), d_{kj}) + t_j) \tag{5}$$

where $\forall m \in M \; X_{kjm} > 0$ finds the latest arrival time to task j out of the assigned agents in X_{kjm} and agent k. Adding together all the costs for an agents path gives us S_k^{pk} the total score for agent k with path p. We can describe the score for slotting task j into position n as $S_j^{pi \varnothing n\{j\}}$. The bundle algorithm for task assignment is taken from the CBBA [5] with the exception of the costing function and assignment limits (3).

3.4   Consensus

Phase 2 of the algorithm takes communications received from all nearby agents and analyses their knowledge on assignments to come to a consensus on which agents are doing which tasks. Using $\sum r_{pj} = 1 \; \forall p$ where r_{pj} is the number of agents with equipment p required for task j we determine whether a task requires multiple agents. Tasks requiring a single agent will require the same consensus algorithm as found in the CBBA [8]. This papers work focuses on tasks that require more than one agent and thus require a different algorithm to form a consensus between agents knowledge.

The multi-agent consensus algorithm is split into two phases; the first equates the receiver's current information with that of the sender using lines 4 to 9 of the algorithm as seen in fig. 1. Secondly the receiver takes new information from the sender and merges it with its own data through lines 10 to 24. The CBBA used a look up table for determining whether to update, leave or reset information; with the adapted problem this becomes problematic. When another agent has differing data to you it doesn't necessarily come down to leave or update the data, the information can potentially merge causing both agents to be correct. Further complications come when equipment requirements are taken into account. The algorithm is split into two phases to best handle merging the incoming data by not having to account for mistakes in our own information according to the sender's knowledge.

The first phase compares all the information the receiver i knows on assignments and compares how up to date that information is with the sender k. When $X_{ijm} > 0$ agent i believes an assignment is taking place between agent m and task j. Comparing $s_{km} > s_{im}$ we can see if the k has had newer information from the assigned agent m to task j. If k has had more recent communications then its data will be more up to date, this could be either a better bid or that the agent is no longer assigned to task j.

During the second phase we update the receiver's information with new information from the sender. From phase 1 we know that all of the receiver's data, according to the sender, is currently up to date. Following this we check every agent m that the sender k believes is assigned to each task j at line 14 in fig. 1. Whenever an agent complies with line 15 such that agent m is not the receiver or currently assigned to the task by the receiver we should compare assignments and potentially update the receiver's information. For tasks that contain a space for the agent m with equipment r_p the assignment matrix can be directly updated with the value from the sender such that $X_{ijm} = X_{kjm}$.

Algorithm 1: Conflict Resolution for Agent i

1:   **send** X_i and s_i to agent k with $g_{ik}(t) = 1$
2:   **receive** X_k and s_k from agent k with $g_{ik}(t) = 1$
3:   **for** X_{ij} $\forall j \in J$ **do**
4:       **if** $X_{ijm} > 0$ and $m \neq i$ $\forall m$ **then**
5:           **if** $s_{km} > s_{im}$ or $m = k$ **then**
6:               $X_{ijm} = X_{kjm}$
7:           **end**
8:       **end**
9:   **end**
10: **for** X_{ijm} $\forall j \in J$ $\forall m \in A$ **do**
11:     **if** $m \neq i$ and $X_{ijm} = 0$ and $X_{kjm} > 0$ **then**
12:         **if** $(\sum (X_{ijn} > 0)$ $\forall r_{pn}) < \sum r_{pj}$ $\forall p \in r_{pm}$ **then**
13:             $X_{ijm} = X_{kjm}$
14:         **elseif** $(\min X_{ijn}$ $\forall n) < X_{kjm}$ and $r_{pn} = r_{pm}$ **then**
15:             $X_{ijn} = 0$
16:             $X_{ijm} = X_{kjm}$
17:         **elseif** $(\min X_{ijn}$ $\forall n) = X_{kjm}$ and $r_{pn} = r_{pm}$ **then**
18:             **if** $(a_m > a_n)$ **then**
19:                 $X_{ijn} = 0$
20:                 $X_{ijm} = X_{kjm}$
21:             **end**
22:         **end**
23:     **end**
24: **end**

Fig. 1. Algorithm for multi-agent multi-task conflict resolution in the CBGA

When a task is full or all the relevant equipment is available, we can replace an agent for the task j given a better bid. We must replace an agent that is carrying at least one piece of identical equipment. Lines 14 and 17 in fig. 1 find an agent with a piece of matching equipment that either has a lower score or an identical score. If this condition is met then we can replace the lower scored agent n by setting $X_{ijn} = 0$ and update the new assignment with $X_{ijm} = X_{kjm}$.

To avoid any chance of deadlocking where agents alternate between who is assigned due to equal scores, the agent with the highest ID gets priority. It's a simple systematic system to guarantee a winner despite equal scores.

4   Performance

4.1   Test Scenario

Each simulation will contain 20 tasks with a varying number of agents. The overall score of the simulation is the sum of all rewards for completed tasks minus the cost of agents traveling to that task. Multi-agent tasks are defined as task requiring more than one agent and will reward an increased score over single-agent tasks. Task complexity will vary throughout the experiments as we modify both the amount of agents required to complete a task and the set of equipment required to complete a task. As we modify each we will look at the overall impact on the score and the amount of communications required. A single instance of communication for an agent is counted as receiving the assignment information another agent. Each experiment was run 200 times and the average data for all agents recorded.

4.2   Results

A task becomes a multi-agent task when it requires more than one agent to complete it. To test the effect of multi-agent tasks we must compare the difference to that of the CBBA where each task requires a single agent.

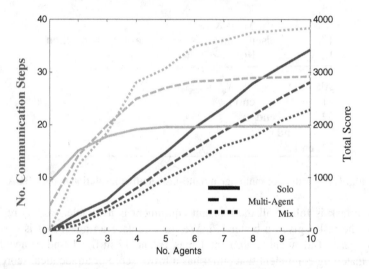

Fig. 2. Total Score and Communication steps between individual tasks, multi-agent tasks and both

In figure 2 we have 3 experiments plotted, in each experiment we used 20 tasks and increased the number of agents in each simulation. The first experiment 'solo' functioned with the CBBA where each task required a single agent. The 'multi-agent' experiment required 2 agents for every task and the 'mixed' experiment had a split between multi-agent tasks and regular tasks. Whilst initially we are inclined to think that complicating the simulation with higher requirements on tasks would produce an increase in communication, we actually find the opposite to be true. As multi-agent tasks are introduced we see that the average number of communications per agent decreases.

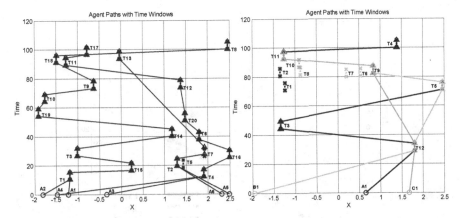

Fig. 3. Agent movement through time and the X axis. Multi-Agent tasks (left) and Multi-Agent, Multi-Equipment tasks for Exp. 3 (right).

Using figure 3 we can evaluate why this communication drop is observed with multi-agent targets. Between $t = 0$ and 20 each agent moves towards its initial task with another agent to help complete it. After completing the first task it makes sense for agents to stay together for the next task with the closest task yielding the best reward, therefore neither agent needs to dispute this choice. Occasionally two groups may attempt the same task which will then require consensus but on average each self-made group continues through the simulation effectively as one entity.

As we expand the tasks further we now introduce heterogeneous agents. In figure 4 we have experiment 1 where two agent types A and B complete solo tasks half requiring agent type A and half requiring agent B. Experiment 2 contains the same scenario as found in experiment 1 except now we've added another type of task that requires both A and B. Finally experiment 3 contains three equipped agents and three different tasks requiring agents A, AB and ABC. What we notice again is a significant reduction in the communication required to meet a consensus noticeably more so once we introduce a task requiring all three equipped agent types. These results might be derived from the time constraints on the tasks which will limit the available options from the maximum 20, tasks down to a much easier to manage set of the earliest obtainable. In Figure 3, for Multi-Agent, Multi-Equipment tasks, we can see how agent C has very little choice in his direction. Therefore it must depend on its teammates to arrive and aid at its tasks. Agent A freely moves between its tasks and when required aid its teammates. The reduced options for each agent greatly reduces the need for communicating between team mates, assuming they're all formulating decisions the same way.

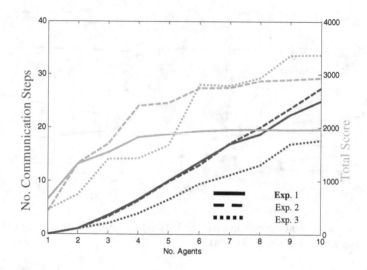

Fig. 4. Comparison of Total Score and No. Communication Steps.

5   Conclusions

This paper presents an extension of the CBBA that solves the multi-agent multi-task assignment problem with group and equipment based dependencies. A new data storage system is proposed to allow agents to deal with multi-agent multi-task assignments, progressing towards dealing with a dynamic environment. Without such a design consensus between agents becomes difficult and storage of data inconsistent with the varying number of agents per tasks.

Communication increases were expected with more complicated tasks that require more information to reach consensus. However as found in practice, a general decrease in communications is made; though the actual size of data messages have increased. These results might be derived from the time constraints on tasks which limit the available options from the maximum 20 tasks down to a more manageable set of the earliest obtainable tasks. Added to this with constraints on which agents can perform each task further reduces the set of achievable tasks for any one agent. The necessity to bid over tasks and form consensus gradually disappears as we tighten the restrictions on each task. In some cases agents do nothing as they're not required, although it's far better for the group as a whole if they don't move, therefore causing no cost on travel or an increase in communications. This situation potentially brings the overall average communication down, where the agents working towards tasks are communicating more than the averages would suggest.

As we increase the number of agents in the simulation the overall score maxes out as agents complete all the available tasks. Eventually the only increase in score is caused by a greater chance of an agent starting near a task than is the case with fewer agents. For multi-agent problems agents group up and stick together in some cases for the entire task. With increasingly complicated group and equipment requirements we found groups continue to work together where possible but often an agent will leave to complete another task and merge back again at a later task. In some respects we can say that when cooperation is a requirement it in fact simplifies the problem rather than complicates it.

References

1. Schneiderman, R.: Unmanned Drones are Flying High in the Military/Aerospace Sector (Special Reports). IEEE Signal Processing Magazine 29(1), 8–11 (2012)
2. Martinez-Val, R., Perez, E.: Aeronautics and astronautics: recent progress and future trends. Part C: Journal of Mechanical Engineering Science 223(12), 2767–2820 (2009)
3. Jones, D.: Power line inspection - a UAV concept. The IEE Forum on Autonomous Systems (Ref. No. 2005/11271), 8–28 (November 2005)
4. Maza, I., Caballero, F., Capitan, J., de Dios, J.M., Ollero, A.: Experimental results in multi-UAV coordination for disaster management and civil security applications. Journal of Intelligent and Robotic Systems 61, 563–585 (2011)
5. Lin, L., Goodrich, M.A.: UAV intelligent path planning for Wilderness Search and Rescue. In: IEEE/RSJ International Conference on Intelligent Robots and Systems, IROS 2009, October 10-15, pp. 709–714 (2009)
6. Sujit, P.B., Beard, R.: Multiple MAV Task Allocation using Distributed Auctions. In: Proc. of the AIAA Guidance, Navigation and Control Conf. (2007)
7. Hoeing, M., Dasgupta, P., Petrov, P., O'Hara, S.: Auction-Based Multi-Robot Task Allocation in COMSTAR. In: Proc. Int. Conf. on Autonomous Agents and Multiagent Systems (2007)
8. Choi, H., Brunet, L., How, J.P.: Consensus-Based Decentralized Auctions for Robust Task Allocation. IEEE Transactions on Robotics 25(4), 912–926 (2009)
9. Lo, V.M.: Heuristic Algorithms for Task Assignment in Distributed Systems. IEEE Transactions on Computers 37(11), 1384–1397 (1998)
10. Kopřiva, Š., Šišlák, D., Pěchouček, M.: Sense and Avoid Concepts: Vehicle-Based SAA Systems (Vehicle-to-Vehicle). In: Angelov, P. (ed.) Sense and Avoid in UAS: Research and Applications. John Wiley & Sons, Ltd., Chichester (2012)
11. Pastor, E., Lopez, J., Royo, P.: UAV Payload and Mission Control Hardware/Software Architecture. IEEE Aerospace and Electronic Systems Magazine 22(6), 3–8 (2007)

From e-Learning to m-Learning:
Context-Aware CBR System

Henda Ouertani Chorfi[1], Aise Zülal Sevkli[2], and Fatiha Bousbahi[1]

[1] King Saud University, College of Computer and Information Sciences,
Information Technology Department, Riyadh, Saudi Arabia
{houertani,fbousbahi}@ksu.edu.sa
[2] Fatih University, Department of Computer Engineering, Büyükçekmece, Istanbul, Turkiye
zsevkli@fatih.edu.tr

Abstract. Mobile learning extends e-learning, from indoors to outdoors by giv-ing learners opportunities to improve their skills when and where needed. By this way, the mobile device can be a powerful tool for learners to acquire in-formation and knowledge. However, one of the biggest challenges in mobile learning is addressing the needs of a varied learner type across a wider variety of devices in different contexts. These new needs faces us to take into account not only users' preferences and devices capabilities but also environmental cha-racteristics. Our main focus in this article is to discuss issues related to e-learning versus m-learning and the design of a mobile learning system based on Case-Based Reasoning approach taking into consideration context-aware of de-vice added to users' preferences and devices' capabilities.

Keywords: m-learning, mobile device, Case-Based Reasoning, content adaption.

1   Introduction

In the e-learning systems developed along the last two decades, developers focused on the user characteristics in order to adapt the proposed content to the user needs and preferences. All the information about learner preferences, knowledge and behavior is accumulated and treated in a user model which is a kind of repository about the user and forms the heart of a learner centric and adaptive system. User model issued to drive instructional decisions in order to make an adaptive e-learning system for indi-vidual students [5].

When coming with the mobile learning, the wide variety of technical characteristic and standards of devices (notebook computers, cellular phones, Personal Communica-tion System (PCS), Personal Digital Assistants (PDAs)...) leads us to take into ac-count new features in the adaption process: the device "preferences". So, delivering tailored contents tend to adapt to not only learner's needs and preferences, but also to mobile device used.

Moreover, mobile learning is not only addressing the needs of a varied learner type across a wider variety of devices but also is dealing with different contexts. These

T. Huang et al. (Eds.): ICONIP 2012, Part IV, LNCS 7666, pp. 528–534, 2012.
© Springer-Verlag Berlin Heidelberg 2012

face us to take into account not only users' preferences and devices capabilities but also environmental characteristics. We talk about context awareness applications.

With the mobility of applications, context awareness gets attraction by researchers. Although there is no unified definition of context , most of researchers agree on definitions of [13] and [7] who first introduced the term of context as any information that can be used to characterize the situation of entities that are relevant to interaction between a user and an application. The entity can be a person, place or object. If application senses changing of any information relevant to behavior or attribute of entities, the application is called context aware application. The sensibility of changing entities can be categorized into three types of context:

— Computing context: available processors, user devices, network capacity, communication cost and communication bandwidth
— User context: user profile , location, collection of nearby people and social situation
— Physical context: lighting and noise, traffic condition, temperature, time.

In this context, to achieve adaptive and smart systems, our hypothesis is that Case-Based Reasoning (CBR) is a promising way. Many works have clearly shown the potential of CBR. Ma et al. [10] use CBR for adapting the behavior of smart homes to users' preferences. In [6], using CBR techniques is twofold: on one hand, it allows to minimize the number of questions to ask to the student. On the other hand, it minimizes the time for finding a new solution (personalized course) by adapting previous ones.

In this work, we study how CBR can be a suitable method for achieving a context aware adaptive system. We propose architecture for mobile learning where we integrate the user model, the device characteristics and the environmental features.

The paper is structured as follows: in the second section, we carry out a comparison between e-learning and m-learning. The third section presents the system architecture. In fourth section, details on modules of proposed system are given. Finally some conclusions and future work are remarked.

2   Differentiating Electronic Learning from Mobile Learning

In the literature, the most recurrent definitions of mobile learning are: "mobile learning is the intersection of mobile computing and e-learning: accessible resources wherever you are, strong search capabilities, rich interaction, powerful support for effective learning, and performance-based assessment. E-learning independent of location in time or space" [12] and "Because mobile devices have the power to make learning even more widely available and accessible, mobile devices are considered by many to be a natural extension of e-learning" [4].

But we believe that it makes more sense when we consider that m-learning is complementary to e-learning as it was e-learning for traditional learning. Indeed, m-learning can enhance e-learning. For example, without an access to computers and Internet, m-learning permits learners improving their already learned courses (in an e-learning environment) with additional features such as alerts, personalized aids and dedicated training. To better situate each learning mode, we summarize in the table below the differences between e-learning and m-learning.

Table 1. Comparisons between e-learning and m-learning

e-learning	m-learning
Learning in classroom or internet labs	Learning anywhere and anytime
Private location	No geographic limitations
Learning can be personalized	Learning has to be personalized
Large units of information	Small units of information
connected to something, tethered	Untethered
Can be real-time or self-paced,	Often self-paced
Presented in a formal and structured manner	Often presented in an informal manner

3   System Architecture

We propose a Mobile Learning Adaption System (MLAS) which applies CBR approach to determine the appropriate content for the learner depending on the whole context (leaner profile, device characteristics and environment state). The advantage of using CBR techniques is principally to minimize the time for finding a new solution (personalized content) by adapting or re-using previous ones. Figure 1 shows the general architecture of MLAS.

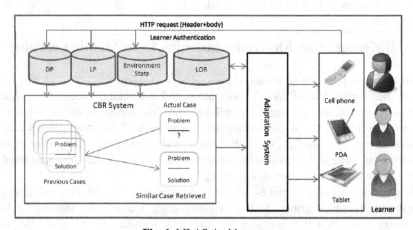

Fig. 1. MLAS Architecture

In this system, we manage four types of data: data about the user (Learner Profile (LP)), data about the device (Device Profile (DP)), data about the environment (Environment state) and the content that will be delivered to the learner (Learning Objects Repository (LOP)).

The MLAS is based on two systems: the CBR system and the adaption system. In the following sections, we give a brief description of each MLAS component.

3.1 Learner Profile

Students achieve higher learning performance if the learning content can be customized and offered according to their diverse learning needs [3]. In MLAS the learning content is generated according to the learner profile (LP) and his feedback taking into consideration the device capabilities.

In the learner profile, we store data about the learner's personnel information (name, gender, level etc.), cognitive information (score, time taken, date of last access etc.) and the learner's preferences in term of multimedia (the desired maximum delivery time, image format, presentation style, the ratio of audio to picture, animation, etc.) [8].

3.2 Device Profile

In the market there are many different types of wireless devices and each type has different features and capabilities. When the learner requests content via wireless device, the MLAS should detect learner device and send appropriate content according to device features and capabilities. Our system achieves device detection problem by using some information such as user-agent, profile headers, etc. in the header of HTTP request send by learner.

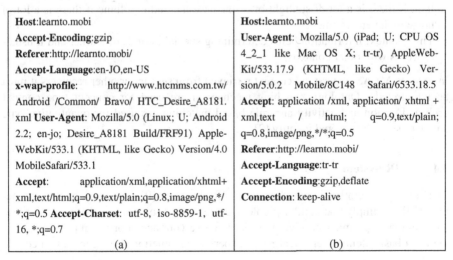

Host:learnto.mobi	**Host**:learnto.mobi
Accept-Encoding:gzip	**User-Agent**: Mozilla/5.0 (iPad; U; CPU OS
Referer:http://learnto.mobi/	4_2_1 like Mac OS X; tr-tr) AppleWeb-
Accept-Language:en-JO,en-US	Kit/533.17.9 (KHTML, like Gecko) Ver-
x-wap-profile: http://www.htcmms.com.tw/	sion/5.0.2 Mobile/8C148 Safari/6533.18.5
Android /Common/ Bravo/ HTC_Desire_A8181.	**Accept**: application /xml, application/ xhtml +
xml **User-Agent**: Mozilla/5.0 (Linux; U; Android	xml,text / html; q=0.9,text/plain;
2.2; en-jo; Desire_A8181 Build/FRF91) Apple-	q=0.8,image/png,*/*;q=0.5
WebKit/533.1 (KHTML, like Gecko) Version/4.0	**Referer**:http://learnto.mobi/
MobileSafari/533.1	**Accept-Language**:tr-tr
Accept: application/xml,application/xhtml+	**Accept-Encoding**:gzip,deflate
xml,text/html;q=0.9,text/plain;q=0.8,image/png,*/	**Connection**: keep-alive
*;q=0.5 **Accept-Charset**: utf-8, iso-8859-1, utf-	
16, *;q=0.7	
(a)	(b)

Fig. 2. Samples of the HTTP headers (a) send by HTC Desire smart phone (b) send by iPod 1.0

The HTTP headers of two wireless devices received at the server-side (learn-to.mobi) are shown in Figure 2. Although the headers provide basis information about device such as device model, manufacturer, client device's OS version, browser version, Java capabilities, etc. MLAS may need different features than basis ones such as screen size, storage, streaming, sound format, and image format, etc.

As a result, the device profile in MLAS should support all information about device capabilities and features for a variety of wireless devices. There are some projects which collect all device features in one file. Wireless Universal Resource File (WURFL) is one of these projects. It presents a set of proprietary APIs and XML configuration file. The detail information about WURFL is given in [11].

In our proposed MLAS, the component of device profile first analyzes HTTP request to detect device model then uses WURFL to access and retrieve more features of user device.

3.3   Environment State

In the Environment state we collect data about four characteristics of context information as in [7]: identity, location, activity (status) and time.

— Identity provides uniqueness of entity that is relevant to the application.
— Location is the position information of entity in 2D space. It can be received from GPS receiver. Location information is used to tell the place where the learner / learner's friends exactly is/are located or is used to list nearby places according to learners interest. For example, if the learner moves from outdoors to indoors, a different network technology may be selected.
— Activity (status) identifies fundamental characteristics of the entity that can be sensed. MLAS stores temperature and light or noise level for places and calendar activities for learners. For example it helps to adjust the screen to not reflect the light if there is a lot of sunlight or to increase the audio volume if there is a lot of noise in the environment.
— Time context information helps determining special periods of learner and it usually works with other contexts.

In order to MLAS understand context retrieved from sensors or other receivers it must be interpreted. Ontology matching mechanism are used for concept type context such as location and activity and rule based matching are used for quantitative type context   such as time and identity [2].

3.4   CBR System

The content generation is based on the CBR approach [9]. The main hypothesis behind CBR is simply that similar problems have similar solutions or that you can reuse the solution of a similar problem in order to solve your actual problem [14]. A case is the most basic element representing an experienced situation. It is, generally a couple of (problem, solution). In our MLAS, the problem corresponds to the learner profile, the device characteristics and the environment state. The solution consists on the rules to be applied to adapt the delivered content.

When a new learner logs in and in order to construct a new problem, the system gets device, environment and learner information from the learner request, the applications detecting the environment state, the device profile and the learner profile.

When a new case is constructed, the CBR system retrieves the most similar problem among the existing problems in the case base. For that, the system calculates its similarity with the cases of the case base. The algorithm to compute similarity uses the Nearest Neighbour Algorithm [1].

We consider that two cases are similar if their similarity is lower than a certain threshold. This similarity is calculated to permit the system to minimize time of constructing a new solution for the current problem by using or adapting the solution of an old and similar case stored in the case base.

Different forms of adaptation exist, such as null adaptation, transformational adaptation (including substitutional and structural adaptation), and generative adaptation [14]. Null adaptation simply applies the solution from the retrieved case to the target case. Since users are not categorized in MLAS, null adaptation is used.

When the case is constructed (problem and solution), it is sent to the adaptation system in order to construct the adaptive content based on the rules in the solution.

3.5   Adaptation System

The adaptation system takes the responsibility of adapting the content to be delivered to the learner. It creates the adaptive contents based on rules. The rules are constructed by the CBR system in the solution depending on learner profile, device characteristics and environment state. Each rule is associated to a conversion function or a filtering process. For example, to provide each mobile device with the markup language that supports (WML, CHTML, XHTML), we use the WALL library [11]. Also, if the device supports GIF format and the LOR contains, only JPG format, the system should create a new GIF image based on original JPG image dynamically (the LOR will be consequently updated). The different conversion rules will be described in future work.

4   Conclusion

In this paper, we propose a new architecture for mobile learning system based on CBR approach. CBR strenghts the m-learning system by storing previous cases to be reused when solving the actual one. Our system shortens the retrieval time of content by reusing similar cases in an intelligent way. Another strong aspect of this system is that it is deals with different data to adapt the content to the learner : characteristics of learner and device and the environment state.

The system is under implementation and we would like in future work to test it on real problem.

References

1. Aha, D.W.: Case-based learning algorithms. In: Proceedings of the DARPA Case- Based Reasoning Workshop, pp. 147–158. Morgan Kaufmann, Washington, D.C. (1991)
2. Al-Mekhlafi, K., Hu, X., Zheng, Z.: An Approach to Context-aware Mobile Chinese Language Learning for Foreign Students. In: Proceeding of Eighth International Conference on Mobile Business (2009)
3. Beekhoven, S., Jong, U.D., Hout, H.V.: Different courses, different students, same results: An examination of differences in study progress of students in different courses. High. Educ. 46(1), 37–59 (2003)
4. Brown, T.H.: Towards a model for m-learning in Africa. International Journal of E-Learning 4(3), 299–315 (2005)
5. Chorfi, H., Jemni, M.: A CBR Adaptive Hypermedia for Edcuation based on XML. In: 6th IEEE International Conference onAdvanced Learning Technologies, ICALT, Kerkrade, The Netherlands, pp. 1092–1096 (2006)
6. Chorfi, H., Jemni, M.: PERSO: Towards an Adaptive e-Learning System. Journal of Interactive Learning Research 15(4), 433–447 (2004)
7. Dey, A.K., Abowd, G.D.: Towards a better understanding of context and context-awareness. In: Proceedings of the Workshop on the What, Who, Where, When and How of Context-Awareness, affiliated with the CHI 2000 Conference on Human Factors in Computer Systems. ACM Press, New York (2000)
8. Hassan, M., Al-Sadi, J.: A New Mo-bile Learning Adaptation Model. International Journal of Interactive Mobile Technologies (iJIM) 3(4) (2009)
9. Kolodner, J.L.: Case-based reasoning. Morgan Kaufmann, San Mateo (1993)
10. Ma, T., Kim, Y.D., Ma, Q., Tang, M., Zhou, W.: Context-aware implementation based on cbr for smart home. In: IEEE Wireless and Mobile Computing, Networking and Communications, WiMob 2005, pp. 112–115. IEEE Computer Society (2005)
11. Passani, L., Kamerman, S.: WURFL Project (2011),
http://wurfl.sourceforge.net/backgroundinfo.-php
12. Quinn, C.: mLearning. Mobile, Wireless, In-Your-Pocket Learning. Linezine (Fall 2000),
http://www.linezine.com/2.1/features/cqmmwiyp.htm
13. Schilit, B., Theimer, M.: Disseminating active map information to mobile hosts. IEEE Network 8(5), 22–32 (1994)
14. Wilke, W., Bergmann, R.: Techniques and Knowledge Used for Adaptation During Case-based Problem Solving. In: Mira, J., Moonis, A., de Pobil, A.P. (eds.) IEA/AIE 1998. LNCS (LNAI), vol. 1416, pp. 497–505. Springer, Heidelberg (1998)

Extreme Learning Machines
for Intrusion Detection Systems

Gilles Paiva M. de Farias[1], Adriano L.I. de Oliveira[1], and George G. Cabral[2]

[1] Federal University of Pernambuco, Recife PE 50740-560, Brazil,
{gpmf,alio}@cin.ufpe.br
[2] Federal Rural University of Pernambuco, Recife PE 52171-900, Brazil,
ggc2@cin.ufpe.br

Abstract. Information is a powerful tool that can be used as a competitive advantage to increase market shares, competitiveness and keep products up-to-date. Protecting the information is a difficult task; intrusion detection systems is one of the tools of great importance for the protection of computer network infrastructures. IDSs (Intrusion Detection Systems) are tools that help users and network administrators to keep safe from intruders and attacks of various natures. Machine learning techniques are one of the most popular techniques for IDSs proposed and investigated in the literature. This paper focuses on the use of ELM (Extreme Learning Machine) and OS-ELM (Online Sequential ELM) techniques applied to IDSs. Some features of these methods that motivate their use for building IDSs are: (i) easy assignment of parameters; (ii) good generalization; and (iii) fast and online training. The results show that the methods can be easily applied to a huge amount of data without a significant generalization loss.

Keywords: Intrusion Detection Systems, Extreme Learning Machines, Online Learning.

1 Introduction

Many business and government organizations use computer networks for sharing information. In this scenario, security is an important aspect. Over the years, computer networks, as well as malicious intrusion techniques, have been improved. Performing only human audit in data is no longer feasible due to the huge amount of information. Thus, a number of algorithms have been proposed for intrusion detection, including AI (Artificial Intelligence) techniques, which were proposed to improve the detection accuracy. The most adopted intelligent methods are Artificial Neural Networks (ANNs) [11], but other techniques are also used, such as genetic algorithms [10], Agents and support vector machines (SVMs) [12]. According to Kemmerer and Vigna [4], system auditing is useless unless the resulting information is analyzed. The approach used to analyze the collected data is also important and there are basically two directions: anomaly detection and misuse detection.

T. Huang et al. (Eds.): ICONIP 2012, Part IV, LNCS 7666, pp. 535–543, 2012.

The *anomaly detection* approach attempts to model the user or application ordinary behavior; next, significant deviations from this behavior are regardedas possible intrusion events. In this case, if a user or application changes their routine of network activities, the system must alarm this changing as a possible intrusion. An advantage of this approach is that previously unknown attacks can be detect. As n disadvantage, this approach may provide relatively high false-positive rates(non malicious behavior detected as an intrusion) . Furthermore, it is a hard task to model what is a normal behavior since users can gradually modify their activities without characterizing an abnormal behavior.

In the *misuse detection* approach, attacks are represented by patterns or signatures, according to Mukkamala et. al. [7]. Misuse detection matches these patterns against the data to be audited looking for evidences of known attacks [4]. Misuse detection yields lower error rates than anomaly detection [8]. However, this approach is not able to detect unknown attack methods. The main issue is how to build permanent signatures having all the possible variations to avoid false-negatives and how to build signatures of non-intrusive activities to avoid false-positives [7].

In this paper, recent machine learning methods, namely Extreme Learning Machine (ELM) [3] and its online version (OS-ELM) [6] are investigated for the intrusion detection task. ELM has as major advantages its fast convergence and the generalization power, compared to more popular techniques such as MLP neural networks [2]. The online nature of the data stream of the networks suggests the use of the OS-ELM. In this case, training is performed online with the data continuously arriving and being incorporated to the model in blocks, referred to as chunks. Therefore, the main motivation behind the present work is not just to seek for the smallest classification error, but also to try to find a model able to fast incorporate new data keeping a good generalization power.

This paper is organized as follows. Next Section briefly presents some related works. Section 3 details the ELM and OS-ELM methods. Section 4 presents the experiments conducted and the analysis of the results obtained. Finally, in Section 5 conclusions are presented.

2   Related Works

An IDS (Intrusion Detection System) is a security system that monitors a computer network aiming at preventing unauthorized access and threats (which may be internal or external to the network).

One of the pioneer works published in the subject of security systems was conducted by James P. Anderson in 1980 [1]. Despite using techniques considered archaic nowadays, a fairly complete overview of information was raised, addressing issues such as internal and external penetrations, types of illegal users, countermeasures, monitoring actions, and so on.

A wide range of works in the area can be found in the literature, addressing topics aimed at specific attacks, as shown by Li and Lee [5], where a specialized solution based on wavelets was used in identification of DDoS attacks

(Distributed Deny of Service). Surveys like the one conducted by Tavallaee et al. [9] present studies which also focuses on a variety of known attacks.

Using the *misuse detection* approach, Mukkamala et al. [7] have used various techniques, such as SVM, RP (Resilient Propagation), SCG (Scaled Conjugated Algorithm), OSS (One-Step Secant Algorithm) and MARS (Multivariate Adaptive Regression Splines) to address the problem. The authors then present a solution that involves various combinations of these techniques resulting in a committee of experts called "An ensemble of ANN, SVM and MARS". They reported good results reaching rates that exceed 99% of correct detection. The authors mention that adjustments must be done in their method so that it can be feasible to be applied in real world, since the method is very complex and time consuming. The work presented some of the highest hit rates being a valuable reference regarding the IDSs subject [7].

The related works present a wide variety of techniques, as well as results and ways of assemble the experiments. A remarkable problem is the lack of a standard experimental methodology that would enable a more fair and accurate comparison of machine learning methods.

3   Extreme Learning Machines

This Section presents the two techniques investigated in this paper for the problem of Intrusion Detection.

3.1   ELM

ELM (Extreme Learning Machine) uses the Moore-Penrose inverse and the smallest norm least-square solution of a general linear system $Ax = y$ [3]. ELM is an extremely fast learning algorithm to single hidden layer feedforward networks (SLFNs), with $\tilde{N}$ hidden neurons, where $\tilde{N} \leq N$, being N the number of training samples.

Given N arbitrary distinct samples in a problem with n attributes and m classes, (x_i, t_i), where $x_i = [x_{i1}, x_{i2}, .., x_{in}]^T \in R^n$ and $t_i = [t_{i1}, t_{i2}, .., t_{im}]^T \in R^m$, an SLFN with $\tilde{N}$ hidden neurons and activation function $g(x)$ is mathematically modeled as:

$$f_{\tilde{N}}(x_j) = \sum_{i=1}^{\tilde{N}} \beta_i g(w_i \cdot x_j + b_i) = o_j, \quad j = 1..N \tag{1}$$

where $w_i = [w_{i1}, w_{i2}, .., w_{in}]^T$ is the weight vector connecting the i_{th} hidden neurons to the input neurons, $\beta_i = [\beta_{i1}, \beta_{i2}, .., \beta_{im}]^T$ is the weight vector connecting the i_{th} hidden neuron and the output neurons, and b_i is the threshold of the i_{th} hidden neuron. $w_i \cdot x_j$ is the inner product of w_i and x_j. SLFNs with $\tilde{N}$ hidden neurons and $g(x)$ activation function can approximate N samples with zero error means that exists β_i such that:

$$f_{\tilde{N}}(x_j) = \sum_{i=1}^{\tilde{N}} \beta_i g(w_i \cdot x_j + b_i) = t_j, \quad j = 1..N \tag{2}$$

Equation 2 can be written as:

$$H\beta = T, \quad \text{where} \quad H = \begin{bmatrix} g(w_1 \cdot x_1 + b_1) & \cdots & g(w_{\tilde{N}} \cdot x_j + b_{\tilde{N}}) \\ \cdots & \cdots & \cdots \\ g(w_1 \cdot x_N + b_1) & \cdots & g(w_{\tilde{N}} \cdot x_N + b_{\tilde{N}}) \end{bmatrix}_{N \times \tilde{N}} \tag{3}$$

$$\beta = \begin{bmatrix} \beta_1^T \\ \vdots \\ \beta_{\tilde{N}}^T \end{bmatrix}_{\tilde{N} \times m} \tag{4} \qquad\qquad T = \begin{bmatrix} t_1^T \\ \vdots \\ t_{\tilde{N}}^T \end{bmatrix}_{N \times m} \tag{5}$$

Equation 4 represents the weights between the hidden neurons and the output layer while in Equation 5, T is the vector containing the outputs of the network. In [3], the authors shown that the smallest norm least-square solution of Equation 3 is given by $\hat{\beta} = H^\dagger T$ ($H^\dagger$ stands for the Moore-Penrose generalized inverse [2] of matrix H).

ELM has the following interesting properties: (i) minimum training error. ELM uses one of the least-square solutions of a general linear system $H\beta = T$ to reach the smallest training error; (ii) smallest norm of weights and best generalization performance; and (iii) uses the minimum norm least-square solution of $H\beta = T$.

3.2   OS-ELM (Online Sequential ELM)

In the real world, training data may arrive chunk-by-chunk or one-by-one (a special case of chunk). In this case, the original ELM algorithm has to be modified, to become online sequential. The output weight matrix $\hat{\beta}$ is a least-square solution of Equation 3. Here, we considere that $rank(H) = \tilde{N}$ the number of hidden nodes. In this case, $H^\dagger$ is given by:

$$H^\dagger = (H^T H)^{-1} \tag{6}$$

This is also called left pseudoinverse of H. If $H^T H$ tends to be singular, it can become nonsingular by choosing a smaller network size $\tilde{N}$ or increasing the data amount N in the initialization phase of OS-ELM. Now, substituting Eq. 6 in $\hat{\beta} = H^\dagger T$:

$$\hat{\beta} = (H^T H)^{-1} H^T H \tag{7}$$

Equation 7 is the least-square solution to $H\beta = T$. The sequential implementation of the least-square solution of 7 is the OS-ELM algorithm. Given a chunk of the initial training set given by:

$$\aleph_0 = (x_i, t_i)_{i=1}^{N_0} \tag{8}$$

Considering $N_0 \geq \tilde{N}$, means that Equation 7 can be written as:

$$\beta_0 = K_0^{-1} H_0^T T_0 \quad \text{where} \quad K_0 = H_0^T H_0 \tag{9}$$

New chunks are represented as:

$$\aleph_1 = (x_i, t_i)_{i=N_0+1}^{N_0+N_1} \tag{10}$$

Where N_1 denotes the number of observations in this chunk. Now, considering H_0 and H_1, the solution of Equation 7 is presented as folows:

$$\beta^{(1)} = K_1^{-1} \begin{bmatrix} H_0 \\ H_1 \end{bmatrix}^T \begin{bmatrix} T_0 \\ T_1 \end{bmatrix}, \quad \text{where} \quad K_1 = \begin{bmatrix} H_0 \\ H_1 \end{bmatrix}^T \begin{bmatrix} H_0 \\ H_1 \end{bmatrix} \tag{11}$$

To perform a sequential learning it is necessary to express $\beta^{(1)}$ as a function of $\beta^{(0)}$, K_1, H_1 and T_1 instead as a function of the dataset $\aleph_0$. All the steps to perform this process are presented in detail in [6]. $\beta^{(1)}$ is:

$$\beta^{(1)} = \beta^{(0)} + K_1^{-1} H_1^T (T_1 - H_1 \beta^{(0)}), \quad \text{where} \quad K_1 = K_0 + H_1^T H_1 \tag{12}$$

Now, generalizing the solution to new arrival, we have the following:

$$\begin{aligned}
\beta^{(k+1)} &= \beta^{(k)} + K_{k+1}^{-1} H_{k+1}^T (T_{k+1} - H_{k+1}\beta^{(k)}) \\
K_{k+1}^{-1} &= (K_k + H_{k+1}^T H_{k+1})^{-1} \\
&= K_k^{-1} - K_k^{-1} H_{k+1}^T \\
&\quad \times (I + H_{k+1} K_k^{-1} H_{k+1}^T)^{-1} H_{k+1} K_k^{-1}
\end{aligned} \tag{13}$$

4 Experiments

The experiments were conducted using the KDD Cup 1999 data set[9]. After the competition, the database became a reference for research in Intrusion Detection, being used by several studies in the literature. KDD Cup99 attacks have 4 categories: (i) DoS (Denial of Service); (ii) R2L (Remote to Local); (iii) U2R (User to Root); and (iv) Probing.

To carry out the experiments, two new datasets derived from the original were created. The first one, referred to as $subset_1$, was created based on the experiments reported by Mukkamala et. al. [7] aiming at comparing the results to [7]. This dataset contains 11982 samples of five classes (normal, probe, DoS, R2L and U2R). The second used dataset, $subset_2$, represents 10% of KDD Cup original dataset and is available at the homepage of the competition. The second dataset contains 494021 samples; the best parameters found for $subset_1$ where also employed for the experiments in $subset_2$. For each dataset, 66% of the data was used for modeling and the remainder 34% was used for testing purposes.

To pick the best parameters configuration, an empirical search was performed and the neighborhood of the best configuration was explored. It was observed that for the same parameters, different models were achieved due to the random initialization of the weights. Thus, for each parameter configuration, thirty executions were performed and the results shown in the following tables represent the average of these executions. The OS-ELM needs one more parameter than the standard ELM, the chunk (which stands for the size of the block of data to be processed at each step).

Table 1 shows the results of the standard ELM. According to these results we can conclude that the number of neurons was not critical for the generalization performance of the ELM. The results for the training and test set show that the method generalizes well for both sets with a small standard deviation.

Table 1. Results for the experiments of the ELM applied to $subset_1$

Neurons	Training acc	Test acc	Training time
20	97.16 ($\pm$ 0.13)	97.10 ($\pm$ 0.15)	36.75 ($\pm$ 0.49)
25	97.30 ($\pm$ 0.07)	97.42 ($\pm$ 0.09)	35.78 ($\pm$ 0.47)
30	97.59 ($\pm$ 0.04)	97.37 ($\pm$ 0.03)	44.13 ($\pm$ 0.49)
35	97.58 ($\pm$ 0.01)	97.40 ($\pm$ 0.03)	40.08 ($\pm$ 0.48)
40	97.61 ($\pm$ 0.02)	97.79 ($\pm$ 0.01)	35.89 ($\pm$ 0.47)
45	97.53 ($\pm$ 0.02)	97.66 ($\pm$ 0.02)	35.16 ($\pm$ 0.50)
50	97.60 ($\pm$ 0.03)	97.76 ($\pm$ 0.02)	36.26 ($\pm$ 0.49)

Table 2 presents the results of the OS-ELM for the $subset_1$. According to Table 2, the number of hidden neurons has a higher influence in the results than the chunk size. A small chunk size makes the modeling phase faster as we can see in Table 3. Therefore, a configuration with a small chunk size and a higher number of hidden neurons yields the best OS-ELM model in terms of the combined training time and accuracy.

Table 2. Accuracy for the experiments of the OS-ELM applied to test portion of $subset_1$

Neurons\Chunk	250	500	1000	2000	4000
20	79.12 ($\pm$ 0.12)	79.68 ($\pm$ 0.12)	79.99 ($\pm$ 0.11)	80.33 ($\pm$ 0.12)	80.38 ($\pm$ 0.11)
25	78.01 ($\pm$ 0.10)	80.31 ($\pm$ 0.09)	80.02 ($\pm$ 0.09)	80.30 ($\pm$ 0.09)	82.99 ($\pm$ 0.08)
30	82.78 ($\pm$ 0.08)	85.22 ($\pm$ 0.07)	85.32 ($\pm$ 0.07)	87.40 ($\pm$ 0.06)	86.99 ($\pm$ 0.06)
35	87.91 ($\pm$ 0.08)	88.36 ($\pm$ 0.06)	93.58 ($\pm$ 0.05)	92.90 ($\pm$ 0.05)	92.55 ($\pm$ 0.04)
40	90.93 ($\pm$ 0.08)	87.93 ($\pm$ 0.07)	89.00 ($\pm$ 0.07)	89.90 ($\pm$ 0.07)	90.01 ($\pm$ 0.06)
45	89.07 ($\pm$ 0.07)	91.12 ($\pm$ 0.07)	89.92 ($\pm$ 0.06)	90.72 ($\pm$ 0.05)	91.96 ($\pm$ 0.05)
50	91.71 ($\pm$ 0.06)	91.01 ($\pm$ 0.05)	90.01 ($\pm$ 0.06)	90.01 ($\pm$ 0.06)	90.13 ($\pm$ 0.07)

Table 4 compares the results obtained by the methods investigated in this paper and other results reported in the literature for this same dataset ($subset_1$). Note that the experiments cannot be fairly compared as an identical reproduction of the same data used by other works is not possible. The results show that the methods investigated in the present work were outperformed by some

Table 3. Training time (seconds) for all the experiments depicted in Table 2

Neurons\Chunk	250	500	1000	2000	4000
20	6.86 (± 0.45)	8.02 (± 0.50)	11.03 (± 0.66)	63.00 (± 0.85)	185.07 (± 1.03)
25	6.85 (± 0.5)	7.03 (± 0.48)	11.02 (± 0.68)	65 (± 1.05)	168.14 (± 2.30)
30	7.6 (± 0.48)	7.05 (± 0.5)	11.03 (± 0.4)	70.01 (± 2.15)	173.02 (± 4.68)
35	6.63 (± 0.52)	7.06 (± 0.51)	12.00 (± 0.67)	63.03 (± 4.14)	207.04 (± 6.08)
40	6.79 (± 0.48)	8.06 (± 0.5)	23.09 (± 0.48)	62.00 (± 4.10)	287.00 (± 9.2)
45	7.45 (± 0.54)	7.98 (± 0.53)	16.12 (± 0.95)	66.1 (± 4.06)	286.03 (± 12.3)
50	7.43 (± 0.6)	7.9 (± 0.52)	16.22 (± 0.7)	64.15 (± 4.30)	285.03 (± 15.43)

of the other works, however, the methods compared cannot online incorporate new data, which is a remarkable advantage of OS-ELM. Furthermore, some of the methods of Table 4 are quite complex which make it impossible to use them in the real world.

Table 4. Comparisons among the investigated methods and other works in literature for the $subset_1$

Algorithm	SVM	RP	SCG	OSS	Ensemble of ANN [7]	MARS [7]	Ensemble of ANN, MARS and SVM [7]	ELM	OS-ELM
Sucess rate	98.85	97.09	80.89	93.64	99.30	92.75	99.82	97.79	93.58

Based on the best configuration obtained on experiments carried out on $subset_1$, the following configurations were used to perform the experiments on $subset_2$: ELM (Neurons = 40) and OS-ELM (Neurons = 35; Chunk = 1000).

Table 5 depicts the results for $subset_2$. This Table shows a small difference in the generalization quality of the results between the methods. However, the OS-ELM performs the training phase more than twice faster than the standard ELM. It is important to emphasize the size of this dataset, 494021 samples (much larger than subset1, which has only 11982 samples). Probably none of the other methods presented in Table 4 can process a dataset of this magnitude.

Table 5. Results for both classifiers applied on $subset_2$

Algorithm	Training	Test	Elapsed time (seconds)
ELM	98.82 (± 0.01)	98.84 (± 0.01)	737.99 (± 3.60) -
Os-ELM	97.37 (± 0.75)	97.34 (± 0.78)	340.070 (± 5.80)

5   Conclusion

This study aimed at investigated the use of the ELM and OS-ELM networks for Intrusion Detection Systems. The main motivations behind employing ELM networks for this problem are its fast convergence and online training mode (in the case of the OS-ELM version). Furthermore, these features do not significantly affect the quality (accuracy) of the classification. Another positive characteristic (regarding the standard ELM) is the existence of just one critical, but easy to assign, parameter: the number of hidden neurons. In the case of the OS-ELM version, it was noted that the parameter chunk was also easy to assign.

The nature of the problem addressed in this work (where the data arrives continuously) suggests the use of an online classifier. The OS-ELM has shown to be effective since a bunch of data is discarded in each iteration and new characteristics are incorporated to the model. Regarding generalization power, the online version of the ELM was outperformed by ELM, however, the training time was much smaller, which is important in practice. Furthermore, its operation mode is quite compatible with the addressed problem.

The results show that both ELM methods achieved a good classification rate, however, this work is not mainly concerned in the best classification rates, since a feasible solution consists in a trade-off between the classification rate and the time spent for modeling.

References

1. Anderson, J.P.: Computer security threat monitoring and surveillance. In: Fort Washington, Technical relatory (1980)
2. Huang, G.-B., Zhou, H., Ding, X., Zhang, R.: Extreme learning machine for regression and multiclass classification. IEEE Transactions on Systems, Man, and Cybernetics 42, 513–519 (2012)
3. Huang, G.-B., Zhu, Q.-Y., Siew, C.-K.: Extreme learning machine: a new learning scheme of feedforward neural networks. In: Proceeding of International Joint Conference on Neural Networks, vol. 2, pp. 985–990. IEEE (2004)
4. Kemmerer, R.A., Vigna, G.: Intrusion detection: a brief history and overview. Computer 35, 27–30 (2002)
5. Li, L., Lee, G.: Ddos attack detection and wavelets. In: Proceedings of The 12th International Conference on Computer Communications and Networks, ICCCN 2003, pp. 421–427. IEEE (2003)
6. Liang, N.Y., Huang, G.B., Saratchandran, P., Sundararajan, N.: A fast and accurate online sequential learning algorithm for feedforward networks. IEEE Transactions on Neural Networks 17, 1411–1423 (2006)
7. Mukkamala, S., Sung, A.H., Abraham, A.: Intrusion detection using an ensemble of intelligent paradigms. Journal of Network and Computer Applications 28, 167, Science Direct (2005)
8. Shiri, F.I., Shanmugam, B., Idri, N.B.: A parallel technique for improving the performance of signature-based network intrusion detection system. In: International Conference on Communication Software and Networks, pp. 692–696. IEEE (2011)
9. Tavallaee, M., Bagheri, E., Lu, W., Ghorbani, A.A.: A detailed analysis of the kdd cup 99 data set. In: IEEE 2009 Symposium on computational Intelligence in Security and Defense Applications (CISDA 2009), pp. 1–6. IEEE (2009)

10. Tian, J., Gao, M.: Network intrusion detection method based on high speed and precise genetic algorithm neural network. In: International Conference on Networks Security, Wireless Communications and Trusted Computing, vol. 2, pp. 619–622. IEEE (2009)

11. Tian, J., Gao, M., Zhang, F.: Network intrusion detection method based on radial basic function neural network. In: International Conference on E-Business and Information System Security (EBISS 2009), pp. 1–4. IEEE (2009)

12. Xiaoqing, G., Hebin, G., Luyi, C.: Network intrusion detection method based on agent and svm. In: The 2nd IEEE International Conference on Information Management and Engineering, pp. 399–402. IEEE (2010)

A Self-Organizing Maps Multivariate Spatio-temporal Approach for the Classification of Atmospheric Conditions

Kostas Philippopoulos and Despina Deligiorgi

National and Kapodistrian University of Athens, Department of Physics,
Division of Environmental Physics and Meteorology, Greece
despo@phys.uoa.gr

Abstract. This work demonstrates the potential of Self-Organizing Maps (SOM) as a multivariate clustering approach of spatio-temporal datasets in atmospheric physics. A comprehensive framework is proposed and the method is applied and assessed for its performance in the field of synoptic climatology within a specific region at southeastern Mediterranean. The results indicate that the SOM can be a powerful tool for the identification and classification of atmospheric conditions, allowing an analytical description of the principal atmospheric states. The coupling of sea level pressure (SLP) and 500hPa geopotential (Φ500) in a synoptic-scale domain with the wind, the specific humidity and the air and dew point temperature in the chosen mesoscale subdomain, allows the SOM algorithm to define the relevant atmospheric circulation patterns. The corresponding patterns are well documented and the method accounts for their seasonality. Furthermore, in the resulting two-dimensional lattice the similar patterns are mapped closer to each other, compared to more dissimilar ones.

Keywords: self-organizing maps, atmospheric circulation, synoptic climatology, pattern recognition.

1 Introduction

Synoptic climatology is defined as the linkage of atmospheric circulation and environmental response [1] as it examines the relationship of large-scale circulation with the regional and local scale climate. An important aspect of climatological research is the classification of atmospheric variables into distinct patterns and relating this information with a wide range of meteorological phenomena. Atmospheric circulation classification is principally associated with the grouping of SLP and geopotential height in a small number of discrete circulation types for analyzing the variability of atmospheric circulation in terms of their frequency changes on different temporal and spatial scales [2]. The classification schemes can be divided into subjective and automated methods, depending on the procedure that is used to assign atmospheric fields into the resulting classes. The subjective schemes employ the expert's knowledge for identifying and allocating each day to the

T. Huang et al. (Eds.): ICONIP 2012, Part IV, LNCS 7666, pp. 544–551, 2012.

circulation types and are typically based on the visual analysis of daily weather maps. On the contrary, the automated classification schemes essentially employ statistical methods for analyzing atmospheric data, with the objective of generating groups of cases with increased internal similarity and at the same time increased external separability. An extensive review on the classifications of atmospheric circulation patterns is presented in [3] and a database of weather and circulation type classification schemes for the European continent in [4]. The SOMs in synoptic climatology are proposed as an alternative method for classifying atmospheric conditions and according to [5] they differ from traditional clustering algorithms from the way the circulation patterns are identified and that the SOM presents an effective means of visualizing the relationships between the resulting patterns. A recent review on SOM applications in meteorology and oceanography is presented in [6] and in synoptic climatology in [7]. Atmospheric circulation constitutes more a continuum than a system with clearly defined and separated states [3] and for its classification, ideally multiple atmospheric variables from various atmospheric levels and spatial domains are required. The novelty of this work is the simultaneous classification of atmospheric fields from two domains and for multiple atmospheric levels, aiming to produce a comprehensive classification of atmospheric conditions, focused on the eastern Mediterranean. In this paper the results of the classification procedure will be discussed in terms of the corresponding atmospheric circulation patterns.

2   Methodology

The aim of this work is to propose an analytical neural network based clustering framework and to examine its performance in the field of synoptic climatology. The SOMs are mainly used for classifying the spatial distribution of a single atmospheric variable (SLP or geopotential height) in a synoptic-scale domain. The proposed framework extends the SOM methodology to multivariate fields, which can be extracted for different spatial domains, following a four-step approach:

- Definition of the relevant variables and corresponding domains
- Data-preprocessing and dimension reduction
- SOM pattern recognition and classification
- Visualization and interpretation of the results.

The definition of the relevant variables and the corresponding domains involves the identification of the appropriate spatial and temporal scales, depending on the atmospheric process under study. Upon the selection of the initial spatio-temporal dataset, the time-series are standardized separately via the calculation of the z-scores, using the corresponding mean and standard deviation values. Subsequently the Principal Components Analysis (PCA) in S-mode is recommended as a pre-processing tool for data reduction purposes. The PCA analysis performs an orthogonal linear transformation, aiming to reduce the dimension of the dataset by finding the linear combinations of the initial variables (principal components - PCs) with the largest variance. The SOM classification procedure is performed on the PC scores of the first PCs that explain more than a predefined percent of the initial variance. The SOM algorithm, proposed by Kohonen [8], is a neural network used for clustering, feature extraction and data visualization. It is an especially powerful visualization tool, as it converts complex, non-linear statistical relations between high-dimensional data into simple geometric relations at a low-dimensional display [9]. The SOM neural network model consists of an input layer and a two- dimensional lattice of neurons (the output or competitive layer), which is fully connected to the input space. Initially the number of neurons is selected and their weight vectors are initialized randomly. Subsequently a training vector is presented to the network and the Euclidean distances between the training vector and the neurons' weight vectors are calculated. The neuron that produces the smallest distance is called the Best Matching Unit (BMU) and its weight vectors along with its neighboring neurons weight vectors are updated towards the input vector. The input vectors are presented sequentially in the network and by using iterative training the neurons are adjusted in a way that different parts of the SOM respond similarly to certain input patterns. The final part of the SOM method is the visualization of the results, where each training vector is associated with one neuron, which represents the resulting patterns of the classification process. The inherent drawback of nonhierarchical clustering algorithms, such as the SOM, is the requirement of predefined number of nodes. Under this framework and for overcoming this weakness, multiple SOM configurations should be examined and the optimum architecture can be determined using quantitative or/and qualitative criteria.

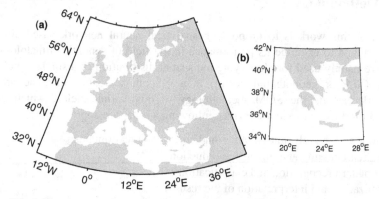

Fig. 1. European continent domain – Domain1 (a) and Greece subdomain (b)

3   Application to Atmospheric Circulation

The application of the methodology is focused at the Eastern Mediterranean and Greece (Fig. 1) and employs a subset from the full resolution ERA-Interim Reanalysis dataset, produced by the European Centre for Medium-Range Weather Forecasts (ECMWF) [10]. SLP and geopotential at 500hPa data are extracted for the European continent domain (Domain1), while the zonal and meridional wind components at 10m and at 850hPa, the specific humidity at 700hPa and the air and dew-point temperature at 2m are extracted for the Greece subdomain (Domain2). The two domains are presented in Fig. 1 and the selected variables are obtained on a daily temporal scale at 12Z, for a 31-year period (1979-2009) with 0.75°x0.75° spatial resolution. The values of the seven atmospheric variables at each grid point form a 9036x1 daily vector and the classification scheme treats each of the 11323 days as a different object. For the purposes of this work a two-dimensional display (lattice) is selected and various hexagonal SOM topologies are examined with the number of nodes ranging from 16 to 40, with varying number of row and column neurons. The optimum SOM architecture is selected using qualitative subjective criteria by examining according to [11] in each case the corresponding results and deciding the optimal number of classes.

4   Results and Discussion

According to the PCA of the initial 11323x9036 dataset, the 99% of the variance is explained from the first 113 PCs and the SOM classification was performed on the corresponding PCs score matrix (11323x113). The optimum configuration consists of 32 nodes, organized into an 8x4 array (Fig. 2). Some general remarks regarding the resulting SOM classification is the ability of the method to produce physical meaningful states, with distinct atmospheric circulation patterns (Fig. 3). The results are in agreement with other atmospheric circulation classifications, focused on the same geographical region [12-14], regardless of the employed classification methodology. An important advantage of the method is their mapping, where closely related patterns are mapped together and the dissimilar ones further apart. The optimum configuration is found to reproduce the expected atmospheric circulation patterns with four anticyclonic patterns, six cyclonic, thirteen mixed and eight smooth fields. Furthermore, the method accounts for the seasonality of the meteorological conditions and produces six summer patterns, organized at the left part of the array, seventeen winter patterns, organized at the right part of the array and nine transient patterns which are observed at the middle part of the array (Fig. 2). It should be noted that in some cases winter patterns are also observed during early spring and late autumn and the summer patterns during September and May, in accordance with the meteorological definition of seasons for the geographical under study area.

Fig. 2. SOM classification topology, absolute frequency of occurrence and seasonality of patterns (light grey corresponds to summer, grey to transient and black to winter patterns)

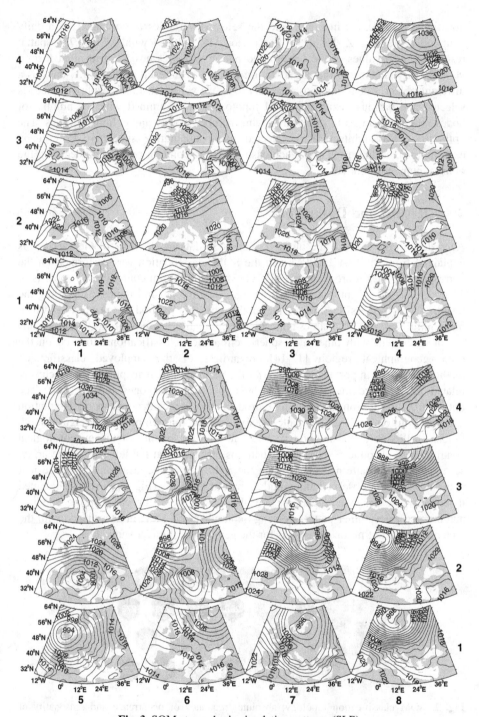

Fig. 3. SOM atmospheric circulation patterns (SLP)

4.1   Winter Circulation Patterns

During winter the study area is influenced from cyclonic (W5.1, W7.1, W8.1, W5.2, W6.2 and W6.3), anticyclonic (W5.3, W5.4, W7.4 and W8.4), smooth field (W6.1, W8.2 and W8.3) and combination of high and low pressure field patterns (W7.2, W7.3, W4.4 and W6.4). In detail, according to W5.1, W7.1 and W8.1 patterns, which are located at the lower-right part of the SOM matrix, the area is influenced from deep low-pressure centers located at the British Isles, the Baltic countries and the Scandinavia respectively. The flow in Eastern Mediterranean according to W5.1 pattern is southwesterly, whereas for the W7.1 and W8.1 patterns a zonal distribution of SLP is observed. The aforementioned pressure systems, according to the W8.2, W6.1 and W8.3 patterns, are displaced in higher latitudes and therefore their effect in Eastern Mediterranean and Greece is attenuated, leading to a smooth SLP distribution. During the cold period of the year Mediterranean cyclogenesis is a frequent phenomenon and the associated W5.2, W6.2 and W6.3 patterns differ in the location of the corresponding low-pressure centers. The W5.2 pattern is associated with depressions formed in the Gulf of Genoa and the Ligurian Sea, while for the W6.2 pattern the low-pressure center is located in central or south Italy. The above-mentioned patterns result the advection of maritime air masses in Greece. The low-pressure center of the W6.3 pattern is over Greece and the area under study is mainly located at the cold sector of a depression. The winter anticyclonic patterns are mainly related with either the Siberian or with the Azores anticyclone and they are located at the upper-right part of the SOM matrix. According to W5.3 and W8.4 patterns, the high-pressure center is located at the Eastern Europe and its influence extends over the Balkans. The patterns are connected with the Siberian anticyclone and establish a northerly flow over Greece. An analogous SLP distribution is observed for the W4.4 pattern, differing in the existence of a low-pressure center at the Gulf of Sidra. The W5.4 is characterized from the existence of an extended high-pressure system in central Europe, which causes an intense pressure gradient over Greece, whereas for the W6.4 pattern the center of high-pressure system is located more westerly and the study area is affected from its combination with a low-pressure center located over Cyprus. Regarding the W7.4 pattern the Azores anticyclone extends over the Iberian Peninsula and the western Mediterranean, resulting into a meridional distribution of SLP over the eastern Mediterranean. According to the W7.2 pattern, the anticyclone penetrates to a lesser extent into the European continent and in combination with the low-pressure system located at the northeastern Europe it leads to an almost meridional SLP distribution at the eastern Mediterranean. The synoptic condition of the W7.3 pattern is also associated with the extension of the Azores anticyclone over central Europe and its interaction with a low-pressure center located at the Gulf of Sidra.

4.2   Summer Circulation Patterns

During summer, the atmospheric circulation over the study area mainly alternates between the S1.1 and S2.4 smooth pressure field patterns and the S1.2, S.1.3, S2.3 and S1.4 patterns, which are produced by the interaction of the Middle East thermal low with the high-pressure fields located at the Western Europe and the Mediterranean. In detail, according to the S1.3 pattern the Azores anticyclone is extended throughout the southern Europe and the greater part of the Mediterranean and in combination with the Middle East thermal low it induces the characteristic summertime SLP distribution over the study area. The same distribution is observed for the S2.3 pattern, differing in the fact that the extension of the Azores anticyclone is displaced northwards and in this case it extends throughout the continental Europe, reaching up to the Balkans. The extension of the anticyclone is somewhat attenuated in the S1.2 and S1.4 patterns, where at the northeastern Europe a low-pressure and a high-pressure centers are located respectively, which do not affect the study area. This characteristic well established summertime SLP distribution in southeastern Mediterranean and Greece induces a northerly flow called Etesians. Regarding the smooth S1.1 and S2.4 patterns the Middle East thermal low is attenuated and the pressure gradient in the study area is weak.

4.3   Transient Seasons Circulation Patterns

The spring and autumn patterns have common characteristics and in this work are referred as transitional period patterns. The SLP distributions, due to the fact that the atmospheric circulation during the transitional periods alternates between winter and summer circulation patterns, produce similar atmospheric circulation types. The pressure gradient for the T3.1pattern, where the cyclonic center at the Scandinavia doesn't affect southeastern Mediterranean and for the T3.4 and T4.3 patterns is weak, which favors the development of local-scale flows. According to the T2.1, T3.2 and T3.3 patterns, the study area is affected from the combination of low-pressure field at the east of Greece with the anticyclonic centers located at the central Europe, at northeastern Europe and between the British Isles and the Scandinavia respectively. According to the T4.1 pattern, the low-pressure system in Northern Europe is extended towards the south, while the East Europe high-pressure system is also extended southwards in Turkey, leading to a weak to moderate meridional pressure gradient over the study area. According to the T2.1 pattern an extended high-pressure field covers the western Mediterranean, the south and east of Europe and interacts with the low-pressure field located at the easternmost part of the Mediterranean. The study area according to the T4.2 pattern is influenced from two low-pressure fields located at the east of Greece and at the north of the British Isles and two high-pressure fields that originate from the Azores and from the Siberian anticyclone located at Russia in northeastern Europe.

5   Conclusions

The results of this work suggest that although the SOM methodology is not primarily a cluster analysis algorithm it is a powerful tool for classifying atmospheric conditions and identifying distinct circulation patterns. The coupling of atmospheric fields from multiple levels and domains resulted in a classification with well-established atmospheric circulation patterns. Following the employed subjective criteria, the optimum SOM topology contains 32 nodes, organized in an 8x4 lattice. The method clearly accounts for the seasonality of the atmospheric circulation and identified six summer patterns located at the left part of the lattice, nine transient season patterns located at its middle part and seventeen winter patterns located at the right part. Further work is proposed to assess the physical mechanisms that drive local climatological conditions by examining their relationship with the identified patterns.

References

1. Yarnal, B.: Synoptic Climatology in Environmental Analysis: A primer. Belhaven Press, London (1993)
2. Beck, C., Philipp, A.: Evaluation and comparison of circulation type classifications for the European domain. Phys. Chem. Earth, Parts A/B/C 35, 9–12, 374–387 (2010)
3. Huth, R., Beck, C., Philipp, A., Demuzere, M., Ustrnul, Z., Cahynova, M., Kysely, J., Tveito, O.E.: Classifications of Atmospheric Circulation Patterns. Ann. N.Y. Acad. Sci. 1146, 105–152 (2008)
4. Philipp, A., Bartholy, J., Beck, C., Erpicum, M., Esteban, P., Fettweis, X., Huth, R., James, P., Jourdain, S., Kreienkamp, F., Krennert, T., Lykoudis, S., Michaelides, S.C., Pianko-Kluczynska, K., Post, P., Rasilla Álvarez, D., Schiemann, R., Spekat, A., Tymvios, F.S.: Cost733cat – A database of weather and circulation type classifications. Phys. Chem. Earth, Parts A/B/C 35, 360–373 (2010)
5. Hewitson, B.C., Crane, R.G.: Self-organizing maps: applications to synoptic climatology. Clim. Res. 22, 13–26 (2002)
6. Liu, Y., Weisberg, R.H.: A Review of Self-Organizing Map Applications in Meteorology and Oceanography. In: Mwasiagi, J.I. (ed.) Self-Organizing Maps - Applications and Novel Algorithm, pp. 253–272. InTech Publishers (2011)
7. Sheridan, C.S., Lee, C.C.: The self-organizing map in synoptic climatological research. Prog. Phys. Geog. 35, 109–119 (2011)
8. Kohonen, T.: Self-Organization and Associative Memory. Springer, New York (1984)
9. Sang, H., Gelfand, A.E., Lennard, C., Hegerl, G., Hewitson, B.: Interpreting self-organizing maps through space-time data models. Ann. Appl. Stats. 2, 1194–1216 (2008)
10. Dee, D.P., Uppala, S.M., Simmons, A.J., Berrisford, P., Poli, P., Kobayashi, S., Andrae, U., Balmaseda, M.A., Balsamo, G., Bauer, P., Bechtold, P., Beljaars, A.C.M., van de Berg, L., Bidlot, J., Bormann, N., Delsol, C., Dragani, R., Fuentes, M., Geer, A.J., Haimberger, L., Healy, S.B., Hersbach, H., Hólm, E.V., Isaksen, L., Kållberg, P., Köhler, M., Matricardi, M., McNally, A.P., Monge-Sanz, B.M., Morcrette, J.J., Park, B.-K., Peu-bey, C., de Rosnay, P., Tavolato, C., Thépaut, J.-N., Vitart, F.: The ERA-Interim reanalysis: configuration and performance of the data assimilation system. Q.J.R. Meteorol. Soc. 137, 553–597 (2011)
11. Michaelides, S.C., Liassidou, F., Schizas, C.N.: Synoptic classification and establishment of analogues with artificial neural networks. Pure Appl. Geophys. 164, 1347–1364 (2007)
12. Kassomenos, P.A.: Anatomy of the synoptic conditions occurring over southern Greece during the second half of the 20th century Part I. Winter and summer. Theor. Appl. Climatol. 75, 65–77 (2003)
13. Kassomenos, P.A.: Anatomy of the synoptic conditions occurring over southern Greece during the second half of the 20th century Part II. Autumn and spring. Theor. Appl. Climatol. 75, 79–92 (2003)
14. Kostopoulou, E., Jones, P.D.: Comprehensive analysis of the climate variability in the eastern Mediterranean. Part I: map-pattern classification. Int. J. Climatol. 27, 1189–1214 (2007)

An Online Signature Verification System for Forgery and Disguise Detection

Abdelâali Hassaïne and Somaya Al-Maadeed

Computer Science and Engineering Department
College of Engineering, Qatar University
Doha, Qatar
{hassaine,s_alali}@qu.edu.qa

Abstract. Online signatures are acquired using a digital tablet which provides all the trajectory of the signature as well as the variation in pressure with respect to time. Therefore, online signature verification achieves higher recognition rates than offline signature verification. Nowadays, forensic document examiners distinguish between forgeries, in which an impostor tries to imitate a given signature of another person and disguised signatures, in which the authentic author deliberately tries to hide his/her identity with the purpose of denial at a later stage. The disguised signatures play an important role in real forensic cases but are not considered in recent literature. In this paper, we propose a new system for online signature verification for both forgeries and disguised signatures. This system extract features from both the questioned and the reference signature. The combination of the features is performed using several classifiers and achieves high performances on several signature databases.

Keywords: Online signature verification, Online signature database, Feature extraction, Forgeries, Disguised signatures.

1 Introduction

Signature verification is a very active research field. It consists in comparing a questioned signature with a set of one or several reference signatures. Signature verification systems can assist forensic experts in deliberating about the authenticity of a questioned signature. Such systems can also be used in banks as security check.

Two kinds of modalities are considered when dealing with the verification of signatures, the offline modality, in which scanned copies of the signatures are available for the verification, and the online modality, in which the signatures are acquired through digital tablets. The online modality provides more information about the signatures (trajectory, speed, pressure...ect), therefore, this modality provides generally better verification results than the offline modality.

Nowadays, forensic document examiners distinguish between forgeries, in which an impostor tries to imitate a given signature of another person and

T. Huang et al. (Eds.): ICONIP 2012, Part IV, LNCS 7666, pp. 552–559, 2012.

disguised signatures, in which the authentic author deliberately tries to modify his/her signature with the purpose of denial at a later stage.

A first survey of existing methodologies in signature verification has been done in [13]. A second survey has been done some years later [8]. A recent survey has also been conducted some years ago [7].

Out of these surveys, we can say that signature verification systems range into two categories of methods: 1)Global methods, in which features are extracted from both the questioned and the reference signatures and then compared. According to the difference between these features a decision is taken on the authenticity of a certain signature. 2)Local methods, in which features are also extracted but they are considered locally, in order to allow a localized comparison. Local methods generally achieve better results than the global methods.

Note that most of the work which has been done in the field of signature verification deal with the detection of forgeries. A recent study by Malik et al. suggests that local features achieve better performances in detecting disguised signatures than global features [11]. This study however only considered the offline modality.

Furthermore, several international competitions on signature verification have been organized. A first one in 2004 dealing only with online signatures [14]. Since 2009, the international competition on signature verification has been organized on a yearly basis [1,10,9]. Note that only the 2010 edition competition dealt with disguised signatures [10], however this competition only considered offline signatures.

In this paper, we describe a new system for online signature verification which is able to detect forgeries but also disguised signatures. We study the performance of several online signature verification features and we propose several classifiers for combining those features.

The reminder of this paper is organized as follows: section 2 gives the details of the acquisition of online signatures. Section 3 presents details of the method and the proposed features. Section 4 presents results of individual features as well as their combinations. Section 5 concludes this paper and draws some future work perspectives.

2   Data Acquisition

Online signatures contain a set of samples, each sample corresponds to the point coordinates on the digitizing tablet along with the corresponding pressure (X_t, Y_t, P_t) where t correspond to time (cf. figure 1).

In this study, we used online signatures acquired using a WACOM Intuos4 digitizing tablet with a sampling rate of 200 Hz, a resolution of 2000 lines/cm and a precision of 0.25 mm. The pressure information is available in 1024 levels.

194 volunteers participated in the data collection process. They were instructed to provide occurrences of their natural signatures, and then to change their signatures in order to deny their identity at a later stage. Other volunteers were then asked to produce a simulation of the genuine signatures that they could see. Figure 2 shows a screenshot of the software which has been developed for the acquisition of online signatures.

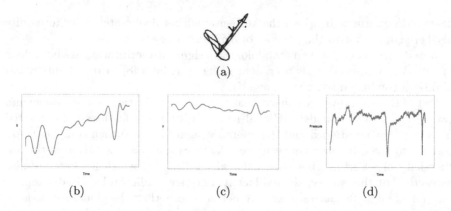

(a)

(b)　　　　　　　　　(c)　　　　　　　　　(d)

Fig. 1. Example of a signature and the corresponding X, Y and Pressure signals

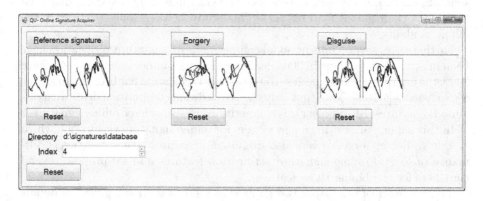

Fig. 2. Acquiring genuine, forgeries and disguised signatures

3　Method Description

Similarly to any automated verification process, signature verification involves two main steps: 1) Feature extraction: In this step, features are extracted from the questioned signature as well as the reference signature. A feature is said to be discriminative if its intra-writer variability is smaller than its inter-writer variability, 2) Matching: In this step, a decision or a score indicating how probable it is that the questioned signature is genuine. This score is generally obtained by a combination of several features. Each of these steps is detailed below.

3.1　Feature Extraction

From the three basic signals which are X, Y and Pressure, several features (or signals) can be extracted from the online signatures [12]. The following list gives a description of the features used in this study.

- **Distances:** The euclidian distance is computed between each successive X and Y coordinates of signature $d_t = \sqrt{(x_t - x_{t-1})^2 + (y_t - y_{t-1})^2}$ (cf. figure 3(a)).
- **Angles:** The angle between the X axis and the line formed with the first signature point and the current point $\alpha_t = atan\frac{Y_t - Y_0}{X_t - X_0}$ (cf. figure 3(b)).
- **Speeds:** The difference between successive distances $S_t = d_t - d_{t-1}$.
- **Angular Speeds:** The difference between successive angles $AS_t = \alpha_t - \alpha_{t-1}$.

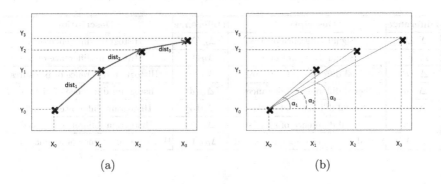

(a)   (b)

Fig. 3. Computing distances and angles

3.2 Matching

In order to compare the questioned signature with the reference signature, we compare the differences between each pair of features. For the sake of simplicity, this is illustrated for the distance feature in figure 4. Two differences are considered: the difference at the signal level Δ_S and the difference at the histogram level Δ_H.

Δ_S is computed after resizing (using a cubic interpolation) the two signals to the same number of samples which has been empirically set to 100 in this study.

$$\Delta_S(d_1, d_2) = \Sigma_{100}^{i=1} |d_1(i) - d_2(i)|.$$

Δ_H is computed after computing the distribution histograms of the two signals with the same number of bins (which has also been set empirically to 100).

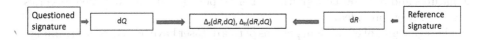

Fig. 4. Comparing the distance feature of the questioned and the reference signature

$$\Delta_H(d_1, d_2) = \Sigma_{100}^{i=1} |Histogram_{d_1}(i) - Histogram_{d_2}(i)| .$$

Note that distribution histograms are widely used in the field of writer identification [5].

Those formulas apply as well on the other features, resulting in the list of differences shown in table 1 for each comparison.

Table 1. List of differences for each comparison

Difference	Description	Difference	Description
$\Delta_S(X)$	Signal difference of X	$\Delta_H(X)$	Histogram difference of X
$\Delta_S(Y)$	Signal difference of Y	$\Delta_H(Y)$	Histogram difference of Y
$\Delta_S(P)$	Signal difference of pressures	$\Delta_H(P)$	Histogram difference of pressures
$\Delta_S(d)$	Signal difference of distances	$\Delta_H(d)$	Histogram difference of distances
$\Delta_S(\alpha)$	Signal difference of angles	$\Delta_H(\alpha)$	Histogram difference of angles
$\Delta_S(S)$	Signal difference of speeds	$\Delta_H(S)$	Histogram difference of speeds
$\Delta_S(AS)$	Signal difference of angular speeds	$\Delta_H(AS)$	Histogram difference of angular speeds

All the above differences have a different discriminative power. In the next section, we present the performance of each of them separately and propose several ways of combining them.

4   Results and Discussion

Two online signature databases have been considered in this study:

ICDAR2009 Signature Verification Competition Dataset. This dataset [1] contains two separate sets, one for training and one for evaluation.

The training set contains signatures of 12 authentic persons, each of them provided 5 signatures. Moreover, 32 persons provided 5 forgeries of the genuine signatures. In sum, each signature has 5 genuine occurences and 620 forgeries. This set has also been used for training purposes.

The evaluation set contains signatures of 77 persons, each of them provided 12 genuine signatures. This set contains a total of 640 forgeries.

QU Online Signature Database. This dataset was acquired as described in 2. It contains signatures of 194 persons, each of them provided at least 3 reference signatures and 3 disguised signatures. For 50 persons, at least 3 skilled forgeries have also been provided.

The signatures of 50 random persons of this dataset have been used for training.

Evaluation on this dataset has been performed by considering the full testing set in a first step, by disregarding disguised signatures in a second step and by disregarding forgeries in a third step.

Table 2 shows the Area Under the ROC Curve (AUC) and the Equal Error Rate (EER) of the presented features on these databases.

Table 2. AUC and EER measures for the presented features

Feature	ICDAR2009		QU all		QU forgeries		QU disguised	
	AUC	EER	AUC	EER	AUC	EER	AUC	EER
$\Delta_H(X)$	0.8365	0.2546	0.6561	0.3959	0.8730	0.2088	0.6299	0.4097
$\Delta_S(X)$	0.8266	0.2152	0.7148	0.3192	0.8155	0.2598	0.7027	0.3305
$\Delta_H(Y)$	0.8347	0.2551	0.5832	0.4337	0.7490	0.3295	0.5631	0.4504
$\Delta_S(Y)$	0.6776	0.3917	0.6690	0.3909	0.6885	0.3790	0.6666	0.3924
$\Delta_H(P)$	**0.9705**	**0.0866**	0.7315	0.3242	0.9526	0.1188	0.7048	0.3480
$\Delta_S(P)$	0.8499	0.1969	**0.7690**	**0.2920**	**0.9860**	**0.0471**	**0.7428**	**0.3094**
$\Delta_H(d)$	0.8464	0.2345	0.5959	0.4182	0.7232	0.3316	0.5805	0.4271
$\Delta_S(d)$	0.6687	0.3957	0.5406	0.4782	0.5495	0.4737	0.5395	0.4787
$\Delta_H(\alpha)$	0.6449	0.3851	0.5641	0.4525	0.6008	0.4343	0.5596	0.4562
$\Delta_S(\alpha)$	0.6478	0.4069	0.5458	0.4752	0.5643	0.4650	0.5435	0.4765
$\Delta_H(S)$	0.6800	0.3519	0.6435	0.4102	0.6083	0.4409	0.6477	0.4063
$\Delta_S(S)$	0.6794	0.3627	0.5351	0.4777	0.5570	0.4642	0.5324	0.4794
$\Delta_H(AS)$	0.5527	0.4842	0.6023	0.4388	0.6818	0.3759	0.5927	0.4462
$\Delta_S(AS)$	0.5857	0.4453	0.5115	0.4936	0.5407	0.4776	0.5079	0.4956

Several classifiers have been tried in order to combine the presented features, including Random Forest [2] with 2000, 5000 and 10000 random trees, logistic regression [6], linear regression, Multivariate Adaptive Regression Splines (MARS) [4] and neural networks with 2, 5 and 10 hidden neurons using a logistic output unit [3]. Table 3 shows the performance of those classifiers on the presented features.

Table 3. AUC and EER measures of combining features using several models

	ICDAR2009		QU all		QU forgeries		QU disguised	
Random Forest (2000 tree)	0.9685	0.0924	0.8602	0.2243	0.9847	0.0459	0.8451	0.2359
Random Forest (5000 tree)	0.9688	0.0932	**0.8604**	**0.2233**	0.9845	0.0454	**0.8454**	**0.2359**
Random Forest (10000 tree)	0.9699	0.0905	0.8604	0.2250	**0.9849**	**0.0449**	0.8453	0.2368
Logistic Regression	0.9523	0.1149	0.8375	0.2334	0.9728	0.0570	0.8211	0.2490
Linear Regression	0.9470	0.1108	0.8424	0.2323	0.9743	0.0576	0.8265	0.2461
MARS	0.9472	0.1153	0.8590	0.2261	0.9829	0.0523	0.8440	0.2368
Neural Nets (2 nodes)	0.9650	0.0784	0.8486	0.2359	0.9761	0.0505	0.8332	0.2460
Neural Nets (5 nodes)	0.9588	0.0912	0.8525	0.2301	0.9791	0.0507	0.8372	0.2425
Neural Nets (10 nodes)	**0.9733**	**0.0782**	0.8499	0.2355	0.9739	0.0544	0.8349	0.2482

Notice that all the classifiers achieve comparable results. However, neural networks and random forests slightly outperform the other classifiers, the latters are specifically prefered in the case of disguised signatures. Finally, although the performance in detecting disguised signatures is acceptable, the performance in detecting forgeries is much more accurate which suggests that further research is needed for detecting disguised signatures.

5   Conclusion

We presented a new system for online signature verification which deals with both forgeries and disguised signatures. This system extracts several features of the signatures and compares them at the histogram level and the signal level. Several classifiers have been tested for combining these features, with neural networks and random forests generally prefered.

It is planned to study new methods of matching these features using dynamic time warping as well as extending this method for the case of multiple reference signatures.

Acknowledgments. This publication was made possible by a grant from the Qatar National Research Fund under its Undergraduate Research Experience Program and Qatar University under Student fund QUST-CENG-FALL-11/12-10. Its contents are solely the responsibility of the authors and do not necessarily represent the official views of the Qatar National Research Fund or Qatar University.

References

1. Blankers, V.L., van den Heuvel, C.E., Franke, K.Y., Vuurpijl, L.G.: Icdar 2009 signature verification competition. In: Proceedings of the 2009 10th International Conference on Document Analysis and Recognition, ICDAR 2009, pp. 1403–1407. IEEE Computer Society, Washington, DC (2009)
2. Breiman, L.: Random forests. Machine Learning 45, 5–32 (2001)
3. Brierley, P.: Tiberius, predictive modelling software (2011), http://www.tiberius.biz
4. Friedman, J.H.: Multivariate adaptive regression splines. Annals of Statistics 19(1), 1–67 (1991)
5. Hassaïne, A., Al-Maadeed, S., Alja'am, J., Jaoua, A., Bouridane, A.: The ICDAR2011 Arabic Writer Identification Contest. In: Proceedings of the Eleventh International Conference on Document Analysis and Recognition, Beijing, China (September 2011)
6. Hosmer, D.W., Lemeshow, S.: Applied Logistic Regression, 2nd edn. Wiley (October 2000)
7. Impedovo, D., Pirlo, G.: Automatic Signature Verification: The State of the Art. IEEE Transactions on Systems, Man, and Cybernetics, Part C: Applications and Reviews 38(5), 609–635 (2008)

8. Leclerc, S., Plamondon, R.: Automatic signature verification: the state of the art-1989-1993. Intl. Journal of Pattern Recognition and Artificial Intelligence 8(3), 643–660 (1994)

9. Liwicki, M., Malik, M., van den Heuvel, C., Chen, X., Berger, C., Stoel, R., Blumenstein, M., Found, B.: Signature Verification Competition for Online and Offline Skilled Forgeries (SigComp2011). In: 2011 International Conference on Document Analysis and Recognition (ICDAR), pp. 1480–1484 (September 2011)

10. Liwicki, M., van den Heuvel, C.E., Found, B., Malik, M.I.: Forensic signature verification competition 4nsigcomp2010 - detection of simulated and disguised signatures. In: Proceedings of the 2010 12th International Conference on Frontiers in Handwriting Recognition, ICFHR 2010, pp. 715–720. IEEE Computer Society, Washington, DC (2010)

11. Malik, M.I., Liwicki, M., Dengel, A.: Evaluation of Local and Global Features for Offline Signature Verification, pp. 26–30. CEUR (2011)

12. Nalwa, V.: Automatic on-line signature verification. Kluwer International Series in Engineering and Computer Science, pp. 143–164 (1997)

13. Plamondon, R., Lorette, G.: Automatic signature verification and writer identification– the state of the art. Pattern Recognition 22(2), 107–131 (1989)

14. Yeung, D.-Y., Chang, H., Xiong, Y., George, S.E., Kashi, R.S., Matsumoto, T., Rigoll, G.: SVC2004: First International Signature Verification Competition. In: Zhang, D., Jain, A.K. (eds.) ICBA 2004. LNCS, vol. 3072, pp. 16–22. Springer, Heidelberg (2004)

A Modular Approach to Support the Multidisciplinary Design of Educational Game Experiences

Telmo Zarraonandia, Paloma Díaz, and Ignacio Aedo

Computer Science Department, Universidad Carlos III de Madrid,
Leganés (Madrid), Spain
{1tzarraon,2pdp}@inf.uc3m.es, 3aedo@ia.uc3m.es

Abstract. The design of a computer game based educational experience is a multidisciplinary task that requires the collaboration of experts in game design, education and in the field of instruction. In this paper we present an educational game (EG) design method which tackles this multiplicity of backgrounds allowing each of the experts to focus on the aspects of the game design related to his/her area of expertise. The method is supported by a modular conceptual model that provides a common framework for describing an EG and facilitates the reuse of design components and the description of new game variants. In order to illustrate the proposal we describe the process of design and implementation of 'Maz-E-nglish', an EG for mobile devices which aims to help children learning English grammar.

Keywords: serious games, education, game based learning, modelling languages, design method, model driven design.

1   Introduction

During the last few years we have witnessed a growing interest in exploring the benefits of integrating computer games in educational processes. There is widespread agreement about the value of computer games to motivate learners [1, 2], but such motivational benefits can only be fully exploited if the game is designed to target and meet some instructional objectives. If not, learners may spend their time learning to be successful at the game without getting any instructional benefits [3]. The design of a successful educational game (EG) is therefore a challenging task involving a variety of knowledge and skills. EG designers not only have to deal with the inherent technical complexity of games design, but also have to interleave learning activities that support the attainment of the learning objectives in a subtle way. In this paper we focus on games that support education and do not diminish the characteristics of games that make them intrinsically motivating, including playability, engagement or enjoyment. Such different and complementary perspectives of an EG leads on to dealing with their design as a multidisciplinary task that requires the collaboration of different roles, such as game designers, educators and domain experts. Therefore, a model supporting EG design should provide an adequate level of abstraction to be easily managed and understood by each of the roles involved in this process.

T. Huang et al. (Eds.): ICONIP 2012, Part IV, LNCS 7666, pp. 560–567, 2012.

In this paper we propose a method for guiding the design of a computer EG which tackles the multiplicity of backgrounds of the roles involved in the task and facilitates that each of them focus on the aspects and perspectives related to his/her area of expertise. The design method is supported by the EG model presented in [4], which serves as a communication tool between the different roles involved in the design process, and promotes reuse of design components, simplifying the adaptation of existing games and the production of variants.

The rest of the paper is organized as follows. First, we introduce some relevant related work on the subject and briefly summarize the EG conceptual model. Next, we present the proposed design method by describing the processes followed to produce an EG for helping children practice English grammar, 'Maz-E-nglish'. Finally, some conclusions and current lines of work are presented.

2   Related Work

Several attempts have been made at analysing and understanding the enjoyment derived from computer games. Among them, Csikszentmihlayi's theory of flow [5] has been extensively used to analyse the components of a game that can increase the degree of player involvement [6, 7]. For instance, the GameFlow model [6] proposes eight elements for achieving enjoyment: concentration, challenge, skills, control, clear goals, feedback, immersion, and social interaction. Another popular approach of analysing games is to focus in on the factors that promote players' motivation. Malone [8] proposes a framework for designing intrinsically motivating game experiences that puts the stress on factors like challenge, curiosity, control and fantasy. Following these ideas, and after an exhaustive review of the literature on the subject, Garris [1] concludes that game characteristics can be described in terms of six broad dimensions or categories: fantasy, rules/goals, sensory stimuli, challenge, mystery, and control. In addition the same author proposes describing the game experience through an interative game cycle of judgment-behaviour-feedback which includes a specific phase of debriefing. Indeed the importance of including some sort of debriefing activity, which allows the learner to connect with what is learnt in the game with the real world has been highlighted by several authors [9, 10, 11]. As Fabricatore states in [12], focusing only on the motivational aspects of the game might derive on EG which lack cohesion between the cognitive task and the game-play. Despite the valuable contribution of these pieces of work in understanding educational game experiences, all these design heuristics provide little support for their practical application. Indeed, notations and models filling the gap between theoretical models and technical designs are somewhat lacking.

3   An Adaptive and Reusable EG Design Model

Figure 1 depicts the latest version of the conceptual model to support the design of adaptive and reusable EGs presented in [4]. The model organizes the game features most often regarded in the literature as significant in producing an engaging, fun and educational game experience. For each feature a set of configurable elements and a basic vocabulary is provided. In order to facilitate the reuse of parts of an EG design, the features and the game elements are arranged in two different and independent sub-models: the game rules model and the scenario model. The idea is to be able to play a game in different scenarios, and to use the same scenario to play different games. Furthermore, in order to facilitate their reuse, the elements of the model are also organized in different layers and classified by the game feature they support. This separation also makes it possible to provide some support to multidisciplinary teams in as much as each member can focus on the elements related to his/her area of expertise. Additionally, since a common vocabulary is provided, made up of the elements of the model, the communication problems usually associated with the use of expert jargon can be reduced.

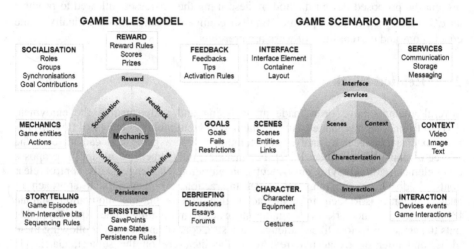

Fig. 1. Model to define the game rules (left) and game scenarios (right)

3.1   Game Rules Model

As depicted on the left hand side of fig. 1, the elements of this model are spread over different levels, so that the definitions of the elements of a specific level are based on the definitions of the elements of the innermost levels. In order to facilitate their reuse, game rules are defined without making any reference to specific resources. The two innermost levels of the model (*Mechanics and Goals*) provide the elements to describe the game mechanics and challenges. Once the basic logic of the game has been established, the designer can use the elements of level 3 to expand the game definition to provide support to social interaction (adding elements such as groups, rules or synchronizing rules); to narrate the story behind the game; to include debriefing activities (such as group discussions or essays) that might facilitate the connection of the lessons learned in the virtual world with their application in real life and to specify elements and feedback mechanisms for increasing immersion. Finally, level 4 makes it possible to add rewarding and persistence mechanisms frequently used in computer games, such as the accumulation of points, the opportunity to explore secret areas or allowing access to complementary games, the possibility to customize an avatar or to define save points, for instance. The use of elements from levels 3 and 4 is optional, as not all the games need to include the features supported by those levels to the same extent.

3.2   Game Scenario Model

The right hand side of Fig 1 depicts the model for the scenario definition. Again, the definitions of the elements on a specific level are based on the definitions of the elements of the innermost levels. The definition of a *Scenario* is composed of a group of interconnected scenes through links, with each scene representing a different physical environment or situation. The designer will define a scene-entity for each object, individual or area of the scene in which she wants to provide the possibility of interacting with. These scene-entities will group a set of graphical and sound resources to represent the different states of that object. In addition, the designers can use the *Characterization* elements, in order to define the appearance and characteristics of the characters associated with the scenario, and *Context* items to describe elements that might be useful in setting the context of situations that will take place in the scenario. Optionally the designer may enrich the scenario definition including the description of *Services* or tools to support and increase the possibilities of the different games that could be played in it. Finally, the last layer of the model allows defining, in an abstract way, the layout in which the representation elements and the services will be organized and presented to the player in each device, and the type of interactions s/he will be able to perform through the corresponding input/output devices.

4   EG Design Process Based on the Model

In this section we describe the steps to be followed in an EG design process supported by the proposed model. To better illustrate our approach and its advantages, we will make use of a specific example: the process of design of 'Maz-E-nglish', an EG for mobile devices to help young learners learning English.

4.1   Game Design Definition Process

Fig 2 depicts the process followed to design the EG using the proposed model. The design process is organized in four steps that are described below.

Step 1: Game Components Initial Definition. The first step of the process is to outline an initial design of the EG that captures the main features to be implemented. Each of the different roles in the design team can tackle this task independently, focusing on the design of the components he/she is expert in. In our case, we will need to involve at least one game designer with experience in creating engaging and fun game experiences. This expert would be best placed to propose an initial definition of the mechanics and goals of the game. On another hand, an educator or pedagogical expert would also be able to provide advice on which game interactions would be more appropriate for teaching children this type of knowledge. He/she can focus on designing adequate feedback, rewarding and debriefing mechanisms. Finally, it will also be necessary to collaborate with an English language specialist who would select the most appropriate resources to be included in the scenario in order support the learning of this specific subject.

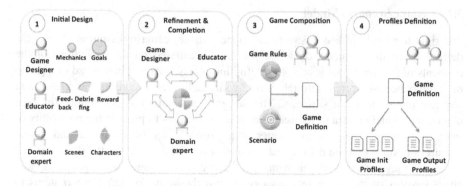

Fig. 2. The four steps of the game design process

Step 2. Refinement and Completion. The next step is to agree on the game components designs and to assemble them to produce a set of game rules and a scenario definition. This will require reviewing and refining the initial designs until the solution satisfies the requirements of the different experts in the team. In this process the original definitions can also be enriched with other game components not initially considered. Following on with our example, the output of this step could be a set of rules named 'Maze' and a scenario called 'Brick & Grammar Labyrinth'. The former describes a game that consists of moving an avatar around a maze in which we obstacles to avoid are found, prizes to collect and language grammar challenges to solve. The objective is to solve and collect as many challenges and coins as possible, and find the exit to the maze before the time limit expires. The scenario contains the description of two scenes. One scene depicts a maze made up of brick walls and rocks, and where prizes and challenges are represented by coins and shields respectively. The other scene depicts a multiple-choice question that includes the entities 'statement', 'answers', and 'solutions'. With regards to the interface design, and as the game is intended to be played on mobile devices, all the space of the screen is used to allocate the scenes representations, one of them at a time. Finally, two possible interaction mechanism to control the avatar were defined: direction buttons and 'point and click' through screen touch.

Step 3. Game Composition. The next step is to carry out a matching process in which the entities of the two designs are associated. As in the games that children play in real life, playing a game in two different scenarios (for instance, outdoors and indoors) may require modifying and adjusting some of the rules of the game. Besides not all games played in a particular scenario have to make use of all the scenario features. Therefore, in this step designers must choose which entities, characters, interactions and services are going to be used and match them with the corresponding entities, operations and elements of the rules perspective. In our case, after merging the set of rules and the scenario, the designers obtain a game in which the player controls an avatar that moves around a brick maze. Every time the avatar interacts with an object that depicts a shield, the player is required to answer a question related to English grammar.

Step 4. Game Profiles Definition. Once the game has been defined, designers can specify different game initialization and output profiles. The former are used to vary parameters for some of the game components so the game experience is adapted to different learner profiles or different levels of difficulty, for instance. This way, in our example a game initialization profile has been defined to allow the educator to set up values for the time limit, number of prizes and challenges, and amount of feedback. On the other hand, the output profiles are used to specify the results or outputs of the game an educator may require to observe. For example, in our game we have defined an output profile focused on the game activity of the learner and another one focused on the educational purpose of the experience. When making use of the former the game engine would retrieve information such as the number of attempts, mazes solved, coins collected or average time taken to completed the maze, whereas when using the latter only the total number of challenges solved by the player will be retrieved.

4.2   Game Implementation

Fig 3 depicts two screenshots of an implementation of the 'Maz-e-nglish' EG. This implementation has been carried out using the game engine AndEngine [13], a 2D Engine OpenGL for the Android platform. In order to play the game the students install the EG package on their mobile devices and upload to the installation one configuration zip file that the educator should provide him/her with. These configuration files contain descriptions of topography of the various labyrinths and the questions to be used in the grammar challenges, expressed in XML language. This allows educators to update the students' game installations with new questions as they progress through the course. The current version of the game does not generate output and log files of the game activity in accordance with the output game profiles.

Fig. 3. Screenshots of the "Maz-E-nglish" EG for Android platforms

4.3   Reuse and Game Variant Production

The game designs obtained following this approach can be easily adapted and modified. For instance, in order to train the language listening ability of young learners a new version of the 'Maz-E-nglish' game was produced. This version made use of an extended version of the 'Brick & Grammar Labyrinth' scenario which includes a new scene for representing listening questions and the description of a service able to play the required audio files in them. On the other hand, by modifying the definition of the interface and interaction layer it could be possible to apply the design to game implementations for desktop computers.

Furthermore, game component designs can be used to accelerate the design of the new game experiences. This way, the rules 'Maze Solving' rules could be combined with a new scenario definition to create a 3D game experience, which could serve to train children in which evacuation routes to follow in an emergency, for instance. On the other hand, the scenario 'Brick & Grammar Labyrinth' could easily be reused in a game in which players have to look for and capture each other around the labyrinth.

5   Conclusions

In order to fulfil the benefits that game based learning can provide it is necessary to provide the means to reduce the high cost associated with the production of EG and to facilitate the design process. A good game design harmonizes both enjoyment and instructional objectives, and therefore a successful design team should integrate experts from at least game design and pedagogy. This work tries to provide support to the members of a multidisciplinary team of these characteristics by providing them with a common framework for describing EG. To help reduce the development costs, these designs might later be implemented using some of the different authoring tools and game engines that provide a set of base services to support the execution of the games, such as AndEngine [13], OpenSpace [14] or RedDwarf Server [15], for instance. Furthermore the modular description of model facilitates the association of different design components with different parts of the game implementation, which in turn facilitates their reuse. This feature will also open the door to the design of a system which allows educators to define their own game based educational experience by exchanging and introducing slight modifications in pre-defined game templates.

Current lines of work include the implementation of a new version of the game able to read and interpret XML files containing descriptions of game output profiles provided by the educators. The new game will retrieve information about the plays accordingly to what it is specified in these files, and send the information to a web repository so that the educator can access it and consult it.

Acknowledgements. This work is supported by the project urThey funded by the Spanish Ministry of Science and Innovation (TIN2009-09687). We also wish to thank the student Eugenio Pérez Martinez for carrying out the implementation work.

References

1. Garris, R., Ahlers, R., Driskell, J.: Games, motivation, and learning: a research and practice model. Simul Gaming 33, 441–467 (2002)
2. Druckman, D.: The Educational Effectiveness of Interactive Games. Simulation and Gaming Across Disciplines and Cultures, 178–187 (1995)
3. Hays, R.T.: The effectiveness of instructional games: A literature review and discussion. Technical report (November 2005),
   http://handle.dtic.mil/100.2/ADA441935 (last accessed May 30, 2012)
4. Zarraonandia, T., Díaz, P., Aedo, I., Ruíz, M.R.: Seeking Reusability of Computer Games Designs for Informal Learning. In: Proceedings of the 2011 11th IEEE International Conference on Advanced Learning Technologies, ICALT (2011)
5. Csikszentmihalyi, M.: Beyond boredom and anxiety. Jossey-Bass, San Francisco (1975)
6. Sweetser, P., Wyeth, P.: GameFlow: A model for evaluating player enjoyment in games. ACM Computers in Entertainment 3, 3 (2005)
7. Kiili, K.: Digital game-based learning: Towards an experiential gaming model. The Internet and Higher Education 8(1), 13–24 (2005)
8. Malone, T.W.: Toward a theory of intrinsically motivating instruction. Cognitive Science 4, 333–369 (1981)
9. Crookall, D.: A guide to the literature on simulation/gaming. Simulation and Gaming Across Disciplines and Cultures, 151–178 (1995)
10. Lederman, L.C.: Debriefing: Toward a Systematic Assessment of Theory and Practice. Simulation & Gaming 23(2), 145–160 (1992)
11. Leemkuil, H.H.: Is it all in the game? Learner support in an educational knowledge management simulation game, University of Twente (2006)
12. Fabricatore, C.: Learning and videogames: An unexploited synergy, In: 2000 AECT National Convention, Long Beach (2000)
13. AndEngine, http://www.andengine.org/ (last accessed May 30, 2012)
14. OpenSpace, http://www.openspace-engine.com (last accessed May 30, 2012)
15. RedDwarf Server, http://www.reddwarfserver.org (last accessed May 30, 2012)

Touch-Based Mobile Phone Interface Guidelines and Design Recommendations for Elderly People: A Survey of the Literature

Muna S. Al-Razgan, Hend S. Al-Khalifa, Mona D. Al-Shahrani,
and Hessah H. AlAjmi

Information Technology Department,
College of Computer and Information Sciences King Saud University
{malrazgan,hendk}@ksu.edu.sa,
{m.n_n.o,hessah.alajmi}@hotmail.com

Abstract. Mobile phones are becoming a great necessity for elderly people; the features they provide supported by rich functionality made them one of the indispensable gadgets used in their daily life. However, as mobile phones get more advanced and their interfaces become more complicated, new design recommendations and guidelines need to be developed to serve the elderly needs. In this paper we present a set of guidelines and design recommendations for touch based mobile phones targeted towards elderly people. These guidelines were distilled and consolidated after a comprehensive review of the literature. We hope that these compiled guidelines will serve as an information base for future designers/developers to use while designing touch based mobile interfaces for elderly people.

Keywords: Elderly people, Mobile phones, Accessibility, Design Recommendations, Guidelines.

1 Introduction

The number of elder people has increased rapidly in recent years [1-3]. Increasing ageing societies has appeared the most among developed countries such Japan, Europe, and North America. The United Nations have projected that by 2050, elderly above the age of 60 will reach a percentage up to 21% of the world population. This increase in elderly numbers represents a recurring need for establishing suitable market of computing technology devices for elderly people [5].

As people get older some of their physical and mental capabilities start to decline [3]. In order to overcome these losses of capabilities for elderly people, recent studies have focused on the development and adaption of technological tools to help the increased number of elderly in society [1][4]. Among these technologies is the mobile phone.

Mobile phones are becoming one of the more utilized technology items for older people. Some of the reasons why older people possess mobile phones are because they are used as memory aids, let them feel safe and secure, keep them related to

T. Huang et al. (Eds.): ICONIP 2012, Part IV, LNCS 7666, pp. 568–574, 2012.

social activities and the most important reason is enabling them to perform their daily life activities independently [2] .

Recent studies showed that the main concern among older people with regard of using mobile phones resides in their complex interfaces and overrated features [4]. The elderly in these studies were related to overly complex interfaces and excessive functionality. This problem might hinder the ability of older people to send text messages or call their relatives because they are afraid of making mistakes. In addition, previous studies have found that mobile phones have built-in features that cannot be modified according to elderly needs.

However, as of today's market, we can see an increase in the number of touch-based mobile phones such as (iPhone, and Android) [2]. Touch-screen mobile phones offer a suitable screen size for older people with many features that can be adjusted according to their needs. Among these features are bigger buttons, larger text message, and spoken interfaces. However, this diversity in touch-based mobile phone market reveals the need for sound design recommendations and guidelines for the design of mobile interfaces for elderly people. Therefore, the overarching objective of this paper is to distill and consolidate guidelines and design recommendations for touch-based mobile phone interfaces appropriate for elderly people. This will be the first step towards providing an information base for future designers/developers intending to design touch based interfaces for elderly people.

The rest of the paper is organized as follows: section 2 reviews the literature related to mobile phone and elder people. Section 3, provides a compiled list of design recommendations and guidelines from previous studies put into an ad hoc framework. Section 4, concludes the paper with future research directions.

2   Literature Survey

There have been many research studies in the use of mobile phones among elderly people. We will present in this section some of the related work that has started from 2002 up to date. An early study presented in [6] concentrated on identifying the important wireless devices and services needed for older people. They articulated the user feedback into a concept scenario to generate design ideas for new products and services; then prioritized the ideas based on the elder's feedback. The result of the study showed a positive opinion about additional values of the services. The study found that the important criteria for senior persons were the ease of use and true need of the device for facilitating independent living. In another study presented in [7] the author pointed out that standard interface techniques are not appropriate for elder users. Designers have to think of the functionality of the output message to design suitable system for older people. Nevertheless the study performed by the authors in [8] addressed touch-based PDA usage by older people and their performance with respect to younger people. Their findings suggest that there are no major differences in performance between older and younger users while physical interaction with PDAs.

Additional study of the use of mobile phones by older persons was discussed in [9], it used a mixed method of quantitative and qualitative approaches to discover the design improvements for the elderly. Older people have strong opinion about some of the advanced features presented in mobile phones such as: the use of one button to lock the cell phone, a panic button for emergency, a screen with only four menus (voice call, text, alarm, and calendar), and button to place unwanted people in the blacklist. Along the same line of research, the author in [10] reported that mobile interfaces had too many menus, and the functions in mobile phones were complicated and difficult to understand. The author indicated from her interviews with elders, the three most-desired functions in mobile devices, which are: address book, diary, and an alarm clock.

The authors in [11] indicated senior users did not purchase their phone; instead they were given to them by relatives (children) or friends. From their observations, it was clear that the ease of use does not influence the intention to purchase a cell phone; instead it prevents senior people from utilizing the functions available in the mobile phone. Their findings related the factors of acceptance with adaption of the mobile phone for elder users.

A comparison study conducted by [12] found that the communication media used by both senior and younger people was the cell phone. An opposite study in [13] revealed that older people fear the use of a new technology and preferred mobile phone with aid features to support their declining abilities.

Touch-screen devices have recently increased in the market. To make these devices more effective and easy to use, the introduction of multimodal feedback from more than one sensory modality had taken place. This kind of feedback is more suited for elder users. The study in [14] demonstrated the enhanced performance for older adults when presented by multimodal feedback with auditory signals through touch-screen devices. Also, to help elderly better deal with touch-screen phones, the study in [15] provided a guideline to icon feedback design for elderly to make the touch screen friendlier.

Investigation of special touch-screen tablets' designs for elderly users was presented in [16, 17]. The authors addressed the important features for seniors. To help seniors remain socially connected and reduce their loneliness, the building bridges project in [16] suggested involving elderly people at every stage of the design process of the technology to be part of their life. Also, in [17] they investigated the optimal number of blocks and targets for a touch-screen tablet intended for seniors. The results do not recommend designing an interface that require the use of two-hands for elderly; and to improve the performance of elder adults, the designed interface has to reduce the cognitive overload for the elderly.

Moreover, the authors in [18] introduced a new tabletop device with touch-based gestural interface to help the elderly communicate with their social network. Some of their design ideas were: elderly people prefer tap gestures because it is easy to understand and remember; also when the touch on the screen is lost during dragging objects, the object should stay where it has been left. Lately, the authors in [19] presented an evaluation of (iPad) current model. Their results showed high acceptance and satisfaction rate among senior users. This finding was clear for both seniors who had experience with PCs and those who had not any experience.

Recent studies have focused on identifying important interface features requested by elder people. The authors in [20] argued about blindly reducing the functionalities that are not effective for elders' mobile-phones. Due to the fact that existing mobile phones do not address the needs of elder people. The authors proposed a worth-driven mobile phone design process to meet the needs of aged people. In addition, the authors in [4] identified the important functions for elderly. They preferred a larger display, a touch-screen, labeled icons and larger fonts. Other features can be eliminated such as the ones related to entertainment.

In a different research where the authors tried to tailor mobile phones to specific user needs, the work in [21] introduced the idea of portable mobile-user interface. This interface can be installed in any mobile-phone regardless of its model or brand. Their prototype addressed the usability problems in existing mobile -phones. As well as the user needs to learn only one user interface and be able to install it in any new phone. Moreover, the author in [22] presented some guidelines for the design of interfaces for elderly people, which was used to develop Mylife application. Mylife is a flexible application that can be presented in touch-screen devices.

Finally, the authors in [23] conducted an experiment to assess standard usage of mobile touch-screen interfaces when used by elderly. Their results showed that elderly are interested in using touch-screen devices, and when provided with one week of training, their performance has increased.

From the above survey, we can can conclude the acceptance of touch-screen devices among elder adults, and suggest a future focus on the interface and applications development for seniors.

3   Consolidated Guidelines and Design Recommendations

Guidelines and design recommendations for mobile phones' interfaces for the elderly is not a new research topic, Kurniawan [10] has already compiled a list of design recommendations targeted for feature phones (this term is used to describe low-end devices with physical keypad). However, to the best of our knowledge we have not seen similar papers targeting touch-based mobile phones.

In this section, a list of design recommendations will be presented based on our extensive literature review. The recommendations are compiled from [4, 5, 9, 14, 16, 17, 18, 20 and 22] and classified into three dimensions, namely: (1) Look and Feel, (2) Functionality and (3) Interaction.

1) Look and Feel
This design requirement includes the following:

- Larger size of mobile phone that consists of three-dimensional appearance button for touch-screens,
- Separate keypads for numbers and letters,
- Good spacing between buttons,
- Larger font for text, and labeled icons.
- In addition, the most important feature should be available directly via a labeled button and not via menu navigation.

2) Interaction

This design requirement includes the following:

- Easy zoom in and out and pinching.
- Tapping with audio confirmation to help elderly with reduced vision.
- Also, the elderly prefers tapping but not drag and drop actions, voice call and slow motion interface.
- The interface should also clearly express where the user is in the dialogue, and which "tasks" are active.
- Moreover, the designer should avoid the following for elderly interfaces:

  — Avoid slide-out keyboard because it bothers the elderly,
  — When the touch is lost during dragging, the object should stay where it has been left,
  — Do not overload the same object with actions performed by tapping and by dragging gestures, and
  — Finally, the screen should not turn off when being idle to avoid confusion. The elder might think that the mobile is not working.

3) Functionality

The most desired functions for elderly are the following:

- Address book linked with caller identification number along with a picture of the caller and on-screen number selections (e.g., press 1 for calendar), i.e. functionality of the same type should be grouped together,
- The main navigation should be placed identically on all "pages", and critical functions should never disappear, and important functions should be placed at the top of the screen to avoid mistake touches.
- Additional request by elderly to have specific buttons for the following actions: single button to return to the home state, a locking button to prevent accidental dialing, a panic button for emergencies, and a button to place a caller/number into the blacklist.
- On the other hand designers should carefully consider naming programs and commands; not too many or too less features for mobile phone interfaces.

Using the above design recommendations will help developers to design the appropriate user-friendly mobile phone interfaces for elderly people and help them use and enjoy this type of technology.

4   Conclusion and Future Work

In this paper, we have conducted a thorough literature survey of the usage of touch-screen devices among elderly people. Based on our extensive study we were able to distill and consolidate a set of design recommendations and guidelines classified into three dimensions, namely: (1) Look and Feel, (2) Functionality and (3) Interaction. This framework of design recommendations can serve as an information base for designer to use when designing touch-based interfaces for elderly people. Our next step will be to test whether these recommendations are applicable for Arab elderly people with no or minimum modifications.

Acknowledgment. This research project was supported by a grant awarded to the Data Management and Engineering Research Group from the College of Computer and Information Sciences, King Saud University.

References

1. Li, G., Zhao, Y., Jiao, B., Korhonen, T.: Design of Easy Access Internet Browsing System for Elderly People Based on Android. In: Rautiainen, M., Korhonen, T., Mutafungwa, E., Ovaska, E., Katasonov, A., Evesti, A., Ailisto, H., Quigley, A., Häkkilä, J., Milic-Frayling, N., Riekki, J. (eds.) GPC 2011. LNCS, vol. 7096, pp. 64–72. Springer, Heidelberg (2012)
2. Plaza, I., Martín, L., Martin, S., Medrano, C.: Mobile applications in an aging society: Status and trends. Journal of Systems and Software 84(11), 1977–1988 (2011)
3. Phiriyapokanon, T.: Is a big button interface enough for elderly users (2011)
4. Sulaiman, S., Sohaimi, I.S.: An investigation to obtain a simple mobile phone interface for older adults. In: 2010 International Conference on Intelligent and Advanced Systems, pp. 1–4 (2010)
5. Massimi, M., Baecker, R.M., Wu, M.: Using Participatory Activities with Seniors to Critique, Build, and Evaluate Mobile Phones. In: 9th International ACM SIGACCESS Conference on Computers and Accessibility, vol. 6185, pp. 155–162 (2007)
6. Mikkonen, M., Väyrynen, S., Ikonen, V., Heikkila, M.O.: User and Concept Studies as Tools in Developing Mobile Communication Services for the Elderly. Personal and Ubiquitous Computing 6(2), 113–124 (2002)
7. Zajicek, M.: Passing on Good Practice: Interface Design for Older Users, pp. 636–640 (2004)
8. Siek, K.A., Rogers, Y., Connelly, K.H.: Fat Finger Worries: How Older and Younger Users Physically Interact with PDAs. In: Proc. of Interact, pp. 267–280 (2005)
9. Kurniawan, S., et al.: A Study of the Use of Mobile Phones by Older Persons, pp. 989–994 (2006)
10. Kurniawan, S.: Mobile Phone Design for Older Persons. Interactions 14, 24–25 (2007)
11. Biljon, J.V., Renaud, K.: A Qualitative Study of the Applicability of Technology Acceptance Models to Senior Mobile Phone Users. In: ER 2008 Workshops (CMLSA, ECDM, FP-UML, M2AS, RIGiM, SeCoGIS, WISM) on Advances in Conceptual Modeling: Challenges and Opportunities, pp. 228–237 (2008)
12. Hashizume, A., Kurosu, M., Kaneko, T.: The Choice of Communication Media and the Use of Mobile Phone among Senior Users and Young Users. In: Lee, S., Choo, H., Ha, S., Shin, I.C. (eds.) APCHI 2008. LNCS, vol. 5068, pp. 427–436. Springer, Heidelberg (2008)
13. Kurniawan, S.: Older people and mobile phones: A multi-method investigation. International Journal of Human-Computer Studies 66(12), 889–901 (2008)
14. Lee, J.-H., Poliakoff, E., Spence, C.: The Effect of Multimodal Feedback Presented via a Touch Screen on the Performance of Older Adults. In: Altinsoy, M.E., Jekosch, U., Brewster, S. (eds.) HAID 2009. LNCS, vol. 5763, pp. 128–135. Springer, Heidelberg (2009)
15. Tsai, W.-C., Lee, C.-F.: A Study on the Icon Feedback Types of Small Touch Screen for the Elderly. In: Stephanidis, C. (ed.) UAHCI 2009, Part II. LNCS, vol. 5615, pp. 422–431. Springer, Heidelberg (2009)

16. Wherton, J., Prendergast, D.: The Building Bridges Project: Involving Older Adults in the Design of a Communication Technology to Support Peer-to-Peer Social Engagement. In: 5th Symposium of the Workgroup Human-Computer Interaction and Usability Engineering of the Austrian Computer Society on HCI and Usability for e-Inclusion, pp. 111–134 (2009)

17. Lepicard, G., Vigouroux, N.: Touch Screen User Interfaces for Older Subjects Effect of the Targets Number and the Two Hands Use, pp. 592–599. Springer, Heidelberg (2010)

18. Leonardi, C., Albertini, A., Pianesi, F., Zancanaro, M.: An Exploratory Study of a Touch-based Gestural Interface for Elderly. In: the 6th Nordic Conference on Human-Computer Interaction: Extending Boundaries, pp. 845–850 (2010)

19. Werner, F., Werner, K., Oberzaucher, J.: Tablets for Seniors – An Evaluation of a Current Model (iPad). In: Ambient Assisted Living, pp. 177–184. Springer, Berlin (2012)

20. Renaud, K., Biljon, J.: Worth-centred mobile phone design for older users. Universal Access in the Information Society 9(4), 387–403 (2010)

21. Olwal, A.: OldGen: Mobile Phone Personalization for Older Adults. In: CHI 2011, Vancouver, BC, Canada, May 7–12, pp. 3393–3396 (2011)

22. Hellman, R.: Usable User Interfaces for Persons with Memory Impairments. In: AAL (2012)

23. Kobayashi, M., Hiyama, A., Miura, T., Asakawa, C., Hirose, M., Ifukube, T.: Elderly User Evaluation of Mobile Touchscreen Interactions. In: Campos, P., Graham, N., Jorge, J., Nunes, N., Palanque, P., Winckler, M. (eds.) INTERACT 2011, Part I. LNCS, vol. 6946, pp. 83–99. Springer, Heidelberg (2011)

A Novel Paradigm for Mining Cell Phenotypes in Multi-tag Bioimages Using a Locality Preserving Nonlinear Embedding

Adnan Mujahid Khan[1], Ahmad Humayun[2], Shan-e-Ahmad Raza[1],
Michael Khan[3], and Nasir M. Rajpoot[1]

[1] Department of Computer Science, University of Warwick, UK
[2] Georgia Institute of Technology, Atlanta, USA
[3] Department of Life Sciences, University of Warwick, UK
{amkhan,nasir}@dcs.warwick.ac.uk

Abstract. Multi-tag bioimaging systems such as the toponome imaging system (TIS) require sophisticated analytical methods to extract molecular signatures of various types of cells. In this paper, we present a novel paradigm for mining cell phenotypes based on their high-dimensional co-expression profiles contained within the images generated by the robotically controlled TIS microscope installed at Warwick. The proposed paradigm employs a refined cell segmentation algorithm followed by a locality preserving nonlinear embedding algorithm which is shown to produce significantly better cell classification and phenotype distribution results as compared to its linear counterpart.

Keywords: Multivariate fluorescence microscopy, Nonlinear embedding, Cancer biology.

1 Introduction

Bioimage computing is rapidly emerging as a new branch of computational biology which deals with the processing and analysis of bioimages as well as the mining and exploration of useful information present in the vast amounts of image data generated regularly in biology labs around the world. Image based systems biology promises to provide functional localization in space and time [1]. Recent advances in single-molecule detection using fluorescence microscopy imaging technologies allow image analysis to provide access to invisible yet reproducible information extracted from bioimages [2]. Highly multiplexed fluorescence imaging techniques such as MELC or toponome imaging system (TIS) [3] generate massive amounts of multi-channel image data, where each individual channel can provide information about the abundance level of a specific protein molecule localized within an individual cell using the corresponding tag. Such high-dimensional representation of multiple co-localized protein expression levels demands for sophisticated analytical methods to extract molecular signatures of diseases such as cancer to not only enable us understand the biological processes behind cancer development but also aid us in early diagnosis

T. Huang et al. (Eds.): ICONIP 2012, Part IV, LNCS 7666, pp. 575–583, 2012.

and appropriate treatment of cancer. In this paper, we address the problem of mining cell phenotypes based on their high-dimensional protein co-expression profiles contained within the TIS images generated by a robotically controlled microscope installed at Warwick. We make three important contributions: First, we perform our analysis at the cell level marking a departure from the existing approaches employing pixel-level analysis [3–5]. Second, we show that the raw protein co-expression vectors have a nonlinear high-dimensional structure which can be effectively visualized using a symmetric neighborhood embedding approach. Third, we demonstrate the effectiveness of the nonlinear embedding coordinates for (a) classifying the tissue type at cellular level as compared to principal component analysis (PCA), its linear embedding counterpart, and (b) mining the cell phenotypes in an exploratory clustering setup using affinity propagation [6].

2   The Mining Framework

The framework presented in this work consists of three stages: pre-processing involves alignment and cell segmentation, non-linear low-dimensional embedding, and unsupervised clustering.

2.1   Pre-processing

Raza *et al.* [7] show that the multi-tag images obtained from TIS possess slight mis-alignment, which can potentially introduce noise when finding functional protein complexes in cancerous and normal tissue samples. In line with this argument, the RAMTaB (Robust Alignment of Multi-Tag Bioimages) [7] algorithm is used for aligning multi-tag fluorescent microscopy images.

Cell segmentation is required in order to restrict the analysis to cellular areas only. For nuclei segmentation, we used the multi-step framework proposed by Al-Kofahi *et al.* [8] on DAPI channel which highlights all the nuclei in the image. Initially, an image is binarized using graph-cut based alogorithm to extract foreground. Next, seed points are detected on the foreground of the binarized image by using multiscale Laplacian of Gaussian (LoG) filter, to perform an initial segmentation. Finally, this initial segmentation is refined by using a second graph-cut based algorithm. Nuclei segmentation results obtained using [8] are further post-processed to cater for very small nuclei, often produced as a result of segmentation errors, by either merging with the nearby nuclei or eliminating them altogether (see Figure 1). This further ensures that analysis is restricted to significantly distinguishable cell nuclei only. We refer to this complete process as cell segmentation in the following text.

2.2   Raw Expression Vector (REV)

We compute mean intensity value for each cell across K antibodies ($K = 12$) and build a $L_i \in \mathbb{R}^K$ vector for each cell i, which we call the Raw Expression

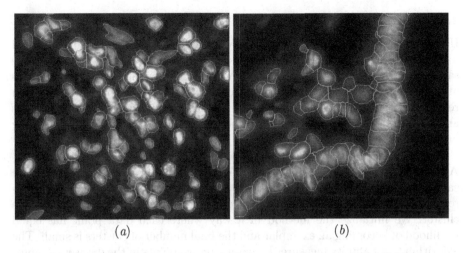

(a)   (b)

Fig. 1. Results of cell segmentation overlaid on top of the two original DAPI images. (a) Cancer; (b) Normal.

Vector (REV). Let N be the total number of cells found in all the stacks, then the data structure can be represented by an $N \times K$ matrix. We normalize the mean intensity value in each column to the range $[0, 1]$ before performing any further analysis.

2.3 Locality Preserving Non-linear Embedding

Most real-world datasets, regardless of their original dimensionality, contain some structure which should be representable in its intrinsic dimensions. We map all $\boldsymbol{L}_i \in \mathbb{R}^K$ REVs to $\boldsymbol{M}_i \in \mathbb{R}^L$, where $L < K$. t-Distributed Stochastic Neighbor Embedding (t-SNE) [9] is a method which provides such mapping through an optimization aiming to retain the original global and local structure. To achieve this, it define a similarity measure between any two points i and j in the original $\mathbb{R}^K$ space $(p_{j|i})$ and another in the lower dimensional $\mathbb{R}^L$ space $(q_{j|i})$ as,

$$p_{j|i} \propto \exp\left(-\|\boldsymbol{L}_i - \boldsymbol{L}_j\|_2^2 / 2\sigma_i^2\right) \ , \quad q_{j|i} \propto \left(1 + \|\boldsymbol{M}_i - \boldsymbol{M}_j\|_2^2\right)^{-1} \ , \quad (1)$$

where σ_i^2 is the variance of the Gaussian centered on $\boldsymbol{L}_i$, and $\|\cdot\|_2$ is the Euclidean norm. The user can control this variance, in turn specifying the number of neighbors affecting $p_{j|i}$. In order to keep the inherent structure of the data, t-SNE constrains the similarity measures for any two points to be roughly equivalent between the high and low dimensional space i.e. $p_{j|i} \approx q_{j|i}$. The Kullback-Leibler divergence is a natural fit to impose this contraint. Hence, the cost function to optimize is $\sum_i \sum_j \mathrm{KL}(p_{j|i} \| q_{j|i})$. To make the cost symmetric, all similarity measures are replaced by,

$$p_{j|i} \xrightarrow[\text{replace by}]{} p_{i,j} = \left(p_{j|i} + p_{i|j}\right) / 2N \ . \quad (2)$$

2.4   Clustering

Using either the original ($\{L_i\}$) or the lower dimensional data ($\{M_i\}$) we would like to observe the different phenotypes in both cancerous and normal tissue samples. Since each dimension in REV encodes the difference in expression levels after adding a particular anti-body, it can be used to cluster pixels based on responses to K anti-bodies. To observe the discrimination between cancerous and normal tissue responses, we experimented with two unsupervised clustering methods briefly described below:

Affinity Propagation Clustering (APC). APC [6] is an approach where each data samples elects another data point within the dataset to act as its representative or exemplar. The points electing a common exemplar form a single cluster. We initialize the method in a way where each data point has equal likelihood of becoming an exemplar and the final number of clusters is small. The algorithm takes affinity measures between any two points in the dataset as input, which is used in each iteration to find which data points are good exemplars for what samples. To achieve this goal, two kinds of messages are shared between every pair of data points i and j. The *responsibility* $r(i, k)$ reflects the suitability point k to represent point i as its exemplar. The reverse message, *availability* $a(i, k)$ defines how much point k thinks it is suited to act as the exemplar of point i. Each iteration updates $r(i, k)$ and $a(i, k)$ in a data-driven fashion.

The affinity measure we use between points i and j is

$$K(i, j) = \exp\left(-\|M_i - M_j\|_2^2 / 2\sigma^2\right), \tag{3}$$

where $\sigma = max(\|M_i - M_j\|_2)/3$. We denote the number of clusters resulting from this approach by $\hat{C}$. It is worth noting that the APC algorithm determines $\hat{C}$ in an unsupervised manner.

Agglomerative Hierarchical Clustering (AHC). This is a bottom-up clustering method [10], which starts with each of the N REVs as being a single cluster, and merges two clusters in each iteration. This process can be better represented as a dendrogram tree structure, where cutting across the tree at level k would give $N - k$ clusters[1]. We aim to get the same number of clusters returned by APC, hence we cut the tree at level $\hat{k} = N - \hat{C}$.

The criterion we employ to select the two clusters to merge aims to minimize the increase in the variance of clusters [11]. Mathematically, at each iteration level k^* we have clusters $S_j = \{M_{(j,1)}, \ldots, M_{(j,n_j)}\}$ where $n_j = |S_j|$ and $j \in \{1, \ldots N - k^*\}$. To make clusters for level $k^* + 1$, we seek clusters $\hat{u}$ and $\hat{v}$ such that

$$\hat{u}, \hat{v} = \operatorname*{arg\,min}_{u,v \in \{1, \ldots N-k^*\}} n_u n_v \left(\|\bar{S}_u - \bar{S}_v\|_2\right)^2 / (n_u + n_v), \tag{4}$$

where $\bar{S}_j$ is the mean vector for set S_j. This step will result in a new cluster formed by merging $S_{\hat{u}}$ and $S_{\hat{v}}$, hence reducing the number of clusters by 1.

---

[1] The algorithm starts at level $k = 0$, where there are N clusters. Cutting the tree at level k means truncating the tree *after* level k.

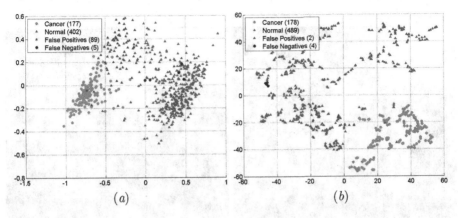

Fig. 2. Illustration of low dimensional embedding of Raw Expression Vectors belonging to Cancer and Normal Cells using two different dimensionality reduction techniques: (a) PCA; (b) t-SNE

3   Experimental Results and Discussion

The data used in this study consists of 3 colon tissue samples. 2 out of the 3 selected tissue samples are taken from healthy colon tissues while 1 taken from cancerous colon tissue. The tissue samples are verified to be normal or cancerous by independent expert pathologists. A library of 26 antibody tags is used on all 3 colon tissues to generate 3 stacks of multi-tag microscopic bioimages (each stack having 26 tags) using TIS [3]. The library of antibody tags used in this study comprise mainly of nuclei, stem-cell and tumor markers as reported in [12]. Out of these 26 antibodies, some antibodies are ignored because their function is not reflective of the cell activity, while others are discarded because the of the poor quality of its image. Subsequently, only 12 antibody tags are used in the subsequent analysis: CD36,CD44, CD57, CD133, CD166, CK19, CK20, Cyclin-A, Cyclin-D1, CEA, Muc2, EpCAM. Each image is of size 1056×1026 with pixel resolution of 206×206 nm/pixel.

All 3 fluorescent microscopy image stacks in the dataset are processed in a similar manner, with image alignment and cell segmentation as described in section 2.1, and finally REV generation as described in section 2.2. For image registration, the default parameters as detailed in [7] are used. For cell segmentation, we tuned for parameters of algorithm in [8] to suit our imaging conditions.

Experiment 1: Linear (PCA) vs. Non-linear (t-SNE) Dimensionality Reduction for Cell Classification. For dimensionality reduction, we used two frameworks: one linear (PCA) and other non-linear (t-SNE). Here we show that PCA fails to preserve pairwise relationship between REVs in high dimensional space, whereas t-SNE not only preserves the pairwise relationships but also provides a much superior visual representation of the protein expression vectors. REVs are reduced to 3 dimensions using PCA and t-SNE and *k-means*

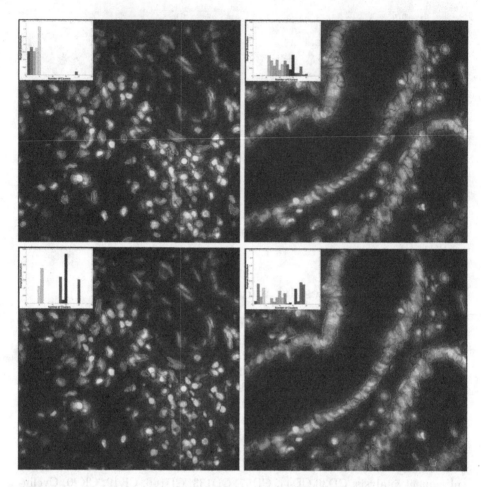

Fig. 3. Visual Overlay of 20 phenotypes found using APC(first row) and AHC(second row) on top of DAPI images for a Caner(first column) and Normal(second column) tissue sample. Marginal distributions of cell phenotypes are shown on the top-left corner of each image. Note the difference in distribution of phenotypes in cancer and normal sample.

clustering (with $k = 2$) is applied on these low-dimensional representations to yield the results. Figure 2 shows the visual comparison of clustering results. The results obtained above are evaluated on three quantitative accuracy measures: Sensitivity (Sen), Specificity (Spec), and Positive Predictive Value (PPV). Let TP denotes the number of true positive(cancer cells correctly classified as cancerous), FP the number of false positive (normal cells incorrectly classified as cancerous), TN the number of true negatives (cancer cells incorrectly classified as normal), and FN the number of false negatives (normal cells correctly classified as normal), then Sen is defined as $TP/(TP+FN)$, Spec as $TN/(TN+FP)$ and

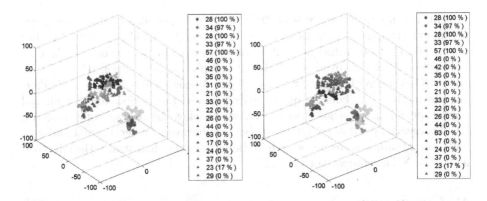

Fig. 4. Scatter plot of t-SNE embedding of REVs, where cancer (circles) and normal cells (triangles) are colored using 20 different phenotypes identified using APC(left) and AHC(right). The legend displays the number of cells present in the cluster and percentage of cancerous cells respectively, identified in the corresponding phenotype. Note that phenotypes are marked on the basis of majority; i.e. a given phenotype is marked as cancer (circle) if majority of its cells belong to cancer and viceversa.

Table 1. Quantitative Comparison of Classification Results using: 12−dimensional REV; 3−dimensional PCA; 3−dimensional t-SNE. Values marked in bold show best results. On all 3 scales, t-SNE embedding gives superior performance.

	Sen	Spec	PPV
REV	0.9725	0.8187	0.6654
PCA	0.9725	0.8187	0.6654
t-SNE	0.9780	0.9959	**0.9889**

Table 2. Quantitative comparison of inter- and intra-class symmetric KL-divergence of cell phenotype distributions for 12−dimensional REV, 3−dimensional PCA, and 3−dimensional t-SNE using APC and AHC

Cluster		Inter-class	Intra-class
REV	AP	29.8565	2.3992
	HC	33.7972	2.9332
PCA	AP	22.9484	**0.2841**
	HC	26.0002	0.697
t-SNE	AP	**41.896**	0.6915
	HC	41.7355	0.8345

PPV as $TP/(TP+FP)$. Table 1 shows quantitative comparison of classification (using k-means, with $k = 2$) when (1) 12−dimensional REVs; (2) 3−dimensional PCA; and (3) 3−dimensional t-SNE data are used for classification. Note that the PPV for t-SNE is approximately 32% higher than those of the original data and PCA.

Experiment 2: Cell Phenotype Analysis Using Unsupervised Clustering: Given the promising cell classification results obtained above, we are further interested in finding different phenotypes present in normal and cancer samples. Unsupervised clustering can be used for such type of analysis. Here, we present a comparative analysis of different phenotypes identified by using two popular clustering frameworks described in section 2.4: APC and AHC. Number of clusters $(\hat{C})$ is identified using APC and the same number is used in AHC.

Figure 3 shows the overlay of different cell phenotypes identified using APC and AHC over DAPI Images, whereas Figure 4 shows the scatter plot of different phenotypes. In order to quantitatively assess the performance of all cell phenotype mining methods used here, we employ the average of symmetric KL-divergence of cell phenotype distributions between cancer/normal and normal/normal samples. Results shown in Table 2 again demonstrates the effectiveness of t-SNE as compared to PCA.

4   Conclusions

We presented a paradigm for cell-level mining of molecular signatures in multi-tag bioimages using a nonlinear embedding approach. This approach is a marked departure from the traditional pixel-level approaches. We showed that the symmetric neighborhood embedding outperforms the original high-dimensional raw protein expression vectors in terms of its ability to discriminate between normal and cancer tissue samples on the basis of their phenotypic distributions. Our future work will employ this paradigm in a large-scale validation for extracting biologically plausible molecular signatures of various cell phenotypes found in cancer specimens.

References

1. Megason, S.G., Fraser, S.E.: Imaging in systems biology. Cell 130(5), 784–795 (2007)
2. Danuser, G.: Computer vision in cell biology. Cell 147(5), 973–978 (2011)
3. Schubert, W., Bonnekoh, B., Pommer, A.J., Philipsen, L., Böckelmann, R., Malykh, Y., Gollnick, H., Friedenberger, M., Bode, M., Dress, A.W.M.: Analyzing proteome topology and function by automated multidimensional fluorescence microscopy. Nature Biotechnology 24(10), 1270–1278 (2006)
4. Humayun, A., Raza, S.E.A., Waddington, C., Abouna, S., Khan, M., Rajpoot, N.M.: A Framework for Molecular Co-Expression Pattern Analysis in Multi-Channel Toponome Fluorescence Images. In: Microscopy Image Analysis with Applications in Biology, MIAAB (2011)
5. Kölling, J., Langenkämper, D., Abouna, S., Khan, M., Nattkemper, T.: Whide – a web tool for visual data mining colocation patterns in multivariate bioimages. Bioinformatics (2012)
6. Frey, B.J., Dueck, D.: Clustering by passing messages between data points. Science 315(5814), 972–976 (2007)
7. Raza, S.: e.A., Humayun, A., Abouna, S., Nattkemper, T.W., Epstein, D.B.A., Khan, M., Rajpoot, N.M.: Ramtab: Robust alignment of multi-tag bioimages. PLoS ONE 7(2) (2012)
8. Al-Kofahi, Y., Lassoued, W., Lee, W., Roysam, B.: Improved automatic detection and segmentation of cell nuclei in histopathology images. IEEE Transactions on Biomedical Engineering 57(4), 841–852 (2010)

9. Van der Maaten, L., Hinton, G.: Visualizing data using t-sne. Journal of Machine Learning Research 9, 2579–2605 (2008)
10. Jain, A., Murty, M., Flynn, P.: Data clustering: a review. ACM Computing Surveys (CSUR) 31(3), 264–323 (1999)
11. Ward Jr, J.: Hierarchical grouping to optimize an objective function. Journal of the American Statistical Association, 236–244 (1963)
12. Bhattacharya, S., Mathew, G., Ruban, E., Epstein, D., Krusche, A., Hillert, R., Schubert, W., Khan, M.: Toponome imaging system: in situ protein network mapping in normal and cancerous colon from the same patient reveals more than five-thousand cancer specific protein clusters and their sub-cellular annotation by using a three symbol code. Journal of Proteome Research (2010)

SNEOM: A Sanger Network Based Extended Over-Sampling Method.
Application to Imbalanced Biomedical Datasets

José Manuel Martínez-García[1],
Carmen Paz Suárez-Araujo[1], and Patricio García Báez[2]

[1] Instituto Universitario de Ciencias y Tecnologías Cibernéticas, Universidad de Las Palmas de Gran Canaria, Las Palmas de Gran Canaria, Canary Islands, Spain
jose.martinez104@alu.ulpgc.es, cpsuarez@dis.ulpgc.es
[2] Departamento de Estadística, Investigación Operativa y Computación, Universidad de La Laguna, La Laguna, Canary Islands, Spain
pgarcia@ull.es

Abstract. In this work we introduce a novel over-sampling method to face the problem of imbalanced classes' classification. This method, based on the Sanger neural network, is capable of dealing with high-dimensional datasets. Moreover, it extends the capability of over-sampling methods and allows generating samples from both minority and majority classes. We have validated it in real medical applications where the involved datasets present an un-even representation among the classes and it has been obtained high sensitivities identifying minority classes. Therefore, by means of this method it is possible to accomplish the design of systems for the medical diagnosis with a high reliability.

Keywords: imbalanced distributions, over-sampling, classification, Sanger network, neural networks.

1 Introduction

Imbalanced datasets involve a great issue for classifiers since there is not an even class representation. Most of standard learning algorithms are not able to deal with imbalanced distributions because they assume an equal representation among the different classes, causing a decline of the classifier performance. This inconvenience is pervasive in a multitude of different domains in which the collection of positive samples (relatives to the minority or interest class) is less than the negative samples collection (majority or popular class).

In the medical field, imbalanced datasets suppose a critical problem because of the cost of misclassifying a minority class (e.g. an infrequent disease) as a majority class (e.g. non-suffering that strange illness), are especially high. In that case, a subject who really suffers the disease could not be properly treated medically.

To face this problem there are mainly two different approaches. The first one, focused on the learning algorithm, attempts to modify the learning strategies of the classifiers with the aim of make up for the lack of minority samples. For this, two

T. Huang et al. (Eds.): ICONIP 2012, Part IV, LNCS 7666, pp. 584–592, 2012.

techniques related to this approach are known, cost-sensitive learning [1,2] and recognition-based learning [3,4].

On the other hand, the second approach seeks to solve the gap between classes from the point of view of the data. Two main strategies are based on the restructuring [5,6] or modification of the dataset [7,8].

Regarding to the dataset modification strategy, there are two options: cutting down the majority class set (under-sampling) or increasing the minority class set (over-sampling). Previous related works revealed the advantages of using sampling techniques for improving the performance of classifiers [9,10] and even combinations of over-sampling the minority class and simultaneously under-sampling the majority class [4].

In this paper we present an extended over-sampling method based on Sanger neural network (SNEOM) for improving the classification of imbalanced datasets with two classes. This method allows obtaining correlated synthetic samples with the input data from an n-dimensional space. Performing a dimensionality reduction using a projective method based on the Sanger network, the input multivariate data is transformed into a less-dimensional space where it is possible to add perturbations over the data, such as a Gaussian noise, in order to generate similar samples to the original ones within the acceptance regions for each variable. This method is termed *extended* because it could be configured for only minority class generation (pure over-sampling approach) and for both minority and majority class generation.

2   SNEOM: A Sanger Network Based Extended Over-Sampling Method

Imbalanced distributions entail one of the most challenging matters in machine learning due to the drawbacks that class under-representation involve for most of classifiers which suppose an even distribution among the different classes belong to the training data set. When a classifier is trained using a dataset that contains a predominant samples' representation of one class with respect to the other classes, it will be over-specialized in majority class recognition, provoking a misclassification of the minority class samples.

In this section we introduce an over-sampling method for both minority and majority classes' generation within imbalanced datasets. Hence it is called extended. Our proposal is intended not only to create minority class samples to produce a balanced class distribution, but to increase the proportion of both class in order to produce a more extensive, diverse and balanced training data set.

SNEOM is a projective neural method for multivariate and relational synthetic data generation. The main feature of this method is that re-sampling is performed on the transformed space of the input data. Thus, SNEOM could be applied on multivariate data, providing the high dimensional data visualization in two or three dimensions. Another important characteristic of our proposal is its capacity for providing balanced classes even with data missing. This attribute is endowed with the neural architecture that implements SNEOM, the Sanger network, specifically the Sanger Network Extension for Missing Data Treatment [11]. This feature will not be evaluated in this paper.

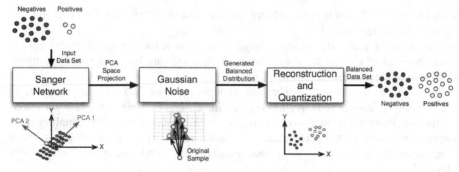

Fig. 1. Schema of the SNEOM method

SNEOM schema is shown in Fig. 1. The first step carried out is data pre-processing. Before performing the projection of the input data it is necessary to standardize the data. This is an essential task because the projective method used is sensitive to the relative scaling of the original variables. Standardization is performed transforming the data to a Gaussian distribution centered at zero and unitary standard deviation.

After pre-processing, a feature extraction process will be performed using a neural projective method. It is generated a transformation of the input space into a new feature space, concretely the Principal Component Analysis (PCA) space, with a lower dimensionality, preserving the maximum possible amount of information. The proposed method implements this transformation process using a neural approach whose processes converge into PCA, the Sanger network, also called Generalized Hebbian Algorithm (GHA) [11].

Sanger network is totally interconnected with the inputs. We have adapted it to be able to face the processing of missing data. To accomplish it we have followed a scheme similar to the found in [12], causing these missing values not to contribute the output nor modify the weights of the networks:

$$y_i = \sum_{j \in P_t}^n w_{ij} x_j \ .$$

(1)

Where P_t is the set of input units, j, whose values x_j are available at time t.

The learning rule that progressively fits this matrix of weights is given by the following expression:

$$\Delta w_{ij} = \begin{cases} \eta(t) y_i \left(x_j - \sum_{k=1}^i y_k w_{kj} \right) & \textit{if } j \in P_t \\ 0 & \textit{otherwise} \end{cases} \ .$$

(2)

In the expression given by Equation 2, the learning ratio $\eta(t)$ is a linearly decreasing function in time t. This formula causes the Sanger weights to converge to the main training data set components.

Once we have performed PCA by means of the Sanger network, we obtain the scores (input dataset projected onto the feature space) and the weights (correlation coefficients between the scores and the input dataset).

The selected PCA number depends on the input variance percentage obtained after the Sanger training. It is important to emphasize that the goodness of the subsequent reconstruction of data to the initial space is strongly related to this parameter, so it is recommendable selecting a quantity of PCA that provides around 90% of the input data variance.

Then we apply our over-sampling procedure on the feature space. Firstly, we set the factors of the minority and majority classes through two parameters entered by the user, which indicate the amount of synthetic samples that the algorithm will generate for each class. If the majority factor is nonzero, it will be considered as extended over-sampling approach. Next, the procedure calculates the acceptance area for each projected sample. It determines the boundaries within new samples could be generated, using the variances of the data distribution. These boundaries are calculated as the standard deviations for each class variable. Once calculated the acceptance area for every class, it is centered at the coordinates of each projected sample. Afterwards, the method takes iteratively each sample and adds it a Gaussian noise with a unitary bell width to perturb the original sample and initializes a new nearby sample. If the new sample falls within the acceptance area, a new synthetic sample is generated and more new synthetic samples will be generated until the factor for that class is reached. If any synthetic sample falls outside the acceptance area, it will not be accepted and the bell width of the Gaussian noise will be decremented 0.1 units with the aim of generating new samples closest to the original one and being more likely to fall inside the acceptance area.

Data projected onto the PCA space preserve the statistics features of the data distribution for each class. Perturbing samples on this transformed space allows us modifying fewer dimensions instead of the whole set of variables in the input space. In this way, the number of dimensions perturbed is reduced and it also facilitates the visualization of the entire process.

After the generation stage, we have the reconstruction phase, which consist of transforming the feature space into the input space again. The original input vector (x) can be reconstructed (Rx) with a minimum loss of information from the m-outputs vector (y) and the Sanger weights (W) (decompression procedure). The equation that carries out the decompression is the following one:

$$Rx_j = \sum_{j=1}^{m} w_{ij} y_i .$$ (3)

Once data is projected onto the input space, the inverse process to the standardization is done and it is performed a quantization on the data. This last step is crucial in order to adjust the generated data to the range of the input data.

3   Applications of SNEOM in Non-balanced Medical Classification Problems

3.1   Datasets

In order to validate SNEOM we have carried out different experiments about imbalanced datasets in the medical field. We have analyzed the results for bimodal classifications, using neural and not neural paradigms, concretely - Naïve Bayes (NB), C4.5 classification tree and Multilayer Perceptron (MLP) - embedded into the data mining software WEKA [13]. The studied classifications have been discriminating between Mild Cognitive Impairment (MCI) and non-MCI subjects, classification of benign and malignant breast tumors, and pathology detection in the vertebral column, distinguishing between normal and pathological explorations. The two latest medical datasets have been taken from the UCI repository [14] whereas the first one dataset have been extracted from real medical consultations using EDEVITALZH clinical environment [15]. These datasets differ on their sizes and imbalance rates, resulting in assorted scenarios where testing our method. With the aim of carry out a homogeneous study of the method, with high reliability in the reconstruction stage and easy visualization we have selected a subset of attributes of the sets taking as starting dataset for the analysis the data related to real MCI consultations from EDEVITALZH environment. Thus, we have used the following datasets:

1. The Mild Cognitive Impairment dataset contains 116 consultations relatives to a cohort of subjects recruited from the dependence care unit for the elderly of Santander (Cantabria, Spain). Within this database there are 92 consultations relative to non-MCI subjects the 24 remaining correspond to consultations where the patient was revealed a MCI. The variables included in this database are: the Mini-Mental State Examination score, the Barthel index score and the educational level.
2. Breast Cancer Wisconsin dataset [14]. This dataset is composed by 683 samples, 444 of them corresponding to benign tumours and the other 239 belongs to malignant tumours. Each instance is represented by four variables regarding to breast cytology information: clump thickness, uniformity of cell size, uniformity of cell shape and marginal adhesion.
3. Vertebral Columns dataset [14]. The total amount of samples contained in this dataset is 310, divided into 210 samples representing abnormal exploration and 100 samples according to a normal exploration. Each patient is represented in the dataset by four biomechanical attributes derived from the shape and orientation of the pelvis and lumbar spine: pelvic incidence, pelvic tilt, lumbar lordosis angle and sacral slope.

3.2   Results and Discussion

Results have been obtained training three different classifying paradigms with three different data distributions, an imbalanced (original) dataset and two SNEOM-balanced datasets using both the standard (OS) and extended (EOS) over-sampling approaches. For the first distribution, classifiers were trained using only original samples from the imbalanced dataset and regarding to the two latter distributions, classifiers were trained using generated samples from our SNEOM method. All the datasets were split into training, test and validation sets, taking a 75% of the samples for an even distribution between the two first sets and the 25% remain for the validation set.

We have been taken into account a variety of measures that allows us to obtain stable and reliable evaluations in order to validate and demonstrate the effectiveness of our proposed method SNEOM. The imbalanced input dataset used for training is denoted in tables as Original. The datasets composed by synthetic samples using the SNEOM method are indicated as $SNEOM_{OS}$ for pure over-sampling and $SNEOM_{EOS}$ for referring to extended over-sampling approach. Best results are in bold with an asterisk.

For the MCI dataset, the obtained results are shown in table 1. In this case, training a Naïve Bayes and a C4.5 classifier using the imbalanced dataset, all the samples were predicted as belonging to the majority class, so that all the minority instances were incorrectly classified (sensitivity = 0%). Training a MLP with the same dataset, only the half of minority samples was hit (sensitivity 50%) and the whole majority samples were fairly classified (100%). Because of the imbalanced validation dataset is composed by roughly 4 times majority samples (23 non-MCI subjects) than minority samples (6 MCI subjects), the Area Under the Curve (AUC) value is the highest in the table (0.993). Regarding to the classifiers trained with the OS dataset, the sensitivities are significantly higher than training the classifiers with the unbalanced dataset but the specificities have been decreased because the classifier is performing a fairer classification that in the previous scenario (previously both positive and negative samples were identified as majority class). Training with the OS dataset, the accuracy is always greater than the accuracy values obtained with the imbalanced training data except for the C4.5 classifier that is equal (79.31%). When the EOS dataset is used, it is achieved a 100% sensitivity rate for all the classifiers. Specificity rates are less than the ones obtained with the imbalanced dataset, due to the strongly influence of classifying all the validation samples as the majority class, but in almost all the cases, it is over the OS rates (except for MLP classifier). Although the AUC value for a MLP classifier trained with the imbalanced dataset is the highest, this value is not reliable for indicating a good classifying performance because only the 50% of the minority samples were correctly classified. The best combination of all for this problem was the NB classifier trained with EOS, obtaining the best accuracy and sensitivity rates (93.1% and 100%, respectively) and a high specificity rate (91.3%).

Table 1. Mild Cognitive Impairment dataset performance results

	Accuracy	Sensitivity	Specificity	AUC
Original+MLP	89.67%	50.00%	100%	0.993
$SNEOM_{OS}$+MLP	90.00%	83.33%	91.30%	0.964
$SNEOM_{EOS}$+MLP	82.76%	100.00%	78.26%	0.986
Original+NB	79.31%	0%	100%	0.500
$SNEOM_{OS}$+NB	82.76%	66.67%	86.96%	0.808
$SNEOM_{EOS}$+NB*	93.10%	100%	91.30%	0.928
Original+C4.5	79.31%	0%	100%	0.500
$SNEOM_{OS}$+C4.5	79.31%	100%	73.91%	0.870
$SNEOM_{EOS}$+C4.5	82.76%	100%	78.26%	0.906

In relation to the Breast Cancer dataset, whose results are represented in Table 2, the sensitivity and AUC values are always better for OS and EOS datasets than for the imbalanced one, so that, using SNEOM there is more likely to correctly identify the minority instances. The accuracy is greater for OS and EOS balanced datasets with respect to the imbalanced dataset in the most of all cases except for the C4.5 classifier that is slightly smaller. The specificity values are very close between the balanced and the imbalanced distributions using a MLP or NB classifiers, getting the greatest value (96.64%) for a NB classifier trained with the OS dataset. This configuration also provides the greatest value for accuracy, so it is one of the best combinations together with the MLP classifier trained with the EOS dataset. The latter achieves the highest sensitivity (96.61%), which is the best rate for recognizing minority class samples.

Table 2. Breast Cancer dataset performance results

	Accuracy	Sensitivity	Specificity	AUC
Original+MLP	94.12%	89.80%	96.40%	0.991
SNEOM$_{OS}$+MLP	95.29%	94.92%	95.50%	0.993
SNEOM$_{EOS}$+MLP*	95.29%	96.61%	94.50%	0.993
Original+NB	95.29%	93.22%	96.40%	0.992
SNEOM$_{OS}$+NB*	95.88%	95.00%	96.40%	0.993
SNEOM$_{EOS}$+NB	95.29%	95.00%	95.50%	0.994
Original+C4.5	94.12%	93.22%	94.59%	0.928
SNEOM$_{OS}$+C4.5	93.53%	96.61%	91.89%	0.961
SNEOM$_{EOS}$+C4.5	93.53%	95.00%	92.79%	0.939

Results for Vertebral Column dataset are depicted in Table 3. It is observable that it is very difficult for classifiers to correctly discriminate among classes due to the lower obtained percentage for all the measures. Here, accuracy is very similar for all the cases, except for the C4.5 classifier with the imbalanced dataset that is higher than the rest of configurations because all the samples were classified as majority class, and consequently all the minority classes were incorrectly identified (sensitivity = 0%). Therefore, for this problem, accuracy is not a reliable measure to compare between different configurations. Sensitivity values are always significantly higher for OS and EOS, and also the AUC values are better when using SNEOM. Specificities are slightly higher for imbalanced datasets, because of the classifiers tend to forecast in most of the cases that the validation samples belong to the majority class. MLP and NB provided good results again. Using a MLP classifier trained with the OS dataset we obtained the best sensitivity and AUC rates (85% and 0.743, respectively), indicating that this classifier trained with the OS training set is capable of identify more minority samples than the others. NB classifier with the EOS training dataset performed the best accuracy, obtaining the best trade-off between the classification of majority and minority classes (sensitivity = 72%, specificity = 61.54%).

Table 3. Vertebral Column dataset performance results

	Accuracy	Sensitivity	Specificity	AUC
Original+MLP	64.94%	40.00%	67.31%	0.700
SNEOM$_{OS}$+MLP*	63.64%	85.00%	55.77%	0.743
SNEOM$_{EOS}$+MLP	63.64%	72.00%	59.62%	0.700
Original+NB	63.64%	64.00%	63.46%	0.708
SNEOM$_{OS}$+NB	62.34%	68.00%	59.62%	0.718
SNEOM$_{EOS}$+NB*	64.94%	72.00%	61.54%	0.739
Original+C4.5	67.53%	0%	100%	0.500
SNEOM$_{OS}$+C4.5	63.64%	72.00%	59.62%	0.671
SNEOM$_{EOS}$+C4.5	62.34%	85.00%	53.85%	0.672

Results obtained for three biomedical datasets with different unbalanced levels and dataset sizes have validated the SNEOM method and justified its goodness for using it in classification tasks with imbalanced classes. In all the experiments, the best results were achieved by means training the classifiers with the SNEOM-balanced datasets and using both pure over-sampling and our novel proposed extended over-sampling approaches.

4   Conclusions

In this work we have presented a novel over-sampling method with the possibility to extend the synthetic samples generation to the majority class in addition to the minority class. This is a neural projective method that is able to deal with high dimensional data and missing values in any variable. In order to validate our method and state the goodness of using it in imbalanced classification, we have applied it to three biomedical imbalanced datasets to train three different classifiers with the original imbalanced dataset and comparing the results with two balanced datasets composed by synthetic samples obtained after applying SNEOM, using both pure over-sampling and extended over-sampling approaches. Several metrics have been calculated in order to evaluate the performances obtained among different classifiers trained with each of data distributions. The best performances were obtained when using SNEOM for balancing the classes' representation into the datasets. We have validated our method and its goodness to apply it on imbalanced scenarios. In addition, we have demonstrated the effectiveness of our extended approach getting better performances in some cases than the standard over-sampling technique. Therefore, by means of this method it is possible to accomplish the design of systems for the medical diagnosis with a high reliability.

Acknowledgments. Authors would like to thank the Canary Islands Government and EU Funds (FEDER) for their support under Research Project "Sol-SubC200801000347". We also sincerely thank the anonymous reviewers for their helpful comments which allowed improving the quality of the paper.

References

1. Domingos, P.: Metacost: A general method for making classifiers cost–sensitive. In: Proceedings of the Fifth International Conference on Knowledge Discovery and Data Mining, pp. 155–164 (1999)
2. Zhou, Z.H., Liu, X.Y.: Training cost–sensitive neural networks with methods addressing the class imbalance problem. IEEE Transactions on Knowledge and Data Engineering 18(1), 63–77 (2006)
3. Manevitz, L., Yousef, M.: One-class SVMs for document classification. Journal of Machine Learning Research 2, 139–154 (2001)
4. Chawla, N.V., Bowyer, K.W., Hall, L.O., Kegelmeyer, W.P.: SMOTE: Synthetic minority over-sampling technique. Journal of Artificial Intelligence Research 16, 321–357 (2002)
5. Argamon-Engelson, S., Dagan, I.: Committee-based sample selection for probabilistic classifiers. Journal of Artificial Intelligence Research (JAIR) 11, 335–360 (1999)
6. Freund, Y., Sebastian Seung, H., Shamir, E., Tishby, N.: Selective sampling using the query by committee algorithm. Machine Learning 28(2-3), 133–168 (1997)
7. Chawla, N.V., Lazarevic, A., Hall, L.O., Bowyer, K.W.: Smoteboost: Improving prediction of the minority class in boosting. In: Proceedings of the 7th European Conference on Principles and Practice of Knowledge Discovery in Databases, pp. 107–119 (2003)
8. Chyi, Y.-M.: Classification analysis techniques for skewed class distribution problems. Master Thesis, Department of Information Management, National Sun Yat-Sen University (2003)
9. Van Hulse, J., Khoshgoftaar, T.M., Napolitano, A.: Experimental perspectives on learning from imbalanced data. In: International Conference on Machine Learning, pp. 935–942 (2007)
10. Batista, G.E.A.P.A., Prati, R.C., Monard, M.C.: A Study of the Behavior of Several Methods for Balancing Machine Learning Training Data. SIGKDD Explorations 6(1), 20–29 (2004)
11. García Báez, P., Suárez Araujo, C.P., Fernández Viadero, C., Regidor García, J.: Automatic Prognostic Determination and Evolution of Cognitive Decline Using Artificial Neural Networks. In: Yin, H., Tino, P., Corchado, E., Byrne, W., Yao, X. (eds.) IDEAL 2007. LNCS, vol. 4881, pp. 898–907. Springer, Heidelberg (2007)
12. Samad, T., Harp, S.A.: Self-organization with partial data. Network 3, 205–212 (1992)
13. Witten, I.H., Frank, E.: Data Mining: Practical machine learning tools and techniques, 2nd edn. Morgan Kaufmann, San Francisco (2005)
14. UCI Machine Learning Repository, http://archive.ics.uci.edu/ml
15. Suarez Araujo, C.P., Perez-del-Pino, M.A., Garcia Baez, P., Fernandez Lopez, P.: Clinical Web Environment to Assist the Diagnosis of Alzheimers Disease and other Dementias. WSEAS Transactions on Computers 6(3), 2083–2088 (2004)

Abductive Neural Network Modeling
for Hand Recognition Using Geometric Features

El-Sayed M. El-Alfy, Radwan E. Abdel-Aal, and Zubair A. Baig

College of Computer Sciences and Engineering
King Fahd University of Petroleum and Minerals
Dhahran 31261, Saudi Arabia
{alfy,radwan,zbaig}@kfupm.edu.sa

Abstract. Hand recognition has received wide acceptance in many applications for automatic personal identification or verification in low to medium security systems. In this paper, we present a new approach for hand recognition based on abductive machine learning and hand geometric features. This approach is evaluated and compared to other learning algorithms including decision trees, support vector machines, and rule-based classifiers. Unlike other algorithms, the abductive learning approach builds simple polynomial neural network models by automatically selecting the most relevant features for each case. It also has acceptable accuracy with low false acceptance and false rejection rates. For the adopted dataset, the abductive learning approach has more than 98% overall accuracy, 1.67% average false rejection rate, and 0.088% average false acceptance rate.

Keywords: Pattern Recognition, Hand Recognition, Biometric Authentication, Geometric Features, Polynomial Neural Networks, Abductive Learning.

1   Introduction

Recently hand recognition has drawn considerable attention and several approaches have been proposed in the literature [1], [2]. It has gained wide acceptance in several low to moderate security systems for personal authentication and identification [3]. Hand geometry recognition uses features extracted from hand images such as hand and finger lengths, widths, thicknesses, and curvatures; and relative locations of joint points [2], [3]. Before a person is granted access to certain services or facilities, the system needs to verify or determine the person's identity. Nowadays, there is a technological shift toward biometric authentication systems as being more reliable alternative or complementary to traditional approaches based on passwords, smart cards, tokens, etc. In contrast to other biometric systems (such as face, iris and fingerprint), hand recognition has many advantages [4]. It is relatively simple and affordable since it uses inexpensive and less complex devices and processing techniques. Moreover, hand geometric features can be easily extracted from low or medium resolution hand images, and hand geometry is non-intrusive, more convenient and publicly accepted. It also requires low amount of data to identify an individual, has

T. Huang et al. (Eds.): ICONIP 2012, Part IV, LNCS 7666, pp. 593–602, 2012.

low failure to enroll (*FTE*) rate, and can be fused easily with other hand-based biometrics (e.g. palmprint and fingerprint) using same or fewer scanners [5].

The aim of this paper is to develop a new system for hand recognition based on abductive machine learning and investigate its performance as compared other baseline learning algorithms. We consider decision trees, support vector machines and rule-based classifiers for comparison.

The rest of the paper is organized as follows. Section 2 describes the components of the proposed approach. Section 3 presents the experimental work and discusses the results. Finally, Section 4 concludes the paper and highlights future work.

2   Hand Recognition System Overview

Like other biometric-based security systems, a hand geometry system works in two phases: enrollment phase and matching or recognition phase. During the first phase, potential users of the system are enrolled by taking one or more hand images for each user. For the purpose of hand geometry, surface (front), dorsal (back) and/or lateral (side) images can be captured for the hand. The hand images are pre-processed and distinguishing features are extracted to create user templates (template = feature set + user's identity) which are stored in a database. During the second phase, a single image is entered and processed to determine the identity of the user (identification mode). In verification mode, a claimed identity is entered along with the image and the system role is to accept or deny the claimed identity [2], [3]. The outline of the main steps in a hand-geometry based authentication system is illustrated in Fig. 1.

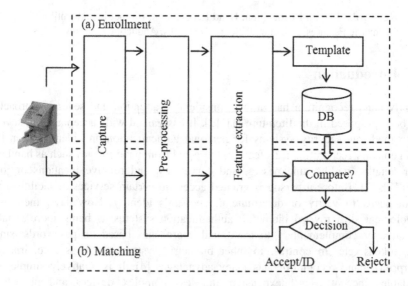

Fig. 1. An overview of the proposed hand recognition system

To extract hand geometry features, acquired hand images need to pass through the following steps. First the hand image is pre-processed to adjust orientation and light of the hand and reduce noise before feature extraction. The hand silhouette is then determined using different thresholding, filtering, segmentation and edge detection techniques. However, these techniques are relatively less complex than those used with other biometrics such as face, fingerprint and iris. The silhouette is used to determine or locate specific points in the hand image called major points such as finger tips, fingers in-between valleys, hand end point located at the wrist, etc. These points are mainly selected based on the design aspects of the authentication system. Once these major points are determined, the geometric features can be extracted by measuring the distances between the valley points and fingertips. Moreover, finger widths can be measured at different positions. Hand geometry features may include fingers lengths and widths, palm length and width, and hand length. Other geometric features can be used such as hand contour coordinates, angles, moments, etc. [4]. The next step is to compare the extracted feature vector against templates stored in the database and compute matching scores depending on identification or verification mode. For identification, the system performs one-versus-all comparison of the feature vector for the given image with all entries in the database using a similarity or distance measure to determine the closest match and report its associated identity. If the goal is verification, the system compares only with the pre-stored templates whose ID's are the same as the claimed ID. Then, based on the imposter and genuine score distributions for the pre-stored templates, the system compares the matching score against a threshold value to decide match (accept) or non-match (reject).

3   Abductive Learning Based Hand Recognition

The problem of hand recognition can be casted as a multi-class classification problem. Given an observed feature vector $\mathbf{x} = (x_1, x_2, ..., x_m)$, it is required to predict a label $y \in \{1, 2, ..., N\}$ where N is the number of classes (subjects). The proposed approach uses N recognizers whose outputs y_i, where $0 \leq y_i \leq 1$ and $i = 1, ..., N$, are then merged using the maximum rule. A schematic diagram of the proposed hand recognition system is shown in Fig. 2.

The construction of each of the N recognizers is based on the abductory inductive mechanism (AIM) [6], which is a self-organizing supervised machine-learning algorithm for automatically synthesizing abductive network models from a database. The automation of model synthesis not only lessens the burden on the analyst but also safeguards the model generated from being influenced by human biases and misjudgments. The process is 'evolutionary' in nature, using initially simple myopic regression relationships to derive more accurate representations in the successive iterations. To prevent exponential growth and limit model complexity, the algorithm selects only relationships having good predicting powers within each phase. Iteration is stopped when the new generation regression equations start to have poorer prediction performance than those of the previous generation, at which point the model starts to become overspecialized and therefore unlikely to perform well with new data.

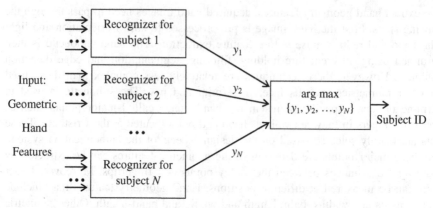

Fig. 2. Schematic diagram of the proposed hand recognition system

The algorithm has three main elements (representation, selection, and stopping) and applies abduction heuristics for making decisions concerning some or all of these three aspects. The relationship between any inputs and output can be approximated by a Kolmogorov-Gabor polynomial of the form [7]:

$$f(\mathbf{x}) = w_0 + \sum_{i=1}^{m} w_i x_i + \sum_{i=1}^{m}\sum_{j=1}^{m} w_{ij} x_i x_j + \sum_{i=1}^{m}\sum_{i=1}^{m}\sum_{k=1}^{m} w_{ijk} x_i x_j x_k \qquad (1)$$

where, $\mathbf{x} = (x_1, x_2, \ldots, x_m)$ is the input vector and w_i, w_{ij}, and w_{ijk} $\forall i, j, k$ are the corresponding weights. Fig. 3 shows a typical example of an abductive model showing various types of processing elements. These elements include:

- White elements: each consists of a constant value plus the linear weighted sum of all outputs of the previous layer. Assume a white element has the n inputs: $f_1, f_2, \ldots, f_n$, then the output will be given by,

$$y = w_0 + \sum_{i=1}^{n} w_i f_i \qquad (2)$$

where w_i is the corresponding weight for the ith input.

- Single, double, and triple elements: each implements a third-degree polynomial expression with all possible cross-terms for one, two, and three inputs respectively, e.g. a double element with two inputs x_1 and x_2 will have the output computed from,

$$f(x_1, x_2) = w_0 + w_1 x_1 + w_2 x_2 + w_3 x_1^2 + w_4 x_2^2 + w_5 x_1 x_2 + w_6 x_1^3 + w_7 x_2^3 \qquad (3)$$

In addition to the main layers of the network, an input layer of normalizers convert the input variables into an internal representation as z scores with zero mean and unity variance, and an output unitizer unit restores the results to the original problem space. Elements in the first layer operate on various combinations of the independent input variables (x's) and the element in the final layer produces the predicted output for the dependent variable y.

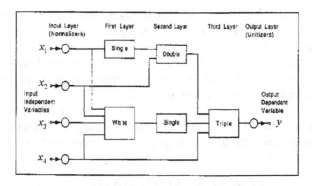

Fig. 3. A typical example of an abductive model showing various types of processing elements

One limitation of this approach is that a large number of samples in the training dataset would be required to calculate all the coefficients. Ivakhnenko has described the classical GMDH approach where it was shown that the Kolmogorov-Gabor polynomial can be constructed iteratively as a combined polynomial of second order polynomials. To illustrate the steps of the classical GMDH approach, consider an estimation data base of n_e observations (rows) and $m+1$ columns for m independent variables (x_1, x_2, ..., x_m) and one dependent variable y. In the first iteration, we assume that our predictors are the actual input variables. The initial rough prediction equations are derived by taking each pair of input variables (x_i, x_j; $\forall i, j \in \{1, 2, ..., m\}$) together with the output y and computing the quadratic regression polynomial [8]:

$$y = f(x_i, x_j) = A + Bx_i + Cx_j + Dx_i^2 + Ex_j^2 + Fx_ix_j \tag{4}$$

Each of the resulting $m(m-1)/2$ polynomials are evaluated using data for the pair of input variables used to generate it, thus producing new estimation variables (z_1, z_2, ..., $z_{m(m-1)/2}$) which would be expected to describe y in better manner than the original variables. The resulting z variables are screened according to some selection criterion and only those having good predicting power are kept. The original GMDH algorithm employs an additional and independent selection set of n_s observations for this purpose and uses the regularity selection criterion based on the root mean squared error r_k over the data set, where,

$$r_k^2 = \sum_{\ell=1}^{n_s}(y_\ell - z_{k\ell})^2 \bigg/ \sum_{\ell=1}^{n_s} y_\ell^2 \quad ; k = 1, 2, ..., m(m-1)/2 \tag{5}$$

Only those polynomials (and associated z variables) that have r_k below a prescribed limit are kept and the minimum value, r_{min}, obtained for r_k is also saved. The selected z variables represent a new dataset for repeating the estimation and selection steps in the next iteration to derive a set of higher-level variables. At each iteration, r_{min} is compared with its previous value and the process is continued as long as r_{min} decreases or until a given complexity is reached. An increasing r_{min} is an indication of the model becoming overly complex, thus over-fitting the estimation data and performing poorly in predicting the new selection data. Keeping an eye on the model complexity is an important aspect of GMDH-based algorithms, so as to achieve the final objective of constructing the model, i.e. using it with new data previously unseen during training. The best model for this purpose is the one which provides the shortest description for the data available [9]. Computationally, the resulting GMDH model can be seen as a layered network of partial quadratic descriptor polynomials, with each layer representing the results of an iteration.

The predicted squared error (*PSE*) criterion [9] is used for selection and stopping, to avoid model overfitting. This criterion minimizes the expected squared error that would be obtained when the network is used for predicting new data. AIM expresses the *PSE* error as follows:

$$PSE = FSE + CPM(2K/N)\sigma_p^2 \qquad (6)$$

where *FSE* is the fitting squared error on the training data, *CPM* is a complexity penalty multiplier selected by the user, K is the number of model coefficients, N is the number of samples in the training set, and σ_p^2 is a prior estimate for the variance of the error obtained with the unknown model. This estimate does not depend on the model being evaluated and is usually taken as half the variance of the dependent variable y. As the model becomes more complex relative to the size of the training set, the second term increases linearly while the first term decreases. *PSE* goes through a minimum at the optimum model size that strikes a balance between accuracy and simplicity (exactness and generality). The user may optionally control this trade-off using the *CPM* parameter. Larger values than the default value of 1 lead to the generation of simpler models that are less accurate but may generalize well with previously unseen data, while smaller values produce more complex networks that may overfit the training data and degrade actual prediction performance.

4   Evaluation

In this section we describe the adopted dataset and discuss the experimental work and results to evaluate the performance of the proposed method using various metrics.

4.1   Dataset Description

The adopted hand dataset has 200 images collected from 20 subjects (persons) using a digital camera and peg-free mode. Each person has 10 images of his/her left hand. This dataset is the same as that for the first 20 subjects in the dataset available at [10]. This dataset is used in [11], [12] as well. Seventeen distinguishing hand-geometric features, as shown in Fig. 4, are computed and each vector is labeled with the person's identification (PID) number. There are four features representing the lengths of four fingers (little, fourth, middle and index); referred to by FL_1, FL_2, FL_3, and FL_4, respectively. Also there are eight features representing the widths of the same four fingers (in the same order) measured at two different locations for each finger. These features are denoted, in order, as FW_{1a}, FW_{1b}, ..., FW_{4a}, and Fw_{4b}. In addition, there are two features representing the length and width of the palm; denoted as PL and PW, respectively. Finally, the last two features are the hand contour length and hand area and are denoted as HCL and HA, respectively.

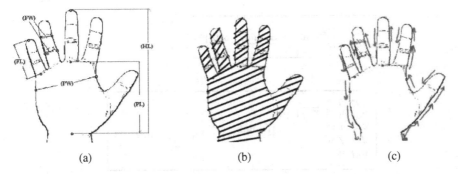

(a)   (b)   (c)

Fig. 4. Hand geometric features [10]: (a) finger length (FL) and width (FW), palm length (PL) and width (PW), hand length (HL) (b) hand area (HA) (c) hand contour length (HCL)

4.2   Experiments and Performance Analysis

Using the above mentioned dataset, we conducted a number of experiments to evaluate and compare the effectiveness of the presented abductive learning approach. The dataset was first randomized and then split into a training set containing 140 cases and an evaluation set containing 60 cases, *i.e.* a training-to-evaluation ratio of 70:30. The split was performed such that this ratio was maintained for each of the 20 subjects.

The first experiment involved the training and evaluation of the proposed system. There are 20 one-versus-all binary classification models, one for each of the 20 subjects. Each model was built using the training dataset and tested on the evaluation dataset. Then, the output of the 20 models was merged using the maximum rule, i.e. the overall predicted class will be the same as that of the model giving the maximum output among the 20 models; see Fig. 2. In all experiments, each of the 20 models was individually optimized by selecting the CPM parameter that gives the minimum absolute error for that model class on the evaluation set. To limit the computational time, only 5 values of CPM were tested for each model: CPM = 0.2, 0.5, 1.0, 2.0, and 5.0. Performance results are given for the optimum set of 20 models obtained in this way. Optimum models generated for the 20 subjects varied in complexity, both in terms of the number of input features selected and the number of model layers. Fig. 5 shows the simplest and most complex optimum models, which happened to correspond to subject 1 (uses only 2 relevant features) and subject 11 (uses only 5 relevant features), respectively. The predictor variables Var_1, Var_2, ..., and Var_17 refer to the 17 features in the same order as they exist in the dataset. Table 1 shows the corresponding feature for each of these predictor variables. The automatic selection of model inputs through abductive learning achieved a dimensionality reduction ranging from 70.6% to 88.2% for individual models.

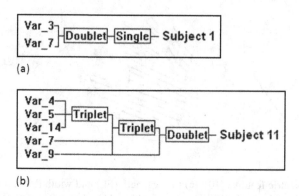

(a)

(b)

Fig. 5. Model structures for: (a) the simplest model (subject 1), (b) the most complex model (subject 11)

All 17 features were selected by the 20 models, collectively - which indicates that all the features are relevant to this classification problem. However, their significance to different models varies. Insight into the relative importance of the various input features can be gained from the frequency each feature is utilized by the 20 models. Table 1 lists the number of times each of the 17 features is used by the 20 models collectively. Based on these results, the input features can be tentatively ranked – with features Var_4, Var_3, Var_12, Var_14, Var_17, and Var_2 at the top of the list while features Var_15, Var_13, and Var_6 at the bottom.

Table 1. Frequency of use of the various input features by the 20 optimal classification models

Feature		Num. of times used by the 20 models	Feature		Num. of times used by the 20 models
ID	Name		ID	Name	
Var_1	FL_1	3	Var_10	FW_{3b}	4
Var_2	FL_2	6	Var_11	FW_{4a}	3
Var_3	FL_3	7	Var_12	FW_{4b}	7
Var_4	FL_4	8	Var_13	PW	2
Var_5	FW_{1a}	4	Var_14	PL	7
Var_6	FW_{1b}	2	Var_15	HL	1
Var_7	FW_{2a}	5	Var_16	HCL	5
Var_8	FW_{2b}	3	Var_17	HA	7
Var_9	FW_{3a}	3			

Evaluation of the 20 optimum models on the evaluation dataset of 60 cases gave only a single misclassification (subject 15 misclassified as subject 2). Table 2 shows the percentage classification error rate on the evaluation set as well as the percentage false rejection rate (*FRR*) and false acceptance rate (*FAR*) for the models. These performance metrics were calculated as follows:

$$Error\ Rate\ (\%) = 100\left(1 - \frac{1}{N}\sum_{i=1}^{20} TP_i\right)$$ (7)

$$FRR_{avg}(\%) = 100\left(\frac{1}{20}\sum_{i=1}^{20}\frac{FN_i}{AP_i}\right) \qquad (8)$$

$$FAR_{avg}(\%) = 100\left(\frac{1}{20}\sum_{i=1}^{20}\frac{FP_i}{AN_i}\right) \qquad (9)$$

where,

N : Number of cases in the evaluation set, $N = 60$.

AP_i: Number of cases in the evaluation set belonging to subject i (i.e. actual +ve)

AN_i: Number of cases in the evaluation set belonging to all subjects other than subject i, i.e. $N - AP_i$ (i.e. actual -ve)

TP_i: Number of cases among the AP_i subset that were classified as subject i (i.e. true +ve)

FN_i: Number of cases among the AP_i subset that were classified as any subject other than subject i (i.e. false -ve)

TN_i: Number of cases among the AN_i subset that were classified as any subject other than subject i (i.e. true +ve)

FP_i: Number of cases among the AN_i subset that were classified as subject i (i.e. false -ve)

To compare the performance with other algorithms, we used the same training and evaluation datasets with three other classifiers [13]. The first algorithm is J48 which is an implementation of a decision tree based classifier. The second algorithm is JRip which is an example of a rule based classifier. The last algorithm is SMO which builds an optimized support vector machine (SVM) using a polynomial kernel. The results are shown in Table 2. By investigating these results, it is clear that the abductive learning approach, compared to the other three algorithms, has higher accuracy and lower error rate, average *FRR* and average *FAR*. In addition, abductive learning automatically selects only most relevant features to build much simpler models.

Table 2. Classification performance and comparison

Approach	Accuracy (%)	Error Rate (%)	Average FRR (%)	Average FAR (%)
Abductive	98.3333	1.6667	1.6667	0.0877
Decision Tree	93.3333	6.6667	6.6667	0.3509
Rule Based	81.6667	18.3333	18.3333	0.9649
SVM	96.6667	3.3333	3.3333	0.1754

5   Conclusion

This paper presented a hand recognition system using abductive learning and hand geometric features. The effectiveness of this approach is evaluated and compared to other machine learning algorithms including decision trees, support vector machines, and rule-based classifiers. Based on the conducted experimental work on the adopted dataset, the abductive learning approach can yield more than 98% overall accuracy, 1.67% average false rejection rate, and 0.088% average false acceptance rate. In addition, it can build simpler polynomial network models by automatically selecting the most relevant features for each subject. For future work, we intend to test the system for other datasets and to enhance the results further by performing collective optimization on all the 20 models simultaneously.

Acknowledgements. The authors would like to thank King Fahd University of Petroleum & Minerals (KFUPM), Saudi Arabia, for funding and support during this work under Grant no. RG1106-1&2.

References

1. Yörük, E., Konukoglu, E., Sankur, B., Darbon, J.: Shape-based hand recognition. IEEE Transactions on Image Processing 15(7), 1803–1815 (2006)
2. Sidlauskas, D.P., Tamer, S.: Hand geometry recognition. In: Jain, A.K., Flynn, P., Ross, A.A. (eds.) Handbook of Biometrics, pp. 91–107. Springer US, Boston (2008)
3. Kumar, A., Wong, D.C.M., Shen, H.C., Jain, A.K.: Personal authentication using hand images. Pattern Recognition Letters 27(13), 1478–1486 (2006)
4. Nicolae, D.: A survey of biometric technology based on hand shape. Pattern Recognition 42(11), 2797–2806 (2009)
5. Pavesic, N., Ribaric, S., Ribaric, D.: Personal authentication using hand-geometry and palmprint features: the state of the art. In: Proc. Workshop on Biometrics, ICPR 2004 (2004)
6. AIM User's Manual. AbTech Corporation, Charlottesville, VA (1990)
7. Anastasakis, L., Mort, N.: The development of self-organization techniques in modeling: A review of the group method of data handling (GMDH). In: Research Report, Department of Automatic Control and Systems Engineering, University of Sheffield (2001)
8. Farlow, S.J.: The GMDH algorithm. In: Farlow, S.J. (ed.) Self-Organizing Methods in Modeling: GMDH Type Algorithms, pp. 1–24. Marcel-Dekker, New York (1984)
9. Barron, A.R.: Predicted squared error: A criterion for automatic model selection. In: Farlow, S.J. (ed.) Self-Organizing Methods in Modeling: GMDH Type Algorithms, pp. 87–103. Marcel-Dekker, New York (1984)
10. http://www.cse.ust.hk/~helens
11. Selvarajan, S., Palanisamy, V., Mathivanan, B.: Human identification and recognition system using more significant hand attributes. In: Proc. of International Conference on Computer and Communication Engineering, ICCCE (2008)
12. Kumar, A., Wong, D., Shen, H., Jain, A.: Personal verification using palmprint and hand geometry biometric. In: Proc. of 4th International Conference on Audio- and Video-Based Biometric Person Authentication (2003)
13. Witten, I.H., Frank, E., Hall, M.A.: Data Mining: Practical Machine Learning Tools and Techniques, 3 edn. Morgan Kaufmann (2011)

Designing Serious Games for Adult Students with Cognitive Disabilities

Javier Torrente, Ángel del Blanco, Pablo Moreno-Ger,
and Baltasar Fernández-Manjón

e-UCM Research Group, Department of Software Engineering and Artificial Intelligence
Complutense University of Madrid
C/Profesor José García Santesmases sn, 28040 Madrid, Spain
{jtorrente,angel.dba,pablom,balta}@fdi.ucm.es

Abstract. Digital games have a great potential to improve education of people with cognitive disabilities. However, this target audience has attracted little attention from industry and academia, compared to other segments of the population. As a consequence, there is little knowledge available about how to design games that are usable and enjoyable by people with cognitive disabilities. In this paper we discuss how the eAdventure game platform can support their special needs. This tool has been used to develop two games to improve professional education of people with cognitive disabilities. Lessons learnt from these experiences are presented to serve as a first step to support further research in this field.

Keywords: accessibility, digital games, eAdventure, education, e-Learning, Game-Based Learning, social inclusion.

1   Introduction

The educational potential of digital games is rapidly being accepted within the academic community, as more experimental research that proves the effectiveness of this paradigm has became available recently. This body of research validate, at least partially, the hypothesis of academics who discussed unique characteristics of games that make them interesting for education [1].

Some of these features could be especially advantageous for students with cognitive disabilities. For example, digital games provide a virtual world that can be used as a safe test environment that students can freely explore, at their own pace, trying out hypothesis and receiving immediate feedback. Students get immersed in this virtual world, where they can rehearse and improve their abilities and knowledge but without taking any risk. In addition, digital games are able to capture students' attention more effectively than other contents, keeping them in the zone of optimal flow for knowledge creation. This characteristic may be especially beneficial for students with intellectual disabilities, as they usually suffer from attention deficit, which is a significant drawback for learning.

T. Huang et al. (Eds.): ICONIP 2012, Part IV, LNCS 7666, pp. 603–610, 2012.

Despite this potential, there are few games directed to people with cognitive disabilities [2, 3]. This deficiency is motivated by multiple causes, but one of the most important is that designing serious games for people with cognitive disabilities is an extraordinary challenge. Making games is always a complex activity requiring wide doses of creativity and highly specialized technical skills. Cognitive disabilities are complex and heterogeneous, difficult to categorize and model, requiring an individualized approach in many cases. Therefore, when the game design must also cope with these special needs the difficulty of the task increases, involving an additional development cost.

Specialized authoring tools can facilitate game development for this target population. High level tools like Unity or Game Maker facilitate the creation of games by providing code abstraction, automation of frequent tasks, built-in modules and game parts that are ready to use. However, it is necessary that these tools accommodate the special needs of people with disabilities to be really effective. But for this to be feasible, it is necessary to understand what are the requirements of this understudied population for interacting with games.

In this paper we discuss how the eAdventure game authoring tool can be used to create games for people with cognitive disabilities. We present two case studies of developing games to educate adults with cognitive disabilities:"My first day at work" and "The big party". The goal in both games is to improve their education as means to increase their opportunities for employment. Finally we discuss lessons learnt for designing games for this target population.

2   Digital Games for People with Cognitive Disabilities

Despite the ever-growing expansion of digital games, the collective of people with cognitive disabilities has not attracted too much attention yet. As recent literature reviews on accessibility in games reveal, there are few games available that cater for the needs of people with cognitive disabilities [2, 3]. Still, some interesting examples can be found, like Ilbo [4], where players navigate through a 3D maze by using their weight while sitting on a chair. Other games are oriented to facilitate collaboration among peers and improve social and communication skills [5], although in some cases the presence of game elements is limited to a 3D virtual world [6].

Most of the limited research reported on games for cognitive disabilities is concentrated on rehabilitation and therapy, usually combined with virtual reality techniques. For example, in [7] virtual reality games developed for the Nintendo™ Wii® are used to improve motor and cognitive skills of children with a diagnosis of Down Syndrome. Despite of research done, this field is also considered to be in its infancy, lacking of proper understanding of what causes the effectiveness of computer and virtual reality games for rehabilitation [8].

Some studies have addressed the potential of digital games to improve education of people with cognitive disabilities. For instance, in [9] computer games are used to teach safety knowledge to children with cognitive disabilities. This study also demonstrates that knowledge constructed in the virtual world can be transferred to persistent skills in the real world. In [10], a puzzle game for training children with autism is described. A relevant study for the topic of this paper is the GOET project, whereby several games were developed to educate students with cognitive disabilities to improve their chances for employment [11].

Generally speaking, research on serious games for people with cognitive disabilities is still in its infancy, compared to other types of disabilities. It is necessary to conduct a deeper analysis of how game design can be tuned to cater for the special needs of this audience.

3   Point and Click Adventure games. eAdventure

Choosing a right type of game is important to minimize the number of accessibility barriers that must be dealt with. *Point-and-click* adventure games is a genre where many of the most frequent accessibility issues are not present. Besides, this genre has been signaled by academics for having significant educational potential. It is a genre where reflection predominates over action. In fact, time pressure is rarely used to get players engaged. Other elements are used instead, as an appealing story or puzzles that players must solve by applying reasoning and problem solving skills. As a consequence, these games are usually low-paced, which is a desirable characteristic for people with cognitive disabilities [13]. Besides, *point-and-click* interaction is usually simple, requiring a minimum amount of input as controls are mouse clicks that could also be replaced by one-switch devices [3].

eAdventure is a game authoring tool especially devised for educational applications [12]. It is oriented to teachers as end users, providing a simple interface and educational features such as a tracking and assessment system. eAdventure supports the development of games accessible for people with cognitive disabilities in several ways. First, eAdventure is focused on *point-and-click* adventure games. Second, eAdventure includes an adaptation engine that adds personalization and flexibility to the game experience. This system can be used to adapt content and puzzles, reducing complexity and the number of objects as needed [14]. Besides, experimental development to improve the accessibility of the platform has been conducted.

4   My First Day at Work

The educational game "My first day at work" aims to facilitate the incorporation of a worker with a cognitive disability to a new company. The game assumes the player has already got his/her first job, and it covers competences and skills needed for daily work and achieve a successful integration into the team:

- Usage of standard equipment and materials used in the office: computer, printer, fax and a copy printer.
- Fundamentals of the e-mail system used in the company: how to access incoming messages, how to compose and send new messages, download files and use attachments.

Besides, the game covers transversal competencies that people with cognitive disabilities have problems to develop frequently:

- Basic social interaction skills, such as how to address colleagues with respect, ask for help when needed, etc.
- Structure of the company and the physical distribution of its headquarters.

The game has the form of an adventure quest where the player must complete different tasks that are assigned by the company's management board. To complete these tasks he/she must interact with different objects and characters.

Additionally, the game "My first day at work" includes accessibility features oriented to overcome potential barriers for students with a visual disability or limited mobility in hands. Therefore the game can be played using the mouse, the keyboard or speech commands, and the return of information is produced either visually or by audio. The game also includes a high contrast mode for people with limited vision. This visualization mode applies an alternative rendering mode to backgrounds and interactive elements, with the purpose of increasing the contrast of such a highly graphical application.

The game was developed in collaboration with Technosite, a company that belongs to the ONCE group (Spanish National Organization for the Blind). Experts in game accessibility, therapists and social workers were involved in the development of the game. A usability evaluation was performed with 15 users that were exposed to the game for one hour. Participants with the slightest disabilities were able to complete the game without further guidance or intervention from researchers. However, participants with severe disabilities had problems to remember short-term goals, which suggested the need for a "task list" feature that could be accessed at all times. Participants showed interest in the game and considered it a good asset to improve their education.

Fig. 1. Screenshots of the game "My first day at work"

5   The Big Party

The game "The Big Party" was designed to train a specific set of social and self autonomy skills and concepts in adults with a cognitive disability.

The topic of the game is to attend a social dinner organized by the company the player works for. The game covers a wide range of issues, from personal hygiene and choosing appropriate clothes for the occasion to addressing other colleagues.

When the game starts, the player chooses his/her gender on the game. This choice will be used by the game to adapt configuration of the resources, clothes, and hygiene habits displayed. The game covers the next specific competences:

- Personal hygiene: processes related to hygiene including showering, brushing teeth, applying cologne and deodorant, combing one's hair, etc. Tasks related to personal care must be executed in a specific right order (for example, cologne should not be applied before taking a shower).
- Preparation before leaving home: adequate dressing for the event.
- Take public transport to reach the event and dealing with unexpected issues (e.g. request help from underground's staff).
- Use of common resources and items in public places and transport vehicles (ticket vendor machines, control points, automatic elevator, etc.).
- Correct use of language in formal occasions.
- Basic rules of behaviour in public places, including interaction with peers, like give greetings, say good bye, bringing up conversation topics that may be of interest for other people or resolution of conflicts (e.g. stepping a colleague by accident). Aspects related to self control and moderate eating and drinking are also considered.

The game is linear, with a specific number of tasks to be completed in a specific order. Thus, completing the game implies succeeding in all game tasks. For that purpose, the player is provided with convenient feedback when he/she fails to complete a task. The player is allowed as many retries as needed.

The game has been developed in collaboration with the Prodis foundation, whose mission is to prepare adult students with cognitive disabilities for professional development. The game has been evaluated in two Living Labs with teachers of special education and also with students with Down Syndrome. The purpose of this evaluation was to identify potential improvements or modifications for enhancing its usability and guarantee usefulness for this particular educational context.

Fig. 2. Pictures of evaluation sessions during development of "The Big Party" (living lab with educators on the left, usability evaluation with students on the right)

6   Lessons Learnt from the Case Studies

Having a flexible and highly configurable game experience was very important in these cases. This is also a requisite identified by previous work in this field [11]. The one-size-fits-all principle does not usually fit games, where players have different motivations and even play styles. In special education this requisite is even more important, as each user is unique and requires personalized attention. In this sense digital games are more suitable than other kind of contents as digital games are flexible and easy to configure.

A good example is the high contrast mode developed for the game "My First Day at Work". Although this mode was developed in collaboration with people that normally use high contrast settings to interact with technology, not all people that participated in the evaluation felt comfortable with the interface. Through the feedback participants provided, researchers noticed that each user had a particular way to interact with the computer. In the case of "The Big Party", diverse aspects were added a posteriori to facilitate understanding and use by people with intellectual disabilities, like allowing multiple retries to complete a task, indication of possible solutions after a failure or mistake, etc. In this manner students could play the games and learn at their own pace.

Another problem found was that many people with intellectual disabilities have difficulties to identify themselves in the games [11]. Finding a solution to this problem is essential or many students would not be able to play as they would not understand what is going on in the game. In this sense, the ideal solution would be to use students' own image to set up a virtual avatar, but from a technical perspective this is quite complex to implement. In the case of "My First Day at Work", the workaround was to provide the player with a finite set of avatars with varied abilities and characteristics to choose from. Hence players could choose the avatar that was more close to their own characteristics and abilities. In "The big party" game students experienced the game in first person, limiting their choices to a simple selection of gender. The preliminary evaluation proved that any improvement in this aspect would be beneficial for the overall usability of the game.

Broadly speaking, design guidelines followed in the development of both games can be repurposed and applied to effectively develop other games for students with intellectual disability. For example, language style should be simple, clear and direct. It is also highly desirable to provide information using multiple modalities (e.g. complementing visual feedback with descriptive sounds, using subtitles but also speech recorded by actors. This feature will also make the games more accessible for students with other disabilities. It is important to gauge game's pace to ensure that players have enough time to read all dialogues, analyze all information provided by the game and take decisions according to options available. The eAdventure platform that was used to develop the games provided ready-to-use solutions that facilitated dealing with this issue (e.g. management of timing, progress in dialogs and interactions).

Reaching the highest level of realism possible is also a recommended practice. This facilitates the acquisition of new knowledge and abilities by students with abstract reasoning deficiencies. For this reason both "My First Day at Work" and "The

Big Party" have been developed combining photos and videos from real environments with cartoon-like designs. This also helps to limit the number of graphic assets required, which reduces the production cost.

Both games were developed following a user-centered methodology, using living labs to identify potential barriers. This methodology allowed for a rapid detection of poor design strategies and supported an agile requirements capture process, which facilitated development and reduced the overall cost. This aspect was crucial for success as how this target population interacts with games is rather unknown and therefore it cannot be anticipated.

These case studies were useful to identify potential improvements in the eAdventure authoring tool. For example, people with Down Syndrome are slower at executing goal-directed tasks/activities compared to typically developing peers. Games usually set out a number of primary goals to entice the player that have to be completed in the long term (e.g. defeat the master boss of a level or unlock all possible levels) and are not prone to change frequently. These are complemented with secondary goals, whose completion is required to progress in the game and achieve the primary goals (e.g. unlock a certain weapon to beat the master boss). Secondary goals are set out frequently, and are used to keep the player challenged and engaged at all times. This structure of primary and secondary goals was also present in both case studies, and resulted to be too complex for some users with Down Syndrome as they were unable to remember short-term goals and had problems to distinguish between primary and secondary goals. This problem could be addressed by developing configurable tasks lists in eAdventure that could be accessed by the player at all times.

7   Conclusions and Future Work

The field of digital games has reached a considerable status of maturity and stability, both in its recreational and serious forms. However, there are areas that have not been thoroughly explored yet. This is the case of games for people with cognitive disabilities. The design of games for this audience is a challenge as classic solutions may not be applicable, given the diversity of this understudied target group that brings together multiple disability profiles with heterogeneous needs. Besides, the potential of games to improve the lives of people with cognitive disabilities remains almost unexplored. Research on digital games should address both issues systematically in the next years.

In this paper we have discussed how the eAdventure game authoring tool can support the needs of students with cognitive disabilities. We have presented the main lessons learnt from designing and developing two games for this purpose with eAdventure, in the aim that they may be useful for other serious games developers. However, this is just a first step. The guidelines discussed in this paper are still general and superficial, based on two examples. It is necessary to carry out a deep analysis of the successful strategies found in these games and others in the literature to produce more concrete guidelines that could be applied in the development of new games but also to improve eAdventure and other game authoring platforms.

Acknowledgments. The next sponsors have partially supported this work: the Spanish Ministry of Science and Innovation (grant no. TIN2010-21735-C02-02); the European Commission, through the Lifelong Learning Programme (projects "SEGAN Network of Excellence in Serious Games" - 519332-LLP-1-2011-1-PT-KA3-KA3NW and "CHERMUG" - 519023-LLP-1-2011-1-UK-KA3-KA3MP) and the 7th Framework Programme (project "GALA - Network of Excellence in Serious Games" - FP7-ICT-2009-5-258169); the Complutense University of Madrid (research group number GR35/10-A-921340) and the Regional Government of Madrid (eMadrid Network - S2009/TIC-1650).

References

1. Gee, J.P.: What video games have to teach us about learning and literacy. Palgrave Macmillan, New York (2003)
2. Westin, T., Bierre, K., Gramenos, D., Hinn, M.: Advances in Game Accessibility from 2005 to 2010. In: Stephanidis, C. (ed.) HCII 2011 and UAHCI 2011, Part II. LNCS, vol. 6766, pp. 400–409. Springer, Heidelberg (2011)
3. Yuan, B., Folmer, E., Harris, F.C.: Game accessibility: a survey. Universal Access in the Information Society 10, 81–100 (2011)
4. Kwekkeboom, B., van Well, I.: Ilbo,
   http://www.game-accessibility.com/index.php?pagefile=ilbo
5. Ohring, P.: Web-based multi-player games to encourage flexibility and social interaction in high-functioning children with autism spectrum disorder. In: Proceedings of the 7th International Conference on Interaction Design and Children, pp. 171–172. ACM, New York (2008)
6. Gaggioli, A., Gorini, A., Riva, G.: Prospects for the Use of Multiplayer Online Games in Psychological Rehabilitation. In: Virtual Rehabilitation, pp. 131–137 (2007)
7. Wuang, Y.-P., Chiang, C.-S., Su, C.-Y., Wang, C.-C.: Effectiveness of virtual reality using Wii gaming technology in children with Down syndrome. Research in Developmental Disabilities 32, 312–321 (2011)
8. Levac, D., Rivard, L., Missiuna, C.: Defining the active ingredients of interactive computer play interventions for children with neuromotor impairments: a scoping review. Research in Developmental Disabilities 33, 214–223 (2012)
9. Coles, C.D., Strickland, D.C., Padgett, L., Bellmoff, L.: Games that "work": using computer games to teach alcohol-affected children about fire and street safety. Research in Developmental Disabilities 28, 518–530 (2007)
10. Sehaba, K., Estraillier, P., Lambert, D.: Interactive Educational Games for Autistic Children with Agent-Based System. In: Kishino, F., Kitamura, Y., Kato, H., Nagata, N. (eds.) ICEC 2005. LNCS, vol. 3711, pp. 422–432. Springer, Heidelberg (2005)
11. Lanyi, C.S., Brown, D.J.: Design of Serious Games for Students with Intellectual Disability. In: Joshi, A., Dearden, A. (eds.) Proceedings of the 2010 International Conference on Interaction Design & International Development, IHCI 2010, pp. 44–54. British Computer Society Swinton, UK (2010)
12. Torrente, J., Del Blanco, Á., Marchiori, E.J., Moreno-Ger, P., Fernández-Manjón, B.: <e-Adventure>: Introducing Educational Games in the Learning Process. In: IEEE Education En-gineering (EDUCON) 2010 Conference, pp. 1121–1126. IEEE, Madrid (2010)
13. IGDA: Accessibility in Games: Motivations and Approaches (2004)
14. Torrente, J., Del Blanco, Á., Moreno-Ger, P., Martínez-Ortiz, I., Fernández-Manjón, B.: Implementing Accessibility in Educational Videogames with <e-Adventure> (2009)

Neural Network Based Approach
for Automotive Brake Light Parameter Estimation

Antonio Vanderlei Ortega and Ivan Nunes da Silva

University of São Paulo, Department of Electrical Engineering, CP 359
CEP 13566.590, São Carlos, SP, Brazil
{avortega,insilva}@sc.usp.br

Abstract. The advantages offered by the electronic component LED (Light Emitting Diode) have caused a quick and wide application of this device in replacement of incandescent lights. However, in its combined application, the relationship between the design variables and the desired effect or result is very complex and it becomes difficult to model by conventional techniques. This work consists of the development of a technique, through artificial neural networks, to make possible to obtain the luminous intensity values of brake lights using SMD (Surface Mounted Device) LEDs from design data. Such technique can be used to design any automotive device that uses groups of SMD LEDs. Results of industrial applications, using SMD LED, are presented to validate the proposed technique.

Keywords: Brake light, SMD LED, neural networks, intelligent systems.

1   Introduction

The LED device is an electronic semiconductor component that emits light. At present time, it has been used in replacement of incandescent lights because of its advantages, such as longer useful life (around 100,000 hours), larger mechanic resistance to vibrations, lesser heating, lower electric current consumption and high fidelity of emitted light color [1].

However, in designs where incandescent lights are replaced by LEDs, some of their important characteristics must be considered, such as direct current, reverse current, vision angle and luminous intensity [2].

In automobile industry, incandescent lights have been replaced by LEDs in the brake lights, which are a third light of brakes [3]. In these brake lights are used sets of SMD LEDs usually organized in a straight line. The approval of brake light prototypes is made through measurements of luminous intensity in different angles, and the minimum value of luminous intensity for each angle is defined according to the application [4].

The main difficulty found in the development of brake lights is in finding the existent relationship between the following parameters: luminous intensity (I_V) of the SMD LED, distance between SMD LEDs (d) and number of SMD LEDs (n), with the desired effect or result, i.e., there is a complexity in making a model by

T. Huang et al. (Eds.): ICONIP 2012, Part IV, LNCS 7666, pp. 611–618, 2012.

conventional techniques of modeling, which are capable to identify properly the relationship between such variables. The prototype designs of brake lights have been made through trials and errors, causing increasing costs of implementation due to time spent in this stage. Moreover, the prototype approved from this system cannot represent the best relationship cost/benefit, since few variations are obtained from configurations of approved prototypes. The artificial neural networks are applied in cases like this one, where the traditional mathematic modeling becomes complex due to nonlinear characteristic of the system.

2   Overview of Automotive Applications Using LEDs

Modern automotive vehicles use incandescent lamps for parking, turning, and brake lights. However, incandescent bulbs consume a disproportionately large amount of energy for the amount of colored light they project from the vehicle's lighting fixture.

In [4] is demonstrated that the conversion of a turn signal from an incandescent light to LED is possible with the latest advancements in LED designs. In [1] is proposed a system based on LEDs for vehicle traffic control applications. From geometric considerations, the system requires a cluster of 200 red, amber, and green or 200 multicolor LEDs for a single three-light system. In [3] is presented a vehicle that uses LEDs in its headlights.

In [8] is described a light-emitting diode brake-light messaging (LEDBM) system that can be used to avoid rear collisions. In [11] is proposed a robust vehicle detection method that uses vision to extract bright regions brake lights. In [12] is presented a vision system dedicated to the detection of vehicles in reduced visibility conditions.

Several more technical studies are also proposed in the literature involving LED applications in automotive industry. In [13] is presented a smart driver for LEDs, particularly for automotive lighting applications, which avoid ringing and overshoot phenomena. In [14] is accomplished a comparison among the most efficient red, amber/yellow and white power LEDs, offering results mainly in terms of luminous efficiency in nominal test conditions. Some thermal considerations related to the LED automotive headlights and components can be also found in [15, 16].

This paper presents an industrial application using artificial neural networks to estimate values of brake light luminous intensity from design data. Although this study is aimed at the application of LED in brake lights, the methods developed and described here can also be used in other applications, such as headlights, turn lights, traffic lights, or any other application where SMD LEDs can be used in groups.

3   SMD LEDs Applied in Brake Lights

LED is an electronic device composed by a chip of semiconductor junction that when traversed by an electric current provides a recombination of electrons and holes. However, this recombination demands that the energy produced by free electrons can be transferred to another state. In semiconductor junctions, this energy is released in form of heat and by emission of photons, i.e., light emission [5]. In silicon and germanium the largest energy emission occurs in form of heat, with insignificant light emission. However, in other materials, such as GaAsP or GaP, the number of light photons emitted is sufficient to build a source of quite visible light [6]. This process of light emission, which is intrinsic characteristic of the LEDs, is called electroluminescence [7].

In brake lights the SMD LEDs are applied in set and generally organized in a straight line on a printed circuit board (PCB). In this PCB, besides the SMD LEDs, there are electronic components, basically resistors, which are responsible for the limitation of electric current that circulates through the SMD LEDs.

The main parameters used in brake lights designs are given by: SMD LED luminous intensity (I_V), distance between SMD LEDs (d) and number of SMD LEDs (n). In Fig. 1 is illustrated a basic representation of a brake light.

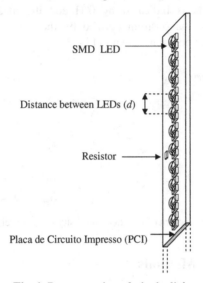

SMD LED

Distance between LEDs (d)

Resistor

Placa de Circuito Impresso (PCI)

Fig. 1. Representation of a brake light

The main function of the brake light is to increase the safety of the vehicle (acting as a prevention system) and to reduce the risk of back collisions. Recent studies show the development of brake lights equipped with modulated signal transmitters containing information about the vehicle in which it is installed. Other vehicles that have the respective reception system of those modulated signals receive them, and their information's have been used to prevent back collisions [8]. In Fig. 2 is illustrated a brake light installed.

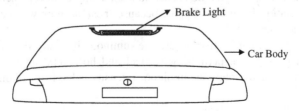

Brake Light

Car Body

Fig. 2. Representation of a brake light installed

At the moment there is no model or technique for designing brake lights and the prototypes are elaborated according to the common sense of designers, i.e., through trial and error methods. This occurs because the relationship between the variables involved with the light emission process of brake lights is completely nonlinear.

After elaboration of the brake light prototype, it is necessary an approval of the sample. The process for the prototype validation is made by measuring the luminous intensity of the brake light in 18 positions or different angles (Fig. 3). After this process, the values obtained in each angle are compared with those values established by governmental rules. The minimum value of luminous intensity (I_{VBL}) in each angle varies according to the application. In Fig. 3 is shown a representation of a generic distribution diagram of brake light luminous intensity (I_{VBL}) in relation to angle. The mean horizontal position is indicated by 0°H and the mean vertical position is indicated by 0°V. Thus, the position defined by the pair of angles (0°V, 5°L) is represented by the shaded position shown in Fig. 3.

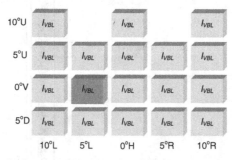

Fig. 3. Generic diagram of luminous intensity (I_{VBL}) in relation to angle

4   Materials and Methods

For this study, 45 samples of brake lights were constructed with the following parameter variations:

- Distance between SMD LEDs (d): 4.5 mm, 5.5 mm and 6.5 mm.
- Number of SMD LEDs (n): 16, 22 and 28.
- Luminous intensity of SMD LED (I_v): 600 mcd, 800 mcd, 1200 mcd, 1500 mcd and 1800 mcd.

It is important to remember that the minimum and maximum values of each parameter in the designed samples must be chosen in such a way as to represent the domain for parameter variation in future designs, because these designs will be made using the proposed neural network.

A photometer was used to measure the luminous intensity of the samples, and it was coupled to a device permitting vertical and horizontal angle variation. In this way, it was possible to obtain the luminous intensity value from 18 different angles.

Initially, the first sample was positioned relative to a screen representing the luminous intensity diagram illustrated in Fig. 3. The photometer was placed at the first angle, and the measurement of the luminous intensity was registered. This procedure was repeated until the luminous intensity value referring to last angle of the sample was registered. Figure 4 illustrates this procedure for the pair of angles (0°H, 5°U) shown in Fig. 3.

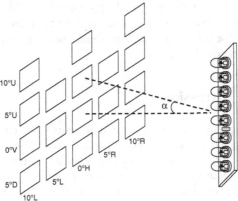

Fig. 4. Luminous intensity diagram in relation to the brake light prototype

The sample was then removed from the device, and a new sample was attached in order to measure the luminous intensity; all procedures are repeated until the value of the last angle of the last sample was registered.

From the design data and measurement results of luminous intensity of brake light samples from different angles, a multilayer perceptron network was trained as previously described. During this stage, a variation of the main network parameters was achieved. The number of layers, number of neurons per layer, activation function for each layer, and type of training were changed in order to obtain a neural network topology that could generate an acceptable mean squared error and ensure an efficient generalization. The best neural architecture for the simulations was selected by means of a cross-validation technique [10].

The topology chosen consisted of two hidden layers with 5 neurons in the first layer and 10 neurons in the second layer. The training algorithm was Levenberg-Marquardt [9]. The main advantage of this algorithm arises from its ability to accelerate the neural network convergence process, and it is considered to be the fastest method for training moderate-sized perceptron networks. In comparative terms, the Levenberg-Marquardt algorithm is about 100 times faster than the backpropagation method. For our application, the network inputs were defined by the 3 main parameters involved in brake light design, i.e.

- Distance between SMD LEDs → d (mm).
- Number of SMD LEDs → n .
- Luminous intensity of SMD LED → I_V (mcd).\

The network output is composed by a unique signal which provides what is the intensity level produced by the brake light in a particular angle, i.e.

- Luminous intensity of brake light → I_{VBL} (cd).

After training, using the 18 different angles, one training for each angle, it was possible to estimate the total luminous intensity produced by the brake light in different angles. To validate the proposed approach are used data coming from samples not used in the network training. A comparison between the estimated values by the network and those provided by experimental tests is accomplished to analyze the efficiency of the proposed approach.

5   Results and Discussion

The computational implementations of the neural networks used in this application were carried out using the software Matlab/Simulink. After the training process, the neural modeling was used to obtain luminous intensity values of brake lights, as previously described. Figure 5 illustrates a comparison between luminous intensity values (I_{VBL}) obtained by experimental tests (ET) and those estimated by the artificial neural network (ANN). In this configuration (Situation I), the used sample presents distance (d) between SMD LEDs equal to 5.5 mm, the number of SMD LEDs (n) is equal to 28 and the luminous intensity of each SMD LED (I_v) has a value equal to 800 mcd.

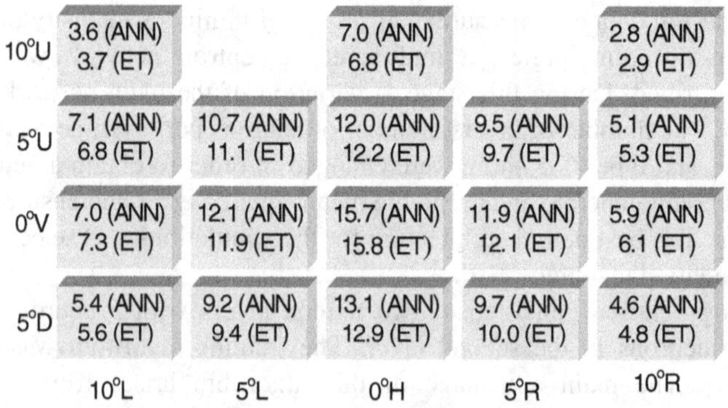

	10°L	5°L	0°H	5°R	10°R
10°U	3.6 (ANN) / 3.7 (ET)		7.0 (ANN) / 6.8 (ET)		2.8 (ANN) / 2.9 (ET)
5°U	7.1 (ANN) / 6.8 (ET)	10.7 (ANN) / 11.1 (ET)	12.0 (ANN) / 12.2 (ET)	9.5 (ANN) / 9.7 (ET)	5.1 (ANN) / 5.3 (ET)
0°V	7.0 (ANN) / 7.3 (ET)	12.1 (ANN) / 11.9 (ET)	15.7 (ANN) / 15.8 (ET)	11.9 (ANN) / 12.1 (ET)	5.9 (ANN) / 6.1 (ET)
5°D	5.4 (ANN) / 5.6 (ET)	9.2 (ANN) / 9.4 (ET)	13.1 (ANN) / 12.9 (ET)	9.7 (ANN) / 10.0 (ET)	4.6 (ANN) / 4.8 (ET)

Fig. 5. Comparative illustration (Situation I)

From Fig. 5 it is observed that the generalization produced by the network to estimate values of luminous intensity in several angles is very satisfactory. In this case, the mean relative errors calculated were around 2.8% and with variance of 1.19%. Figure 6 illustrates another comparison between luminous intensity values (I_{VBL}) obtained by experimental tests (ET) and those estimated by the artificial neural network (ANN). For this configuration (Situation II), the used sample presented the same distance (d) and the same SMD LEDs number of the previous situation; but, the luminous intensity of each SMD LED (I_v) has a value equal to 1200 mcd.

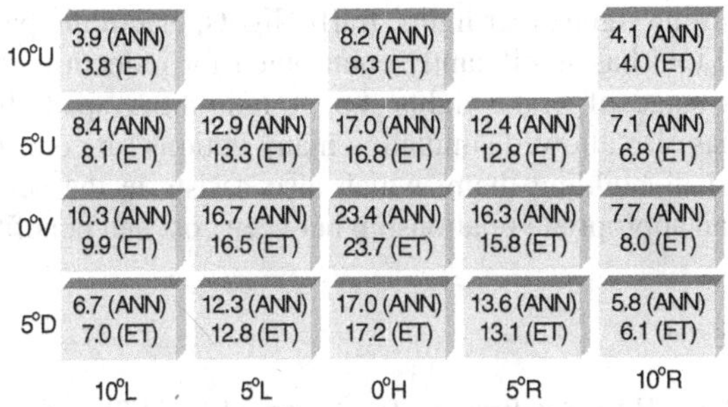

10°U	3.9 (ANN) 3.8 (ET)		8.2 (ANN) 8.3 (ET)		4.1 (ANN) 4.0 (ET)
5°U	8.4 (ANN) 8.1 (ET)	12.9 (ANN) 13.3 (ET)	17.0 (ANN) 16.8 (ET)	12.4 (ANN) 12.8 (ET)	7.1 (ANN) 6.8 (ET)
0°V	10.3 (ANN) 9.9 (ET)	16.7 (ANN) 16.5 (ET)	23.4 (ANN) 23.7 (ET)	16.3 (ANN) 15.8 (ET)	7.7 (ANN) 8.0 (ET)
5°D	6.7 (ANN) 7.0 (ET)	12.3 (ANN) 12.8 (ET)	17.0 (ANN) 17.2 (ET)	13.6 (ANN) 13.1 (ET)	5.8 (ANN) 6.1 (ET)
	10°L	5°L	0°H	5°R	10°R

Fig. 6. Comparative illustration (Situation II)

In this case (Situation II), the mean relative errors calculated were around 3.0% and with variance of 1.61%. Figure 7 illustrates another comparison between luminous intensity values (I_{VBL}) obtained by experimental tests (ET) and those estimated by the artificial neural network (ANN). In this configuration (Situation III), the used sample presents distance (d) between SMD LEDs equal to 6.5 mm, the number of SMD LEDs (n) is equal to 22 and the luminous intensity of each SMD LED (I_v) has a value equal to 1200 mcd.

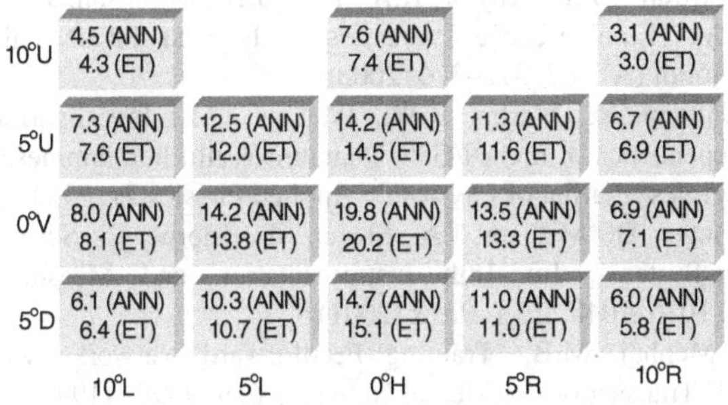

10°U	4.5 (ANN) 4.3 (ET)		7.6 (ANN) 7.4 (ET)		3.1 (ANN) 3.0 (ET)
5°U	7.3 (ANN) 7.6 (ET)	12.5 (ANN) 12.0 (ET)	14.2 (ANN) 14.5 (ET)	11.3 (ANN) 11.6 (ET)	6.7 (ANN) 6.9 (ET)
0°V	8.0 (ANN) 8.1 (ET)	14.2 (ANN) 13.8 (ET)	19.8 (ANN) 20.2 (ET)	13.5 (ANN) 13.3 (ET)	6.9 (ANN) 7.1 (ET)
5°D	6.1 (ANN) 6.4 (ET)	10.3 (ANN) 10.7 (ET)	14.7 (ANN) 15.1 (ET)	11.0 (ANN) 11.0 (ET)	6.0 (ANN) 5.8 (ET)
	10°L	5°L	0°H	5°R	10°R

Fig. 7. Comparative illustration (Situation III)

In this case (Situation III), the mean relative errors calculated were around 2.9% and with variance of 1.47%.

Through these results it is possible to infer that the network presented efficient results for estimation of luminous intensity values of brake lights. It should be taken into account that the proposed neural network has considered the main parameters involved with the design of brake lights. In the selection process of the best neural architecture used in simulations was adopted the cross-validation technique [10].

6   Conclusions

This work presents a technique based on use of artificial neural networks for determination of luminous intensity values for brake lights, in which are considered the main design characteristics. Therefore, the developed tool constitutes a new technique that can efficiently be applied in this type of problem. The developed methodology can also be generalized and used in other applications that use groups of SMD LEDs, such as in headlights, turning lights, rear lights, traffic lights, electronic panels of messages, etc. The developed tool has significantly contributed for reduction of costs in relation to implementation stage of brake lights, i.e, it minimizes spent time in prototype designs. The tool has also allowed simulating many options for configurations of brake lights without need of building them, which also assists in the selection process of sample that offers an appropriate relationship between cost and benefit.

References

1. Peralta, S.B., Ruda, H.E.: Applications for Advanced Solid-State Lamps. IEEE Industry Applications Magazine 4, 31–42 (1998)
2. Edwards, P.R., Martin, R.W., Watson, I.M., Liu, C., Taylor, R.A., Rice, J.H., Robinson, J.W., Smith, J.D.: Quantum Dot Emission from Site-Controlled InGaN/GaN Micropyramid Arrays. Applied Physics Letters 85, 4281–4283 (2004)
3. Voelcher, J.: Top 10 Tech Cars. IEEE Spectrum 41, 20–27 (2004)
4. Young, W.R., Wilson, W.: Efficient Electric Vehicle Lighting Using LEDs. In: Southcon, pp. 276–280. IEEE Press, New York (1996)
5. Streetman, B.G., Banerjee, S.: Solid State Electronic Devices. Prentice Hall, Englewood Cliffs (1999)
6. Martin, R.W., Edwards, P.R., Taylor, R.A., Rice, J.H., Robinson, J.W., Smith, J.D., Liu, C., Watson, I.M.: Luminescence Properties of Isolated InGaN/GaN Quantum Dots. Physica Status Solidi (A) 202, 372–376 (2005)
7. Pecharroman-Gallego, R., Martin, R.W., Watson, I.M.: Investigation of the Unusual Temperature Dependence of InGaN/GaN Quantum Well Photoluminescence over a Range of Emission Energies. Journal of Physics D: Applied Physics 21, 2954–2961 (2004)
8. Griffiths, P., Langer, D., Misener, J.A., Siegel, M., Thorpe, C.: Sensor-Friendly Vehicle and Roadway Systems. In: 18th Instrumentation and Measurement Technology Conference, pp. 1036–1040. IEEE Press (2001)
9. Hagan, M.T., Menhaj, M.B.: Training Feedforward Networks with the Marquardt Algorithm. IEEE Transactions on Neural Networks 6, 989–993 (1994)
10. Haykin, S.: Neural Networks - A Comprehensive Foundation. Prentice-Hall, Upper Saddle River (1999)
11. Kim, S., Oh, S.-Y., Kang, J., Ryu, Y., Kim, K., Park, S.-C., Park, K.: Front and Rear Vehicle Detection and Tracking in the Day and Night Times Using Vision and Sonar Sensor Fusion. In: IEEE/RSJ International Conference on Intelligent Robots and Systems, pp. 2173–2178. IEEE Press (2005)
12. Cabani, I., Toulminet, G., Bensrhair, A.: Color-Based Detection of Vehicle Lights. In: IEEE Intelligent Vehicles Symposium, pp. 278–283. IEEE Press (2005)
13. Pasetti, G., Costantino, N., Tinfena, F., D'Abramo, P., Fanucci, L.: A Flexible LED Driver for Automotive Lighting Applications: IC Design and Experimental Characterization. IEEE Transactions on Power Electronics 27, 1071–1075 (2012)
14. Gacio, D., Cardesin, J., Corominas, E.L., Alonso, J.M., Dalla-Costa, M., Calleja, A.J.: Comparison among Power LEDs for Automotive Lighting Applications. In: IEEE Ind. Appl. Soc. Ann. Meeting (IAS), pp. 1–5. IEEE Press (2008)
15. Donahoe, D.N.: Thermal Aspects of LED Automotive Headlights. In: IEEE Vehicle Power and Propulsion Conference, pp. 1193–1199. IEEE Press (2009)
16. Bielecki, J., Jwania, A.S., El Khatib, F., Poorman, T.: Thermal Considerations for LED Components in an Automotive Lamp. In: 23rd Annual IEEE Semiconductor Thermal Measurement and Management Symposium, pp. 37–43. IEEE Press (2007)

A Single Neuron Model for Pattern Classification

B. Chandra[1] and K.V. Naresh Babu[2]

[1] Department of Mathematics,
Indian Institute of Technology Delhi, Delhi-110016
bchandra104@yahoo.co.in
[2] Department of Mathematics,
Indian Institute of Technology Delhi, Delhi-110016
vnareshiitd@gmail.com

Abstract. A biologically realistic non linear integrate and fire model is proposed in this paper. Its complete solution is derived and used for the construction of aggregation function in Multi layer perceptron model for classification of UCI Machine learning datasets. It is found that a single neuron in the conventional neural network is sufficient for the classification datasets. It has been observed that the proposed neuron model is far superior in terms of classification accuracy when compared with single integrate and fire neuron model. It is observed that biological phenomenon makes artificial neural network efficient for the classification.

Keywords: Biological neuron models, Back Propagation, Single integrate and fire neuron model.

1   Introduction

The mammalian brain consists of billions of neurons which are interconnected and exchange information through neural networks. Several mathematical models have been proposed for the efficient simulation of a biological neuron as well as mathematical explanation for the generation of action potential. Louis et al [2] designed Integrate-and-fire neuron model using an electric circuit consists of a parallel capacitor and resistor. A leak term was added to the membrane potential in the Integrate and Fire neuron model by stein [11]. Hodgkin et al [6] proposed a set of nonlinear ordinary differential equations that approximates the electrical characteristics of excitable cells such as neurons and explains the ionic mechanisms of action potentials. The four dimensional Hodgkin-Huxley model is reduced into a two dimensional model called Fitz-Nagumo model by Nagumo et al [4].It describes regenerative self excitation of a neuron. Later Morris et al [9] taken the best aspects of Hodgkin-Huxley and Fitz-Hugh-Nagumo models and developed a Morris-Lucar model. It is a voltage-gated calcium channel model with a delayed-rectifier potassium channel. A three dimensional model named Hindmarsh-Rose model was proposed by Hindmarsh et al [5] which allows dynamic behavior for the membrane potential. A generalized non linear integrate and fire neuron model was developed by Abbott et al [1]. Leaky integrate

T. Huang et al. (Eds.): ICONIP 2012, Part IV, LNCS 7666, pp. 619–625, 2012.

and fire neuron model, quadratic model [7] and Exponential integrate and fire neuron model [3] were derived as special instances of generalized non linear integrate and fire neuron model. Exponential integrate and fire neuron model is equivalent to leaky integrate fire neuron model for a spike with very sharp intention.

A biologically plausible artificial neuron model has been proposed whose aggregation function is obtained from the complete solution of a proposed non linear integrate-Fire model. The timing of the spikes contains neuronal information rather than the geometrical shape of the action potential. Hence the aggregation function is rendered from the relationship between injected current and Interspike interval of the proposed non linear Integrate-and-Fire neuron model. Back-propagation [10] has been used as learning algorithm without using any hidden layer. A single neuron has been used for the network architecture. The effectiveness of the proposed neuron model has been shown by comparing with the performance of single Integrate-fire neuron model [12] on UCI machine learning datasets. It has been observed that the proposed model reduces the computational time and increases the classification accuracy. It is observed that the inclusion of biological phenomenon in an artificial neural network can make it robust for classification.

Rest of the paper is organized as follows. Section 2 contains an overview of existing neuron models. Section 3 presents the derivation of proposed neuron model and spikes obtained from it. Section4 deals with the derivation of complete solution of proposed model and weight updating equations for proposed neuron model. In Section 5 results obtained from proposed neuron model and single integrate and fire neuron model on various benchmark datasets are discussed. Conclusions are given in section 6.

2   Biological Neuron Models

Various mathematical models for biological neurons have been developed till now in the literature to represent biological activities of a neuron. This section explains some of the important spiking models as follows.

2.1   Integrate and Fire Neuron Model

The electronic circuit of an integrate-and-fire model contains a capacitor C in parallel with a resistor R driven by input current I(t).The input current I(t) divided into two components I_R and I_C.Using ohm's law, I_R can be computed as $\frac{u(t)}{R}$ where $u(t)$ denotes voltage across the resistor. .Since capacity $C = \frac{q}{u}$ where q is the charge and u denotes the voltage, I_C is computed as $I_C = C\frac{du}{dt}$. Hence the input current I(t) is given by

$$I(t) = \frac{u(t)}{R} + C\frac{du}{dt} \qquad (1)$$

Multiply equation (1) by R and substitute the time constant $\tau_m = RC$ of the leaky integrator to obtain the standard form of (1).

$$\tau_m \frac{du}{dt} = -u(t) + RI(t) \tag{2}$$

Whenever the membrane potential u(t) increases with respect to time, It crosses a threshold u_θ at that point a spike is produced and the voltage resets to the resting potential u_{rest}. This reset condition and equation (2) defines the basic integrate and fire neuron model. Generalization of non linear integrates and fire model is discussed in the next section.

2.2 Non Linear Integrate and Fire Neuron Model

In general a non linear integrate and fire neuron model [1] equation (2) is replaced by

$$\tau_m \frac{du}{dt} = F(u) + G(u)I^{ext} \tag{3}$$

The dynamics of membrane potential u are stopped if u reaches the threshold u_θ then reinitialized at $u = u_{rest}$. A comparison with equation (2) makes G(u) as a voltage-dependent input resistance while $-\frac{F(u)}{u - u_{rest}}$ as voltage-dependent decay constant.

In this paper, a new non linear integrate and fire neuron model is developed. Its solution has been used for pattern classification.

3   Proposed Non Linear IFN Model

The proposed non linear integrate and fire neuron model is represented by

$$\tau \frac{du}{dt} = a\frac{cosec(Tan^{-1}u)}{\sqrt{1+u^2}} + RI(t) \tag{4}$$

Where u denotes the membrane potential, I denotes the external current, τ denotes membrane time constant, R denotes the input resistance $\frac{cosec(Tan^{-1}u)}{\sqrt{1+u^2}}$ denotes a non-linear function of membrane potential. The dynamics of membrane potential u are interrupted if u reaches the threshold v then reinitialized at $u = u_{rest}$.

Equation (4) is solved by using variable and separable method as given by

$$\frac{du}{a\frac{cosec(Tan^{-1}u)}{\sqrt{1+u^2}} + RI(t)} = \frac{dt}{\tau} \tag{5}$$

The solution is given by

$$\frac{u}{RI} - \frac{a}{RI^2}\log(a + RI(t)u) = \frac{t}{\tau} + Cons \tag{6}$$

Equation (6) represents the Interspike interval and the frequency f is given by

$$f = \frac{1}{\tau\left(\frac{u}{RI} - \frac{a}{RI^2}\log(a + RIu) - Const\right)} \tag{7}$$

Spikes can be observed in response to the proposed non linear integrates fire neuron model. Simulated spikes are shown in Fig-1. It is drawn using $\tau = 10$, Threshold = 4 and $I_0 = 1$.

4   Proposed Neuron Model

Biological neurons exchange neuronal information [8] based on the timing of the spikes generated by the action potential. Motivated from the non-linear map between injected current and Interspike interval of proposed integrate-and-fire neuron model as shown in equation (6), following aggregation function is assumed instead of the weighted sum of a traditional neuron

$$net = \prod_{i=1}^{n}\left(p_i\left(x_i - \log(q_i x_i)\right) + r_i\right) \tag{8}$$

Where x_i denotes i^{th} input variable, n denotes number of input neurons and q_i, p_i and c_i denotes weight parameters.

In comparison with equation (6), $\frac{1}{RI}$ is denoted as p_i , $\frac{a}{RI}$ is denoted as q_i and the constant term is denoted by r_i.

Proposed neuron model contains an input layer and a single neuron in the output layer without a hidden layer. Weighted product of input neurons is used as aggregation to the single output neuron as shown in equation (8). To minimize the error function, Gradient descent rule is applied for the patterns in the training dataset. If t denotes the target and y denotes the actual output of the neuron then the error function is given by equation (9).

$$E = \frac{1}{2}(y - t)^2 \tag{9}$$

In gradient descent, weight parameters are modified in proportion to the negative of the error derivative with respect to each parameter p_i, q_i and r_i as shown in equation (10), (11) and (12)

$$\Delta p_i = -\epsilon \frac{\partial E}{\partial p_i} \tag{10}$$

$$\Delta q_i = -\epsilon \frac{\partial E}{\partial q_i} \tag{11}$$

$$\Delta r_i = -\epsilon \frac{\partial E}{\partial r_i} \tag{12}$$

Partial derivatives of the Error function with respect to each parameter q_i , p_i and c_i is given by the following equations (13), (14) and (15)

$$\frac{\partial E}{\partial q_i} = -(t - y)y(1 - y)(net).\left(\frac{1}{(p_i(x_i - \log(q_i x_i)) + r_i)}\right)(x_i - \log(q_i x_i)) \tag{13}$$

$$\frac{\partial E}{\partial p_i} = (t - y)y(1 - y)(net).\left(\frac{1}{(p_i(x_i - \log(q_i x_i)) + r_i)}\right)\left(\frac{p_i}{q_i}\right) \tag{14}$$

$$\frac{\partial E}{\partial r_i} = -(t - y)y(1 - y)(net)\left(\frac{1}{(p_i(x_i - \log(q_i x_i)) + r_i)}\right) \tag{15}$$

Parameters are updated using the following equations (16), (17) and (18)

$$p_i(k + 1) = p_i(k) + \Delta p_i \tag{16}$$

$$q_i(k + 1) = q_i(k) + \Delta q_i \tag{17}$$

$$r_i(k + 1) = r_i(k) + \Delta c_i \tag{18}$$

5   Results

This section explains the performance evaluation of the proposed neuron model and compared with the single integrate neuron model on various standard datasets. Performance evaluation was carried out on Benchmark datasets taken from UCI machine learning repository. Datasets used are listed in Table 1. Ten cross validation is performed on all the datasets. The number of iterations required to achieve mean square error of the order of 0.0001 using single integrate fire neuron model and proposed neuron model is given in Table 1. p_i, q_i, r_i are random values drawn from uniform distribution on the open interval $(0, 1)$.

It is observed that the number of iterations needed to obtain MSE of order 0.0001 is lesser for the proposed neuron model when compared with Single Integrate- Fire neuron model for all the datasets. It can be inferred that the proposed neuron model outperforms single integrate and fire neuron model even In case of high dimensional datasets like Mushroom and Page block. It is found that a single neuron in output layer for the proposed neuron model is sufficient for the applications that need a number of neurons in different hidden layers of a traditional neural network. The superiority of the performance of the proposed Neuron model is also depicted graphically in Figure 2.

Table 1. No. of epochs required for achieving M.S.E of order 0.0001

S.No	Datasets	No.of features	No.of Records	IFN	Proposed Model
1	Balanced Scale	4	625	14	10
2	Iris	4	150	96	70
3	Wine	13	178	22	14
4	Breast cancer	9	286	72	64
5	credit	15	690	59	40
6	lymphocytes	18	148	69	30
7	mushroom	22	8124	46	28
8	Page block	10	5473	26	15
9	Hayes-Roth	5	160	95	10
10	Dermatology	33	366	90	76
11	Diabetes	20	691	70	10
12	Monk	7	432	80	72
13	Voting	16	435	70	52
14	Ecoli	8	336	60	14

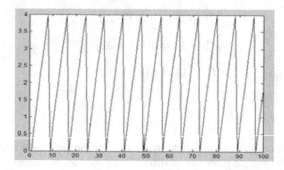

Fig. 1. Spikes generated using proposed non linear IFN model

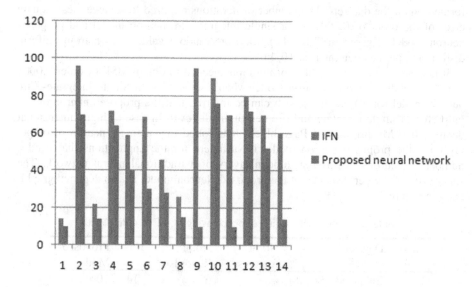

Fig. 2. Number of epochs required to achieve M.S.E of order 0.0001

6   Conclusions

This paper proposes a new neuron model whose aggregation function is designed from the complete solution of a proposed neuron model. The concept of neuronal information exchanged from one neuron to other depending on the timing of the spikes generated has been used to design the proposed neuron model. Training of UCI Machine learning datasets shows that a single neuron in the output layer is capable of performing classification task faster in terms of number of epochs as compared with single integrate and fire neuron model.

References

1. Abbott, L.F., van Vreeswijk, C.: Asynchronous states in a network of pulse-coupled oscillators. Phys. Rev. E 48, 1483–1490 (1993)
2. Abbott, L.F.: Lapique's introduction of the integrate-and-fire model neuron. Brain Research Bulletin 50(5/6), 303–304 (1907)
3. Brette, R., Gerstner, W.: Adaptive exponential integrate-and-fire model as an effective description of neuronal activity. J. Neurophysiol. 94, 3637–3642 (2005)
4. FitzHugh, R.: Impulses and physiological states in theoretical models of nerve membrane. Biophysical J. 1, 445–466 (1961)
5. Hindmarsh, J.L., Rose, R.M.: A model of neuronal bursting using three coupled first order differential equations. Proc. R. Soc. Lond. B. 221, 87–102 (1984)
6. Hodgkin, A., Huxley, A.: A quantitative description of membrane current and its application to conduction and excitation in nerve. J. Physiol. 117, 500–544 (1952)
7. Ermentrout, G.B., Kopell, N.: Parabolic bursting in an excitable system coupled with a slow oscillation. SIAM J. Appl. Math. 46, 233–253 (1986)
8. Maass, W.: Computation with spiking neurons. In: The Handbook of Brain Theory and Neural Networks, 2nd edn., pp. 1080–1083. MIT Press, Cambridge (2003)
9. Morris, C., Lecar, H.: Voltage Oscillations in the barnacle giant muscle fiber. Biophys. J. 35, 193–213 (1981)
10. Rumelhart, D.E., Hinton, G.E., Williams, R.J.: Learning representations by back-propagating errors. Nature 323, 533–536 (1986)
11. Stein, R.B.: A theoretical analysis of neuronal variability. Biophys. J. 5, 173–194 (1965)
12. Yadav, A., Mishra, D., Ray, S., Yadav, R.N., Kalra, P.K.: Learning with Single Integrate-and-Fire Neuron. In: Proceedings of IEEE International Joint Conference on Neural Networks, vol. 4, pp. 2156–2161 (2005)

Continuous Classification of Spatio-temporal Data Streams Using Liquid State Machines

Stefan Schliebs and Doug Hunt

Auckland University of Technology, New Zealand
{sschlieb,dphunt}@aut.ac.nz

Abstract. This paper proposes to use a Liquid State Machine (LSM) to classify inertial sensor data collected from horse riders into activities of interest. LSM was shown to be an effective classifier for spatio-temporal data and efficient hardware implementations on custom chips have been presented in literature that would enable relative easy integration into wearable technologies. We explore here the general method of applying LSM technology to domain constrained activity recognition using a synthetic data set. The aim of this study is to provide a proof of concept illustrating the applicability of LSM for the chosen problem domain.

Keywords: Liquid State Machine, Spatio-temporal data processing, Data mining, Equestrian sport

1 Introduction

Spatio-temporal data from inertial sensors (accelerometers, gyroscopes, magnetometers) worn on the human body have recently been used as the input to gesture recognition systems [5], orientation tracking systems [14] and activity recognition systems [1]. Unconstrained activity recognition presents a number of challenges including a general lack of overall context in many situations that makes it difficult to distinguish two similar movements (e.g. turning a door knob to open a door and turning a key to start a vehicle). Domain constrained activity recognition, however, has been shown to be more achievable and reliable [6,13]. Constraining the domain to a particular sport (such as equestrian sport) is also potentially useful, particularly if the rules or traditions of that sport add further activity and style constraints. We propose to construct a domain constrained, activity classification system to classify the inertial data collected from horse riders to recognise activities of interest within equestrian sport. If successful this classification system will become part of a larger riding coaching system.

In this study, we investigate the suitability of a Liquid State Machine (LSM) [7] as the core of this classification system. LSM technology has been shown to be effective at continuous speech recognition [12] and is implementable on custom chips that enable relative easy integration into wearable technologies. Sensor data streams for speech have some aspects in common with inertial data streams (spatio-temporal, continuous data stream, real time requirements, digitally encoded analogue data). In addition the ability to possibly implement the LSM on chip for wearable applications has further potential benefits.

T. Huang et al. (Eds.): ICONIP 2012, Part IV, LNCS 7666, pp. 626–633, 2012.

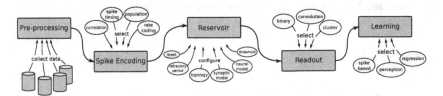

Fig. 1. Proposed framework to classify inertial sensor data into activities of interest

The aim of this study is to provide a simple proof of concept illustrating the applicability of LSM for the chosen problem domain. In the next sections, we first explain the proposed framework and then provide some experimental results on the continuous classification of a synthetic spatio-temporal data set.

2 Framework

An LSM consists of two main components, a "liquid" (also called reservoir) in the form of a recurrent Spiking Neural Network (SNN) [4] and a trainable read-out function. The liquid is stimulated by spatio-temporal input signals causing neural activity in the SNN that is further propagated through the network due to its recurrent topology. Therefore, a snapshot of the neural activity in the reservoir contains information about the current and past inputs to the system. The function of the liquid is to accumulate the temporal and spatial information of all input signals into a single high-dimensional intermediate state in order to enhance the separability between network inputs. The readout function is then trained to transform this intermediate state into a desired system output.

The framework proposed in this paper is illustrated in Figure 1. In the first step, the collected raw sensory time series data is pre-processed. Typical data cleaning procedures are normalization, feature selection and outlier detection, but also Fourier or wavelet transformations may be considered here. The cleaned signal is then encoded into an input compatible with the reservoir. Numerous encoding algorithms have been proposed in literature and we will consider two of them in our experiments presented in the next section. Feeding the encoded input into the reservoir results in a temporal change of neural activity in the SNN which is read out periodically. A machine learning algorithm learns the mapping from the extracted readouts to a desired class label.

Several design and configuration choices have to be made for each component of the framework. The ellipses in Figure 1 show some possible options for these decisions. In the remaining part of this section, we explain each framework component in greater detail and indicate some of our design decisions for the experiments presented in section 3.

2.1 Data

For the experimental analysis presented in this paper, we decided to generate a synthetic multi-sensor time series data set so that we have control over aspects such as the number of sensor signals, problem difficulty and signal-to-noise ratio. Furthermore, this data set provides us with an ideal solution, since the exact differentiation between signal and noise is known a priori, and it allows the analysis of the proposed method for a large variety of testing scenarios.

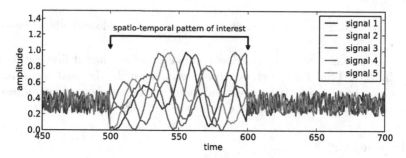

Fig. 2. Synthetic multi-sensor data investigated in this study

The constructed N synthetic time series are generated as a superposition of C sine waves. Signal $s_n(t)$ is described as

$$s_n(t) = \sum_{i=1}^{C} a_n^{(i)} \sin(\omega_n^{(i)} t + \phi_n^{(i)}) \tag{1}$$

where $C \in \mathbb{N}$ is the number of superimposed functions and the parameters $a_n^{(i)} \in \mathbb{R}$, $\omega_n^{(i)} \in \mathbb{R}$, $\phi_n^{(i)} \in \mathbb{R}$ represent the amplitudes, frequencies and phases of the individual sine functions, respectively. For our study, we chose $C = 5$ superimposed sine waves that were parametrized with a set of random amplitudes, frequencies and phases. The $N = 5$ time series $s_n(t)$ were "pasted" into a uniformly distributed noise sequence at various locations and a small Gaussian noise ($\sigma = 0.01$) was added to the resulting signal. Finally, we normalized the signal to be between $[0, 1]$. Figure 2 depicts part of the generated time series.

2.2 Encoding

The time series data obtained from the sensors are presented to the reservoir in form of an ordered sequence of real-valued data vectors. In order to compute an input compatible with the SNN, each real value of a data vector is transformed into a spike train using a spike encoding. We explore two different encoding schemes in this paper, namely Ben's Spike Algorithm (BSA) [11] and a population encoding technique.

BSA assumes the analog signal to be a convolution of a spike train. The algorithm attempts to estimate the corresponding spike train responsible for this convolution by reversing the convolution process. The method has a threshold parameter which we set to 0.955 and we use the discrete linear filter presented in [3]. More detailed explanations of this encoding along with pseudo code can be found in [11].

In contrast to BSA, the population encoding uses more than one input neuron to encode a single time series. The idea is to distribute a single input to multiple neurons, each of them being sensitive to a different range of values. Our implementation is based on arrays of receptive fields with overlapping sensitivity profiles as described in [2,9]. We refer to the mentioned references for further details and examples of this encoding algorithm.

As a result of the encoding, input neurons emit spikes at predefined times according to the presented data vectors.

2.3   Reservoir

For the reservoir, we employ the Leaky Integrate-and-Fire neuron which is arguably one of the best known models for simulating SNN. This neural model is based on the idea of an electrical circuit containing a capacitor with capacitance C and a resistor with a resistance R, where both C and R are assumed to be constant. The dynamics of a neuron i are then described by the following differential equations:

$$\tau_m \frac{dV_i}{dt} = -V_i(t) + R \; I_i^{\text{syn}}(t) \tag{2}$$

$$\tau_s \frac{dI_i^{\text{syn}}}{dt} = -I_i^{\text{syn}}(t) \tag{3}$$

The constant $\tau_m = RC$ is called the membrane time constant of the neuron. Whenever the membrane potential V_i crosses a threshold ϑ_i from below, the neuron fires a spike and its potential is reset to a reset potential V_r. We use an exponential synaptic current I_i^{syn} for a neuron i modelled by Eq. 3 with τ_s being a synaptic time constant.

Similar to [10], we define a dynamic firing threshold as a separate differential equation:

$$\tau_\vartheta \frac{d\vartheta_i}{dt} = \theta - \vartheta_i(t) \tag{4}$$

where θ is the minimum firing threshold of the neuron and τ_ϑ is the time constant for the dynamic threshold. Whenever the neuron i emits a spike, its threshold ϑ_i is increased by a constant $\Delta\theta_i$, i.e. $\vartheta_i \leftarrow \vartheta_i + \Delta\theta_i$. A dynamic synapse model based on the short-term plasticity (STP) proposed by Markram et al. [8] is used to exchange information between connected neurons.

We construct a reservoir having a small-world inter-connectivity pattern as described in [7]. A recurrent SNN is generated by aligning 500 neurons in a three-dimensional grid of size $10 \times 10 \times 5$. In this grid, two neurons A and B are connected with a connection probability

$$P(A, B) = C \times e^{\frac{-d(A,B)}{\lambda^2}} \tag{5}$$

where $d(A, B)$ denotes the Euclidean distance between two neurons and λ corresponds to the density of connections which was set to $\lambda = 3$ in our simulations.

Parameter C depends on the type of the neurons. We discriminate into excitatory (ex) and inhibitory (inh) neural types resulting in the following parameters for C: $C_{ex-ex} = 0.3$, $C_{ex-inh} = 0.2$, $C_{inh-ex} = 0.4$ and $C_{inh-inh} = 0.1$. The network contained 70% excitatory and 30% inhibitory neurons.

2.4 Readout and Learning

In this study, we use the typical analog readout function in which every spike is convolved by a kernel function that transforms the spike train of each neuron in the reservoir into a continuous analog signal. We use an exponential kernel with a time constant of $\tau = 50$ms. The convolved spike trains are then sampled using a time step of 10ms resulting in 500 time series – one for each neuron in the reservoir. In these series, the data points at time t represent the readout for the presented input sample. A very similar readout was used in many other studies, e.g. in [12] for a speech recognition problem.

Readouts were labeled according to their readout time t. If the readout occurred at the time when a sensor signal of interest (e.g. mounting/dismounting the horse) was fed into the reservoir, then the corresponding readout is labeled as a class-1 or class-2 sample. Consequently, a readout belongs to class 0, if it was obtained during the presentation of a noise part of the input signal.

The final step of the LSM framework consists of a mapping from a readout sample to a class label. The general approach is to employ a machine learning algorithm to learn the correct mapping from the readout data. In fact, since the readout samples are expected to be linearly separable with regard to their class label [7], a comparably simple learning method can be applied for this task. From the labeled readouts, we obtained a linear regression model mapping a reservoir readout sample to the corresponding class label, i.e. either 0 (noise) or 2 (pattern A) or −2 (pattern B).

3   Results

Here we present our preliminary findings on the classification of the synthetic time series data introduced in section 2.1. Figure 3 shows the outputs obtained from each of the individual processing steps of the LSM framework. A set of five synthetic time series was generated over a time window of 4200ms which included ten occurrences of two alternating temporal patterns, cf. Figure 3A. The encoded spike trains derived from the given time series are depicted in Figure 3B (BSA encoding) and 3C (population encoding), respectively. The figures show a raster plot of the neural activity of the input neurons over time. A point in these plots indicates a spike fired by a particular neuron at a given time.

The obtained spike trains were then fed into a reservoir resulting in characteristic response patterns for each encoding type, cf. Figure 3D and 3F. The small number of input neurons employed by the BSA encoding results in a rather sparse reservoir response. Clearly, a repeating pattern of neural activity is observed, however, the relevance for the detection of the patterns of interest is less

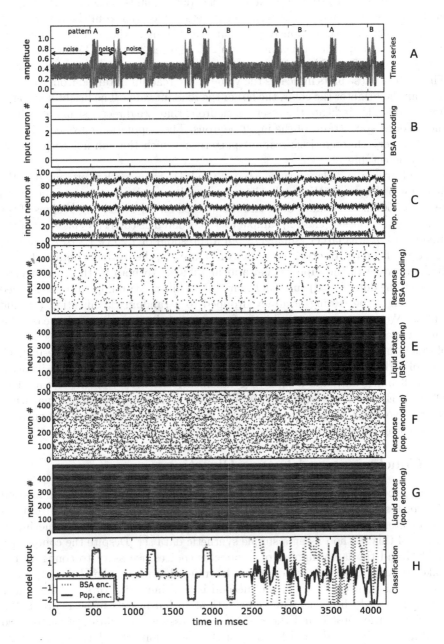

Fig. 3. Experimental results obtained from the continuous classification of the synthetic data set. See text for detailed explanations on the figure.

obvious. The response of the population-encoded signal is denser and the impact of the non-noise signal on the reservoir is detectable even from the raster plot.

The reservoir is continuously read out every 10ms of the network simulation using the technique described in section 2.4. Figures 3E and 3G show the readouts over time for BSA and population-encoded reservoir inputs, respectively. The color in these plots indicates the value of the readout obtained from a certain neuron; the brighter the color, the larger the readout value. The bright horizontal lines in this plot indicate the reservoir neurons that are directly stimulated from the encoded spike trains of the input neurons. The population-encoded stimulus causes a very characteristic readout pattern in which the non-noise signals are clearly detectable.

The learning and classification step of the LSM framework is presented in the last plot of Figure 3. We used a linear regression model that was trained with the first 2500ms of readout data and then tested on the remaining 1700ms of the simulation. A class-1 (class-2) sample was considered as correctly classified, if the model output was larger (smaller) than a threshold of 1.25 (-1.25). As already suspected from the reservoir responses of Figure 3D, the readouts obtained using BSA encoding of the input signal appear difficult to map to the correct class label, cf. the dotted (red) curve in Figure 3G. While the training data was learned well enough, the testing accuracy dropped to a random classification (around 24.7%). We assume an unsuitable configuration of the reservoir as the reason for this low performance. On the other hand, the model responses for the readouts obtained using population encoding showed much more promising results, cf. the solid (blue) curve in Figure 3G. The model reported perfect classification on the training data and a satisfying classification accuracy on the testing data (93.5%). Errors usually occurred immediately after the onset of a signal when the reservoir had not yet accumulated sufficient information about the pattern. Considering the short training period, this result is very encouraging.

4   Conclusions and Future Directions

In this paper, we have proposed an LSM based, continuous classification method to detect spatio-temporal patterns in inertial sensor data obtained from accelerometers, gyroscopes and magnetometers. We have presented some initial results of this method using a simple synthetic data set. Despite the complexity of LSM and the large number of parameter choices necessary to configure the spike encoding, reservoir, readout and learning algorithms, the obtained results are promising. A core part of the method is the simulation of a spiking neural networks for which efficient implementations in hardware can be realized [12]. The low energy profile and the real-time processing capabilities of customized hardware are highly desired features of the continuous classification of inertial sensor readouts.

A future study will investigate the suitability of the proposed framework for classifying the collected real-world data from horse riders. We are specifically interested in a robust configuration of the reservoir and an efficient implementation on a mobile platform. Contemporary bluetooth enabled mobiles may have sufficient computational resources to allow the simulation of the reservoir and the continuous classification of streamed inertial sensor data in real time.

References

1. Bao, L., Intille, S.S.: Activity Recognition from User-Annotated Acceleration Data. In: Ferscha, A., Mattern, F. (eds.) PERVASIVE 2004. LNCS, vol. 3001, pp. 1–17. Springer, Heidelberg (2004)
2. Bohte, S.M., Kok, J.N., Poutré, J.A.L.: Error-backpropagation in temporally encoded networks of spiking neurons. Neurocomputing 48(1-4), 17–37 (2002)
3. de Garis, H., Nawa, N., Hough, M., Korkin, M.: Evolving an optimal de/convolution function for the neural net modules of atr's artificial brain project. In: International Joint Conference on Neural Networks, IJCNN 1999, vol. 1, pp. 438–443 (1999)
4. Gerstner, W., Kistler, W.M.: Spiking Neuron Models: Single Neurons, Populations, Plasticity. Cambridge University Press, Cambridge (2002)
5. Junker, H., Amft, O., Lukowicz, P., Tröster, G.: Gesture spotting with body-worn inertial sensors to detect user activities. Pattern Recognition 41(6), 2010–2024 (2008)
6. Lukowicz, P., Ward, J.A., Junker, H., Stäger, M., Tröster, G., Atrash, A., Starner, T.: Recognizing Workshop Activity Using Body Worn Microphones and Accelerometers. In: Ferscha, A., Mattern, F. (eds.) PERVASIVE 2004. LNCS, vol. 3001, pp. 18–32. Springer, Heidelberg (2004)
7. Maass, W., Natschläger, T., Markram, H.: Real-time computing without stable states: A new framework for neural computation based on perturbations. Neural Computation 14(11), 2531–2560 (2002)
8. Markram, H., Wang, Y., Tsodyks, M.: Differential signaling via the same axon of neocortical pyramidal neurons. Proceedings of the National Academy of Sciences 95(9), 5323–5328 (1998)
9. Schliebs, S., Defoin-Platel, M., Kasabov, N.: Integrated Feature and Parameter Optimization for an Evolving Spiking Neural Network. In: Köppen, M., Kasabov, N., Coghill, G. (eds.) ICONIP 2008, Part I. LNCS, vol. 5506, pp. 1229–1236. Springer, Heidelberg (2009)
10. Schliebs, S., Fiasché, M., Kasabov, N.: Constructing Robust Liquid State Machines to Process Highly Variable Data Streams. In: Villa, A.E.P., Duch, W., Érdi, P., Masulli, F., Palm, G. (eds.) ICANN 2012, Part I. LNCS, vol. 7552, pp. 604–611. Springer, Heidelberg (2012)
11. Schrauwen, B., Van Campenhout, J.: BSA, a fast and accurate spike train encoding scheme. In: Proceedings of the International Joint Conference on Neural Networks, vol. 4, pp. 2825–2830 (July 2003)
12. Schrauwen, B., D'Haene, M., Verstraeten, D., Campenhout, J.V.: Compact hardware liquid state machines on fpga for real-time speech recognition. Neural Networks 21(2-3), 511–523 (2008)
13. Stiefmeier, T., Ogris, G., Junker, H., Lukowicz, P., Troster, G.: Combining motion sensors and ultrasonic hands tracking for continuous activity recognition in a maintenance scenario. In: 2006 10th IEEE International Symposium on Wearable Computers, pp. 97–104. IEEE (2006)
14. Zhu, R., Zhou, Z.: A real-time articulated human motion tracking using tri-axis inertial/magnetic sensors package. IEEE Transactions on Neural Systems and Rehabilitation Engineering 12(2), 295–302 (2004)

On the Selection of Time-Frequency Features for Improving the Detection and Classification of Newborn EEG Seizure Signals and Other Abnormalities

Boualem Boashash[1,2] and Larbi Boubchir[1,*]

[1] Electrical Engineering Department
Qatar University College of Engineering, P.O. Box 2713, Doha, Qatar
[2] University of Queensland, Centre for Clinical Research
Royal Brisbane & Womens Hospital, Herston, QLD 4029, Australia
{boualem,larbi}@qu.edu.qa

Abstract. This paper presents new time-frequency features for seizure detection in newborn EEG signals. These features are obtained by translating some relevant time features or frequency features to the joint time-frequency domain. A calibration procedure is then used for verification. The relevant translated features are ranked and selected according to maximal-relevance and minimal-redundancy criteria. The selected features improve the performance of newborn EEG seizure detection and classification systems by up to 4% for 100 real newborn EEG segments.

Keywords: Time-frequency analysis, instantaneous frequency, features translation, time-frequency features, features selection, seizure, newborn EEG, detection, classification.

1 Introduction

Newborn neurological abnormalities such as seizures can lead to permanent brain damage or even fatalities if not detected and treated early. Analyzing the Electroencephalogram (EEG) is a proven approach for detecting newborn seizures, but their manual detection is time consuming and subjective. It is therefore desired to develop automated techniques by processing newborn EEG signals and extracting features which can then be used for classification [1, 7–9]. As newborn EEG signals are non-stationary signals with time-varying frequency content, a time-frequency signal analysis is used to characterize the different seizure patterns present in these signals [1].

The proposed method for EEG seizure includes three steps: (1) Signal analysis, (2) features extraction to characterize the different abnormalities and (3) classification of these features. Previous techniques for detecting EEG seizure include using EEG features in time [2–5], frequency [3, 4], time-scale [6] or T-F domains [7–9], as well as nonlinear characteristics of EEG signals [1]. The

---

* Corresponding author.

T. Huang et al. (Eds.): ICONIP 2012, Part IV, LNCS 7666, pp. 634–643, 2012.
© Springer-Verlag Berlin Heidelberg 2012

time-domain features extracted from EEG signals include: zero-crossing rates, average amplitude, derivatives of the signal's amplitude and entropy-based features. Frequency-domain features extracted from the spectrum of EEG signals include: spectrum normalized power, sub-band powers, intensity weighted mean frequency, and intensity weighted bandwidth. The T-F features are extracted from the T-F representation of the EEG signal and include the instantaneous frequency (IF) and other non-stationary features, such as those recent and new T-F features based on signal and image processing techniques proposed in [7–9].

This study aims to extend this previous work and define new T-F features by translating some relevant time features or frequency features to the joint T-F domain in order to improve the performance of newborn EEG seizure detection and classification systems. Then, the relevant translated features along with the signal & image-related features proposed recently in [7–9] are used to define a new vector of T-F features for characterizing and classifying different newborn EEG seizures and other abnormalities.

2   Newborn EEG Seizure Detection and Classification

2.1   Time-Frequency Distributions

In order to develop seizure detection methods in the T-F domain, it is necessary to select a suitable T-F Distribution (TFD) to represent the EEG signals. The most common are quadratic TFDs (QTFDs) [1] such as the Wigner-Ville distribution (WVD), smoothed WVD (SWVD), Gaussian kernel distribution (GKD), modified-B distribution (MBD), and spectrogram (SPEC).

The discrete version of a QTFD for a given analytic signal $z[n]$ associated with the real discrete time signal $x[n], n = 0, 1, \ldots, N-1$ is given by [1] (p. 240)

$$\rho[n, k] = 2 \underset{n \to k}{\mathrm{DFT}} \left\{ G[n, m] \underset{n}{*} (z[n + m] z^*[n - m]) \right\} \tag{1}$$

where G is the time-lag kernel of the TFD and $\underset{n}{*}$ stands for convolution in time. $\rho[n, k]$ is represented by an $N \times M$ matrix ρ where M is the number of FFT points used $(N \geqslant M)$ in calculating the TFD. Note that $n = t.f_s$ and $k = \frac{2M}{f_s} f$ where t and f are, respectively, the continuous time and frequency variables, and f_s is the sampling frequency of the signal [1]. The time-lag kernel $G[n, m]$ of the selected QTFDs used in this study can be found in [1](pp. 71-76).

2.2   T-F Approach for Newborn EEG Seizure Detection

As newborn EEG signals are non-stationary, they are best represented by a TFD, which is intended to describe how the energy of the signal is distributed over the two-dimensional T-F space [1]. The TFD shows the start and stop times of signal components and their frequency range, as well as the component variation in frequency with time (described by the IF) [11, 12]. The IF can be estimated using a peak detector in the T-F domain that selects the frequency with the maximum value in the T-F representation as a function of time.

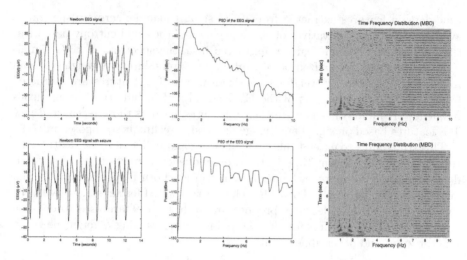

Fig. 1. An example of newborn EEG signal with seizure (2nd row) and non-seizure (1st row) in the time (1st column), frequency (2nd column) and joint T-F domains (3rd column). The T-F representation was generated using the MBD with the smoothing parameter $\beta = 0.01$ and the lag window length of 127.

Figure 1 shows an example of seizure and non-seizure newborn EEG signals in the time, frequency and joint T-F domains, in order to illustrate the difference between them and show how the TFD plot can provide more information about the IF, non-stationary nature and multi-component characteristics of the signals than the time or the frequency representations [1, 7–9].

This T-F approach for automatic classification of newborn EEG seizure includes pre-processing of EEG signals, finding their optimal TFDs, extracting features from the TFDs, and finally allocating the T-F features to the relevant class. In order to classify EEG signals, features are extracted from the TFDs of the EEG segments. The extracted features need have the ability to discriminate between different classes. Based on the EEG classification system described in [7–9], the vector containing the selected T-F features is defined and then used to train a multi-class SVM classifier.

3   Feature Extraction and T-F Translation

Relevant time features or frequency features are selected [2–5, 10] and then translated in the T-F domain. Tables 1 and 2 show the time features and frequency features that have been translated to the T-F domain.

Note that the mappings of frequency features in Table 2 involve summation along the time axis of T-F representations so as to calibrate them by transforming them with a frequency domain representation (assuming that marginals are satisfied). The same applies to Table 1, swapping time and frequency. This is a first step to show the relationships between frequency only features and T-F

Table 1. Time features and their T-F translation. *(Given a real discrete time EEG signal x, then z and ρ are respectively the analytical signal (obtained using the Hilbert transform) and the TFD of z. Also,* **T** *and* **Tt** *stand respectively for the features in the time domain and the translated features in the T-F domain).*

Time features (T)	T-F translation (Tt)
Mean and variance of the EEG segment [2, 3]	
$T_1 = \mu = \frac{1}{N}\sum_{n=1}^{N}|z[n]|$ $\quad$ $Tt_1 = \mu_{(t,f)} = \frac{1}{NM}\sum_{k=1}^{M}\sum_{n=1}^{N}\rho[n,k]$	
$T_2 = \sigma^2 = \frac{1}{N}\sum_{n=1}^{N}(\mu-|z[n]|)^2$ $Tt_2 = \sigma^2_{(t,f)} = \frac{1}{MN}\sum_{k=1}^{M}\sum_{n=1}^{N}(\mu_{(t,f)}-\rho[n,k])^2$	
First and second derivatives of the raw amplitude (skewness, kurtosis) [2, 3]	
$T_3 = \frac{1}{N\sigma^3}\sum_{n=1}^{N}(|z[n]|-\mu)^3$ $\quad$ $Tt_3 = \frac{1}{(NM-1)\sigma^3_{(t,f)}}\sum_{k=1}^{M}\sum_{n=1}^{N}(\rho[n,k]-\mu_{(t,f)})^3$	
$T_4 = \frac{1}{N\sigma^4}\sum_{n=1}^{N}(|z[n]|-\mu)^4$ $\quad$ $Tt_4 = \frac{1}{(NM-1)\sigma^4_{(t,f)}}\sum_{k=1}^{M}\sum_{n=1}^{N}(\rho[n,k]-\mu_{(t,f)})^4$	
Median absolute deviation of the amplitude [3]	
$T_5 = \frac{1}{N}\sum_{n=1}^{N}(|z[n]|-\mu)$ $\quad$ $Tt_5 = \frac{1}{NM}\sum_{k=1}^{M}\sum_{n=1}^{N}|\rho[n,k]-\mu_{(t,f)}|$	
Coefficient of variation of the EEG segment [2]	
$T_6 = \frac{\sigma}{\mu}$ $\quad$ $Tt_6 = \frac{\sigma_{(t,f)}}{\mu_{(t,f)}}$	
RMS amplitude [4]	
$T_7 = \sqrt{\frac{\sum_{n=1}^{N}z[n]^2}{N}}$ $\quad$ $Tt_7 = \sqrt{\frac{\sum_{n=1}^{N}\sum_{k=1}^{M}\rho[n,k]}{NM}}$	
Amplitude based feature: inter-quartile range [5]	
$T_8 = z[\frac{3(N+1)}{4}] - z[\frac{N+1}{4}]$ $\quad$ $Tt_8 = \frac{1}{M}\sum_{k=1}^{M}(\rho[\frac{3(N+1)}{4},k] - \rho[\frac{N+1}{4},k])$	
Shannon entropy [3, 4]	
$T_9 = -\sum_{n=1}^{N}z[n]\log_2(z[n])$ $\quad$ $Tt_9 = -\sum_{n=1}^{N}\sum_{k=1}^{M}\rho[n,k]\log_2(\rho[n,k])$	

features for verification purposes so that the extension from frequency domain to T-F domain is validated.

4 T-F Translated Features Selection

4.1 T-F Features Definition

The relevant translated features listed in Tables 1 and 2 can be ranked and selected according to the *minimum redundancy* and *maximum relevance* criteria defined below in (2) and (4), using the mutual information defined in [13].

Maximum Relevance means to find features satisfying (2), which approximates the maximum dependency with the mean value of all mutual information values between individual feature F_i and class h:

$$\max V(S,h), \quad V = \frac{1}{|S|}\sum_{F_i \in S} I(F_i; h) \tag{2}$$

Table 2. Frequency features and their T-F translation. *(For a real discrete time EEG signal x, then Z and ρ are respectively the Fourier transform of the analytic signal z and the TFD of z. Also, **F** and **Ft** stand for the features in the frequency domain and the translated features in the T-F domain, respectively).*

Frequency features (F)	T-F translation (Ft)
Spectral flux: difference between normalized spectra magnitudes [3] $F_1 = \sum_{k=1}^{M} \left(Z^{(l)}[k] - Z^{(l-1)}[k] \right)^2$ $Z^{(l)}$ and $Z^{(l-1)}$ are normalized magnitude of the Fourier transform at l and $l-1$ frames	$Ft_1 = \sum_{n=1}^{L} \sum_{k=1}^{P} (\rho[n,k] - \rho[n+L,k])^2$ L is a predetermined lag and P is the total of sub-bands
Spectral centroid: average signal frequency weighted by magnitude [3] $F_2 = \frac{\sum_{k=1}^{M} k|Z[k]|}{\sum_{k=1}^{M} |Z[k]|}$	Instantaneous Frequency (IF) *IF is considered here as the T-F extension of spectral centroid* $Ft_2 = \frac{\sum_{k=1}^{M} k\rho[n,k]}{\sum_{k=1}^{M} \rho[n,k]}$
Spectral Roll-Off i.e. spectral concentration below threshold [3] $F_3 = \lambda \sum_{k=1}^{M} |Z[k]|$ λ is chosen to be 0.85 ($\simeq$ frequency under which 85% of the signal power resides)	$Ft_3 = \lambda \sum_{n=1}^{N} \sum_{k=1}^{M} \rho[n,k]$
Spectral flatness indicates whether the distribution is smooth or spiky [3] $F_4 = \left(\prod_{k=1}^{M} Z[k] \right)^{\frac{1}{M}} \left(\sum_{k=1}^{M} Z[k] \right)^{-1}$	Energy localization (i.e. energy of spikes) $Ft_4 = \frac{\left(\prod_{n=1}^{N} \prod_{k=1}^{M} \rho[n,k] \right)^{\frac{1}{NM}}}{\sum_{k=1}^{M} \sum_{n=1}^{N} \rho[n,k]}$
Maximum power of the frequency bands [2, 4] $F_5 = \sum_{k=1}^{\delta} |Z[k]|^2$ $F_6 = \sum_{k=\delta+1}^{M} |Z[k]|^2$	Sub-bands energies $Ft_5 = \sum_{n=1}^{N} \sum_{k=1}^{M_\delta} \rho[n,k]$ $Ft_6 = \sum_{n=1}^{N} \sum_{k=M_\delta+1}^{M} \rho[n,k]$ where $M_\delta = \lfloor M/f_s \rfloor$
Spectral entropy measure the regularity of the power spectrum of the EEG signal [4] $F_7 = \frac{1}{\log(M)} \sum_{k=1}^{M} P(Z[k]) \log P(Z[k])$	Rényi entropy $Ft_7 = \frac{1}{1-\alpha} \log_2 \left(\sum_{n=1}^{N} \sum_{k=1}^{M} \rho^\alpha[n,k] \right)$

where h represents the target classes (e.g. seizure or non-seizure classes) and $I(S;h)$ is the mutual information between the features set $S = \{F_1, F_2, \ldots, F_m\}$ and the target h defined as follows

$$I(S;h) = \int \int p(S;h) \log \frac{p(S;h)}{p(S)p(h)} dS dh \qquad (3)$$

Minimum Redundancy means to select relevant features with less redundancy

$$\min W(S), \quad W = \frac{1}{|S|^2} \sum_{F_i, F_j \in S} I(F_i; F_j) \qquad (4)$$

where S is the set of features and $I(F_i; F_j)$ is the mutual information between features F_i and F_j defined as follows

$$I(F_i; F_j) = \int \int p(F_i; F_j) \log \frac{p(F_i; F_j)}{p(F_i)p(F_j)} dF_i dF_j \qquad (5)$$

where $p(F_i; F_j)$ is the joint probability density function (PDF) of F_i and F_j, and $p(F_i)$ and $p(F_j)$ are the marginal PDFs of F_i and F_j, respectively [13].

This study uses the feature selection method called mRMR (minimum-Redundancy-Maximum-Relevance), proposed recently in [13]. This allows to select the maximum relevant features with a minimum redundancy using a mutual information measure. The combination of these criteria can be made using an additive combination ($\max(V - W)$) which was used in this study, or a multiplicative combination ($\max(V/W)$).

4.2   T-F Features Translation Verification by Calibration

To verify that the translated features are meaningful, we selected a middle way between the T-F domain and the time or frequency domains to calibrate the new T-F features with the time or frequency domain only features. This is done by calculating the power of the analytic signal associated with the EEG signal in time or frequency. The TFD time marginal (TM) and frequency marginal (FM) are estimated by taking the average of $\rho[n, k]$ over the frequency or time samples.

$$\sum_k \rho[n, k] = |z[n]|^2 \ , \quad \sum_n \rho[n, k] = |Z[k]|^2 \qquad (6)$$

Using these two power estimates, we compare the results of classification using the features defined earlier either in time domain, frequency domain or joint T-F domain in terms of the total accuracy obtained.

5   Performance Evaluation, Results and Discussion

This study uses a database described in [8] which includes 14-channel EEG recordings of nine neonates using mono-polar montages. The recordings were marked by a neurologist for seizures [8]. From the data set, we extracted two sets of seizure and non-seizure EEG epochs referred to as sets **S** and **N** respectively. Each set contains 50 segments. Each segment was band-pass filtered in the range 0.5-10 Hz and down-sampled from 256 to 20 Hz as neonatal EEG seizures have spectral activities mostly below 12 Hz. The length of each selected segment is 12.8 s with 256 samples [7–9].

The translated features $\{\mathbf{Tt}, \mathbf{Ft}\}$ were extracted from the TFD of each EEG segment of length T seconds. Only five TFDs are chosen in this simulation, for practical considerations in terms of relevance: MBD, SPEC, SWVD, GKD and WVD. The parameters of the MBD and GKD were respectively chosen as $\beta = 0.05$ and $\sigma = 0.9$ with lag window length of 127. The selected window $w[n]$

Table 3. Total classification accuracy results for real newborn EEG data using the T-F translated features with multi-class SVM classifier, for each QTFD. *(The T-F translated features set {**Tt**, **Ft**} given in Tables 1 and 2 are used. The results between parentheses are the classification results using both original time only and original frequency only features {**T**, **F**}).*

TFD	Class	Total	Classifier outputs		Statistical parameters (%)		
			N	S	Sensitivity	Specificity	Total accuracy
WVD	N	50	49(47)	1(3)	98(94)	92(90)	95(92)
	S	50	4(5)	46(45)	92(90)	98(94)	
SWVD	N	50	48(47)	2(3)	96(94)	88(90)	92(92)
	S	50	6(5)	44(45)	88(90)	96(94)	
CWD	N	50	48(47)	2(3)	96(94)	92(90)	94(92)
	S	50	4(5)	46(45)	92(90)	96(94)	
MBD	N	50	**49**(47)	**1**(3)	**98**(94)	**94**(90)	**96**(92)
	S	50	**3**(5)	**47**(45)	**94**(90)	**98**(94)	
SPEC	N	50	48(47)	2(3)	96(94)	92(90)	94(92)
	S	50	4(5)	46(45)	92(90)	96(94)	

Table 4. Ranking of the T-F translated features based on the combination of the minimum-redundancy and maximum-relevance criteria. *(The proposed T-F features {**Tt**, **Ft**} are ranked by order of maximum-relevance and minimum-redundancy criterion, for each TFDs. The score values are the entropy score according to the minimum redundancy-maximum-relevance criteria).*

WVD		SWVD		CWD		MBD		SPEC	
Feature	Score	Feature	Score	Feature	Score	Feature	Score	Feature	Score
Tt_6	0.506	Tt_3	0.588	Tt_6	0.506	Tt_6	0.692	Tt_3	0.666
Ft_2	0.158	Ft_2	0.115	Ft_2	0.158	Ft_2	0.012	Ft_2	0.105
Tt_4	0.052	Tt_4	0.030	Tt_4	0.052	Tt_3	0.154	Tt_6	0.148
Ft_7	-0.003	Ft_7	0.009	Ft_7	-0.003	Ft_7	-0.033	Ft_7	-0.007
Ft_4	-0.009	Tt_6	-0.001	Tt_3	-0.024	Tt_2	-0.032	Tt_4	0.057
Tt_3	-0.041	Tt_2	-0.041	Tt_9	-0.041	Ft_4	-0.028	Ft_4	-0.005
Ft_6	-0.050	Ft_4	-0.034	Ft_4	-0.059	Tt_4	-0.049	Ft_1	-0.023
Tt_2	-0.092	Tt_9	-0.066	Ft_1	-0.079	Tt_9	-0.067	Tt_9	-0.063
Ft_3	-0.103	Ft_3	-0.095	Ft_3	-0.087	Ft_3	-0.094	Tt_8	-0.087
Ft_1	-0.143	Ft_1	-0.118	Tt_2	-0.129	Ft_1	-0.118	Tt_2	-0.105
Ft_5	-0.158	Ft_6	-0.133	Ft_5	-0.141	Tt_5	-0.128	Ft_3	-0.113
Tt_9	-0.171	Tt_8	-0.154	Ft_6	-0.157	Ft_6	-0.146	Ft_6	-0.135
Tt_5	-0.185	Ft_5	-0.184	Tt_7	-0.178	Tt_8	-0.165	Ft_5	-0.162
Tt_8	-0.206	Tt_5	-0.199	Tt_8	-0.200	Tt_1	-0.192	Tt_5	-0.183
Tt_7	-0.227	Tt_1	-0.218	Tt_5	-0.220	Tt_7	-0.216	Tt_7	-0.207
Tt_1	-0.234	Tt_7	-0.241	Tt_1	-0.230	Ft_5	-0.228	Tt_1	-0.218

for the SWVD and SPEC distributions is a Hanning window of length $\lfloor N/4 \rfloor$ samples. The simulations were carried out in Matlab. For performance evaluation, a multi-class SVM classifier was trained using the T-F features extracted from the newborn EEG signals in the database. The newborn EEG database was split in two parts; 60% of the data were used for training and 40% for testing the classifier.

Table 5. Total accuracy classification obtained using MBD with relevant selected features based on minimum-redundancy-maximum-relevant selection and the signal- & image-related features proposed in [8]. *(The combination of the selected features with the T-F features based on signal and image processing techniques improves total accuracy classification. Classification results were obtained using MBD).*

TFD	T-F features	Total accuracy
MBD	$\{Tt_6, Tt_2\}$	96%
	$\{$signal & image-related features$\}$	96%
	$\{$signal & image-related features$\}+\{Tt_6, Tt_2\}$	97%

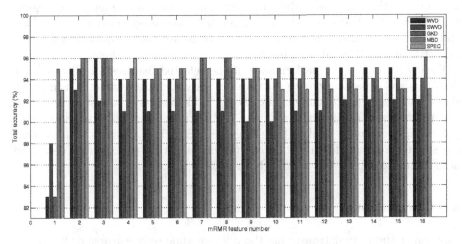

Fig. 2. Accuracy of selected features in classification. *(The total classification accuracy results obtained on the real newborn EEG data using the selected T-F features set according to the minimum-redundancy-maximum-relevant selection. The feature numbers corresponds to the top-ranked features given in Table 4).*

Table 3 shows the classification total accuracy results using the translated features extracted from different TFDs of newborn EEG segments. The results between parentheses are the classification results using the original time only and frequency only features $\{T, F\}$. These results are obtained using the multi-class SVM classifier. Each row shows, for a particular TFD, the total number of EEG segments correctly classified as well as those misclassified as other classes, and also the statistical parameters: sensitivity, specificity and total classification accuracy. Table 3 shows that the use of the translated features improve significatively the classification results compared to the use of both original time and frequency features $\{T, F\}$ by up to 4% for 100 segments. This is confirmed by the *total classification accuracy* calculated for each TFD. The best total classification accuracy is obtained using the MBD; and is 96% for 100 segments. This can be improved by increasing the number of EEG segments in the training-step.

The results shown in Table 3 confirm that including T-F features improves the classification performance.

To assess the redundancy and the relevance of the proposed T-F features in the EEG seizure classification system, the proposed T-F features were ranked according to the criteria defined in Section 4. Table 4 shows the ranking of the translated features based on minimum-redundancy and maximum-relevance criteria. In particular, we observe that the feature Tt_6 corresponding to the translation of *coefficient of variations* time feature is the most relevant feature of the MBD as confirmed by the entropy score (first row).

We have calculated the total accuracy classification using the n top-ranked features where $n = 1, \ldots, 16$, for each TFD. Figure 2 shows the results obtained according to the minimum-redundancy-maximum-relevance criteria. We observe that the best total accuracy classification is 96%; and is obtained with the MBD and SPEC using only 2 features. These features are the 2 top-ranked features for MBD or SPEC listed in Table 4.

Finally, to assess the performance of the selected features for EEG seizure detection and classification, we use two T-F feature classes extracted using both signal and image processing techniques [8]. Table 5 shows the results using the selected relevant T-F features with the T-F features based on signal and image processing techniques used in [8], using the MBD. These results indicate that the selected relevant features improve the classification performance described in [8] by up to 1% for 100 real newborn EEG segments.

6   Conclusion

This paper shows that translating the relevant time only features or frequency only features into the joint T-F domain allows to define new T-F features with better performance in EEG classification. The selection of a minimum set of relevant translated T-F features according to a combined minimal-redundancy and maximal-relevance criterion can improve significantly the performance of the newborn seizure classification system, including reducing computation cost. The improvement obtained with the T-F features may be explained by the use of the non-stationarity characteristics of the signals such as the IF (feature Ft_2). Finally, the proposed T-F features can be applied to detect other newborn EEG abnormalities. The extraction of these features can then be improved using specially defined TFDs such as [14].

Acknowledgments. The authors thanks Dr. T. Ben-Jabeur and Dr. N. Khan for valuable feedback, and Ms. Y. Bahnasy and Ms. N. Saad for contributions to the calibration work. This publication was made possible by a grant from the Qatar National Research Fund under its National Priorities Research Program award number NPRP 09-465-2-174.

References

1. Boashash, B.: Time-Frequency Signal Analysis and Processing: A Comprehensive Reference. Elsevier, Oxford (2003)
2. Aarabi, A., Wallois, F., Grebe, R.: Automated neonatal seizure detection: A multistage classification system through feature selection based on relevance and redundancy analysis. Journal of Clinical Neurophysiology 147(2), 328–340 (2006)
3. Löfhede, J., et al.: Automatic classification of background EEG activity in healthy and sick neonates. Journal of Neural Engineering 7(1), 016007 (2010)
4. Greene, B.R., et al.: A comparison of quantitative EEG features for neonatal seizure detection. Journal of Clinical Neurophysiology 119(6), 1248–1261 (2008)
5. Mitra, J., Glover, J.R., Ktonas, P.Y., et al.: A multistage system for the automated detection of Epileptic seizures in neonatale Electroencephalography. Journal of Clinical Neurophysiology 26(4), 218–226 (2006)
6. Subasi, A.: EEG signal classification using wavelet feature extraction and a mixture of expert model. Expert Systems with Applications 32(4), 1084–1093 (2007)
7. Boashash, B., Boubchir, L., Azemi, G.: Time-frequency signal and image processing of non-stationary signals with application to the classification of newborn EEG abnormalities. In: 11th IEEE International Symposium on Signal Processing and Information Technology, Bilbao, Spain, December 14–17, vol. 1, pp. 120–129 (2011)
8. Boashash, B., Boubchir, L., Azemi, G.: A methodology for time-frequency image processing applied to the classification of non-stationary multichannel signals using instantaneous frequency descriptors with application to newborn EEG signals. EURASIP Journal on Advances in Signal Processing, 117 (2012), doi:10.1186/1687-6180-2012-117
9. Boashash, B., Boubchir, L., Azemi, G.: Improving the classification of newborn EEG time-frequency representations using a combined time-frequency signal and image approach. In: 11th International Conference on Information Sciences, Signal Processing and their Applications, Montreal, Quebec, Canada, July 3-5, pp. 316–321 (2012)
10. Bahnasy, Y., Saad, N., Boubchir, L., Boashash, B.: Calibration of time features and frequency features in the time-frequency domain for improved detection and classification of seizure in newborn EEG signals. In: 11th International Conference on Information Sciences, Signal Processing and their Applications, Montreal, Quebec, Canada, July 3-5, pp. 1483–1484 (2012)
11. Boashash, B.: Estimation and interpreting the instantaneous frequency of a signal - I. Fundamentals. Proc. IEEE 80(4), 520–538 (1992)
12. Boashash, B.: Estimation and interpreting the instantaneous frequency of a signal - II. Algorithms and applications. Proc. IEEE 80(4), 540–568 (1992)
13. Peng, H., Long, F., Ding, C.: Feature selection based on mutual information: criteria of max-dependency, max-relevance, and min-redundancy. IEEE Transactions on Pattern Analysis and Machine Intelligence 27(8), 1226–1238 (2005)
14. Abed, M., Belouchrani, A., Cheriet, M., Boashash, B.: Time-frequency distributions based on compact support kernels: properties and performance evaluation. IEEE Transactions on Signal Processing 60(6), 2814–2827 (2012)

Turf Grass Irrigation Using Neuro-Fuzzy System

Azlinah Mohamed, Nur Fharah Anuar, Sofianita Mutalib,
Marina Yusoff, and Shuzlina Abdul-Rahman

Faculty of Computer & Mathematical Sciences, Universiti Teknologi MARA,
40450 Shah Alam, Selangor, Malaysia
{azlinah,sofi,marinay,shuzlina}@tmsk.uitm.edu.my

Abstract. Turf grass needs water to survive and stay green, but too much water
can really damage it. Turf grass irrigation processes can lead to excess water
consumption. The irrigation process is based on several factors, which are evapotranspiration rate, grass evapotranspiration rate and tensiometer reading.
This study proposes an irrigation process system using Neuro-Fuzzy method
that was experimented on real meteorology data. Both the backpropagation and
resilient backpropagation were explored and compared. The system with the
resilient backpropagation method has achieved higher accuracy rate compared
to the backpropagation method with an average of 10% of reduction in water
usage.

Keywords: Neuro-Fuzzy, Resilient Backpropagation, Turf Grass, Water Irrigation System.

1   Introduction

Turf grass has been widely used in golf courses all around the world and sometimes
as house lawn, usually refers to lawn grass. Maintaining turf grass health and beauty
of its greens have been one of the top considerations for every lawn owner. One of
the important elements for turf grass health and beauty is the sufficiency of water
however, water supply is gradually scarce. Therefore, it is important to use the resources wisely and this can be solved by irrigating periodically based on several de-
terminant factors. The process of watering the turf grass is call irrigation process.
According to Normas [1], irrigation is simply the act of watering the lawn, plants,
flower or garden. There are many ways to determine when the irrigation process is
needed, for example to irrigate at the first sign of moisture stress. When turf grass is
under moisture stress, it becomes dull and bluish-green, the leaf blades fold or roll
and footprints remain after you walk over the area. If dry conditions persist, the grass
wilts [1] and usually irrigation will be exercised on that portion of the lawn that first
exhibits these signs. Timely irrigation is required for effective and efficient water
consumption. Automated irrigation systems are a convenient and cost-effective solution. This is because, once a watering schedule is set, the landscape is watered based
on the weather accordingly. It is a consistent watering schedule resulted into healthier
plant and greener. However, the weather in many parts of the country changes and
constant change of schedule is needed.

T. Huang et al. (Eds.): ICONIP 2012, Part IV, LNCS 7666, pp. 644–651, 2012.
© Springer-Verlag Berlin Heidelberg 2012

The highly managed turf grass such as in golf courses requires managers to maintain green, lush turf grass regardless of weather or other environmental conditions, which sometimes results in frequent over watering or inefficient use of irrigation water. Often, the decision to irrigate is based on incomplete information.   Several factors such as evapotranspiration rates and water level in the root zone determine the health of turf grass. Evapotranspiration (ET) is a combination of transpiration and evaporation process that transport the water into the atmosphere from surfaces [2]. Since water scarcity is increasingly a problem, there is a need for an effort to irrigate effectively, because effective irrigation is to water precisely and consistently [3]. This requires an irrigation system that can make accurate decisions based on uncertain or approximate inputs. The key contributions of our work are the proposal of a Neuro-Fuzzy system on irrigation problem applied on real meteorology data. This work is an extension of previous work [7][8] in which Fuzzy method was applied with Sugeno and Mamdani approaches. The approach is further evaluated on fuzzy system and several neural networks (NNs) algorithms. The remainder of this paper is organised as follows. Section 2 discusses the related works and section 3 presents the methodology. Next, section 4 presents the results and discussion. In section 5, we conclude the paper.

2   Related Works

The total amount of irrigation water depends on several factors such as grass species, soil type, ET process, grass ET process, weather condition such as rainy or dry season, fertilizer process and soil moisture. Generally, all turf grass planted in Malaysia is from the warm-season turf grasses, which are Bermudagrass, Cowgrass, Zoysiagrass and Paspalumgrass [1]. There are three main water levels in the soil: saturation, field capacity and wilting point. The saturation level is referred to as the soil moisture and completely filled with water. Field capacity occurs when the excess water from the saturated condition of the soil has been drained due to the gravitation pull. The permanent wilting point is the level where the grass is unable to extract water for its need and the plant might wilt and die. Between the level of field capacity and the permanent wilting point is the available soil moisture. The available soil moisture can be divided into two categories: the available water with no stress and the available water with stress [4]. The available water with no stress area is where the maximum soil water deficit is been applied. This maximum soil water deficit is the amount of water stored in the plant's root zone that is readily available to the plant. It is the amount of water, which allowed to deficit before the plant reaches its wilting point. The available water with stress area is where the wilting point starts. This is happened when the soil moisture content is lower than the maximum soil water deficit [5]. If the moisture level is lower than the maximum soil water deficit, the plant will start to wilt. If the moisture level reaches the permanent wilting point, the plant will die.

The study by Kazuhiro et al. [6] developed a Fuzzy Expert System for the melon cultivation in greenhouse. In their study, the fuzzy control system was developed for the on-off control irrigation system. The fuzzy control system is programmed to take the soil moisture content from the various climate sensors. The aims were to save water resources and preserve the melon quality [6]. The other example is the use of

Decision Support System to optimize the use of irrigation water fertilizer in farming [8]. The intelligent methods that applied in the optimization process are the combination of backpropagation multilayer perceptron, radial basis function neural networks (NNs), ANFSI NN and fuzzy logic. Meanwhile, the study by [7] was restricted to the Bermuda Turf grass characteristic, the loam soil characteristic, 92.9m^2 lawn sizes, pop-up spray head sprinkler, ET, the climatology factor and the soil moisture reading from tensiometer. The results showed that the fuzzy expert irrigation system performed better based on the lower annual average water usage for the whole year recorded. These literatures suggest the use of Neuro-Fuzzy that could benefit our study.

2.1   Evapotranspiration

Water can lose through the process of transpiration and evaporation. Transpiration is referred to the transfer of water into the atmosphere. While evaporation is a process of the return of water back to the atmosphere through direct evaporative loss from the soil surface, standing water, and water on surfaces such as leaves or roofs [2]. Daily ET rate can be calculated by using the Hargreaves-Samani equation [10]. This equation used minimum climatologically data, which are the maximum and minimum value of temperature data, evaporation data and solar radiance data [11]. Grass evaporation or crop evaporation refers to the loss of water to the atmosphere from the grass surface. Different grasses with different characteristic will result in different ET level. The grass evaporation process also takes into consideration of climatologically aspect such as wind, temperature humidity, soil characteristic and water availability. Grass ET can be calculated using simple equation. The grass coefficient rate is required prior to obtaining the grass ET. In general, the grass coefficient for the warm turf grass season is 0.6 [12]. Table 1 shows the percentage of moisture deficiency in the soil to determine the availability of water for the three different soil textures. Zero percentage of moisture deficiency shows that the soil is in the field capacity structure and the water is filling the soil. Meanwhile 75% to 100% of moisture deficiency shows that, the soil is in the completely dry structure and no water at all in the soil.

Table 1. Soil Moisture Interpretation Chart [20]

Soil moisture deficiency	Moderately coarse texture	Medium texture	Fine and very fine texture
0% (field capacity)	Upon squeezing, no free water appears on soil but wet outline of ball is left on hand.		
0-25%	Forms weak ball, breaks easily when bounced in hand.*	Forms ball, very pliable, slicks readily.*	Easily ribbons out between thumb and forefinger.*
25-50%	Will form ball, but falls apart when bounced in hand.*	Forms ball, slicks under pressure.*	Forms ball, will ribbon out between thumb and forefinger.*
50-75%	Appears dry, will not form ball with pressure.*	Crumbly, holds together from pressure.*	Somewhat pliable, will ball under pressure.*
75-100%	Dry, loose, flows through fingers.	Powdery, crumbles easily.	Hard, difficult to break into powder.

Squeeze a handful of soil firmly to make ball test.

2.2   Neuro-Fuzzy System

Neuro-Fuzzy system is a control system that uses fuzzy logic and NNs. The system is widely used by other industries and may be well suited for controlling irrigation in turf grass. Such systems can make accurate decisions based on uncertain or approximate inputs [3][13]. Neuro-Fuzzy can provide approximation to real world problem and the approaches are known to be robust alternatives to conventional deterministic and programmed models. In Fuzzy method, human experts are needed to provide the knowledge while in NNs the knowledge is learned through data via the training process. In short, a fuzzy system does not have any learning ability while, NNs does not have the ability to represent the knowledge. Usually, the common Mamdani-style fuzzy rule is used and it is in the form If-Then Else Rules [9][16]. Therefore, in this study, we employed Mamdani method which is performed in four steps: fuzzification of the input variable, rule evaluation, aggregation, and defuzzification to obtain the crisp value. One of the NNs methods is resilient backprogation. It differs from standard backpropagation because the resilient backpropagation performs supervised batch learning. Resilient backpropagation method can eliminate the harmful influence of the size of the partial derivative on the weight step [14]. It uses the update-value method, where a weight-specific apply to determine the size of the weight change [15]. By combining fuzzy and resilient backpropagation systems, each of their advantages can be achieved in one powerful system.

3   Methodology

The climatological data are collected for two consecutive years: 2007-2008 from the Department of Meteorology Malaysia. Three main factors are contributed to the irrigation process: ET, grass ET and tensiometer reading. In calculating these input factors, data such as solar radiation, temperature, rainfall and evaporation were collected. The algorithm was implemented using JAVA. The logical design is divided into two phases, which are the Fuzzy logic phase, and NNs phase. Table 2 shows the characteristic of four different turf grasses collected from the expert in one of the golf clubs in Malaysia [17]. All information in this table is used in calculating the irrigation process in the system.

Table 2. Turf grass Association

Turf grass	Root Depth (ft)	Coefficient Rate (Kc)	Available water in sandy soil (5.2 in/ft)	50% of MAD (water that can be depleted)
Bermuda	4	0.6	20.8 inches	10.4 inches
Zoysia	2	0.6	10.4 inches	5.2 inches
Paspulum	3	0.6	15.6 inches	7.8 inches
Cow	1.5	0.6	7.8 inches	3.9 inches

There are three formulas involved in calculating the irrigation process. The first formula used to obtain the data of ET rate, which is calculated based on the temperature rate and solar radiance rate using the Hargreaves-Samani equation [18]. The ET rate equation is written as Equation (1):

$$ETo = 0.0023 \, (Tc+17.8) \, Ra \, (Td) \, \frac{1}{2} \qquad (1)$$

- Tc is the monthly mean temperature (degrees centigrade)
- Ra is the extraterrestrial solar radiation expressed in mm/month
- Td is the difference between the mean minimum and mean maximum temperatures (oC).

The second formula used is to get the grass evaporation data where the calculation concerned of ET rate and grass coefficient rate. The grass coefficient rate used in the system is 0.6 which is the coefficient rate used for all warm turf grasses [18]. The grass ET equation is written as:

$$ETc = Kc * ETo \qquad (2)$$

- Kc is the grass coefficient factor
- ETo is the ET rate

Finally the third formula is used in order to get data for tensiometer reading in centibars. The reproduction of tensiometer data was based on the total amount of soil moisture loss from the saturated level until the wilting point in the soil. The total amount of water in 1 ft depth of soil for loam soil is 5.2 in/ft [4]. Assuming the root of the turf grass reaches the maximum growth, 4 ft deep. Thus, we can assume that the amount of water in the soil is 20.8in/ft (5.2in/ft * 4). From the initial amount of water, we deduce the ET rate. Then, the deduced amount is calculated again by the ET rate of the next day. The process is repeated until the end of one-year data. Besides that, the amount of external moisture is also considered. This external moisture was provided by rainfall. The expert and the tutorial for managing the home garden from RainBird's website [19] aided the creation of the data. Next, the tensiometer reproduction data were created from the moisture reproduction data by calculating the percentage of water loss. The reproduction moisture was taken from the generated reproduction moisture. The tensiometer reproduction formula is as follows:

$$100 - ((reproduction \; moisture/initial \; reproduction \; moisture)*100) \qquad (3)$$

Dataset of year 2007 is used to train the system. In training stage, the precise weight in NNs for the system was obtained. We applied the weight in the testing stage and used dataset of 2008. The system was trained on a regular basis to make sure the system is stable to provide unvarying result each time the dataset is trained.

4   Result and Findings

The findings are discussed based on several parameters and the result comparison between the Neuro-Fuzzy with Backpropagation method and Neuro-Fuzzy with

Resilient Backpropagation method. Parameters involved in determining the amount of water used in irrigation process are varied in the length of the turf root, tensiometer reading or soil moisture, available water storage, ET rate to grass coefficient rate or grass ET rate. Some experts mentioned that all parameters are required to determine the exact time to irrigate the turf grass and others might use only the ET and grass coefficient rate. It is always depending on the expert's opinion, knowledge and experiences. In most cases the irrigation process is determined using 'wait for it' method that is Footprinting. The method is to see how fast the turf grass will back to its original state after it has been stepped on. The faster means that plant cells have not been losing their water content yet. By using this method, the parameter used to determine the amount of water is completely ignored. On the other hands, they also used timetabling manner in performing the irrigation process by using the sprinkler.

Both Fuzzy and Neuro-Fuzzy methods in this system gave the amount of water used to perform the irrigation process on specific day. The water amount gave by the Fuzzy is the amount of water that have depleted. While the amount of water gave by the Neuro-Fuzzy is the amount of water that have depleted that considers the turf grass ability to stay alive without 100 percent of its available water. The result gave by Fuzzy for both NNs used in the system are not the same value. This is due to the system architecture where the amount of water provided by both NN methods was calculated back in Fuzzy separately between the two Neuro-Fuzzy systems. The results showed that Fuzzy methods require more amount of water usage compared to Neuro-Fuzzy System. We also developed independent Fuzzy system for each NN method, so that we could compare result produced by NN methods to fuzzy method. For Resilient backpropagation, the Neuro-Fuzzy requires only $756169.3l$ compared to $810332.2l$ with Fuzzy method. A lot of water could be saved during one-year duration in a wide area of lawn garden and it is better from backpropagation method. For both NN methods, the Neuro-Fuzzy could save the water less than 10% (in the range 6.7%-7.3%). The result of the calculation is shown in Table 3. The turf grass used in this study is Bermudagrass which is located at Glenmarie Golf Courses. An analysis was also performed on the NNs accuracy for both methods as shown in Table 4. On average, the accuracy rate for backpropagation is 93.8% while for resilient backpropagation is 95.5%. This shows the superiority of the latter method.

Table 3. Comparison of Required Water for Irrigation Between Neuro-Fuzzy with BP and Neuro-Fuzzy with RBP for Glenmarie Golf Courses

NNs Methods	Turf grasses	Fuzzy (l)	Neuro-Fuzzy (l)	Water Saved (l)
Backpropagation (BP)	Bermudagrass	796193.6	737880.6	58313.0 (7.3%)
Resilient BP	Bermudagrass	810332.2	756169.3	54162.9 (6.7%)

Table 4. Comparison Between Backpropagation and Resilient Backpropagation

NNs Method	MSE	Epoch	Accuracy Rate (%)
Backpropagation	0.001	127	89
	0.0001	249	92
	0.00001	789	96
	0.000001	3270	98
Average			93.8
Resilient Backpropagation	0.001	99	92
	0.0001	192	94
	0.00001	562	97
	0.000001	2599	99
Average			95.5

5   Conclusion

Neuro-Fuzzy method is one of the possible solutions that can be used in the irrigation process besides Fuzzy method. Both the resilient backpropagation and backpropagation NN methods are suitable for the irrigation process. However, our results showed that resilient backpropagation is better than the backpropagation in term of reducing the occurrence of stress in turf grass while backpropagation is slightly better in term of reducing the utilization of water in one year. Based on the result of the system, it is concluded that the system achieved higher accuracy rate for the resilient backpropagation compared to the standard backpropagation. The study also showed that the use of Neuro-Fuzzy system required 10% less water amount compared to Fuzzy.

Acknowledgements. This work has been made possible with the support of the, Universiti Teknologi MARA, Shah Alam, Selangor, Malaysia.

References

1. Normas, Y.: Malaysian Turf Reference Management,
   http://www.mynormas.com (retrieved November 15, 2008)
2. Burba, G., Pidwirny, M.: The Encyclopedia of Earth (2007)
3. Bremer, D.: Soil-Moisture Sensors Can Help Regulate Irrigation. Turf grass Trends (2003)
4. Ball, J.: Soil and Water Relationship. The Samuel Robert Nobel Foundation (2001)
5. British Columbia Ministry of Agriculture, Food and Fisheries: Soil Water Storage Capacity and Available Soil Moisture. Water Conservation FactSheet. 619.000-1 (2002)
6. Kazuhiro, N., Takako, A., Yoshitaka, M.: A Study on Development of Intelligent Irrigation Systems for Melon Cultivation in Greenhouse. Laboratory of Agricultural System Engineering, Faculty of Agriculture, Niigata University, Ikarashi, Japan (nd)
7. Shuzlina, A.R., Izham, F.A.J., Ku Shairah, J., Azlinah, M.: Intelligent Water Dispersal Controller Using Mamdani Approach. In: Proceedings of the 8th WSEAS 2007, International Conference on Fuzzy Systems, Vancouver, Canada, June 18-20 (2007)
8. Mutalib, S., Rahman, S.A., Yusoff, M., Mohamed, A.: Fuzzy Water Dispersal Controller Using Sugeno Approach. In: Gervasi, O., Gavrilova, M.L. (eds.) ICCSA 2007, Part I. LNCS, vol. 4705, pp. 576–588. Springer, Heidelberg (2007)

9. Ajith, A.: Adaptation of Fuzzy Inference System Using Neural Learning. In: Nedjah, N., et al. (eds.) Fuzzy System Engineering: Theory and Practice. STUDFUZZ, ch. 3, pp. 53–83. Springer, Germany (2005)

10. Samani, Z.: Estimating Solar Radiation and Evapotranspiration Using Minimum Climatological Data (nd)

11. Orang, M.N., Snyder, R.L., Matyac, J.S.: Consumptive Use Program Model. The California Water Plan (nd)

12. Richie, W.E., Green, R.L., Gibeault, V.A.: California Turf grass Culture: Using ETo (Reference Evapotranspiration) for Turf grass Irrigation Efficiency, University of California, vol. 47, pp. 3–4 (1997)

13. Kasabov, N.: Evolving Neuro-Fuzzy Inference System. Evolving Connectionist System (2002)

14. Riedmiller, M.: Rprop-Description and Implementation Details. University of Karlsruhe Technical Report (1994)

15. Lambros, S., Panos, I., Spiridon, L.: Coding potential prediction in Wolbachia using artificial NNs. In: Silico Biology 7, Bioinformation Systems e.V. 0010 (2007)

16. Aznarte, J.L., Benitez Sanchez, J.M.: Forecasting Airbone Pollen Concentration Time Series with Neural and Neuro-Fuzzy Models. Department of Computer Science and Artificial Intelligence, Department of Botany. University of Granada (2005)

17. Malaysian Meteorology Department (2009) (retrieved March 4, 2009 from Meteorology Department)

18. Christian, N.: Fundamental of Turfgrass Management, Iowa State University. John Wiley & Sons Inc., USA (2004)

19. RainBird: Water Saving Tips (retrieved January 3, 2009),
http://www.rainbird.com

20. Miles, D.L., Broner, I.: Estimating Soil Moisture. Colorado State University Extension, Agriculture 4, 700 (1998)

A Target-Reaching Controller for Mobile Robots Using Spiking Neural Networks

Xiuqing Wang[1], Zeng-Guang Hou[2], Feng Lv[1],
Min Tan[2], and Yongji Wang[3]

[1] Vocational & Technical Institute, Hebei Normal University,
Shijiazhuang 050031, Hebei, China
xiuqingwang2004@yahoo.com.cn, lvfeng@mail.tsinghua.edu.cn
[2] State Key Laboratory of Management and Control for Complex Systems,
Institute of Automation, The Chinese Academy of Sciences, Beijing 100090, China
{zengguang.hou,min.tan}@ia.ac.cn
[3] State Key Laboratory of Computer Science, Institute of Software,
The Chinese Academy of Sciences, Beijing 100190, China
ywang@itechs.iscas.ac.cn

Abstract. Autonomous navigation plays important role in mobile robots. In this paper, a navigation controller based on spiking neural networks (SNNs) for mobile robots is presented. The proposed target-reaching navigation controller, in which the reactive architecture is used, is composed of three sub-controllers: the obstacle-avoidance controller and the wall-following controller using spiking neural networks (SNNs), and the goal-reaching controller. The experimental results show that the navigation controller can control the mobile robot to reach the target successfully while avoiding the obstacle and following the wall to get rid of the deadlock caused by local minimum. The proposed navigation controller does not require accurate mathematical models of the environment, and is suitable to unknown and unstructured environment.

Keywords: Mobile robot, spiking neural networks, navigation controller, target reaching, obstacle-avoidance, wall-following.

1 Introduction

Navigation of mobile robots refers to planning a path leading to a specified goal, following the planned path, and avoiding obstacles based on sensor readings and deduction in unknown, uncertain and unstructured environment. The autonomous navigation plays important role in mobile robots for fulfillment of given tasks.

Traditionally, the navigation control of mobile robots requires accurate mathematical models of environment, and is effective only in structured environment. Besides, the traditional navigation controller can only fulfill some simple and repetitive tasks, and the robots are usually controlled to follow planned paths. It is very difficult to build the mathematical models of unknown and unstructured environment. Since neural networks are useful tools for modeling and control of nonlinear systems, some neural network based controllers for mobile robots have been developed successfully [1]-[5].

T. Huang et al. (Eds.): ICONIP 2012, Part IV, LNCS 7666, pp. 652–659, 2012.

Till now artificial neural networks (ANNs) have developed from the first generation to the third generation. The first generation of neural networks consists of McCulloch-Pitts threshold neurons, and the second generation neurons use continuous activation functions to compute their output signals. Spiking neural networks (SNNs) are considered as the third generation [8].

Spiking neural networks have many advantages over the other two generations of neural networks: SNNs represent more plausible models of real biological neurons than those classical ones. SNNs, which convey information by individual spike times, have stronger computational power than other types of neural networks (NNs)[9]. Besides the computation capability, SNNs also show their capabilities in pattern recognition and classification [10]-[16]. Compared with the classical neural networks (NNs), spikes are conveyed in SNNs. Thus, SNNs have better robustness to noise, and this also leads to the possibility of fast and efficient implementations. Moreover, spikes can be modeled relatively easily by digital circuits.

Because SNNs use temporal and spatial information, they can be used for "real" dynamic environments. Since mobile robots often work in the unstructured and dynamic environments, SNNs are more suitable for robots' controller design than the traditional ANNs.

SNNs have been employed in the robotic area successfully, such as path planning [21], environment perception [15], [16], and behavior controller [6], [7], [17]-[22].

In this paper, a behavior based target-reaching controller using spiking neural networks is designed. The navigation controller is based on the previous designed obstacle-avoidance controller [6] and the wall-following controller [7].

The paper is organized as follows: Section 2 gives the kinematic model and the sensory system of the mobile robot CASIA-I. Section 3 discusses the proposed navigation controller based on spiking neural networks. Section 4 presents the simulation results. The paper concludes in Section 5.

2   Kinematic Model and Sensory System of the Mobile Robot

2.1   The Kinematic Model

In this study, the mobile robot as shown in Fig. 1 is a system satisfying the nonholonomic constraints. In 2-dimensional Cartesian space, the pose of the mobile robot is represented by

$$q = (x, y, \alpha)^T \tag{1}$$

where $(x, y)^T$ is the position of the robot in the reference coordinate system XOY, and the heading direction α is taken counterclockwise from the positive direction of X-axis. $X_r O_r Y_r$ is the coordinate for the robot system. (x_t, y_t) is the coordination of the target for the mobile robot. ϕ is taken counterclockwise from the positive direction of X_r-axis to xx_t. More details can be found in [6].

2.2   The Mobile Robot's Sonar Sensory System

Ultrasonic sensor has been widely used in mobile robots due to cheapness and reliability. The mobile robot CASIA-I in our experiment has a peripheral ring of sixteen evenly distributed Polaroid ultrasonic sensors which are denoted by S1 to S16 [6].

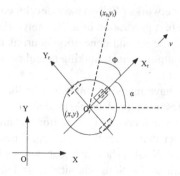

Fig. 1. Pose of the mobile robot

3 Navigation Controller Based on Spiking Neural Networks

There are many different kinds of behavior-based controllers. According to the various given tasks, the entire task module can be divided into goal task module and the sub-goal task module. By various classification standards, the behaviors of the mobile robots can not only be classified into the planned behaviors and the reactive behaviors, but also be classified into combined behaviors and basic behaviors. In this paper, the combined behavior or the target-reaching behavior has been divided into several sub-goal behaviors, which include obstacle-avoidance behavior, the wall-following behavior, and the goal-reaching behavior. In this paper, target-reaching navigation controller based on SNNs is designed and the block diagram of the controller is shown in Fig. 2. The proposed navigation controller is composed of three modules: obstacle-avoidance module, wall-following module and goal-reaching module. The different control modules can transit from one to another under appropriate transition conditions.

The reactive architecture, which was first proposed by Arkin [23], is used in the proposed mobile robot navigation controller. In such behavior-based architecture, environmental model is not necessary any more. The mobile robots' activities can be divided into series of basic behaviors. In the reactive architecture, the relationship between the sensors and the executors is built directly, which is the mapping between the

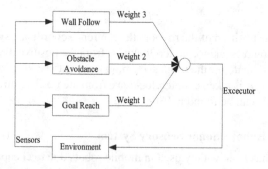

Fig. 2. Block diagram for the mobile robots' navigation controller

patterns of the sensory information and that of the mobile robots' activities. By the reactive architecture, the robots' capability to respond to environment is enhanced greatly. In the proposed navigation controller, the weight of the wall-following module is the largest, and that of the goal-reaching module is the smallest. The weight of the obstacle-avoidance module is less than that of the wall-following module and larger than that of the goal-reaching module.

3.1 Control Strategy for the Goal-Reaching Module

The goal-reaching controller module includes such sub-modules: position/ orientation adjustment module and the moving forward directly to the goal module. The different modules can transit from one module to another under appropriate transition conditions. Angle ϕ in Fig. 1 which denotes the orientation of the robot is calculated by

$$\phi = \begin{cases} \arctan(\frac{y_t-y}{x_t-x}) & x_t - x > 0 \\ \pi + \arctan(\frac{y_t-y}{x_t-x}) & x_t - x < 0, y_t - y > 0 \\ -\pi + \arctan(\frac{y_t-y}{x_t-x}) & x_t - x < 0, y_t - y < 0 \\ \pi/2 & x_t - x = 0, y_t - y > 0 \\ -\pi/2 & x_t - x = 0, y_t - y < 0. \end{cases} \quad (2)$$

When ϕ is larger than a given threshold, the position/ orientation adjustment module will work. In the position/ orientation adjustment module, the robots' position/ orientation should be adjusted as follows: control the robot's left and right wheels to turn at the same speed but in the opposite directions. By such way, the robot will rotate by ϕ. When ϕ is less than a given threshold, position/ orientation adjustment module is turned to the moving forward directly to the goal module; that is, the robot will move forward directly to the goal.

3.2 Module Transition Conditions for the Navigation Controller

Suppose that the reading of the ith sonar sensor is $S_r(i)$ (The sonar sensor model can be referred in [6]). When $S_r(i) > d_{far}$ (d_{far} is the threshold), it is supposed that there is no obstacle in the direction of sensor $S_r(i)$. The obstacles in different directions affect the robot differently, thus the threshold values are different for the sensors in different directions. The threshold value for the front obstacle of the sensor is $d_{thr_{obstacle}}$, and the threshold value for the obstacles at two sides is d_{far}. The distance between the center of the robot and the target is denoted by d_{obj}.

Wall-Following Module. Wall-following behavior is a commonly used control strategy in mobile robot control. Wall-following behavior can help the robot get rid of the deadlocking caused by the local minimum. Besides, the robot can navigate smoothly in the structured corridor scene by the wall-following strategy.

The transition condition for starting the wall-following module is A: When the robot moves forwards and the front sensors $S_r(i)$ ($i = 4, 5, 6$) satisfy the following inequality

$$S_r(i) < d_{thr_{obstacle}}, \quad (3)$$

the wall-following module will be activated.

The transition condition for exiting the wall-following module is B:

$$(\phi \in [-\frac{\pi}{2}, \frac{\pi}{2}]) \cap_{i=4,5,6} (S_r(i) > d_{far}).$$ (4)

Obstacle Avoidance Module. The transition condition for starting the obstacle avoidance module is C:

$$\cap (S_r(i) < d_{far})(i = 1, 2, 3, 7, 8, 9, 10, 16) \cap \overline{A}$$ (5)

Goal Reaching Module. The transition condition for starting the goal-reaching module is D:

$$((sr(i) > d_{far}, i = 1 - 10, 16) \cap \overline{A} \cap \overline{C}) \cup ((sr(i), i = 1 - 9) > d_{obj})$$ (6)

The performance of the controller depends on the sub-module of the navigation controller and also the transition conditions.

3.3   Principle of Obstacle-Avoidance Module

In the obstacle-avoidance controller based on SNN, the readings of the sonar sensors from S1 to S9 are employed. The block diagram of the SNN obstacle-avoidance controller is shown in Fig.3. The sensors are divided into three groups: Group 1, including sensors from S1 to S3; Group 2, from S4 to S6; Group 3, from S7 to S9.

$$S(m) = \min(S_r(i))$$ (7)

where $m = 1, 2, 3$ and i is the number of the sensors in the divided groups. In each group, the smallest reading $S(m)$ will be the testing result. Then $S(m)$ is encoded by spiking frequency coding. The information of the sonar sensors are used as the input to the corresponding sensory neuron. The approximate neuron is used to judge whether the opposite obstacle is so close that the robot need to turn around or not. The turning neurons decide which direction the mobile robot would turn to when the opposite

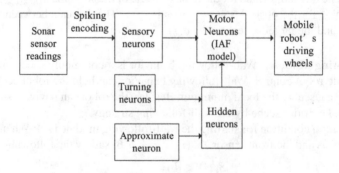

Fig. 3. Block diagram of obstacle-avoidance controller based on SNNs

obstacle is very close. Spiking coincidence detecting coding is used for the hidden neurons. Spiking integrated-and-fire (IAF) model and spiking frequency coding are used for the motor neurons in the SNNs controller. Hebbian learning is used for the obstacle-avoidance SNN controller. The frequency of the output spikes of the motor neurons controls the rotating speed of the mobile robot's driving wheels.

The simulation results and more details of the obstacle-avoidance controller can be referred in [6].

3.4   Principle of Wall-Following Module

The principle of the wall-follow controller is similar to that of the obstacle-avoidance controller. In the wall-following controller, too far and too close detection neurons are used to judge whether the robot is too far away or too close to the wall. If it is true, the corresponding spiking neurons will fire. Hebbian learning is also used for the SNNs wall-following controller. The wall-following controllers can control the robot with the varying initial pose to follow the wall clockwise and counter-clockwise. And with the wall-following controller, the robot can get rid of the deadlock problem caused by local minimum. The structure of the clockwise wall-following controller, the simulation results and the other details of the wall-following controller can be referred in [7].

4   Simulation for Target-Reaching Navigation

The robot could explore the unknown environment to get the obstacle's location by sensors in real time, and then use this information to do local path planning. The simulation is carried out supposing that the robot can be localized accurately, and the simulation results are shown in Fig. 4.

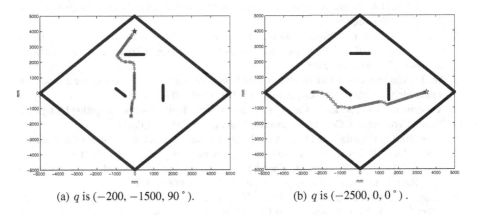

(a) q is $(-200, -1500, 90°)$.   (b) q is $(-2500, 0, 0°)$.

Fig. 4. Navigation results for the robot with varying initial poses q

In Fig. 4 (a), "*" represents the starting position. The mobile robot went through the following processes by use of the navigation transition conditions:

Goal-reaching module → obstacle avoidance module → goal-reaching module → wall-following module → goal-reaching module, and at last the robot arrived at the target point denoted by "⋆".

Similarly, in Fig. 4 (b), the robot went through the following process: Goal-reaching module → wall-following module → obstacle avoidance module → goal-reaching module → obstacle avoidance module → obstacle avoidance module → goal reaching module, and then the robot arrived at the target point. In the simulation, if it is believed that the robot has arrived at the target point, and the robot stops walking. From the simulation results it can be seen that the navigation controller can successfully control the robot with varying initial positions to the target.

5   Conclusion

The proposed navigation controller based on SNNs can control the robot with varying initial positions to the target successfully. The proposed transition conditions of the sub-modules in the navigation controller are feasible. The controller has simpler structure and can be implemented easily.

Acknowledgments. The authors would like to thankful Prof. N. Kasabov of Auckland University of Technology for helpful comments and suggestions. This work was supported in part by the National Natural Science Foundation of China (Grants 61175059, 60974063, 61005009) the Natural Science Foundation of Hebei (Grant F2010000437), and the key foundation of Hebei Normal University (Grant L2010Z08).

References

1. Pomerleau, D.A.: Neural Network Perception for Mobile Robot Guidance. Kluwer Academic Publishers, Norwell (1993)
2. Na, Y.K., Oh, S.Y.: Hybrid Control for Autonomous Mobile Robot Navigation Using Neural Network Based Behavior Modules and Environment Classification. Autonomous Robots 15(2), 193–206 (2003)
3. Ye, J.: Adaptive Control of Nonlinear PID-based Analog Neural Networks for a Nonholonomic Mobile Robot. Neurocomputing 71, 1561–1565 (2008)
4. Rossomando, F.G., Soria, C., Carelli, R.: Autonomous Mobile Robots Navigation Using RBF Neural Compensator. Control Engineering Practice 19, 215–222 (2011)
5. Mohareri, O., Dhaouadi, R., Rad, A.B.: Indirect Adaptive Tracking Control of a Nonholonomic Mobile Robot via Neural Networks. Neurocomputing 88, 54–66 (2012)
6. Wang, X.Q., Hou, Z.-G., Tan, M., Cheng, L.: A Behavior Controller for Mobile Robot Based on Spiking Neural Networks. Neurocomputing 71, 655–666 (2008)
7. Wang, X.Q., Hou, Z.-G., Tan, M., Wang, Y.: The Wall-Following Controller for the Mobile Robot Using Spiking Neurons. In: IEEE 2009 ICAICI, pp. 194–199. IEEE Press, New York (2009)
8. Vreeken J.: Spiking Neural Networks, An Introduction. Technical Report UU-CS-2003-008, Institute for Information and Computing Sciences, Utrecht University (2002),
http://ailab.cs.uu.nl/pubs/SNNVreekenIntroduction.pdf
9. Maass, W., Bishop, C.M. (eds.): Pulsed Neural Networks. MIT-Press, Cambridge (1999)

10. Schliebs, S., Defoin-Platel, M., Worner, S., Kasabov, N.: Integrated Feature and Parameter Optimization for Evolving Spiking Neural Networks: Exploring Heterogeneous Probabilistic Models. Neural Networks 22(5-6), 623–632 (2009)
11. Wysoski, S., Benuskova, L., Kasabov, N.: Fast and adaptive network of Spiking neurons for multi-view visual pattern recognition. Neurocomputing 71(13-15), 2563–2575 (2008)
12. Kasabov, N.: Integrative Connectionist Learning Systems Inspired by Nature: Current Models, Future Trends and Challenges. Natural Computing 8(2), 199–218 (2009)
13. Kasabov, N.: To Spike or Not to Spike: A Probabilistic Spiking Neuron Model. Neural Networks 23(1), 16–19 (2010)
14. Kasabov, N.: Evolving Spiking Neural Networks and Neurogenetic Systems for Spatio- and Spectro-Temporal Data Modelling and Pattern Recognition. In: Liu, J., Alippi, C., Bouchon-Meunier, B., Greenwood, G.W., Abbass, H.A. (eds.) WCCI 2012. LNCS, vol. 7311, pp. 234–260. Springer, Heidelberg (2012)
15. Wang, X.Q., Hou, Z.G., Tan, M., Wang, Y.: Corridor-Scene Classification for Mobile Robot Using Spiking Neurons. In: IEEE Fourth International Conference on Natural Computation, vol. 4, pp. 125–129. IEEE Press, New York (2008)
16. Wang, X.Q., Hou, Z.G., Tan, M.: Improved Mobile Robot's Corridor-Scene Classifier Based on Probabilistic Spiking Neuron Model. In: IEEE International Conference on Cognitive Informatics and Cognitive Computing, pp. 348–355. IEEE Press, New York (2011)
17. Floreano, D., Mattiussi, C.: Evolution of Spiking Neural Controllers for Autonomous Vision-Based Robots. In: Gomi, T. (ed.) ER-EvoRob 2001. LNCS, vol. 2217, pp. 38–61. Springer, Heidelberg (2001)
18. Floreano, D., Zufferey, J.C., Nicoud, J.D.: From Wheels to Wings with Evolutionary Spiking Neurons. Artificial Life 11, 121–138 (2005)
19. Hagras, H., Pounds-Cornish, A., Colley, M.: Evolving Spiking Neural Network Controllers for Autonomous Robots. In: IEEE International Conference on Robotics & Automation 2004, pp. 4620–4626. IEEE Press, New York (2004)
20. Soula, H., Alwan, A., Beslon, G.: Learning at the Edge of Chaos: Temporal Coupling of Spiking Neurons Controller for Autonomous Robotic. In: American Association for Artificial Intelligence (AAAI) Spring Symposia on Developmental Robotic 2005, p. 6 (2005)
21. Alamdari, R.S.A.: Unknown Environment Representation for Mobile Robot Using Spiking Neural Networks. Transactions on Engineering, Computing and Technology 6, 49–52 (2005)
22. Wang, X.Q.: Research on Eenvironment Perception and Behavior Control for Mobile Robots Based on Spiking Neural Networks. Ph.D. Thesis, Chinese Academy of Sciences, China (2007)
23. Arkin, R.C.: Integrating Behavioral, Perceptual, and World Knowledge in Reactive Navigation. Robotics and Autonomous Systems 6(1-2), 105 (1990)

Assessing Reliability of Substation Spare Current Transformer System

Cristiano G. de Melo,
Renata Maria Cardoso R. de Souza, and Liliane R.B. Salgado

Center of Informatics,
Federal University of Pernambuco (UFPE), Recife, Brazil
{cgm,rmcrs,liliane,lncs}@cin.ufpe.br
http://www.cin.ufpe.br

Abstract. This paper relates a case study involving Current Transformers (CTs) from the electrical system of the Companhia Hidro Elétrica do São Francisco (CHESF) in order to estimate the optimum number of spare. Two models have been considered in this work. The first and widely used in sparing analyses consists of an application of the Poisson distribution to calculate the reliability of a system. The second is based on Monte Carlo simulation, in which any probability distribution and stochastic times can be used. The comparison between models has shown that the Monte Carlo model is more efficient due to its stochastic nature.

Keywords: Current transformers, Monte Carlo simulation, reliability assessment.

1 Introduction

Generally considered as very simple equipments which cannot be the origin of an electrical transmission incident, Current Transformers (CTs) are often neglected. However, the CTs are vital equipments for the reliability of the system power. Its role is to provide accurate inputs to protection, control and metering systems including revenue metering. The main tasks of the CTs are: to transform currents from a usually high value to a value easy to handle for relays and instruments, and to insulate the metering circuit from the primary high voltage system.

Failures of these CTs are a concern to utility management because the long outage duration might cause loss for the companies, besides turn the most vulnerable system, or still cause interruptions of power supply in case of catastrophic failure. Therefore, the number of spare CTs should be adequate to replace failed equipments, which are removed from service, because the time for renewal of a failed unit by repair or by procuring a new unit could last up to 12 months depending on the nature of damage. The failure duration can be considerably reduced if there are spare CTs as the replacement with a spare takes only a day or two. However, the adequate calculation of the number of spare equipment is a hard task since an exceeding number of spare equipment may jeopardize

T. Huang et al. (Eds.): ICONIP 2012, Part IV, LNCS 7666, pp. 660–667, 2012.

the budget of the companies, while inventory limitation may put in risk the reliability of the system [1]-[2].

Since the reliability study for CTs is still an open research topic, the major aim of this paper is to discuss the application of probabilistic models to assess the reliability of a substation group that shares the same spare inventory. The first model has been used in many works [4]-[5] and calculates the reliability of a system, in a period of time during which the inventory cannot be replenished. The second one consists of a Monte Carlo simulation model based on an approach introduced in [1]-[2]. The proposed simulation model is a modification of the model introduced in [1]-[2] to consider the repair time of each failed CT in a given period when computing the system failure time in this period. Monte Carlo models are much more flexible since exponential and non-exponential times can be used. Several examples with a real set of CTs of an important generation/transmission company of the northeast region of Brazil are considered to demonstrate the useful of the approach.

This paper is organized as follow. Section 2 shows the current transform data set considered in this study. Sections 3 and 4 describes the Poisson and Monte Carlo simulation models, respectively. Section 5 describes the application to estimate the reliability and presents a comparative study between the models. Section 6 concludes the paper.

2   Substation Current Transformer Data

The data set of the Companhia Hidro Elétrica do São Francisco (CHESF), responsible for the generation and transmission of energy in high and extra high voltage in the northeast region in Brazil, was used to estimate the number of spare CTs. The system under consideration covers a geographical area with six Regional Centers of maintenance. The sample involves a set of 1537 CTs with voltages of 69, 230 and 500-kV installed in 26 East Regional Center substations, the major Regional Center of the CHESF. Fig. 1 illustrates the geographical area in which the substations are found in the East Regional Center. Here, CTs with voltage of 138-kV are not considered because the quantity is very small and historical occurrences for this equipment have not been found.

The system data are summarized in Table 1, where N represents the number of CTs in operation and n the number of spares in the inventory. The data concerning to historical occurrences of frequency failures were extracted from maintenance service orders occurred from January 2005 to August 2010.

Table 1. System Data

Voltage (kV)	N	n
69	832	63
230	566	31
500	139	2

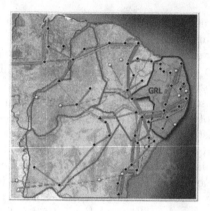

Fig. 1. East Regional Center area

3   Poisson Distribution Based Model

The probability of occurrence of exact x failures of a unit during time interval t is given by the Poisson probability distribution [6]

$$P_x(t) = \frac{(\lambda t)^x}{x!} \exp(-\lambda t), t > 0 \tag{1}$$

where λ is the failure rate of the component, i.e.

$$\lambda = \frac{NumberOfFailuresPerUnitTime}{NumberOfComponentsExposedToFailures} \tag{2}$$

Given a value of N, the mean number of failures per year is equal to $N\lambda$. To determine the probable number of failed units out of a given set of N units the following formula is used [3], [5]:

$$P_x(t) = \frac{(N\lambda t)^x}{x!} \exp(-N\lambda t) \tag{3}$$

Reliability is the probability that a system will perform a required function without failure under stated conditions for a stated period of time. In standby systems, where the spare component starts operating immediately after the failure of the main component, the reliability is given by [1]-[2]

$$R(t) = P_0(t) + P_1(t) \tag{4}$$

where $P_0(t)$ and $P_1(t)$ represent, respectively, the probability of occurring 0 and 1 failure in t. Combining (3) and (4)

$$R(t) = \exp(-\lambda t)(1 + \lambda t). \tag{5}$$

Generalizing for N main components and n spares

$$R(t) = \exp(-\lambda t) \sum_{k=0}^{n} \frac{(N\lambda t)^k}{k!} \tag{6}$$

4   Monte Carlo Simulation Based Model

The Monte Carlo method is a simulation model that uses random number generation for any probability distribution to assign values to variable to be investigated. The Monte Carlo simulation used in [2] it was applied in the group of equipments of this paper to evaluate reliability of a system composed with N field CTs and an inventory with n spares. The performance indices are calculated based on the analysis of a great number of operation years, simulated by a chronological sampling process. During this process, probability distributions associated with the operating and repair/replacement times of each CT of the system are simulated.

In the Monte Carlo simulation, the following assumptions are considered:

1. The operation and repair/replacement times are stochastic and they may follow any probability distribution and
2. The failure times of the equipments are independents and the system failure time in a given period is computed takes into account the number of equipments that are in failure in this period.

Considering the exponential distribution, the operation and repair/replacement times for each CT are sampled, respectively, by [1]–[2]

$$t_{op} = -\frac{1}{\lambda} \times \ln(U_{01}) \tag{7}$$

$$t_r = -\frac{1}{\mu} \times \ln(U_{01}) \tag{8}$$

where λ is the failure rate of a CT, μ is the replenishment rate and U_{01} is a pseudo-random number with uniform distribution between 0 and 1.

As in the model based on the Poisson distribution, it is possible to determine the system reliability during the interval between replenishment. According to [1], this index can be calculated by

$$R_{MC} = \frac{NumberOfPeriodsWithoutFailures}{TotalNumberOfObservationPeriods} \tag{9}$$

This index represents the probability of the initial inventory to be able to meet all equipment failures occurred in the period.

4.1   Other Indices

The following indices used in [1] and [2] can be calculated by the Monte Carlo simulation:

- *Failure probability*: The failure probability represents the chance of finding the system in states with less than the number of CTs in operation, being calculated by

$$P_{failure} = \frac{TotalFailureDuration}{SimulationTime} \tag{10}$$

- *Mean failure frequency*: The mean failure frequency represents the expected number of failures in the system per time unit, i.e.,

$$F_{failure} = \frac{NumberOfSystemFailure}{SimulationTime} \tag{11}$$

- *Mean failure duration*: It corresponds to the mean time during which the system will remain without operating/revenue metering and/or without protection indication, partial or totally, each time a failure occurs. In this case,

$$D_{failure} = \frac{TotalFailureDuration}{NumberOfSystemFailure} \tag{12}$$

4.2   Algorithm

The Monte Carlo simulation model developed in this paper is based on an algorithm which checks the state of each CT of the system at each time. Three states are considered: run (in operation), stop (in the process of repair or replacement), and spare (available in the inventory). The algorithm proceeds as follows.

Algorithm 1. Modeling algorithm

1: **Consider**: values for N, n, λ and μ;
2: **Compute**: the operation and repair/replacement times of all CTs in operation according to (7) and (8);
3: **Find**: the CT with minimum value of t_{op} out of state *run* and go to state *stop*. The simulation time takes t_{op};
4: **For**: each occurred failure, the CTs under repair are organized in order from lowest to highest according to the expression $(t_{op} + t_r)$;
5: **For**: each equipment under repair, verify if the state *stop* was concluded. If there is a deficit in the field, the equipment goes into the state *run* and compute t_{op} and t_r, otherwise goes into the state *spare*;
6: **If**: there is a deficit in the field and the number spare in the inventory ia equal to zero, compute the failure time of system, otherwise a CT of the inventory goes into the state *run* computing t_{op} and t_r;
7: **Repeat**: step 3 until to finish the total simulation time;
8: **Compute**: the reliability indices.

---

5   Application and Results

There are two failure modes for a CT: 1) field-repairable failure and 2) non-field-repairable. The installation and field repairable times of a CT is one to two days, which is shorter than the replacement or procurement time of one year of buying or rebuilding an equipment in the case of a catastrophic failure and non-available spare. In relation to CTs, failures that are not field-repairable in field can not be catastrophic. However, in some cases it is better not to repair the equipment in the field due to low transportation costs. Therefore, non-field-repairable failures have been considered to calculate the failure rate in the reliability analysis.

The Poisson and Monte Carlo model is utilized (the latter with the simulation of approximately 100.000 years) for calculating the system reliability for the group of voltage 69-kV CTs (see Table 1). Consider the minimum reliability of 0.9950, a number typically used in the electric utility industry [5], the reliability and corresponding number of spare requirements for the group of voltage 69-kV CTs are presented in Table 2. It can be seen in this table that the number of spares should be 20 for the Poisson model and 23 for the Monte Carlo model. The difference between the two results is due to the fact that in the Poisson model is assumes that the inventory be completely replenished at the beginning of the period, and that in the Monte Carlo model the repair or the procurement time of a new CT is treated as a stochastic rather than as a deterministic variable.

Table 2. System Reliability for the Group of Voltage 69-KV Current Transformer-Poisson and Monte Carlo Model

N	Reliability	
	Poisson	Monte Carlo
7	0.1626216	0.1626216
8	0.2583486	0.0409400
...	...	...
20	**0.9965032**	0.9772100
21	0.9983555	0.9879700
22	0.9992576	0.9941000
23	0.9996778	**0.9970300**
24	0.9998654	0.9988100

It is observed a difference for small n. This difference is due to the fact that the Poisson model admits that there are N CTs continually in operation, once the Poisson model uses a constant total failure rate $N\lambda$. As the number of spares increases, it is possible to note an approximation among the model results. The results for remaining voltages using the reliability criterion of 0.9950 for both models (i.e., Poisson and Monte Carlo) are summarized in Table 3.

Table 4 shows the system unavailability (hours/year) and the mean failure frequency (failures/year), for n varying from 20 to 24 for the group of voltage 69-kV CTs (Monte Carlo model), while in Table 5 the estimate number of CTs and

Table 3. Estimated Number of Current Transformer Spares for Different Voltage-Poisson and Monte Carlo Model

Voltage (kV)	Estimate Number of Spare (n)	
	Poisson	Monte Carlo
69	20	23
230	6	8
500	3	4

Table 4. Unavailability, Mean Failure Frequency and MTBF for the Group of Voltage 69-kV Current Transformer-Monte Carlo Model

n	Unavailability (h/y)	Frequency (f/y)
20	56.6124	0.07551
21	28.3782	0.04181
22	10.9101	0.01622
23	4.3371	0.00834
24	1.7496	0.00352

Table 5. Estimated Inventory

Index	69-kV	230-kV	500-kV
N	23	8	4
$R(t)$ (Poisson)	0.9996778	0.9998177	0.9994622
R_{MC}	0.9969200	0.9988700	0.9973100
$P_{failure}$	0.000495	0.0001997	0.0006039
$F_{failure}$	0.008339	0.0016499	0.0028499
$D_{failure}$	0.059365	0.1210083	0.2119101

the result of the index are shown. To achieve the optimal point, the inventory of 69, 230, and 500-kV CT groups should have, respectively, 23, 8, and 4 units of equipment, against the current 63, 31, and 2.

The results obtained by applying the Poisson and Monte Carlos models has shown that the estimates obtained by the Poisson distribution are optimistic concerning the availability performance of the set of CTs, since this model assumes that the inventory be completely replenished at the beginning of the period. Thus, the number of spares estimated by the Poisson model is less than that estimated by the Monte Carlo one.

The estimate number of spare CTs for a system is a function of input parameters such as failure rates, repair times, inventory information, and the reliability level demanded from the system. The higher the equipment failure rate and the repair time, the higher the system unavailability and the number of spare requirements for maintaining an adequate level of system reliability would be. Therefore, is important that utility non-neglect the failures records in such equipment.

6   Conclusion

This paper discusses an application to estimate the optimal number of spare CTs in transmission system using two probabilistic models. The first model, used in many works, consists of an application of the Poisson distribution, and the second is based on the Monte Carlo simulation that considers the repair/replacement time as a stochastic variable. A numerical example using 69, 230, and 500-kV CT systems illustrated the application of the proposed models. Both models estimated the number of spare CTs and, as expected, the Monte Carlo simulation provided the best result.

Acknowledgments. The authors would like to thank CHESF for information provided and supporting the development this work. The authors also thank CNPq and Facepe (Brazilian Agencies) for its financial support.

References

1. Costa, J.G.C., Silva, A.M.L.: Monte Carlo simulation to assess the optimum number of distribution spare transformers. In: 10th Int. Conf. Probability Methods Applied to Power Systems (PMAPS), Rincón, PR, pp. 25–29 (May 2008)
2. Costa, J.G.C., Silva, A.M.L., Chowdhuryl, A.A.: Probabilistic methodologies for determining the optimal number of substation spare transformers. IEEE Transaction on Power System 25, 68–77 (2010)
3. Nahman, J.M., Tanaskovic, M.R.: Probability models for optimal sparing of distribution network transformers. IEEE Transaction on Power Delivey 24, 758–763 (2009)
4. Kogan, V.I., Roeger, C.J., Tipton, D.E.: Substation distribution transformers failures and spares. IEEE Transaction on Power System 11, 1906–1912 (1996)
5. Chowdhury, A.A., Koval, D.O.: Development of probabilistic models for computing optimal distribution substation spare trans formers. IEEE Transaction on Industry Application 41, 1493–1498 (2005)
6. Mayer, P.L.: Introductory Probability and Statistical Application, 2nd edn. Addison-Wesley Pub. Co. (1970)
7. Li, W., Vaahedi, E., Mansour, Y.: Determining number and timing of substation spare transformers using a probabilistic cost analysis approach. IEEE Transactions on Power Delivery 14, 934–939 (1999)

Study on Supply Disruption Management of Supply Chain Based on Case-Based Reasoning

Zhang Daohai[1,2]

[1] School of Management and Engineering, Nanjing University, Nanjing, China
[2] Computational Experiment Center for Social Science, Jiangsu University, Zhenjiang, China
zdh@ujs.edu.cn

Abstract. Emergency is an important factor resulting in supply disruption risk. From the angle of risk management, we build supply disruption management model of supply chain, and discuss whether decision-making mechanism of the case-based reasoning, in this paper, can bring the better effect by using computational experiment for supply disruption. The main results show that risk assessment, risk identify, risk control and risk evaluation mechanism based on the case-based reasoning can effectively deal with supply disruption risk, bringing more profit and service level for enterprises.

Keywords: Case-based Reasoning, Supply Disruption, Disruption Management, Computational Experiment.

1 Introduction

Over the past few years many types of unpredictable disasters such as terrorist acts, accidents, natural calamities etc. have occurred, which indicates that our world is increasingly more uncertain and vulnerable. Moreover, global sourcing, lean production, increased reliance on outsourcing, produced number of suppliers, and focusing on reducing inventory etc., have formed now new trends and practices. This new kind of interdependent relationship of supply chains, makes the enterprise operation efficiency, but also brings a more risk factors, and shows clearly that supply chains seem to be more fragile today than in the past. Although these trends and practices have reduced the normal costs in supply chains, they have also made supply chains more susceptible to disruptions and have created longer and more complex supply chains in which the domino effects of disruptions have been exacerbated. For example, a fire in the Phillips Semiconductor plant in Albuquerque, New Mexico in March 2000 caused its major customer, Ericsson, to lose $400 million in potential revenues. The one-month-long brutal winter weather caused by heavy snowfalls that occurred in large tracts of China in January 2008 caused transport chaos and disrupted supplies of energy and food.

Supply disruption management has received increasing attention from both industry and academia. Firms are starting to realize that supply disruption severely affects their ability to successfully manage their supply chains. The academia has devoted

T. Huang et al. (Eds.): ICONIP 2012, Part IV, LNCS 7666, pp. 668–676, 2012.

much research effort to studying this issue. Many papers have been published with advising on how to manage their supply chains in the presence of supply disruption. For example, Hui Yu and Caihong Sun et al. (2007) study the impacts of supply disruption on the supply chain system by using simulation approach in which two different distribution function of random variable are used to express the supply disruption. Comparison between these two simulation results and possible coordination mechanism under the supply disruption are proposed. Zhang Heng and Hua Xinglai et al.(2008) analyse the main sources and results of equipment supply chain risk and provide a model of supplier selection, which can make the risk of supply disruption minimization. Jian Li and Shouyang Wang et al. (2009) investigate the sourcing strategy of a retailer and the pricing strategies of two suppliers in a supply chain under an environment of supply disruption and devise a coordination mechanism to maximize the profits of both suppliers. Xiangguo Chen and Jianhua Ji et al. (2009) establish a shared saving model, the optimal proportion of shared saving for maximizing supplier, manufacturer and the whole supply chain profit has been calculated and the results show that the optimal proportion of shared saving is different on maximizing profit of supplier, manufacturer and the whole supply chain, and the proportion is decided by the relative importance of contribution to the cost reduction of partners and their bargain ability. Jing Hou and Zeng Amy Z et al. (2010) study a buy-back contract between a buyer and a backup supplier when the buyer's main supplier experiences disruptions.

These research results have riched disruption management content of supply chains. On quantitative side, the researches mostly based on some contract mechanism to coordinate and optimize supply chains, make them have some flexibility and achieve integral ability to resist risks, but mostly based on some theory research with simplifying, which may make research results have great distance with realities, impacting the operability of scheme(Leiming Li and Bingquan Liu, 2010). Supply chain emergency is typical of uncertain type of risk with much occasionality. It is especially important that how to assess and identify risk, then to make correctly emergency, after emergency how to evaluate emergency effect in real-time and adjust emergency strategy. By using the computer simulation with multi-discipline method, can effectively represent the typical scene with more rapid computation, and count up a large number of operation performance indicators, to provide the necessary tools for risk management and emergency response plan, but few literature studies these questions by using computational experiment of Multi-agent evolution modeling. So, this paper focuses on the supply chain caused by emergencies in the supply disruption, trying to use computational experiment method, combining research paradigm of risk management, with research of risk assessment, risk identification, risk control and risk evaluation.

2   Disruption Management Model Based on Case-Based Reasoning

This paper is based on the case-based reasoning, using computational experiment method to monitor supply system of supply chain in real-time, putting forward supply disruption management framework, model and algorithm so as to provide new train of thought and method for enterprise to resist an emergency risk.

2.1   Supply Disruption Management Framework

The core idea of case-based reasoning method is through the visit of the past similar case's optimal solution, then to get the optimal coping strategies. It fully uses the implicit knowledge and information of the past similar cases, that is, through the key indicators of the monitoring, looking for similar cases, assessing and estimating the possible risk at the first time, then getting the optimal emergency strategy, and through the later evaluation updating the case, so as to make emergency strategy continuously optimize. Thus it can provide a more superior solution to deal with continuing possible similar incidents risk.

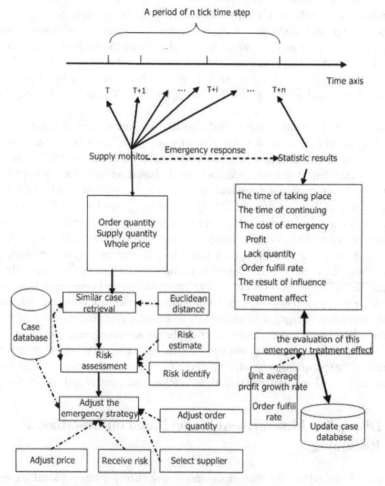

Fig. 1. Supply Disruption Management Framework on Case-based Reasoning

Overall research framework as shown in Fig. 1., considering a period of n tick time step, the supply chain's member can monitor current order quantity, supply quantity of upstream and wholesale price of upstream each period. Then it uses these monitored results to find the similar case from the case database by the similar case retrieval algorithm and according to the statistical results of case history, assesses this accident risk to identify whether it is conventional risk or sudden risk. Once more according to the historical case emergency treatment evaluation result, it adjusts the emergency strategy, including receiving risk, adjusting order quantity, adjusting price, selecting supplier; Finally according to the evaluation of this emergency treatment effect with unit average profit growth rate and order fulfill rate, it updates the case database.

2.2   Research Question and Hypothesis

Because of the complexity of the supply chain system itself, so we can't go to depict the supply chain system all the influence factors, can only focus on the most important factors to test whether the risk assessment, identification, control and evaluation mechanism based on case-based reasoning, in this paper, is better for supply chain enterprises to a supply disruption risk. We made some basic hypothesis model as follows:

(1) The experiment model assumes the second stage supply chain is composed of four suppliers and two manufacturers with relationship of competition. Any manufacturer can adopt the case-based reasoning or not, and supply disruption in any supplier occurs by a certain probability. In this scene we see the supply disruption transfers and effects and the agent's adaptive ability.

(2) We take durable products as study object, assuming that the normal market demand is subject to uniform distribution. Each period the surpluses turn into the next period inventory products and the supply chain only provides one type products. Similar agents compete on the price and the service level of the products.

(3) We set the scene as follows: m1 and m2 are as two manufacturers, m1 uses the case-based reasoning decision, m2 not. They all have four suppliers s1, s2, s3 and s4. Structure model describes as shown in Fig. 2, we respectively take Y as case-based reasoning decision, N as no.

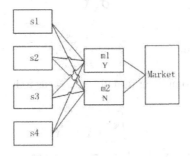

Fig. 2. Structure Model

2.3   Model Represent

Supply Disruption Model. Considering a supplier by the effect of the emergency, the supply capacity coefficient should be experience from the four stages with early happened, to down, then to up, last to recovery, we summarize the several stages, with type (1) model to describe.

$$p(t) = C - A \bullet (t - k) \bullet e^{-B(t-k)} \tag{1}$$

$p(t)$ is supply capacity coefficient of t ($t \geq k$) tick after the emergency; $C > 0$, for adjustment coefficient; $A > 0$, for disruption fluctuation coefficient; $B > 0$, for urgent degree coefficient of disruption period; $k > 0$, for happened point of disruption period. In this experiment, we assume $C = 1, A = 0.5, B = 0.2$.

Case Retrieval Mode. Assume $P = \{P_1, P_2, \cdots, P_m\}$ as source cases set, $P_i (i = 1, 2, \cdots, m)$ is the i source case. Assume $Z = \{Z_1, Z_2, \cdots, Z_n\}$ as case attributes set, $Z_j (j = 1, 2, \cdots, n)$ is the j attribute. Constitute type (2) matrix as follow.

$$X = \begin{bmatrix} X_{11} & X_{12} & \cdots & X_{1n} \\ X_{21} & X_{22} & \cdots & X_{2n} \\ \vdots & \vdots & \vdots & \vdots \\ X_{m1} & X_{m2} & \cdots & X_{mn} \end{bmatrix} \tag{2}$$

X_{ij} represents the j attribute values of the i case.

Index value dimension of different factors is different. In order to eliminate the influence of dimension, we dispose the original data to standardization. Standardization formula type (3) as follow.

$$M_{ij} = \frac{X_{ij} - \overline{X}_j}{\sigma(X_j)} \tag{3}$$

M_{ij} is the standard score of X_{ij}, $\overline{X}_j = \dfrac{1}{m} \sum_{i=1}^{m} X_{ij}$, $\sigma(X_j)$ is the standard deviation of the factor Z_j, computing formula type (4) as follow.

$$\sigma(X_j) = \sqrt{\frac{\sum_{i=1}^{m} (X_{ij} - \overline{X}_j)^2}{m}} \tag{4}$$

Take w_j as the weight of the factor Z_j. In order to eliminate the influence of subjective, we use the machine learning to weight distribution, computing formula type (5) as follow.

$$w_j = \frac{\sigma(X_j)}{\sigma(X_1) + \sigma(X_2) + \cdots + \sigma(X_n)} = \frac{\sigma(X_j)}{\sum_{j=1}^{n} \sigma(X_j)} \tag{5}$$

Take $T = \{T_1, T_2, \cdots, T_n\}$ as the target case, which has properties such as type (6) matrix as follow.

$$X = \begin{bmatrix} X_{11} & X_{12} & \cdots & X_{1n} \\ X_{21} & X_{22} & \cdots & M_{2n} \\ \vdots & \vdots & \vdots & \vdots \\ X_{m1} & X_{m2} & \cdots & X_{mn} \\ T_1 & T_2 & \cdots & T_n \end{bmatrix} \tag{6}$$

After computing, get standard score matrix such as type (7) as follow.

$$X' = \begin{bmatrix} M_{11} & M_{12} & \cdots & M_{1n} \\ M_{21} & M_{22} & \cdots & M_{2n} \\ \vdots & \vdots & \vdots & \vdots \\ M_{m1} & M_{m2} & \cdots & M_{mn} \\ M_{(m+1)1} & M_{(m+1)2} & \cdots & M_{(m+1)n} \end{bmatrix} \tag{7}$$

Through the standard score matrix computing with Euclidean distance, get the similarity, computing formula type (8) as follow.

$$D(T, P_i) = \sqrt{\left(\sum_{j=1}^{n} w_j (M_{ij} - M_{(m+1)j})^2 \right)} \tag{8}$$

Trough type (8) computing, get the similarity matrix D of the target cases with the source case P_i.

3   Experiment Result and Analysis

3.1   Initial Parameter Setting

We mainly adopt Fig. 1 framework and Fig. 2 model, use the computational method to realize the construction of the participating members, set up corresponding environmental parameters, flow and rules, through the control of the corresponding input parameters, get the results we're interested in, and reveals whether the case-based reasoning decision-making mechanism can bring good effect for manufacturer. The initial main parameters setting is shown as in Table 1.

Table 1. Initial Parameter Setting

manufacturer		supplier	
Whole price	20	Whole price	10
Inventory cost	0.2	Max capacity of each period	10000
Fixed order cost	100	Lead time	1
Transportation cost	1	Wait time	2
Product inventory	1000	Material inventory	1000
Safety inventory coefficient	1.65	Safety inventory coefficient	1.65
Inventory strategy	(S,s)	Form of distribution	Order proportion
Service level	5	Inventory strategy	(S,s)
Lead time	1	Service level	5
Wait time	2	Disruption prob.	0.2
Number of supplier choice	2	Disruption period	50

3.2   Experiment Result and Analysis

As Table 1 shown, manufacturer m1 is set to the decision-making of case-based reasoning, while manufacturer m2 is set to the decision-making based on historical comprehensive utility (price, service level, lead time and order fulfill rate), other decision mode and parameter is same, As Fig. 3, Fig. 4 and Fig. 5 shown, m1 and m2 is facing with the fluctuations of lack quantity, its own profit level and delivery quantity at tick 36. In the beginning, owing to the lack of historical records, m1 case study is not obvious, with the passage of time, as can be seen from the figures: early time, m1 and m2 operating curve fluctuations are similar, along with supply disruption, as Fig. 3 shown, the shortage of m1 is much smaller than m2, and as Fig. 5 shown, delivery quantity of m1 also keeps higher level in the period of disruption, and delivery of m2 often occurs broken goods phenomenon. From the fluctuations of profit level as shown in Fig. 4, we see that m1 has maintained a higher level of profit than m2 in supply disruption period.

From the contrast experiment analysis of the same level and ability of the manufacturers, the case-based reasoning in this paper can effectively deal with supply disruption, and help enterprises get more profit and service level.

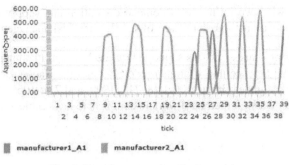

Fig. 3. Experiment result of lack quantity

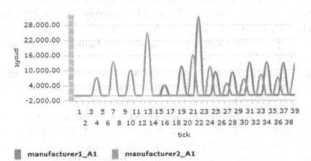

Fig. 4. Experiment result of profit level

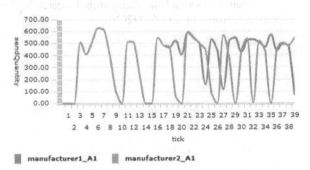

Fig. 5. Experiment result of delivery quantity

4   Conclusion

We build the supply chain risk management model with case-based reasoning, the manufacturers of the supply chain may decide whether to use the case-based reasoning to deal with supply disruption. We contrast that the two identical manufacturers in the same market environment with the different ways of decision-making will bring what influence. The main results show that the case-based reasoning mechanism, in this paper, can better bring to supply disruption influence. But we see emergency not only influence supply, and at the same time also can affect demand, next we will further research with two influence of demand disruption and supply disruption, exploring more practical method and means of supply chain risk management.

Acknowledgement. This work was supported by the National Planning Office of Philosophy and Social Science under Grant 11&ZD169 and the National Natural Science Foundation of China under Grant 71001028, 71171099, and 71101067. We are grateful to all the team members for their hard work to complete today's achievements.

References

1. Yu, H., Sun, C., Chen, J.: Simulating the Supply Disruption For the Coordinated Supply Chain. Journal of Systems Science and Systems Engineering 16, 323–335 (2007)
2. Zhang, H., Hua, X., Liu, G.: Risk Factors Analysis and Decision-making Model Research of Equipment Supply Chain. Journal of the Academy of Equipment Command & Technology 19, 43–46 (2008)
3. Li, J., Wang, S., Cheng, T.C.E.: Competition and Cooperation in a Single-retailer Two-supplier Supply Chain with Supply Disruption. International Journal of Production Economics 124, 137–150 (2010)
4. Chen, X.G., Ji, J.H., Zhu, C.B.: Optimal Proportion of Shared Saving Between Supply Chain Partners. Journal of Harbin Institute of Technology 41, 230–233 (2009)
5. Hou, J., Zeng, A.Z., Zhao, L.: Coordination with a Backup Supplier Through Buy-back Contract under Supply Disruption. Transportation Research Part E-Logistics and Transportation Review 46, 881–895 (2010)
6. Li, L., Liu, B.: Supply Chain Disruption Risk Management Research Review. Science and Technology Management Research 14, 236–239 (2010)

Global Minimizer of Large Scale Stochastic Rosenbrock Function: Canonical Duality Approach

Chaojie Li and David Yang Gao

School of Science, Information Technology and Engineering,
University of Ballarat, Mt Helen, VIC 3350, Australia
cjlee.cqu@163.com

Abstract. Canonical duality theory for solving the well-known bench-mark test problem of stochastic Rosenbrock function is explored by two canonical transformations. Global optimality criterion is analytically obtained, which shows that the stochastic disturbance of these parameters could be eliminated by a proper canonical dual transformation. Numerical simulations illustrate the canonical duality theory is potentially powerful for solving this benchmark test problem and many other challenging problems in global optimization and complex network systems.

Keywords: Global Optimation, Canonical Duality Theory, Stochastic Rosenbrock Function.

1 Preliminary

Almost all of benchmark test problems in previous literature are deterministic by parameters. However, it is usually more difficult for algorithms to deal with stochastic functions. In Yang's work [2], a stochastic parameter is introduced in Rosenbrock's function such that this well-known benchmark test problem can be proposed as

$$(\mathcal{P}): \quad \min \left\{ P(X) = \sum_{i=1}^{n-1} [(x_i - 1)^2 + 100\epsilon_i(x_{i+1} - x_i^2)^2] \mid X \in \mathcal{X} \right\}, \quad (1)$$

where $X = \{x_i\} \in \mathcal{X} = \mathbb{R}^n$ is a real unknown vector, and the random parameters $\{\epsilon_i\}$ are drawn from a uniform distribution in $[0, 1]$. For stochastic functions, most deterministic algorithms such as hill climbing and Nelder-Mead downhill simplex method would simply fail.

2 Canonical Dual Approach

Following the standard procedures in the canonical dual transformation (see Gao, 2009), we first introduce the so-called geometrical operator $\xi = \Lambda(X)$: $\mathcal{X} \to \mathcal{E}_a \subset \mathbb{R}^{n-1}$

$$\xi = \{\xi_k\} = \epsilon_k^{\frac{1}{2}}(x_k^2 - x_{k+1}), \quad (2)$$

T. Huang et al. (Eds.): ICONIP 2012, Part IV, LNCS 7666, pp. 677–682, 2012.

and a canonical function $V(\xi) = 100 \sum_{k=1}^{n-1} \xi_k^2$ such that the duality relation

$$\varsigma = \{\varsigma_k\} = \left\{ \frac{\partial V(\xi_k)}{\partial \xi_k} \right\} = \{200\xi_k\} \tag{3}$$

is invertible. Thus, we have

$$\xi_k = \frac{1}{200}\varsigma_k \quad \forall k = 1, \ldots, n-1, \tag{4}$$

and the conjugate function of $V(\xi_k)$ can be obtained uniquely by the Legendre transformation

$$V^*(\varsigma) = \sum_{k=1}^{n-1} \xi_k \varsigma_k - V(\xi)$$

$$= \sum_{k=1}^{n-1} \{\xi_k \varsigma_k - 100\xi_k^2\} = \sum_{k=1}^{n-1} \frac{1}{400}\varsigma_k^2. \tag{5}$$

Then, the total complementary function can be defined as

$$\Xi(X,\varsigma) = \sum_{k=1}^{n-1} (x_k - 1)^2 + \Lambda(X)^T \varsigma - V^*(\varsigma)$$

$$= \sum_{k=1}^{n-1} \left[(x_k - 1)^2 + \epsilon_k^{\frac{1}{2}}(x_k^2 - x_{k+1})\varsigma_k - \frac{1}{400}\varsigma_k^2 \right]. \tag{6}$$

For a fixed ς in the canonical dual feasible space $\mathcal{S}_a \subset \mathbb{R}^{n-1}$ defined by

$$\mathcal{S}_a = \left\{ \varsigma \in \mathcal{S} \mid \epsilon_k^{\frac{1}{2}}\varsigma_k + 1 \neq 0, \quad \forall k = 1, \ldots, n-2, \quad \varsigma_{n-1} = 0 \right\},$$

the criticality condition $\nabla_X \Xi(X,\varsigma) = 0$ leads to the following analytical solution

$$X = \{x_k\} = \left\{ \frac{\epsilon_{k-1}^{\frac{1}{2}}\varsigma_{k-1} + 2}{2(\epsilon_k^{\frac{1}{2}}\varsigma_k + 1)} \right\}. \tag{7}$$

Substituting this result into the total complementary function $\Xi(X,\varsigma)$, the canonical dual problem can be finally formulated as

$$(\mathcal{P}^d): \quad P^d(\varsigma) = \max \left\{ n - 1 - \sum_{k=1}^{n-1} \left[\frac{(\epsilon_{k-1}^{\frac{1}{2}}\varsigma_{k-1} + 2)^2}{4(\epsilon_k^{\frac{1}{2}}\varsigma_k + 1)} + \frac{1}{400}\varsigma_k^2 \right] \mid \varsigma \in \mathcal{S}_a^+ \right\}, \tag{8}$$

where

$$\mathcal{S}_a^+ = \{\varsigma \in \mathcal{S}_a \mid \epsilon_k^{\frac{1}{2}}\varsigma_k + 1 > 0, \quad \forall k = 1, \ldots, n-2\}. \tag{9}$$

By introducing $G(\varsigma), F(\varsigma)$ and $\mathcal{S}_a^+$ such that

$$
G(\varsigma) = \begin{bmatrix} \epsilon_1^{\frac{1}{2}}\varsigma_1 + 1 & & & & \\ & 2(\epsilon_2^{\frac{1}{2}}\varsigma_2 + 1) & & & \\ & & \ddots & & \\ & & & 2(\epsilon_{n-2}^{\frac{1}{2}}\varsigma_{n-2} + 1) & \\ & & & & 2 \end{bmatrix} \tag{10}
$$

$$
F(\varsigma) = \begin{bmatrix} 1 \\ \epsilon_1^{\frac{1}{2}}\varsigma_1 + 2 \\ \vdots \\ \epsilon_{n-3}^{\frac{1}{2}}\varsigma_{n-3} + 2 \\ \epsilon_{n-2}^{\frac{1}{2}}\varsigma_{n-2} + 2 \end{bmatrix}. \tag{11}
$$

We have the following theorem (see Gao, 2009)

Theorem 1. *If $\bar{\varsigma}$ is a critical point of $(\mathcal{P}^d)$, then the vector*

$$
\bar{X} = G^{-1}(\bar{\varsigma})F(\bar{\varsigma}) \tag{12}
$$

is a critical point of $(\mathcal{P})$ and

$$
P(\bar{X}) = P^d(\bar{\varsigma}). \tag{13}
$$

If $\bar{\varsigma} \in \mathcal{S}_a^+$, then $\bar{\varsigma}$ is the global maximizer of the canonical dual problem $(\mathcal{P}^d)$ on $\mathcal{S}_{a+}$. The vector $\bar{X}$ is a global minimal to the primal problem, and

$$
P(\bar{X}) = \min_{X \in \mathcal{X}} P(X) = \max_{\varsigma \in \mathcal{S}_a^+} P^d(\varsigma) = P^d(\bar{\varsigma}). \tag{14}
$$

The proof of this Theorem can be intuitively derived from the paper by Gao (2003).

3   An Alternative Transformation

In this section, we choose an alternative canonical dual transformation for stochastic function, which shows analytically that the stochastic perturbation of this problem would never change the global minimal elements.

Let $\xi = \{\xi_k\} = \{x_k^2 - x_{k+1}\} \in \mathbf{R}^{n-1}$. The canonical function $V(\xi)$ has the form of

$$
V(\xi) = 100 \sum_{k=1}^{n-1} \epsilon_k \xi_k^2. \tag{15}
$$

Thus, the associated canonical dual variable $\varsigma = \{\varsigma_k\} = \nabla V(\xi) = \{200\epsilon_k\xi_k\}$ and

$$
V^*(\varsigma) = \sum_{k=1}^{n-1} \frac{1}{400\epsilon_k}\varsigma_k^2. \tag{16}
$$

Correspondingly, the total complementary function can be written as

$$\Xi(X, \varsigma) = \sum_{k=1}^{n-1} \left[(x_k - 1)^2 + (x_k^2 - x_{k+1})\varsigma_k - \frac{1}{400\epsilon_k}\varsigma_k^2 \right]. \tag{17}$$

By which, the second type of the canonical dual problem can be formulated as

$$(\mathcal{P}^d): \quad P^d(\varsigma) = \max \left\{ \{n - 1 - \frac{1}{2}F(\varsigma)^T G^{-1}(\varsigma)F(\varsigma) - \frac{1}{400\epsilon}\varsigma^T\varsigma \ | \ \varsigma \in \mathcal{S}_a^+ \} \right\} \tag{18}$$

where

$$G(\varsigma) = \begin{bmatrix} \varsigma_1 + 1 & & & \\ & 2(\varsigma_2 + 1) & & \\ & & \ddots & \\ & & & 2(\varsigma_{n-2} + 1) \\ & & & & 2 \end{bmatrix} \tag{19}$$

$$F(\varsigma) = \begin{bmatrix} 1 \\ \varsigma_1 + 2 \\ \vdots \\ \varsigma_{n-3} + 2 \\ \varsigma_{n-2} + 2 \end{bmatrix} \tag{20}$$

$$\mathcal{S}_a^+ = \{\varsigma \in \mathbb{R}^{n-1} \ | \ \varsigma_k > -1, \quad \forall k = 1, ..., n - 2, \quad \varsigma_{n-1} = 0\}. \tag{21}$$

Theorem 1 still holds for this second canonical dual problem. However, by numerical experiments we can see that the stochastic perturbation does not have any impact on the global optimal solution.

4   Illustration

In this section we list some numerical examples with different dimensions, which are more general than normal Rosenbrock function.

Example 1. Consider

$$(\mathcal{P}): \quad \min \left\{ P(X) = \sum_{i=1}^{3} [(x_i - 1)^2 + 100\epsilon_i(x_{i+1} - x_i^2)^2] \ | \ X \in \mathcal{X} \right\}. \tag{22}$$

This problem has the global minimum of all ones and a local minimum near $(x_1, x_2, x_3, x_4) = (-1, 1, 1, 1)$. Correspondingly, the canonical dual problem is

$$\max \left\{ P^d(\varsigma) = 3 - \sum_{k=1}^{3} \left[\frac{(\epsilon_{k-1}^{\frac{1}{2}}\varsigma_{k-1} + 2)^2}{4(\epsilon_k^{\frac{1}{2}}\varsigma_k + 1)} + \frac{1}{400}\varsigma_k^2 \right] \ | \ \varsigma \in \mathcal{S}_a^+ \right\} \tag{23}$$

where $\mathcal{S}_a^+$ is defined by (9).

And an alternative canonical dual problem is

$$\max\left\{ \ P^d(\varsigma) = 3 - \sum_{k=1}^{3}\left[\frac{(\varsigma_{k-1}+2)^2}{4(\varsigma_k+1)} + \frac{1}{400\epsilon_k}\varsigma_k^2\right] \ \Big| \ \varsigma \in \mathcal{S}_a^+ \right\} \qquad (24)$$

where $\mathcal{S}_a^+$ is defined by (21). Obviously, it is easy to find results by Matlab optimization tools FMINCON. Note that $\varsigma_3 = 0$, the contour of dual problem can be obtained directly(see Fig. 1). Thus, the global maximum of dual problem is $(\varsigma_1,\varsigma_2,\varsigma_3)=(0,\ 0,\ 0)$ and the global minimum of primal problem is $(x_1,x_2,x_3,x_4)=(1,\ 1\ ,\ 1,\ 1)$.

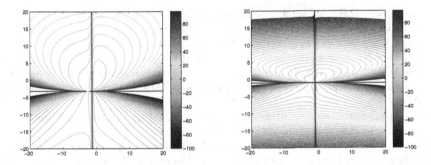

Fig. 1. (a) Dual problem n=3 determined by (23) ; (b) Dual problem n=3 determined by (24)

Example 2. Consider N= 1000, 3000, 5000, 10000, 20000. The global minimum is inside a long, narrow, banana shaped flat valley. In this case, it is difficult to solve exactly the primal problem ($\mathcal{P}$) by gradient methods. Fortunately, the canonical dual problem is concave maximization over a cone, which can be solved easily, fast and exactly by gradient method.

For (23) and (24), the initial points are chosen randomly from -5 to 5 with the constraints (9) and (21), respectively. With these numerical computation settings, L-BFGS method can quickly solve all these test problems and accurately converge to the global maximizer $\varsigma = (0, 0, ..., 0)$ with the optimal value $P^d(\varsigma) = 0(10^{-8})$.

All of numerical experiments have been carried out on personal notebook computer with Intel(R)Core i5-2430M @2.40GHz Windows 7 Home Prem.

5   Conclusion

This paper illustrates that the well-known benchmark test problem of Rosenbrock function with stochastic parameters can be easily solved by the canonical duality theory. Numerical examples show that even though the nonconvex primal problem has been disturbed by stochastic parameters, the canonical duality theory can avoid defective influence to achieve global optimal solution stably. The canonical duality theory can be used for solving some more challenging problems in global optimization and complex network systems.

Acknowledgements. This research is supported by US Air Force Office of Scientific Research under the grant AFOSR FA9550-10-1-0487 and by Australia Government grant through the Collaborative Research Network(CRN) to the university of Ballarat.

References

1. Rosenbrock, H.H.: An automatic method for finding the greatest or least value of a function. The Computer Journal 3, 175–184 (1960)
2. Yang, X.S., Deb, S.: Engineering optimization by cuckoo search. Int. J. Math. Modelling Num. Optimisation 1, 330–343 (2010)
3. Gao, D.Y.: Duality, triality and complementary extremum principles in nonconvex parametric variational problems with applications. IMA J. Appl. Math. 61, 199–235 (1998)
4. Gao, D.Y.: Pure complementary energy principle and triality theory in finite elasticity. Mech. Res. Commun. 26(1), 31–37 (1999)
5. Gao, D.Y.: General analytic solutions and complementary variational principles for large deformation nonsmooth mechanics. Meccanica 34, 169–198 (1999)
6. Gao, D.Y.: Duality Principles in Nonconvex Systems: Theory, Methods and Applications. Kluwer Academic, Dordrecht (2000)
7. Gao, D.Y.: Canonical dual transformation method and generalized triality theory in nonsmooth global optimization. J. Glob. Optim. 17, 127–160 (2000)
8. Gao, D.Y.: Perfect duality theory and complete solutions to a class of global optimization problems. Optim. 52, 467–493 (2003)
9. Gao, D.Y.: Perfect duality theory and complete solutions to a class of global optimization problems. Optimization 52, 467–493 (2003)
10. Gao, D.Y.: Nonconvex semi-linear problems and canonical dual solutions. In: Gao, D.Y., Ogden, R.W. (eds.) Advances in Mechanics and Mathematics, vol. II, pp. 261–312. Kluwer Academic, Dordrecht (2003)
11. Gao, D.Y.: Canonical duality theory and solutions to constrained nonconvex quadratic programming. J. Glob. Optim. 29, 377–399 (2004)
12. Gao, D.Y.: Solutions and optimality to box constrained nonconvex minimization problems. J. Ind. Manag. Optim. 3(2), 293–304 (2007)
13. Gao, D.Y.: Canonical duality theory: theory, method, and applications in global optimization. Comput. Chem. 33, 1964–1972 (2009)
14. Gao, D.Y., Wu, C.Z.: On the triality theory in global optimization. J. Global Optimization (2010)
15. Gao, D.Y., Zhang, J.P.: Canonical Duality Theory for Solving Minimization Problem of Rosenbrock Function. arxiv.org/abs/1108.1241v1

Study on Landslide Deformation Prediction
Based on Recurrent Neural Network
under the Function of Rainfall

Huangqiong Chen[1], Zhigang Zeng[1], and Huiming Tang[2]

[1] Department of Control Science and Engineering, Huazhong University of Science
and Technology, Wuhan 430074, China
[1] Key Laboratory of Image Processing and Intelligent Control of Education Ministry
of China, Wuhan 430074, China
chqkhs@163.com
[2] Faculty of Engineering, China University of Geosciences, Wuhan, Hubei 430074, China

Abstract. Landslide deformation prediction has significant practical value that
can provide guidance for preventing the disaster and guarantee the safety of
people's life and property. In this paper, a method based on recurrent neural
network (RNN) for landslide prediction is presented. The results show that the
prediction accuracy of RNN model is much higher than the feedforward neural
network model for Baishuihehe landslide. Therefore, the RNN model is an effective and feasible method to further improve accuracy for landslide displacement prediction.

Keywords: landslide, deformation prediction, RNN, Jordan network, rainfall.

1   Introduction

In recent years, China has suffered an extensive loss of life, great economic damage
and negative environmental impacts as a result of geo-hazards such as earthquake,
floods, and landslides. Landslide can be defined as a geological phenomenon under
the influence of gravity, which can occur in offshore, coastal and onshore environments. Landslide includes a wide range of ground movements, such as rockfalls, deep
failure of slopes and shallow debris flows.

China is one of the areas where suffer the most serious landslide hazard in Asia as
well as the world. Especially, in West China the large-scale landslides are notable for
their scale complex formation mechanism and serious destruction which are typical
and representative in the world [1]. The last decades have witnessed an increase in the
magnitude and frequency of these catastrophes. This is probably due to the rapid economic development and high population growth of China, which results in overex-
ploitation of the environment including increased deforestation and occupation of
flood plains and hillside areas. In September 1991, for example, Touzhai landslide
occurred in Zhaotong city, Yunnan Province, resulting in a death toll of 216 people
and direct economic losses up to 12 million Chinese yuan [1]. Another case was
Zhouqu country mudslide, which occurred in August 2010 in Gansu, China. It killed
more than 1765 people and also caused tremendous damage and losse [2].

T. Huang et al. (Eds.): ICONIP 2012, Part IV, LNCS 7666, pp. 683–690, 2012.

Landslides have represented 4.89% of the natural disasters that occurred world-wide during the years 1990 to 2005[3]. This trend will continue in future under increased unplanned urbanization and development, continued deforestation and increased regional precipitation in landslide prone areas. Therefore, the prediction of landslide is essential for carrying out quicker and safer mitigation programs, as well as future planning of the area. However, the prediction of landslide is a difficult task to tackle and requires a thorough study of past activities using a complete range of investigative methods to determine the change condition. Most of the slop movements or potential failures may be predicted if proper investigations of parameters are performed [4].

Since 1960s, many approaches have been used to model landslides and make prediction based on the deformation or displacement. They can be roughly classified into three types [5]: i) Deterministic models, such as Saito model [6], inverse-velocity model [7], Voight model [8], deformation power model [9], slide-clock friction models [10-11] and so on. ii) Statistics models, such as grey system model [12] and regression model [13]. iii) Nonlinear models, such as nonlinear dynamic model [14], neural network models [4, 15-16] and support vector machine models [17-18]. The availability of a wide range of methods makes it difficult to choose the optimal method. So emphasis should be given to the validation of the model results and on the estimation of the model's validation.

Artificial neural networks (ANNs) are generic nonlinear function approximators, which are presented as a black-box in general. They are extensively used for pattern recognition and classification [19] which have also a wide applicability in system identification of landslides. The artificial neural network approach has many advantages compared to other statistical methods, it is an effective way for forecasting in complex nonlinear dynamic systems. From a structural point of view, ANNs can be classified into two main types: feedforward neural networks (FNNs) and recurrent neural networks (RNNs). In recent years, the ANNs have been widely used in modeling the evolution or deformation of landslides. However, the previous works mainly focus on the FNNs, the RNNs have not been applied in the area of landslide prediction so far. Based on the previous research works [20-22], we know that RNNs are more appropriate for presenting nonlinear dynamic systems than FNNs. Since the landslides are essentially nonlinear dynamical systems, using RNNs for modeling can anticipate their behaviors more accurately.

For the above purpose, this paper presents a framework to establish a landslide prediction model based on RNNs. The method was implemented for the prediction in the Sichuan province in China.

2   Methodology

2.1   Network Structure

The Jordan's architecture is chosen for the time series prediction, the network consists of a context layer, an input layer, a hidden layer and an output layer. It has been shown theoretically that the original Jordan net is not capable of representing arbitrary dynamic systems [23]. So we adopt the modified Jordan net which introduces self-feedback connections for the context units. This neural network also has four layers, with the main feedback connections taken from the output layer to the context layer. The modified Jordan net has better dynamic capabilities than the original Jordan network, and its structure is shown in Fig. 1.

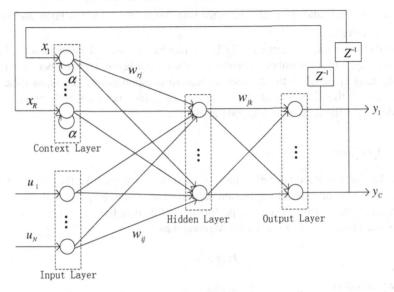

Fig. 1. Structure of modified Jordan network

$$O_r^t = O_{k}^{t-1} + \alpha O_r^{t-1} \tag{1}$$

$$net_j^t = \sum_{i=1}^{N} u_i^t w_{ij} + \sum_{r=1}^{R} O_r^t w_{rj} \tag{2}$$

$$O_j^t = f(net_j^t) \tag{3}$$

$$net_k^t = \sum_{j=1}^{M} O_j^t w_{jk} \tag{4}$$

$$O_k^t = g(net_k^t) \tag{5}$$

where $0 \le \alpha \le 1$. $r = 1,2,...,R$. $r = 1,2,...,R$. R is the number of nodes in context layer. $i = 1,2,...,N$. N is the number of input nodes. $j = 1,2,...,M$. M is the number of hidden nodes. $k = 1,2,...,C$. C is the number of nodes in output layer. O_r^t and O_r^{t-1} are the outputs of the context node r at time t and $t-1$, respectively.

O_k^t and O'^{-1}_k are the outputs of the output node k at time t and $t-1$, respectively. O_j^t is the output of the hidden node j at time t. w_{ij} is the weight of the connection that connects the external input node i to the hidden node j. w_{rj} is the weight of the connection that connects the context node r to the hidden node j. w_{jk} is the weight of the connection that connects node j in the hidden layer to node k in the output layer. $f(\cdot)$ and $g(\cdot)$ are the activation functions of hidden layer and output layer, respectively.

In Jordan network, as shown in Fig.1, the number of internal input (context layer) neurons is equal to the number of neurons in the output layer. The number of external input neurons is equal to the number of feature components of the input data. The second layer is the hidden layer. The nodes in the hidden layer are fully connected to the nodes in the input layer and the output layer.

2.2   Algorithm

We use Levenberg-Marquardt algorithm to train the network. Like the Quasi-Newton algorithm, it was designed to approach second-order training speed without computing the Hessian matrix. When the performance function has the form of a sum of squares, the Hessian matrix can be approximated as

$$H = J^T J \tag{6}$$

and the gradient is

$$g = J^T e \tag{7}$$

where J is the Jacobian matrix which contains the first derivatives of the network errors with respect to the weights and biases, e is a vector of network errors.

The Levenberg-Marquardt algorithm use the Hessian matrix as in equation (6) to do correction in the following Newton-like update

$$x(k+1) = x(k) - [J^T J + \mu I]^{-1} J^T e \tag{8}$$

When the coefficient μ is zero, it is just Newton's method. When μ is large, this becomes gradient descent algorithm with a small step size.

3   Case Study

3.1   Area Description

In order to test the validity of the above prediction modeling, now this paper uses Hubei Baishuihe landslide as an example. Baishuihe landslide is located on the south bank of Yantze River, which belongs to the Three Gorges reservoir and it is 56 km away from the Three Gorges Dam. The bedrock geology of the study area mainly consists of sandstone and mudstone, which is an easy slip stratum. The slope is of the category of bedding slopes. Slippage and deformation of the bedding slope are easily to occur under external forces such as tectonic joint cutting, incised unloading and edge collapse loading of the Yangtze River, and rainfall and so on. The warning area of Baishuihe landslide is large in scale and is the category of forward slopes. In August 2004, as the landslide had significant deformation, 85 people of 21 households in this area moved out. Currently, there are massive farmlands and citrus orchards in the slope. Once the slope failure of the warning area occurs, the safety of the inhabitant in reservoir area and shipping will be under serious threat, and the riverside highway will be destroyed. Considering the morphological and geological conditions of the Baishuihe landslide, it is evident that slope stability processes are the most relevant problem for public safety and land use.

3.2   Analysis and Results

Deformation observed data are shown in Fig. 2. There are 47 groups of measurement data from May 1, 2005 to June 1, 2007, each time step represents a month.

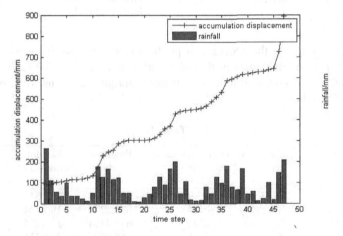

Fig. 2. The accumulation displacement data collected at Bashuihe Landslide during May 1, 2005 to June 1, 2007. Each time step represents a month.

This example uses the model proposed in section 2. Here, the input is rainfall, and the output is accumulation displacement of Baishuihe landslide. According to the investigation [24], most landslides occurred with lag effect with respect to the rainfall. Hence, to predict the accumulation displacement at time step $t+1$, we use the rainfall in time steps t and $t+1$. The anterior data of 37 time steps is used to build up model and the latter 10 time steps for prediction. In this model, we let α equal 1. We continuously increase both the number of neurons in the hidden layer until the network performed well in terms of the mean square error (MSE) and the error autocorrelation function. After several trials, the best number of hidden neurons is determined to be 10.

We use BP neural network and Jordan network to model, and the error performance curves are shown in Fig. 3 and Fig. 4 respectively. According to the figures, we can see that the Jordan net has quicker speed than the BP network.

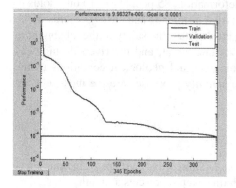

Fig. 3. Error performance curve trained by BP network

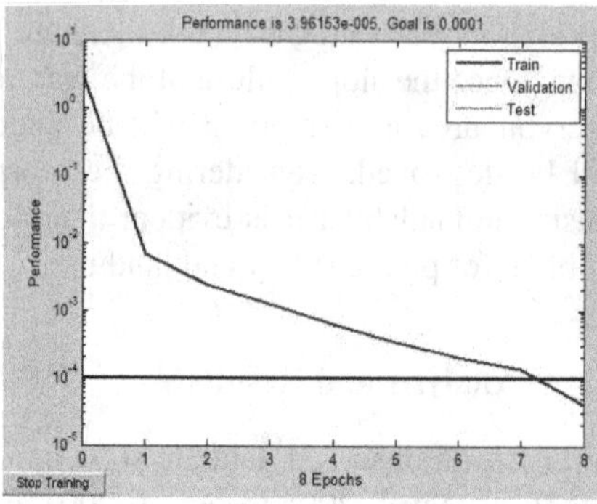

Fig. 4. Error performance curve trained by Jordan network

In order to improve the accuracy of prediction, we process the accumulation displacement data with taking logarithm first, and then normalize. The calculation result of Baishuihe landslide that achieved by using Jordan neural network modeling is shown in Table 1.

Table 1.

Time (year.month.day)	time step	Actual measurement value (mm)	Predicted value (mm)	Absolute error (mm)	Relative error (%)
2006.9.1	38	603.9	623.1644	19.2644	3.19
2006.10.1	39	615.4	635.2774	19.8774	3.23
2006.11.1	40	618	635.4894	17.4894	2.83
2006.12.1	41	623.1	645.0331	21.9331	3.52
2007.1.1	42	628.7	653.1564	24.4564	3.89
2007.2.1	43	630.8	655.9058	25.1058	3.98
2007.3.1	44	637	657.2566	20.2566	3.18
2007.4.1	45	643.1	664.7725	21.6725	3.37
2007.5.1	46	724.8	733.5701	8.7701	1.21
2007.6.1	47	892.3	891.7646	-0.5354	-0.06

As shown in Table 1, the predicting values and actual measurement values are very close for every calculation, and the relative error falls into 4 percent, the predicting precision is high enough which can satisfy the request of deformation prediction of landslide in medium-term and long-term.

4   Conclusion

The displacement prediction of landslides is a complex and nonlinear question. The RNN modeling which is very good in learning and processing information can work out the complex nonlinear relation by using the present data to learn models. This paper has investigated the use RNN for identification of landslide dynamic systems. The neural network employed was Jordan net. Identification results obtained show that the RNN approach is effective at predicting landslide displacement. Therefore, this method has a good perspective in application and further development.

Acknowledgments. The work is supported by the Natural Science Foundation of China under Grant 60974021, the 973 Program of China under Grant 2011CB710606, the Fund for Distinguished Young Scholars of Hubei Province under Grant 2010CDA081, and the Specialized Research Fund for the Doctoral Program of Higher Education of China under Grant 20100142110021.

References

1. Huang, R.Q.: Large-scale Landslides and their Sliding Mechanisms in China since the 20th Century. Chinese Journal of Rock Mechanics and Engineering 26, 433–454 (2007)
2. Tang, C., Rengers, N., van Asch, T.W.J., et al.: Triggering Conditions and Depositional Characteristics of a Disastrous Debris Flow Event in Zhouqu City, Gansu Province, Northwestern China. Natural Hazards and Earth System Sciences 11, 2903–2912 (2011)
3. Kanungo, D.P., Arora, M.K., Sarkar, S., Gupta, R.P.: A Comparative Sstudy of Conventional, ANN Black Box, Fuzzy and Combined Neural and Fuzzy Weighting Procedures for Landslide Susceptibility Zonation in Darjeeling Himalayas. Eng. Geol. 85, 347–366 (2006)
4. Neaupane, K.M., Achet, S.H.: Use of Backpropagation Neural Network for Landslide Monitoring: a Case Study in the Higher Himalaya. Eng. Geol. 74, 213–226 (2004)
5. Li, X.Z., Kong, J.M., Wang, Z.Y.: Landslide Displacement Prediction Based on Combining Method with Optimal Weight. Nat. Hazards 61, 635–646 (2012)
6. Saito, M.: Forecasting the Time of Occurrence of a Slope Failure. In: Proceedings of the 6th International Conference on Soil Mechanics and Foundation Engineering, Montréal, Que, pp. 537–541. Pergamon Press, Oxford (1965)
7. Fukuzono, T.: A new Method for Predicting the Failure Time of a Slope. In: Proceedings of the Fourth International Conference on Landslides, pp. 145–150. Japan Landslide Society, Tokyo (1985)
8. Voight, B.: A Relation to Describe Rate-dependent Material Failure. Science 243, 200–203 (1989)
9. Xu, J.L., Liao, X.P.: Prediction for Huangci Landslide and its Theory and Method. Chin. J. Geol. Hazard. Control 7, 18–25 (1996)
10. Helmstetter, A., Sornette, D., Grasso, J.R., Andersen, J.V., Gluzman, S., Pisarenko, V.: Slider-block Friction Model for Landslides: Application to Vaiont and La Clapiere Landslides. J. Geophys. Res. 109(B02409), 1–15 (2004)
11. Sornette, D., Helmstetter, A., Andersen, J.V., Gluzman, S., Grasso, J.R., Pisarenko, V.: Towards Landslide Predictions: Two Case Studies. Physica A: Stat. Mech. Appl. 338, 605–632 (2004)

12. Lu, P., Rosenbaum, M.S.: Artificial Neural Networks and Grey Systems for the Prediction of Slope Stability. Nat. Hazards 30, 383–398 (2003)
13. Randall, W.J.: Regression Models for Estimating Coseismic Landslide Displacement. Eng. Geol. 91, 209–218 (2007)
14. Long, H., Qin, S.Q., Zhu, S.P., Wan, Z.Q.: Nonlinear Dynamic Model and Catastrophe Analysis of Slope Evolution. J. Eng. Geol. 9, 331–335 (2001) (in Chinese)
15. Wu, Y.P., Teng, W.F., Li, Y.W.: Application of Grey-neural Network Model to Landslide Deformation Prediction. Chin. J. Rock Mech. Eng. 26, 632–636 (2007) (in Chinese)
16. Ran, Y.F., Xiong, G.C.: Study on deformation prediction of landslide based on genetic algorithm and improved BP neural network. Kybernetes 39, 1245–1254 (2010)
17. Feng, X.T., Zhao, H.B., Li, S.J.: Modeling Non-linear Displacement Time Series of Geomaterials Using Evolutionary Support Vector Machines. Int. J. Rock Mech. Min. Sci. 41, 1087–1107 (2004)
18. Dong, H., Fu, H.L., Leng, W.M.: Nonlinear Combination Predicting Based on Support Vector Machines for Landslide Deformation. J. China Railw. Soc. 29, 132–136 (2007) (in Chinese)
19. Melchiorre, C., Castellanos Abella, E.A., Westen van, C.J., Matteucci, M.: Evaluation of Prediction Capability, Robustness, and Sensitivity in Non-linear Landslide Susceptibility Models, Guantanamo, Cuba. Computers & Geosciences 37, 410–425 (2011)
20. Lee, H., Park, Y.: Nonlinear System Identification Using Recurrent Networks. In: Proceedings of the 1991 IEEE International Joint Conference on Neural Networks, pp. 2410–2415 (1991)
21. Karaboga, D., Kalinli, A.: Training Recurrent Neural Networks for Dynamic System Identification Using Parallel Tabu Search Algorithm. In: Proceedings of the 12th IEEE International Symposium on Intelligent Control, pp. 113–117 (1997)
22. Yu, W.: Nonlinear System Identification Using Discrete-time Recurrent Neural Networks with Stable Learning Algorithms. Inf. Sci. 158, 131–147 (2004)
23. Pham, D.T., Oh, S.J.: A Recurrent Backpropagation Neural Network for Dynamic System Identification. Journal of Systems Engineering 2, 213–223 (1992)
24. Zhang, M.S., Li, T.L.: Triggering Factors and Forming Mechanism of Loss Landslides. Journal of Engineering Geology 19, 530–540 (2011)

Vehicle Image Classification Based on Edge: Features and Distances Comparison

Fabrízia Medeiros de S. Matos[1] and Renata Maria Cardoso R. de Souza[2]

[1] Federal Institute of Education, Science and Technology of Paraíba, João Pessoa, Brazil
fabrizia@ifpb.edu.br
[2] Computing Center (CIN), Federal University of Pernambuco, Recife, Brazil
rmcrs@cin.ufpe

Abstract. Automatic vehicle classification is an important task in Intelligent Transport System (ITS) because it allows the traffic parameter, called vehicles count by category, to be obtained. In terrestrial public roads, variants sources of information for vehicles counter by category have been used such as video, magnetic induction coil, sound sensors, temperature sensors and microwave. The use of video has increased support for traffic management due to the advantages of installation cost and a wide range of information it contains. This paper presents comparison of vehicle image classification based on edge features. Contour points number, height, width and fractal dimension are used like features. Nearest neighbor, adaptive nearest neighbor and adaptive distance are used in classification. The experimental platform is built on Matlab R2009a.

Keywords: vehicle image classification, traffic video, edge, fractal, adaptive-NN, adaptive distance.

1 Introduction

Problems related to traffic, such as congestion, mobility and low fatalities in accidents happen in urban centers around the world and interfere negatively in their inhabitants and services. It is not easy to increase the road traffic infrastructure. An alternative to problems associated with traffic has been to use existing road traffic infrastructure in a controlled manner [1].

An Intelligent Transportation System (ITS) is defined as an application that incorporates electronic, computer and communication technologies into vehicles and roadways for monitoring traffic conditions, reducing congestion, enhancing mobility, and so on. Automatic vehicle classification is an important task in ITS because it allows the traffic parameter called vehicles count by category to be obtained, which is used to control and manage the roadways traffic [2].

Methods based on different sensors to get information about vehicles count by category, including video, magnetic induction coil, sound sensors, temperature sensors and microwave have been proposed. Magnetic induction coil installation and maintenance cause damage to the road. The device has a short life; it obtains little traffic information and is also sensitive to weather and traffic speed [1-2].

T. Huang et al. (Eds.): ICONIP 2012, Part IV, LNCS 7666, pp. 691–698, 2012.

Methods for obtaining automatic traffic parameters from video images have become known for being easier to be maintained and then they became popular techniques in ITS. Further advantages are related to the wide range of information they contain [2]. Different methods have been presented for vehicles classification based on road video images [1-7]. Some methods use size vehicle feature while others introduce new descriptors, such as texture, linearity, PCA (Principal Components Analysis) and LDA (Linear Discriminant Analysis). In the classification phase, methods make use of known techniques, such as Nearest Neighbor (NN), neural networks, Support Vector Machine (SVM) and Hidden Markov Models (HMM). However, the great dynamics of real scenes consists in overcoming challenges. Lighting and climatic variations, shadows away from the camera, noise, image quality, dynamic positioning of vehicles on roads and heavy reliance on detection techniques are considered aspects that hinder the automatic methods for collecting traffic parameters based on image processing.

The method and comparison proposed in this paper, considering vehicle image classification, are based on edge features and NN. The method classified a test image into four classes (motorcycle, car, bus and truck). These classes are defined based on reports of the Brazilian Ministry of Transport [8]. Comparison has the objective of to evaluate the potential of classification based on edge features with NN.

This paper is organized as follows: related works are presented in Section 2; details of the proposed methods are in Section 3; Section 4 reports experimental results, and a conclusion is presented in Section 5.

2   Related Works

Some methods proposed for classification of vehicle images make use of vehicle dimension as a feature [3-4],[6-7]. Other methods make use of transformation for dimensional reduction [1-2],[5]. Proposed methods do not have uniformity about the number of classes. Some of them are only classified into two categories, but defined in different ways, such as car and not a car [4], taxi and bus [5]. Among the methods that use a greater number of classes, there is also no consensus with, for example, the use of three [3],[6], five [7] and eight classes [1].

2.1   Methods Based on Dimension of the Vehicle Image Edge

One of the first proposed methods estimates width, length and height of the vehicle [6]. However, firstly it fulfills removal of shadow, background subtraction, vehicle modeling (phase that makes a 3D cuboid wire frame projected on a 2D plane) and vehicle mask (phase that builds a block to estimate dimensions). Considering three classes (taxi, minibus and double-decker), the error results in dimension estimation achieved 10,85%. No results of classification are presented.

By using an analysis of Time-Spatial Image (TSI), seven features (width, area, compactness, length-width ratio, major axis, minor rectangularity and solidity) are obtained from segmented image for vehicle classification [7]. Five classes are considered and error rate ranges from 6,9%   up to 14,29%, depending on controlled time conditions (normal, summer and clouds).

3   Proposed Method and Comparison

The steps of the proposed method for vehicle image classification are similar to a traditional classification process, including pre-processing, training and classification. The features are based on image edge dimensions and three distances are used in the NN classification. The proposed comparison associates these features and distances.

3.1   Pre-processing

In Figure 1 pre-processing block diagram is presented. It is composed of five sequential steps, each one generating a new image which is the input of the next stage.

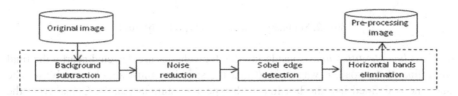

Fig. 1. Pre-processing block diagram

Background Subtraction: In order to remove non-relevant objects of the scene, this phase accomplishes a subtraction arithmetic operation, pixel by pixel (matrix elements), between an image and its background image.

Noise Reduction: It was made by applying the spatial wiener filter [9-10].   The used neighborhood size (mask) was 5 x 5.

Edge Detection: Sobel approximation edge detection [9] is used to build the image edge. The output is a binary image with the same dimension of the original image.

Horizontal Bands Elimination: This phase removes horizontal lines after identification of a horizontal line block without information. The objective is to identify the starting and stopping lines that seem to be the most relevant scene limits of the image edge, and to remove the horizontal lines which are outside this interval (lines that seem to be not part of the analyzed vehicle). This process is carried out in the four following steps:

(i) Summary table: Let us assume that matrix M (dimension r x c) is the image edge. Starting from the first M line (line 1) up to the last line (line r), the number of lines, with and without information, is counted and stored in sequence. New lines in summary table must always be added whenever a change regarding the line type occurs from without information to with information or inversely.

In Figure 2 two summary tables are presented, respectively based on a motorcycle (a) and a truck image (b). For each example, the original image, the image resultant from the edge detection step and the summary table are, respectively, presented. In summary table, the Type columns specify if it is a line block with information (type=1) or not (type=0). The Number column sets the consecutive lines number of the line type that is being specified in the summary. The sum of the Number column is equal to r. Col1 and Col2 columns, defined only for line type equal to 1, represent the first and last columns considering all lines summarized in a specific summary line.

a)

Summary Table			
Type	Number	Col1	Col2
1	2	22	26
0	3	0	0
1	5	21	30
0	3	0	0
1	7	10	28
0	13	0	0
1	6	32	35	start / stop
0	22	0	0

b)

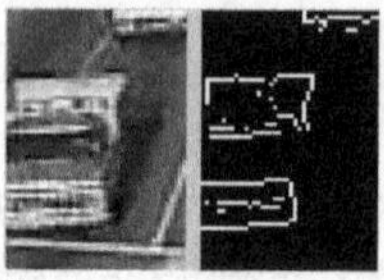

Summary Table			
Type	Number	Col1	Col2
1	5	26	43
0	10	0	0
1	15	3	28	stop
0	9	0	0
1	12	2	24	start
0	10	0	0

Fig. 2. Summary table: (a) motorcycle, (b) truck

(ii) Start line identification: Considering that the vehicles are detected nearby the final white line of the road, it was defined that the start line must also be nearby it. The start line is defined like the last line with type 1 in summary table. Considering the two summary tables in Figure 2, the start lines are, respectively, the seventh and fifth.

(iii) Stop line search process: The search process of the stop line is based on the existence and finding of line blocks, without information, of large, medium and small dimension. Let us assume that *nlin* is the number of image lines (r), k is the start line, the predecessor line is j ($j=k-1$) and that *summ* is the summary table. Part of the stop line search process is defined according to the following algorithms:

```
%--- horizontal stop line search process
bd_large  = fix(nlin * 0.4);
bd_median = fix(nlin * 0.2);
bd_small  = fix(nlin * 0.1);
tol= 0.5;    %Tolerance
band1 = k;   band2 = k;
while (k >= 3)& (j >= 1)
   %----- Large dimension block
   if summ(j,2) >=  bd_large
      break
   end
   %----- Median dimension block
   if (summ(j,2) >= bd_median) & (summ(j,2) < bd_large)
      if ~(h_similar_median(summ, band1, band2, tol))
         break
      end
```

```
end
%----- Small dimension block
if (summ(j,2) >= bd_small) & (summ(j,2) < bd_median)
    pos_band1 = h_calc_band1(summ, band1);
    if pos_band1 < nlin*0.8
        if ~( h_similar_small(summ, band1, band2, tol) );
            break
        end
    end
end
    band1 = k-2;     k=k-2;     j=j-2;
end %while
```

The *h_similar_median* and *h_similar_small* functions are based on width and position similarities. Each function returns 1 (one) if the upper block width (defined between the number 1 lines and the *band1* number) and the bottom block width (between the number *band2* lines and the start line number) present some similarity.

Considering the summary tables presented in Figure 2, the stop lines are, respectively, the seventh and third lines. The stop line search process, for this motorcycle and bus, finishes in the algorithms blocks called median and small, respectively.

(iv) Lines deletion: All lines out of start and stop lines have their matrix elements set to zero.

In Figure 3 four pre-processing steps results are showed. An image sequence of a motorcycle (a) and of a truck (b) is shown. The first two images of each class are, respectively, the original image and the background respectively (both acquired from the video). The four remaining images are, respectively, background subtraction, noise reduction, edge detection and horizontal bands elimination results.

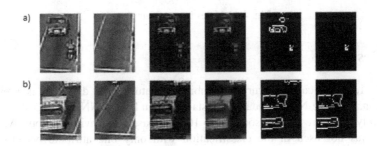

Fig. 3. Pre-processing steps results: (a) bus, (b) truck

3.2   Features and Classification

From each final pre-processing image, following features extractions are performed:

Edge Points Number: The matrix elements with value equal to one are counted.

Height and Width: Width is calculated as the difference between the larger and smaller edge column. Height is calculated as the difference between the larger and smaller edge line. According to these calculations, lines and columns are considered where there is information (matrix element with value 1).

Fractal Dimension: The fractal dimension is computed according to [11].

The NN classification is used with Euclidean distance, adaptive-NN [12] and adaptive distance [13]. Changes about k parameter (neighboring numbers) referring to NN were neither accomplished nor evaluated. The tests are made with $k=1$.

3.3   Test Methodology

The used data set image was manually obtained from a real video of an urban road, with three lanes. The dimension selected to vehicle images was 61 x 43, because it allows long vehicles (bus and truck) to be seen. For each of the four classes, 25 examples were acquired, totalizing 100 vehicles images. The vehicles images were obtained from any lanes of the road and it occurred that in a vehicle image part of another vehicle was shown. In Figure 4, two random images of each of the four classes were presented in sequence (motorcycle, car, bus and truck). There is neither centralization, nor illumination normalization.

By leaving one image out, it was meant to divide the dataset into training and test. Being all data set $X = \{x_1, x_2, ..., x_n\}$, n =100, when the test image is x_i ($1 \leq i \leq n$), the training set is $X' = X - \{x_i\}$ (x_i is removed from the training set)

Fig. 4. Vehicles image examples

4   Results

In Table 1, classification accuracy combining features and distances is presented (Euclidean, adaptive-NN and adaptive distance are in Euc, A.NN and A.Dist columns, respectively). Each result is associated to 100 test based on leaving one out. Adaptive distance was not used in the classification with only one attribute because such distance can only be applied to more than one attribute.

In Figure 5, distribution graphs referring to width and fractal dimension are, respectively, presented. Both graphs include the features of all data images. It is possible to notice that these features, in isolated form (only width or fractal), do not define clearly a decision limit, because there are value overlaps of different classes.

Table 1. Classification Accuracy

Features	NN Without Bands Elimination			NN With Bands Elimination		
	Euc	A. NN	A. Dist	Euc	A. NN	A. Dist
Points Number	46	**52**	-	31	34	-
Heigth	18	19	-	39	41	-
Width	28	28	-	34	36	-
Fractal	43	51	-	40	50	-
P.Number + Heigth	45	45	46	40	39	49
P.Number + Width	48	48	47	41	37	42
P.Number + Fractal	**51**	49	50	34	**44**	**55**
Heigth + Width	35	41	35	40	42	41
Heigth + Fractal	36	43	**51**	**46**	45	53
Width + Fractal	50	42	49	40	36	53

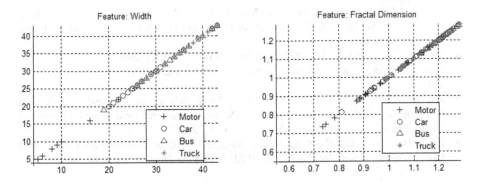

Fig. 5. Width and fractal distributions graph

5   Conclusion

The proposed comparison based on edge features is feasible. By applying different features and distances, the accuracy was low and it rarely changed. Low accuracy can be associated to attributes with low meaningfulness, because in image edge some characteristics, of the original image, are lost. Furthermore, these attributes, in regard to image classification with unwanted objects in the scene (another vehicle, for instance), are unfit. It is also unfeasible the use of linear classifiers over these features, because they do not define a good separation border of classes.

The best accuracy was obtained with horizontal bands elimination step and adaptive distance. Bands elimination seems to be appropriated to pre-processing because the edge dimensions become more useful. The results obtained by means of the use of adaptive distance, even better than the ones obtained with the use of adaptive-NN, suggest that this distance (adaptive distance) is promising when the separation border of classes is confusing.

Future works aim at applying an algorithm of hierarchical and non-linear classification as well as adding a routine of vertical bands elimination.

Acknowledgment. The authors thank Transportation and Traffic Superintendence of João Pessoa for providing videos; and Brazilian agencies CNPq and CAPES for their financial support.

References

1. Morris, B.T., Trivedi, M.M.: Learning, Modeling and Classification of Vehicle Track Patterns from Live Video. IEEE Transaction on Intelligent Transportation Systems 9(3), 425–437 (2008)
2. Wang, W., Shang, Y., Guo, J., Qian, Z.: Real-Time Vehicle Classification Based on Eigenface. In: International Conference on Consumer Electronics, Communications and Networks, pp. 4292–4295 (2011)
3. Ranga, H.T.P., Ravi, K.M., Raja, S.S., Naveen, K.S.K.: Vehicle Detection and Classification Based on Morphological Technique. In: International Conference on Signal and Image Processing, pp. 45–48 (2010)
4. Pumrin, S., Dailey, D.J.: Vehicle Image classification via Expectation-Maximization Algorithm. In: Proc. of the International Symposium on Circuits and Systems, vol. 2, pp. 468–471 (2003)
5. Zhu-Yu, Z., Tian-Min, D., Xian-Yang: Study for Vehicle Recognition and Classification Based on Gabor Wavelets Transform & HMM. In: International Conference on Consumer Electronics, Communications and Networks, pp. 5272–5275 (2011)
6. Lai, A.H.S., Fung, G.S.K., Yung, N.H.S.: Vehicle Type Classification from Visual-Based Dimension Estimation. In: Proc. of the IEEE Intelligent Transportation System, pp. 201–206 (2001)
7. Rashid, N.U., Mithun, N.C., Joy, B.R., Rahman, S.M.M.: Detection and Classification Using Time-Spacial Image. In: International Conf. and Computer Engineering, pp. 502–505 (2010)
8. Ministry Transport of Brazil,
   http://www.transportes.gov.br/conteudo/53887
9. Gonzalez, R.C., Woods, R.E., Eddins, S.L.: Digital Image Processing Using Matlab. Pearson Prentice Hall (2004)
10. Khireddine, A., Benmahammed, K., Puech, W.: Digital Image Restoration by Wiener Filter in 2D Case. Advances in Engineering Software 38, 513–516 (2007)
11. Ebrahimpour-Komleh, H., Chandran, V., Sridharan, S.: Face Recognition Using Fractal Codes. In: International Conference on Image Processing, vol. 3, pp. 58–61 (2001)
12. Wang, J., Neskovic, P., Cooper, L.N.: Improving Nearest Neighbor Rule with a Simple Adaptive Distance Measure. In: Jiao, L., Wang, L., Gao, X.-b., Liu, J., Wu, F. (eds.) ICNC 2006. LNCS, vol. 4221, pp. 43–46. Springer, Heidelberg (2006)
13. Carvalho, F.A.T., Souza, R.M.C.R.: Unsupervised Pattern Recognition Models for Mixed Feature-Type Symbolic Data. Pattern Recognition Letter 31, 430–443 (2010)

Author Index